HANDBOOK OF VISUAL OPTICS
Fundamentals and Eye Optics

VOLUME I

edited by
Pablo Artal

CRC Press
Taylor & Francis Group
Boca Raton London New York

CRC Press is an imprint of the
Taylor & Francis Group, an **informa** business

MATLAB® is a trademark of The MathWorks, Inc. and is used with permission. The MathWorks does not warrant the accuracy of the text or exercises in this book. This book's use or discussion of MATLAB® software or related products does not constitute endorsement or sponsorship by The MathWorks of a particular pedagogical approach or particular use of the MATLAB® software.

CRC Press
Taylor & Francis Group
6000 Broken Sound Parkway NW, Suite 300
Boca Raton, FL 33487-2742

First issued in paperback 2019

© 2017 by Taylor & Francis Group, LLC
CRC Press is an imprint of Taylor & Francis Group, an Informa business

No claim to original U.S. Government works

ISBN-13: 978-0-4822-3785-6 (hbk)
ISBN-13: 978-0-367-86992-2 (pbk)

Library of Congress Cataloging-in-Publication Data

Names: Artal, Pablo, editor.
Title: Handbook of visual optics / [edited by] Pablo Artal.
Description: Boca Raton : Taylor & Francis, [2017] | Includes bibliographical references.
Identifiers: LCCN 2016030030| ISBN 9781482237856 (hbk : alk. paper) | ISBN 9781315373034 (ebk) | ISBN 9781315355726 (epub) | ISBN 9781315336664 (mobi/Kindle) | ISBN 9781482237863 (web PDF)
Subjects: | MESH: Vision, Ocular--physiology | Optical Phenomena | Vision Tests--instrumentation | Eye Diseases--therapy
Classification: LCC QP475 | NLM WW 103 | DDC 612.8/4--dc23
LC record available at https://lccn.loc.gov/2016030030

Visit the Taylor & Francis Web site at
http://www.taylorandfrancis.com

and the CRC Press Web site at
http://www.crcpress.com

Contents

Preface

For many years, first as a student and later as a senior researcher in the area of physiological optics, I have wanted a comprehensive resource for frequently arising questions. Although the situation in today's Internet era is different than before, still I believe there is need for a reliable single source of encyclopedic knowledge. Finally, the dream of my youth—a handbook in visual optics—is a reality and in your hands (or on your screen). I hope this will help interested readers for a long time to come.

At the beginning of this adventure of compiling the handbook, I wanted to accomplish a number of goals (probably, too many!). Among others, I wanted to provide general useful information for beginners, or for those approaching the field from other disciplines, and the latest research presented from the most recent experiments in laboratories. As with most activities in life, success depends on the quality of individuals involved. In this regard, I was tremendously fortunate to have such an exceptional group of contributors. If we can apply the optical equivalence, this handbook is the result of a *coherent superposition* of exceptional expertise.

This handbook builds from the fundamentals to the current state of the art of the field of visual optics. The eye as an optical instrument plays a limiting role in the quality of our vision. A better understanding of the optics of the eye is required both for ophthalmic instrumentation and vision correction. The handbook covers the physics and engineering of instruments together with procedures to correct the ocular optics and its impact on visual perception. The field of physiological, or visual, optics is a classic area in science, an arena where many new practical technologies have been tested and perfected. Many of the most brilliant scientists in history were interested in the eye. Based in well-established physical and physiological principles, the area was described as nearly completed in the second part of the twentieth century. However, from the 1980s onward, a tremendous new interest in this field appeared. This was driven in part by new technology, such as lasers and electronic cameras, which allowed the introduction of new instrumentation. For example, the use of wave-front sensors and adaptive optics concepts on the eye completely changed the field. In relatively few years, these ideas expanded to the clinical areas of ophthalmology and optometry. Today, research in new aspects of vision correction and instruments is extremely active, with many groups working on it around the world. This area is a mixture of fundamentals and applications, and is at the crossroad of many disciplines: physics, medicine, biology, psychology, and engineering. I tried to find an equilibrium among the different approaches and sensibilities to serve all tastes. This book can be accessed sequentially, but also by individual parts whenever a particular topic is required.

The handbook is organized in two volumes, with five total parts. Volume One begins with an introductory part that gives an exceptional appetizer by two giants of the field: Gerald Westheimer presents an historical account of the field, and David Williams explores the near past and the future. Part II covers background and fundamental information on optical principles, ocular anatomy and physiology, and the eye and ophthalmic

instruments. Each chapter is self-contained but oriented to provide the proper background for the rest of the handbook. Basic optics is covered by Schwiegerling (geometrical optics), Malacara (wave optics), and Sasián (aberrations). The concepts of photometry and colorimetry are summarized in Chapter 6 (Ohno). The basics and limits of the generation of visual stimuli are described in Chapter 7 (Farrell et al.). Furlan provides a complete revision on the main ophthalmic instruments, and Dainty an introduction on adaptive optics. While the first chapters of this part are devoted to the more technical aspects, the three next chapters have a different orientation to provide the physiological basis for the eye and the visual system. Choh and Sivak describe the anatomy and embryology of the eye in Chapter 10. Freed reviews the retina, and Winawer the architecture of the visual system. In the final chapter in this part, Pelli and Solomon describe psychophysical methods. Part II sets the foundation for the various principles that follow in the rest of the handbook.

Part III covers the current state of the art on the understanding of the optics of the eye and the retina. Collins et al. and Manns describe, respectively, what we know today about the optical properties of the cornea and the lens. Atchison reviews in Chapters 16 and 17 the different schematics eyes and the definitions and implications of the axes and angles in ocular optics. The optics of the retina is detailed in Chapter 18 (Vohnsen). Once the different components are evaluated, the next chapters concentrate on the impact of optical quality. Refractive errors (Wilson) and monochromatic (Marcos et al.) aberrations are described. Although traditionally most attention has been paid to optical characteristics of the eye in the fovea, the important role of peripheral optics is described in Chapter 21 (Lundström and Rosén). Tabernero describes personalized eye models in Chapter 22. Beyond refractive errors and aberrations, scattering in the eye affects image quality. van den Berg exhaustively reviews the state of the art of the impact and measurements of this phenomenon (Chapter 23). The eye in young subjects has the ability to focus objects placed at different distances efficiently. Bharadwaj provides a review of the accommodative mechanism (Chapter 24), and Winn and Gray describe its dynamics (Chapter 25). The eyes are continually moving to place the fovea on the area of interest. This dynamic behavior has important implications described in Chapter 26 (Anderson). Although the human eye is very robust, serving us over many years, aging obviously affects its optics. In Chapter 27, Charman reviews how the eye changes with age. Several species are able to detect the state of polarization of light. Although our visual system is not capable of something similar, polarization plays a role in optical properties as described in Chapter 28 (Bueno).

Volume Two focuses on the important topics of instrumentation and vision correction. Part I is dedicated to novel ophthalmic instrumentation for imaging, including the anterior segment and the retina, and for visual testing. An introductory chapter is dedicated to reviewing the concepts of light safety (Barat). Molebny presents a complete description of different wavefront sensors and aberrometers in Chapter 2. Hitzenberger reviews the principle of low-coherence interferometry (Chapter 3). This was the basis for one of the most successful techniques in ophthalmology: optical coherence tomography (OCT). Grulkowski concentrates on the current state of the art in OCT applied to the anterior segment (Chapter 4). Popovic (Chapter 5) and Doble (Chapter 6) present how adaptive optics implemented in ophthalmoscopes has changed the field in recent years. A different application of adaptive optics is its use for visual testing. Fernandez (Chapter 7) shows the history, present, and future potential of this technology. Imaging of the ocular media using multiphoton microscopy is a recent scientific frontier. Jester (Chapter 8) and Hunter (Chapter 9) cover, respectively, the applications of this emerging technology for the cornea and the retina.

Part II describes the different devices and techniques for surgical and nonsurgical visual correction procedures, from traditional to futuristic approaches. Ophthalmic lenses are still the most widely used approach and clearly deserve to be well recognized. Malacara (Chapter 10) presents a complete overview of this topic. Contact lenses are described in depth in Chapter 11 (Cox). The specific case of correcting highly aberrated eyes is addressed in Chapter 12 (Marsack and Applegate). A particularly relevant type of correcting devices is intraocular lenses (IOLs), implanted to substitute the crystalline lens after cataract surgery. Two emerging types of IOLs, accommodating and adjustable, are reported in Chapters 13 (Findl and Himschall) and 14 (Sandstedt). Chapter 15 (Alio and El Bahrawy) presents a review of refractive surgical approaches for the cornea. The potential for nonlinear manipulation of the ocular tissues may open the door to new reversible future treatments. Chapter 17 (van de Pol) presents the state of the art of using corneal onlays and inlays for vision correction.

Part III reviews the relationship between the ocular optics and visual perception. Aspects related to optical visual metrics (Chapter 18, Guirao) and the prediction of visual acuity (Chapter 19, Navarro) are included. Adaptation is a key element in vision and may have significant clinical implications. Chapters 20 (Webster and Marcos) and 21 (Shaeffel) describe adaptation to blur and contrast. Visual functions change with age. A description of these characteristics is a useful resource for those interested in any practical application. Chapter 22 (Wood) reviews age-related aspects of vision. Finally, Chapter 23 (Jimenez) explores the impact of the eye's optics in stereovision.

I thank the many people who contributed to this handbook: of course, all the authors for providing accurate and up-to-date chapters; Carmen Martinez for helping me with secretarial work, and Luna Han from Taylor & Francis Group for her guidance and patience. I am also indebted to the financial help received by my lab, which allowed dedication to this endeavor: the European Research Council, the Spanish Ministry of Science, and the Fundacion Seneca, Murcia region, Spain.

Pablo Artal
Universidad de Murcia
Murcia, Spain

Editor

Pablo Artal was born in Zaragoza (Spain) in 1961. He studied Physics at the University of Zaragoza. In 1984, he moved to Madrid with a predoctoral fellowship to work at the CSIC "Instituto de Optica." He was a postdoctoral research fellow, first at Cambridge University (UK) and later at the Institut d'Optique in Orsay, France. After his return to Spain, he obtained a permanent researcher position at the CSIC in Madrid. In 1994 he became the first full professor of optics at the University of Murcia, Spain, where he founded his "Laboratorio de Optica."

Prof. Artal was secretary of the Spanish Optical Society from 1990 to 1994, associated dean of the University of Murcia Science Faculty from 1994 to 2000, and director of the Physics Department at Murcia University from 2001 to 2003. From 2004 to 2007 he was in charge of the reviewing grants panel in physics at the Spanish Ministry of Science in Madrid. Since 2006 he is the founding director of the Center for Research in Optics and Nanophysics at Murcia University. He was president of the Academy of Science of the Murcia Region from 2010 to 2015. From 2015 he is the president of the "Fundación de Estudios Medicos," an outreach organization dedicated to promote science. During his career he often spent periods doing collaborative research in laboratories in Europe, Australia, Latin America, and the United States. This included two sabbatical years in Rochester (USA) and Sydney (Australia).

Dr. Artal's research interests are centered on the optics of the eye and the retina and the development of optical and electronic imaging techniques to be applied in vision, ophthalmology, and biomedicine. He has pioneered a number of highly innovative and significant advances in the methods for studying the optics of the eye and has contributed substantially to our understanding of the factors that limit human visual resolution. In addition, several of his results and ideas in the area of ophthalmic instrumentation over the last years have been introduced in instruments and devices currently in use in clinical ophthalmology.

He has published more than 200 reviewed papers that received more than 7600 citations with an H-index of 45 (12700 and 60 in Google scholar) and presented more than 200 invited talks in international meetings and around 150 seminars in research institutions around the world. He was elected fellow member of the Optical Society of America (OSA) in 1999, fellow of the Association for Research in Vision and Ophthalmology in 2009 and 2013 (gold class), and fellow of the European Optical Society in 2014.

In 2013, he received the prestigious award "Edwin H. Land Medal" for his scientific contributions to the advancement of diagnostic and correction alternatives in visual optics. This award was established by the OSA and the Society for Imaging Science and Technology to honor Edwin H. Land. This medal recognizes pioneering work empowered by scientific research to create inventions, technologies, and products. In 2014, he was awarded with a prestigious "Advanced Grant" of the European Research Council. In 2015, he received the "King Jaime I Award on New Technologies" (applied research). This is one of the most prestigious awards for researchers in all areas in Spain. It consists of a medal, mention, and 100000€ cash prize.

He is a coinventor of 22 international patents in the field of optics and ophthalmology. Twelve of them extended to different countries and in some cases expanded to complete families of patents covering the world. Several of his proposed solutions and instruments are currently in use in the clinical practice. Dr. Artal is the cofounder of three spin-off companies developing his concepts and ideas.

He has been the mentor of many graduate and postdoctoral students. His personal science blog is followed by readers, mostly graduate students and fellow researchers, from around the world. He has been editor of the *Journal of the Optical Society of America A* and the *Journal of Vision.*

Contributors

Andrew J. Anderson
Department of Optometry and Vision Sciences
The University of Melbourne
Melbourne, Victoria, Australia

David A. Atchison
School of Optometry and Vision Science
Queensland University of Technology
Brisbane, Queensland, Australia

Shrikant R. Bharadwaj
Brien Holden Institute of Optometry and Vision Sciences
Bausch and Lomb School of Optometry
and
Hyderabad Eye Research Foundation
L V Prasad Eye Institute
Hyderabad, India

Juan M. Bueno
Laboratorio de Óptica
Universidad de Murcia
Murcia, Spain

W. Neil Charman
Division of Pharmacy and Optometry, Faculty of Biology,
    Medicine and Health
University of Manchester
Manchester, United Kingdom

Vivian Choh
School of Optometry and Vision Science
University of Waterloo
Waterloo, Ontario, Canada

Michael Collins
School of Optometry and Vision Science
Queensland University of Technology
Brisbane, Queensland, Australia

Chris Dainty
Institute of Ophthalmology
University College London
London, United Kingdom

Carlos Dorronsoro
Instituto de Optica
Consejo Superior de Investigaciones Científicas
Madrid, Spain

Joyce E. Farrell
Department of Electrical Engineering
Stanford University
Stanford, California

Michael A. Freed
Department of Neuroscience
University of Pennsylvania
Philadelphia, Pennsylvania

Walter D. Furlan
Diffractive Optics Group
Universidad de Valencia
Valencia, Spain

Lyle S. Gray
Department of Life Sciences
Glasgow Caledonian University
Glasgow, United Kingdom

Arthur Ho
Brien Holden Vision Institute
Sydney, Australia

and

Miller School of Medicine
University of Miami
Miami, Florida

and

School of Optometry and Vision Science
University of New South Wales
Sydney, Australia

Hiroshi Horiguchi
Department of Ophthalmology
Jikei University School of Medicine
Tokyo, Japan

Haomiao Jiang
Department of Psychology
Stanford University
Stanford, California

Linda Lundström
Department of Applied Physics
KTH Royal Institute of Technology
Stockholm, Sweden

Daniel Malacara
Centro de Investigación en Optica
León, Mexico

Fabrice Manns
College of Engineering
and
Bascom Palmer Eye Institute
University of Miami
Coral Gables, Florida

Susana Marcos
Instituto de Optica
Consejo Superior de Investigaciones Científicas
Madrid, Spain

Yoshi Ohno
Sensor Science Division
National Institute of Standards and Technology
Gaithersburg, Maryland

Jean-Marie Parel
Bascom Palmer Eye Institute
and
College of Engineering
University of Miami
Miami, Florida

Denis G. Pelli
Department of Psychology
Center for Neural Science
New York University
New York, New York

Pablo Pérez-Merino
Instituto de Optica
Consejo Superior de Investigaciones Científicas
Madrid, Spain

Scott Read
School of Optometry and Vision Science
Queensland University of Technology
Brisbane, Queensland, Australia

Robert Rosén
Applied Research
AMO Groningen BV
Groningen, the Netherlands

José Sasián
College of Optical Sciences
University of Arizona
Tucson, Arizona

Jim Schwiegerling
College of Optical Sciences
University of Arizona
Tucson, Arizona

Jacob G. Sivak
School of Optometry and Vision Science
University of Waterloo
Waterloo, Ontario, Canada

Joshua A. Solomon
Centre for Applied Vision Research
City University London
London, United Kingdom

Juan Tabernero
Laboratorio de Óptica
Universidad de Murcia
Murcia, Spain

Thomas J.T.P. van den Berg
Netherlands Institute for Neuroscience
Royal Academy
Amsterdam, the Netherlands

Stephen Vincent
School of Optometry and Vision Science
Queensland University of Technology
Brisbane, Queensland, Australia

Brian Vohnsen
School of Physics
University College Dublin
Dublin, Ireland

Sarah Walters
Center for Visual Science
University of Rochester
Rochester, New York

Brian A. Wandell
Department of Electrical Engineering
and
Psychology Department
Stanford University
Stanford, California

Gerald Westheimer
Department of Molecular and Cell Biology
University of California, Berkeley
Berkeley, California

David R. Williams
Center for Visual Science
University of Rochester
Rochester, New York

David A. Wilson
Brien Holden Vision Institute
and
School of Optometry and Vision Science
University of New South Wales
Sydney, New South Wales, Australia

Jonathan Winawer
Department of Psychology
Center for Neural Science
New York University
New York, New York

Barry Winn
Sohar University
Sohar, Sultanate of Oman

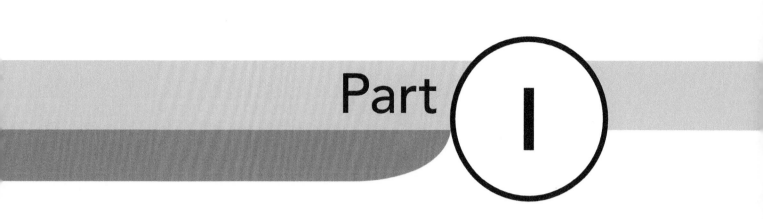

Part I

Introduction

1

History of physiological optics in the twentieth century

Gerald Westheimer

Contents

1.1 STATUS AT THE BEGINNING OF THE CENTURY

Physiological Optics, as confirmed by its central manifestation, Helmholtz's three-volume handbook, was understood at the time to be synonymous with the current *Vision Science*. But nomenclature has to go along with the explosive expansion of scientific knowledge. Hence the more optical components are now subsumed under *Visual Optics*, and even here further subdivision is needed. Optical imagery in the living eye is continually conditioned on factors arising from being embedded in motor apparatuses, specifically those controlling the pupil aperture and the ciliary muscle. Hence a division into *structural visual optics*, relating to the image-forming properties of the static normal eye, and *functional visual optics*, which would fold in accommodative and aperture size factors, seems indicated. Though it is not recognized as a distinct discipline, one can identify a branch of research as *histological optics*. Insofar as it transmits light unimpeded, eye tissue, such as the cornea and the crystalline lens, needs to have unusual biological structure. This became more evident and constituted a challenge around the middle of the century when electron microscopy began to reveal the subcellular makeup of corneal and lenticular layers.

The pioneering study by David Maurice (1957) on the cornea was influential here.

The final biological stage of light capture resides in the receptors. Starting with an observation by Ernst Brücke in the 1840s, there have been consistent attempts to assign to them special light-gathering properties. The directional sensitivity and wave guide nature of retinal receptors has been an active area now for the last 80 years. It may be noted parenthetically that the optics of invertebrate eyes, left out of consideration in this review, has deservedly been given much attention (Exner 1891, Snyder and Menzel 1975, Land and Nilsson 2002).

1.2 THE FOUNDATIONS

The eye's image-forming properties were well understood in the middle of the eighteenth century as shown in the classic treatise by Robert Smith (1689–1768) (Smith 1738). Through the efforts of astronomers, for example, Bessel, Seidel, and Airy (1801–1892), optics as a discipline was thoroughly established in the nineteenth century. Maxwell's (1831–1879) electromagnetic theory took command of the subject in 1861 and has never needed superseding. The giants of physiological optics, Thomas Young (1773–1839), Jan Purkinje (1787–1869), Listing (1808–1882), and

Helmholtz (1821–1894), laid and cemented the foundations, and ophthalmologists Donders (1818–1889), Landolt (1846–1926), and Snellen (1834–1908) developed clinical applications. Thus, a century ago, at the time of the beginning of the First World War, a student had available comprehensive compendia containing the available knowledge, specifically Helmholtz's *Physiological Optics* in the new edition updated in particular by Gullstrand (1862–1930), the *Graefe-Saemisch Handbuch der Augenheilkunde* in its many volumes and several editions, followed in the next couple of decades by important chapters in Vol XII of Bethe et al., *Handbuch der normalen und pathologischen Physiologie* (1932) and *in Abderhalden's Handbuch* (1920), and in Vol 1 of Duke-Elder's *Textbook of Ophthalmology* (1932). When the author entered optometry school in 1940, the assigned textbook, the second edition of Emsley's *Visual Optics* (Emsley 1939), contained a treatment of the subject that would rival current accounts and in some respects exceed their scope. By that time, too, quantum theory insofar as its characterization of the photon was concerned had solid footing and quite soon thereafter gripped the vision community when used to underpin our understanding of the absolute visual threshold (Hecht et al. 1942). The vision community has been well served by the authoritative treatment of the subject by Yves LeGrand (1908–1986) in various versions of his textbook (LeGrand 1949), a model of clarity.

Exhaustive literature surveys of vision science up to the beginning of the twentieth century had been provided in the encyclopedic scholarship displayed in the appendixes by A. Koenig to the second edition of Helmholtz's *Physiological Optics* (almost 8000 references up to 1894) and F. Hofmann in his *Graefe-Saemisch Handbuch* chapters (almost 1500 references on spatial vision alone) and by A. Tschermak in the voluminous footnotes of his chapters in *Bethe et al. Handbuch* (Tschermak 1931).

As will be seen, developments in visual optics during the second half of the twentieth century required an expanded view beyond geometrical optics and the simple application of diffraction theory in Airy's disk, yet the foundations for it were well in place. Abbe (1873) and Rayleigh (1896) in their treatment of microscope resolution used a framework that contained, almost explicitly, all the elements that were to become mainstream in the Fourier theory of optics that has since become dominant. Though it was more of academic than practical interest at the time, the theory applicable to coherent light (and what is more challenging, partially coherent light) was put on the table by van Cittert (1934), and so were the celebrated polynomials of Zernike (1934).

The upswing in the growth of optics, specifically as it plays a role in vision, in the middle of the twentieth century thus had their origin elsewhere. Most prominently, it was the harnessing of scientific and industrial resources in the conduct of the Second World War that ended in 1945. The scientific community virtually unanimously rallied behind the war effort, contributing insight and inventiveness to a heady mixture that also included technological innovation and industrial prowess. After the war, this continued in university and some corporate laboratories, blossoming into a research enterprise of unprecedented magnitude and productivity. The sequence into cybernetics (Wiener 1948), information theory (Shannon and Weaver 1949), and the linear systems approach (Trimmer 1950) and control theory was seamless and

so was the progress toward labs equipped with oscilloscopes, photomultipliers, digital computers, transistor devices, in time followed by lasers, LCD, and deformable mirrors. For decades, vision laboratories thrived on "war surplus" lenses, mirrors, prisms, and filters. Light that was once generated by candles, and had its intensity controlled by the inverse square law and its duration by episcotister disks, was produced by lamps with specific filaments and then by high-pressure mercury arcs with wavelength range restricted by narrowband interference filters and intensity adjusted by neutral density wedges. To achieve high retinal illuminance, the filaments were imaged in the pupil in Maxwellian view well before its optical subtleties had been realized (Westheimer 1966).

1.3 STRUCTURAL OPTICS OF THE EYE

1.3.1 EYE DIMENSION AND AXIAL LENGTH

The ingenious method of measuring the eye's axial length by utilizing x-ray phosphenes (Goldmann and Hagen 1942) soon gave way to sonography that in the form of corneal pachometry (Molinari 1982) is in clinical use and now has become a reliable means of evaluating the refractive needs associated with cataract extraction (Hoffer 1981).

1.3.2 CORNEA

Because it is the principal source of the eye's refractive power and because, unlike the other refractive surfaces, it is immediately accessible, the cornea has always attracted much attention. Gullstrand in his appendix to the Third Edition of Helmholtz went into much detail about the shape of the corneal surface and the various means of measuring it. Keratoscopy and keratometry formed a strong chapter in Abderhalden's *Handbuch*. As contact lenses became ubiquitous and their fitting needed good information of the corneal surface on which they rested and whose optical properties they largely preempted, rapid and accurate measurements of corneal curvature could be performed in the clinic by cleverly designed electro-optical apparatuses, the subject of continued attention and technical innovation (Fowler and Dave 1994). Polarization effects, which can be made visible, have been ascribed to the cornea (Stanworth and Naylor 1950).

1.3.3 THE CRYSTALLINE LENS

The anterior and posterior surfaces of the crystalline lens are of critical relevance in how the eye accommodates, that is, changes focus under neural control by contraction of the ciliary muscle. E.F. Fincham (1937a), in part using optical means, made the major contribution to this topic in his monograph. It became clear early in the optical modeling of the eye that the anterior and posterior curvature of the lens could not fully account for its total refractive power for a biologically realistic refractive index, and giving the eye a solid interior core was not supported by evidence. Hence the proposition that a remarkable proportion of its total refractive power is provided by a refractive index gradient had gained acceptance by the time Gullstrand wrote his 1911 appendix. More recent approaches show just how challenging a topic this is (Campbell 1984, Pierscionek and Chan 1989).

1.3.4 TRANSMISSION OF THE OCULAR MEDIA

For a variety of reasons, the transmissivity of the ocular media has been of interest, more recently because of concern for possible damage from exposure to intense sources. For decades the data accepted as authoritative came from the study by Ludvigh and McCarthy (1938). It formed the basis for the confirmation that the energy exchange at absolute visual threshold involved only a handful of photons (Hecht et al. 1942). Wavelength dependency of light absorption in the media became an issue in the characterization of retinal photopigments, one of the most important research enterprises of vision science in the middle of the twentieth century (Rushton 1959, Wald 1964).

1.3.5 RETINAL OPTICS

In the vertebrate, before it reaches the receptors, light has to traverse several retinal layers that therefore have to be essentially transparent, not necessarily a quality automatically associated with active neural tissue. Haidinger's brushes, an entoptic phenomenon, have their origin in retinal optical structure (Naylor and Stanworth 1954). Myelination of the ganglion cell axons, helpful in enhancing velocity of action potentials, does not start till they exit the eyeball at the optic disk. The vascular tree of Purkinje, a prominent feature of the fundus, is somehow compensated for and made visible only by special tricks. The central region of the retina in the primate is suffused by a pigment, selectively absorbing light of some wavelengths. It seems to have a role in Haidinger's brushes (Bone and Landrun 1984, Mission 1993), but whether it is the origin of the entoptic phenomenon known as Maxwell's spot (Maxwell 1890/1965, p. 278) has been subject of an interesting debate (Polyak 1941, Walls and Mathews 1952). Of great significance are its possible protective properties (Snodderly 1995).

Optics becomes critical however, in the operation of receptor cells, whose diameter is of the order of the wavelength of light. A start was made in the 1840s by Ernst Brücke, at the time Helmholtz's fellow student in Johannes Müller's Institute in Berlin, who made the observation that rodlike retinal receptor cells acted like light guides. He seems not to have published it; all we have is Helmholtz's (1866) report and the comment that light once it had entered a receptor and impinged on the cylindrical boundary separating media of high from low refractive index would undergo total reflection and proceed further along the receptor and not leave it.

Receptor optics became mainstream with the discovery of the retinal directional sensitivity by Stiles and Crawford (1933) and the conjecture by Toraldo (1949) of retinal cones being wave guides. This set into motion extensive research activity, still ongoing. The state of the subject is well captured in the contributions to Enoch and Tobey's *Vertebrate Receptor Optics* (1981).

Whereas rods and their rhodopsin photopigment had been fully identified with scotopic vision, the same could not be said about cones and the cone pigments till the 1960s. Before that, because there was no firsthand knowledge of the phototransduction that underlies color vision, the possibility remained open that there was only a single cone pigment and that wavelength analysis came about through an intracellular filtering process, as indeed is the case with oil drops in birds (Walls 1942). Of historical interest, therefore, is the attempt by Ingelstam (1956) to show that the ultramicroscopic structure of receptors, which had just been discovered, might allow wavelength-dependent differential energy concentration. Most of these conjectures were laid to rest by Brindley and Rushton's (1959) demonstration that to the human observer colors looked the same whether light entered the retina from the front or the back. The effect of the concentration of photopigments on their absorption spectrum—called self-screening—must, however, be considered in color vision theory (Brindley 1960) and probably plays a role in the Stiles–Crawford effect of the second kind (Stiles 1939), color changes associated with direction of incident light.

1.4 THE RETINAL IMAGE

1.4.1 ABERRATIONS OF THE EYE

A theoretical approach to the monochromatic aberrations in an optical system requires adequate knowledge of the optical parameters, position and curvature of the surfaces, and refractive index of the media. Because the precision needed to estimate image quality by ray tracing was lacking, this topic of visual optics was largely unattended until it was, so to speak, turned upside down quite recently by nulling out the aberrations. There was a brief flurry of activity centered on spherical aberration, when it was fingered to account for night myopia (Koomen et al. 1951), of practical importance during the Second World War. The enlarged pupil in the scotopic state allows light to enter into the eye through regions manifesting spherical aberration, but the more likely explanation of accommodation activity in empty fields (Otero and Aguilar 1951) won out. In a curious interlude, a quite adequate experimental determination of the eye's spherical aberration (Ivanoff 1953) was marred by inclusion of a point derived from the wrong supposition that the eye was always focused precisely on the target plane. When this is corrected (Westheimer 1955), outlines of spherical aberration across the pupil looked regular.

Because it needs to be factored into the stimulus situation in color vision research, vision researchers throughout the twentieth century remained aware of the eye's chromatic aberration (Hartridge 1918, Ames and Proctor 1921). Axial chromatic aberration (Wald and Griffin 1947) was mostly seen consonant with that of eye media with the dispersion of water. To obviate possible effects of chromatic aberration in color vision research, an "achromatizing lens" with the reverse of the eye's chromatic aberration was designed (Bedford and Wyszecki 1957). During the same period, the role of lateral chromatic aberration of the eye in engendering spurious stereoscopic disparity was given due consideration (Vos 1960), but some inadequacies in the explanation remain.

1.4.2 QUALITY OF THE RETINAL IMAGE

Helmholtz was fully aware that the central issue and best descriptor in the specification of the quality of the retinal image is the light distribution at a sharp target edge, though it took a little while for the realization that the point-spread function is even more basic. The most influential contributor at the beginning of the century was Hamilton Hartridge (1922), and it is hard to imagine anyone doing better at a time before the idea of direct measurement took hold. Assuming that the shape of the point-spread function was Gaussian and making rather good guesses of the parameter, Fry and Cobb (1935) were able to achieve some synthesis between image light spread and thresholds for simple line targets.

In retrospect, the direction of future development was clearly foreshadowed by exceedingly insightful indirect approaches to retinal image quality by LeGrand (1935) using interference and Shlaer (1937) employing what is tantamount to Abbe's theory of microscope resolution. But, at the start of my career in vision science just after the Second World War, these were not adequately understood or appreciated.

In their place, the interest was in direct measurements, first in an approachable if not particularly informative animal preparation, the excised steer eye, expeditiously brought from the abattoir to the lab, as Jay Enoch, one of the collaborators explained to me (Boynton et al. 1954). Needless to say, light spread was very extensive, making one the researchers wonder why, if the image is so bad, visual acuity is so good (DeMott 1959). As the research during the remainder of the century, and continuing to the present, has made clear, such a proposition was ill-posed—the need instead was to pursue the question of how good, in the end, the retinal image might actually be, with all the experimental prowess that can be marshaled.

The most interesting and productive laboratory of the time was at the Institut d'Optique in Paris where Arnulf and his students were in daily contact with the change in approach to the theory of optics that began with Duffieux's (1946) paperback. To call the turn to Fourier optics revolutionary would be an exaggeration, because it is implicit in the resolution formulations of Rayleigh and Abbe and *Fourier's Lehrsatz* is explicitly used on p. 185 of Born (1933). In the single most significant paper in the subject of retinal imagery of the twentieth century, Francoise Flamant (1955) used the principle of the ophthalmoscope to measure the width of the reflected image of a narrow slit. Being familiar with the theorem that convolution becomes multiplication in the Fourier domain, she undid the double convolution due to the light traversing the eye media twice by taking the square root of its Fourier transform. Needless to say, Flamant's results were much closer to the human optics than those on the excised steer eye. In a sequence of more and more sophisticated experiments using photomultiplier tubes in place of Flamant's grainy photographic film (Westheimer and Campbell 1962, Campbell and Gubisch 1966), objective data were accumulated on the optical image quality of the normal human eye that proved quite compatible with psychophysical ones employing the principle of interference fringes (Westheimer 1960, Campbell and Green 1965). Conjectures on hypothetical image sharpening mechanisms with their improbable information-theoretical basis could be discounted (Gubisch 1967).

1.4.3 OPTICAL TRANSFER FUNCTION

Nowhere is the complementarity of the traditional spatial and the modern spatial-frequency descriptions of imagery more evident than in counterposing the image light spread and the optical transfer functions. They are Fourier transforms of each other

and therefore have their x-axes point in opposite directions: light spread over extensive regions tends toward infinite distances in the image plane but toward zero (the DC point) in the spatial-frequency spectrum. In principle, the diffraction image of a point source with a round pupil never stays at zero intensity, though its central lobe, the Airy disk, has a well-defined diameter. But the spatial-frequency spectrum has a distinct cutoff point beyond which there is no representation of grating targets. When this became understood and increasingly popular from the middle of the twentieth century on, the resolution limit of optical devices was better appreciated and could be related to the electrical circuits to which they were increasingly being coupled. Yet the fundamental distinction always needs pointing out: on the one hand, a firm cutoff spatial frequency in optical imagery and, on the other, the sloping transfer function, in principle never quite reaching zero transmission in electronics.

The eye's actual optical transfer function exemplified by the original one provided in the Campbell and Green study of 1965 included the effect of aberrations and the pupil diameter, but still needed extension to include not only amplitude but also phase, discarded in the power spectrum. For many years, from the seminal paper by Schade (1956) and the widely quoted data of van Nes and Bouman (1967) on, the majority of interpretations of the modulation transfer function of the whole visual system failed to stress that it lacked phase information and hence did not allow a unique description of light spread from the power spectrum. It took almost a couple of decades from the first enunciation of the Fourier theory of vision till the explicit demonstration that phase was more important than amplitude (Piotrovsky and Campbell 1982). Yet it has been shown by Hopkins (1955) and Steel (1956) that defocus manifests itself prominently in the phase of the optical transfer function.

1.4.4 STREHL RATIO

Attempts at capturing image quality in a single number go back to Strehl (1895), a high school teacher with an abiding interest in telescope design, who suggested the ratio of the height of the actual point-spread function at its center to that of the diffraction image, generally the Airy disk, defined by the instrument's aperture and the wavelength of light. It was conservatively estimated at 0.2 (Gubisch 1967) in a good eye, but in practice may be much less, because even a miniscule level of stray light (see below) at outlying image distances, covering as it does large retinal areas, would be integrated in the constant volume of light involved in computing the Strehl ratio. Areal summation of light probably makes a low value of the Strehl ratio not as severe a visual handicap as it may appear.

1.4.5 STRAY LIGHT

Whereas the shape of the central lobe of the point-spread function is an important factor in visual acuity, its long tail plays a role in a different visual phenomenon, glare. The veiling effect of bright sources in quite remote retinal areas can have deleterious influence on some visual tasks and early on in the twentieth century attempts were made to distinguish between optical and physiological causes (Holladay 1927, Stiles and Crawford 1937). This required the estimation of the retinal illuminance distribution caused by light scatter in the eye, which also, depending on the wavelength and the red reflectance of the fundus, could act as an integrating sphere. The visual system has been used as a null detector to calibrate the threshold-raising effect of a uniform field of known luminance against that of distant outer zones of bright glare sources. This has yielded useful contributions to our knowledge of the quality of the retinal image (Fry and Alpern 1953, Vos 1962).

1.5 OPHTHALMIC INSTRUMENTATION

1.5.1 OPHTHALMOSCOPY

The introduction of the ophthalmoscope by Helmholtz led to an unsurpassed blossoming in the diagnosis of eye diseases and, when the optical industry was ready at the turn of the twentieth century, to the development of high-performing instruments. Successively versions were self-luminous, reflex-free, and stereoscopic. The Thorner design made by Busch was pitted against the Gullstrand version made by Zeiss.

1.5.2 OPTOMETERS AND AUTOMATIC OBJECTIVE REFRACTOMETERS

In order to clearly visualize the fundus in ophthalmoscopy, the patient's refractive error needs to be compensated. Schmidt-Rimpler in 1877 used this phenomenon to obtain an objective measure of the refractive error. Since then there have been many versions of what are called *optometers* or *refractometers* or *refractionometers*. E.F. Fincham's design of a *Coincidence Optometer* is perhaps the highlight of this trend (Fincham 1937b) early in the century. They depended on an observer detecting either the sharpness of an image or, as in Fincham's instrument, the alignment of two lines each carried by a separate beam through a different region of the eye's pupil.

The automatic recording infrared optometer of Campbell and Robson (1959) put an end to an era when records of the eye's accommodative changes were secured by cinematography of the Purkinje image from the anterior surface of the lens (Allen 1949) and the emphasis shifted to using light reflected from the fundus.

With the advent of modern optical and electronic components, automated objective refractometers became compact and user-friendly and by the turn of the twenty-first century had established themselves firmly in the eye clinic. Documentation of this development can be found elsewhere.

1.6 SPURT AT THE END OF THE TWENTIETH CENTURY

The narrative so far, covering developments in physiological optics narrowly defined to include the optical properties of human eye in the major portion of twentieth century, was informed by the author's personal experience: undergraduate training based on the state of knowledge prior to the outbreak of the Second World War, his active involvement in the discipline over the rest of the century, and his personal acquaintance with all the major participants in the story.

Much of the groundwork was laid in the British Isles by Smith and Porterfield in the eighteenth century and by Thomas Young,

J.C. Maxwell, Lord Rayleigh, Airy, and others in the nineteenth century and preserved in ophthalmological (Duke-Elder 1932) and optometric texts (Laurence 1926, Emsley 1939). Specifically deserving of mention as upholding and furthering the tradition in the twentieth century are E.F. Fincham (1893–1963) and Arthur G. Bennett (1912–1994).

On the continent, where Kepler, Descartes, and Scheiner had earlier clarified the image-forming properties of the eye, the work of the giants, Gauss and Listing (1808–1882), Purkinje, and Helmholtz, soon percolated down to the clinic and then to productive collaboration with optical industry. Von Graefe (1828–1870), Donders, Landolt, Snellen, and Gullstrand held on and maintained the tradition, and so did, at least in the realm of scholarship, Moritz von Rohr (1868–1940) and Armin von Tschermak (1870–1952) in the next generation. Emblematic of what followed is Max Born (1882–1970) and his magisterial textbook *Optik*. Removed in 1933 from their native habitat, they reemerged in another, more welcoming language and environment and with immensely augmented success and influence. *Principles of Optics* by Born and Wolf is now in its seventh edition.

Important laboratories in the middle of the twentieth century were located in the Netherlands, sparked by Maarten Bouman (1919–2011), and at the Istituto Nazionale di Ottica in Arcetri-Florence under the auspices of the Ronchi family. In Paris at the Institut d'Optique, where Marechal and others (Fleury et al. 1949) dug deeply into the fundamentals of image formation, diffraction, Fourier filtering, and apodization (Dossier 1954), Albert Arnulf (1898–1984) led a group of investigators who in the 1950s were unmatched in the point of attack and skill in physiological optics experiments. Fergus Campbell once told me that whenever he started a research project, he found that Arnulf had been there before. Cambridge, England, had been the site of Thomas Young's major discovery. In the middle of the twentieth century, it saw an extraordinary blossoming of vision research and, as the host of innumerable students and visitors, predominantly from the United States, had a lasting international impact. Of the several centers in the United States, mention should be made of the Dartmouth Eye Institute (Burian 1948) where collaboration with a research arm of the American Optical Company resulted in the design of ophthalmic diagnostic and corrective devices. The work of scientists Paul Boeder and Kenneth N. Ogle (1902–1968), later at the Mayo Clinic, helped give it an optical basis. In the same period, vision research in general, and often physiological-optical in substance, was prominent at Columbia University, where Selig Hecht (1882–1947) in Biophysics and C.H. Graham (1906–1971) in Psychology operated well-supported laboratories and their many students spread a research culture characterized by up-to-date methodology and experimental rigor. The same applied to Lorrin Riggs (1912–2008) at the Psychology Department of Brown University and Glenn A. Fry (1908–1996) at the Ohio State University School of Optometry.

The dramatic transformation that took place in the last third of the twentieth century had its origin less in any theoretical or conceptual changes than in the prodigious advances in the materials from which optical and electronic components are constructed: optical fibers, crystals, transistors, integrated circuits, CRTs, LEDs, LCDs, and the list goes on. Right from its inception a couple of centuries ago, the study of the eye's image-forming properties has always been a prologue and necessary preliminary to vision as a perceptual process, it is what would now be called the front end. Adherent of idealist or materialist philosophy alike, the student of physiological optics was interested in the rules imposed by the laws of physics on what reaches the brain. Expressing electromagnetic disturbance distributions in Fourier—rather than position—space was not a revolutionary step, but generating such disturbance with lasers, orders of magnitude higher in intensity and coherence than other light sources, was. Quantitative change in sufficient measure, so F. Engels interpreted G.F. Hegel, becomes a qualitative one. When the intensity and coherence of light sources, the sensitivity of light detection, its temporal and spatial resolution, the storage capacity for the resultant signals, and the speed and power of analysis increase by a factor of 10^3, in some cases even 10^6, the character of the whole enterprise changes. This has been the case within a single generation for the entire armamentarium used by the vision scientists in his—now, of course, his or her—laboratory. Unimagined ease of generating visual stimuli with devices controlled by fast and powerful computers with virtually unlimited memory has relegated to the historical dust heap the metal and wood—more recently even the electronic—shops of just 50 years ago. Flexible and versatile optical components made of novel materials are allowing experimental forays that Selig Hecht, let alone Helmholtz or Lord Rayleigh, could not have dreamt of. The evidence of this development is provided in the following chapters, detailing stunning advances, yet built on the strong knowledge base erected over the course of many previous generations and solidified by the twentieth-century scientific work sketched here.

REFERENCES

Abbe, E. Beiträge zur Theorie des Mikroskops und der mikroskopischen Wahrnehmung. *Arch Mikroskop Anatomie* 9 (1873): 413–468.

Abderhalden, E. (Ed.). *Handbuch der biologischen Arbeitsmethoden*. Berlin, Germany: Urban & Schwarzenberg, 1920.

Allen, M.J. An objective high speed photographic technique for simultaneously recording changes in accommodation and convergence. *Am J Optom Arch Am Acad Optom* 26 (1949): 279–289.

Ames, A. and C.A. Proctor. Dioptrics of the eye. *J Opt Soc Am* 5 (1921): 22–84.

Bedford, R.E. and G. Wyszecki. Axial chromatic aberration of the human eye. *J Opt Soc Am* 47 (1957): 565.

Brindley, G.S. *Physiology of the Retina and Visual Pathway*, 1st edn. London, U.K.: Edward Arnold, 1960.

Brindley, G.S. and W.A.H. Rushton. The colour of monochromatic light when passed into the human retina from behind. *J Physiol* 147 (1959): 204–208.

Bone, R.A. and J.T. Landrum. Macular pigment in Henle fiber membranes: A model for Haidinger's brushes. *Vision Res* 24 (1984): 103–108.

Born, M. *Optik*. Berlin, Germany: J. Springer, 1933.

Boynton, R.M., J.M. Enoch, and W.R. Bush. Physical measures of stray light in excised eyes. *J Opt Soc Am* 44 (1954): 879–886.

Burian, H.M. The history of the Darmouth Eye Institute. *AMA Arch Ophthalmol* 40 (1948): 163–175.

Campbell, F.W. and D.G. Green. Optical and retinal factors affecting visual resolution. *J Physiol* 181 (1965): 576–593.

Campbell, F.W. and R.W. Gubisch. Optical quality of the human eye. *J Physiol* 186 (1966): 558–578.

Campbell, F.W. and J.G. Robson. High-speed infrared optometer. *J Opt Soc Am* 49 (1959): 268–272.

Campbell, M.C. Measurement of refractive index in an intact crystalline lens. *Vision Res* 24(1984): 409–415.

DeMott, D.W. Direct measures of the retinal image. *J Opt Soc Am* 49 (1959): 571–579.

Dossier, B. Recherches sur l'Apodisation des Images Optiques. *Revue d'Optique* 33 (1954): 257–267.

Duffieux, P.M. *L'integrale de Fourier et ses applications a l'optique*. Imprimeries Oberthur, Rennes, France, 1946.

Duke-Elder, W.S. *Text-Book of Ophthalmology*, Vol. I. London, U.K.: Henry Kimpton, 1932.

Emsley, H.H. *Visual Optics*, 2nd edn. London, U.K.: Hatton Press, 1939.

Enoch, J.M. and F.L. Tobey. *Vertebrate Photoreceptor Optics*. Berlin, Germany: Springer-Verlag, 1981.

Exner, S. *Die Physiologie der Facettirten Augen*. Leipzig, Germany: Deuticke, 1891.

Fincham, E.F. The mechanism of accommodation. *Br J Ophthalmol* (1937a), Monograph Supplement VIII.

Fincham, E.F. The coincidence optometer. *Proc Phys Soc* 49 (1937b): 456–468.

Flamant, F. Étude de la répartition de lumière dans l'image rétinienne d'une fente. *Revue d'Optique* 34 (1955): 433–459.

Fleury, P., A. Maréchal, and C. Anglade. (Eds.). *La Théorie des Images Optique, Colloques Internationaux du CNRS*. Paris, France: Éditions de la Revue d'Optique, 1949.

Fowler, C.W. and T.N. Dave. Review of past and present techniques of measuring corneal topography. *Ophthal Physiol Opt* 14 (1994): 49–58.

Fry, G.A. The image-forming mechanism of the eye. In *Handbook of Physiology*, Vol. 1, J. Field (Ed.). Washington, DC: American Physiological Society, 1959, pp. 647–670, XXVII.

Fry, G.A. and M. Alpern. The effect of a peripheral glare source upon the apparent brightness of an object. *J Opt Soc Am A* 43 (1953): 189–195.

Fry, G.A. and P.W. Cobb. A new method for determining the blurredness of the retinal image. *Trans Am Acad Ophthalmol Otolaryngol* (1935): 432–448.

Goldmann, H. and R. Hagen. Zur direkten Messung der Totalbrechkraft des lebenden menschlichen Auges. *Ophthalmol Basel* 104 (1942): 15–22.

Goodman, J.W. *Introduction to Fourier Optics*. San Francisco, CA: McGraw-Hill, 1968.

Gubisch, R. Optical performance of the human eye. *J Opt Soc Am A* 57 (1967): 407–415.

Gullstrand, A. Appendix to Part I. In *Handbuch der Physiologischen Optik*, H.V. Helmholtz (Ed.). Hamburg, Germany: Voss, 1909.

Hartridge, H. Chromatic aberration and the resolving power of the eye. *J Physiol* 52 (1918): 175–246.

Hartridge, H. Visual acuity and the resolving power of the eye. *J Physiol* 57 (1922): 52–67.

Hecht, S., S. Shlaer, and M.H. Pirenne. Energy, quanta and vision. *J Gen Physiol* 25 (1942): 819–840.

Helmholtz, H.V. *Handbuch der Physiologischen Optik*, 3rd edn. Hamburg, Germany: Voss, 1909.

Hoffer, K.J. Accuracy of ultrasound intraocular lens calculation. *AMA Arch Ophthalmol* 99 (1981): 1819–1823.

Hofmann, F.B. Die Lehre vom Raumsinn. In *Handbuch der gesamten Augenheilkunde*, Vol. 1, A. Graefe and T. Saemisch (Eds.). Berlin, Germany: Julius Springer, 1920, pp. 1–213.

Holladay, L.L. Action of a light source in the field of view in lowering visibility. *J Opt Soc Am* 14 (1927): 1–15.

Hopkins, H.H. The frequency response of a defocused optical system. *Proc R Soc (Lond) A* 231 (1955): 91–103.

Ingelstam, E. Possible interpretation of the ultrastructure of retinal receptors including an electromagnetic theory of the primary mechanism of colour vision. In *Problems in Contemporary Optics. Proceedings of the Florence Meeting,* September 10–15, 1954. Firenze, Italy: Istituto Nazionale di Ottica, 1956, pp. 640–667.

Ivanoff, A. *Les Aberrations de L'Oeil.* Paris, France: Revue d'Optique, 1953.

Koomen, M., R. Scolnik, and R. Tousey. A study of night myopia. *J Opt Soc Am* 41 (1951): 80–83.

Land, M.F. and D.E. Nilsson. *Animal Eyes.* New York: Oxford University Press, 2002.

Laurence, L. *Visual Optics and Sight Testing*, 3rd edn. London, U.K.: School of Optics, 1926.

Le Grand, Y. Sur la mesure de l'acuité visuelle au moyen de franges d'interférence. *Compte rendus de l'Académie des Sciences* 200 (1935): 490–491.

Le Grand, Y. *Optique Physiologique.* Paris, France: Editions de la revue d'optique, 1949.

Ludvigh, E. and E.F. McCarthy. Absorption of visible light by the refractive media of the human eye. *AMA Arch Ophthalmol* 20 (1938): 37–51.

Maurice, D.M. The structure and transparency of the cornea. *J Physiol* 136 (1957): 263–286.

Maxwell, J.C. *The Scientific Papers of James Clerk Maxwell.* New York: Dover, 1890/1965, Vol. II.

Misson, G.P. The form and behaviour of Haidinger's brushes. *Ophthal Physiol Opt* 13 (1993): 392–396.

Molinari, J.F. A review of pachometry. *Am J Optom Physiol Opt* 59 (1982): 912–917.

Naylor, E.J. and A. Stanworth. The measurement of the Haidinger effect. *J Physiol* 123 (1954): 30-1P.

Otero, J.M. and M. Aguilar. Accommodation and night myopia. *J Opt Soc Am* 41 (1951): 1061–1962.

Pierscionek, B.K. and D.Y. Chan. Refractive index gradient of human lenses. *Optom Vis Sci* 66 (1989): 822–829.

Piotrowski, L.N. and F.W. Campbell. A demonstration of the visual importance and flexibility of spatial-frequency amplitude and phase. *Perception* 11 (1982): 337–346.

Polyak, S.L. *The Retina.* Chicago, IL: University of Chicago Press, 1941.

Purkinje, J. *Beobachtungen und Versuche zur Physiologie der Sinne.* Prague, Czech Republic: Calve, 1823.

Rayleigh, L. On the theory of optical images, with special reference to the microscope. *Philos Mag* 42 (1896): 167–195.

Rushton, W.A.H. Visual pigments in man and animals and their relation to seeing. *Prog Biophys Biophys Chem* 9 (1959): 240–283.

Schade, O.H. Optical and photoelectric analog of the eye. *J Opt Soc Am* 46 (1956): 721–739.

Shannon, C.E. and W. Weaver. *The Mathematical Theory of Communication.* Urbana, IL: U. Illinois Press, 1949.

Shlaer, S. The relation between visual acuity and illumination. *J Gen Physiol* 21 (1937): 167–188.

Snodderly, D.M. Evidence for protection against age-related macular degeneration by carotenoids and antioxidant vitamins. *Am J Clin Nutr* 62 (1995): 1448S–1461S.

Snyder, A.W. and R. Menzel. (Eds.). *Photoreceptor Optics.* New York: Springer-Verlag, 1975.

Southall, J.P.C. (Ed.). *Helmholtz's Treatise on Physiological Optics.* New York: Dover, 1924/1962.

Stanworth, A. and E.J. Naylor. The polarization optics of the isolated cornea. *Br J Ophthalmol* 34 (1950): 201–211.

Steel, W.H. The defocused image of sinusoidal gratings. *Opt Acta* 3 (1956): 49–55.

Stiles, W.S. The directional sensitivity of the retina and the spectral sensitivities of the rods and cones. *Proc R Soc Lond B Biol Sci* 127 (1939): 64–105.

Stiles, W.S. and B.H. Crawford. The luminous efficiency of rays entering the eye pupil at different points. *Proc R Soc (Lond) B* 112 (1933): 428–450.

Stiles, W.S. and B.H. Crawford. The effect of light glaring source on extrafoveal vision. *Proc R Soc (Lond) B* 122 (1937): 255–280.

Strehl, K. Aplanatische und fehlerhaft Abbildung im Fernrohr. *Zeitschrift für Instrumentenkunde* 15 (1895): 362–370.

Toraldo di Francia, G. Retina cones as dielectric antennas. *J Opt Soc Am* 39 (1949): 324.

Trimmer, J.D. *The Response of Physical Systems.* New York: John Wiley, 1950.

Tschermak, A. Optischer Raumsinn. In *Handbuch der normalen und pathologischen Physiologie*, A. Bethe et al. (Eds.). Berlin, Germany: Julius Springer, 1931.

van Cittert, P.H. Die Wahrscheinliche Schwingungsverteilung in einer von einer Lichtquelle direkt oder mittels einer Linse beleuchteten Ebene. *Physica* 1(1934): 201–210.

van Nes, F.L. and M.A. Bouman. Spatial modulation transfer in the human eye. *J Opt Soc Am* 57 (1967): 401–406.

Vos, J.J. Some aspects of color stereoscopy. *J Opt Soc Am* 50 (1960): 785–790.

Vos, J.J. *On Mechanisms of Glare.* Utrecht, the Netherlands: University of Utrecht, 1962.

Wald, G. The receptors of human color vision. *Science* 145 (1964): 1007–1016.

Wald, G. and D.R. Griffin. The change in refractive power of the human eye in dim and bright light. *J Opt Soc Am* 37 (1947): 321–336.

Walls, G.L. The vertebrate eye and its adaptive radiation. Detroit, MI: Cranbrook Institute, 1942.

Walls, G.L. and R.W. Mathews. New means of studying color blindness and normal foveal color vision. *Univ Calif Publ Psychol* 7 (1952): 1–172.

Westheimer, G. Spherical aberration of the eye. *Opt Acta* 2 (1955): 151.

Westheimer, G. Modulation thresholds for sinusoidal light distributions on the retina. *J Physiol Lond* 152 (1960): 67–74.

Westheimer, G. and F.W. Campbell. Light distribution in the image formed by the living human eye. *J Opt Soc Am* 52 (1962): 1040–1045.

Westheimer, G. The Maxwellian view. *Vision Res* 6 (1966): 669–682.

Westheimer, G. Optical properties of vertebrate eyes. In *Handbook of Sensory Physiology*, Vol. VII/2, M.G.F. Fuortes (Eds.). Berlin, Germany: Springer-Verlag, 1972, pp. 449–482.

Wiener, N. *Cybernetics: Or, Control and Communication in the Animal and the Machine.* Cambridge, MA: Technology Press, 1948.

Zernike, F. Beugungstheorie des Schneidenverfahrens und Seiner Verbesserten Form, der Phasenkontrastmethode. *Physica* 1 (1934): 689–704.

2 Possibilities in physiological optics

David R. Williams and Sarah Walters

Contents

2.1 INTRODUCTION

This chapter speculates about future directions of physiological optics, identifying a few of the grand challenges that we think might offer the richest rewards, though they may also rank among the most difficult to achieve. The cliché that prediction is hard, especially when it is about the future, bears repeating here. Nonetheless, we have decided to plunge ahead, driven by the conviction that short-term planning for the next experiment always benefits from a longer-term vision for the larger scientific and technological goals our field could eventually realize. Optimized investment decisions in science and engineering require a risk–benefit analysis, whether undertaken by those who provide resources or those who use them. Deciding what criteria to deploy to define benefit is by itself controversial, ranging from the value of advancing basic science to improving eye care to realizing commercial success. We have tended to favor challenges here where fundamental advances in vision science are most likely to result in eye care improvements. Risk is equally difficult to calibrate. At the time of this writing, the pendulum has swung in the direction of risk aversion, at least in many Western countries, with increasing resource competition in science and engineering demanding increasingly compelling evidence that each new endeavor will succeed. One of the motivations for writing this chapter is to remind ourselves that there are many potentially transformative benefits from investing our energy and resources in physiological optics, if only we are willing to accept the risks required to secure them. While the list of challenges we have selected is idiosyncratic, we hope that it will be useful, especially to young scientists who are pondering where they might have

the biggest impact. We feel that the chapter will have succeeded if a single independent thinker concludes that our crystal ball is murky and is inspired to innovate in an entirely different and productive direction.

2.2 DISRUPTIVE TECHNOLOGIES FOR REFRACTING THE EYE

Over the last 400 years, the evolution of technology to measure and correct the eye's optical defects has sprung as much from experts in other fields, especially optics and astronomy, as it has from within. Examples abound. In the early seventeenth century, Johannes Kepler (1604), the astronomer best known for describing the laws of planetary motion, arguably provided the first clear articulation of the fact that the retinal image is inverted. Shortly afterward, Christoph Scheiner (1619), an astronomer well known for his bitter dispute with Galileo about the discovery of sunspots, was the first to measure the eye's plane of focus. He used a simple subjective method, now referred to as Scheiner's disc, that consisted of two laterally displaced pinholes in an opaque plate held close to the pupil. Light passing through the holes formed a double image on the retina for objects at distances nearer or further than the focal plane. Only for objects lying at the focal plane did these double images merge into one, uniquely identifying the focal plane. Scheiner's disc was arguably the first wavefront sensor ever deployed in the human eye, the fact that it could characterize only the single, lowest-order Zernike mode notwithstanding. Examples from the nineteenth century of the dominant role played by astronomers in vision correction include Sir John Herschel's concept for the first contact lens to correct the optical defects of the cornea (Herschel 1845). Thomas Young (1801) first described the astigmatism of the eye, though he was much better known for his contributions to the wave theory of light, among other things. The migration of technology from optics and astronomy to vision continues today, the best example being the application of the Hartmann–Shack wavefront sensor to measure the eye's wave aberration, which was borrowed from the optical metrology of astronomical telescopes (Liang et al. 1994). There is no reason we can think of to suppose that this historical trend will diminish, and it seems likely that advances in optics, astronomy, and increasingly microscopy will continue to fuel future generations of ophthalmic technology.

The explosive development of computing power over the last 50 years has already transformed ophthalmic instrumentation, and we speculate that its impact is destined to accelerate. A telling example comes from Smirnov's landmark paper on the aberrations of the human eye in 1961, published before the potential of computation was clear (Smirnov 1961). Smirnov described a subjective method, similar to but much more sophisticated than Scheiner's, to characterize not only the eye's defocus but also astigmatism and higher-order aberrations. Smirnov did not believe that his approach would ever find practical application, stating the following:

> The method applied in the present work of determining the wave aberration is quite laborious; although the measurements can be taken in 1–2 hours, the calculations take 10–12 hours… Therefore, it is unlikely that such detailed measurements will ever be adopted by practitioner-ophthalmologists.

Smirnov could not have foreseen the digital revolution that ultimately made possible automated computation of the wave aberration in a small fraction of a second (Hofer et al. 2001). The marriage of optical metrology with modern computational methods presaged the widespread use of Hartmann–Shack and related wavefront sensing methods in the eye. These methods provided a much more complete description of the eye's monochromatic aberrations than was possible before. Moreover, the wave aberration specifies how light passing through each and every point in the pupil must be advanced or delayed to achieve perfect imaging, thereby providing a map in the pupil plane that indicates where modifications are required to improve vision correction technology. The introduction of accurate wavefront sensing methods raised the possibility of achieving supernormal vision through the correction of higher-order aberrations. Enthusiasm for this idea has waned, at least as applied to the normal population who typically have relatively minor loss in image quality due to aberrations beyond the defocus and astigmatism. Moreover, technologies for vision correction that are commercially available are sufficiently imprecise that higher-order aberration correction is difficult to achieve in any case. Nonetheless, for some patients, such as those with keratoconus or penetrating keratoplasty, the correction of higher-order aberrations remains a very exciting challenge. Yoon et al. have demonstrated the ability to do this with contact lenses, but the technology is not yet available for the typical patient (Sabesan et al. 2007, 2013).

Despite the plethora of devices now available to autorefract the eye, the wavefront sensor being just one especially sophisticated example, none have replaced the subjective phoropter-based refraction, the ubiquitous procedure in the optometry clinic. A purely automatic refraction could greatly accelerate patient flow through the clinic and likely achieve a more reliable refraction than the slow traditional method that requires the patient's response. Evidence for this comes from the fact that a wavefront sensor combined with a deformable mirror in an adaptive optics retinal imaging system can track the optical defects of the eye automatically in real time with sufficient accuracy to resolve the smallest rods and cones in the living human eye. The tolerances required for this level of resolution are considerably tighter than those required to refract the eye to within a quarter or even an eighth of a diopter, essentially because we can now build cameras with considerably higher resolution than a patient's visual system. Indeed, wavefront aberrometry has already demonstrated reliable refractive measurement and has been used to greatly enhance understanding of clinically challenging conditions, ultimately leading to expanded treatment options (Applegate et al. 2014). A full adaptive optics phoropter (Jaeken et al. 2014), though expensive, could offer even more functionality, its closed-loop control could provide a more accurate correction than an open-loop wavefront sensor, and it would offer the opportunity for the patient to view various corrections prior to delivery of spectacles, contact lenses, or some other modality for vision correction.

The most common argument against abandoning the subjective refraction is that, unlike the objective autorefraction, it engages the patient's nervous system in the final judgment of optimum image quality. There is no question that the optical correction that optimizes subjective image quality is not necessarily the same as that which delivers the best objective optical quality (Chen et al. 2005), and that the neural visual system is adapted to the particular pattern of aberrations with which it has prior experience (Artal et al. 2004). This is the basis for the standard, time-honored practice of undercorrecting astigmatism. The patient presenting with a large amount of astigmatism is likely to reject, at least initially, a complete correction of astigmatism because the brain's understanding of what the world should look like has been learned with that aberration present. On the other hand, the mystery surrounding the brain's contribution to the best refraction is rapidly falling away, to the point where its contribution could probably be faithfully incorporated in an automatic refraction. There is mounting evidence that the brain is plastic in the face of changes in the wave aberration (Mon-Williams et al. 1998; Sabesan and Yoon 2010), and some reason to suspect that providing the best optical correction will, provided adequate time is allowed for adaptation, eventually provide the best visual performance. It seems likely, and would also be open to empirical validation, that the optimum subjective refraction lies on a line in aberration space between the native wave aberration and the best correction in purely optical terms. In that case, a single scalar multiplier could be selected by the clinician to set the appropriate balance between the two. Admittedly, there are additional, second-order factors that can influence the optimal refraction as well. For example, the eye's Stiles–Crawford effect, which apodizes the generalized pupil function, has a minor influence on refraction (Atchison et al. 1998), but this could also be incorporated into an automated computation of the optimal refraction.

It seems possible that the refraction could be achieved objectively with a very simple camera, perhaps even a smartphone camera, equipped with the appropriate computational algorithm. Methods to refract the eye with a cell phone already exist, using the subjective Hartmann–Shack approach (Pamplona et al. 2010). Others have obtained an objective refraction using the light reflected from the retina in a manner consistent with eccentric photorefraction. However, the accuracy is poor to date, and it is typically used only as a screening tool (Arnold and Armitage 2014). In principle, it should be possible to compute the lower-order wave aberration (sphere, cylinder, and axis) simply from the evolving intensity distribution of light returning from the retina, without the need for a Hartmann–Shack wavefront sensor.

The inertia of the optometry profession, which long ago established adequate albeit tedious methods to refract the eye, not to mention the risk that optometrists might be disenfranchised by automation, may actually pose a larger hurdle than the technical challenges that remain. Indeed, a sufficiently sophisticated autorefractor could provide a refraction without the need for human intervention at all, at kiosks in grocery stores and possibly even at home. We speculate that the time will come when completely automated refraction will be routine, and perhaps even coupled to a 3-D printer that could fabricate personalized vision correction technology on the spot.

2.3 A TRANSFORMATIONAL TECHNOLOGY FOR VISION CORRECTION

The history of the major technologies for vision correction has marched inexorably toward approaches that are increasingly proximal to the eye itself. The first approach was a crude lens held before the eye, followed by spectacles fixed to the head, then followed by contact lenses resting on the surface of the cornea. Next came refractive surgery in which the cornea itself is modified. Here we will not review the myriad alternative vision correction technologies that are under development today but instead briefly describe an approach that we think, though it is in an early stage of development, shows unusual promise because it could correct vision but leave the patient in as close to the native state as possible. In refractive surgery, the patient's refractive error is corrected in a one-time procedure that does not require the patient to wear an additional optical device. But despite the impressive improvements that have been made in refractive surgery, especially since the early days of radial keratotomy, even the best refractive surgical techniques today carry some risk because the corneal epithelium must be debrided, as in PRK, or a corneal flap cut as in LASIK. In either case, vision correction requires the removal of corneal tissue with an excimer laser. The application of wavefront sensing to refractive surgery made it possible to discover where in the pupil plane refractive surgery had failed to correct the wavefront, providing the critical feedback for continuous improvement in patient outcomes (Krueger et al. 2004). Perhaps the most important contribution of wavefront sensing in refractive surgery has been not the correction of higher-order aberrations in the patient, but correction of iatrogenic aberrations induced by the procedure, especially spherical aberration (Yoon et al. 2005).

It may eventually be possible to correct the eye's optics without the need to debride, cut, or ablate the cornea (Savage et al. 2014).

They have shown that a femtosecond laser can produce local changes in refractive index in living cornea at exposures that are below those that produce scattering changes caused by tissue damage. This approach, known as laser-induced refractive index change (LIRIC), has the potential to be a much less invasive treatment for refractive errors, since there is no flap-cutting step in which focused femtosecond laser pulses are applied above the damage threshold in corneal tissue. Interestingly, the range of exposures that modify the refractive index had never been discovered before, presumably because changes in refractive index unaccompanied by changes in transparency are not typically seen with casual inspection. LIRIC follows conceptually from earlier efforts to adjust refractive state by light modification of the refractive index of an intraocular lens (Sandstedt et al. 2006), but with the added advantage of modifying the native collagen of the anterior optics of the eye.

To date, the refractive index change produced with a single exposure regimen in the living cat eye has remained stable for upward of 12 months (Savage et al. 2014). It remains to be seen how large a refractive correction can be produced with this method relative to the range of refractive errors in the human population, though it is estimated that ±3.00 D is feasible (Krueger et al. 2013). Although initial in situ studies were performed in cat cornea, histological studies have indicated LIRIC could be used for noninvasive refractive correction in human corneal tissue. Should this method be translated to safe use in humans, it could transform how vision is corrected, obviating the need for tissue-removing refractive surgery, contact lenses, or spectacles in many cases.

This approach does not directly address presbyopia, the loss of accommodation with age. A variety of creative approaches have been developed to combat this problem, including multifocal lenses (which could also be realized with LIRIC), variable focus spectacles, and corneal inlays to name a few (Charman 2014a), and it seems certain that innovations in this space will continue. Although these approaches certainly make the inevitable loss of accommodation manageable, the true restoration of accommodative ability would be preferable. So far, however, this has proven resistant to successful implementation (Charman 2014b). Many of the currently available surgical approaches to correcting presbyopia require the lens to be removed, which is typically only necessitated in the case of a cataractous lens. One such example is that of capsular refilling, in which the dysfunctional crystalline lens is removed and the capsular bag is filled with a material that returns deformability (Nishi et al. 2009; Hirnschall and Findl 2012). IOLs that accommodate are a propitious concept, though there is a history of overpromising functionality in this area (Findl and Leydolt 2007). Nevertheless, advances are being made to restore accommodative function to the native lens noninvasively. Myers and Krueger (1998) first proposed using a femtosecond laser to restore some deformability to the lens, and this method is now under development (Lubatschowski et al. 2010; Krueger et al. 2013). Regardless of the approach taken in attempts to restore accommodation, none have achieved a full restoration of accommodative ability to date, and this will likely require further innovation. One could imagine that regeneration of the crystalline lens to a more youthful state may one day be possible through the use of stem cells (Charman 2014b). Efforts to correct presbyopia would certainly benefit from a more complete understanding of the mechanisms leading to loss of

Introduction

accommodation, and in fact, may pave the way to realizing a therapy that prevents presbyopia altogether.

2.4 CAN ELUCIDATING THE FUNDAMENTAL MECHANISM OF EMMETROPIZATION HELP US PREVENT REFRACTIVE ERROR?

If we had the capacity to prevent refractive errors before they occur, it could ultimately obviate the need for more cumbersome and/or invasive corrective solutions such as those discussed earlier. Over the last three decades, a number of elegant experiments using animal models have greatly clarified the role of visual experience on the development of refractive error (Wallman and Winawer 2004; Smith et al. 2014; Chakraborty and Pardue 2015). It has been shown that viewing the world through positive or negative lenses slows or accelerates eye growth respectively in a number of species including the primate, firmly establishing a feedback loop that controls eye growth based on visual experience (Hung et al.1995). We also know from experiments in which the optic nerve is sectioned that a control loop resides in the eye that does not require input from extraretinal stages of the visual system (Wildsoet and Wallman 1995; Wildsoet 2003). This control loop is even more localized since different regions of the retina that experience different refractive states grow at different rates (Wallman and Adams 1987; Wallman and Winawer 2004; Smith et al. 2010). This machinery must contain a sensor that generates an error signal indicating the sign of defocus in the retinal image, and then this signal must somehow modulate local eye growth.

Neither of these components of the control loop has been thoroughly characterized, though tantalizing clues are beginning to appear (Zhong et al. 2004; Ho et al. 2012). Given the remarkable diversity of cell classes in the mammalian retina (Dacey 1999; Masland 2012), it seems likely that there is a neural circuit specialized to sense the sign of defocus in the retinal image. A conceivable way this could be achieved is through measurement of the contrast of the retinal image simultaneously in two axially distinct focal planes. The difference between the signals from these two planes could provide an opponent error signal sufficient to guide eye growth. Whether evolution has actually generated such a circuit is anyone's guess, but it may not be so far-fetched given that the retina has at least two different planes, axially displaced, for absorbing light, albeit for a different purpose. Intrinsically photosensitive ganglion cells (Berson et al. 2002) in the inner retina respond directly to the incoming light, controlling pupil diameter and the body's circadian clock. Ho et al. (2012) reported a differential response in the human multifocal ERG to the sign of defocus, though this method does not reveal the cellular basis for the response. Diedrich and Schaeffel (2009) failed to find a subset of cells sensitive to defocus recording from chick retinal ganglion cells with a multielectrode array, but the signal must exist somewhere in the retina, waiting to be discovered.

This discovery would not only be a major scientific advance, it might also point the way to new methods to prevent refractive error from developing in childhood. It is conceivable that pharmacological intervention, perhaps in the form of eye drops, could someday mitigate the development of refractive error. In fact, the idea of a pharmacological solution to myopia was proposed as early as the nineteenth century (Saw et al. 2002). Topically administered atropine appears to show some success in inhibiting the progression of axial eye growth, but the underlying cause of this effect remains elusive (Tong et al. 2014). The molecular pathways underlying emmetropization have a genetic basis that could provide a route to identifying them (Wojciechowski 2011; Zhang 2015). If and when an effective pharmacological agent is identified, we speculate that it is highly likely that a fundamental understanding of the mechanisms of emmetropization, established with basic scientific research using animal models, will have been of critical importance in guiding the search.

Perhaps the circuit that provides the error signal for eye growth, once discovered, could eventually be controlled through the genetic expression of a channelrhodopsin combined with spectacles with a spectral transmittance designed to increase or decrease the rate of eye growth as required. On the other hand, someone may eventually devise an effective environmental solution. There has been some excitement about the discovery that myopia is reduced by outdoor activity and exposure to sunlight during development (Rose et al. 2008; Wang et al. 2015). But so far, despite the fact that nearly 2.5 billion humans are myopic (Asbell 2016) and despite everything we have learned about the factors that control emmetropization, a robust lifestyle regimen for avoiding refractive error has proven frustratingly difficult to identify. A solution based on prevention that is less cumbersome than the methods currently available for correction could be transformative.

2.5 VIRTUAL AND AUGMENTED REALITY: A RENAISSANCE FOR APPLIED VISUAL PSYCHOPHYSICS

The remarkable developments over the last 50 years in computational power and digital connectivity are now transforming communication so rapidly that their impact seems destined to be profound. Just the ability to copy text more efficiently than could a scribe, enabled by Gutenberg's invention of moveable type in the mid-1400s, is often credited with fomenting the Renaissance and the Scientific Revolution (Man 2002). Our new found ability to move enormous amounts of information around the planet essentially instantly, not to mention the burgeoning capacity with which computers can make intelligent decisions about that information, will surely have an impact at least as significant as moveable type. Whereas the telephone of just a few years ago delivered information only via voice, today's smartphone now makes accessible a substantial fraction of the cumulative wisdom (and folly) of the human species with a remarkably small number of key strokes, delivering information through both auditory and visual sensory modalities. We suggest that we are now at the same point in delivering digital information to the eyes that we were a half a millennium ago in delivering focused images to them. At that time, before the advent of spectacles, there was limited availability of handheld lenses to correct vision. The convenience of methods to fix these lenses in place with respect to the eyes, either by pinching them onto the nose or with the hinged extensions that wrap around the ears we now use, predated the widespread adoption of spectacles (Rubin 1986). The smartphone, for all its technical sophistication, is the modern-day equivalent of the handheld lens, an awkward technology from an ergonomic perspective that requires at least one and often both hands to use. A recent survey by Verizon and KRC Research stated that smartphone users drop their device, on average, upward of once per week (Burrows 2016), and another survey determined that 26% of iPhone owners had damaged the screen (SquareTrade 2014). Freeing the hands by attaching the technology to the head is the next logical step just as it was for spectacles.

The heads up displays developed by the military beginning in the mid-twentieth century were among the first attempts to augment reality with supplementary analog and then digital information. The Google Glass is a recent commercial innovation in this vein, and while its rollout has been controversial (Efrati 2013; Rosenberger 2014), it seems inevitable to us that future designs will become ubiquitous, especially as improvements in voice recognition also help to free up the hands for other tasks. The growing commercial interest in augmented reality technologies parallels that in virtual reality systems, where visual information about the local environment is completely replaced by that of a remote environment or one generated by computer. Although virtual reality systems have enormous commercial potential in the gaming industry, applications for the technology span a wide gamut of industries that is ever growing as commercial systems improve. As one could imagine, virtual reality offers a unique ability to simulate emergency situations for training purposes, whether that be for respondents to fires or natural disasters, or medical personnel treating injuries in the emergency room or on the battlefield (Brownstone 2014). In the medical sphere, a number of virtual reality simulations have been developed for practice of medical procedures (Ruthenbeck and Reynolds 2015). On the complete opposite end of the spectrum, corporations in the design and manufacturing space are making use of virtual reality to assist engineers with product development and testing (Reis 2014). Other applications are as diverse as providing children with practice crossing the street safely (Schwebel et al. 2016) and assisting those who are autistic with immersion in a variety of social situations (Lorenzo et al. 2016).

While the study of human factors of vision has had a long and valuable history of informing the design of everything from television to road signs to aircraft cockpits, the commercial push for augmented and virtual reality is now reinvigorating the practical relevance of the sometimes arcane field of visual psychophysics. The goal of designing both virtual and augmented reality systems is to deliver imagery to the eye as seamlessly as it would appear in native viewing. Failure to achieve this can have, and historically often has had, significant negative consequences for the user. Visual fatigue and even nausea continue to pose formidable problems for the design of such technology. The visual system maintains an exquisite calibration between sources of information about its own eye and head movements, such as the vestibular system, and the consequences they produce for the retinal image. This calibration is critical in normal viewing

for distinguishing motions of the retinal image generated by self-motion and motion in the external world. When head movements, for example, are sensed by a virtual reality system and then used to displace the virtual field of view accordingly, small errors in gain and timing can be very disruptive. The virtual field of view can depart from a natural view in many other ways. For example, motion parallax and binocular disparity cues must be captured faithfully, as well as distance-dependent blur (Held et al. 2010, 2012). Ideally, the virtual reality device would create a light field that is physically indistinguishable from that arriving at the eye from the full 3-D scene. Holograms have this property but can pose problems with speckle. Approaches with roots dating back to the integral photography methods of Lippmann (1908) may be more promising (Okano et al. 1999). In the case of augmented reality technology, the need to merge digital information with a clear view of the world raises a number of issues that visual psychophysicists are especially well equipped to answer. For example, issues of visual performance as a function of retinal eccentricity, visual attention, and visual crowding are all critical to making the best engineering decisions about how to integrate digital information without compromising performance in the real world. Of course, a premium will be placed on designs that are unobtrusive, lightweight, and compact, while also providing high resolution over a large field of view. Conventional rotationally symmetric optics constrain such parameters unduly and the best designs will likely abandon that constraint in favor of free-form optic designs (Bauer and Rolland 2014).

Just as methods to correct vision have historically migrated toward closer proximity to the eye, it is natural to ponder whether contact lenses might someday deliver digital imagery to the retina. If such a contact lens or IOL could replicate the wavefront of the light that emanated from a visual scene, it would be possible to create a retinal image without the need for external projection optics to deliver the image to the retina. The technical challenges of achieving this are indeed formidable and a problem exacerbated by the fact that an eye tracker would be required to enable the observer to scan the scene through natural eye movements.

2.6 HOW GOOD CAN THE OPHTHALMOSCOPE GET?

The last quarter century has witnessed striking improvements in our ability to visualize the interior of the living eye, especially through increases in resolution. Perhaps the most successful of these innovations has been optical coherence tomography (OCT), which can image both the anterior and posterior segment of the eye with an axial resolution as small as 3 microns (Fercher et al. 2003; Drexler and Fujimoto 2015; Huang et al. 2015). Prior to OCT, we had essentially no axial imagery of the eye so that its introduction provided access to an entirely new spatial dimension in which diseases of the eye could be observed. The use of adaptive optics to correct the eye's aberrations, reviewed in Williams 2011 and Roorda and Duncan 2015, provides a transverse resolution in two spatial dimensions of roughly 2 μm, exceeding that of conventional ophthalmoscopy including OCT. Since then, a host of additional imaging modalities have been combined successfully with adaptive optics such as fluorescence imaging, first of the single-photon variety, and subsequently two-photon imaging. Fluorescence imaging offers an exciting opportunity to not only image structure but also track functional activity of retinal neurons. In particular, two-photon imaging allows for monitoring of molecular species that are excited in the ultraviolet, previously inaccessible with one-photon methods due to the transmission window of the ocular media. Recently, the demonstration of split detection and other nonconfocal imaging methodologies have shown considerable promise, enabling visualization of structures and cell classes that have until now been elusive in confocal imaging because of the transparency of the retina (Elsner et al. 2000; Chui et al. 2012; Scoles et al. 2014; Sulai et al. 2014; Guevara-Torres et al. 2015).

Since the fundamental work of Abbe (1873), the diffraction limit has long been presumed to pose a fundamental barrier to the resolution of optical instruments. Overcoming the diffraction limit could allow imaging features at even smaller spatial scales in retinal imaging. The field of microscopy has recently witnessed a remarkable transformation in which not one but several new methods have emerged to overcome the diffraction limit, including structured illumination (Gustafsson 2000), stimulated emission depletion (Hell and Wichmann 1994), stochastic optical reconstruction microscopy (Rust et al. 2006), and photoactivated localization microscopy (Betzig et al. 2006). None of these methods has yet been applied successfully to retinal imaging, though efforts have begun (Shroff et al. 2009, 2010).

Though the first closed-loop adaptive optics systems for imaging the eye were demonstrated nearly 20 years ago (Liang et al. 1997), their widespread use has been slow to develop. To date, their cost, their complexity, and the time required for postprocessing the images they generate have tended to make them better suited for the research laboratory than the clinic. The quality of the images they produce has systematically and significantly improved since their introduction and methods to accelerate image processing have been developed (Yang et al. 2014), but hurdles remain. For example, a natural clinical application of the adaptive optics scanning light ophthalmoscope (AOSLO) would be to count the fraction of individual receptors missing as a quantitative measure of the severity of a retinal degeneration or the efficacy of therapy in rescuing receptors. But, as pointed out by Roorda and Duncan (2015), the quantitative interpretation of such images is complicated by fact that the failure to image cells at any particular location does not guarantee that they are absent. Other factors within either the system itself or the eye can often obscure cells, confounding an accurate count.

Tracking of eye motion to produce a stabilized retinal image shortens image acquisition time and improves quality, especially in patients whose retinal degeneration precludes them from achieving good fixation (Ferguson et al. 2010; Yang et al. 2014; Sheehy et al. 2015; Zhang et al. 2015). Stabilization enables testing the response of single photoreceptors to a visual stimulus (Harmening et al. 2014) and holds great promise for studying retinal disease as well as spatially localized visual psychophysics (Bruce et al. 2015). Regardless, in vivo adaptive optics ophthalmoscopy has given us a new microscopic view of a host of retinal diseases (Carroll et al. 2013) that was only rarely accessible before from the postmortem histology of retinas fortuitously obtained but often compromised by an artifact. In fact, adaptive

optics–based measures, largely focused on the photoreceptors to date, are already being deployed in clinical trials (Roorda and Duncan 2015). No doubt the number of papers in that literature will continue to grow that provide new adaptive optics–based descriptions of the impact of specific retinal diseases, especially with the incorporation of different imaging modalities and functions such as tracking and microstimulation.

2.7 OUTSTANDING ISSUES ABOUT HOW THE RETINA CATCHES PHOTONS

Understanding how the optics of photoreceptors guide photons to the photopigment has been a revered problem among the most dedicated disciples of physiological optics ever since the discovery of the Stiles–Crawford effect (Stiles and Crawford 1933). For cones outside the foveal center, light entering near the pupil center can be absorbed with nearly an order of magnitude more efficiency than light entering through the pupil margin. This subjective phenomenon, along with its objective correlates observed when light exits the pupil after two transits through the receptors (Marcos and Burns 1999), reflects the waveguide properties of receptors. This waveguiding improves the retinal image in at least two ways. First, by funneling light entering the photoreceptor aperture at the lip of the inner segment into the smaller diameter outer segment, it increases the photon yield and reduces photon noise in the retinal image (Packer and Williams 2003). Without this funneling effect, receptors would need to support more photopigment molecules to sustain the same photon capture rate they presently enjoy. Not only would there be metabolic costs of this, but also the more photopigment molecules the receptors need to maintain, the higher the receptor noise due to random thermal isomerization, which also presumably encourages optical concentration of light onto a smaller reservoir of photopigment. Second, waveguiding, along with other strategies such as absorbing uncaught photons via the rich concentration of melanin in the RPE, saves the precious photopigment molecules for the imaging-forming photons that arrive directly from the pupil while effectively rejecting photons scattered by the sclera or the retina that would otherwise reduce image contrast.

Perhaps the most interesting mystery remaining in the realm of photoreceptor optics is how the receptors manage to orient themselves toward the pupil, as several studies have shown they are equipped to do (Bonds and MacLeod 1978; Applegate and Bonds 1981; Smallman et al. 2001). Smallman et al. (2001) showed that a shift in the entry point of light in the pupil caused by cataract surgery was followed over the next several days by a systematic shift in the Stiles–Crawford maximum toward the postsurgical pupil. Neither the error signal that captures the initial misalignment nor the motor that corrects for the misalignment has been discovered. Speculation about both error signal and motor has largely focused on the receptors themselves (e.g., Laties and Burnside 1979). On the other hand, it is unclear how a photoreceptor would directly sense its own misalignment. Variations in the distribution of intensity within the receptor might be a cue, perhaps created by changes in the modes propagating within the waveguide, and the motor might involve the cilium within the receptor. But it seems more likely to us that the job of the cilium is to keep the receptor rigid and that the structure that holds the outer segment tips, the microvilli of the retinal pigment epithelium, may steer the receptor toward the incoming light. In that case, the retinal pigment epithelium may have the capacity to detect the alignment of the receptors via the shadows that the inner and outer segments cast on them, with the microvilli remodeling so as to steer the receptors. Of course, for this speculation to be true, the pigment epithelium would need to have its own intrinsic light sensitivity, and such a mechanism has not been reported. Sun et al. (1997) have identified at least one photopigment in RPE, peropsin, but its function is unknown. Experiments that might put this speculation to rest presumably must manipulate receptor orientation, perhaps with a displaced artificial pupil, while simultaneously observing the phototropic machinery in single receptors, ideally all in the living human eye. We may now be close to meeting this technical challenge since it is now possible to measure the orientation of single receptors in the living eye (Roorda and Williams 2002; Morris et al. 2015a,b; Walker et al. 2015) and increasingly sophisticated imaging tools are becoming available, such as split detection adaptive optics imaging (Scoles et al. 2014) and fluorescence adaptive optics imaging (Gray et al. 2006), which have the ability to visualize individual receptor inner segments and retinal pigment epithelial cells in the primate eye.

One dangling controversy begging to be settled revolves around the claim that Müller cells are also waveguides, giving light entering the peripheral retina access to photoreceptors without scattering by the overlying neural retina and vasculature (Franze et al. 2007; Labin and Ribak 2010; Labin et al. 2014). The foveal excavation obviates the need for such a mechanism in central vision so the controversy lies in peripheral retina where the inner retina is thick. There are a number of observations that cast doubt on the Müller cell waveguide theory even there. While the beacon-like brightness of single rods and cones viewed in reflected light with adaptive optics scanning laser ophthalmoscopy is an unmistakable signature of photoreceptor waveguiding, the Müller cells are remarkably invisible at all depth planes throughout the retina. If they were light guides, one would expect that they would reveal themselves in reflected light as the receptors do. And, as Austin Roorda has pointed out (personal communication), the clear visibility of the shadows of one's own retinal vasculature when the entry point of light in the pupil is rapidly changed is inconsistent with a waveguide role for Müller cells. If they were effective guides, the photoreceptors would have direct access to the light distribution near the retinal–vitreal boundary, and no shadows of blood vessels would be seen. A careful histological study could help resolve this controversy by determining definitively whether Müller cells are oriented toward the pupil center as are the receptors throughout the retina. Laties and Enoch developed specialized histological methods to preserve receptor orientation, showing that primate rods and cones align themselves dutifully toward the pupil no matter where they are located throughout peripheral retina, just as one would expect of a well-aligned array of waveguides (Laties and Alan 1969; Laties and Enoch 1971). While the classic textbook description (e.g., Polyak 1957) suggests that the Müller cells have a radial orientation, implying that they point to the center of the globe rather than the pupil, this is no substitute for a careful histological examination designed to directly address this question. The Müller cell waveguide hypothesis also predicts that the best subjective focus in peripheral vision should occur in a plane coincident with the apertures of the putative waveguides in the inner retina rather than at the photoreceptors, a striking departure from conventional thinking that careful psychophysical observations coupled with adaptive optics imaging of the retina could put to experimental verification.

2.8 CAN OPTICAL TECHNOLOGY HELP US DISENTANGLE THE NEURAL CIRCUITRY OF THE RETINA?

The retina tiles the visual field with about 5 million cones and 120 million rods the signals from which converge on only about 1.2 million ganglion cells that convey the retinal image from the eye to the brain. This number of ganglion cells sounds large until one recalls that a 12 megapixel digital camera in a typical smartphone today has an order of magnitude more pixels. The number seems smaller still when one considers that ganglion cells must cover a visual field with a solid angle on the order of 4 steradians, which is more than 4 times larger than that of a typical cell phone camera. Therefore, the 2-D sampling density in samples/steradian of a typical smartphone camera is about 40 times greater than the average sampling density of ganglion cells in the visual field of the human eye. The visual system can cope with such a small number of ganglion cells in part because the sampling density is so highly nonuniform, with the fovea having a sampling density that is roughly two orders of magnitude higher than peripheral retina. This combined with a saccadic eye movement system for high-speed rotation of the eye allows the brain to quickly position the highly resolving fovea where it is needed.

However, this simple analysis ignores an important feature of the mosaic of ganglion cells, which is that it is not just one array but as many as two-dozen independent arrays of morphologically distinct subclasses of ganglion cells, each completely tiling the retina and necessarily sampling the retinal image more coarsely than the aggregate array. Every time the function of a new class of cell is characterized, we learn it has an important role for the organism. For example, the recent excitement about intrinsically photosensitive ganglion cells has revealed one circuit specifically dedicated to controlling the circadian rhythm of the organism and another for controlling pupil size (Berson et al. 2002; Gamlin et al. 2007). We have known for some time that there is complex opponent circuitry for comparing the outputs of the three classes of cones (Dacey and Packer 2003). These discoveries shift our conception of the retina away from the notion that it is a passive conduit for the brain to access the retinal image, and toward the notion that it may have considerably more autonomy in processing the retinal image that we might have imagined. Nothing excites a retinal physiologist more than the possibility that the tissue he has devoted his life to, which after all has the consistency of a single ply of wet toilet paper, is in fact a sophisticated, intelligent array processor essential for processing the retinal image so exquisitely that even the extraocular ganglia can make sense of it. Many questions remain about the functional role of classes of ganglion cell for vision. For example, it seems likely, though the definitive evidence does not yet exist, that the primate retina has circuitry for processing motion (Dacey 2004) just as do its mammalian relatives, the mouse and the rabbit (Vaney et al. 2012). The taxonomy of retinal circuits for color and luminance processing no doubt demands elaboration. And as discussed in an earlier section, lurking somewhere in the retinal circuitry is the capability to sense the sign of defocus, just to mention a few examples of opportunities for discovery.

So while there remains much to discover about what the retina is capable of computing and the circuitry that is responsible, we could accelerate the rate of discovery with more efficient tools for recording from it. Since the beginning of retinal physiology, the microelectrode has been the technology of choice, but sampling 1.2 million ganglion cells, one at a time, to characterize a population of two-dozen anatomically distinct classes of cells is a daunting task at best. Arrays of microelectrodes have helped greatly to overcome this limitation, but there remains a sampling limitation imposed by the fixed locations of the microelectrodes in the array imbedded in the complex geometry of the neural circuitry from which it records. This poses a rate-limiting factor in our effort to sift through the large numbers of ganglion cells to discern all their functional roles, especially those cells that are rare in the population.

The recent development of technology to record optically from neurons (Lütcke et al. 2010) may ultimately challenge the microelectrode as the best means to record the individual activity of large numbers of neurons (London et al. 1987). This method relies on a virus such as adeno-associated virus (AAV) that has been modified to deliver to the neuron, and it infects the genetic machinery to express fluorescent reporter molecules. For example, a common strategy is to express a calcium indicator such as GCaMP6, the amount of fluorescence of which informs us about the electrical response of the neuron (Chen et al. 2013). In the case of the retina, the virus is delivered via intravitreal or, in some case, intraretinal injection. After retinal neurons begin to express the genetically encoded calcium indicator, visual stimuli can be delivered, and changes in fluorescence can be measured in all the cells within the imaging field of view. This simultaneously provides a high-resolution image of the neurons under study as well as a detailed map of the response of different parts of the circuitry to visual stimulation. By recording potentially from hundreds and perhaps thousands of cells at the same time, the method may accelerate the search for cells with new functional roles. Moreover, the development of high resolution, fluorescence adaptive optics imaging has made it possible to image retinal neurons expressing fluorescence markers in the living eye (Gray et al. 2006; Geng et al. 2012). Yin et al. (2013) have shown that it is possible to optically record from mouse ganglion cells in the living mouse eye, and Yin et al. (2014) have extended this approach to the living monkey eye. In vivo imaging of the neural responses of retinal cells offers the advantage that the retina is not disturbed by removing it from the eye to make the recordings. Moreover, because it is possible to record repeatedly over days, weeks, and months from the same cells, a whole host of experiments, such as studies of retinal development or the efficacy of therapy for a retinal disease, are now possible. While challenges remain in the development of this technology, such

as the inconsistency of expression, sluggish temporal response of calcium indicators, and the stability of the preparation over time, we speculate that adaptive optics coupled with in vivo recording methods will greatly expand the questions about retinal function that can be addressed.

2.9 CAN OPTICAL TECHNOLOGY ACCELERATE THE NEXT GENERATION OF CURES FOR BLINDNESS?

Despite the enormous societal cost of retinal disease as well as the vast investment to date in developing therapies for it, the number of effective therapies is frustratingly small. Nonetheless, there is a growing sense that we are on the verge of a revolution in new treatments for retinal disease, engendered by novel strategies that have emerged from fields as diverse as molecular biology and photonics. These strategies include optoelectronic, gene therapy, optogenetic, and stem cell approaches (Theogarajan 2012; Garg and Federman 2013). In the optoelectronic approach (O'Brien et al. 2012), an electrode array is inserted either onto the surface of the retina or behind it, designed to stimulate surviving retinal neurons electrically. There is either an array of light detectors inside the eye or an external camera that, following appropriate signal conditioning, drives the electrodes that are in contact with the retina. Though progress to date has not yet produced detailed imagery of the world remotely close to what a sighted person enjoys, this approach has been successful at generating visual phosphenes reliably in the blind visual field in some patients. These phosphenes have proven useful for accomplishing simple visual tasks such as motion detection, object recognition, and letter identification (the Lasker/IRRF Initiative for Innovation in Vision Science 2014). The long-term promise of the optoelectronic approach is diminished by the invasive surgery required to implant the electrode array, the challenge of establishing a stable interface between electrodes and retinal neurons, and the need for a separate light sensor or camera.

A commonality of the remaining approaches described here is that they directly modify or restore the existing biological machinery for sight. Gene therapy offers particular promise in restoring vision in cases where the specific genetic mutation causing retinal degeneration is known, such as in Leber congenital amaurosis. In that case, a modified AAV vector delivers a gene to defective photoreceptors that corrects the mutation, ideally restoring normal photoreceptor function. Several clinical trials are underway, addressing a variety of inherited retinal degenerations as well as age-related macular degeneration (Carvalho and Vandenberghe 2015). Results from RPE65 Leber congenital amaurosis trials have shown that vision improvement is possible with no adverse effects from AAV-mediated gene delivery (Boye et al. 2013). One of the challenges of this approach is that there are a remarkably large number of genetic disorders that compromise outer retinal function, at present knowledge numbering 240 (Daiger et al. 2016), and it may well be impractical to develop a therapy individualized for each of these. An alternative gene therapy approach is to deploy the same delivery method but to insert genes that improve the microenvironment in the outer retina, for example by reducing oxidative stress, thereby forestalling the demise of receptors (Cepko 2012).

In the optogenetic approach, a modified AAV vector is injected either into the vitreous or the subretinal space. The virus payload causes each infected cell to produce channelrhodopsin, a light-gated ion channel, throughout the cell membrane. These channels respond directly to incoming light, conferring light sensitivity on neurons that would normally be inherently blind. Vision restoration then involves the direct stimulation of such retinal neurons that have survived retinal degeneration, whether photoreceptors, bipolar cells, or ganglion cells. Encouraging results have shown that mice with optogenetically restored vision can correctly identify the direction of a light source (Doroudchi et al. 2011). One of the key challenges currently is the poor quantum efficiency of channelrhodopsin approaches. The density of light sensitive molecules is far less than in photoreceptors and the optics for concentrating light described earlier in the chapter in normal photoreceptors is not available to channelrhodopsin. So a viable optogenetic approach will likely require a camera and goggles to deliver a bright image to the retina optimized spectrally and spatiotemporally for the specific light-gated channels deployed (Pan et al. 2015).

Finally, stem cell therapy involves the injection of progenitor cells into the subretinal space where outer retinal degeneration is occurring that are designed to replace RPE or photoreceptor cells that have been damaged or lost. Ideally these cells would establish functional connections with neighboring cells in the retina, restoring visual function. Transplantation of stem cell derived retinal tissue into nonhuman primate models of retinal degeneration has provided evidence of differentiation into both rod and cone photoreceptors, as well as some anatomical evidence of neural connectivity (Shirai et al. 2016). Nevertheless, immunological rejection remains a challenge and must be overcome before retinal stem cell transplantation is a routine clinical procedure.

The progress in developing all these exciting new vision restoration methods is fundamentally limited by the inability to track the efficacy of therapeutic intervention in the living eye at a microscopic spatial scale. While a rich battery of visual tests is readily available, none directly measures retinal function at a cellular spatial scale in the living eye. The gold standard for measuring success is inevitably psychophysical of course, and while such measures are eventually required, it is time consuming, especially in the animal models that are needed for rigorous evaluation of any new approach. Moreover, because the psychophysical response depends on the entire pathway including both eye and brain, it provides little insight into the status of the specific retinal circuitry targeted for restoration. Whole animal electrophysiological methods such as the electroretinogram (ERG) and visual evoked potential (VEP) are less informative than they might be for the same reason. Single unit and multielectrode physiology can provide functional assessment at a cellular scale, but the number of cells that can be monitored is often limited and experiments are typically performed in the excised retina at a single time point, precluding longitudinal measurements. Incorporating direct in vivo microscopic observation with an AOSLO of the retinal location where a therapy has been introduced should accelerate the development cycle for any of these vision restoration approaches. Consider, for example, the benefit of AOSLO imaging for the stem cell approach.

Instead of injecting a bolus of cells into the eye and waiting for several months to determine if vision is restored, one might be able to track these cells, suitably labeled, over time, assessing their viability. It may be possible to obtain in vivo structural evidence that such cells have established neural connections. But even more compelling would be the use of the optical recording technique that uses a genetically encoded calcium indicator and adaptive optics, described earlier, that could establish whether functional connectivity has been established. The ability to image repeatedly in the living eye would avoid piecing together the progression of therapeutic intervention from many different animals sacrificed at different time points, which increases variability and is expensive in both time and money.

Note that the optogenetic technologies described earlier would allow us to control neurons as well as record from them, creating a two-way, read–write optical interface with the nervous system. A robust interface between computers and the central nervous system is already providing unprecedented insight into the fundamental principles upon which the brain operates. From a medical perspective, it is also enabling the precise delivery of therapeutics and communication with advanced prosthetic devices, such as artificial limbs. What we learn from a focused effort on an optogenetic retinal interface, where the natural window of the pupil provides direct optical access to the central nervous system, could conceivably benefit these other efforts to link computers to other parts of the brain. The potential of a truly reliable, parallel interface with the central nervous system is tremendous. Do our species have a future in which our brains have hardwired access to the increasingly vast store of information available from the future Internet, obviating the need for virtual and augmented reality goggles entirely? No one knows, but, with the introduction of optogenetics, it seems more likely now than ever before.

REFERENCES

Abbe, E. 1873. Beiträge zur Theorie des Mikroskops und der mikroskopischen Wahrnehmung. [Contributions to the Theory of the Microscope and of Microscopic Perception.] *Archiv für mikroskopische Anatomie* 9 (1):413–418. doi: 10.1007/bf02956173.

Applegate, R, D Atchison, A Bradley, A Bruce, M Collins, J Marsack, S Read, L N Thibos, and G Yoon. 2014. Wavefront refraction and correction. *Optometry and Vision Science* 91 (10):1154–1155. doi: 10.1097/opx.0000000000000373.

Applegate, RA and AB Bonds. 1981. Induced movement of receptor alignment toward a new pupillary aperture. *Investigative Ophthalmology and Visual Science* 21 (6):869.

Arnold, RW and MD Armitage. 2014. Performance of four new photoscreeners on pediatric patients with high risk amblyopia. *Journal of Pediatric Ophthalmology and Strabismus* 51 (1):46–52.

Artal, P, L Chen, EJ Fernández, B Singer, S Manzanera, and DR Williams. 2004. Neural compensation for the eye's optical aberrations. *Journal of Vision* 4 (4):4–4. doi: 10.1167/4.4.4.

Atchison, DA, A Joblin, and G Smith. 1998. Influence of Stiles–Crawford effect apodization on spatial visual performance. *Journal of the Optical Society of America A* 15 (9):2545–2551. doi: 10.1364/JOSAA.15.002545.

Asbell, PA. 2016. Myopia, just a refractive error? *Eye and Contact Lens* 42 (1):1.

Bauer, A and JP Rolland. 2014. Visual space assessment of two all-reflective, freeform, optical see-through head-worn displays. *Optics Express* 22 (11):13155–13163. doi: 10.1364/OE.22.013155.

Berson, DM, FA Dunn, and M Takao. 2002. Phototransduction by retinal ganglion cells that set the circadian clock. *Science* 295 (5557):1070–1073.

Betzig, E, GH Patterson, R Sougrat, O.W Lindwasser, S Olenych, JS Bonifacino, MW Davidson, J Lippincott-Schwartz, and HF Hess. 2006. Imaging intracellular fluorescent proteins at nanometer resolution. *Science* 313 (5793):1642–1645. doi: 10.1126/science.1127344.

Bonds, AB and DI MacLeod. 1978. A displaced stiles-crawford effect associated with an eccentric pupil. *Investigative Ophthalmology and Visual Science* 17 (8):754.

Bonora, S and RJ Zawadzki. 2013. Wavefront sensorless modal deformable mirror correction in adaptive optics: Optical coherence tomography. *Optics Letters* 38 (22):4801–4804.

Boye, SE, SL Boye, AS Lewin, and WW Hauswirth. 2013. A comprehensive review of retinal gene therapy. *Molecuar Therapy* 21 (3):509–519.

Brownstone, S. 2014. Medics prepare for battlefield trauma in oculus rift, without leaving their chair. Accessed April 12, 2016. http://www.fastcoexist.com/3032243/medics-prepare-for-battlefield-trauma-in-oculus-rift-without-leaving-their-chair.

Bruce, KS, WM Harmening, BR Langston, WS Tuten, A Roorda, and LC Sincich. 2015. Normal perceptual sensitivity arising from weakly reflective cone photoreceptors. *Investigative Ophthalmology and Visual Science* 56 (8):4431–4438. doi: 10.1167/iovs.15-16547.

Burrows, P. 2016. New Verizon survey identifies the worst phone owners in America. Accessed April 12, 2016. http://www.verizonwireless.com/news/article/2016/03/new-verizon-survey-identifies-the-worst-phone-owners-in-america.html.

Carroll, J, DB Kay, D Scoles, A Dubra, and M Lombardo. 2013. Adaptive optics retinal imaging—Clinical opportunities and challenges. *Current Eye Research* 38 (7):709–721. doi: 10.3109/02713683.2013.784792.

Carvalho, LS and LH Vandenberghe. 2015. Promising and delivering gene therapies for vision loss. *Vision Research* 111, Part B:124–133. doi: 10.1016/j.visres.2014.07.013.

Cepko, CL. 2012. Emerging gene therapies for retinal degenerations. *The Journal of Neuroscience* 32 (19):6415–6420. doi: 10.1523/jneurosci.0295–12.2012.

Chakraborty, R and MT Pardue. 2015. Molecular and biochemical aspects of the retina on refraction. *Progress in Molecular Biology and Translational Science* 134:249–267. doi: 10.1016/bs.pmbts.2015.06.013.

Charman, WN. 2014a. Developments in the correction of presbyopia I: Spectacle and contact lenses. *Ophthalmic and Physiologic Optics* 34 (1):8–29. doi: 10.1111/opo.12091.

Charman, WN. 2014b. Developments in the correction of presbyopia II: Surgical approaches. *Ophthalmic and Physiologic Optics* 34 (4):397–426. doi: 10.1111/opo.12129.

Chen, L, B Singer, A Guirao, J Porter, and DR Williams. 2005. Image metrics for predicting subjective image quality. *Optometry and Vision Science* 82 (5):358–369. doi: 10.1097/01.OPX.0000162647.80768.7F.

Chen, T-W, TJ Wardill, Y Sun, SR Pulver, SL Renninger, A Baohan, ER Schreiter et al. 2013. Ultrasensitive fluorescent proteins for imaging neuronal activity. *Nature* 499 (7458):295–300. doi: 10.1038/nature12354.

Chui, TYP, DA VanNasdale, and SA Burns. 2012. The use of forward scatter to improve retinal vascular imaging with an adaptive optics scanning laser ophthalmoscope. *Biomedical Optics Express* 3 (10):2537–2549. doi: 10.1364/BOE.3.002537.

Dacey, DM. 1999. Primate retina: Cell types, circuits and color opponency. *Progress in Retinal and Eye Research* 18 (6):737–763.

Dacey, DM. 2004. Origins of perception: Retinal ganglion cell diversity and the creation of parallel visual pathways. In *The Cognitive Neurosciences III*, eds. MS Gazzaniga. Cambridge, MA: The MIT Press.

Dacey, DM and OS Packer. 2003. Colour coding in the primate retina: Diverse cell types and cone-specific circuitry. *Current Opinion in Neurobiology* 13 (4):421–427. doi: 10.1016/S0959-4388(03)00103-X.

Daiger, SP, LS Sullivan, and SJ Bowne. 2016. "RetNet: Retinal Information Network." The University of Texas Health Science Center, Last modified March 9, 2016. Accessed April 12. https://sph.uth.edu/retnet/.

Diedrich, E and F Schaeffel. 2009. Spatial resolution, contrast sensitivity, and sensitivity to defocus of chicken retinal ganglion cells in vitro. *Visual Neuroscience* 26 (5–6):467–476.

Doroudchi, MM, KP Greenberg, J Liu, KA Silka, ES Boyden, JA Lockridge, AC Arman et al. 2011. Virally delivered channelrhodopsin-2 safely and effectively restores visual function in multiple mouse models of blindness. *Molecular Therapy* 19 (7):1220–1229.

Drexler, W and JG Fujimoto. 2015. Retinal optical coherence tomography imaging. In *Optical Coherence Tomography: Technology and Applications*, eds. Wolfgang, D and GJ Fujimoto, pp. 1685–1735. Cham, Switzerland: Springer International Publishing.

Efrati, A. 2013. Google glass privacy worries lawmakers. *Wall Street Journal* (Online), Accessed April 12, 2016. http://www.wsj.com/articles/SB10001424127887324767004578487661143483672.

Elsner, AE, M Miura, SA Burns, E Beausencourt, C Kunze, LM Kelley, JP Walker et al. 2000. Multiply scattered light tomography and confocal imaging: Detecting neovascularization in age-related macular degeneration. *Optics Express* 7 (2):95–106. doi: 10.1364/OE.7.000095.

Fercher, AF, W Drexler, CK Hitzenberger, and T Lasser. 2003. Optical coherence tomography—Principles and applications. *Reports on Progress in Physics* 66 (2):239.

Ferguson, RD, Z Zhong, DX Hammer, M Mujat, AH Patel, C Deng, W Zou, and SA Burns. 2010. Adaptive optics scanning laser ophthalmoscope with integrated wide-field retinal imaging and tracking. *Journal of the Optical Society of America A* 27 (11):A265–A277. doi: 10.1364/JOSAA.27.00A265.

Findl, O and C Leydolt. 2007. Meta-analysis of accommodating intraocular lenses. *Journal of Cataract and Refractive Surgery* 33 (3):522–527. doi: 10.1016/j.jcrs.2006.11.020.

Franze, K, J Grosche, SN Skatchkov, S Schinkinger, C Foja, D Schild, O Uckermann, K Travis, A Reichenbach, and J Guck. 2007. Muller cells are living optical fibers in the vertebrate retina. *Proceedings of the National Academy of Sciences of the United States of America* 104 (20):8287–8292. doi: 10.1073/pnas.0611180104.

Gamlin, PDR, DH McDougal, J Pokorny, VC Smith, K-W Yau, and DM Dacey. 2007. Human and macaque pupil responses driven by melanopsin-containing retinal ganglion cells. *Vision Research* 47 (7):946–954. doi: 10.1016/j.visres.2006.12.015.

Garg, SJ and J Federman. 2013. Optogenetics, visual prosthesis and electrostimulation for retinal dystrophies. *Current Opinion in Ophthalmology* 24 (5):407–414.

Geng, Y, A Dubra, L Yin, WH Merigan, R Sharma, RT Libby, and DR Williams. 2012. Adaptive optics retinal imaging in the living mouse eye. *Biomedical Optics Express* 3 (4):715–734. doi: 10.1364/BOE.3.000715.

Gibson, W. The Science in Science Fiction" on Talk of the Nation. National Public Radio. Accessed November 30, 1999. Available at http://www.npr.org/templates/story/story.php?storyId=1067220.

Gray, DC, W Merigan, JI Wolfing, BP Gee, J Porter, A Dubra, TH Twietmeyer et al. 2006. In vivo fluorescence imaging of primate retinal ganglion cells and retinal pigment epithelial cells. *Optics Express* 14 (16):7144–7158.

Guevara-Torres, A, DR Williams, and JB Schallek. 2015. Imaging translucent cell bodies in the living mouse retina without contrast agents. *Biomedical Optics Express* 6 (6):2106–2119. doi: 10.1364/BOE.6.002106.

Gustafsson, MGL. 2000. Surpassing the lateral resolution limit by a factor of two using structured illumination microscopy. *Journal of Microscopy* 198 (2):82–87. doi: 10.1046/j.1365-2818.2000.00710.x.

Harmening, WM, WS Tuten, A Roorda, and LC Sincich. 2014. Mapping the perceptual grain of the human retina. *The Journal of Neuroscience* 34 (16):5667–5677. doi: 10.1523/jneurosci.5191-13.2014.

Held, RT, EA Cooper, and MS Banks. 2012. Blur and disparity are complementary cues to depth. *Current Biology* 22 (5):426–431. doi: 10.1016/j.cub.2012.01.033.

Held, RT, EA Cooper, JF O'Brien, and MS Banks. 2010. Using blur to affect perceived distance and size. *ACM Transactions on Graphics* 29 (2):19. doi: 10.1145/1731047.1731057.

Hell, SW and J Wichmann. 1994. Breaking the diffraction resolution limit by stimulated emission: Stimulated-emission-depletion fluorescence microscopy. *Optics Letters* 19 (11):780–782. doi: 10.1364/OL.19.000780.

Herschel, JFW. 1845. Light. In *Encyclopaedia Metropolitana*, eds. Edward, S, HJ Rose, and HJ Rose. London, U.K.: B. Fellowes et al.

Hirnschall, N and O Findl. 2012. Lens refilling. In *Presbyopia: Origins, Effects and Treatment*, eds. Pallikaris, I, S Plainis, and WN Charman. Thorofare, NJ: Slack.

Ho, W-C, O-Y Wong, Y-C Chan, S-W Wong, C-S Kee, and HHL Chan. 2012. Sign-dependent changes in retinal electrical activity with positive and negative defocus in the human eye. *Vision Research* 52 (1):47–53.

Hofer, H, P Artal, B Singer, JL Aragón, and DR Williams. 2001. Dynamics of the eye's wave aberration. *Journal of the Optical Society of America A* 18 (3):497–506. doi: 10.1364/JOSAA.18.000497.

Hofer, H, N Sredar, H Queener, CH Li, and J Porter. 2011. Wavefront sensorless adaptive optics ophthalmoscopy in the human eye. *Optics Express* 19 (15):14160–14171.

Huang, D, Y Li, and M Tang. 2015. Anterior eye imaging with optical coherence tomography. In *Optical Coherence Tomography: Technology and Applications*, eds. Wolfgang, D and GJ Fujimoto, pp. 1649–1683. Cham, Switzerland: Springer International Publishing.

Hung, LF, MLJ Crawford, and EL Smith. 1995. Spectacle lenses alter eye growth and the refractive status of young monkeys. *Nature Medicine* 1 (8):761–765.

Jaeken, B, L Hervella, GM Perez, JM Marín, and P Artal. 2014. Clinical validation of adaptive-optics guided refraction. *Investigative Ophthalmology & Visual Science* (Meeting Abstract) 55 (13):2114–2114.

Kepler, J. 1604. *Ad Vitellionem paralipomena, quibus astronomiae pars optica traditur.* [The Optical Part of Astronomy.] Francofurti: apud Claudium Marnium & haeredes Ioannis Aubrii.

Krueger, RR, SM MacRae, and RA Applegate. 2004. The future of customization. In *Wavefront Customized Visual Correction*, eds. Ronald, RK, RA Applegate, and SM MacRae. Thorofare, NJ: Slack.

Krueger, RR, J-M Parel, KR Huxlin, WH Knox, and K Hohla. 2013. The furture of ReLACS and femtosecond ocular surgery. In *Textbook of Refractive Laser Assisted Cataract Surgery*, eds. Ronald, RK, JH Talamo, and RL Lindstrom. New York: Springer.

Labin, AM and EN Ribak. 2010. Retinal glial cells enhance human vision acuity. *Physical Review Letters* 104 (15):158102. doi: 10.1103/PhysRevLett.104.158102.

Labin, AM, SK Safuri, EN Ribak, and I Perlman. 2014. Muller cells separate between wavelengths to improve day vision with minimal effect upon night vision. *Nature Communications* 5:4319. doi: 10.1038/ncomms5319.

Laties, AM and B Burnside. 1979. The maintenance of photoreceptor orientation. In *Motility in Cell Function: Proceedings of the First John M. Marshall Symposium on Cell Biology*, eds. Pepe, F, V Nachmias, and JW Sanger. New York: Academic Press.

Laties, AM and JM Enoch. 1971. An analysis of retinal receptor orientation. I. angular relationship of neighboring photoreceptors. *Investigative Ophthalmology* 10 (1):69.

Laties, MD and M Alan. 1969. Histological techniques for study of photoreceptor orientation. *Tissue and Cell* 1 (1):63–81.

Liang, J, B Grimm, S Goelz, and JF Bille. 1994. Objective measurement of wave aberrations of the human eye with the use of a Hartmann–Shack wave-front sensor. *Journal of the Optical Society of America A* 11 (7):1949–1957. doi: 10.1364/JOSAA.11.001949.

Liang, J, DR Williams, and DT Miller. 1997. Supernormal vision and high-resolution retinal imaging through adaptive optics. *Journal of the Optical Society of America A* 14 (11):2884–2892.

Lippmann, G. 1908. Épreuves réversibles. Photographies intégrales. [Reversible Prints. Integral Photographs.] *Comptes Rendus de l'Académie des Sciences* 146 (9):446–451.

London, JA, D Zecevic, and LB Cohen. 1987. Simultaneous optical recording of activity from many neurons during feeding in navanax. *Journal of Neuroscience* 7 (3):649.

Lorenzo , G, A Lledó, J Pomares, and R Roig. 2016. Design and application of an immersive virtual reality system to enhance emotional skills for children with autism spectrum disorders. *Computers and Education* 98: 192–205.

Lubatschowski, H, S Schumacher, M Fromm, A Wegener, H Hoffmann, U Oberheide, and G Gerten. 2010. Femtosecond lentotomy: Generating gliding planes inside the crystalline lens to regain accommodation ability. *Journal of Biophotonics* 3 (5–6):265–268.

Lütcke, H, M Murayama, T Hahn, DJ Margolis, S Astori, S Meyer, W Göbel et al. 2010. Optical recording of neuronal activity with a genetically-encoded calcium indicator in anesthetized and freely moving mice. *Frontiers in Neural Circuits* 4: 9.

Man, J. 2002. *Gutenberg: How One Man Remade the World with Words.* New York: Wiley.

Marcos, S and SA Burns. 1999. Cone spacing and waveguide properties from cone directionality measurements. *Journal of the Optical Society of America A* 16 (5):995–1004.

Masland, RH. 2012. The neuronal organization of the retina. *Neuron* 76 (2):266–280.

Mon-Williams, M, JR Tresilian, NC Strang, P Kochhar, and JP Wann. 1998. Improving vision: Neural compensation for optical defocus. *Proceedings of the Royal Society of London. Series B: Biological Sciences* 265 (1390):71–77. doi: 10.1098/rspb.1998.0266.

Morris, HJ, L Blanco, JL Codona, SL Li, SS Choi, and N Doble. 2015a. Directionality of individual cone photoreceptors in the parafoveal region. *Vision Research* 117:67–80.

Morris, HJ, JL Codona, L Blanco, and N Doble. 2015b. Rapid measurement of individual cone photoreceptor pointing using focus diversity. *Optics Letters* 40 (17):3982–3985.

Myers, RI and RR Krueger. 1998. Novel approaches to correction of presbyopia with laser modification of the crystalline lens. *Journal of Refractive Surgery* 14 (2):136–139.

Nishi, Y, K Mireskandari, P Khaw, and O Findl. 2009. Lens refilling to restore accommodation. *Journal of Cataract and Refractive Surgery* 35 (2):374–382.

O'Brien, EE, U Greferath, KA Vessey, AI Jobling, and EL Fletcher. 2012. Electronic restoration of vision in those with photoreceptor degenerations. *Clinical and Experimental Optometry* 95 (5):473–483. doi: 10.1111/j.1444-0938.2012.00783.x.

Okano, F, J Arai, H Hoshino, and I Yuyama. 1999. Three-dimensional video system based on integral photography. *Optical Engineering* 38 (6):1072–1077. doi: 10.1117/1.602152.

Packer, O and DR Williams. 2003. Light, the retinal image, and photoreceptors. In *The Science of Color*, ed. Shevell, SK. New York: Elsevier.

Pamplona, VF, A Mohan, MM Oliveira, and R Raskar. 2010. NETRA: Interactive display for estimating refractive errors and focal range. *ACM Transactions on Graphics* 29 (4):1.

Pan, Z-H, Q Lu, A Bi, AM Dizhoor, and GW Abrams. 2015. Optogenetic approaches to restoring vision. *Annual Review of Vision Science* 1 (1):185–210. doi: 10.1146/annurev-vision-082114-035532.

Polyak, SL. 1957. *The Vertebrate Visual System*. Chicago, IL: University of Chicago Press.

Reis, M. 2014. Could virtual reality be the next big thing in education? Accessed April 12, 2016. http://www.forbes.com/sites/ptc/2014/08/27/could-virtual-reality-be-the-next-big-thing-in-education/#36bd98382aa3.

Roorda, A and JL Duncan. 2015. Adaptive optics ophthalmoscopy. *Annual Review of Vision Science* 1 (1):19–50. doi: 10.1146/annurev-vision-082114-035357.

Roorda, A and DR Williams. 2002. Optical fiber properties of individual human cones. *Journal of Vision* 2 (5):404-U1.

Rose, KA, IG Morgan, J Ip, A Kifley, S Huynh, W Smith, and P Mitchell. 2008. Outdoor activity reduces the prevalence of myopia in children. *Ophthalmology* 115 (8):1279–1285.

Rosenberger, R. 2014. Google glass and highway safety—Messy choices. *IEEE Technology and Society Magazine* 33 (2):23–25. doi: 10.1109/MTS.2014.2319931.

Rubin, ML. 1986. Spectacles: Past, present, and future. *Survey of Ophthalmology* 30 (5):321.

Rust, MJ, M Bates, and X Zhuang. 2006. Sub-diffraction-limit imaging by stochastic optical reconstruction microscopy (STORM). *Nature Methods* 3 (10):793–796.

Ruthenbeck, GS and KJ Reynolds. 2015. Virtual reality for medical training: The state-of-the-art. *Journal of Simulation* 9 (1):16–26. doi: 10.1057/jos.2014.14.

Sabesan, R, TM Jeong, L Carvalho, IG Cox, DR Williams, and G Yoon. 2007. Vision improvement by correcting higher-order aberrations with customized soft contact lenses in keratoconic eyes. *Optics Letters* 32 (8):1000–1002. doi: 10.1364/OL.32.001000.

Sabesan, R, L Johns, O Tomashevskaya, DS Jacobs, P Rosenthal, and G Yoon. 2013. Wavefront-guided scleral lens prosthetic device for keratoconus. *Optometry and Vision Science* 90 (4):314–323. doi: 10.1097/OPX.0b013e318288d19c.

Sabesan, R and G Yoon. 2010. Neural compensation for long-term asymmetric optical blur to improve visual performance in keratoconic eyes. *Investigative Ophthalmology & Visual Science* 51 (7):3835–3839. doi: 10.1167/iovs.09-4558.

Sandstedt, CA, SH Chang, RH Grubbs, and DM Schwartz. 2006. Light-adjustable lens: Customizing correction for multifocality and higher-order aberrations. *Transactions of the American Ophthalmological Society* 104:29–39.

Savage, DE, DR Brooks, M DeMagistris, LS Xu, S MacRae, JD Ellis, WH Knox, and KR Huxlin. 2014. First demonstration of ocular refractive change using blue-IRIS in live cats. *Investigative Ophthalmology and Visual Science* 55 (7):4603–4612.

Saw, SM, G Gazzard, K-G Au Eong, and DTH Tan. 2002. Myopia: Attempts to arrest progression. *The British Journal of Ophthalmology* 86 (11):1306–1311.

Scheiner, C. 1619. Oculus. Oeniponti, Innsbruck, Austria: Apud Danielem Agricolam.

Schwebel, DC, T Combs, D Rodriguez, J Severson, and V Sisiopiku. 2016. Community-based pedestrian safety training in virtual reality: A pragmatic trial. *Accident Analysis and Prevention* 86:9–15. doi: 10.1016/j.aap.2015.10.002.

Scoles, D, YN Sulai, CS Langlo, GA Fishman, CA Curcio, J Carroll, and A Dubra. 2014. In *Vivo Imaging of Human Cone Photoreceptor Inner Segments. Investigative Ophthalmology and Visual Science* 55 (7):4244–4251. doi: 10.1167/iovs.14-14542.

Sheehy, CK, P Tiruveedhula, R Sabesan, and A Roorda. 2015. Active eye-tracking for an adaptive optics scanning laser ophthalmoscope. *Biomedical Optics Express* 6 (7):2412–2423. doi: 10.1364/BOE.6.002412.

Shirai, H, M Mandai, K Matsushita, A Kuwahara, S Yonemura, T Nakano, J Assawachananont et al. 2016. Transplantation of human embryonic stem cell-derived retinal tissue in two primate models of retinal degeneration. *Proceedings of the National Academy of Sciences* 113 (1):E81–E90. doi: 10.1073/pnas.1512590113.

Shroff, SA, JR Fienup, and DR Williams. 2009. Phase-shift estimation in sinusoidally illuminated images for lateral superresolution. *Journal of the Optical Society of America A* 26 (2):413–424. doi: 10.1364/JOSAA.26.000413.

Shroff, SA, JR Fienup, and DR Williams. 2010. Lateral superresolution using a posteriori phase shift estimation for a moving object: Experimental results. *Journal of the Optical Society of America A* 27 (8):1770–1782.

Smallman, HS, DIA MacLeod, and P Doyle. 2001. Vision: Realignment of cones after cataract removal. *Nature* 412, 604–605 doi:10.1038/35088126

Smirnov, MS. 1961. Measurement of the wave aberration of the human eye. *Biofizika* 6:687–703.

Smith, EL, LF Hung, and B Arumugam. 2014. Visual regulation of refractive development: Insights from animal studies. *Eye* 28 (2):180–188.

Smith, EL 3rd, L-F Hung, J Huang, TL Blasdel, TL Humbird, and KH Bockhorst. 2010. Effects of optical defocus on refractive development in monkeys: Evidence for local, regionally selective mechanisms. *Investigative Ophthalmology and Visual Science* 51 (8):3864–3873.

SquareTrade. 2014. New Study Shows Damaged iPhones Cost Americans $10.7 Billion, $4.8B in the Last Two Years Alone. Accessed April 12, 2016. http://www.squaretrade.com/press/new-study-shows-damaged-iphones-cost-americans-10.7billion-4.8b-in-the-last-two-years-alone.

Stiles, WS and BH Crawford. 1933. The luminous efficiency of rays entering the eye pupil at different points. *Proceedings of the Royal Society of London. Series B, Containing Papers of a Biological Character* 112 (778):428–450.

Sulai, YN, D Scoles, Z Harvey, and A Dubra. 2014. Visualization of retinal vascular structure and perfusion with a nonconfocal adaptive optics scanning light ophthalmoscope. *Journal of the Optical Society of America A* 31 (3):569–579.

Sun, H, DJ Gilbert, NG Copeland, NA Jenkins, and J Nathans. 1997. Peropsin, a novel visual pigment-like protein located in the apical microvilli of the retinal pigment epithelium. *Proceedings of the National Academy of Sciences* 94 (18):9893–9898.

The Lasker/IRRF Initiative for Innovation in Vision Science. 2014. Chapter 1—Restoring vision to the blind: The new age of implanted visual prostheses. *Translational Vision Science and Technology* 3 (7):3–3. doi: 10.1167/tvst.3.7.3.

Theogarajan, L. 2012. Strategies for restoring vision to the blind: Current and emerging technologies. *Neuroscience Letters* 519 (2):129–133. doi: 10.1016/j.neulet.2012.02.001.

Tong, LMG, VA Barathi, and RW Beuerman. 2014. Atropine and other pharmacological approaches to prevent myopia. In *Myopia*, eds. Beuerman, RW, S-M Saw, and DTH Tan. Princeton, NJ: World Scientific.

Vaney, DI, B Sivyer, and WR Taylor. 2012. Direction selectivity in the retina: Symmetry and asymmetry in structure and function. *Natural Review Neuroscience* 13 (3):194–208.

Walker, MK, L Blanco, R Kivlin, SS Choi, and N Doble. 2015. Measurement of the photoreceptor pointing in the living chick eye. *Vision Research* 109 (Pt A):59–67.

Wallman, J and JI Adams. 1987. Developmental aspects of experimental myopia in chicks: Susceptibility, recovery and relation to emmetropization. *Vision Research* 27 (7):1139–1163.

Wallman, J and J Winawer. 2004. Homeostasis of eye growth and the question of myopia. *Neuron* 43 (4):447–468. doi: 10.1016/j.neuron.2004.08.008.

Wang, Y, H Ding, WK Stell, L Liu, S Li, H Liu, and X Zhong. 2015. Exposure to sunlight reduces the risk of myopia in rhesus monkeys. *PloS One* 10 (6): e0127863.

Wildsoet, C and J Wallman. 1995. Choroidal and scleral mechanisms of compensation for spectacle lenses in chicks. *Vision Research* 35 (9):1175–1194.

Wildsoet, CF. 2003. Neural pathways subserving negative lens-induced emmetropization in chicks—insights from selective lesions of the optic nerve and ciliary nerve. *Current Eye Research* 27 (6):371–385.

Williams, DR. 2011. Imaging single cells in the living retina. *Vision Research* 51 (13):1379–1396. doi: 10.1016/j.visres.2011.05.002.

Wojciechowski, R. 2011. Nature and nurture: The complex genetics of myopia and refractive error. *Clinical Genetics* 79 (4):301–320. doi: 10.1111/j.1399-0004.2010.01592.x.

Yang, Q, J Zhang, K Nozato, K Saito, DR Williams, A Roorda, and EA Rossi. 2014. Closed-loop optical stabilization and digital image registration in adaptive optics scanning light ophthalmoscopy. *Biomedical Optics Express* 5 (9):3174–3191. doi: 10.1364/BOE.5.003174.

Yin, L, Y Geng, F Osakada, R Sharma, AH Cetin, EM Callaway, DR Williams, and WH Merigan. 2013. Imaging light responses of retinal ganglion cells in the living mouse eye. *Journal of Neurophysiology* 109 (9):2415–2421.

Yin, L, B Masella, D Dalkara, J Zhang, JG Flannery, DV Schaffer, DR Williams, and WH Merigan. 2014. Imaging light responses of foveal ganglion cells in the living macaque eye. *The Journal of Neuroscience* 34 (19):6596–6605.

Yoon, G, S MacRae, DR Williams, and IG Cox. 2005. Causes of spherical aberration induced by laser refractive surgery. *Journal of Cataract and Refractive Surgery* 31 (1):127–135. doi: 10.1016/j.jcrs.2004.10.046.

Young, T. 1801. The Bakerian lecture: On the mechanism of the eye. *Philosophical Transactions of the Royal Society of London* 91:23–88. doi: 10.1098/rstl.1801.0004.

Zhang, J, Q Yang, K Saito, K Nozato, DR Williams, and EA Rossi. 2015. An adaptive optics imaging system designed for clinical use. *Biomedical Optics Express* 6 (6):2120–2137. doi: 10.1364/BOE.6.002120.

Zhang, Q. 2015. Genetics of refraction and myopia. *Progress in Molecular Biology and Translational Science* 134:269–279. doi: 10.1016/bs.pmbts.2015.05.007.

Zhong, X, J Ge, EL Smith III, and WK Stell. 2004. Image defocus modulates activity of bipolar and amacrine cells in macaque retina. *Investigative Ophthalmology and Visual Science* 45 (7):2065.

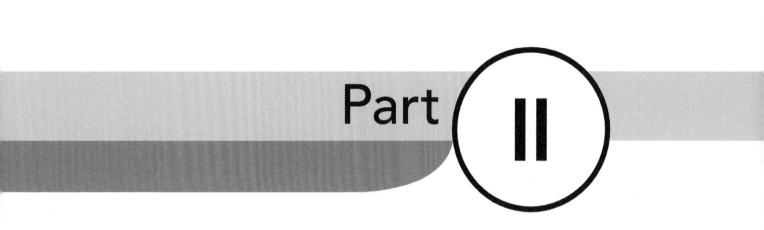

Part II

Fundamentals

3 Geometrical optics

Jim Schwiegerling

Contents

3.1 INTRODUCTION

3.1.1 WHAT IS GEOMETRICAL OPTICS?

Geometrical optics is a branch of optics that typically examines the transfer of light from a source to some destination via an optical system. These systems can be composed of multiple lenses, mirrors, prisms, and windows. Consequently, geometrical optics concerns itself with the refraction and reflection of light at interfaces and its propagation through various media. Light exhibits wave phenomenon such as interference and diffraction (treated in Chapter 5). However, the description of refraction, reflection, and propagation of waves is mathematically intensive. Geometrical optics makes a series of approximations that vastly simplifies the analysis. These approximations are valid for incoherent light. The main approximation of geometrical optics is to replace the wavefront with a series of rays and analyze how the rays move through space and interact with objects. Rays represent the local normals to wavefronts and illustrate the direction of propagation of the wave at a given point. In homogeneous materials, rays will travel in a straight line until they reach a boundary. At the boundary, the rays will refract and reflect and then continue their straight-line propagation in the ensuing material. This simplification enables optical systems to be designed and analyzed in a straightforward manner leading to systems for illumination or imaging having the desired properties of the designer. This chapter examines the basic description of the properties of rays and their interaction with optical elements such as lenses and prisms. In addition, definitions for the various properties of optical systems are provided.

3.1.2 SIGN CONVENTION

Prior to developing the foundations of geometrical optics, a coordinate system and a consistent sign convention need to be defined. The axis for an optical system will, in general, be taken as the z-axis. Light will typically travel from left (more negative values of z) to right (more positive values of z) with regard to this axis, unless the light is reflected from a surface. The vertical axis will be denoted as the positive y-axis. The coordinate system will be considered right-handed, meaning that the positive x-axis will be into the plane of the paper in the ensuing figures. Additional quantities such as distances, radii of curvature, and angles and their respective signs are defined with respect to this coordinate system. Distances are measured from a reference point and are signed to be consistent with the coordinate system. For example, if one optical surface is located at $z = 0$ and a second optical surface is located at $z = +10$, then the distance from the first surface to the second surface is positive. Similarly, distances measured in the $-z$ direction are negative. The same convention for distances holds for the x and y directions. An object with its base on the z-axis and its top at $y = 5$ would have a positive height, while its image may be upside down with its base still on the z-axis, but its "top" now at $y = -3$. In this latter case, the image height is negative. Angles are measured with regard to a reference line such as the z-axis or a local normal to an optical surface. Counterclockwise angles are considered positive and clockwise angles are negative. For example, a ray that starts on the z-axis ($y = 0$) at an object may propagate in the positive z direction to an optical surface and strike that surface at a height of $y = 2$. This ray would have a positive angle with respect to the z-axis. Finally, the radius of curvature of a spherical surface has a positive value if its center of curvature is to the right of the surface. Similarly, surfaces with negative radii have their center of curvatures to the left of the surface. Arrowheads are used in the figures to help illustrate the sign of the quantities depicted. Figure 3.1 summarizes the various sign and coordinate conventions described earlier.

3.1.3 WAVELENGTH, SPEED OF LIGHT, AND REFRACTIVE INDEX

The electromagnetic spectrum represents a continuum of radiation that propagates as transverse waves. The distinguishing feature between various elements of the spectrum is the wavelength, λ, or the distance between the peaks of the propagating waves. The electromagnetic spectrum describes everything from gamma rays with a wavelength comparable to the size of atomic nuclei to radio waves with wavelengths comparable to the size of a skyscraper. All of these waves have further in common their speed in vacuum, $c \cong 3 \times 10^8$ m/s. In visual optics, typically only a small subset of the electromagnetic spectrum is considered where the wavelengths interact with the components of the human visual system to form images and/or enable diagnostic and therapeutic treatment of the eye. In the ensuing discussion, the spectrum will be restricted to the ultraviolet, visible, and near-infrared wavelengths. This range corresponds to wavelengths of roughly

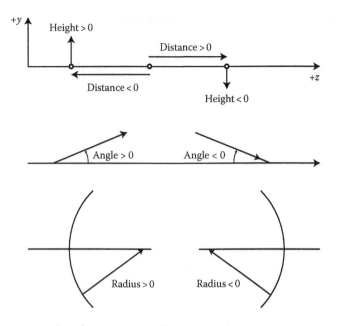

Figure 3.1 Coordinate system and sign convention.

$0.1 \leq \lambda \leq 1.0$ µm. Finally, when these waves enter a material such as water or glass, they slow down. For example, the speed of visible light in glass is typically reduced to about 2×10^8 m/s. This speed change in various materials can be captured by defining the material's index of refraction, n, as

$$n = \frac{\text{Speed in vacuum}}{\text{Speed in material}}. \tag{3.1}$$

For the given glass example, the refractive index would be $n = 1.5$. Note, when the electromagnetic waves travel in air, their speed is essentially the same as when they are in vacuum. Consequently, the refractive index of air is typically taken as $n = 1.0$.

3.2 WAVES, RAYS

3.2.1 VERGENCE

If a stone is dropped into a calm pool of water, circular ripples will propagate outward from the point where the stone entered the water. In a similar fashion, a point source of light will radiate spherical wavefronts in all directions. These wavefronts will remain perfectly spherical until they interact with some object in the environment. The properties of this spherical wavefront at any point in space can be defined by its *vergence*. Vergence is defined as

$$\text{Vergence} = \frac{n}{z}, \tag{3.2}$$

where

n is the refractive index of the material in which the spherical wave is propagating

z is the distance between the measurement point in space and the location of the point source

In visual optics, it is common to measure this distance in units of meters. The units of vergence, therefore, are reciprocal meters or *diopters* (D). As the distance between the point source and the measurement point increases, the radius of the spherical wavefront will continue to become larger. At extreme distances, z will approach infinity and the vergence will become zero. The preceding description of vergence considers the case of diverging spherical wavefronts. By convention, diverging spherical waves have a negative vergence, which requires that z be negative. Converging spherical waves, where the spherical wavefronts collapse to a single point, can also be described with vergence. In this case, the distance z is taken as a positive value. To summarize, a negative vergence describes a diverging spherical wave, a positive vergence describes a converging spherical wave, and a flat or plane wave is equivalent to a vergence of zero.

3.2.2 RAYS AND WAVEFRONTS

Rays are a mathematical simplification that describes the local direction a wavefront is propagating. Consequently, rays are always perpendicular to the wavefront. For the spherical waves described in the preceding section, a wave with positive vergence is converging. The rays associated with this wave all point to the single point to which the spherical wave collapses. A wave with negative vergence is diverging. The rays associated with this wave all appear to be emanating from the same point. Finally, the rays associated with a plane wave are all parallel. Figure 3.2 illustrates the various spherical wavefronts, their vergences, and associated rays.

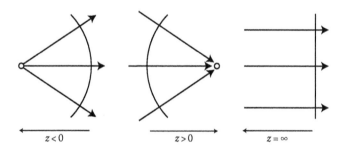
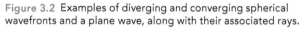

Figure 3.2 Examples of diverging and converging spherical wavefronts and a plane wave, along with their associated rays.

3.3 LAWS OF REFRACTION AND REFLECTION

3.3.1 REFLECTION FROM A PLANAR SURFACE

The law of reflection governs how a ray reflects from a surface. In much the same fashion as an elastic collision, a ray incident on a planar surface will reflect from the surface at the same angle. Figure 3.3a shows the reflective surface and its normal. A ray incident at an angle i with respect to this normal will reflect from the surface at an angle i'. The law of reflection, in conjunction with the sign convention, states that $i = -i'$. As drawn in the figure, the angle i is positive and the angle i' is negative. A further consequence of the law of reflection is that the incident and emerging rays and the surface normal must all lie in the same plane.

3.3.2 SNELL'S LAW AT AN INTERFACE

When an incident ray reaches a boundary between two transparent media, the transmitted ray is bent or refracted. Figure 3.3b illustrates the refracting of a ray at the boundary between two regions of indices n and n', respectively. Snell's law, as shown in the following equation, describes the degree of refraction:

$$n \sin i = n' \sin i', \tag{3.3}$$

where i and i' are the angles the incident and emerging rays form with respect to the surface normal. The typical convention of using unprimed variables for values before the interface and primed variables for values after the interface will be used in the ensuing discussion. A ray in a medium of refractive index n that is incident on a surface at an angle i will refract to leave the surface at angle i'. The newly refracted ray will then continue in a straight line until it intercepts a new boundary. As with the law of reflection, the incident and emerging rays and the surface normal must all lie in the same plane.

As seen in the figure, the incident and refracted angles are measured counterclockwise from the surface normal and are consequently both positive. Finally, it is convenient to combine Snell's law and the law of reflection into a single mathematical expression. This can be easily accomplished in the reflective

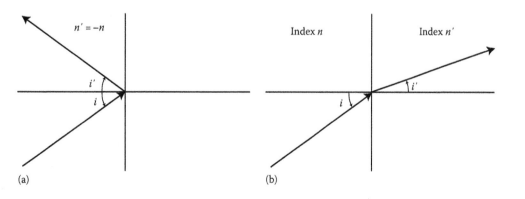

(a)   (b)

Figure 3.3 (a) For a reflective surface, the angle of incidence equals minus the angle of reflection. (b) For a refractive surface, the direction of the refracted rays is governed by Snell's law.

case by negating the index of refraction after the surface such that $n' = -n$. With this condition, Snell's law becomes

$$n \sin i = n' \sin i'$$

$$n \sin i = -n \sin i'$$

$$\sin i = \sin(-i')$$ (3.4)

$$i = -i',$$

which is just the law of refraction. For systems with multiple reflective surfaces, the refractive index simply toggles between a positive and a negative value following each reflection.

3.3.3 TOTAL INTERNAL REFLECTION

Snell's law can also be rewritten in the form

$$\sin i = \frac{n'}{n} \sin i'. \quad (3.5)$$

The maximum value of the sine function is one. In cases where $n' > n$, meaning the incident ray is in a higher refractive index material than the refracted ray, the incident angle i cannot exceed to so-called critical angle, i_c. Otherwise, the maximum value of the sine function is exceeded. The critical angle occurs when $\sin i' = 1$ or

$$i_c = \sin^{-1}\left(\frac{n'}{n}\right). \quad (3.6)$$

For angles of incidence less than the critical angle, the incident ray will refract as governed by Snell's law. If the incident ray meets or exceeds the critical angle, the ray will reflect as governed by the law of reflection. This reflective process is known as total internal reflection and is often exploited to provide reflection without the need to mirror coat a surface.

3.3.4 PRISMS

Prisms are optical elements, often triangular in shape, which are used to invert, revert, or deviate the direction of light beams within an optical system. Prisms can combine the effects of total internal reflection, conventional reflection (with a mirrored surface), and refraction to cause these changes in beam direction. A common prism used in visual optics is a thin triangular prism used to treat strabismus or misalignment between the two eyes. The prism is placed in front of the misaligned eye so that the line of sight through that eye is deviated such that it becomes parallel with the line of sight of the fellow eye. In this manner, the two eyes are looking at the same point in the scene and the images from each eye can be fused. Figure 3.4 shows a typical thin triangular prism. The magnitude of the deviation is dictated by the apex angle of the triangular prism and is usually given in units of prism diopters (Δ). The definition of one *prism diopter* is the prismatic deviation of a beam of light by 1 cm at a distance of 1 m.

The orientation of a prism is defined by its base (the wide end of the prism). Base up and base down orientations cause vertical deviations, while base out and base in cause horizontal deviations. By rotating the prism about the line of sight, deviations in different directions can be achieved. Base out orients the prism base oriented toward the temple, while base in orients the prism nasally. Figure 3.5 shows the orientation of the prism with respect to the misaligned eye for different types of strabismus.

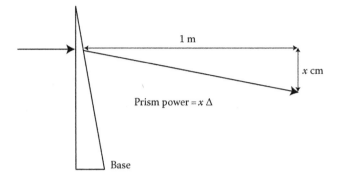

Figure 3.4 The power of a prism is given by the deviation in cm at a distance of 1 m. The orientation of a prism is described by the direction of its base.

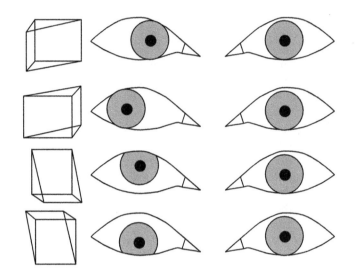

Figure 3.5 Rotating the orientation of the prism corrects for different types of ocular misalignment.

3.4 REFRACTION AND REFLECTION FROM A SPHERICAL SURFACE

3.4.1 REFRACTION FROM A SPHERICAL SURFACE

Prisms and other optical elements with flat surfaces only deviate the direction of a beam passing through them, but do not change the beam's vergence. To change vergence, a surface must have curvature. Snell's law, as defined in Equation 3.3, holds for spherical surfaces as well. The angle of incidence and refraction are measured with respect to the surface normal. However, since the surface is curved, the orientation of the normal now depends upon where the ray strikes the surface. The spherical refracting surface shown in Figure 3.6 separates two optical spaces with refractive indices of n and n', respectively. The optical axis is the line passing through the center of curvature of the spherical surface. The intersection of the optical axis with the spherical surface is called the surface vertex. This surface has a radius of curvature R, which is the distance from a point on the surface to the center of curvature. This radius defines the shape of the spherical surface. From the previously defined sign convention, R is a positive value in this case since the center of curvature lies

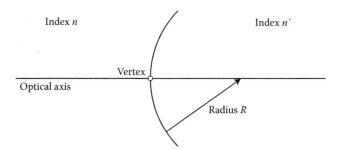

Figure 3.6 Spherical refractive surface of radius R, separating spaces with refractive index n and n'.

to the right of the surface. Alternatively, the spherical surface can also be defined by its curvature, $C = 1/R$. The power ϕ of the spherical surface describes its ability to modify the vergence of an incident wave. The power is defined as

$$\phi = (n' - n) \cdot C = \frac{n' - n}{R}, \tag{3.7}$$

which has units of reciprocal distance. When this distance is measured in units of meters, the power of the surface is in units of diopters. In cases such as that shown in Figure 3.6, when $n' > n$, rays parallel to the optical axis will converge as they pass through the surface. This situation is associated with a positive power. When $n' < n$, the same incident rays would diverge after refracting at the surface. The surface would have negative power. For example, the average cornea has a radius of curvature of $R = 7.8\,\text{mm}$ and an index of refraction $n' = 1.376$. In air ($n = 1.0$), the cornea would have a power

$$\phi = \frac{1.376 - 1.0}{0.0078\,\text{m}} = 48.21\,\text{D}. \tag{3.8}$$

Rays traveling parallel to the optical axis are said to be collimated and on-axis. When these rays intercept a surface with positive power, they will refract and converge to a point F'. Similarly, there exists a point F in which rays that diverge from F intercept the positive powered surface and are refracted so that they become parallel to the optical axis. The points F and F' are called the front and rear focal points of the surface. The focal length is defined as the reciprocal of the power

$$f = \frac{1}{\phi}. \tag{3.9}$$

The distance from the surface to F' is called rear focal length, f_R', and is defined as

$$f_R' = \frac{n'}{\phi} = n'f. \tag{3.10}$$

Similarly, the distance from the surface to F is called the front focal length, f_F, and is defined as

$$f_F = -\frac{n}{\phi} = -nf. \tag{3.11}$$

The focal length f describes the equivalent or effective focal length of the surface in air, whereas the rear focal length f_R' and the front focal length f_F are scaled by their respective refractive indices. The front and rear focal lengths describe the physical length from the surface to their respective focal points. The concepts for a single spherical refracting surface can be extended to a thick lens. A thick lens is an optical element made of a material with refractive index n_{lens}, which has, in general, two spherical refracting surfaces separated by a distance t. The power Φ of a thick lens is given by

$$\Phi = \phi_1 + \phi_2 - \frac{t}{n_{lens}}\phi_1\phi_2, \tag{3.12}$$

where ϕ_1 and ϕ_2 are the powers of the individual refracting surfaces of the thick lens. Often, lenses can be approximated as "thin" lenses when the thickness t is much less than the magnitude of the radii of curvature of either of the refracting surfaces. In this case, the power of the thin lens is given by the sum of the powers of the surfaces, or

$$\Phi = \phi_1 + \phi_2. \tag{3.13}$$

Similar definitions for the focal length of the lens now hold, but the total power Φ is used instead of the surface power ϕ. Based on similar arguments, two thin lenses in air, separated by some distance t, would have a net power of

$$\Phi_{net} = \Phi_1 + \Phi_2 - t\Phi_1\Phi_2, \tag{3.14}$$

where

Φ_1 is the power of the first thin lens
Φ_2 is the power of the second thin lens

If the two thin lenses are placed in contact, then $t = 0$ and

$$\Phi_{net} = \Phi_1 + \Phi_2. \tag{3.15}$$

In other words, the powers of two thin lenses in contact add to create the net power.

3.4.2 REFLECTION FROM A SPHERICAL MIRROR

Reflective spherical surfaces can also modify the vergence of an incident beam. Convex and concave mirrors are treated in much the same way as the spherical refracting surfaces described earlier. When examining the law of reflection, it was convenient to assume that the refractive index after reflection was the negative of the index of refraction prior to refraction, or $n' = -n$. Under this assumption, Snell's law collapsed to the law of reflection. Using the same assumption, the power of a spherical mirror is given by

$$\phi = \frac{(n' - n)}{R} = -\frac{2n}{R}. \tag{3.16}$$

As in Equation 3.9, the focal length of a spherical mirror is

$$f = \frac{1}{\phi} = -\frac{R}{2n}, \tag{3.17}$$

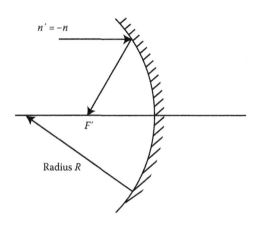

Figure 3.7 Collimated light focuses to a point halfway to the center of curvature of the mirror.

and the front and rear focal lengths are

$$f_F = f'_R = -\frac{n}{\phi} = -nf = \frac{R}{2} = \frac{1}{2C}. \qquad (3.18)$$

This last result means that the front and rear focal points for a mirror lie halfway between the center of curvature and the vertex of the surface. Figure 3.7 shows a concave mirror. In visual optics, concave mirrors are often used in illumination systems for surgical microscopes to concentrate light onto the surgical field. The properties of convex mirrors are used in keratometry to measure the radius of curvature of the anterior corneal surface.

3.5 GAUSSIAN IMAGING EQUATION

The Gaussian imaging equation allows the determination of the object and image locations for a surface of power ϕ. The distance from the surface vertex to the object will be denoted by z and the distance from the surface to the image plane is denoted by z'. In keeping with the sign convention, these distances are negative if the distance is measured to the left of the surface and positive if it is to the right of the surface. The Gaussian imaging equation relates these distances to the surface power as

$$\frac{n'}{z'} = \phi + \frac{n}{z}. \qquad (3.19)$$

The planes located at z and z' are said to be conjugate since they satisfy the Gaussian imaging equation. Figure 3.8 shows an object point a distance z from the refracting surface. This point is imaged to a point a distance z' from the single refracting surface. In examining Equation 3.19 and comparing it to the definition of vergence in Equation 3.2, it is evident that n/z is the vergence of the object point as measured from the plane of the refracting surface. Similarly, n'/z' is the vergence of the image point as measured from the plane of the refracting surface. Consequently, the image vergence is the sum of the object vergence and the power of the refracting surface. The refracting surface modifies the object vergence to produce the image vergence. The Gaussian

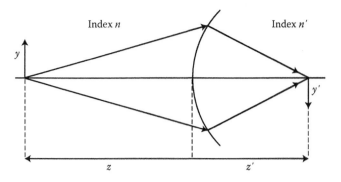

Figure 3.8 An object located at a distance z from a single refracting surface of power ϕ will be imaged to a point located at z'. By convention $z < 0$ and $z' > 0$ in this figure.

imaging equation holds for mirror surfaces as well, again with the assumption that $n' = -n$.

The transverse (or lateral) magnification, m, of the system for an extended object can also be defined by the object and image distances. Figure 3.8 shows an extended object of height y, located a distance z to the left of the surface. The image of this object has a height y'. The sign convention says that positive heights are above the optical axis and negative heights are below the optical axis. The transverse magnification is the ratio of the image height to the object height and is related to the object and image distances by

$$m = \frac{y'}{y} = \frac{nz'}{n'z}. \qquad (3.20)$$

Figure 3.8 shows a case where $m < 0$, which means that the image is inverted relative to the object.

3.5.1 THICK LENSES AND GAUSSIAN IMAGING

In the previous section, only a single refracting surface was considered. In a similar manner, the imaging equation will hold for a thin lens with its power defined as the sum of the surface powers as in Equation 3.13. In applying the Gaussian imaging equation to thick lenses with power Φ as defined in Equation 3.12, a slight modification is required. For the thick lens, the object and image distances z and z' need to be measured relative to the front and rear principal planes of the lens, P and P'. The principal planes can be considered the planes of effective refraction for the lens. The object wave with its object vergence entering the front principal plane is mapped to the rear principal plane, where it is then converted to the image vergence by adding the power of the lens. In the cases of a single refracting (or reflecting) surface and a thin lens, both the principal planes were located at the surface. For a thick lens, the principal planes are in general displaced from the surface vertices and are separated from one another. The front principal plane is located at a distance d from the first surface vertex, where this distance is given by

$$d = \frac{\phi_2}{\Phi} \frac{t}{n_{lens}}. \qquad (3.21)$$

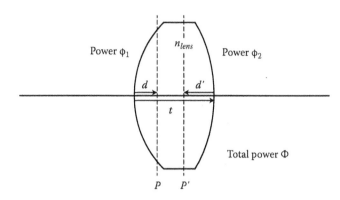

Figure 3.9 A thick lens and its principal planes.

Similarly, the rear principal plane is located a distance d' from the second surface vertex, where this distance is given by

$$d' = -\frac{\phi_1}{\Phi}\frac{t}{n_{lens}}, \tag{3.22}$$

where

ϕ_1 and ϕ_2 are the powers of the individual refracting surfaces of the thick lens

t is the thickness of the lens

n_{lens} is the refractive index of the lens material

The total power of the thick lens Φ is defined in Equation 3.12. Figure 3.9 shows the locations of the principal planes for a thick lens.

3.5.2 CARDINAL POINTS

In general, optical systems will consist of multiple thick lens elements whose surface shapes and materials are carefully chosen to provide the imaging requirements and quality desired for the application. These complex systems can be reduced to a series of six "special" points known as the cardinal points that fully define the Gaussian imaging and magnification properties of the system. Four of these six points have already been encountered in the preceding discussion. The front and rear focal points, F and F', arise when collimated light enters or leaves the optical system. For collimated incident rays, the optical system converts the vergence of the bundle so that the

exiting vergence appears to emanate from the rear focal point. Similarly, an object point located on the front focal point will result in a beam emerging from the optical system that has zero vergence or, in other words, is collimated. The second set of points associated with the optical system is the front and rear principal points. The principal points are located where their respective principal planes cross the optical axis. For a thick lens, the locations of the principle planes (points) were defined by Equations 3.21 and 3.22. For a general optical system, the locations can be identified by ray tracing the system. Once located, the object and image distance are measured with respect to the principal points and the Gaussian imaging formula holds. The principal planes are said to be planes of unit magnification. A ray striking the front principal plane at a given location is mapped to an identical location on the rear principal plane and appears to emerge from that point. In general, the emerging ray will be traveling in a different direction than the incident ray. The last pair of cardinal points is the nodal points, N and N'. A ray traveling at an angle to the optical axis and passing through the front nodal point emerges from rear nodal point at the same angle. The nodal points can be found relative to the principal points. In general, the nodal points of a system are shifted relative to the corresponding principal points such that

$$\overline{PN} = \overline{P'N'} = f_F + f'_R = (n'-n)f, \tag{3.23}$$

where

n is the object space index

n' is the image space index

In cases where the object and image space indices are the same (e.g., a camera lens in air), the nodal points are located at the principal planes. In cases where the object and image refractive indices differ (e.g., the eye where $n = 1.0$ and the image space contains vitreous humor with refractive index $n' = 1.336$), the nodal points are shifted relative to the principal points.

With knowledge of these six cardinal points, a multielement optical system can be reduced to a "black box" and properties of the rays entering and leaving the system can be easily determined without knowledge of all of the surfaces, spacing, and materials within the box. Figure 3.10 shows the properties of the cardinal points of a generalized optical system.

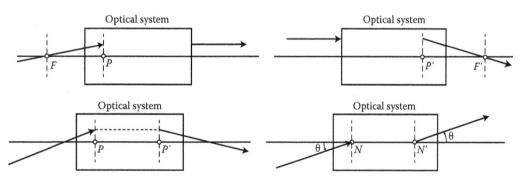

Figure 3.10 Cardinal points of an optical system.

3.5.3 APERTURE STOP AND PUPILS

The aperture stop is a mask within the system that limits the size of the bundle of rays that passes through an optical system. This mask can be a separate mask or the clear opening of one of the optical elements. The aperture stop is typically circular, but this is not a requirement. Each point source on an object radiates light into all directions. The aperture stop in effect blocks most of these rays and only allows a limited cone of light from the point source to pass through the optical system. The object and image locations, as well as the system magnification, are independent of the aperture stop location. However, the aperture stop dimensions and location affects the amount of light that reaches the image, as well as how well the rays come to a focus. A judicious choice of the location of the aperture stop can optimize the quality of an optical system.

The entrance pupil is the image of the aperture stop formed by all of the lens (optical) surfaces preceding the stop, and the exit pupil is the image of the stop formed by all of the surfaces following the stop. If the entrance pupil is the object for the entire system, its image is formed at the exit pupil. For a well-corrected system, the exit pupil is a 1:1 mapping of the entrance pupil, which is to say that the scaling of rays passing through the entrance pupil will be mapped to the same relative position in the exit pupil. The entrance pupil can be thought of as a port that captures light from the object scene. The light that gets into the entrance pupil makes it to the exit pupil (assuming no loss due to reflections and/or vignetting) and contributes to the image. In the eye, for example, the iris acts as the aperture stop. The cornea acts as lens and forms an image of the iris located approximately 3 mm posterior to the corneal vertex. This image is the entrance pupil of the eye and it is approximately 10% larger than the physical size of the opening in the iris.

3.5.4 CHIEF AND MARGINAL RAYS

There are several special rays that provide useful information regarding the properties of an optical system. One set of rays are called marginal rays. These rays start at the object on the optical axis and pass through the edge of the aperture stop. In an ideal system, the marginal rays will propagate through the optical system and ultimately converge to the optical axis at the image plane. Furthermore, if the marginal rays intersect the optical axis at some location between the object and image, then an intermediate image is formed. These intermediate image planes are useful in that a mask or reticle can be placed at the location, having the effect of superimposing the reticle pattern onto the final image. This technique is often used in microscope eyepieces, for example, to allow scaled rulings to be placed over the image to measure features.

A second special ray defined with regard to the aperture stop is the chief ray (principal ray). The chief ray is defined as the ray that starts at the edge of the object and passes through the center of the aperture stop. When the chief ray crosses the optical axis, it is called a pupil plane. The size of this pupil is defined by the height of the marginal ray in the pupil plane.

The positions of the entrance and exit pupil are determined by where the chief ray appears to cross the optical axis in object and image space. The chief ray incident on the first surface is projected to determine the point it crosses the optical axis. This point defines the location of the entrance pupil. The size of the entrance pupil is determined by projecting the incident marginal ray onto the plane of the entrance pupil. The marginal ray will define the boundary of the entrance pupil. Similar definitions hold for the exit pupil, where the emerging chief and marginal rays are used to define the location and size of the exit pupil.

3.6 CYLINDRICAL AND TORIC SURFACES

Most optical systems are rotationally symmetric. The spherical surfaces of the lenses are chosen to provide the imaging properties and magnification required for a specific task. Multiple elements may be used to optimize image quality. Occasionally, rotationally symmetric aspheric surfaces are used to further improve image quality or reduce the required number of elements. However, some optical systems lack this rotational symmetry. The human eye, for example, often suffers from astigmatism since its optical surfaces are not necessarily rotationally symmetric. Consequently, astigmatic lenses often appear in visual optics since they are used to compensate for astigmatism in the eye. The simplest astigmatic lens is a cylindrical lens.

3.6.1 POWER AND AXIS OF A CYLINDRICAL LENS

A cylindrical lens has one or more surfaces that are a section of a cylinder. Figure 3.11a shows an example of a positive powered cylindrical lens. A cross section through one meridian of the cylindrical lens shows a circular surface of radius R_s, meaning the power in this meridian is given by Equation 3.7. However, in a perpendicular meridian, the cross section is flat, meaning there is no power in this direction. Consequently, a cylindrical lens is a lens with power along one meridian and no power along an orthogonal meridian. The cylindrical lens can also be rotated about the optical axis. To describe this orientation, the *cylinder axis* θ_f is defined as the angle of the zero-power meridian, measured counterclockwise from the horizontal axis. The power and orientation of a cylindrical lens can be described by $\phi_s \times \theta_f$, where $\phi_s = (n' - n)/R_s$ is the cylindrical lens power. Note that due to the shape of the cylindrical lens, there is a redundancy in the definition of the cylinder axis. If the zero-power axis of cylindrical lens is oriented along the horizontal meridian, then $\theta_f = 0°$ or $\theta_f = 180°$. By convention, the cylinder axis is uniquely defined as being in the range $0° < \theta_f \leq 180°$. With this definition, the horizontally oriented cylindrical lens has a cylinder axis $\theta_f = 180°$.

3.6.2 TORIC AND SPHEROCYLINDRICAL SURFACES

A more complicated astigmatic lens introduces both spherical and cylindrical power simultaneously. Examples of refractive surfaces that exhibit power variation are toric and spherocylindrical surfaces. These surfaces have a short radius of curvature R_s along one meridian (called the steep meridian) and a longer radius of curvature R_f along the orthogonal meridian (called the flat meridian). The shape of toric and spherocylindrical surfaces match along these principal meridia, but in general the surface shapes are slightly different away from these meridia. The powers

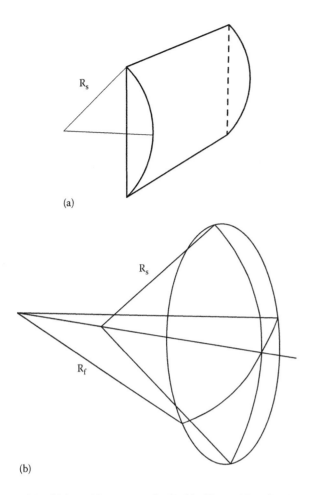

(a)

(b)

Figure 3.11 (a) A positive powered cylindrical lens with radius R_s. (b) A spherocylindrical surface with radii R_f and R_s.

along these meridia can be calculated in much the same manner as a spherical surface, with

$$\phi_s = \frac{n' - n}{R_s} \quad \text{and} \quad \phi_f = \frac{n' - n}{R_f}. \qquad (3.24)$$

The toric or spherocylindrical surfaces can be thought of as a surface with high power ϕ_s along the steep meridian. Rotating through the meridia, the power gradually reduces in a sinusoidal fashion and reaches a minimum ϕ_f along the flat meridian. The power then sinusoidally increases back to ϕ_s at 180° away from the starting point. A cylindrical lens is a special case of these types of lenses where the power along the flat meridian is zero. Figure 3.11b shows an example of a spherocylindrical surface.

As with the cylindrical lens, the lens can be rotated about the optical axis so that one of the lens' principal meridia is at an angle measured counterclockwise from the horizontal meridian. Again, there is redundancy in the description of this orientation, so the angles between 1° and 180° are used to uniquely describe the orientation. One example where toric surfaces are analyzed is in keratometry. Keratometry is the measure of the flat and steep radii of the anterior corneal surface. Typically, keratometry is measured by aligning an illuminated ring of known size in front of the cornea. The anterior cornea is treated as a convex mirror and forms an image of the ring behind the cornea. The size of the image is measured and compared to the known ring size to

determine the magnification imparted by the anterior cornea. From the magnification m, the corneal radius can be determined by combining Equations 3.18 through 3.20, such that

$$R = \frac{2mz}{m-1}, \qquad (3.25)$$

where z is the distance between the keratometer target and the eye. Note that in the sign convention described previously, z is a negative quantity. In cases of corneal astigmatism, the ring image will be elliptical with the long axis of the ellipse corresponding to the flat meridian and the short axis of the ellipse corresponding to the steep meridian. Thus, the orientation of the ellipse determines the orientation of the principal meridia of the cornea. The magnification along both the short and long axes of the ellipse can be measured separately and Equation 3.25 is used to determine the radii R_s and R_f along their respective axis. Finally, Equation 3.24 is used to determine corneal power along each direction. If $n' = 1.376$ and $n = 1.0$ in these equations, then the true optical power of the anterior cornea is determined. In keratometry, an artificial refractive index called the "keratometric refractive index" $n' = n_k$ is used in these expressions. The value of n_k has been chosen to reduce the anterior corneal power to account for the negative power imparted by the posterior corneal surface. So keratometry is an estimate of the total corneal power (combined anterior and posterior powers) based solely on the anterior corneal shape. Typical values of the keratometric refractive index are $n_k = 1.3375$ and $n_k = 1.3315$, depending on the device used to measure the cornea.

As an example, suppose the cornea has a radius $R_s = 7.8$ mm along a meridian oriented at 30° and a radius $R_f = 8.0$ mm along a meridian oriented at 120°. Based on Equation 3.23 and using $n' = 1.376$, the true anterior corneal powers are $\phi_s = 48.21$ D and $\phi_f = 47.00$ D. The keratometry values associated with this cornea are given by

$$K_s = \frac{1.3375 - 1}{7.8} = 43.27 \text{ D} \quad \text{and} \quad K_f = \frac{1.3375 - 1}{8.0} = 42.19 \text{ D}, \qquad (3.26)$$

where K_s and K_f have been used to distinguish between the keratometric power and the true anterior corneal power. Note, K_s is typically referred to as the "steep-K value" and K_f is typically referred to as the "flat-K value." Finally, the orientation of the steep and flat meridians are specified along with these keratometric powers and the "@" symbol. In the preceding example, the full keratometry measurement would be described as 43.27 D @ 30°/42.19 D @ 120°. The absolute difference between the keratometric powers suggests that the example cornea has slightly more than 1 D or corneal astigmatism.

Lenses for correcting refractive error also incorporate astigmatic surfaces. From a thin lens standpoint, these types of lenses can be considered as a combination of a spherical lens and a cylindrical lens. In Equation 3.15, the powers of two thin lenses simply added when the separation between them was negligible. For astigmatic thin lenses, the same effect holds, but each of the principal meridia needs to be considered independently. Combining a thin spherical lens with a thin cylindrical lens results in a thin spherocylindrical lens. This resultant spherocylindrical lens has the spherical lens power along one principal meridian and the combined spherical and cylindrical power along the other principal meridian. For example, the combination of a –3.00 D spherical lens with a +1.50 D cylindrical lens with cylinder axis at 180° leads to a spherocylindrical lens with power of –3.00 D along the horizontal axis since the cylindrical lens has no power in this direction. Along the vertical axis, the resultant lens has a power of –1.50 D since along this meridian the powers of the spherical and cylindrical lens add. The prescription for this spherocylindrical lens is typically written as

$$-3.00/+1.50\times180°, \qquad (3.27)$$

where the first component is the spherical lens power and the second component is the cylindrical lens power and its cylinder axis. This prescription is said to be in "plus cylinder form" since the cylindrical lens power is positive.

As a second example, consider a spherical lens with a power of –1.50 D combined with a cylindrical lens of power –1.50 D with cylinder axis at 90°. In this case, the resultant spherocylindrical lens has a power of –1.50 D along the vertical axis since only the spherical lens contributes in this direction. Along the horizontal axis, the power is –3.00 D since both the spherical and cylindrical lens combines in this direction. The prescription for this lens is typically written as

$$-1.50/-1.50\times90°, \qquad (3.28)$$

where again the first component is the spherical lens power and the second component is the cylindrical lens power and its cylinder axis. This prescription is said to be in "minus cylinder form" since the cylindrical lens power is negative. However, comparing the properties of the resultant spherocylindrical lenses from the two examples shows that both lenses have identical power distributions. There are always two ways to achieve a given spherocylindrical lens. One prescription is in plus cylinder form and the other is in minus cylinder form.

It is useful to convert between the two cylinder forms. The following steps perform this conversion:
1. The new spherical component is the sum of the spherical and cylindrical powers of the old form.
2. The new cylindrical component is the negative of the old cylinder component.
3. The new cylinder axis is 90° from the old cylinder axis.
4. If the new axis does not fall within the 1°–180° range, then add or subtract 180° from the new axis to place it in this range.
This technique works for converting from plus to minus cylinder form, as well as from minus to plus cylinder form.

The previous examples are used to illustrate this conversion technique. To convert the plus cylinder form prescription –3.00/ + 1.50 × 180° to minus cylinder form, the first step is to add the spherical and cylindrical powers. The new spherical power is therefore –3.00 + 1.50 = –1.50 D. The second step is to negate the cylindrical power, so the new cylindrical power is –(+1.50) = –1.50 D. The third step is to rotate the axis by 90°, leading to a new cylinder axis of 180° + 90° = 270°. Finally, the new cylinder

axis should be in the range of 1°–180°, so it is observed that the 270° is the same as the 90° axis and the new cylinder axis adjusted accordingly. The final prescription in minus cylinder form based on these steps is given in Equation 3.28. Note, steps 3 and 4 can be combined by simply rotating the axis 90° in the direction that puts it in the desired range. In the preceding example, at step 3, the new cylinder axis can easily be obtained by calculating instead 180°−90° = 90°.

A final concept that often arises with astigmatic lenses is spherical equivalent power (SEP). The SEP is the average power of a spherocylindrical lens. As described previously, the power of a spherocylindrical lens varies sinusoidally from a minimum power along its flat meridian to a maximum power along its steep meridian. The SEP is given by the spherical power plus half the cylindrical power. This definition holds for both the plus and minus cylinder forms. From the previous spherocylindrical examples, the SEP is given by

$$SEP = -3.00 + \frac{+1.50}{2} = -1.50 + \frac{-1.50}{2} = -2.25\,D. \quad (3.29)$$

Finally, the Jackson crossed cylinder is a specialty lens often used in ophthalmic optics. It is a spherocylindrical lens with a SEP of zero. It has a power ϕ in one meridian and a power $-\phi$ in the orthogonal meridian. These lenses introduce pure astigmatism and no spherical component. The prescription for such a lens is given by

$$\phi/-2\phi\times\theta, \qquad (3.30)$$

where θ is the cylinder axis. The Jackson crossed cylinder is formed by combining a plano-convex positive cylindrical lens with a plano-concave negative cylindrical lens. The powers of each lens have equal magnitude, but opposite sign. The axes of the lenses are set to be orthogonal. Figure 3.12 demonstrates a Jackson crossed cylinder.

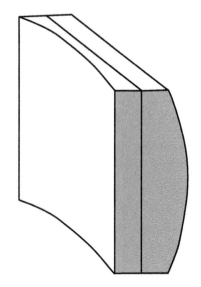

Figure 3.12 The Jackson crossed cylinder can be formed by combining conventional positive and negative cylindrical lenses.

3.7 VISUAL INSTRUMENTS

There are a variety of instruments that are designed to use the eye as the final detector. In many cases, these systems do not form an image per se but, instead, have the emerging light collimated or slightly diverging so that the eye can ultimately form the final image on the retina. In such cases, the lateral magnification of the optical system often becomes ill-defined because the eye needs to be considered as part of the overall system. Visual instruments typically use the angular subtense of the light emerging from the systems relative to an unaided view as a measure of magnification. The systems described in this section follow these specifications.

3.7.1 SIMPLE MAGNIFIER AND MAGNIFYING POWER

Perhaps, the simplest visual instrument is the simple magnifier. This device can be as simple as a single positive lens. The lens forms a virtual image of a given object where the image distance is a comfortable viewing distance from the eye. This virtual image also subtends a larger angle than the original object so that it appears larger to the viewer. One way of increasing the angular subtense of an object is to bring it close to the eye. Figure 3.13a shows an object with height h close to the eye that subtends an angle u_0. The problem with this arrangement is that the eye must accommodate in order to focus on such a near object. This accommodation can produce excessive strain or may not even be possible since people gradually lose their ability to accommodate with age.

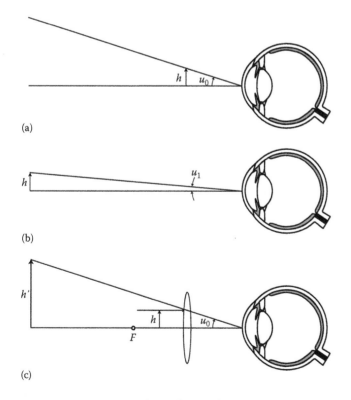

(a)

(b)

(c)

Figure 3.13 (a) Bringing an object close to the eye allows the angular subtense to reach u_0, but the proximity may be difficult to accommodate to. (b) Moving the object to a comfortable viewing distance reduces the angular subtense to u_1. (c) A simple magnifier places a virtual image of the object at a comfortable viewing distance and increases its angular subtense.

Moving the object further away can alleviate this strain by placing the object at a comfortable viewing distance. This distance is typically taken as 250 mm. Figure 3.13b shows the object at this viewing distance. Moving the object away from the eye reduces its angular subtense to u_1. A simple magnifier is used to recover the original angular subtense, but place the image of the object at a comfortable viewing distance. If the object is placed within the focal length of the magnifier, then a virtual image as shown in Figure 3.13c is formed. This virtual image now has an angular subtense of u_0. The magnifying power (MP) is defined as

$$MP = \frac{\text{Angular subtense of the virtual image with magnifier}}{\text{Angular subtense of the object at a comfortable viewing distance}} = \frac{u_0}{u_1}.$$

(3.31)

The MP is a measure of how much larger the virtual image is relative to the object at some finite distance. This concept will appear again when discussing microscopes in Section 3.7.2.

3.7.2 MICROSCOPES

Microscopes extend the capabilities of the simple magnifier, enabling much higher magnifications to be achieved. Most modern microscopes are based on infinity-corrected objective lenses. Figure 3.14 illustrates such a system. The microscope consists of three lens groups: an infinity-corrected objective with power Φ_{obj}, a tube lens with power Φ_{tube}, and an eyepiece with power Φ_{eye}. The corresponding focal lengths are $f_{obj} = 1/\Phi_{obj}$, $f_{tube} = 1/\Phi_{tube}$, and $f_{eye} = 1/\Phi_{eye}$, respectively. The object is placed in the front focal plane of the objective lens. The objective lens collimates light from points on the object. The tube lens collects these collimated beams and creates a magnified intermediate image. The eyepiece then acts much like a simple magnifier and reimages the intermediate image to a comfortable viewing distance with magnifying power MP_{eye}. The magnification of a microscope with an infinity-corrected objective is given by

$$\text{Magnification} = \frac{\Phi_{obj}}{\Phi_{tube}} MP_{eye} = \frac{f_{tube}}{f_{obj}} MP_{eye}.$$

(3.32)

The collimated space between the objective and tube lenses is advantageous over more traditional designs where the objective lens forms the intermediate image directly at a standardized location. In the infinity-corrected objective case, additional

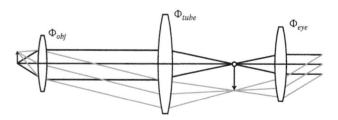

Figure 3.14 A microscope with an infinity-corrected objective.

optical elements such as filters, polarization elements, and beam splitters can be placed in the collimated space without introducing aberrations.

3.7.3 TELESCOPES AND ANGULAR MAGNIFICATION

Telescopes are *afocal* optical systems, or systems with a net power of zero. The net power of a system with two separated thin lenses is given by Equation 3.14. The first thin lens is called the objective lens and has a power Φ_{obj}. The second lens is called the eyepiece and has a power Φ_{eye}. The corresponding focal lengths are $f_{obj} = 1/\Phi_{obj}$ and $f_{eye} = 1/\Phi_{eye}$, respectively. From Equation 3.14, the net power of a telescope is

$$\Phi_{net} = \Phi_{obj} + \Phi_{eye} - t\Phi_{obj}\Phi_{eye}. \tag{3.33}$$

For the telescope to have a net power of zero, the separation between the two thin lenses must satisfy

$$t = \frac{\Phi_{obj} + \Phi_{eye}}{\Phi_{obj}\Phi_{eye}} = f_{obj} + f_{eye}. \tag{3.34}$$

Since telescopic systems do not form an image, the lateral magnification is again ill-defined. For telescopes, angular magnification is used instead, where the angular magnification, m_a, is defined as the ratio of the angular subtense of the object as viewed from the objective to the angular subtense of the object as viewed by the eye through the telescope.

Two common types of telescopes are the Galilean and the Keplerian telescope. These telescopes are shown in Figure 3.15. The Galilean telescope shown in Figure 3.15a consists of a low power positive lens (objective) and a high power negative lens (eyepiece). Since the focal length of the eyepiece is negative, the separation between the object and the eyepiece is less than the focal length of the objective, leading to a compact system. The angular magnification of the Galilean telescope is given by

$$m_a = -\frac{\Phi_{eye}}{\Phi_{obj}} = -\frac{f_{obj}}{f_{eye}}. \tag{3.35}$$

Since the powers of the two lenses have opposite sign, the angular magnification is positive, meaning the image viewed through the Galilean telescope is upright. The primary drawback to the Galilean telescope is that the exit pupil is located in between the two lenses. This means that the eye cannot be placed at the position of the exit pupil and consequently the field of view of the telescope is reduced.

A second common telescope type is the Keplerian telescope shown in Figure 3.15b. This telescope consists of two positive powered lenses. The separation between the two thin lenses is again given by the sum of the focal lengths as in Equation 3.34. In this case, both focal lengths are positive, so the separation between the lenses is larger than the equivalent Galilean telescope. The angular magnification for the Keplerian telescope is again given by Equation 3.35. However, since both lens powers are positive, the angular magnification is negative, leading to an inverted image. One advantage of the Keplerian telescope over the Galilean is that the exit pupil of the Keplerian telescope lies outside the two thin lenses. The eye can be placed at the location of the exit pupil, and the field of view of the system is consequently larger compared to the Galilean system.

Both telescopic systems are routinely used in visual optics. A common application is to provide increased angular magnification to people with low vision. Both systems can be used, but the Keplerian telescope requires additional prisms or lenses to make the perceived image upright. The trade-off between the two systems is a compact Galilean telescope with a small field of view or a longer Keplerian telescope with a wider field of view.

3.8 SUMMARY

Geometrical optics studies how light beams travel through optical systems. For imaging systems, the main goal is to relay the object plane to the image plane. In addition, the size of the image is typically important. The Gaussian imaging equation provides the connection between the object and image, as well as the system magnification. While the techniques demonstrated in this chapter examine simplified cases of thin lenses and systems of a few lenses, these examples are sufficient to understanding the basic mechanisms of the image-forming process. Optical system designers will typically use many more elements to optimize the quality of the image in terms of brightness and sharpness. This chapter also showed that these more complex systems can be reduced to a "black box," where only the cardinal points need to be known to determine the imaging properties. In this manner, the designer can be concerned with the elements within the box, and the user can implement the system without full knowledge of every component within the box. Visual optics employs a wide array of optical systems, from systems for presenting visual stimuli, to heads-up display systems, to diagnostic imaging devices, to low vision aids. The workings of the rich and varied systems can often be understood with the fundamentals presented here.

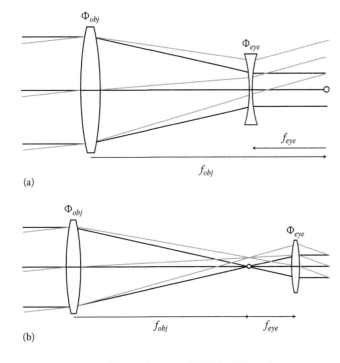

Figure 3.15 (a) A Galilean telescope. (b) A Keplerian telescope.

4 Wave optics

Daniel Malacara

Contents

4.1 WAVE NATURE OF LIGHT

Most light phenomena can be explained by modeling the light as a transverse wave. However, a few phenomenon, like the photoelectric effect or the blackbody radiation, cannot be interpreted, unless the light is considered as a particle, which we call photon. So for most practical purposes, the light is a transverse wave or, more specifically, an electromagnetic wave. A monochromatic light wave has an electric field with sinusoidal oscillation in a perpendicular direction to its traveling path. It also has a sinusoidal magnetic field in a plane perpendicular to the electric field, as illustrated in Figure 4.1.

4.1.1 MATHEMATICAL REPRESENTATION OF WAVES

The electromagnetic waves are transverse propagating in space, where the distance between two consecutive crests at a given time is called the wavelength, as illustrated in Figure 4.2. If an observer is fixed in space, two consecutive crests pass through that point, separated in time by the period T. The frequency ν is the number of crests passing in a unit time, which is given by the inverse of the period as follows:

$$\nu = \frac{1}{T}. \tag{4.1}$$

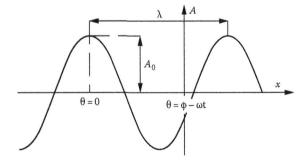

Figure 4.1 Electromagnetic wave with its electric and magnetic fields, in mutually perpendicular planes.

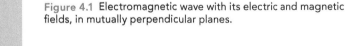

Figure 4.2 Illustration of some parameters for an electromagnetic wave.

If the crest waves propagate with a speed v, the frequency and the wavelength are related by

$$\lambda\nu = v. \tag{4.2}$$

Two additional parameters are defined to specify the wavelength and frequency of the electromagnetic wave. These are the angular frequency, ω, measured in radians per unit time

$$\omega = 2\pi\nu \tag{4.3}$$

and the wave number k

$$k = \frac{2\pi}{\lambda}. \tag{4.4}$$

Thus, Equation 4.2 can be written as

$$\frac{\omega}{k} = v. \tag{4.5}$$

Let us assume that the wave propagates, starting at the origin ($x = 0$) at the time $t = 0$, along the x-axis, in the positive direction. The instantaneous value of the electric field $E(x, t)$ at a given point x in space and at a given time t can be expressed by

$$E(x,t) = A\cos(kx - \omega t + \phi_0). \tag{4.6}$$

This may be called the "instantaneous amplitude," but more frequently, the "disturbance." The maximum value of the disturbance is A, and it is called the "amplitude." The phase of the wave at the origin, at the time $t = 0$, is ϕ_0. Thus, the phase ϕ of the wave at the point x and the time t is given by $kx - \omega t + \phi_0$. Given the amplitude A and a certain phase, the disturbance can be calculated with this expression. The two parameters defining the disturbance are the amplitude and the phase. So, it is desirable to express the disturbance by a 2D vector, where the two components are defined by the amplitude and the phase. Thus, waves can be added as vectors and at the end we have both the final amplitude and the final phase. This representation is quite simple in terms of complex numbers, as follows:

$$E(x,t) = \cos\left(kx - \omega t + \phi_0\right) + i\sin\left(kx - \omega t + \phi_0\right)$$
$$= \cos\phi_0(x,y) + i\sin\phi_0(x,y)$$
$$= R(x,t) + iI(x,t). \tag{4.7}$$

But using Euler's formula

$$E(x,t) = Ae^{i\left(kx - \omega t + \phi_0\right)} = Ae^{i\phi(x,t)}, \tag{4.8}$$

where the disturbance is the real value of this expression. If the numerical or algebraic values of the disturbance of a wave are known, the amplitude A is given by

$$A = \sqrt{R(x,t)^2 + I(x,t)^2} \tag{4.9}$$

and the phase is given by

$$\phi(x,t) = \tan^{-1}\left(\frac{I(x,t)}{R(x,t)}\right). \tag{4.10}$$

4.1.2 ELECTROMAGNETIC WAVES

There is a fixed relation between the electric and magnetic fields that depends on the medium on which they propagate. In a dielectric medium this relation is as follows:

$$H = \frac{n}{c\mu}E, \tag{4.11}$$

where

c is a constant equal to the speed of light in vacuum
n is the refractive index
μ is the medium magnetic permeability, which is almost the same for all dielectrics

Thus, basically the ratio between E and H is a function of the refractive index. The magnitude S of the Poynting vector is the light power (energy per unit time) flowing through a unit area, perpendicular to the direction of propagation of the light beam. The Poynting vector $\mathbf{S}$ can be written in terms of the electric and magnetic fields vectors as

$$\mathbf{S} = \mathbf{E} \times \mathbf{H}. \tag{4.12}$$

If we express the electric and magnetic vector fields of the electromagnetic wave $\mathbf{E}$ and $\mathbf{H}$ in its complex representation, as in Equation 4.8, we get

$$\mathbf{E} = \mathbf{E}_0\, e^{i(kx-\omega t)} \quad \text{and} \quad \mathbf{H} = \mathbf{H}_0 e^{i(kx-\omega t)}. \tag{4.13}$$

Since their instantaneous values in the point z at the time t are their real values, the Poynting vector is

$$\mathbf{S} = \mathrm{Re}(\mathbf{E}) \times \mathrm{Re}(\mathbf{H})$$
$$= \mathbf{E}_0 \times \mathbf{H}_0 \cos^2(kx-\omega t) = \mathbf{S}_0 \cos^2(kx-\omega t), \tag{4.14}$$

where $\mathbf{S}_0$ is the amplitude or the maximum value of the Poynting vector. Since the vectors $\mathbf{E}$ and $\mathbf{H}$ are perpendicular to each other, the first one pointing in the direction of the y-axis and the second in the direction of the z-axis, we can write their instantaneous values E_y and H_z as scalar values, as follows:

$$E_y = E_{y0}e^{i(kx-\omega t)} \quad \text{and} \quad H_z = H_{z0}e^{i(kx-\omega t)}, \tag{4.15}$$

where their scalar amplitudes are E_{y0} and H_{z0}. Then, for dielectrics, the magnitude $\mathbf{S}_0$ of the Poynting vector is given by

$$|\mathbf{S}_0| = \frac{n}{c\mu} E_{y0}^2. \tag{4.16}$$

This means that the flux of energy per second, per unit transverse section area, is directly proportional to the square of the

amplitude of the electric field and to the refractive index. The net flux of energy of an electromagnetic wave is given by the time average of the Poynting vector as follows:

$$\langle \mathbf{S} \rangle = \frac{\mathbf{S}_0}{2}. \tag{4.17}$$

Since the average of $\cos^2(kx-\omega t)$ in a period is equal to 1/2, this value represents the average energy per second, per unit transverse area, in watts per square meter, carried by an electromagnetic wave.

4.1.3 WAVE EQUATION

In a dielectric material with the electrical free charge density $\rho = 0$ and with electrical conductivity σ, the Maxwell equations in the MKS system of units are given by

$$\nabla \times \mathbf{E} = -\mu\frac{\partial \mathbf{H}}{\partial t}$$

$$\nabla \times \mathbf{H} = \sigma\mathbf{E} + \varepsilon\frac{\partial \mathbf{E}}{\partial t} \tag{4.18}$$

$$\nabla \cdot \mathbf{E} = 0$$

$$\nabla \cdot \mathbf{H} = 0.$$

From these equations, after some algebraic manipulation, and using some well-known vector identities, we obtain

$$\nabla^2 \mathbf{E} = \sigma\mu\frac{\partial \mathbf{E}}{\partial t} + \mu\varepsilon\frac{\partial^2 \mathbf{E}}{\partial t^2}, \tag{4.19}$$

which is the wave equation. For the case of dielectrics, this equation reduces to

$$\nabla^2 \mathbf{E} = \mu\varepsilon\frac{\partial^2 \mathbf{E}}{\partial t^2}. \tag{4.20}$$

Solving this equation the behavior of electromagnetic waves in dielectrics and metals can be explained with great detail.

4.1.4 WAVE SUPERPOSITION

Let us assume that we have two electromagnetic waves propagating along the same path on the x-axis. These two waves have the same phase equal to zero at the origin, at the time zero, amplitudes A_1 and A_2, angular frequencies ω_1 and ω_2, and wave numbers k_1 and k_2. Since the two waves are in the same medium, we have

$$\frac{\omega_1}{k_1} = \frac{\omega_2}{k_2} = v. \tag{4.21}$$

The sum of the disturbances is

$$E_T = E_1 + E_2 = A_1 e^{i\theta_1} + A_2 e^{i\theta_2}$$
$$= \left(A_1 + A_2 e^{i(\theta_2 - \theta_1)}\right)e^{i\theta_1}. \tag{4.22}$$

This expression can also be written as

$$E_1 + E_2 = \left(A_1 + A_2 \cos\left(\theta_2 - \theta_1\right) + i\, A_2 \sin\left(\theta_2 - \theta_1\right) \right) e^{i\theta_1}. \quad (4.23)$$

Then, using Euler's formula, we can write this result in the form

$$E_T(x,t) = A_T e^{i\theta_T} \quad (4.24)$$

and we obtain

$$A_T = \left(A_1^2 + A_2^2 + 2A_1 A_2 \cos\left(\theta_2 - \theta_1\right) \right)^{1/2} \quad (4.25)$$

and

$$\theta_T = \theta_1 + \tan^{-1}\left(\frac{A_2 \sin\left(\theta_2 - \theta_1\right)}{A_1 + A_2 \cos\left(\theta_2 - \theta_1\right)} \right), \quad (4.26)$$

which is sometimes written as

$$\theta_T = \theta_1 + \tan^{-1}\left(\frac{2 \sin\left(\dfrac{\theta_2 - \theta_1}{2}\right) \cos\left(\dfrac{\theta_2 - \theta_1}{2}\right)}{\left(\dfrac{A_1}{A_2} - 1\right) + 2\cos^2\left(\dfrac{\theta_2 - \theta_1}{2}\right)} \right). \quad (4.27)$$

This is a general expression for the superposition of two sinusoidal waves with different or equal frequencies and wavelengths.

4.1.4.1 Two waves with the same frequency and wavelength, traveling in the same direction

If the two waves have the same wavelengths and the same frequencies and are traveling in the same direction, we can write the phases as

$$\theta_1 = kx - \omega t \quad \text{and} \quad \theta_2 = kx - \omega t + \phi. \quad (4.28)$$

Thus, the amplitude of the added combination is

$$A_T = \left(A_1^2 + A_2^2 + 2A_1 A_2 \cos\phi \right)^{1/2} \quad (4.29)$$

and the phase

$$\theta_T = kx - \omega t + \tan^{-1}\left(\frac{2 \sin\left(\dfrac{\phi}{2}\right) \cos\left(\dfrac{\phi}{2}\right)}{\left(\dfrac{A_1}{A_2} - 1\right) + 2\cos^2\left(\dfrac{\phi}{2}\right)} \right). \quad (4.30)$$

We can see that the combination of the two waves is another wave with the same wavelength and frequency as the original components. The phase differs from the original waves by just a constant. The amplitude of the combination is also a function of the constant ϕ. If this phase difference between the two waves is zero or a multiple of 2π, the final amplitude is the sum of the amplitudes of the two components.

4.1.4.2 Two waves with different frequency and wavelength, traveling in the same direction

If the wavelength of the two added waves is different, their phases can be written as

$$\theta_1 = k_1 x - \omega_1 t \quad \text{and} \quad \theta_2 = k_2 x - \omega_2 t. \quad (4.31)$$

And thus, the resultant amplitude is

$$A_T = \left(A_1^2 + A_2^2 + 2A_1 A_2 \cos\left(\left(k_2 - k_1\right)x - \left(\omega_2 - \omega_1\right)t\right) \right)^{1/2}. \quad (4.32)$$

This amplitude is not constant. The profile of the square of the amplitude (irradiance) is sinusoidal, as shown in Figure 4.3. The minimum value occurs when the cosine function is equal to –1 and the maximum value when the cosine function is equal to +1. These maximum and minimum values are equal to the sum and the difference of the amplitudes of the two component waves. The wavelength L of this modulating function can be obtained from

$$\frac{1}{L} = \frac{k_2 - k_1}{2\pi} = \frac{1}{\lambda_2} - \frac{1}{\lambda_1}, \quad (4.33)$$

and the frequency f of this modulating function is

$$f = \frac{\omega_2 - \omega_1}{2\pi} = \nu_2 - \nu_1. \quad (4.34)$$

The phase of the inner (or carrier) electromagnetic wave is given by

$$\theta_T = k_1 x - \omega_1 t + \tan^{-1}\left(\frac{2 \sin\left(\dfrac{\left(k_2 - k_1\right)x - \left(\omega_2 - \omega_1\right)t}{2}\right) \times \cos\left(\dfrac{\left(k_2 - k_1\right)x - \left(\omega_2 - \omega_1\right)t}{2}\right)}{\left(\dfrac{A_1}{A_2} - 1\right) + 2\cos^2\left(\dfrac{\left(k_2 - k_1\right)x - \left(\omega_2 - \omega_1\right)t}{2}\right)} \right). \quad (4.35)$$

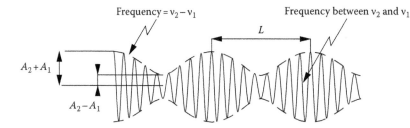

Figure = $\nu_2 - \nu_1$ Frequency between ν_2 and ν_1

L

$A_2 + A_1$

$A_2 - A_1$

Figure 4.3 The profile of the square of the amplitude (irradiance) is sinusoidal in a modulated wave.

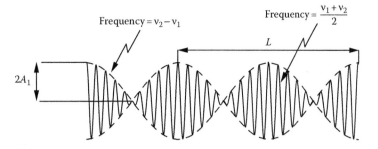

Figure 4.4 The profile of the amplitude modulation envelope is sinusoidal in a modulated wave with equal amplitudes for two components.

An interesting particular case occurs when the amplitudes of the two waves traveling along the same path are equal to A. Then,

$$A_T = \sqrt{2}A\left(1+\cos\left((k_2-k_1)x-(\omega_2-\omega_1)t\right)\right)^{1/2}$$

$$= 2A\cos\left(\frac{(k_2-k_1)x}{2}-\frac{(\omega_2-\omega_1)t}{2}\right). \qquad (4.36)$$

Thus, the profile of the amplitude modulation envelope is sinusoidal, as shown in Figure 4.4. The phase is given by

$$\theta_T = \frac{(k_1x-\omega_1t)+(k_2x-\omega_2t)}{2} = \frac{(k_2+k_1)x-(\omega_2+\omega_1)t}{2}. \qquad (4.37)$$

4.1.4.3 Two waves with the same frequency and wavelength, traveling in opposite directions

Let us now consider the two waves traveling in opposite directions. The phases of these two waves can be written as

$$\theta_1 = kx-\omega t \quad \text{and} \quad \theta_2 = -kx-\omega t. \qquad (4.38)$$

Thus, using Equation 4.16 we have

$$E_T = A_1 e^{i(kx-\omega t)} + A_2 e^{i(-kx-\omega t)}$$

$$= \left(A_1 e^{i(kx)} + A_2 2\, e^{i(-kx)}\right)e^{-i\omega t}$$

$$= \left((A_1-A_2)e^{ikx} + 2A_2(\cos kx)\right)e^{-i\omega t}. \qquad (4.39)$$

The first term in this expression represents a wave traveling in the positive direction of x, with amplitude equal to the difference of the amplitudes of the two waves. If the two amplitudes are equal, this traveling wave does not exist. The second term is a stationary wave oscillating in time, but without traveling along the x-axis.

4.1.4.4 Stationary waves

Stationary waves can be produced in the laboratory in two simple manners. The first one is by reflecting a light beam in a plane mirror. Winner in 1890 registered these stationary waves by means of an inclined photographic plate. In 1891, Gabriel Lippmann recorded the stationary waves by means of a thick photographic emulsion. A series of clear and dark layers were produced in the photographic emulsion (Figure 4.5). Lippmann was awarded the Nobel Prize in Physics in 1908 due to an invention to produce color photographs with this method.

The analysis of the behavior of the electric and magnetic fields and the Poynting vector in stationary electromagnetic waves is interesting, as we will describe. Let us assume that the electric field of the first of two waves with the same amplitude A_0, traveling in opposite directions is

$$E_{1y} = A_0 e^{i(kx-\omega t)} \qquad (4.40)$$

and that the magnetic field is

$$H_{1z} = A_0 \frac{n}{c\mu} e^{i(kx-\omega t)}. \qquad (4.41)$$

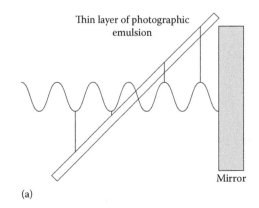

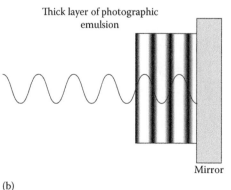

(a)   (b)

Figure 4.5 (a) Wiener's experiment and (b) Lippmann's experiment.

The electric field for the second wave, traveling in opposite direction is

$$E_{2y} = A_0\, e^{i(-kx-\omega t)}, \qquad (4.42)$$

and its magnetic field is

$$H_{2z} = -A_0\, \frac{n}{c\mu}\, e^{i(-kx-\omega t)}. \qquad (4.43)$$

Hence, the total electric field for the combination of the two waves is

$$E_{Ty} = A_0\left(e^{ikx} + e^{-ikx}\right)e^{-i\omega t} = 2A_0 \cos kx\, e^{-i\omega t}, \qquad (4.44)$$

and the corresponding magnetic field is

$$H_{Tz} = A_0 \frac{n}{c\mu}\left(e^{ikx} - e^{-ikx}\right)e^{-i\omega t}$$

$$= 2iA_0 \frac{n}{c\mu}\sin kx\, e^{-i\omega t}$$

$$= -2A_0 \frac{n}{c\mu}\cos\left(kx \pm \pi/2\right)e^{-i(\omega t \mp \pi/2)}. \qquad (4.45)$$

We may observe that while in a single traveling electromagnetic wave the electric and magnetic fields are in phase, in a stationary wave they are out of phase by 90° in both space and time (Figure 4.6). The scalar value S of the Poynting vector has a positive value if this vector points in the positive direction of x and negative otherwise. It is given by

$$S = \mp A_0^2 \left(\frac{n}{c\mu}\right)^2 \sin 2kx \sin 2\omega t. \qquad (4.46)$$

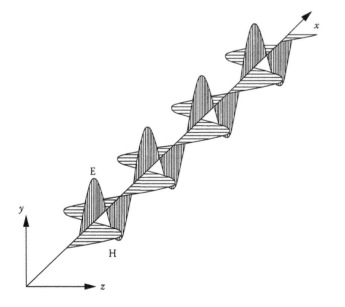

Since the functions sin $2kx$ and sin $2\omega t$ have positive and negative values, depending on the angle, the amplitude of the Poynting vector also has positive and negative values. This means that the energy is oscillating, along the x-axis. The average of the Poynting vector is zero, since the average of sin $2kx$ and cos $2kx$ in one period is equal to zero.

4.2 INTERFERENCES

When two light beams with the same frequency and wavelength travel along the same path, the final amplitude of the combination is a function of the phase difference between them, as in Equation 4.29. If the phase difference is equal to $2\pi m$, where m is an integer, the amplitude is equal to the sum of the amplitudes of the components. If the phase difference is equal to $2\pi(m + 1/2)$, the amplitude of the sum of these two beams is equal to the difference of the two amplitudes. This phenomenon, known as interference, will be described with detail in the next sections.

4.2.1 COHERENCE OF A LIGHT BEAM

The concept of coherence between two light sources is related to their capacity to interfere and form good contrast fringes. To interfere as described in Section 4.1.4.1, it is necessary that the two light waves being added have a constant phase difference. In other words, it should not change with time. This is possible only if the two light waves are perfectly synchronized. This synchronization is almost impossible if the two light beams originate from different light sources or from different point in an extended light source. If the two beams originate at different light sources, their relative phase will change randomly, very fast in time. Then, we say that the two beams are incoherent to each other. In this case, the amplitude of the combination will change quite fast in time. We observe that the combination has irradiance equal to the sum of the irradiances of the two components, since the average of the cosine function in Equation 4.29 is equal to zero.

Let us assume that we have two partially coherent light sources, as illustrated in Figure 4.7, and we wish to calculate the final irradiance at the point P on the observing screen. The light source 1 produces an optical disturbance $U_1(t)$, at the point P, with a light wave that was emitted at the time t. The light source 2 produces an optical disturbance $U_2(t + \tau)$ at the same point P, with a light wave emitted at the time $t + \tau$.

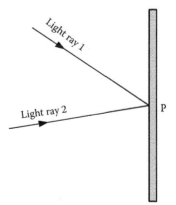

Figure 4.6 Electric and magnetic fields in a stationary electromagnetic wave. The fields are out of phase by 90° in both space and time.

Figure 4.7 Two partially coherent light source illuminating a point **P**.

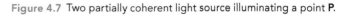

Therefore, the irradiance at this point P is the complex square of the sum of the two disturbances:

$$I(P) = \left[U_1(t) + U_2(t + \tau)\right] \cdot \left[U_1^*(t) + U_2^*(t + \tau)\right], \quad (4.47)$$

where U_1^* and $U_2^*(t + \tau)$ are the complex conjugates of $U_1(t)$ and $U_2(t + \tau)$, respectively.

The irradiances I_1 and I_2 at the point P, due to the illumination of the two light sources 1 and 2, are the following:

$$\begin{aligned} I_1 &= U_1(t)U_1^*(t) \\ I_2 &= U_2(t + \tau)U_2^*(t + \tau). \end{aligned} \quad (4.48)$$

Then, we can prove that the total irradiance at the point P is given by

$$I(P) = I_1 + I_2 + U_1(t) \cdot U_2^*(t + \tau) + U_1^*(t) \cdot U_2(t + \tau). \quad (4.49)$$

Now, if we now define the *mutual coherence function* $\Gamma_{12}(\tau)$ as

$$\Gamma_{12}(\tau) = U_1(t)U_2^*(t + \tau), \quad (4.50)$$

where we may observe that this mutual coherence function has the units of irradiance, then we can write this expression for the irradiance as

$$I(P) = I_1 + I_2 + 2\Gamma_{12}^{(r)}(\tau), \quad (4.51)$$

where $\Gamma_{12}^{(r)}(\tau)$ is the real part of $\Gamma_{12}(\tau)$.

This irradiance value is completely general, independent of the degree of coherence on the light sources. If the light sources are totally incoherent, the real part of the mutual coherence function is zero. On the other hand, if the light sources are fully coherent, we can write the optical disturbances as

$$U_1(t) = A_{01} \, e^{-i\omega t} \quad (4.52)$$

and

$$U_2^*(t + \tau) = A_{02} \, e^{-i\omega(t + \tau)}. \quad (4.53)$$

Hence, the mutual coherence function can be written as

$$\Gamma_{12}(\tau) = A_{01} A_{02} \, e^{i\omega t} \quad (4.54)$$

and its real part as

$$\Gamma_{12}^{(r)}(\tau) = A_{01} A_{02} \cos \omega t. \quad (4.55)$$

If we substitute this expression in Equation 4.51, we obtain the irradiance for the sum of two coherent light waves. Now, we define the *degree of complex coherence* as the following expression without units:

$$\gamma_{12}(\tau) = \frac{\Gamma_{12}(\tau)}{\sqrt{I_1 I_2}}. \quad (4.56)$$

And its real part as the *normalized correlation function* is as follows:

$$\gamma_{12}^{(r)}(\tau) = \mathrm{Re}\left[\gamma_{12}(\tau)\right]. \quad (4.57)$$

Then, we can write the irradiance as

$$\begin{aligned} I(P) &= I_1 + I_2 + 2\sqrt{I_1 I_2}\,\gamma_{12}^{(r)}(\tau) \\ &= I_1 + I_2 + 2\sqrt{I_1 I_2}\,\left|\gamma_{12}(\tau)\right|\cos \alpha, \end{aligned} \quad (4.58)$$

where

$\left|\gamma_{12}(\tau)\right|$ is the modulus or magnitude of the degree of complex coherence

α is the phase of the degree of complex coherence

The normalized correlation function has oscillatory values about zero, whose value depends on the relative positions of the light sources. The amplitude of these variations depends on the degree of coherence of the two light sources, which is between –1 and +1 for perfectly coherent sources and it is constant for noncoherent sources. The amplitude of the oscillation of $\gamma_{12}(\tau)$ is the module of $\gamma_{12}(\tau)$, represented by $\left|\gamma_{12}(\tau)\right|$.

When the two light waves are partially coherent, we represent the minimum and maximum values of the irradiance in the interference pattern by $I_{mín}$ and $I_{máx}$, respectively. If there is perfect coherence, $I_{mín}$ is zero, but $I_{máx}$ is different from zero. For partial coherence, $I_{mín} \neq I_{máx}$, with both being different from zero. Following Michelson's definition of contrast or visibility of the interference fringes, it can be specified by

$$V = \frac{I_{máx} - I_{mín}}{I_{máx} + I_{mín}}, \quad (4.59)$$

with values between zero and one. Thus, there is an alternative manner to write the degree of coherence of two light waves, by means of the visibility of the interference fringes, as we will see. From Equation 4.58, the maximum and minimum irradiances are given by

$$I_{máx} = I_1 + I_2 + 2\sqrt{I_1 I_2}\,\left|\gamma_{12}^{(r)}(\tau)\right| \quad (4.60)$$

when $\cos \alpha = 1$ and

$$I_{mín} = I_1 + I_2 - 2\sqrt{I_1 I_2}\,\left|\gamma_{12}^{(r)}(\tau)\right| \quad (4.61)$$

when $\cos \alpha = -1$. Substituting these values in the equation for the visibility, we obtain

$$V = 2\left|\gamma_{12}^{(r)}(\tau)\right|\frac{\sqrt{I_1 I_2}}{(I_1 + I_2)}. \quad (4.62)$$

Frequently, the magnitude of the degree of complex coherence is determined by measuring the contrast of the interference pattern.

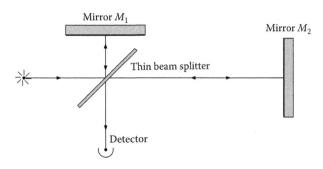

Figure 4.8 Interference with two different light paths, to explain the temporal coherence concept.

4.2.1.1 Temporal coherence

Let us assume that the two interfering light waves arrive to the observing screen along the same path and that also were emitted from the same point and in the same direction from the illuminating extended light source. Then, their phase difference is only due to the traveling time difference between the two light beams. These interference conditions can be applied with the arrangement in Figure 4.8. The beam splitter divides the amplitude of the incident beam and sends one beam to mirror M_1 and another beam to mirror M_2. The two beams come back to the beam splitter, where they recombine and send them to a light detector or observation screen. Since the two mirrors are not necessarily at the same distance from the beam splitter, the two light beams arrive to the detector with different traveling times. Then, the *self-coherence function* is defined as

$$\Gamma_{11}(\tau) = U_1(t) \cdot U_1^*(t + \tau) \qquad (4.63)$$

and the *degree of complex self-coherence* as

$$\gamma_{12}(\tau) = \frac{\Gamma_{11}(\tau)}{I_1}. \qquad (4.64)$$

The values of this function oscillate symmetrically about the zero value when the difference τ between the traveled times for the two light beams changes. This time difference τ can change if any of the two mirrors is displaced along its perpendicular. If the degree of complex self-coherence is zero, we say that light source is temporarily incoherent. This happens when the light beam is not monochromatic. The reason is that, as we described

in Section 4.1.4.2, when light beams of different wavelengths are mixed, the light beam is modulated.

If the light source is partially monochromatic, the oscillations in the value of the degree of complex self-coherence have small amplitude with neither maxima nor minima well defined. There are two equivalent manners to describe this, as we will see next.

a. Each color or frequency emitted by the light source produces its own fringe pattern in the detector, with the fringe positions and spacing, different for each color. When several colors are present, the different fringe patterns will be different, producing a confusing result with low contrast.

b. If the light source is not monochromatic, the wave train will have a short length. To say it in different words, the light beam will have amplitude modulation. If regions of the wave train, with different modulation amplitudes interfere, the fringe visibility will be low. Hence, the fringe contrast will change for different optical path differences.

In conclusion, the more monochromatic a light source is, the larger its wave train is and also its time of coherence. The degree of temporal coherence can be specified by the length of the wave train or by the length of the time of coherence, which is the coherence length divided by the group velocity.

Let us consider now a spectral line with width $\Delta\lambda$, as in Figure 4.9. Then, the wave train is finite in length, not periodical, and has a group length L. If we have a combination of two separated spectral lines with a certain width, the wave train has a shape as illustrated in Figure 4.10. The temporal coherence or wave train length of some common light sources is in Table 4.1.

A case of particular interest is when two light waves with slightly different frequency are added, as described in

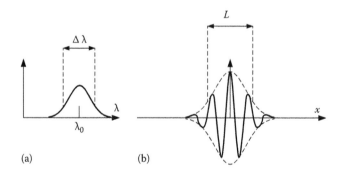

(a)   (b)

Figure 4.9 (a) A spectral line with a certain width and (b) its wave train.

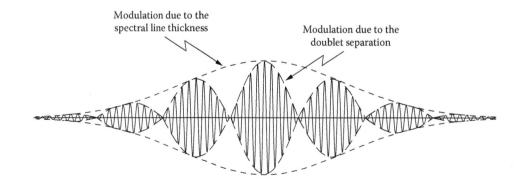

Figure 4.10 Wave train produced by a pair of spectral lines separated by $\Delta\lambda$.

Table 4.1 **Coherence lengths for some common light sources**

LIGHT SOURCE	WAVELENGTH	COHERENCE LENGTH
White light[a]	White light	2 μm
Sodium doublet	589 nm	0.6 mm
Laser diode (multimode)	650 nm	0.8 mm
Ion argon	488/515 nm	20 mm
Gas discharge spectral line	656.3 nm	15–30 cm
He–Ne laser (multimode)	632.8 nm	20–80 cm
GaAlAs laser (single mode)	670–905 nm	3 m
He–Ne laser (single mode)	632.8 nm	20–200 m

[a] The spectral width is the half-width of the sensitivity curve for the human eye.

Section 4.1.4.2, but completely incoherent to each other, because their frequency difference is fast and randomly changing. This is the case of the sodium yellow doublet in Table 4.1. In this case, the modulation pulse separations and lengths also change, so fast, that no detector is able to detect and measure.

4.2.1.2 Spatial coherence

Two interfering light sources are separated, but placed at equal distances from the observation point **P**, so that the traveling time difference τ is zero. Then, we can define the *mutual coherence time* as

$$\Gamma_{12}(0) = U_1(t) \cdot U_2^*(t) \qquad (4.65)$$

and the degree of *complex mutual coherence* as

$$\gamma_{12}(0) = \frac{\Gamma_{12}(0)}{\sqrt{I_1 I_2}}, \qquad (4.66)$$

which changes its magnitude when the separation **S** between the two light sources is modified.

Figure 4.11 shows two slits (or pinholes), illuminated by an extended light source. The two slits or pinholes can be considered as secondary light sources whose spatial coherence decreases when their separation increases. The mutual coherence of the two slits is a function of the angular diameter of the extended light source as observed from the plane of the slits and also of the degree of coherence of the light source. The two slits will have a perfect mutual coherence only if the angular diameter of the light source is zero or if the illuminating light source is perfectly coherent, as we will see next. If the two slits or pinholes have the same area, their irradiances I_1 and I_2 on the observing screen will be the same, with a value equal to I. Hence, the fringe visibility is given by

$$V = \left| \gamma_{12}^{(r)}(0) \right| I. \qquad (4.67)$$

An extended light source is coherent if any two points in the light source are mutually coherent. It is easy to conclude that a sufficient but not necessary condition for an extended light source to be coherent is that it should be monochromatic.

4.2.1.3 Van Cittert–Zernike theorem

This theorem allows us to evaluate the degree of spatial coherence between to separated points in space, illuminated by an extended incoherent light source. Let us assume that we have a plane extended light source with low spatial coherence plane, as shown in Figure 4.12. In an observation plane, parallel to the light source, there are two points S_1 and S_2. Let us now assume that the observation plane is an opaque screen and that there are two pinholes at the points S_1 and S_2, where the light can go through and diffract. Thus, these pinholes can be considered as two point light sources. Then, an observing screen, parallel to the light source and to the plane with the

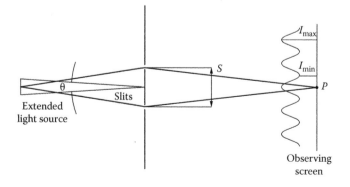

Figure 4.11 A pair of slits (or pinholes) illuminated with as single light source, to explain the spatial coherence concept.

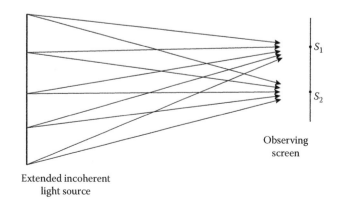

Figure 4.12 Illustration of the van Cittert–Zernike theorem.

pinholes, is placed in front of these two slits. Then, interference fringes appear in the observing screen.

However, since the light source is extended, the light beams passing through the two pinholes S_1 and S_2 are not fully mutually coherent. This means that there is not perfect spatial coherence. Thus, the fringe contrast is not high, unless the extended light source has a small angular diameter, as seen from the plane of the two pinholes S_1 and S_2. The van Cittert–Zernike method allows us to evaluate the degree of coherence between the two pinholes $|\gamma_{12}^{(r)}(0)|$, so that we can evaluate the visibility of the interference fringes:

$$|\gamma_{12}(\tau)| = \left| \frac{\iint_S I(\xi,\eta) e^{ik(\xi\theta_x + \eta\theta_y)} \, d\xi d\eta}{\iint_S I(\xi,\eta) d\xi d\eta} \right|, \tag{4.68}$$

where $k = 2\pi/\lambda$, the coordinates (ξ, η) are in the light source plane, and the irradiance emitted by each of the two pinholes is $I(\xi, \eta)$. The two points S_1 and S_2 have an angular separation θ as seen from the light source. Since the relative positions of the two pinholes can be any pair of points in the plane x–y, this angle θ has two components θ_x and θ_y, along the axes x and y, respectively. The integrals are evaluated in the plane of the light source.

Equation 4.68 is a general expression for any possible configuration, but there are two cases of great practical interest. The first one is when the light source is a long slit with a small width w:

$$|\gamma_{12}(0)| = \left| \mathrm{sinc}\left(\frac{w\theta_x}{\lambda} \right) \right| = \left| \mathrm{sinc}\left(\frac{wa}{\lambda z} \right) \right|, \tag{4.69}$$

where

   w is the light source width
   z is the distance to the plane with the two points S_1 and S_2
   a is the separation between these two points, measured along the x-axis

The function $\mathrm{sin}\,\phi$ is equal to $(\sin\phi)/\phi$.

The second configuration of practical interest is when the light source is a circle with diameter d:

$$|\gamma_{12}(0)| = \left| \frac{2J_1\left(\frac{\pi d\theta_x}{\lambda} \right)}{\left(\frac{\pi d\theta_x}{\lambda} \right)} \right| = \left| \frac{2J_1\left(\frac{\pi da}{\lambda} \right)}{\left(\frac{\pi da}{\lambda} \right)} \right|, \tag{4.70}$$

where

   d is the diameter of the light source
   z is the distance to the plane with the points S_1 and S_2
   a is the separation between these two points, measured along the x-axis

The function $(J_1(\phi)/\phi)$ is the Airy function, where $J_1(\phi)$ is the first-order Bessel function.

4.2.2 YOUNG'S DOUBLE-SLIT EXPERIMENT

If we illuminate a pair of slits with a single light source as illustrated in Figure 4.13, the light passing through each of the slits diffracts, angularly expanding the light and illuminating the observation plane. The distances from the slits to a point P on the screen are not necessarily the same. Thus, the optical path difference OPD is a function of the position of the point P on the observing screen. For a relatively small field, the fringe pattern consists of straight and parallel fringes. The optical path difference and the condition for a bright fringe can be written as

$$OPD = A + B - C - D = m\lambda$$
$$\approx d\sin\theta = m\lambda. \tag{4.71}$$

The lower approximation is made assuming that the light source is at the axis of symmetry of the system and that the observation plane is far from the slits, as compared with the slit separation d. If the two slits have the same narrow width, so that the irradiance of both slit is the same, the irradiance on the observation screen is given by

$$I = 2I_0\left[1 + \cos(kOPD)\right] =$$
$$= 4I_0 \cos^2 \frac{kOPD}{2} \approx 4I_0 \cos^2 \frac{\pi d \sin\theta}{\lambda}. \tag{4.72}$$

The bright fringes are located where the cosine argument is equal to zero or a multiple of π. Then, the value of $(d\sin\theta)/\lambda$ is an integer. Then, the angular separation between two fringes, as seen from the slits, is equal to λ/d. Let us assume that the observation is made from the plane of the slits, or by looking through the slits, with the light source at the front. Since the angular resolution of the human eye is a minute of arc, the maximum slit separation has to be equal to 1.7 mm. If the slits are not extremely narrow, the fringes position is the same, but their irradiance decreases for large values of the angle θ. If the wavelength λ is slightly modified by $\Delta\lambda$, the order of interference m at a point P where a bright fringe was is no longer an integer. Then, the fringe is displaced, and the order of interference at the point P has the new value:

$$\Delta m = -\frac{OPD}{\lambda_0^2}\Delta\lambda. \tag{4.73}$$

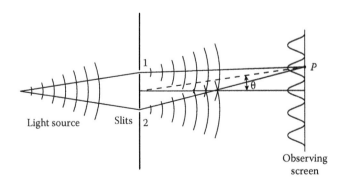

Figure 4.13 Young's double-slit experiment.

From this relation we can see that if the optical path difference *OPD* is large, small wavelength changes Δλ produce large fringe displacements. A good fringe contrast for a large field is obtained only if a highly monochromatic light source is used. A white light source produces fringes in a relatively small field.

If the point light source is displaced in a direction perpendicular to the slits, the optical path difference *OPD* changes, displacing the fringes in the opposite direction. Thus, if the light source is extended, the fringe contrast decreases. To say it with other words, a good contrast is achieved only with a light source with a good spatial coherence is used. The only way to increase the contrast with a light source with low spatial coherence is by reducing the fringe separation, thus increasing the fringe separations. The contrast remains acceptable if the variations in the optical path difference Δ*OPD*, for different points of the extended light source, are smaller than one-fourth of the wavelength of the light.

4.2.3 MICHELSON INTERFEROMETER

This is probably one of the most popular of all interferometers, designed by Albert A. Michelson (1852–1931). Figure 4.14 illustrates this interferometer, where the light from an extended light source goes to a semireflecting beam splitter P_1 that divides the light beam in amplitude. One of the two beams goes to the mirror M_1 and the other goes to the mirror M_2. After reflection on these mirrors, the beams come back to the beam splitter, where they join again and then go to the observing screen or to the eye of the observer. It is important to notice the light beam that gets reflected at the mirror M_1 travels twice through the beam splitter plate, before reaching the observer. However, the other light beam, going to the mirror M_2, travels through the beam splitter plate only once. This instrument is an uncompensated interferometer. If a glass plate P_2 with identical glass and thickness is added to the

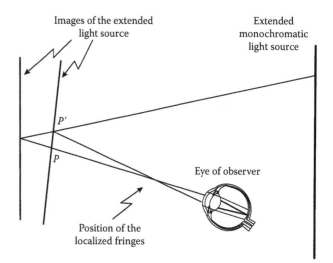

Figure 4.15 Observed two virtual images of the light source in the Michelson interferometer.

interferometer, in the path going to the mirror M_2, then, both paths from the light source to the observer will travel the same amount of glass. This is a compensated interferometer. The observer looking through the instrument will see two virtual images of the light source (Figure 4.15). One of the images is formed by the reflection on the mirror M_1 and the other image is formed by the reflection on the mirror M_2. The longitudinal separation between the two virtual images of the light source is equal to the optical path difference, only if the interferometer is compensated. Otherwise, they are different.

4.2.3.1 Coherence requirements

The temporal and spatial coherence requirements for the light source increase with the optical path difference, mainly if the Michelson interferometer is not compensated. This fact can be explained as follows. First, let us consider a nonmonochromatic light beam traveling a glass plate. Since the light beam is not monochromatic, its wave train is finite, with a modulation function. Due to the glass chromatic dispersion, when the light travels the plate, each one of the monochromatic components has a different phase delay when they exit the plate with refractive index n and thickness t, according to the expression in the following:

$$\theta = \frac{2\pi}{\lambda}(n-1)t. \tag{4.74}$$

So, when the monochromatic light components are out, they will have a different phase relation than when they entered the plate, producing a wave train with a different shape than the one at the entrance. If the interferometer is not compensated, one of the beams travels through the plate three times, while the other beam travels through it only once. This causes that the two interfering light beams have different wave trains with different modulating function. However, the difference is not great if the spectral width is not too large. The interference fringes will have a good visibility only if the two interfering wave trains have the same shape and are not one ahead of the other. If the spectral width is not extremely large, the wave trains can be made to

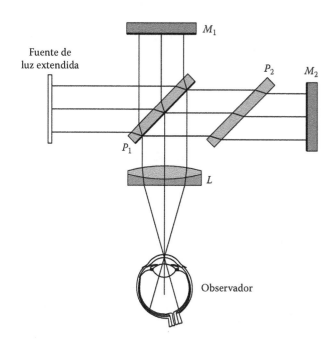

Figure 4.14 Michelson interferometer.

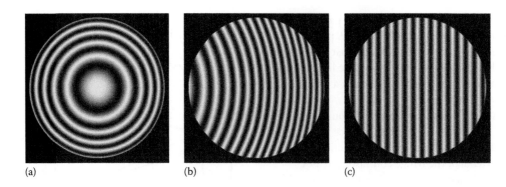

Figure 4.16 Fringes in a Michelson interferometer. (a) Equal inclination fringes, (b) localized fringes, and (c) equal thickness fringes.

coincide by just moving one of the two mirrors along its axis. This means that the difference of optical paths is made equal to zero for the central wavelength. When the interferometer is illuminated with white light or the spectral width is too large, the interferometer has to be compensated.

An alternative equivalent explanation for the temporal coherence requirement is that the optical path difference should be smaller than a quarter wavelength for all wavelengths in the light beam. If the light source is white, it is impossible to satisfy this condition.

4.2.3.2 Interference fringes classification

The two observed virtual images of the extended light source are one behind the other and parallel to each other, only if the two mirrors are exactly perpendicular to the light beams arriving to them. By tilting or axially displacing the two mirrors, the relative positions of the two virtual images of the light source can be modified, producing different types of fringes. These can be equal inclination fringes, equal thickness fringes, and localized fringes. Next, these fringes will be described.

If the two virtual images of the light source are axially separated, but parallel to each other, the optical path difference (*OPD*) is a function of the angle of observation with respect to the optical axis. They will be concentric rings and are called "equal inclination fringes." In uncompensated interferometers, the inclined beam splitter produces slightly elliptical fringes. These fringes are observed at an infinite distance by an unaccommodated eye or with a telescope.

Now, let us assume that the two virtual images of the extended light source are not exactly parallel to each other, but form a small angle between them. Then, the center of the ring fringes in the fringe pattern displaces to one side. Then, the fringes will transform in arcs and may eventually become straight and parallel, when the angle is quite large. Under these conditions the optical path difference is a function of the local separation between the two images of the light source. The interference fringes thus observed are called "equal thickness fringes." Strictly speaking, the fringes are of the equal thickness type only if the eye or observing optical system is focused at an infinite distance. When the eye is not focused at an infinite distance, the observed fringes will be of an intermediate type to the equal inclination and the equal thickness fringes. The interference pattern is formed by localized fringes. The fringes are arcs with its convexity toward the thinner side of the wedge between the virtual images

of the light source. A light ray exits one of the images of the light source at the point *P* on the other from the corresponding point at the other light source and with the same inclination respect to the image. The point where they cross is located near the light sources. For this reason, these are called "localized fringes." Figure 4.16 illustrates the interference fringes that can be obtained with a Michelson interferometer.

4.2.3.3 Complementary interference pattern

In a Michelson interferometer, as in all other interferometers with a dielectric beam splitter (and most with a metal beam splitter), two complementary fringe patterns are formed. One of the two patterns is the one already described. The second fringe pattern would be observed from the light source, looking toward the interferometer. As illustrated in Figure 4.17, the two virtual images of the light source reflected from each of the two mirrors, M_1 and M_2, will be observed from the light source. They form an interference pattern. We can easily find that the optical path difference for both fringe patterns is the same. So, if there is destructive or constructive interference in one of the patterns, it will be identical in the same place in the other pattern. This seems to contradict the principle of conservation of energy. If there is constructive interference, the amplitude is duplicated and the irradiance is multiplied by four. If there is destructive interference, the amplitudes subtract and the irradiance becomes zero. Apparently in one case, energy is created and in the other destroyed.

This paradox is solved if we remember from the electromagnetic theory of light or from the Stokes relations that upon internal and external reflection in a dielectric like glass, there is one and only

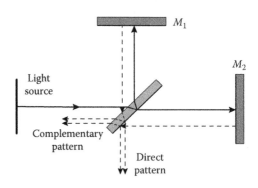

Figure 4.17 Observation of two complementary interference patterns in a Michelson interferometer.

one change of phase equal to 180°. Since the number of external and internal reflection is not the same for both interference patterns, both of them will be complementary. In other words, a bright fringe in one of them corresponds to a dark fringe in the other. If the beam splitter is metallic, the two fringe patterns are not exactly complementary, since part of the energy heats the metal film.

4.3 DIFFRACTION

We have seen in Section 4.1.4.1 that the length of any wave train cannot be shortened without dispersing or widening its wavelength spectrum. In a similar manner, the width of a wavefront cannot be reduced with a diaphragm without angularly dispersing it in a wider angle. The former phenomenon could be named temporal diffraction and the latter spatial diffraction or just diffraction. This phenomenon is illustrated in Figure 4.18, where the light illuminates a point on the screen that is outside of the geometrical limits. Diffraction was first observed in 1665 by Francesco Maria Grimaldi, who was the first to use the term diffraction.

4.3.1 HUYGENS AND HUYGENS–FRESNEL THEORY

To explain diffraction, Christian Huygens proposed that after the diffracting diaphragm, each point in the wavefront acts as a new source of secondary spherical wavelets. This model, illustrated in Figure 4.19, is known as the Huygens principle. According to Huygens, after the diffracting aperture the secondary wavelets add up producing a new wavefront that acts as an envelope for all secondary wavelets. The wavelets add together in such a way that the final wavefront is the envelope. Since at the edge the envelope is curved, the light travels outside of the geometrical paths. Thus, the angular deflection of the light beam near the edges of the aperture is explained, but not the presence of fringes near the shadows.

Augustin Jean Fresnel modified Huygens' theory almost a century later. He assumed that the wavelets not only formed an envelope to produce the new wavefront, but that they interfere among them, as in common interference phenomena. Thus, to find the light pattern after diffraction by an aperture, the wavelets have to be added at every point on the observing screen, taking into account their relative amplitudes and phases. This model is known as the Huygens–Fresnel principle. The success of this theory is so great that the irradiances on the screen can

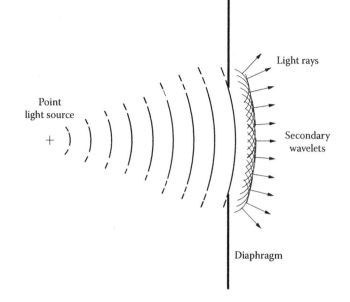

Figure 4.19 Huygens principle to explain diffraction.

be calculated with high accuracy, including the presence of any fringes. However, this theory has two problems. One of them is that it cannot explain why the wavelets are propagated only forward and not backward. The second deficiency is that the phases at the observing screen are not correctly evaluated, since they are off by 90°, with respect to their real measured values. These deficiencies in Huygens–Fresnel theory were corrected by Kirchhoff in 1876. In spite of the great success of Kirchhoff's theory in the calculation of diffraction patterns, this theory can still be improved. Better and more accurate results are obtained using the electromagnetic theory.

4.3.2 KIRCHHOFF THEORY

Kirchhoff proved in 1876 that the intuitive but powerful theory by Huygens and Fresnel could be mathematically justified with an integral theorem based in the wave equation. Let us begin by mentioning, without deriving, the Helmholtz–Kirchhoff theorem. Further details can be found in any advanced book in physical optics and diffraction. This theory begins with the Green theorem. Let us assume that we have a closed surface, as in Figure 4.20. The theorem states that if we know the optical disturbance or its normal values over a closed surface, it is possible to calculate the values of the optical disturbance at any

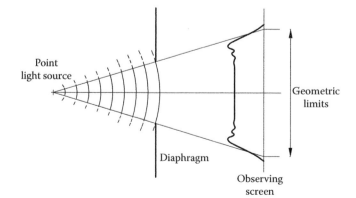

Figure 4.18 Diffraction of a light beam.

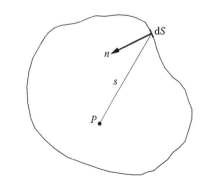

Figure 4.20 Helmholtz–Kirchhoff theorem.

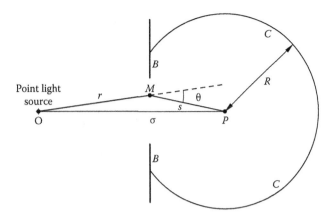

Figure 4.21 Surface to derive Kirchhoff's integral.

point inside this closed volume. The optical disturbance can be the result of the illumination produced by any distribution of light sources, inside as well as outside the closed volume. The theorem allows us the calculation of this optical disturbance $U(P)$ at the observing point P inside the volume, as follows:

$$U(P) = \frac{1}{4\pi} \iint_S \left(U \frac{\partial}{\partial n}\left(\frac{e^{iks}}{s} \right) - \frac{e^{iks}}{s} \frac{\partial U}{\partial n} \right) dS, \qquad (4.75)$$

where the integration should be performed over the whole closed surface S.

Observing Figure 4.21, let us assume that the optical disturbance is produced by a point light source at the point O outside the closed volume. Since the amplitude produced by the light source is inversely proportional to the distance to the light source, the disturbance at any point M on the surface would be given by

$$U_M = \frac{A\,e^{ikr}}{r}. \qquad (4.76)$$

Hence, we can find

$$\frac{\partial U_M}{\partial n} = \cos(n,r)\frac{\partial}{\partial r}\left(\frac{A\,e^{ikr}}{r} \right)$$

$$= \frac{A\,e^{ikr}}{r}\left(ik - \frac{1}{r} \right)\cos(n,r), \qquad (4.77)$$

where (n, r) is the angle between the line r and the normal to the surface. If the distance r is much larger than the wavelength of the light, we have

$$\frac{1}{r} \ll k. \qquad (4.78)$$

Therefore,

$$\frac{\partial U_M}{\partial n} = \frac{iAk\,e^{ikr}}{r}\cos(n,r), \qquad (4.79)$$

and in an analogous manner we can find

$$\frac{\partial}{\partial n}\left(\frac{e^{iks}}{s} \right) = \frac{iAk\,e^{iks}}{s}\cos(n,s), \qquad (4.80)$$

and if we substitute Equations 4.76, 4.89, and 4.80 in Equation 4.75, we find

$$U(P) = \frac{iA}{2\lambda} \iint_S \frac{e^{ik(r+s)}}{rs}\left[\cos(n,s) + \cos(n,r) \right] dS. \qquad (4.81)$$

This expression can be used to solve diffraction problems, by taking the closed surface as in Figure 4.21. It is formed by three parts, which are the area A inside of the diffracting aperture, the area B on the diaphragm, and the spherical fragment C with center at the observation point P.

The optical disturbance U over region B on the diaphragm is zero, since it is opaque and the light does not go through it. Then, the spherical fragment C is made to grow to an infinite, making its contribution to the optical disturbance equal to zero. Thus, the only nonzero contribution comes from the aperture σ. As a result, this expression becomes the well-known Kirchhoff integral, with σ being the diffracting aperture:

$$U(P) = \frac{iA}{2\lambda} \iint_\sigma \frac{e^{ik(r+s)}}{rs}\left[\cos(n,s) + \cos(n,r) \right] dS. \qquad (4.82)$$

If the angle between the incident light ray and the diffracted light ray is θ, as illustrated in Figure 4.21, we can see that

$$\cos(n,s) + \cos(n,r) = 1 + \cos\theta. \qquad (4.83)$$

Hence, the Kirchhoff integral becomes

$$U(P) = \frac{iA}{\lambda} \iint_\sigma \left(\frac{1 + \cos\theta}{2} \right)\frac{e^{ik(r+s)}}{rs}\, dS. \qquad (4.84)$$

Observing this expression, we can see that there are many similarities with the Huygens–Fresnel equation, but also some differences. The first evident difference is the presence of the imaginary letter i in front of the integral. The meaning is that the phase at the observation point P is shifted by 90° with respect to the one obtained with the Huygens–Fresnel theory.

Another characteristic of this expression is the presence of the so-called inclination factor $(1 + \cos\theta)/2$. With this factor, the amplitude of the wavelets decreases when the diffraction angle grows.

A third characteristic is the factor $1/(rs)$, which takes into account the fact that the amplitude decreases inversely with the distance.

4.3.3 FRESNEL DIFFRACTION

For historical reasons, diffraction phenomena are classified as follows: (1) Fresnel diffraction, when the light source, the observation plane, or both are at a finite distance from the diffracting diaphragm, and (2) Fraunhofer diffraction, when both the light source and the observation plane are at an infinite distance from the diffracting aperture.

4.3.3.1 Single slit: Cornu spiral

Using the Huygens–Fresnel theory, we can find the Fresnel diffraction pattern of a slit with width S. The geometry of this problem is in Figure 4.22, where O is a point light source and P is a point on the screen, where the final amplitude is to be calculated. This point P is on the plane of symmetry of the slit. We divide the cylindrical wavefront in horizontal strips with width ds and height s. The optical path difference between the light passing through point Q in the horizontal strip and through the point F in the optical axis is given by OPD. This distance is approximately equal to the sum of the sagittas of the arcs, from point Q to the point R:

$$OPD = \frac{S^2}{2a} + \frac{S^2}{2b} = \frac{a+b}{2ab}s^2.$$ (4.85)

Hence, the phase difference between one light ray that goes through the aperture with a height s and the axial ray is given by

$$\delta = kOPD = \frac{k(a+b)}{\lambda ab}s^2.$$ (4.86)

According to the Huygens–Fresnel principle, we sum with their corresponding phases, the amplitude contributions of all strips with width ds. The contribution of each strip to the amplitude at the point P is directly proportional to its width ds. If the aperture is quite wide compared with its distances to the light source and to the observation plane, we must take into account the obliquity factor, given by

$$\text{Obliquity factor} = \frac{1+\cos\theta}{2}.$$ (4.87)

This obliquity factor appeared in a natural manner in Kirchhoff's theory, but it was previously postulated by Huygens. Here, we assume that the angle θ is so small that it is not necessary to take this factor into account. Now, we define a dimensionless variable v as

$$v = \sqrt{\frac{2(a+b)}{ab\lambda}}s.$$ (4.88)

Hence, the phase difference of the light ray passing through the strip with width ds with respect to the axial light ray becomes

$$\delta = \frac{\pi}{2}v^2$$ (4.89)

and since the contribution of the light wave passing through the whole strip is directly proportional to the width ds and hence also to dv.

When summing the amplitude contributions of all strips, it is necessary to take into account their relative phases δ. This can be done graphically, summing their amplitudes as vectors, with magnitude dv and phase δ (direction of the vector). In this manner, we obtain the graph in Figure 4.23, where the amplitude is directly proportional to the resultant vector $\mathbf{R}$. Using this method, it is possible to generate a curve that can be used to find the amplitude at the point P, for any width and position of the diffracting slit. Using the dimensionless variable v, we can find

$$dx = dv\cos\delta = \cos\left(\frac{\pi v^2}{2}\right)dv,$$ (4.90)

$$dy = dv\sin\delta = \text{sen}\left(\frac{\pi v^2}{2}\right)dv.$$ (4.91)

Integrating these expressions we find the Cornu spiral, as follows:

$$x = \int_0^v \cos\left(\frac{\pi v^2}{2}\right)dv$$ (4.92)

and

$$y = \int_0^v \sin\left(\frac{\pi v^2}{2}\right)dv.$$ (4.93)

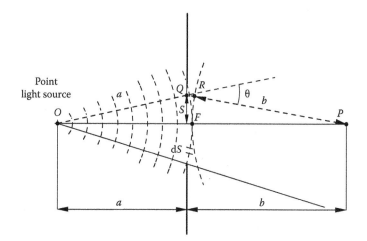

Figure 4.22 Fresnel diffraction for a rectangular aperture.

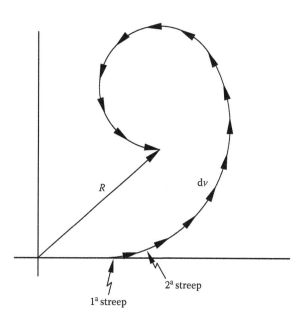

Figure 4.23 Vector sum for the elements of the diffracted wavefront.

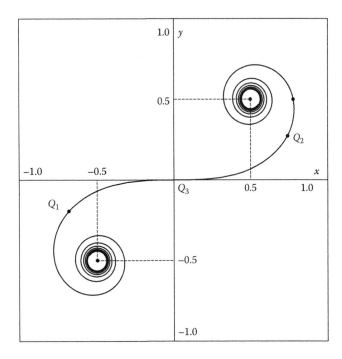

Figure 4.24 Cornu spiral.

These are the Fresnel integrals, whose numerical value can be found in many mathematical tables. The plot of x vs. y produces a curve called Cornu spiral, illustrated in Figure 4.24. If the diffracting slit is centered on the optical axis and has a total width S, the first step is to calculate the total value of v. Then, we place two symmetrical points Q_1 and Q_2 in the Cornu spiral, so that their separation along the spiral is v. The separation between these two points, measured along a straight line, represents the amplitude of the illumination at the point P.

The complete diffraction pattern of a slit can be found by moving up or down the observation point P or, equivalently, moving the slit down or up, with respect to the line joining the light source with the observation point P. To do this, we move the points Q_1 and Q_2 along the Cornu spiral, in such a way that their distance, measured along the curve, remains constant. The distance between these two points, measured in straight line, represents the resultant amplitude at the point P, for that slit position. The angle of the straight line joining the two points Q_1 and Q_2 with respect to the x-axis represents the phase. However, this phase is shifted 90° with respect to the measured phase, as we

explained before. Figure 4.25 shows the irradiance variation along a direction perpendicular to the diffracting slit.

To find the diffraction pattern of a straight edge, we may think in an infinitely wide slit with one of its edges at infinity. Then, we have a fixed point Q_3 at the center of one of the spirals and another point Q_2 on the Cornu spiral. The full diffraction pattern for the straight edge is found by displacing the point Q_2 along the curve. This irradiance variation in the diffraction pattern of a straight edge is illustrated in Figure 4.26.

If the slit is infinitely wide or, in other words, that there is not diffraction slit present, the amplitude in the observation screen is represented by the distance along a straight line, between the points Q_1 and Q_2. We can observe in this figure that the resultant phase has a value that is $\pi/4$ (45°) ahead of the phase that would be produced with an infinitely narrow slit. This is one of the disadvantages of the Huygens–Fresnel model.

The calculation of the Fresnel diffraction patterns of curved obstacles is more complicated, but the physical principle is the same. Figure 4.27 shows some of these diffraction patterns.

4.3.3.2 Circular aperture

The diffraction by a circular aperture can be done in a simple manner only for observation points located on the optical axis. However, even with this restriction, very interesting results can be obtained. We use the same geometry in Figure 4.21, but thinking in a circular aperture instead of a slit. Then, we can calculate the amplitude at the point P on the optical axis in the following manner. We divide the wavefront in narrow rings, concentric with point F. Each ring has a radius s. Then, we sum the amplitude contributions of each zone in order to find the resultant amplitude at the point P.

The phase difference between the light ray passing through the ring with radius s and a light ray along the optical axis is

$$\delta = Ks^2, \tag{4.94}$$

where

$$K = \frac{\pi(a+b)}{ab\lambda}. \tag{4.95}$$

The amplitude contribution of each zone is directly proportional to its area, which is directly proportional to its

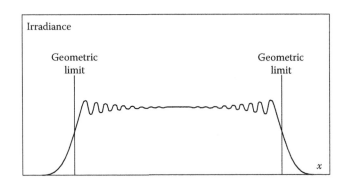

Figure 4.25 Irradiance in the diffraction pattern of a wide slit.

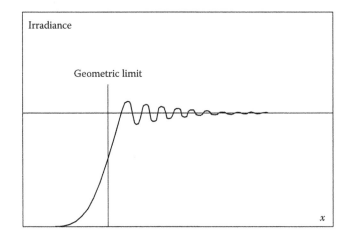

Figure 4.26 Irradiance in the diffraction pattern of straight edge.

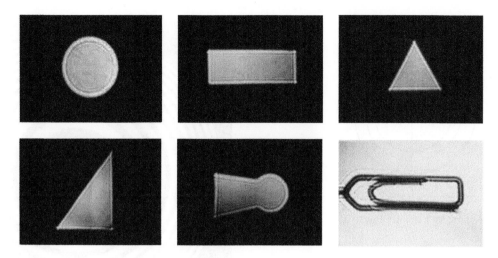

Figure 4.27 Fresnel diffraction patterns for several diaphragms and objects.

radius s and also to its width ds. From this, we can obtain a plot analogous to the Cornu spiral, represented by the following relations:

$$dx = As\, ds \cos \delta \qquad (4.96)$$

and

$$dy = As\, ds \sin \delta, \qquad (4.97)$$

but differentiating Equation 4.94, we may find

$$d\delta = 2Ks\, ds. \qquad (4.98)$$

Hence, by substituting this result in Equations 4.96 and 4.97, we find

$$dx = \frac{A}{2K} \cos \delta\, d\delta \qquad (4.99)$$

and

$$dy = \frac{A}{2K} \sin \delta\, d\delta. \qquad (4.100)$$

Integrating this expression with the limits zero and δ, we obtain

$$x = \frac{A \sin \delta}{2K} \qquad (4.101)$$

and

$$y = \frac{A(1 - \cos \delta)}{2K} \qquad (4.102)$$

from which we get

$$x^2 + \left(y - \frac{A}{2K} \right)^2 = \left(\frac{A}{2K} \right)^2, \qquad (4.103)$$

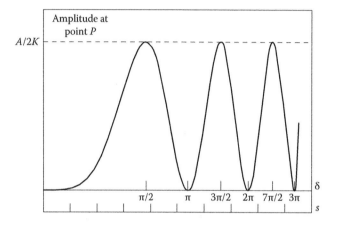

Figure 4.28 Amplitude on the optical axis of a circular aperture.

which represents a circle with center on the y-axis and tangent to the x-axis, with radius equal to $A/2K$. This means that if the radius s of the circular diffracting aperture increases in a continuous manner, the amplitude at the point P oscillates between a value equal to zero and a maximum value, as illustrated in Figure 4.28. The amplitude oscillations continue until the aperture becomes infinitely large. However, physically this is unacceptable, since the amplitude oscillations must be dampening as the aperture grows, to finally reach a constant value when the aperture is extremely large.

The problem is solved when we take into account the inclination factor $(1 + \cos \theta)/2$. In this manner, the diameter of the circles (amplitude A) described in the plot given by Equation 4.103 will decrease as the angle θ increases. Then, the circular plot becomes a spiral, as in Figure 4.29. The amplitude oscillations are damped as in Figure 4.30. They converge to a constant value equal to the amplitude at point P when there is no diffracting aperture.

4.3.3.3 Fresnel zone plate: Pinhole camera

The amplitude at the point P on axis oscillates because some rings contribute in a constructive manner to the amplitude, while others contribute in a destructive manner. If we cover with an

Fundamentals

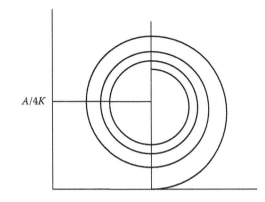

Figure 4.29 Diffraction spiral for a circular aperture.

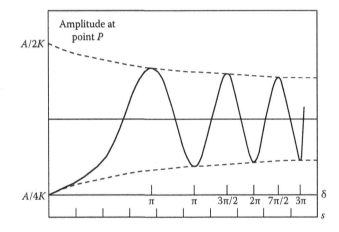

Figure 4.30 Decreasing amplitude oscillation for amplitudes at the optical axis, in Fresnel diffraction of a circular aperture.

Figure 4.31 Fresnel zone plate.

In analogy with convergent lenses, if we define the focal length f of the Fresnel zone plate as

$$f = \frac{2s_0^2}{\lambda}, \qquad (4.106)$$

we can write

$$\frac{1}{f} = \frac{1}{a} + \frac{1}{b}. \qquad (4.107)$$

As a result, the Fresnel zone plate can, as thin lenses, form images. With this expression we can calculate the position of the image for any position of the object, where a is the distance from the object to the zone plate and b is the distance from the zone plate to the image. In this expression the distance a is positive for real objects and the distance b is positive for real images.

It is important to notice that there is not only a single point on the optical axis where the amplitude is strong. There are also other points on the optical axis where the amplitude is strong, almost as at the point P. Let us assume that there is another point P_2 where the phase difference between the ray from the edge of the first clear circular zone and the light ray passing at the edge of this zone is $3\pi/2$. Then, the central zone has a central nucleus with constructive interference plus two rings, the inner one with destructive interference and the outer one with constructive interference. The rest of the transparent rings will be formed by two annular zones with constructive interference (the innermost and the outermost) and one annular zone with destructive interference (the intermediate). For the same point P_2 on axis, the zone plate dark zones are opaque for

opaque ring the zones that contribute destructively, the amplitude contributions of all uncovered ring will always be constructive. This process will strongly increase the final amplitude at the point P. The diffracting aperture so obtained, illustrated in Figure 4.31, is called a Fresnel zone plate.

If the amount of energy at the point P increases, as a consequence of the principle of conservation of energy, the amount of energy outside of the vicinity of point P has to decrease. This effect of concentration of energy on the optical axis is similar to the concentration of energy produced by a convergent lens, but with some important difference, as we will later describe.

The phase at the edge of the central clear zone is $\pi/2$, with respect to the phase at the center of this zone. Hence, the radius s of this central zone can be obtained from Equations 4.94 and 4.95:

$$s_0 = \sqrt{\frac{ab\lambda}{2(a+b)}}. \qquad (4.104)$$

Thus, we can find

$$\frac{\lambda}{2s_0^2} = \frac{1}{a} + \frac{1}{b}. \qquad (4.105)$$

two ring subzones with destructive interference and one ring subzone with constructive interference.

Generalizing this result, we see that there are bright points on axis with diffraction order m wherever a clear ring on the zone plate is transparent for m annular subzones with constructive interference and $(m - 1)$ annular subzones with destructive interference. An important result is that not all bright points are equally strong. The larger the order of diffraction m, the dimmer the point is. The same effect appears for negative diffraction orders.

In conclusion, a Fresnel zone plate acts as a convergent or a divergent lens, with multiple focal lengths, both positive (convergent) and negative (divergent), given by

$$f = \frac{2s_0^2}{(2m-1)\lambda},$$ (4.108)

where m is a positive or negative integer.

It is possible to use a Fresnel zone plate or just a circular hole with its diameter equal to the diameter of the central transparent zone in the Fresnel zone plate, to form images like a lens. For example, we can form images and make a photographic camera with just a pinhole with the proper diameter.

4.3.4 FRAUNHOFER DIFFRACTION: FOURIER TRANSFORMS

When the diffraction phenomenon is performed with both the light source and the observation plane at infinity, we have Fraunhofer diffraction. The light source can truly be at infinity, like in the case of a star, but a frequent arrangement is to place the point light source at the focus of a convergent lens, as in Figure 4.32. Then, the virtual image of the light source is at an infinite distance. The observing screen can also be optically placed at infinity by using a convergent lens with its focus at the observation plane. Since all rays with the same inclination after the diffracting aperture arrive at the same point on the observing screen, we may interpret the Fraunhofer diffraction pattern as a representation of the angular distribution of the light after passing the diffracting aperture.

The inclination factor is not important in Fraunhofer diffraction, since the angles for the diffracted ray are small. Thus, if the illuminating light beam is perpendicular to the diffracting aperture, we may write Kirchhoff's integral as

$$U(P) = \frac{i}{\lambda} \iint_\sigma B \frac{e^{iks}}{s} dS,$$ (4.109)

where B is a constant, directly proportional to the amplitude on the pane of the diffracting aperture. If the diffracting aperture is at the x–y plane and the amplitude in this plane is a function of x and y, we may consider that B is not a constant, but a function $B(x, y)$. The integral is performed over the diffracting aperture. The distance s is the total optical path from the point (x, y) to the observation plane. The factor $1/s$ can be considered as a constant for all points (x, y), included in the function (or constant) $B(x, y)$. Thus, we can write

$$U(P) = \frac{i}{\lambda} \iint_\sigma B(x, y) e^{iks} dx dy.$$ (4.110)

Let us now consider a diffracted light wave in a given direction, as illustrated in Figure 4.33. The total distance from the point (x, y) at the diffracting aperture to the observing screen is infinite for all points. However, there is a phase difference between a light ray from the point (x, y) and the light ray that passes through the center of the aperture. The optical path difference between these two rays is equal to $OPD = s - s_0$.

Hence, we obtain

$$U(P) = \frac{i}{\lambda} \iint_\sigma B(x, y) e^{iks_0} e^{ikOPD} dx dy,$$ (4.111)

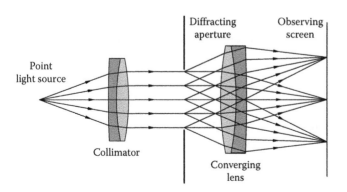

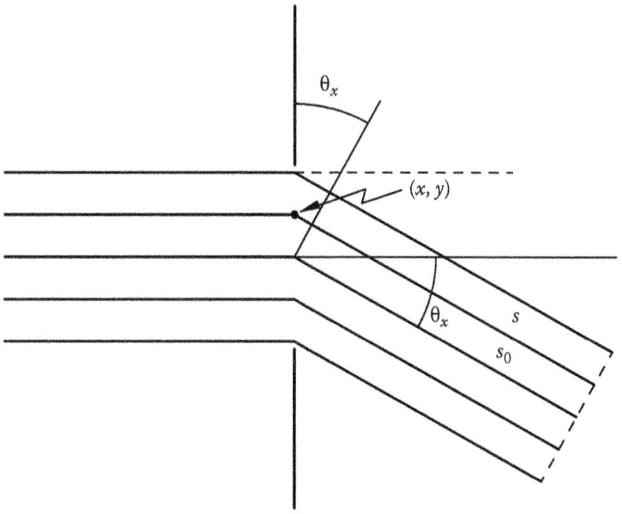

Figure 4.32 Optical arrangement to produce Fraunhofer diffraction.

Figure 4.33 Light rays in Fraunhofer diffraction.

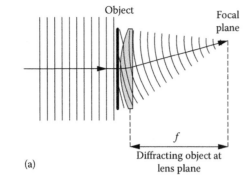

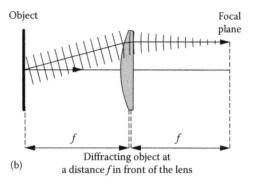

Figure 4.34 Light rays' paths in Fraunhofer diffraction.

but if we make the constant s_0 equal to zero, the first exponential term after $B(x, y)$ in this integral is equal to one, obtaining

$$U(P) = \frac{i}{\lambda} \iint_\sigma B(x, y) e^{ikOPD} \, dxdy. \tag{4.112}$$

The optical path difference *OPD* can be written as a function of the angles θ_x and θ_y in the directions x and y, as

$$OPD = x \sin\theta_x + y \sin\theta_y. \tag{4.113}$$

Thus, if we define

$$k_x = k \sin\theta_x \tag{4.114}$$

and

$$k_y = k \sin\theta_y, \tag{4.115}$$

we can write

$$U(k_x, k_y) = \frac{i}{\lambda} \iint_\sigma B(x, y) e^{i(k_x x + k_y y)} \, dx \, dy. \tag{4.116}$$

We observe that this expression is quite similar to a Fourier transform. So, using the Fourier theorem, if we have a function $B(x, y)$ and we define another function $U(k_x, k_y)$ as in expression 116, it is possible to show that the function $B(x, y)$ can be written as

$$B(x, y) = \frac{i}{\lambda} \int_{-\infty}^{\infty} \int_{-\infty}^{\infty} U(k_x, k_y) e^{-i(k_x x + k_y y)} \, dk_x \, dk_y, \tag{4.117}$$

where the functions $B(x, y)$ and $U(k_x, k_y)$ are the 2D Fourier transforms of each other.

If we now assume that the amplitude $B(x, y)$ defined in all the plane of the diffracting screen, is zero in all the opaque regions, thus, except for a constant phase factor and a multiplicative constant, the Fraunhofer diffraction pattern is the Fourier transform of the amplitude function over the diffracting aperture. It is interesting to notice that this result for Fraunhofer diffraction can be generalized to cases when

the illuminating wavefront is not flat, but it may be slightly spherical or may have small deformations. Then in general, the function $B(x, y)$ can be written as a complex function as follows:

$$\begin{aligned} B(x, y) &= B_0(x, y) e^{-i\phi(x, y)} \\ &= B_0(x, y) e^{-ikW(x, y)}, \end{aligned} \tag{4.118}$$

where
  $B_0(x, y)$ is a real function that describes the amplitude variations over the aperture
  $\phi(x, y)$ represents the phase variations (due to the wavefront deformations)

The wavefront deformations are $W(x, y)$.

We have explained before that in order to observe the Fraunhofer diffraction pattern at an infinite distance, we have to use a convergent lens in front of the aperture and to place the observing plane at the focus as in Figure 4.34a. However, the distance from the diffracting aperture to the convergent lens is also important. Changing this distance changes the phase over the diffraction pattern, but the amplitude distribution remains constant. In order to have a constant phase, from the center of the aperture to the observation plane should be the same for any direction of the diffracted beam. This is true if the diffracting aperture is in front of the converging lens, at its front focal plane, as in Figure 4.34b.

4.3.4.1 Single slit and rectangular aperture

In this section, the Fraunhofer diffraction produced by a slit or a rectangular aperture will be described. Let us assume that a slit is located in the plane x-y, centered and along the y-axis. If the slit has a width $2a$, we can write Equation 4.116 as

$$U(k_x) = A \int_{-\infty}^{\infty} e^{ik_x x} \, dx, \tag{4.119}$$

where the amplitude at the slit is a constant and the constant i/λ is part of this constant A. Then, integrating, we obtain

$$U(k_x) = Aa \left(\frac{e^{ik_x a} - e^{-ik_x a}}{2ik_x a} \right). \tag{4.120}$$

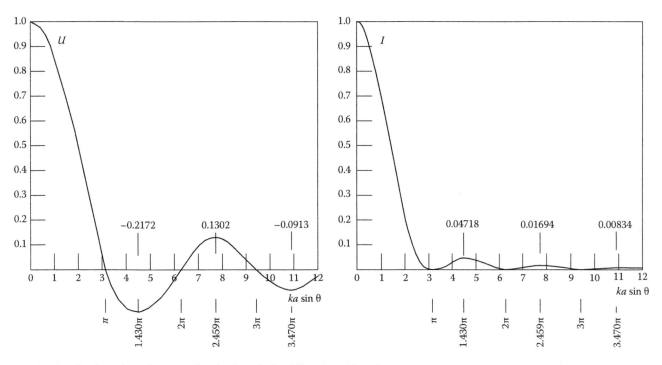

Figure 4.35 Amplitude and irradiance profiles for Fraunhofer diffraction with a slit.

Then, using the Cauchy relation, we have

$$U(k_x) = U_0 \left(\frac{\sin(ka \sin \theta)}{ka \sin \theta} \right), \quad (4.121)$$

where U_0 is a constant. Figure 4.35 shows this amplitude distribution and its corresponding irradiance, as a function of $ka \sin \theta$.

The first dark fringe appears at an angle θ from the optical axis is given by

$$\sin \theta = \frac{\lambda}{2a}. \quad (4.122)$$

We may observe that this diffraction pattern becomes narrower when the slit width $2a$ grows. By analogy with the slit, for a rectangular aperture with width $2a$ and length $2b$, we may obtain

$$U(\theta_x, \theta_y) = U_0 \left(\frac{\sin(ka \sin \theta_x)}{ka \sin \theta_x} \right) \left(\frac{\sin(kb \sin \theta_y)}{kb \sin \theta_y} \right). \quad (4.123)$$

Figure 4.36 shows the Fraunhofer diffraction patterns for circular, rectangular, and triangular apertures. Figure 4.37 shows the gradual transition of the three Fresnel diffraction patterns to Fraunhofer diffraction, as the distance to the observing plane increases.

4.3.4.2 Circular aperture

The circular aperture is another quite common and important type of diffracting aperture. This shape of aperture is found in most optical instruments. The diffraction pattern can be found by integrating Equation 4.116 in a circle. Let us assume that the amplitude $A(x)$ is constant inside the circular aperture and that the constant i/λ is inside of this constant A, obtaining

$$U(k_x, k_y) = A \iint_{\sigma} e^{i(k_x x + k_y y)} \, dx \, dy, \quad (4.124)$$

but, changing to polar coordinates, we have

$$x = \rho \cos \alpha, \quad y = \rho \sin \alpha, \quad (4.125)$$

where ρ and α are the polar coordinates of a point inside the circular aperture. On the other hand, we can write

$$\sin \theta_x = \cos \psi \cos \theta, \quad \sin \theta_y = \sin \psi \sin \theta, \quad (4.126)$$

where, if η and ξ are the Cartesian coordinates of a point in the diffraction pattern, ψ is the angle between the a radial line going from the optical axis to the point in the diffraction pattern and the η axis. On the other hand, θ is the angle between the lines from the center of the diffracting aperture to the point in the diffraction pattern, with the optical axis. Hence, we can write

$$x \sin \theta_x + y \sin \theta_y = \rho \sin \theta \cos (\alpha - \psi). \quad (4.127)$$

Now, substituting these results into Equation 4.124 and transforming the integral to polar coordinates, where σ is the circle with radius a, we have

$$U(k_x, k_y) = A \int_{\rho=0}^{a} \int_{\alpha=0}^{2\pi} e^{ik\rho \, \text{sen} \, \theta \cos(\alpha - \psi)} \, \rho \, d\rho \, d\alpha. \quad (4.128)$$

To perform this integration is relatively simple with a basic knowledge of Bessel functions. Then, the following result can be found:

$$U(\theta) = U_0 \left(\frac{J_1(ka \sin \theta)}{ka \sin \theta} \right), \quad (4.129)$$

where

a is the radius of the diffracting aperture
$J_1(x)$ is the *Bessel function* of the first kind and the first order

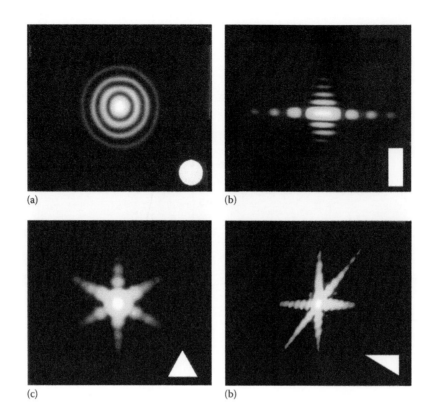

Figure 4.36 Fraunhofer diffraction patterns for a (a) circular aperture, (b) rectangular aperture, (c) equilateral triangular aperture, and (d) rectangular triangular aperture.

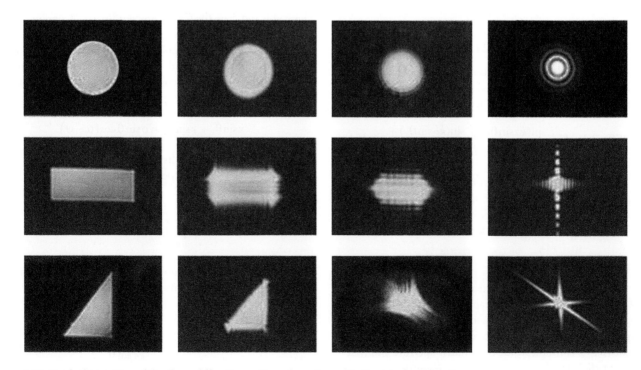

Figure 4.37 Gradual transition of the three diffraction patterns from Fresnel to Fraunhofer diffraction.

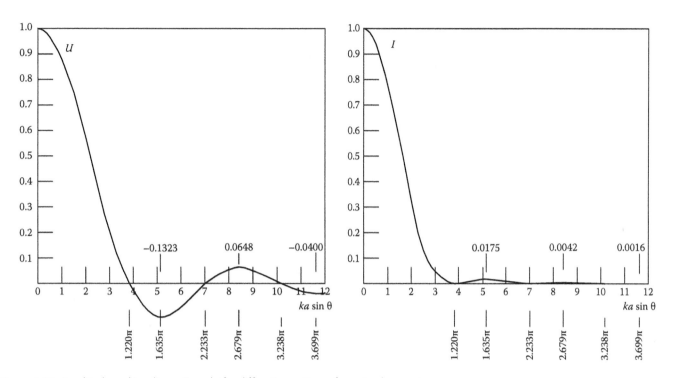

Figure 4.38 Amplitude and irradiance Fraunhofer diffraction patterns for a circular aperture.

This function $U(\theta)$ is called the "Airy function" and its corresponding irradiance is $U^2(\theta)$, illustrated in Figure 4.38.

The circle inside the first diffraction ring is the Airy disc, which has an angular radius θ as observed from the diffracting aperture, equal to

$$\sin\theta = 1.22\frac{\lambda}{2a} = 1.22\frac{\lambda}{D}, \qquad (4.130)$$

where D is the aperture diameter. If the diffraction pattern is in the focal plane of a lens with diameter D and focal length f, the radius r of the Airy disc is given by

$$r = 1.22\frac{\lambda f}{D}. \qquad (4.131)$$

If we assume that the wavelength of the light is 500 nm, the Airy disc diameter becomes

$$2r = 1.22\frac{f}{D}\,\mu m, \qquad (4.132)$$

which is quite useful and easy to remember, since the diameter of the Airy disc in microns is nearly equal to the focal ratio of the optical system or lens.

The finite size of the Airy disc is the reason for the limitation in the resolving power of optical instruments and systems with circular optical lenses and components. Figure 4.39 shows images for one and for two Airy diffraction patterns with two different exposure times and with different separations.

4.4 POLARIZED LIGHT

4.4.1 UNPOLARIZED AND POLARIZED LIGHT: MALUS LAW

To study most optical phenomena, like reflection, refraction interference, and diffraction, it is not necessary to know if the light waves are longitudinal, like sound waves or transverse, like a string or water wave. However, in some materials and under some special conditions, the transmitted or reflected light irradiance shows some dependence on the angular orientation (rotation about the optical axis). This indicates that the light wave is transverse and not longitudinal. From electromagnetic theory, we know that light wave has a transverse nature, with an electric field and a magnetic field mutually perpendicular, as studied at the beginning of this chapter.

However, the electric and magnetic field are not always restricted to oscillate in a plane, as we will now study. When they are in a plane we say that the light beam is linearly polarized. Let us think of many light beams traveling along the same path, but all linearly polarized in many different planes and mutually incoherent. The result is a light beam whose electric (and magnetic) field rotates its plane of oscillation quite fast and at random. This is an unpolarized light beam. Most light beams are of this type. A mixture of linearly polarized and unpolarized light is said to be partially polarized.

A polarization analyzer is an optical device that contains a transverse axis of symmetry. If a light beam of linearly polarized light arrives to this analyzer with its plane of polarization oriented along this axis of symmetry, all the light passes through it. However, if the plane of polarization is perpendicular to the axis, the light cannot pass through the analyzer, as illustrated in Figure 4.40.

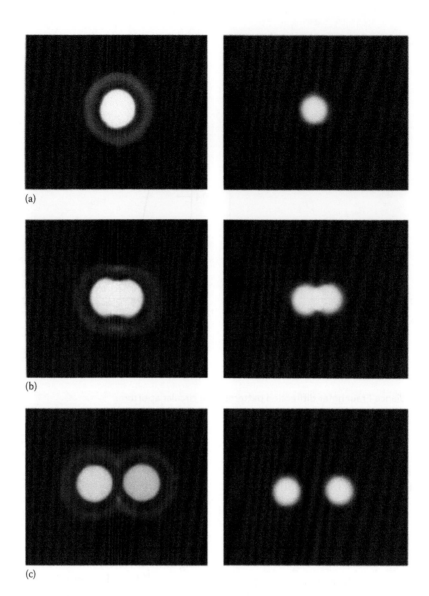

Figure 4.39 Airy diffraction patterns with two different exposure times. (a) A single image, (b) two images just resolved, and (c) two images perfectly separated.

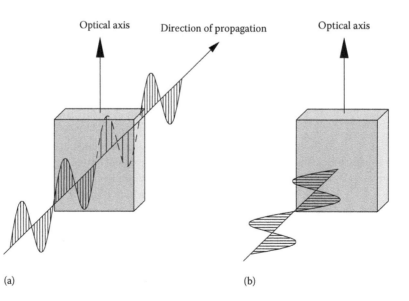

Figure 4.40 Polarized light analyzer. (a) Optical axis parallel to the electric field. (b) Optical axis perpendicular to the electric field.

If the plane of polarization and the axis of the analyzer are at an angle θ, only the projected component of the electric field along the axis of the analyzed can go through it. Thus, if the electrical vector has an amplitude E, the transmitted electric component E_T has the amplitude

$$E_T = E \cos\theta, \tag{4.133}$$

where E is the amplitude of the incident light beam. Hence, if the incident irradiance is I, the transmitted irradiance I_T is

$$I_T = I \cos^2\theta. \tag{4.134}$$

This relation is known as Malus law.

If two mutually coherent polarized light beams are superimposed along the same path, the electrical beams add in space like vectors, producing a light beam with another state of polarization. This is a generalized concept of interference where complete constructive or destructive interference can take place only if the two interfering beams are in the same polarization state. An interesting case is when the two interfering light beams are fully coherent, linearly polarized and with its planes of polarization orthogonal to each other.

4.4.2 ELLIPTICAL AND CIRCULARLY POLARIZED LIGHT

Let us assume that the two linearly polarized waves travel along the z-axis and with a phase difference δ between them. The planes of polarization of the two waves are x-z and y-z. The resultant electrical vector at the point z, and the time t, can be represented in a graphic manner as in Figure 4.41 and in analytic form as follows:

$$E_x = a_1 \cos(kx - \omega t) \tag{4.135}$$

and

$$E_y = a_2 \cos(kx - \omega t + \delta). \tag{4.136}$$

Now, we can rewrite Equation 4.136 as

$$E_y = a_2 \cos(kx - \omega t)\cos\delta - a_2 \left[1 - \cos^2(kx - \omega t)\right]^{1/2} \sin\delta. \tag{4.137}$$

Then, using Equation 4.135, we get

$$E_y = \frac{a_2}{a_1} E_x \cos\delta - a_2 \left(1 - \frac{E_x^2}{a_1^2}\right)^{1/2} \sin\delta, \tag{4.138}$$

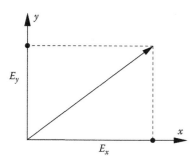

which can be transformed into

$$\frac{E_x^2}{a_1^2} + \frac{E_y^2}{a_2^2} - \frac{2E_x E_y}{a_1 a_2}\cos\delta = \sin^2\delta. \tag{4.139}$$

This expression represents an ellipse whose major semiaxis forms an angle ψ with respect to the x-axis. To prove it, we rotate by an angle ψ to a new coordinate system (η, ξ), as illustrated in Figure 4.42.

We use the transformation equations

$$E_x = E_\eta \cos\psi - E_\xi \sin\psi \tag{4.140}$$

$$E_y = E_\eta \sin\psi + E_\xi \cos\psi \tag{4.141}$$

to obtain

$$\left[a_2^2 \cos^2\psi + a_1^2 \sin^2\psi - 2a_1 a_2 \sin\psi \cos\psi \cos\delta\right]E_\eta^2$$
$$+ \left[a_2^2 \sin^2\psi + a_1^2 \cos^2\psi + 2a_1 a_2 \sin\psi \cos\psi \cos\delta\right]E_\xi^2$$
$$- 2\left[\left(a_2^2 - a_1^2\right)\sin\psi \cos\psi + a_1 a_2 \left(\cos^2\psi - \sin^2\psi\right)\cos\delta\right]E_\eta E_\xi$$
$$= a_1^2 a_2^2 \sin^2\delta. \tag{4.142}$$

Now, we select the angle ψ in such a way that the ellipse semiaxes coincide with the η and ξ axes. This is done by setting the coefficient of $E_\eta E_\xi$ equal to zero. Thus,

$$\left(a_2^2 - a_1^2\right)\sin\psi \cos\psi + a_1 a_2 \left(\cos^2\psi - \sin^2\psi\right)\cos\delta = 0. \tag{4.143}$$

Thus, the ellipse rotation angle is

$$\tan 2\psi = \left(\frac{a_1 a_2}{a_1^2 - a_2^2}\right)\cos\delta. \tag{4.144}$$

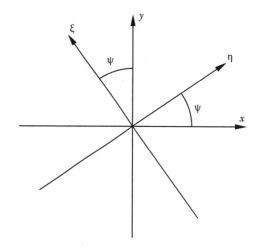

Figure 4.41 Interference between two linearly polarized light waves in mutually perpendicular planes.

Figure 4.42 Rotation to a new system of coordinates.

Finally, the ellipse equation in the new coordinate system can be written as

$$\left[a_2^2 \cos^2 \psi + a_1^2 \sin^2 \psi - 2a_1 a_2 \sin \psi \cos \psi \cos \delta \right] E_\eta^2$$
$$+ \left[a_2^2 \sin^2 \psi + a_1^2 \cos^2 \psi + 2a_1 a_2 \sin \psi \cos \psi \cos \delta \right] E_\xi^2$$
$$= a_1^2 a_2^2 \sin^2 \delta \tag{4.145}$$

that has the general form

$$\frac{E_\eta^2}{a^2} + \frac{E_\xi^2}{b^2} = 1, \tag{4.146}$$

where the constants a and b are the major and minor semiaxis of the ellipse, respectively. In order to find the explicit values of these two constants, it is necessary first to prove the following expression:

$$\left[a_2^2 \cos^2 \psi + a_1^2 \sin^2 \psi - 2a_1 a_2 \sin \psi \cos \psi \cos \delta \right]$$
$$\times \left[a_2^2 \sin^2 \psi + a_1^2 \cos^2 \psi + 2a_1 a_2 \sin \psi \cos \psi \cos \delta \right]$$
$$= a_1^2 a_2^2 \sin^2 \delta \tag{4.147}$$

that can be transformed into

$$\left(a_2^4 - a_1^4 \right) \sin^2 \psi \cos^2 \psi + a_1^2 a_2^2 \left(\sin^4 \psi + \sin^4 \psi \right)$$
$$+ 2a_1^2 a_2^2 \left(a_2^2 - a_1^2 \right) \sin \psi \cos \psi \left(\sin^2 \psi - \sin^2 \psi \right) \cos \delta$$
$$- 4a_1^2 a_2^2 \sin^2 \psi \cos^2 \psi \cos^2 \delta = a_1^2 a_2^2 \sin^2 \delta. \tag{4.148}$$

Now, using the values of $\sin 2\psi$ and $\cos 2\psi$, we can transform this expression into

$$\frac{a_2^4 - a_1^4}{4} \sin^2 2\psi + a_1^2 a_2^2 \left(1 - \frac{\sin^2 \psi}{2} \right)$$
$$+ a_1 a_2 \left(a_1^2 - a_2^2 \right) \sin 2\psi \cos 2\psi \cos \delta - a_1^2 a_2^2 \sin^2 2\psi \cos^2 \delta$$
$$= a_1^2 a_2^2 \sin^2 \delta. \tag{4.149}$$

Finally, substituting here the value of $\cos \delta$ from Equation 4.144, it is possible to prove this equation. If we now divide both terms in Equation 4.145 by $a_1 a_2 \sin 2\delta$ and use Equation 4.147, we can find that

$$a^2 = a_2^2 \sin^2 \psi + a_1^2 \cos^2 \psi + 2a_1 a_2 \sin \psi \cos \psi \cos \delta \tag{4.150}$$

and

$$b^2 = a_2^2 \cos^2 \psi + a_1^2 \sin^2 \psi - 2a_1 a_2 \sin \psi \cos \psi \cos \delta. \tag{4.151}$$

With these two expressions and Equation 4.148, we can find that

$$a^2 b^2 = a_1^2 a_2^2 \sin^2 \delta, \tag{4.152}$$

and summing Equations 4.151 and 4.152, we can find that

$$a^2 + b^2 = a_1^2 + a_2^2. \tag{4.153}$$

Then, by using these two expressions the ellipse semiaxes can be found from a_1, a_2, and δ. Figure 4.43 shows the types of

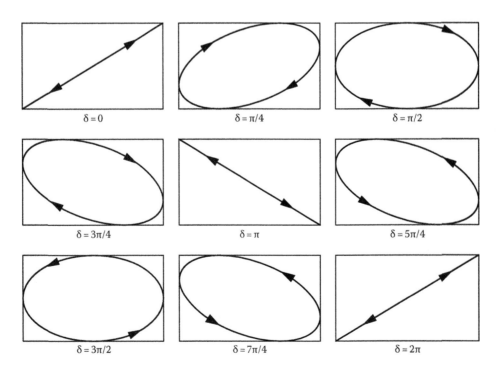

Figure 4.43 Plot of electric vector magnitude and direction along the trajectory, for different phase difference of the orthogonal components. The polarization is right handed if $\sin \delta > 0$ and left handed if $\sin \delta < 0$. The phase increases toward the back of the drawing and decreases toward the front. In other words, the drawings are observed from the light source side.

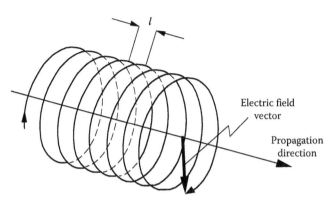

Figure 4.44 Electrical field in elliptically polarized light.

ellipses that can be obtained, for different values of the phase difference between the two orthogonal components. As we may observe, all the ellipses are tangent to the sides of a rectangle with width and height $2a_1$ and $2a_2$, respectively.

In light elliptically polarized, the electrical vector changes its magnitude and direction along its trajectory. If we represent the electrical vector by an arrow, the end of the arrow will follow the shape of a flattened corkscrew, as in Figure 4.44. In right-handed circularly polarized light, the electric vector turns clockwise, as seen from the light source and in left-handed circularly polarized light otherwise.

From Figure 4.43, we see that when $a_1 = a_2$, the light is right-handed circularly polarized, if $\delta = \pi/2$, and left handed if $\delta = 3\pi/2$. We can see that if the two waves have the same phase ($\delta = 0$), the ellipse becomes a straight line, thus, the resultant light is linearly polarized.

4.4.3 POLARIZED LIGHT REPRESENTATION

Completely polarized light is represented by three parameters, which can be, for example, the semiaxes a_1, a_2, and the phase δ, but there are several other ways to represent these parameters, so that combinations of several states of polarized light can be combined and evaluated. Next, we will describe some of these models.

4.4.3.1 Poincaré sphere

As we just pointed out, to determine the complete polarization state of fully polarized light, we need their independent parameters, for example, the semiaxes a and b and the ellipse orientation ψ.

In order to mathematically describe fully polarized light, George G. Stokes introduced in 1852 the four parameters, S_0, S_1, S_2, and S_3, which can completely specify the state of polarization. These Stokes parameters are defined in terms of a_1, a_2, and δ, as follows:

$$S_0 = a_1^2 + a_2^2 \tag{4.154}$$

$$S_1 = a_1^2 - a_2^2 \tag{4.155}$$

$$S_2 = 2a_1 a_2 \cos \delta \tag{4.156}$$

$$S_2 = 2a_1 a_2 \sin \delta, \tag{4.157}$$

where the parameter S_0 represents the irradiance of the light beam. Since the light is fully polarized, only three of these parameters are independent. In this case, we can write

$$S_0^2 = S_1^2 + S_2^2 + S_3^2. \tag{4.158}$$

Generalized Stokes parameters for partially polarized light will be described in the next section. If we define an angle χ as

$$\tan \chi = \frac{b}{a}, \tag{4.159}$$

then

$$\sin 2\chi = \frac{2 \tan \chi}{1 + \tan^2 \chi} = \frac{2ab}{a^2 + b^2}. \tag{4.160}$$

Then using Equations 4.152 and 4.153, we can write

$$\sin 2\chi = \frac{2a_1 a_2 \sin \delta}{a_1^2 + a_2^2}. \tag{4.161}$$

Now, substituting here the values of S_0 and S_3 from Equations 4.154 and 4.155:

$$\sin 2\chi = \frac{S_3}{S_0}. \tag{4.162}$$

On the other hand, using Equations 4.156 and 4.157 in Equation 4.144, we can see that

$$\tan 2\chi = \frac{S_2}{S_1}. \tag{4.163}$$

Observing Equation 4.158 we could conclude that the parameters S_1, S_2, and S_3 can be represented by points in a sphere with radius S_0. With Equations 4.162 and 4.163 we can see that the angles 2χ and 2ψ can be represented as in Figure 4.45. This is

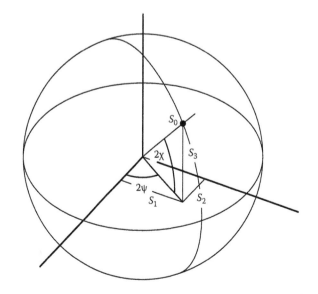

Figure 4.45 Poincaré sphere.

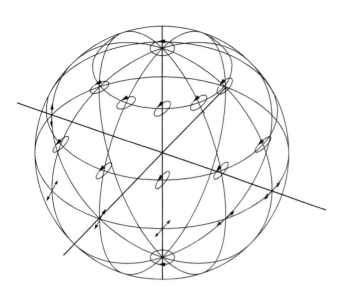

Figure 4.46 Ellipses on the Poincaré sphere.

the Poincaré sphere, from which we can see that the Stokes parameters can be written as follows:

$$S_1 = S_0 \cos 2\chi \cos 2\psi. \tag{4.164}$$

$$S_2 = S_0 \cos 2\chi \,\text{sen}\, 2\psi. \tag{4.165}$$

$$S_3 = S_0 \sin 2\chi. \tag{4.166}$$

Different points on this sphere represent ellipses with different eccentricities and orientations. At the poles of the sphere, there are circles and at the equator they become straight segments. Right-handed ellipses are in the north hemisphere and left handed in the south hemisphere. In this manner, different points in the sphere represent different states of fully polarized light, as illustrated in Figure 4.46.

4.4.3.2 Mueller matrices

A given state of polarization of a light beam with an ellipse with a certain eccentricity and orientation, plus some amount of unpolarized light, could be moved to another state of polarization if some polarization devices, like analyzers, polarizing prisms, or phase plates, are inserted in the light beam. Let us assume that the initial polarization state is defined by the four values of the Stokes parameters, which can be represented by I_1, M_1, C_1, and S_1. If an optical device changes the state of polarization to another with parameters I_2, M_2, C_2, and S_2, we can characterize the new state by a linear transformation with a matrix called the Mueller matrix, whose values depend on the optical device and its orientation, as follows:

$$\begin{bmatrix} I_2 \\ M_2 \\ C_2 \\ S_2 \end{bmatrix} = \begin{bmatrix} m_{00} & m_{01} & m_{02} & m_{03} \\ m_{10} & m_{11} & m_{12} & m_{13} \\ m_{30} & m_{31} & m_{32} & m_{33} \\ m_{40} & m_{41} & m_{42} & m_{43} \end{bmatrix} \begin{bmatrix} I_1 \\ M_1 \\ C_1 \\ S_1 \end{bmatrix}. \tag{4.167}$$

Here, all the elements are real numbers. In a synthetic manner, we write

$$\mathbf{S}_2 = \mathbf{M} \cdot \mathbf{S}_1, \tag{4.168}$$

where

$\mathbf{S}_1$ and $\mathbf{S}_2$ are the input and output Stokes vectors
$\mathbf{M}$ is the system Mueller matrix

The Mueller matrices for some optical elements that alter the state of polarization are as follows:

$$\frac{1}{2}\begin{bmatrix} 1 & 1 & 0 & 0 \\ 1 & 1 & 0 & 0 \\ 0 & 0 & 0 & 0 \\ 0 & 0 & 0 & 0 \end{bmatrix}$$

Horizontal linear polarizer

$$\frac{1}{2}\begin{bmatrix} 1 & -1 & 0 & 0 \\ -1 & 1 & 0 & 0 \\ 0 & 0 & 0 & 0 \\ 0 & 0 & 0 & 0 \end{bmatrix}$$

Vertical linear polarizer

$$\frac{1}{2}\begin{bmatrix} 1 & 0 & 1 & 0 \\ 0 & 0 & 0 & 0 \\ 1 & 0 & 1 & 0 \\ 0 & 0 & 0 & 0 \end{bmatrix}$$

Linear polarizer at 45°

$$\frac{1}{2}\begin{bmatrix} 1 & 0 & -1 & 0 \\ 0 & 0 & 0 & 0 \\ -1 & 0 & 1 & 0 \\ 0 & 0 & 0 & 0 \end{bmatrix}$$

Linear polarizer at − 45°

$$\begin{bmatrix} 1 & 0 & 0 & 0 \\ 0 & 1 & 0 & 0 \\ 0 & 0 & 0 & -1 \\ 0 & 0 & 0 & 0 \end{bmatrix}$$

Quarter wave phase plate
vertical fast axis

$$\begin{bmatrix} 1 & 0 & 0 & 0 \\ 0 & 1 & 0 & 0 \\ 0 & 0 & 0 & 1 \\ 0 & 0 & -1 & 0 \end{bmatrix}$$

Quarter wave phase plate
horizontal fast axis

$$\begin{bmatrix} 1 & 0 & 0 & 0 \\ 0 & 1 & 0 & 0 \\ 0 & 0 & -1 & 0 \\ 0 & 0 & 0 & -1 \end{bmatrix}$$

Half wave phase plate
vertical fast axis

4.4.4 NATURAL LIGHT AND PARTIALLY POLARIZED LIGHT

We have seen that a fully polarized light beam has in general elliptical polarization, which can take the particular states of linear or circularly polarized. It, still remains to formally define *natural light* or *unpolarized* light and *partially polarized light*. Fully or partially polarized light can be found in any of the following seven states:

1. Linear
2. Circular
3. Elliptical
4. Unpolarized

5. Partially linear
6. Partially circular
7. Elliptically partial

The other three possible combinations, that is, linear with circular, linear with elliptical, and circular with elliptical, produce one of the seven states just mentioned.

A fully polarized, partially polarized, or unpolarized light beam can be modeled by two linearly polarized beams in mutually perpendicular planes. When analyzed, this combination with an analyzer has an irradiance $I(\theta)$ different for every orientation θ of the analyzer. To find this irradiance, let us imagine that the two orthogonally polarized beams have amplitudes E_x and E_y and that can be represented by

$$E_x = a_1\, e^{i(kz-\omega t)} \tag{4.169}$$

$$E_y = a_2\, e^{i(kz-\omega t+\delta-\varepsilon)}, \tag{4.170}$$

where δ is the phase difference and ε is a possible phase change in the y component, introduced by a phase plate with its fast axis along the y-axis. Hence, the transmitted amplitude with the analyzer with an orientation θ given by

$$E(\theta,\varepsilon) = E_x \cos\theta + E_y \sin\theta. \tag{4.171}$$

Hence, the irradiance is

$$I(\theta,\varepsilon) = E(\theta)E^*(\theta). \tag{4.172}$$

From Equations 4.169 through 4.172, we may obtain

$$I(\theta,\varepsilon) = \left(a_1\, e^{i(kz-\omega t)} \cos\theta + a_2\, e^{i(kz-\omega t+\delta-\varepsilon)}\sin\theta \right)$$
$$\times \left(a_1\, e^{-i(kz-\omega t)} \cos\theta + a_2\, e^{-i(kz-\omega t+\delta-\varepsilon)}\sin\theta \right), \tag{4.173}$$

which can be reduced to the following expression:

$$I(\theta,\varepsilon) = a_1^2 \cos^2\theta + a_2^2 \sin^2\theta + 2a_1 a_2 \cos(\delta-\varepsilon)\sin\theta\cos\theta. \tag{4.174}$$

This expression gives us the transmitted irradiance for any orientation of the analyzer and for any state of full polarization, including the possible presence of a phase retarder, with its fast axis, parallel to the x-axis. When the polarization is not full but partial, this is a valid expression if we take the temporal averages of each term and represent them using the symbol $\langle\rangle$, to obtain the following:

$$\langle I(\theta,\varepsilon)\rangle = \langle a_1\rangle^2 \cos^2\theta + \langle a_2\rangle^2 \sin^2\theta + 2\langle a_1 a_2 \cos(\delta-\varepsilon)\rangle \sin\theta\cos\theta$$
$$= \frac{1}{2}\left(\langle a_1\rangle^2 + \langle a_2\rangle^2\right) + \frac{1}{2}\left(\langle a_1\rangle^2 - \langle a_2\rangle^2\right)\cos2\theta$$
$$+ \langle a_1 a_2 \cos(\delta-\varepsilon)\rangle \sin2\theta. \tag{4.175}$$

To represent the state of polarization, including partially polarized and unpolarized light, it is necessary to specify four independent parameters. These parameters will determine the irradiance, eccentricity, orientation of the ellipse, and degree of polarization. This method was proposed by G. Stokes, who defined the vector $[I, M, C, S]$. This vector is formed by a set of four numbers called Stokes parameters for partially polarized light or just Stokes parameters. They are defined as the temporal averages of the Stokes parameters for monochromatic light, as follows:

$$I = \langle S_0\rangle = \langle a_1\rangle^2 + \langle a_2\rangle^2. \tag{4.176}$$

$$M = \langle S_1\rangle = \langle a_1\rangle^2 - \langle a_2\rangle^2. \tag{4.177}$$

$$C = \langle S_2\rangle = 2\langle a_1 a_2 \cos\delta\rangle. \tag{4.178}$$

$$S = \langle S_3\rangle = 2\langle a_1 a_2 \sin\delta\rangle. \tag{4.179}$$

These parameters represent the following properties of the partially polarized light beam:

I represents the irradiance of the light beam.

M represents the predominance of the horizontal or vertical component, according to its sign.

C represents the ellipse orientation, given by the angle χ toward $-45°$ or toward $+45°$, according to its sign.

S represents the handiness of the polarization state, being positive if it is right handed and left otherwise.

If the light is fully polarized, the phase difference δ between the two mutually perpendicular polarizing planes and the ratio of the amplitudes a_1/a_2 are constant, so that $\langle a_1\rangle\langle a_2\rangle = \langle a_1 a_2\rangle$. Then, the Stokes parameters, according to Equation 4.158, satisfy the following relation:

$$M^2 + C^2 + S^2 = I^2. \tag{4.180}$$

If the light is not polarized, we can assume that the phase δ has fast-changing random values. Hence, the averages of $\cos\delta$ and $\sin\delta$ are both equal to zero. Also, the amplitudes of a_1 and a_2 are equal. Thus, the Stokes vector components M, C, and S are zero. Thus, we can write

$$M^2 + C^2 + S^2 = 0. \tag{4.181}$$

Then, it is natural to define the degree of polarization as

$$V = \frac{\sqrt{M^2 + C^2 + S^2}}{I}, \tag{4.182}$$

where V is zero for unpolarized light, one for full polarized light, and with intermediate values for partially polarized light.

With the definitions of the Stokes parameters, we can write the irradiance as a function of the angle θ of the analyzer and of the additional phase ε introduced by a phase plate, from Equation 175, as

$$\langle I(\theta,\varepsilon)\rangle = \frac{1}{2}\left(I + M\cos2\theta + C\cos\varepsilon\sin2\theta + S\sin\varepsilon\sin2\theta \right). \tag{4.183}$$

Table 4.2 **States of polarization and the Stokes vector**

POLARIZATION STATE	$(a_1^2 + a_2^2)$	a_2/a_1	δ	I	M	C	S	V
Linear (horizontal)	1	0	—	1	1	0	0	1
Linear (vertical)	1	∞	—	1	–1	0	0	1
Linear at 45°	1	1	0°	1	0	1	0	1
Linear at –45°	1	–1	180°	1	0	–1	0	1
Circular (right)	1	1	90°	1	0	0	1	1
Elliptical (right)	1	0.5	90°	1	0.6	0	0.8	1
Unpolarized	1	1	Random	1	0	0	0	0
Linear (horizontal partial)	1	0.5	Random	1	0.6	0	0	0.6
Lineal a 45° (partial)	1	1	0°	1	0	0.3	0	0.3
Circular (partial)	1	1	0°	1	0	0	0.2	0.2

Table 4.2 shows some examples of Stokes vectors for different types of polarization.

If a light beam is perfectly monochromatic, the two orthogonal components, E_x and E_y, are coherent to each other; they have a constant amplitude and a constant phase difference δ. Hence, the light beam is completely polarized. However, if the light beam is perfectly polarized, the light beam is not necessarily monochromatic.

In interference theory, two separated light sources can be mutually coherent only if they come from the same light source. To say it with different words, two independent light sources cannot be coherent to each other. Polarization phenomena can be considered as a generalized interference, where for a light beam to be completely polarized it is necessary that two beams along the same trajectory are polarized in mutually perpendicular planes and coherent to each other. This is possible only if the two linearly polarized waves come from the same fully polarized light wave. Thus, these two linearly polarized waves can be produced if an unpolarized light beam passes through a polarizer, with an angle different from zero with respect to the vertical or horizontal plane. In this manner, the two waves so obtained are polarized in perpendicular planes and are mutually coherent, even if they are not monochromatic, or it can even be white light.

By means of a quarter-wave phase plate and appropriately selecting the angle of its axis with respect to the polarized light beams, we can obtain a full polarized beam in any desired state. The light beam is fully polarized, since the two conditions for this are fulfilled, that is, that the ratio of the amplitudes of the components and their phase difference are constant.

Only one problem remains, when the spectral width of the light source is quite large, for example, when the light is white. Then, the state of polarization for each color (wavelength) will be quite similar, but not identical, because the phase delay is not exactly the same (in radians or degrees) for all colors. However, perfect full polarization can be achieved by using an achromatic phase plate. Concluding, full polarization can be obtained with nonmonochromatic light, including white light.

We may think of an unpolarized light wave as being produced by two light waves, with the same irradiance, linearly polarized in two orthogonal planes, and fully incoherent to each other. Equivalently, we may think that the state of polarization changes randomly and extremely fast, due to the fast and randomly changing phase difference δ.

The average irradiance of a nonpolarized light wave is a constant for any plane of polarization. If one of the two waves had a greater irradiance, the result would be partially linearly polarized light. Partially linearly polarized light can be generated when an unpolarized light beam is reflected with an incidence angle greater than zero on a dielectric or metal, and also by scattering. We may think of partially polarized light as a mixture of unpolarized and polarized light.

4.4.5 POLARIZATION STATES IDENTIFICATION

The state of polarization can be detected by means of analyzers, also called polarizers and phase retarder plates. Under some special circumstances, polarized light can also be detected with the human eye. For example, if we observe the blue sky in a clear day, without clouds, through a polarizer, we may see a dim but clear image of about 2°–3° of angular diameter, as illustrated in Figure 4.47. This image, known as Haidinger brushes, disappears

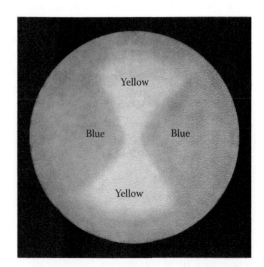

Figure 4.47 Haidinger brushes.

after some seconds; however, if the analyzer is rotated 90°, it appears again with a different orientation. The explanation of this phenomenon is based on the assumption that in the retina of the human eye there are many dichroic small particles oriented in a radial pattern. However, this has not been confirmed. The eyes of bees and some other insects are quite sensible to polarized light. The sensitivity to polarized light is so high that these insects can orient themselves during their flight, by just observing the polarized light in the sky.

The type and degree of polarization of a light beam can be measured with just the help of an analyzer and a quarter-wave retarder phase plate. These phase retarders or phase plates are optical devices that have different optical paths for two mutually perpendicular orientations of the electric vector, as illustrated in Figure 4.48. Any glass or dielectric plate with thickness t and refractive index n delays the phase by

$$\phi = k(n-1)t. \tag{4.184}$$

The phase delay is constructed with an anisotropic material. The maximum and minimum refractive indices in this plate are for electric fields parallel to the plate faces and mutually perpendicular. The fast axis is for the direction of the electric field with the minimum refractive index and the other is the slow axis. Then, the two perpendicular components of the polarization have different phase delays, whose difference is

$$\Delta\phi = k(n_s - n_f)t, \tag{4.185}$$

where the subscripts s and f mean slow and fast, respectively. If the thickness of the phase plate is such that the phase difference is equal to $\pi/2$, we have a quarter-wave phase plate. Many isotropic transparent materials, as those shown in Figure 4.49, can act as phase retarders.

If a fully linearly polarized light beam enters a quarter-wave phase plate, with its plane of polarization at 45° with respect to the slow and fast axis, circularly polarized light will be produced. But if circularly polarized light arrives at the phase plate, linearly polarized light at 45° with respect to the fast and slow axes will come out of the phase plate.

(b)

Figure 4.49 Several phase retarders: (a) between two crossed polarizers and (b) without polarizers.

If the plane of polarization of a fully linearly polarized light beam enters the phase plate is different from zero, 90°, or 45°, the light exiting the phase plate will be elliptically polarized, with its semiaxes parallel to the phase plate axis. However, if light elliptically polarized arrives to the phase plate, the exiting light will be linearly polarized.

If the light entering the phase plate is elliptically polarized, with its semiaxes forming an angle with respect to the axes of the phase plate, the transmitted light will also be elliptically polarized, but not equal to the incident light.

These procedures are very useful to experimentally determine and measure the polarization states of a light beam, as described in Table 4.3.

Another way to specify and measure the state and degree of polarization is by the use of the Stokes parameters. The measurement procedure can be found with the help of Equation 175 for the transmitted irradiance of the polarized beam through the analyzer, as a function of the angle of the axis of the analyzer with the x-axis. If only the analyzer is present, without the phase retarder, by using the Stokes parameter definitions, we can obtain

$$\langle I(\theta,0°) \rangle = \frac{1}{2}\left(I + M\cos 2\theta + C\sin 2\theta\right). \tag{4.186}$$

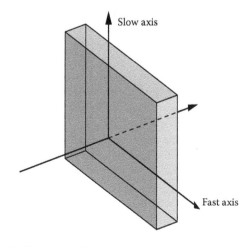

Figure 4.48 Phase retarder.

Table 4.3 **Identification of the state of polarization**

An analyzer is rotated in front of the light beam, and the transmitted irradiance:	A quarter-wave phase plate is located before the analyzer. If there is a minimum of the irradiance, one of the axes of the phase plate must be parallel to the axis of the analyzer. Then, when the analyzer is rotated in front of the phase plate, the irradiance:		State of polarization:
Becomes zero at some angles			Linearly polarized
It has a minimum at some angles	Becomes zero at some angles		Elliptically polarized
	Reaches a minimum, different from zero at some angles	At the same previous orientation of the analyzer	Linearly polarized (partial)
		At a different than the previous orientation of the analyzer	Elliptically polarized (partial)
Remains constant	Becomes zero at some angles		Circularly polarized
	Reaches a minimum, different from zero at some angles		Circularly polarized (partial)
	Remains constant		Unpolarized

If we insert a quarter-wave retarder plate in the path of the light, with an axis orientation such that its fast axis is parallel to the y-axis ($\varepsilon = 90°$), the irradiance after the analyzer would be

$$\langle I(\theta,90°)\rangle = \frac{1}{2}\left(I + M\cos 2\theta + S\sin 2\theta\right). \qquad (4.187)$$

Then, by performing four measurements, with the help of an analyzer and a quarter-wave phase plate, it is possible to determine the Stokes parameters. The four measurements are made with four different angles of the analyzer, with respect to the x-axis, as follows:

1. 0° without phase plate: $\langle I(0°,0°)\rangle = \frac{1}{2}\left(I+M\right)$.

2. 45° without phase plate: $\langle I(45°,0°)\rangle = \frac{1}{2}\left(I+C\right)$.

3. 90° without phase plate: $\langle I(90°,0°)\rangle = \frac{1}{2}\left(I-M\right)$.

4. 45° with the phase plate: $\langle I(45°,90°)\rangle = \frac{1}{2}\left(I+S\right)$.

With these four measurements we can calculate the Stokes parameters, with the following expressions:

$$I = \langle I(0°,0°)\rangle + \langle I(90°,0°)\rangle$$
$$M = \langle I(0°,0°)\rangle - \langle I(90°,0°)\rangle$$
$$C = 2\langle I(45°,0°)\rangle - I \qquad (4.188)$$
$$S = 2\langle I(45°,90°)\rangle + I.$$

With this method, commercial instruments had been constructed to completely measure the state of polarization of a light beam.

4.4.6 GENERATION OF POLARIZED LIGHT

A fully linearly polarized light beam can in practice be obtained by several different devices or procedures that will now be described.

4.4.6.1 By absorption: Polarizers

A full linearly polarized beam can be obtained from an unpolarized beam by means of a filter that is transparent only to the

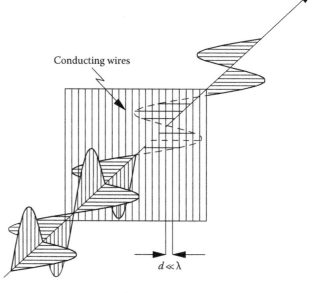

Conducting wires

$d \ll \lambda$

Figure 4.50 Microwaves polarizer.

waves having the desired polarization plane. This kind of filter has been constructed and it is very popular, with the name of polarizer or polaroid. To understand how they work, let us consider a polarizer for microwaves, made with a grating of conducting wires, as illustrated in Figure 4.50. The separation between the wires in this grating is smaller than the wavelength of the microwave, to avoid that it works like a diffraction grating. If the electric field is parallel to the conducting wires, an electrical current will be induced and the energy of the microwave is absorbed. If the electrical field is perpendicular to the wires, no current is induced, and the microwave passes through the grating.

Some natural materials, like tourmaline, have a microscopic structure formed by molecular electrically conducting chains, whose separations are smaller than the wavelength of the visible light. These materials are good to work as polarizers.

The Polaroid Corporation manufactures polarizers with a process invented by its founder, E. Land, in 1938. A large sheet of polyvinyl alcohol is heated and then suddenly stretched. With this operation, the molecules acquire a preferential orientation

Figure 4.51 Two crossed polarizers.

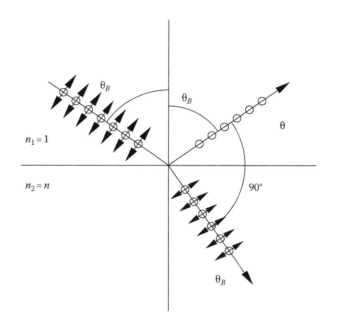

Figure 4.53 Brewster angle.

in the direction of the stretching. Later, the sheet of polyvinyl alcohol is immersed in an iodine solution. This process makes the molecules chain to become conducting. These polarizers are illustrated in Figure 4.51. Figure 4.52 shows the transmission characteristics of a typical polarizer (Polaroid HN-38). We may observe that the maximum efficiency of the polarizer occurs for a wavelength close to 600 nm. This efficiency is quite low in the blue region, close to 400 nm. When the polaroid sheets are crossed, and are illuminated with visible light, centered in the yellow region, the transmission is lower than 0.01%.

4.4.6.2 By reflection or refraction: Polarizing prism

If we think of a ray of light arriving to a flat optical surface, not perpendicularly to the surface, the plane of incidence is that which contains the incident ray and the normal to the surface. Now, let us assume that the ray is fully linearly polarized. By definition we say that the light beam has *s* polarization if the electrical vector is perpendicular to the plane of incidence

(the subscript *s* comes from the German word senkretch, meaning perpendicular). On the other hand, we say that the light beam has *p* polarization if the electrical vector is on the plane of incidence (the subscript *p* comes from the word parallel).

With the electromagnetic theory of light it can be shown that given an angle of incidence θ, the reflectivity and transmissivity of a reflecting surface is in general different for light beams with *p* polarization than for light beams with s polarization. Hence, if a beam of unpolarized light is reflected or refracted on a surface, both beams are reflected with different reflectivity. Hence, if a beam of unpolarized light is reflected in a surface, the light polarizes at least in a partial manner.

Let us now consider a light ray incident upon a dielectric plane surface, where it is reflected and refracted, as in Figure 4.53.

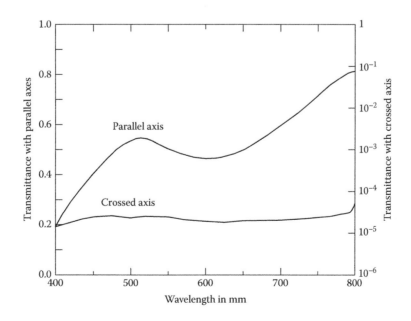

Figure 4.52 Desired and undesired transmittances in a Polaroid HN-38 polarizer.

If the angle between the refracted and the reflected ray is 90°, the reflected ray in plane *s* will be fully polarized and the ray in plane *p* is partially polarized.

The angle of incidence θ that is necessary in order to make the angle formed by the refracted and reflected rays equal to 90° has the name of Brewster angle. It can be calculated by means of Snell's law

$$\frac{\sin \theta_B}{\sin \theta'_B} = n,$$

(4.189)

and the condition

$$\theta + \theta'_B = 90°,$$

(4.190)

in which, after combining, we obtain

$$\tan \theta_B = n.$$

(4.191)

The reflected light is fully linearly polarized if the incidence angle is the Brewster angle. The refracted light is not fully polarized, but the degree of polarization can be increased as much as desired by means of a stack of glass plates, as shown in Figure 4.54. The advantage of this type of device over the polaroid sheets is that it can be used in the violet or infrared regions.

A variant of this type of polarizer is the polarizing beam splitter, as illustrated in Figure 4.55, which splits the light beam into two linearly polarized light beams, in perpendicular planes. This prism was designed by Banning in 1947. The system is formed by two rectangular prisms with a stack of thin films between them, with alternating high refractive indices (n_H)

and low refractive indices (n_L). The optical thickness of each film is made equal to a quarter wave in order to increase the reflectance. Since the reflections in thin films are related to the Brewster angle, we can write

$$\tan \theta_H = \frac{n_L}{n_H},$$

(4.192)

where θ_H is the Brewster angle, with incidence in the medium with high index of refraction. The incidence angle θ_G in the glass should be 45° and it is related with θ_H by means of the Snell law, as follows:

$$n_G \sin \theta_G = n_H \sin \theta_H,$$

(4.193)

where *n* is the refractive index of the glass. Using these conditions and an angle equal to 45°, we obtain the following relation for the refractive indices:

$$\frac{2}{n_G^2} = \frac{1}{n_H^2} + \frac{1}{n_L^2}.$$

(4.194)

Thus, to obtain a good beam splitter, the refractive indices of the glass and the indices of refraction of the thin films should as close to their ideal values as possible.

4.4.6.3 By double refraction

When refracting a light beam, an optical anisotropic material splits it in two rays, linearly polarized, in mutually perpendicular planes, *s* and *p*. However, Snell's refraction law is satisfied only for one of them. Some anisotropic materials are crystals.

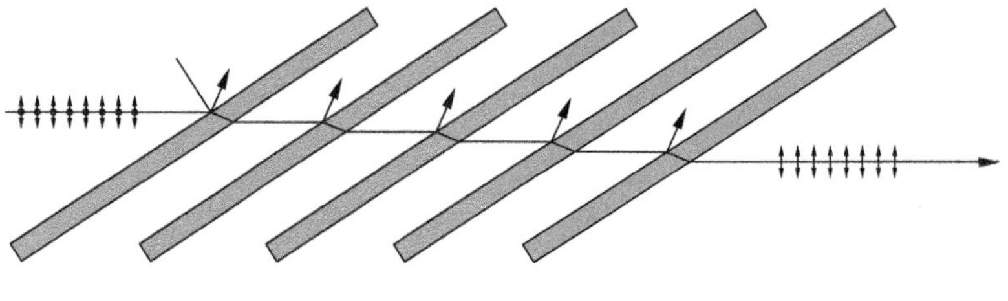

Figure 4.54 Polarizer formed by a stack of inclined glass plates.

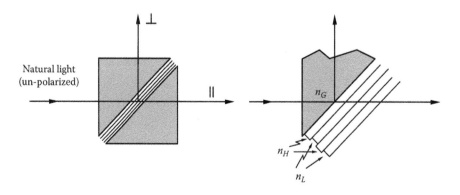

Figure 4.55 Polarizing beam splitter.

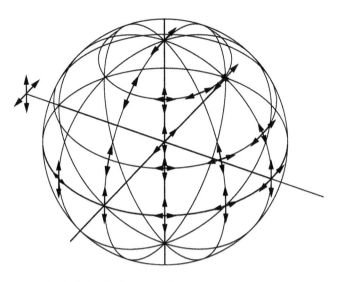

Figure 4.56 Polarization by scattering.

With crystals we can construct devices that allow us to obtain linearly polarized light. One of them is the Nicol prism.

4.4.6.4 By scattering

Let us consider a light beam illuminating an atom, molecule, or particle, with its dimensions much smaller than the wavelength of the light. Then, the light energy is absorbed and then reemitted as a spherical wave. This phenomenon is known with the name of scattering. However, in the reemitted light the electric vector has to have the same direction as in the incident wave. Hence, if the incident light is unpolarized, the emitted wave cannot have any electric field in the direction of propagation of the incident beam but will have electric field vector in all other directions, perpendicular to this trajectory, as illustrated in Figure 4.56. In some directions the exiting light will have partial polarization. This is the reason why the blue sky is partially polarized.

RECOMMENDED READING

American Institute of Physics, *Polarized Light: Selected Reprints*, American Institute of Physics, New York, 1963.

Angela, A.D., M. Lavery, M. Padgett, and A. Forbes, Unraveling Bessel Beams, *Optics and Photonics News*, Vol. 24, Pags. 22–29, June 2013.

Born, M. and E. Wolf, *Principles of Optics*, 7th edn., Cambridge University Press, Cambridge, U.K., 1999.

Chiao, R. and P.W. Milonni, Fast Light, Slow Light, *Optics and Photonics News*, Vol. 13, Pags. 26–30, June 2002.

Donald, R.H., Applications of laser light, in *Scientific American*, septiembre de 1968, reimpreso en *Lasers and Light*, A.L. Schawlow (comp.), W. H. Freeman and Company, San Francisco, CA, 1969.

Goodman, J., *Introduction to Fourier Optics*, 3rd edn., Roberts and Co., Greenwood Village, CO, 2005.

Goldstein, D., *Polarized Light*, 2nd edn., Marcel Decker, New York, 2003.

Lene, V.H., Frozen Light, *Scientific American*, Vol. 265, Pags. 66–70, July 2001.

Ronald, N.B., The Fourier Transform, *Scientific American*, Vol. 260, Pags. 62–69, June 1989.

William, A.S., *Polarized Light*, Harvard University Press, Cambridge, MA, 1962.

William, A.S. and S.S. Ballard, *Polarized Light*, Van Nostrand, Princeton, NJ, 1964.

Yturi, R., O. Kocharovskaya, G.R. Welch, and M.O. Scully, Slow, Ultraslow, Stored and Frozen, *Optics and Photonics News*, Vol. 13, Pags. 44–48, June 2002.

Aberrations in optical systems

José Sasián

Contents

5.1 INTRODUCTION

This chapter provides an introduction to optical aberrations in axially symmetric systems. An understanding of the process of optical image formation is important in a variety of fields. Many systems are modeled to some useful extent as linear shift invariant systems, which are characterized by a point spread function. A remarkable attribute of such linear systems is that the image is given by the convolution of the object with the system's point spread function. Since aberrations influence the point spread function, it is then important to understand the nature of optical aberrations. Sharp imaging requires the absence of aberrations.

The concept of a ray is intuitive from the observation of shadows, columns of light in the sky, and image formation in a camera obscura. The invention and understanding of the magnifier, the microscope, the telescope, the camera obscura, and the human eye led to the discovery of optical aberrations. The explanation of optical aberrations has been based on the geometrical concept of ray (see Chapter 4). Aberrations are understood as departures from an ideal behavior. The ideal behavior is application dependent; it is often defined by a simple imaging model, based on the principle of central projection. In central projection the image is found to be a scaled copy of the object, and ideal imaging is analytically represented by Newtonian or Gaussian imaging equations. Light rays departing from an object point are expected to arrive to the ideal image point. However, in practice, rays may not intersect the ideal image point and the error is then called and characterized as angular, transverse, or longitudinal ray error.

An alternative approach that is fundamental for understanding aberrations is based on wave propagation. In this case, ideal imaging requires that a spherical wavefront as it departs from an object point arrives to an ideal image point as a spherical wavefront. In practice, the wavefront is deformed as it passes through an optical system. This wavefront deformation from a spherical form is called wave aberration. The geometrical wavefront is the surface of constant optical path length taking as a reference the object point.

It must be realized that symmetry plays an important role in describing aberrations. Initially, we are concerned with optical systems that have a rotational axis of symmetry. This symmetry imposes some restrictions in the nature of the aberrations such systems may have. The aberrations themselves have some level of symmetry. Therefore, identifying symmetrical properties is relevant in understanding aberrations as ultimately this provides useful insight.

There are several metrics for describing aberration, namely, wavefront deformation, angular ray aberration, transverse ray aberration, and longitudinal ray aberration, and for visual systems, low-order aberrations might be given in diopters of optical power. Then in speaking of optical aberrations it is critical to specify the metric used to describe them.

The aberrations discussed in this chapter are those of axially symmetric systems. To understand aberrations in optical systems that are nonaxially symmetric, it is desirable to first have a solid understanding of aberrations in axially symmetric systems. The purpose of this chapter is to provide such a solid foundation.

5.2 AXIALLY SYMMETRICAL SYSTEMS

To build an understanding about aberrations, a model of an optical system is necessary. We assume at this time that the optical system has rotational symmetry; this implies the existence of an optical axis about which the system is symmetric. A rotation of the system about the axis cannot be distinguished under rotational symmetry.

The space where the object resides is infinite in extent and is called the object space; similarly, the space where the image resides is called the image space and is infinite in extent. A lens or optical system to be well defined requires of an aperture stop that is assumed to be circular and laying on a plane perpendicular to the optical axis. One key function of the aperture stop is to clearly define light beams that propagate from the stop to the image plane. Another key function of the aperture stop is to determine the amount of light accepted by the system. By definition the entrance pupil is the image of the aperture stop in object space, and the exit pupil is the image of the aperture stop in image space. These system's entities are shown in Figure 5.1.

Aberration theory aims at describing an optical system with ray trace information from two key rays: the chief and marginal ray. The marginal ray is defined as passing by the center of the field of view and the edge of the aperture stop. The chief ray is defined as passing by the edge of the field of view and the center of the aperture stop. These rays behave in an ideal manner and can be traced using Newtonian or Gaussian imaging equations, or the so-called first-order ray tracing equations.

The first-order ray trace equations are refraction $n'u' = nu - \phi y$ and transfer $y' = y + ut$. Here n is the index of refraction, u is the ray slope, y is the ray height, $\phi = (n' - n)c$ is the surface optical power, and c is the surface curvature. Primed quantities are after refraction and unprimed quantities before refraction. Quantities for the chief ray are barred and for the marginal ray are unbarred. First-order ray tracing is performed in an iterative manner, surface after surface, and it requires minimum computation effort.

A propagating ray needs to be defined in an optical system and this is done by two points. One point is defined in the object plane by the field vector $\vec{H}$ and another point is defined in the exit pupil plane by the aperture vector $\vec{\rho}$. These vectors are normalized so their magnitude ranges from zero to unity and are illustrated in Figure 5.2. The maximum field extent is given by the chief ray height $\bar{y}_o$ in the object plane. The maximum

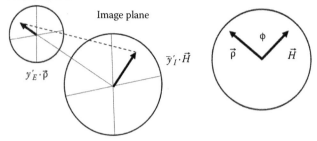

Figure 5.2 The field and aperture vectors and their projection looking down the optical axis. (Reprinted from Sasián, J., *Introduction to Aberrations in Optical Imaging Systems*, Cambridge University Press, Cambridge, U.K., Copyright 2013. With permission of Cambridge University Press.)

aperture extent is given by the marginal ray height y'_E in the exit pupil plane. Thus any ray propagating in the system is clearly defined by the vectors $\bar{y}_o \vec{H}$ and $y'_E \vec{\rho}$. The angle between the field and aperture vectors is ϕ; when $\phi = 0°$ meridional rays are defined and when $\phi = 90°$ sagittal rays are defined.

Figure 5.1 shows an ideal ray as a broken line propagating through an optical system. A real ray is also shown as a solid line. By definition both rays do coincide at the object point $\bar{y}_o \vec{H}$ and at the exit pupil point $y'_E \vec{\rho}$. Elsewhere, they may depart in position due to aberration, and in particular at the image plane they depart by the transverse ray error $\bar{y}_i \Delta \vec{H}$ and at the entrance pupil by the transverse ray error $y_E \Delta \vec{\rho}$.

Thus for understanding a rotationally symmetric system, we have a model that is ideal in that imaging an object is described by the Newtonian or Gaussian equations, and in which we have a stop aperture, and entrance and exit pupil. Rays propagate perfectly and the image is a scaled copy of the object. Then ray or wavefront aberrations are defined as departures from ideal behavior. Rays are represented as line normals to the wavefront.

5.2.1 WAVEFRONT DEFORMATION FROM A SPHERE

While describing optical systems using the concept of ray is intuitive, it is fundamentally more insightful to describe aberrations by the wavefront deformation. The emphasis in this chapter is thus given to wavefront aberration. Figure 5.3 shows that the wavefront deformation is measured with the aid of a reference sphere. This sphere is centered in the ideal image point and passes by the on-axis exit pupil point. The wavefront deformation from a sphere is measured as an optical path difference as it is the product of the index of refraction in image space and the distance between the reference sphere and the wavefront, along the ray.

One significant realization is that the wavefront deformation can be described as a superposition of basic deformation forms of increasing order. That is, the wavefront deformation can be expanded as a polynomial on $\vec{H}$ and $\vec{\rho}$ where each term represents an aberration term. Furthermore, given the axial symmetry of the system only a specific set of terms can result. The axial symmetry requires the wavefront deformation, which is a scalar, to be a function of the rotational invariants $\vec{H} \cdot \vec{H}$, $\vec{H} \cdot \vec{\rho}$, and $\vec{\rho} \cdot \vec{\rho}$.

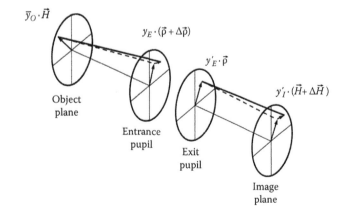

Figure 5.1 Model of an optical system including the optical axis, the object and image planes, the entrance and exit pupil planes, an ideal ray as a broken line, a real ray as a solid line, and the field and aperture vectors. (Reprinted from Sasián, J., *Introduction to Aberrations in Optical Imaging Systems*, Cambridge University Press, Cambridge, U.K., Copyright 2013. With permission of Cambridge University Press.)

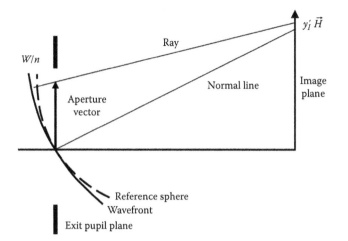

Figure 5.3 Construction to illustrate the wavefront deformation in relation to the reference sphere, the wavefront, and the ray. (Reprinted from Sasián, J., *Introduction to Aberrations in Optical Imaging Systems*, Cambridge University Press, Cambridge, U.K., Copyright 2013. With permission of Cambridge University Press.)

The wavefront deformation $W(\vec{H}, \vec{\rho})$, or aberration function, is expressed as

$$W\left(\vec{H}, \vec{\rho}\right) = W_{000} + W_{200}\left(\vec{H} \cdot \vec{H}\right) + W_{111}\left(\vec{H} \cdot \vec{\rho}\right) + W_{020}\left(\vec{\rho} \cdot \vec{\rho}\right)$$

$$+ W_{040}\left(\vec{\rho} \cdot \vec{\rho}\right)^2 + W_{131}\left(\vec{H} \cdot \vec{\rho}\right)\left(\vec{\rho} \cdot \vec{\rho}\right) + W_{222}\left(\vec{H} \cdot \vec{\rho}\right)^2$$

$$+ W_{220}\left(\vec{H} \cdot \vec{H}\right)\left(\vec{\rho} \cdot \vec{\rho}\right) + W_{311}\left(\vec{H} \cdot \vec{H}\right)\left(\vec{H} \cdot \vec{\rho}\right)$$

$$+ W_{400}\left(\vec{H} \cdot \vec{H}\right)^2 \ldots$$

The expansion of the wavefront deformation assumes that the system properties are smooth in behavior. Table 5.1 presents the aberration terms up to fourth order, and Figure 5.4 illustrates their form. Note that these terms represent a set of basic forms that also depend on the coefficients W.

The zero-order term W_{000} is called piston; it represents a uniform advance or delay of the wavefront and therefore it does not degrade an image. The second-order terms are piston $W_{200}(\vec{H} \cdot \vec{H})$, magnification $W_{111}(\vec{H} \cdot \vec{\rho})$, and focus $W_{020}(\vec{\rho} \cdot \vec{\rho})$. Piston varies quadratic with the field of view but is uniform with respect to the aperture; therefore, it again does not degrade an image and is neglected. Since the reference sphere is centered at the ideal image point, then the magnification and focus terms are equal to zero. The fourth-order terms, which are known as the primary aberrations, are spherical, coma, astigmatism, field curvature, and distortion. The quartic piston term is again neglected; however, piston terms in multiple aperture systems may need to be accounted for whenever light interference is important.

The aberration coefficients provide the aberration amplitude at the edge of the exit pupil. Spherical aberration depends on the fourth order on the system aperture and is uniform across the field of view. Whenever spherical aberration is present, it can be observed at the center of the field of view as other aberrations become zero. Spherical aberration can be considered

Table 5.1 **Wavefront aberrations**

ABERRATION NAME	VECTOR FORM
Zero order	
Uniform piston	W_{000}
Second order	
Quadratic piston	$W_{200}\left(\vec{H} \cdot \vec{H}\right)$
Magnification	$W_{111}\left(\vec{H} \cdot \vec{\rho}\right)$
Focus	$W_{020}\left(\vec{\rho} \cdot \vec{\rho}\right)$
Fourth order	
Spherical aberration	$W_{040}\left(\vec{\rho} \cdot \vec{\rho}\right)^2$
Coma	$W_{131}\left(\vec{H} \cdot \vec{\rho}\right)\left(\vec{\rho} \cdot \vec{\rho}\right)$
Astigmatism	$W_{222}\left(\vec{H} \cdot \vec{\rho}\right)^2$
Field curvature	$W_{220}\left(\vec{H} \cdot \vec{H}\right)\left(\vec{\rho} \cdot \vec{\rho}\right)$
Distortion	$W_{311}\left(\vec{H} \cdot \vec{H}\right)\left(\vec{H} \cdot \vec{\rho}\right)$
Quartic piston	$W_{400}\left(\vec{H} \cdot \vec{H}\right)^2$

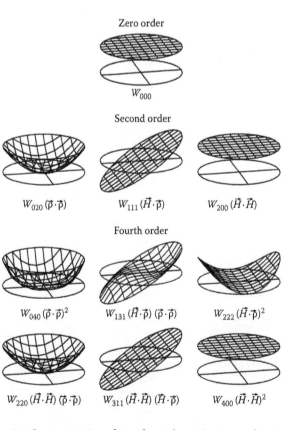

Figure 5.4 Representation of wavefront aberration terms showing the aperture dependence. (Reprinted from Sasián, J., *Introduction to Aberrations in Optical Imaging Systems*, Cambridge University Press, Cambridge, U.K., Copyright 2013. With permission of Cambridge University Press.)

as the variation of focal length with aperture; it has axial symmetry. Coma aberration depends linearly with the field of view and cubically with the system's aperture; it has plane symmetry. Coma can be considered as a variation of magnification with aperture. Astigmatism aberration depends quadratically on the field of view and quadratically on the aperture; it has double plane symmetry. Astigmatism can be considered as a variation of the focal length with the aperture azimuth. Noting the dependence of the aberrations on the field and aperture is important, as well as noting the symmetries of each aberration form. Spherical, coma, and astigmatism degrade the image of a point object as the image, geometrically, will not be a point. Figure 5.5 illustrates the images that are produced in the presence of these aberrations. The amount of aberration is given by the aberration coefficient expressed in waves. If the aberration is a small fraction of a wave, say, 1/10, then the image can be considered as diffraction limited, that is, essentially perfect. As the aberration increases, the image degrades by distributing the energy over a larger region in a light diffracted pattern.

Field curvature and distortion preserve point images but change axially and transversely the position of the image. Field curvature depends quadratically with the field of view and quadratically with the aperture; it has axial symmetry. In the presence of field curvature, the image no longer focuses on a plane but on a curved surface. Distortion aberration varies cubically with the field of view and linearly with the aperture; it has plane symmetry.

The discussion treats each aberration term as separated; however, most often aberrations appear as a combination of them. Nevertheless, recognizing each aberration term is important. It must be understood that the fourth-order aberrations are in practice the most significant. Aberrations of higher order, as a function of the field and aperture, also are present, and the overall superposition of all aberration orders provides the overall wavefront deformation.

As noted earlier, light rays are represented as lines normal to the wavefront. The wavefront is examined at the exit pupil, and if aberrated, it causes the rays to not pass through the ideal image point where the reference sphere is centered.

Figure 5.6 shows the ray trajectory from the exit pupil plane to the image plane under the presence of the primary aberrations. Ray fans are shown in the tangential azimuth (or meridional plane containing the optical axis) and in the sagittal azimuth (perpendicular to the meridional plane). Ray fans are shown for the on-axis, 0.7 field, and full field positions.

The transverse ray aberration $\Delta\vec{H}$, that is, the distance from the actual ray intersection on the image plane to the ideal image point, is proportional to the derivative of the wavefront deformation. We can write for $\Delta\vec{H}$

$$\Delta\vec{H} = -\frac{1}{\text{Ж}}\,\vec{\nabla}_\rho W(\vec{H}, \vec{\rho})$$

The transverse ray error $\Delta\vec{H}$ is a normalized (by the chief ray image height) vector and is expressed as the gradient of the

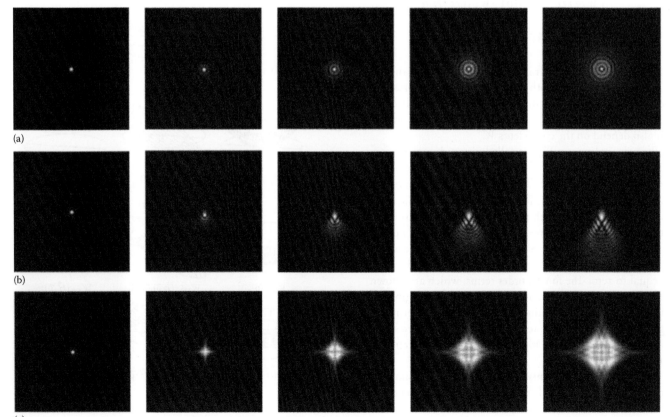

Figure 5.5 Computer-generated images under the presence of aberration. (a) Spherical aberration ranging from 0 to 4 waves. (b) Coma aberration ranging from 0 to 4 waves. (c) Astigmatism aberration ranging from 0 to 4 waves.

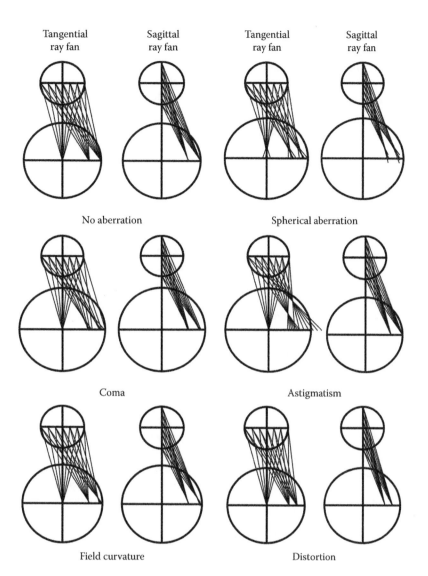

Figure 5.6 Ray fans from the exit pupil to the image plane in the presence of aberration. (Reprinted from Sasián, J., *Introduction to Aberrations in Optical Imaging Systems*, Cambridge University Press, Cambridge, U.K., Copyright 2013. With permission of Cambridge University Press.)

aberration function $W(\vec{H}, \vec{\rho})$. The symbol Ж $= n(\bar{u}y - u\bar{y})$ stands for the Lagrange invariant of the system.

5.2.2 ABERRATION COEFFICIENTS

The discussion earlier provides insight into the nature of the primary wavefront aberrations that an axially symmetric system may exhibit. The next matter is to determine the aberration coefficients for a particular system. The aberration coefficients depend on the constructional parameters of a system such as the radii of curvature, surface distances, and index of refraction. We now discuss the aberration coefficients of a system of surfaces that are perfectly spherical in shape.

Aberration theory aims at providing a substantial amount of information about an optical system with minimum calculation. Notably, from the first-order ray trace of a marginal and chief ray, the aberration coefficients are determined. Table 5.2 provides the aberration coefficients in terms of Seidel sums S_I, S_{II}, S_{III}, S_{IV}, and S_V; these use only quantities associated with the ray trace

Table 5.2 Aberration coefficients in terms of Seidel sums

COEFFICIENT	SEIDEL SUM
$W_{040} = \dfrac{1}{8}S_I$	$S_I = -\sum\limits_{i=1}^{j}\left(A^2 y\Delta\left(\dfrac{Fu}{n}\right)\right)_i$
$W_{131} = \dfrac{1}{2}S_{II}$	$S_{II} = -\sum\limits_{i=1}^{j}\left(A\bar{A}y\Delta\left(\dfrac{u}{n}\right)\right)_i$
$W_{222} = \dfrac{1}{2}S_{III}$	$S_{III} = -\sum\limits_{i=1}^{j}\left(\bar{A}^2 y\Delta\left(\dfrac{u}{n}\right)\right)_i$
$W_{220} = \dfrac{1}{4}\left(S_{IV} + S_{III}\right)$	$S_{IV} = -Ж^2\sum\limits_{i=1}^{j}P_i$
$W_{311} = \dfrac{1}{2}S_V$	$S_V = -\sum\limits_{i=1}^{j}\left(\dfrac{\bar{A}}{A}\left[Ж^2 P + \bar{A}^2 y\Delta\left(\dfrac{u}{n}\right)\right]\right)_i$

of chief and marginal first-order rays. A Seidel sum calculates the argument for each system surface i and then sums up all the arguments for the number of surfaces j to determine the system aberration coefficient.

The quantities used in the arguments are the chief and marginal ray refraction invariants $\bar{A} = n(\bar{u} + \bar{y}c)$ and $A = n(u + yc)$, the marginal ray height y and slope u, and the index of refraction n. The Petzval term is $P = -(1/n' - 1/n)c$, $\Delta(u/n) = u'/n' - u/n$, and the Lagrange invariant is $\text{Ж} = n(\bar{u}y - u\bar{y})$.

5.2.3 COOKE TRIPLET LENS EXAMPLE

To put things in perspective consider the Cooke triplet lens in Figure 5.7 that operates at F/4 for a field of +/–20°. Table 5.3 provides the aberrations contributed by each surface and then the total aberration in the bottom row. Clearly, aberration decomposition like this provides useful information about how aberrations occur and how they are balanced within the lens. The total aberration is not zero as there are higher-order aberrations that are balanced with the primary aberrations.

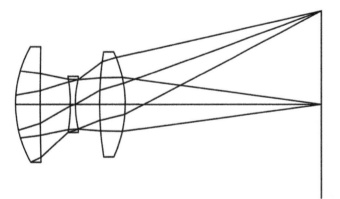

Figure 5.7 Layout of a Cooke triplet lens. A set of three rays from two field points, on-axis and full field, are traced. (Reprinted from Sasián, J., *Introduction to Aberrations in Optical Imaging Systems*, Cambridge University Press, Cambridge, U.K., Copyright 2013. With permission of Cambridge University Press.)

Table 5.3 **Wave aberration coefficients of a Cooke triplet (waves at 587 nm)**

SURFACE	W_{040}	W_{131}	W_{222}	W_{220}	W_{311}
1	6.77	16.16	9.64	44.06	52.59
2	3.78	-44.19	129.24	62.29	-364.36
3	-16.16	96.72	-144.77	-100.68	301.39
4	-8.01	-56.45	-99.48	-92.30	-325.33
5	1.34	20.24	76.6	51.72	391.53
6	14.94	-32.46	17.64	45.68	-49.63
Sum	2.66	0.02	-11.13	10.78	6.19

5.3 PUPIL ABERRATIONS

Since the entrance and exit pupil planes are optically conjugated, aberrations can be defined. These are known as pupil aberrations and are described with the pupil aberration function

$$\bar{W}\left(\vec{H}, \vec{\rho}\right) = \bar{W}_{000} + \bar{W}_{200}\left(\vec{\rho}\cdot\vec{\rho}\right) + \bar{W}_{111}\left(\vec{H}\cdot\vec{\rho}\right) + \bar{W}_{020}\left(\vec{H}\cdot\vec{H}\right)$$

$$+ \bar{W}_{040}\left(\vec{H}\cdot\vec{H}\right)^2 + \bar{W}_{131}\left(\vec{H}\cdot\vec{H}\right)\left(\vec{H}\cdot\vec{\rho}\right) + \bar{W}_{222}\left(\vec{H}\cdot\vec{\rho}\right)^2$$

$$+ \bar{W}_{220}\left(\vec{H}\cdot\vec{H}\right)\left(\vec{\rho}\cdot\vec{\rho}\right) + \bar{W}_{311}\left(\vec{\rho}\cdot\vec{\rho}\right)\left(\vec{H}\cdot\vec{\rho}\right) + \bar{W}_{400}\left(\vec{\rho}\cdot\vec{\rho}\right)^2$$

where to distinguish from the image aberration function the coefficients are barred. In this function, the field and aperture vectors interchange role. It is interesting that the image and pupil aberration coefficients are related as shown in Table 5.4.

An interpretation of the pupil aberrations is found by considering the ray error $\Delta\vec{\rho}$ at the entrance pupil $\Delta\vec{\rho} = -\frac{1}{\text{Ж}}\nabla_H \bar{W}(\vec{H}, \vec{\rho})$.

For a given field point, the cross section of the beam is distorted as shown in Figure 5.8. Each pupil aberration produces a distortion of a square grid at the exit pupil. Pupil spherical aberration produces a beam displacement that is cubic with the field of view.

In some lenses such as those in mobile phones, the amount of pupil spherical aberration is so large that the chief ray slope changes and may have a positive effect on improving the relative illumination. Pupil coma also impacts the relative illumination by changing the size of the entrance pupil as a function of the field of view. A full understanding of optical aberrations includes pupil aberrations as these enter into the computation of irradiance changes, higher-order aberration coefficients, and beam cross section sizes.

5.4 NONAXIALLY SYMMETRIC SYSTEMS

Symmetry is an important attribute that determines the nature of the aberrations that can occur in a given class of optical systems. As the symmetry of a system is reduced, say, by limiting the symmetry to plane symmetry, more aberration terms are possible.

Table 5.4 **Relationship between pupil and image aberration coefficients**

$\bar{W}_{040} = W_{400}$
$\bar{W}_{131} = W_{311} + \frac{1}{2}\text{Ж}\cdot\Delta\left\{\bar{u}^2\right\}$
$\bar{W}_{222} = W_{222} + \frac{1}{2}\text{Ж}\cdot\Delta\left\{u\bar{u}\right\}$
$\bar{W}_{220} = W_{220} + \frac{1}{4}\text{Ж}\cdot\Delta\left\{u\bar{u}\right\}$
$\bar{W}_{311} = W_{131} + \frac{1}{2}\text{Ж}\cdot\Delta\left\{u^2\right\}$
$\bar{W}_{400} = W_{040}$

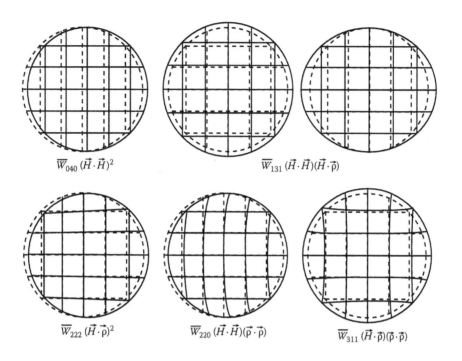

Figure 5.8 Pupil aberrations represent a cross section change at the entrance pupil of an optical system. The broken line represents a uniform object grid at the exit pupil. The solid line represents the grid image at the entrance pupil. (Reprinted from Sasián, J., *Introduction to Aberrations in Optical Imaging Systems*, Cambridge University Press, Cambridge, U.K., Copyright 2013. With permission of Cambridge University Press.)

It turns out that there are no new aberration shapes but the field dependence becomes richer in possibilities. For example, in addition to spherical aberration, uniform coma and astigmatism are possible over the field of view.

The treatment earlier assumes that the surfaces are perfectly spherical. However, fabrication errors, or reduced system symmetry, do call for other wavefront expansions. For example, in describing the aberrations of the eye, the Zernike polynomials are a clear choice. As an actual eye has no exact symmetry, it is necessary to allow a richer possibility for the wavefront deformation. The study of aberration is closely related to the system symmetry and symmetry awareness is important.

The foundation to discuss aberrations in nonaxially symmetric systems is a clear understanding of the aberrations in axially symmetric systems. This chapter is aimed at providing the essential information for understanding aberrations in axially symmetric systems.

FURTHER READING

Buchdahl, H. A., *An Introduction to Hamiltonian Optics*, Cambridge University Press, Cambridge, U.K., 1970.

Greivenkamp, J., *Field Guide to Geometrical Optics*, SPIE Press, Bellingham, WA, 2004.

Hopkins, H. H., *The Wave Theory of Aberrations*, Oxford University Press, Oxford, U.K., 1950.

Sasián, J., *Introduction to Aberrations in Optical Imaging Systems*, Cambridge University Press, Cambridge, U.K., 2013.

Steward, G. C., *The Symmetrical Optical System*, Cambridge University Press, London, U.K., 1928.

Welford, W., *Aberrations of the Symmetrical Optical System*, Academic Press, Academic Press, New York, London, 1974.

6 Photometry

Yoshi Ohno

Contents

6.1 INTRODUCTION

Photometry is the measurement of light, which is electromagnetic radiation detectable by the human visual system in the wavelength range from 360 to 830 nm. Photometry is needed in many applications where light as observed by human eyes is evaluated, for example, in evaluation of products in lighting, displays, and traffic signs. In photometry, optical radiation is measured with spectral weighting by the standardized spectral response of the human eye. Photometry normally uses optical radiation detectors that mimic the spectral response of the eye, or spectroradiometry incorporating appropriate calculations for weighting by the spectral response of the eye. Typical photometric quantities include luminous flux (unit: lumen), luminous intensity (unit: cd), illuminance (unit: lux), and luminance (unit: candela per square meter).

Similar to photometry, the measurement of the entire optical radiation spectrum (and often involves spectrally resolved measurements) is called "radiometry," and a similar set of quantities is used, such as radiant flux, radiant intensity, and irradiance. Photometry and radiometry are closely related; thus, some principles in radiometry are also important for photometry.

Measurement of color is often regarded as a part of photometry, as it also involves measurement of visible radiation. Measurement of color is also important in evaluating many industrial products. Colorimetry is the measurement science used to quantify and describe physically the human color perception. The basis of colorimetry was established by Commission Internationale de l'Éclairage (CIE) in 1931, and it evolved much since then.

In this chapter, the fundamentals of photometric quantities and units, some important principles in photometry and radiometry, and the fundamentals of color quantities are described. The terminology used in this chapter follows international standards and recommendations [1,2].

6.2 BASIS OF PHYSICAL PHOTOMETRY

6.2.1 SPECTRAL LUMINOUS EFFICIENCY FUNCTION

The relative spectral responsivity of the human eye was first standardized by the CIE in 1924 [3] and redefined as part of the colorimetric standard observers in 1931 [4]. It is called "the spectral luminous efficiency function for photopic vision," or the $V(\lambda)$ function, defined in the region from 360 to 830 nm, and is normalized to one at its peak, 555 nm (Figure 6.1). The values were republished by CIE in 1983 [5] and adopted by Comité International des Poids et Mesures (CIPM) in 1983 [6] to supplement the 1979 definition of the candela. The tabulated values of the function at 1 nm increments are republished in a CIE standard [7]. In most cases, the region from 380 to 780 nm suffices for calculation with negligible errors because the value of the $V(\lambda)$ function falls below 10^{-4} outside this region. Thus, a photodetector having a spectral responsivity matched to the $V(\lambda)$ function achieves the role of human eyes in photometry.

The $V(\lambda)$ function is defined for the *CIE standard photometric observer for photopic vision*, which assumes a 2° field of view at relatively high luminance levels (higher than approximately 1 cd/m²). The human vision in this level is called photopic vision.

The spectral response of human vision deviates significantly at very low levels of luminance (less than approximately 10^{-3} cd/m²). This type of vision is called scotopic vision. Its spectral responsivity, peaking at 507 nm, is designated by the $V'(\lambda)$ function, which was defined by CIE in 1951 [8], recognized and republished by CIPM

in 1982 [9]. The human vision in the region between photopic vision and scotopic vision is called mesopic vision. The spectral luminous efficiency functions for mesopic vision was published recently by CIE [10]. However, it has not yet been adopted as official photometric units. In current practice, almost all photometric quantities are given in terms of photopic vision, even at low light levels. Quantities in scotopic vision are seldom used except for special calculations in research purposes.

6.2.2 PHOTOMETRIC BASE UNIT, THE CANDELA

While the $V(\lambda)$ was defined in 1924, the physical standard for photometry evolved later. Until 1948, the flame of a specially fabricated candle or oil lamp was used as a unit of luminous intensity. In 1920, *the international candle* was adopted by the CIE. In 1948, to realize more stable and reproducible standards, the Conférence Générale des Poids et Mesures (CGPM) (the General Conference on Weights and Measures) adopted a definition based on a platinum blackbody at its freezing temperature and adopted the "candela" (Latin name of candle).

The candela was then redefined in 1979 by the CGPM [11], which is still the current definition. The candela is defined in terms of a specific amount of optical power of a monochromatic radiation at a specific wavelength as

> The candela is the luminous intensity, in a given direction, of a source that emits monochromatic radiation of frequency 540×10^{12} hertz and that has a radiant intensity in that direction of (1/683) watt per steradian.

The value of K_m (683 lm/W) was determined based on measurements by several national laboratories in such a way that consistency was maintained with the prior unit so that the magnitude of one candela did not change. Technical details on the redefinition of the candela are available in references [12,13]. The key point of this 1979 redefinition is the establishment of an explicit link between the photometric units and the radiometric units.

It should be noted that the 1979 definition is given only at one wavelength where the $V(\lambda)$ function peaks and did not refer to the $V(\lambda)$ function itself. This was because, at that time, it was considered that the values of $V(\lambda)$ would be changed often and then the definition of the candela would not have to be revised.

To fill in this incompleteness of the definition, the CIPM supplemented the definition of the candela in 1983 [6], in which a photometric quantity X_v is defined in relation to the corresponding radiometric quantity $X_{e,\lambda}$ by the equation:

$$X_v = K_m \int_{360\,nm}^{830\,nm} X_{e,\lambda} V(\lambda) \mathrm{d}\lambda. \qquad (6.1)$$

The constant, K_m, relates the photometric quantities and radiometric quantities and is called the "maximum spectral luminous efficacy (of radiation) for photopic vision." The value of K_m is given by the 1979 definition of candela, which defines the spectral luminous efficacy of light at the frequency 540×10^{12} Hz (at the wavelength 555.016 nm in standard air) to be 683 lm/W. The value of K_m is calculated as $683 \times V(555.000\ nm)/V(555.016\ nm) = 683.002$ lm/W [5]. K_m is normally rounded to 683 lm/W with negligible errors.

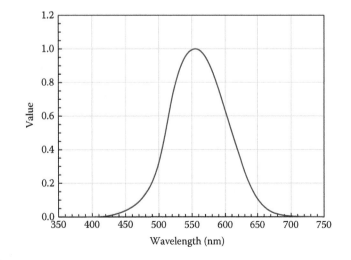

Figure 6.1 CIE $V(\lambda)$ function.

6.3 PHOTOMETRIC QUANTITIES AND UNITS

In 1960, the Système International (SI) was established, and the candela became one of the seven SI base units [14]. For further details on the SI units, Refs. [15–17] can be consulted.

Several quantities and units, defined in different geometries, are used in photometry and radiometry. Table 6.1 lists the photometric quantities and units, along with corresponding quantities and units for radiometry.

While the candela is the SI base unit, the luminous flux (lumen) is perhaps the most fundamental photometric quantity, as the other photometric quantities are defined in terms of lumen with an appropriate geometric factor. The definitions of these photometric quantities are described as follows. The definitions of some terms given here are simplified from those given in Ref. [1], which provides the official, rigorous definitions.

6.3.1 LUMINOUS FLUX

Luminous flux (Φ_v) is the time rate of flow of light as weighted by $V(\lambda)$. The unit of luminous flux is the lumen (lm). It is defined as

$$\Phi_v = K_m \int_\lambda \Phi_{e,\lambda} V(\lambda) d\lambda, \tag{6.2}$$

where $\Phi_{e,\lambda}$ is the spectral concentration of radiant flux as a function of wavelength λ. The term, luminous flux, is often used in the meaning of total luminous flux in photometry.

6.3.2 LUMINOUS INTENSITY

Luminous intensity (I_v) is the luminous flux from a point source emitted per unit solid angle in a given direction, as defined by

$$I = \frac{d\Phi}{d\Omega}, \tag{6.3}$$

where $d\Phi$ is the luminous flux leaving the source and propagating in an element of solid angle $d\Omega$ containing the given direction. The unit of luminous intensity is the candela (cd = lm sr^{-1}) (Figure 6.2).

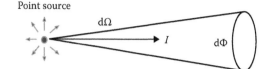

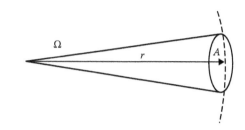

Figure 6.2 Luminous intensity.

Figure 6.3 Solid angle.

6.3.2.1 Solid angle

The solid angle (Ω) of a cone is defined as the ratio of the area (A) cut out on a spherical surface (with its center at the apex of that cone) to the square of the radius (r) of the sphere:

$$\Omega = \frac{A}{r^2}. \tag{6.4}$$

The unit of solid angle is steradian (sr), which is a dimensionless unit (Figure 6.3).

6.3.3 ILLUMINANCE

Illuminance (E_v) is the density of incident luminous flux at a point on a surface and is defined as luminous flux per unit area:

$$E = \frac{d\Phi}{dA}, \tag{6.5}$$

where $d\Phi$ is the luminous flux incident on an element dA of the surface containing the point. The unit of illuminance is lux (lx = lm m^{-2}) (Figure 6.4).

Table 6.1 **Quantities and units used in photometry and radiometry**

PHOTOMETRIC QUANTITY	UNIT	RELATIONSHIP WITH LUMEN	RADIOMETRIC QUANTITY	UNIT
Luminous flux	lm (lumen)		Radiant flux	W (watt)
Luminous intensity	cd (candela)	lm sr^{-1}	Radiant intensity	W sr^{-1}
Illuminance	lx (lux)	lm m^{-2}	Irradiance	W m^{-2}
Luminance	cd m^{-2}	lm sr^{-1} m^{-2}	Radiance	W sr^{-1} m^{-2}
Luminous exitance	lm m^{-2}		Radiant exitance	W m^{-2}
Luminous exposure	lx s		Radiant exposure	W m^{-2}
Luminous energy	lm s		Radiant energy	J (joule)
Total luminous flux	lm (lumen)		Total radiant flux	W (watt)
Color temperature	K (kelvin)		Radiance temperature	K

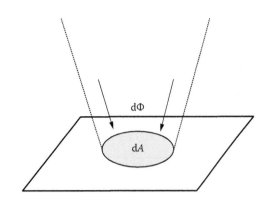

Figure 6.4 Illuminance.

6.3.4 LUMINANCE

Luminance (L_v) is the luminous flux per unit solid angle emitted from a surface element in a given direction, per unit projected area of the surface element perpendicular to the direction. The unit of radiance is W sr^{-1} m^{-2}, and that of luminance is cd/m^2. These quantities are defined by

$$L = \frac{d\Phi}{d\Omega \cdot dA \cdot \cos\theta},$$ (6.6)

where

  $d\Phi$ is the radiant flux (luminous flux) emitted (reflected or transmitted) from the surface element and propagating in the elementary solid angle $d\Omega$ containing the given direction
  dA is the area of the surface element
  θ is the angle between the normal to the surface element and the direction of the beam

The term $dA \cos\theta$ gives the projected area of the surface element perpendicular to the direction of measurement (Figure 6.5).

  Note that the Equations 6.3, 6.5, and 6.6 are not mathematical derivatives but quotients (see note in ILV luminance definition [1]).

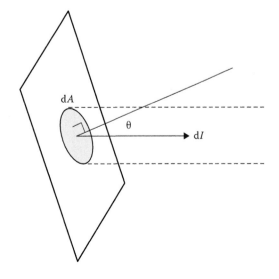

Figure 6.5 Luminance.

6.3.5 LUMINOUS EXPOSURE

Luminous exposure (H_v) is the time integral of illuminance $E_v(t)$ over a given duration Δt:

$$H = \int_{\Delta t} E(t) dt.$$ (6.7)

The unit of luminous exposure is lux·second (lx·s).

6.3.6 LUMINOUS ENERGY

Luminous energy (Q_v) is the time integral of the luminous flux (Φ) over a given duration Δt:

$$Q = \int_{\Delta t} \Phi(t) dt.$$ (6.8)

The unit of luminous energy is lumen·second (lm·s).

6.3.7 TROLAND

Troland is a unit of the retinal illuminance when a surface with a luminance of 1 cd/m^2 is viewed through the eye's entrance pupil with an area of 1 mm^2. The troland value, T, for the luminance, L (cd/m^2), of an external field and the pupil size, p (mm^2), is given by

$$T = L \cdot p$$ (6.9)

This unit is not an SI unit, not used in metrology; however, it is introduced here because this unit is commonly used by vision scientists.

6.4 PRINCIPLES IN PHOTOMETRY

Several important theories in practical photometry and radiometry are introduced in this section.

6.4.1 INVERSE SQUARE LAW

Illuminance E (lx) at a distance d (m) from a point source having luminous intensity I (cd) is given by

$$E = \frac{I}{d^2}.$$ (6.10)

For example, if the luminous intensity of a lamp in a given direction is 1000 cd, the illuminance at 2 m from the lamp in this direction is 250 lx. Note that the inverse square law is valid only when the light source is regarded as a point source. Sufficient distances relative to the size of the source are needed to assume this relationship.

6.4.2 LAMBERT'S COSINE LAW

The luminous intensity of a Lambertian surface element is given by (Figure 6.6)

$$I(\theta) = I_n \cos\theta.$$ (6.11)

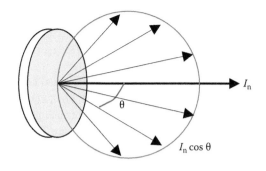

Figure 6.6 Lambert's cosine law.

6.4.2.1 Lambertian surface

A surface whose luminance is the same in all directions of the hemisphere above the surface.

6.4.2.2 Perfect (reflecting/transmitting) diffuser

A Lambertian diffuser with a reflectance (transmittance) equal to 1.

6.4.3 RELATIONSHIP BETWEEN ILLUMINANCE AND LUMINANCE

The luminance L (cd/m²) of a Lambertian surface of reflectance ρ, illuminated by E (lx) is given by (Figure 6.7)

$$L = \frac{\rho \cdot E}{\pi}. \qquad (6.12)$$

6.4.3.1 Reflectance (ρ)

The ratio of the reflected flux to the incident flux in a given condition. The value of ρ can be between 0 and 1.

In the real world, there is no existing perfect diffuser or perfectly Lambertian surfaces, and Equation 6.12 does not apply. For real object surfaces, the following terms apply.

6.4.3.2 Luminance factor (β)

Ratio of the luminance of a surface element in a given direction to that of a perfect reflecting or transmitting diffuser, under specified conditions of illumination. The value of β can be larger than 1. For a Lambertian surface, reflectance is

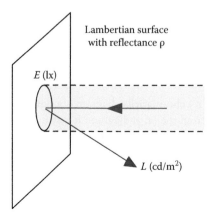

Figure 6.7 Relationship between illuminance and luminance.

equal to the luminance factor. Equation 6.12 for real object is restated using β as

$$L = \frac{\beta \cdot E}{\pi}. \qquad (6.13)$$

6.4.3.3 Luminance coefficient (q)

Quotient of the luminance of a surface element in a given direction by the illuminance on the surface element, under specified conditions of illumination, is

$$q = \frac{L}{E}. \qquad (6.14)$$

Using q, the relationship between luminance and illuminance is thus given by

$$L = q \cdot E. \qquad (6.15)$$

Luminance factor corresponds to radiance factor, and luminance coefficient corresponds to radiance coefficient in radiometry. Bidirectional reflectance distribution function is also used for the same concept as radiance coefficient.

6.4.4 PLANCK'S LAW

The spectral radiance of a blackbody at a temperature T (K) is given by

$$L_e(\lambda, T) = c_1\, n^{-2} \pi^{-1} \lambda^{-5} \left[\exp\left(\frac{c_2}{n\lambda T} \right) - 1 \right]^{-1}, \qquad (6.16)$$

where

$c_1 = 2\pi h c^2 = 3.7417749 \times 10^{-16}$ W·m²
$c_2 = hc/k = 1.438769 \times 10^{-2}$ m·K (1986 CODATA from Ref. [18])
h is Planck's constant
c is the speed of light in vacuum
k is the Boltzmann constant
n (=1.00028) is the refractive index of standard air [19]
λ is the wavelength in standard air

6.5 COLORIMETRIC QUANTITIES

6.5.1 COLOR MATCHING FUNCTIONS AND TRISTIMULUS VALUES

The basis of colorimetry was established by CIE in 1931 based on a number of visual experiments that defined a set of three spectral weighting functions [4]. These functions, shown in Figure 6.8, are called the "CIE 1931 XYZ color matching functions (CMFs)" denoted as $\bar{x}(\lambda)$, $\bar{y}(\lambda)$, $\bar{z}(\lambda)$. These functions were derived from a linear transformation of the original set of color matching functions in such a way that $\bar{y}(\lambda)$ is equal to $V(\lambda)$. The tabulated values of the 1931 CMFs at 5 nm interval are available in Ref. [20] and at 1 nm interval in Ref. [21].

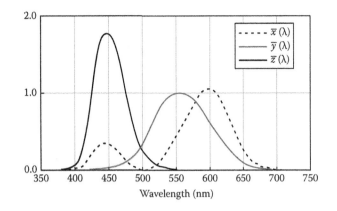

Figure 6.8 CIE 1931 *XYZ* color matching functions.

By using the CMFs, light stimuli having any spectral power distribution $\phi_\lambda(\lambda)$ can be specified for color by three values:

$$X = k \int_\lambda \phi_\lambda(\lambda)\,\bar{x}(\lambda)\mathrm{d}\lambda$$

$$Y = k \int_\lambda \phi_\lambda(\lambda)\,\bar{y}(\lambda)\mathrm{d}\lambda \qquad (6.17)$$

$$Z = k \int_\lambda \phi_\lambda(\lambda)\,\bar{z}(\lambda)\mathrm{d}\lambda,$$

where

$\phi_\lambda(\lambda)$ is the spectral distribution of light stimulus
k is a normalizing constant

These integrated values are called "tristimulus values." Two light stimuli having the same tristimulus values have the same color even if the spectral distributions are different. For light sources and displays, $\phi_\lambda(\lambda)$ is given in quantities such as spectral irradiance and spectral radiance. If $\phi_\lambda(\lambda)$ is given in an absolute unit (such as W m⁻² nm⁻¹, W m⁻² sr⁻¹ nm⁻¹) and $k = 683$ lm/W is chosen, Y yields an absolute photometric quantity such as illuminance (in lux) or luminance (in cd/m²).

For object colors, $\phi_\lambda(\lambda)$ is given as

$$\phi_\lambda(\lambda) = R(\lambda) S(\lambda), \qquad (6.18)$$

where

$R(\lambda)$ is the spectral reflectance factor of the object
$S(\lambda)$ is the relative spectral distribution of the illumination

and

$$k = \frac{100}{\displaystyle\int_\lambda S(\lambda)\,\bar{y}(\lambda)\mathrm{d}\lambda}, \qquad (6.19)$$

so that $Y = 100$ for a perfect diffuser and Y indicates the luminance factor of the object surface. To calculate color of objects from spectral reflectance factor $R(\lambda)$, one of the standard illuminants (see Refs. [20,22]) is used.

Tristimulus values can be obtained either by numerical summation of Equation 6.17 from the spectral data $\phi_\lambda(\lambda)$ obtained by a spectroradiometer or spectrophotometer or by broadband measurements using detectors having relative spectral responsivity matched to the color matching functions. Such a device using three (or four) detector channels is called "tristimulus colorimeter."

When applying colorimetric data for real visual color matching, it should be noted that the $\bar{x}(\lambda)$, $\bar{y}(\lambda)$, $\bar{z}(\lambda)$ color matching functions are based on experiments using 2° field of view and applicable only to narrow fields of view (up to 4°). Such an ideal observer is called the "CIE 1931 standard colorimetric observer." In 1964, the CIE defined a second set of standard color matching functions for a 10° field of view, denoted as $\bar{x}_{10}(\lambda)$, $\bar{y}_{10}(\lambda)$, $\bar{z}_{10}(\lambda)$, to supplement those of the 1931 standard observer. This is called the "CIE 1964 supplementary standard colorimetric observer" and can be used for a field of view greater than 4°. The 2° observer is used in most applications for colorimetry of light sources. The 10° observer is often used in object color measurements. For further details of colorimetry and color science, refer to official CIE publications [20–22] and other general references [23].

6.5.2 CHROMATICITY COORDINATES

While the tristimulus values can specify color, it is difficult to associate what color it is from the three numbers. By projecting the tristimulus values onto a unit plane ($X + Y + Z = 1$), color of light can be expressed on a two-dimensional plane. Such a unit plane is known as the *chromaticity diagram*. The color can be specified by the *chromaticity coordinates* (x, y) defined by

$$x = \frac{X}{X+Y+Z}; \quad y = \frac{Y}{X+Y+Z} \qquad (6.20)$$

The diagram using the chromaticity coordinates (x, y), as shown in Figure 6.9a, is referred to as the *CIE 1931 (x, y) chromaticity diagram*, or the *CIE (x, y) chromaticity diagram*. The boundaries of this horseshoe-shaped diagram are the plots of monochromatic radiation (called the "spectrum locus").

The (x, y) chromaticity diagram is significantly nonuniform in terms of color difference. The minimum perceivable color differences in the CIE (x, y) diagram, known as the "MacAdam ellipses," are shown in Figure 6.9a. To improve this, in 1960, CIE defined an improved diagram—*CIE 1960 (u, v) chromaticity diagram* (now obsolete), and in 1976, a further improved diagram—*CIE 1976 uniform chromaticity scale* (UCS) *diagram*, or the CIE (u', v') diagram, as shown in Figure 6.9b, with its chromaticity coordinate (u', v') given by

$$u' = \frac{4X}{X+15Y+3Z}; \quad v' = \frac{9Y}{X+15Y+3Z}. \qquad (6.21)$$

While the (u', v') chromaticity diagram is a significant improvement from the (x, y) diagram, it is still not satisfactorily uniform. Both of these diagrams are widely used. Note that these chromaticity diagrams are intended to present color of light sources (emitted light) and not color of objects (reflected light). Presentation of object colors requires a three-dimensional color space that incorporates another dimension—lightness (black to white). Refer to Ref. [20] for the details of object color specification.

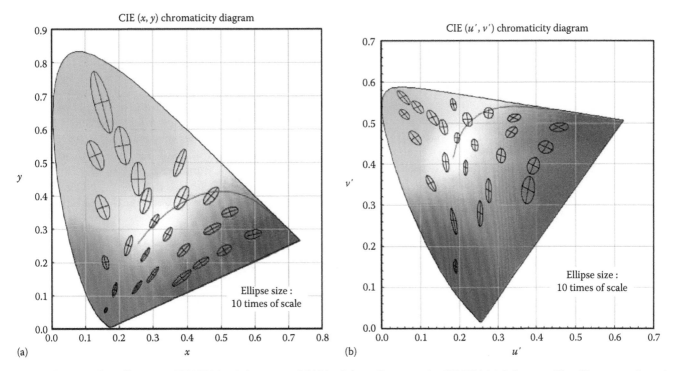

Figure 6.9 (a) MacAdam ellipses on CIE 1931 (x, y) diagram and (b) MacAdam ellipses on the CIE 1976 (u'v') diagram. The ellipses are plotted 10 times their actual size. The curve near the center region is the Planckian locus.

6.5.3 CORRELATED COLOR TEMPERATURE

Figure 6.10 shows the trace of the (x, y) chromaticity coordinate of blackbody radiation (See 6.4.4) at its temperature from 1,600 to 20,000 K. This trace is called the "Planckian locus." The colors on the Planckian locus can be specified by the blackbody temperature in kelvin and is called *color temperature*. The colors around the Planckian locus from about 2,500 to 20,000 K can be regarded as *white*, 2,500 K and lower being reddish white and 20,000 K and higher being bluish white. The point labeled "Illuminant A" is the typical color of an incandescent lamp and

"Illuminant D65" the typical color of daylight, as standardized by the CIE [22] as *CIE Standard Illuminants A and D65*. The colors of most of the traditional lamps for general lighting fall in the region between 2700 and 6500 K.

By definition, color temperature cannot be used for colors away from the Planckian locus, in which case *correlated color temperature* (CCT) is used. CCT is defined as the temperature of the blackbody whose chromaticity is closest to that of the light source in question on the CIE (u', 2/3 v') coordinates [1]. The (u', 2/3 v') coordinate means the CIE 1960 (u, v) diagram, which is now obsolete. Based on the definition, the iso-CCT lines are perpendicular to the Planckian locus on the (u, v) diagram, but not perpendicular on the (x, y) diagram (see Figure 6.10) due to its nonuniformity. To calculate CCT, find the point on the Planckian locus that is at the shortest distance from the given chromaticity point on the (u, v) diagram. CCT is the temperature of the Planck's radiation at that point. Practical methods of computing CCT are available in Ref. [24].

CCT is widely used to specify the chromaticity of general illumination sources. However, CCT provides only one dimension of the chromaticity, which is a two-dimensional quantity. Another important dimension with respect to CCT is the shift of chromaticity from the Planckian locus. Duv (symbol: D_{uv}) is defined as the distance from the chromaticity coordinate of the test light source to the closest point on the Planckian locus on the CIE (u', 2/3 v') coordinates, with a plus sign for above and a minus sign for below the Planckian locus [25]. Duv is important when the color quality of illumination sources is evaluated. Further details and calculation methods for Duv are available in Ref. [24].

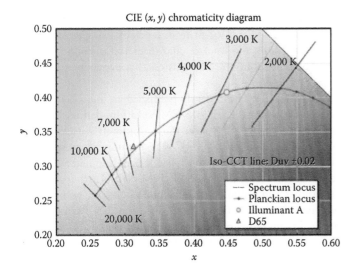

Figure 6.10 Planckian locus on (x, y) chromaticity diagram and iso-CCT lines.

Fundamentals

6.5.4 COLOR QUANTITIES FOR SINGLE-COLOR LIGHTS

In addition to chromaticity coordinates x, y and u', v', the following quantities are used to specify the color and spectrum of single-color sources such as LEDs. The definitions in this section follow Ref. [26].

Peak wavelength λ_p: The wavelength at the maximum of the spectral distribution.

Spectral bandwidth (at half intensity level) $\Delta\lambda_{0.5}$: Calculated as the width between the wavelengths at half of the peak of spectral distribution, as shown in Figure 6.11. It is also denoted as $\Delta\lambda$(FWHM).

Centroid wavelength λ_c: Calculated as the "center of gravity wavelength," according to the equation

$$\lambda_c = \frac{\int_\lambda \lambda \cdot S(\lambda)d\lambda}{\int_\lambda S(\lambda)d\lambda}. \tag{6.22}$$

Dominant wavelength λ_d: Wavelength of the monochromatic stimulus that, when additively mixed in suitable proportions with the specified achromatic stimulus, matches the color stimulus considered. Equal energy spectrum with $(x, y) = (0.3333, 0.3333)$ is used as the achromatic stimulus. See Figure 6.12, where N denotes the achromatic stimulus.

Excitation purity p_e: Defined as the ratio NC/ND in Figure 6.12. The value of excitation purity is unity if the chromaticity of the LED is on the spectrum locus.

6.6 FUTURE PROSPECT FOR PHOTOMETRY

The $V(\lambda)$ function was determined on the basis of experimental studies of photopic vision with a narrow field of view (about 4° or less). For situations where the visual target has an angular subtense larger than 4° or is seen off-axis, the CIE has defined the $V_{10}(\lambda)$ function [27], based on experimental studies for photopic vision with a 10° field of view. However, photometric units based on the $V_{10}(\lambda)$ function have not been adopted by CIPM. Such work is in progress.

It is also known that the $V(\lambda)$ underestimates the visual response in the blue region, and an improved function, known as $V_M(\lambda)$, was published by CIE as a supplement to $V(\lambda)$ [28]. The $V_M(\lambda)$ is not recognized by CIPM and might be used only for research purposes.

In real applications, perceived brightness and measured photometric quantities do not agree in some cases. However, at present there is no agreed photometric quantity that is more satisfactory than luminance or luminous intensity for quantifying the absolute brightness of luminous sources. CIE described a supplementary system of photometry that provides a more perceptually relevant approach for comparative brightness evaluation of lights at any level, including mesopic levels [29]. This system introduces the concept of equivalent luminance and develops a photometric model to calculate brightness-related equivalent luminance by using existing photometric and colorimetric quantities and introducing a chromatic contribution to brightness that depends upon the adaptation level.

In colorimetry, CIE recently published an improved color matching functions (CIE 2015 CMFs) [30] to define chromaticity coordinates that agree more accurately with visual perception. The new set of the CMFs, including new spectral luminous efficiency functions, has not been adopted as CIE standards nor recognized by the CIPM. With further application studies and support by the color and lighting community, these new functions may replace the current CMFs in the future. The proposed spectral luminous efficiency functions also require further application studies for a possible future update of the $V(\lambda)$ function.

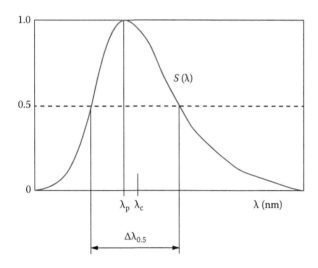

Figure 6.11 Typical relative spectral distribution of an LED.

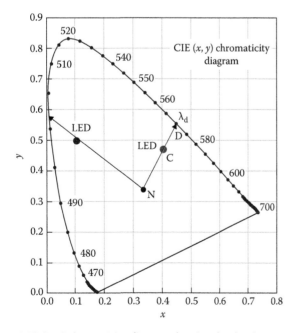

Figure 6.12 (x, y) chromaticity diagram showing the dominant wavelength and excitation purity.

REFERENCES

1. CIE S 017/E:2011, ILV: International Lighting Vocabulary.
2. ISO 80000-7:2008 (E), Quantities and Units, Part 7 Light, First Edition.
3. CIE Compte Rendu, p. 67 (1924).
4. CIE Compte Rendu, Table II, pp. 25–26 (1931).
5. CIE 18.2-1983, *The Basis of Physical Photometry.*
6. BIPM Monographie 1983, *Principles Governing Photometry,* Bureau International Des Poids et Mesures (BIPM), Pavillon Breteuil, F-92310 Sèvres, France (1983).
7. ISO 23539:2005(E)/CIE S 010/E:2004 Photometry—The CIE System of Physical Photometry.
8. CIE Compte Rendu, Vol. 3, Table II, pp. 37–39 (1951).
9. CIPM, Comité Consultatif de Photométrie et Radiométrie 10e Session–1982, BIPM, Pavillon de Breteuil, F-92310, Sèvres, France (1982).
10. CIE 191:2010, Recommended System for Mesopic photometry based on Visual Performance, International Commission on Illumination, Vienna, Austria.
11. CGPM, Comptes Rendus des Séances de la 16e Conférence Générale des Poids et Mesures, Paris 1979, BIPM, F-92310 Sèvres, France (1979).
12. W.R. Blevin and B. Steiner, A radiometric realization of the photometric units, *Metrologia* **11**, 97 (1975).
13. W.R. Blevin, The candela and the watt, *CIE Proc.* P-79-02 (1979).
14. Le Système International d'Unité (SI), *The International System of Units (SI)*, 8th edn, Bur. Intl. Poids et Mesures, Sèvres, France (2006), www.bipm.org/en/publications/si-brochure/.
15. Thompson, A. and B.N. Taylor, *Guide for the Use of the International System of Units (SI)*, Natl. Inst. Stand. Technol. Spec. Publ. 811, Gaithersburg, MD, (2008).
16. Taylor, B.N. and A. Thompson, *The International System of Units (SI)*, Natl. Inst. Stand. Technol. Spec. Publ. 330, Gaithersburg, MD, (2008).
17. ISO 1000:1992, SI units and recommendations for the use of their multiples and of certain other units.
18. E.R. Cohen and B.N. Taylor, The 1986 CODATA recommended values of the fundamental physical constants, *Journal of Research of the National Bureau of Standards*, **92**(2), 85–95 (1987).
19. Blevin, W.R., Corrections in optical pyrometry and photometry for the refractive index of air, *Metrologia* **8**, 146 (1972).
20. CIE 15.2:2004, *Colorimetry*, 3rd edn.
21. CIE S014-1/E: 2006 Colorimetry—Part 1; CIE Standard Colorimetric Observers, International Commission on Illumination, Vienna, Austria.
22. CIE S014-2/E: 2006 Colorimetry—Part 2; CIE Standard Illuminants, International Commission on Illumination, Vienna, Austria.
23. G. Wyszecki and W.S. Stiles, *Color Science: Concepts and Methods, Quantitative Data and Formulae*, John Wiley & Sons, Inc., New York (1982).
24. Y. Ohno, Practical use and calculation of CCT and Duv. *LEUKOS*, **10**(1), 47–55, (2013). DOI: 10.1080/15502724.2014.839020.
25. ANSI_NEMA_ANSLG, C78.377-2015, *Specifications for the Chromaticity of Solid State Lighting Products*, American National Standards Institute, Washington, DC.
26. CIE 127-2007, *Measurement of LEDs*, 2nd edn, International Commission on Illumination, Vienna, Austria.
27. CIE 165:2005, CIE 10 degree Photopic Photometric Observer, International Commission on Illumination, Vienna, Austria.
28. CIE 86-1990, CIE 1988 2° Spectral Luminous Efficiency Function for Photopic vision, International Commission on Illumination, Vienna, Austria.
29. CIE 200:2011, CIE Supplementary System of Photometry, International Commission on Illumination, Vienna, Austria.
30. CIE 170-2:2015, Fundamental chromaticity diagram with physiological axes—Part 2: Spectral luminous efficiency functions and chromaticity diagrams, International Commission on Illumination, Vienna, Austria.

7 Characterization of visual stimuli using the standard display model

Joyce E. Farrell, Haomiao Jiang, and Brian A. Wandell

Contents

7.1 INTRODUCTION

Visual psychophysics advances by experiments that measure how sensations and perceptions arise from carefully controlled visual stimuli. Progress depends in large part on the type of display technology that is available to generate stimuli. In this chapter, we first describe the strengths and limitations of the display technologies that are currently used to study human vision. We then describe a standard display model that guides the calibration and characterization of visual stimuli on these displays (Brainard et al., 2002; Post, 1992). We illustrate how to use the standard display model to specify the spatial–spectral radiance of any stimulus rendered on a calibrated display. This model can be used by engineers to assess the trade-offs in display design and by scientists to specify stimuli so that others can replicate experimental measurements and develop computational models that begin with a physically accurate description of the experimental stimulus.

7.2 DISPLAY TECHNOLOGIES FOR VISION SCIENCE

An ideal display system for science and commerce would deliver the complete spectral, spatial, directional, and temporal distribution of light rays, as if these rays arose from a real 3D scene. The full radiometric description of light rays in the 3D scene is called the "light field" (Gershun, 1939). For vision science, the simplified and related representation is the irradiance the scene produces at the cornea—this is the only part of the scene radiance that the retina encodes. The complete radiometric description of the rays at the cornea, sometimes referred to as the plenoptic function (Adelson and Bergen, 1991), specifies the rate of incident photons from every direction at each point in the pupil plane. To achieve an accurate dynamic reproduction of a scene, the plenoptic function must change as the head and eyes move.

Commonly used scientific displays do not approach this ideal. Instead, most displays emit light rays from a planar surface in a wide range of directions, and the spectral radiance is invariant as the subject changes head and eye position. The displays themselves are limited in various ways; for example, the pixels produce a limited range of spectral power distributions (SPDs), typically being formed as the weighted sum of three spectral primaries. Despite these limitations, modern displays create a very compelling perceptual experience that captures many important elements of the original scene. The ability to program these displays with computers and digital frame buffers has greatly enlarged the range of stimuli used in visual psychophysics compared to the optical benches and tachistoscopes used by previous generations.

The vast majority of modern displays comprise a 2D matrix of picture elements (pixels) at a density of 100–400 pixels per inch (4–16 pixels per mm). Each pixel typically contains three different light sources (subpixels, Figure 7.1). The pixels are intended to be identical across the display surface (spatial homogeneity).

Most displays are designed with three types of subpixels with SPDs that peak in the long-, middle-, and short-wavelength regions of the visible spectrum (Figure 7.2). Each type of subpixel is called a display primary. The relative SPD of each primary is designed to be invariant as its intensity is varied (spectral homogeneity). In normal operation, the three subpixel intensities are controlled to match the color appearance of an experimental stimulus. Three primaries are used because experiments show that subjects can match the color appearance of a wide range of SPDs using the mixture of just three independent light sources

(Wandell, 1995, Chapter 4; Wyszecki and Stiles, 1967). Modern displays effectively comprise a very large number of color-matching experiments, one for each pixel on every frame.

Display architectures are distinguished by (1) the physical process that produces the light and (2) the spatial arrangement of the pixels and subpixels. Key design parameters of commercial displays are energy efficiency, brightness, spatial resolution, darkness, color range, temporal refresh, and update rates. The relative importance of these parameters depends on the application.

The three main display technologies used in vision experiments today are cathode ray tubes (CRTs), liquid crystal displays (LCDs), and organic light-emitting diodes (OLEDs). Color CRTs were developed by RCA in the 1950s (Law, 1976) and were the nearly universal display technology for several decades. They remain an important display technology for vision researchers, although now they are rarely sold as consumer products. Invented at RCA labs in the 1970s (Kawamoto, 2002), LCDs were introduced as small mobile displays in digital watches, calculators, and other handheld devices; later, they enabled the widespread adoption of laptop computers. OLEDs were invented at Kodak in the 1980s (Tang and VanSlyke, 1987) and were first introduced as displays for digital cameras. Large OLED displays are expensive, but they have some advantages over LCDs: they achieve a deeper black and they have better temporal resolution.

Despite the fact that LCDs have displaced CRTs in the market, CRTs are still widely used in vision science. A recent sampling from the *Journal of Vision* suggests that scientists mainly

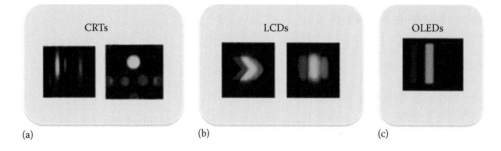

Figure 7.1 Camera images of a white pixel. (a) Shows a white pixel illuminated on a Dell CRT Display Model P1130 (left) and an Hewlett-Packard CRT Display Model Number D2845 (right). (b) Shows a white pixel on a Dell LCD Display Model 1907FPc (left) and a Dell LCD Display Model 1905FP (right). (c) Shows a white pixel on a Sony OLED Display Model PVM-24. (From Farrell, J. et al., *J. Dis. Technol.*, 4, 262, 2008.)

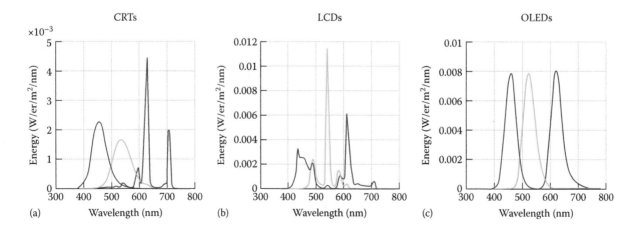

Figure 7.2 (a) Spectral power distribution of blue, green, and red color primaries of a cathode ray tube (CRT) (Dell CRT Display Model P1130), (b) liquid crystal display (LCD) (Dell LCD Display Model 1907FPc), and (c) organic light-emitting diode (OLED) display (Sony PVM-2451).

use CRTs and with some use of LCDs, while the OLEDs are not yet common. One reason CRTs are preferred is that the intensity of each primary can be accurately controlled beyond 10 bits (Brainard et al., 2002). As shown by simulation later, this intensity precision is valuable for visual psychophysical experiments that measure detection or discrimination thresholds.

7.2.1 CATHODE RAY TUBES

CRTs create light by directing an electron beam onto one of three different types of phosphors (Castellano, 1992). When irradiated by electrons, the phosphors emit light with a spectral radiance distribution that is unique to that material. CRT phosphors are painted on a transparent glass surface in a pattern of alternating dots or stripes, and they are selected to emit predominantly in the long (red), middle (green), and short (blue) wavebands (Figures 7.1a and 7.2a). The amount of light from each type of phosphor is controlled by the intensity of the electron beam that is incident on the phosphor. The spatial properties of the display are determined by the size and spacing of the phosphor dots or stripes.

The temporal properties of the display are determined by the frequency with which each phosphor is stimulated by electrons and the rate at which the phosphorescence decays (see Figure 7.3b).

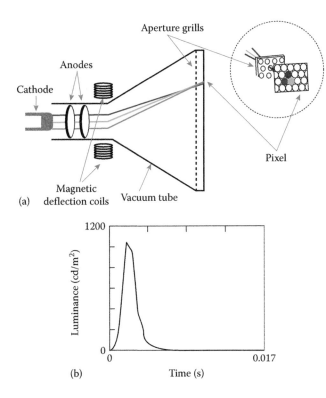

Figure 7.3 (a) Cathode ray tube components. Independent electron beams are created and controlled by using three cathode ray guns that generate the electrons and anodes that attract the electrons. The electron beam is directed by magnetic coils to traverse the display surface. The surface is coated with "red," "green," and "blue" phosphors that emit visible light when an electron is absorbed. An aperture grille (shadow mask) is positioned to such that one of the electron beams strikes the red phosphors, another electron beam strikes the green phosphors, and a third electron beam strikes the blue phosphors. (b) Temporal response of pixel luminance (97.5% of peak) measured during one frame. (From Cooper, E.A. et al., *J. Vis.*, 13, 16, 2013.)

The refresh rate is determined by how fast an electron beam can scan across the many rows of pixels in a display. The more rows there are, the more time it takes for the electron beam to return to the same phosphor dot. When the refresh rate is slow and the phosphor decay is fast, the display appears to flicker. Longer phosphor decay times reduce the visibility of flicker, but increase the visibility of motion blur (Farrell, 1986; Zhang et al., 2008).

In addition to scanning through many rows of pixels, the electron beam intensity modulates as the beam traverses phosphors within each row. The electron beam modulation rate, referred to as slew rate, is not fast enough to change perfectly as the beam moves between adjacent pixels. Consequently, the ability to control the light from adjacent pixels within a row is not perfectly independent (Lyons and Farrell, 1989). We will explain the consequence of this slew rate limitation later in this chapter.

7.2.2 LIQUID CRYSTAL DISPLAYS

LCDs are a large array of light valves that control the amount of light that passes from a backlight, which is constantly on, to the viewer (Figure 7.4). The backlight is usually a fluorescent tube or sometimes a row of LEDs positioned at the edge of the LC array. The photons from the backlight are spread uniformly across the back of the display using diffusing filters. To reach the viewer, the backlight must pass through a polarization filter, a layer of LC material, a second polarization filter, and then a color filter. The ability of photons to traverse this path is controlled by the alignment of the LCs that determines the polarization of the photons and thus how much light passes between the two polarization filters. The state of the LC is determined by an electric field that is controlled by digital values in a frame buffer, under software control. Even when the LC is in a state that permits transmission (open), only a small fraction (about 3%) of the backlight photons pass through the two polarizers, color filter, and electronics.

The spectral radiance of an LCD pixel is determined by the SPD of the backlight and the transmissivity of the optical elements (polarizers, LC, and color filters). The spatial properties of an LCD are determined by the dimensions of a panel of thin-film transistors (TFTs) that control the voltage for each pixel component and the size and arrangement of each individual filter in the color filter array. The temporal properties of an LCD are determined by the modulation rate of the backlight and the temporal response of the LC (Yang and Wu, 2006). LCDs use sample and hold circuitry that keeps the LCs in their "open" or "closed" state (see Figure 7.4b). This means that flicker is not visible, but a negative consequence of the slow dynamics is that LCDs can produce visible motion blur. Furthermore, LCs respond faster to an increase in voltage (changing the alignment of the LCs) than they do to a decrease in voltage (returning toward its natural state). Consequently, a change from white to black is faster than a change from black to white. Some LCD manufacturers have introduced circuitry to "overdrive" and "undershoot" the voltage delivered to each pixel. This additional circuitry reduces the visible motion blur, but it makes it impossible to separately control the spatial and temporal properties of the display. The slow and asymmetric changes in the state of

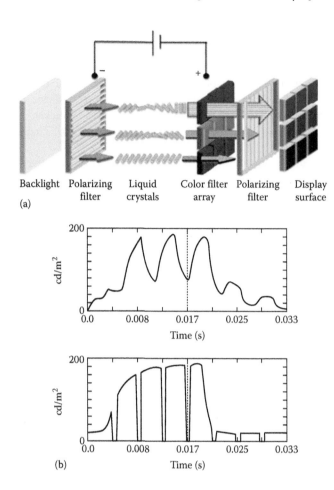

Figure 7.4 Components of a twisted nematic liquid crystal display (LCD). (a) A fluorescent or light-emitting diode (LED) backlight produces light that is passed through a polarizing filter to a layer of liquid crystals. In the absence of electric current, the liquid crystals are in their natural "twisted" state and guide the light through a second polarizer and a color filter. When electric current is applied, the liquid crystals "untwist" and are aligned to be perpendicular to the second polarizing filter, blocking the light. The amount of current varies the orientation of the liquid crystals and consequently the amount of transmitted light. (b) Temporal responses. The graphs plot pixels luminance over two frames. The pixel is set to 97.5% of maximum luminance in the first frame and to 2.5% of maximum in the second. The dashed line delineates the end of the first frame, during which all pixels are on, and the beginning of the second frame, during which all pixels are off. The top figure shows data measured from an LCD with a fluorescent backlight and the bottom figure shows data measured from an LCD with an LED backlight. The responses are slow and asymmetric. (From Cooper, E.A. et al., *J. Vis.*, 13, 16, 2013.)

LCs also make it difficult to have precise control in the timing of visual stimuli (Tobias and Tanner, 2012).

Another limitation of LCDs is that in the "off" state, photons from the backlight find their way through the filters to the viewer. Consequently, LCDs do not achieve a complete black background. Recently, manufacturers introduced LED backlit panels that can be locally dimmed in different regions. In this way, one portion of the image can be much brighter than another, and a portion of the display can be nearly black. This design extends the image dynamic range, but such LCDs are difficult to calibrate because of the complexity of the design, control circuitry, and spatial distribution of the LED back panel.

7.2.3 ORGANIC LIGHT-EMITTING DIODES

OLEDs emit light by applying an electric current to an electroluminescent layer of organic molecules. Each diode (pixel) consists of two layers of organic molecules that are sandwiched between a cathode and an anode (Figure 7.5). There are several ways to produce the different primaries: (1) each diode can be made from a different substance that emits light in a distinct wavelength band, (2) color filters can be placed in front of a single type of diode, or (3) the emissions from a single type of OLED can be used to excite different types of phosphors (Tsujimura, 2012). Since OLEDs do not use a backlight, each pixel can be black, emitting only light that is scattered from nearby pixels.

The spatial properties of an OLED display are determined by the spatial arrangement of OLEDs that are deposited onto glass.

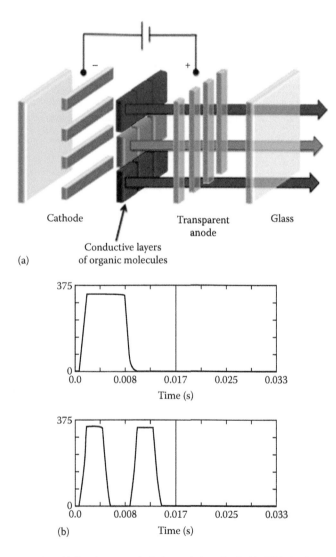

Figure 7.5 (a) Passive organic light-emitting diode (OLED) pixel array: electroluminescent light is generated when current is applied to the conductive layers of organic material sandwiched between a cathode and an anode. An active matrix OLED includes a thin film transistor that is placed on top of the anode to control the electrical signal at each pixel. (b) Temporal response. The luminance time course for two frames when pixels are set to 97.5% of peak and then (dashed line) 2.5% of peak. The bottom graph shows the temporal profile in a "flicker-free" mode that rapidly turns on and off the OLED pixels within a single frame. (From Cooper, E.A. et al., *J. Vis.*, 13, 16, 2013.)

Some types of OLEDs (polymer OLEDs) can be printed onto plastic using a modified inkjet printer (Bale et al, 2006; Carter et al, 2005), but these flexible displays are still experimental and hence will not be considered here.

OLEDs can be rapidly turned on and off; hence, the display dynamics are mainly limited by how often the electronics updates the subpixel intensities. The rate at which the pixel intensities can be changed (update rate) limits the motion velocities that can be represented (Watson et al., 1986). To reduce the visibility of flicker and motion blur, OLEDs can be refreshed at a rate that exceeds the update rate (see Figure 7.5b).

7.2.4 DIGITAL LIGHT PROJECTORS

The digital light projector (DLP) display technology is a micro-electromechanical device consisting of an array of microscopically small mirrors arranged in a matrix on a semiconductor chip—one mirror for each pixel (Florence and Yoder, 1996; Younse, 1993). The system includes a constant backlight, and each mirror can be in one of two states: it either reflects the backlight photons toward or away from the viewer.

The mirrors can alternate state very rapidly, and varying the percentage of time the mirror is directing light toward the viewer controls the light intensity at each pixel. In the single-chip DLP, color is controlled using a rapidly spinning color wheel that interposes different color filters between the light source. The single-chip DLP design uses a color wheel whose rotation is synchronized with the control signals sent to the chip. While most display technologies use subpixel primaries that are adjacent in space, the DLP color primaries are adjacent in time—a technique called field-sequential color. Some DLP devices include only three (red, green, and blue [RGB]) primaries, while others include a fourth (white or clear) primary. The white primary increases the maximum display brightness, but at the highest brightness levels the display has a vanishingly small color gamut (Kelley et al., 2009).

A problem with the single-chip DLP design is that field-sequential color can produce visible color artifacts when the eye moves rapidly across the image. High-speed eye movements cause the sequential RGB images to project to different retinal positions (Zhang and Farrell, 2003). A more expensive three-chip DLP design is often used in home and movie theaters. The three-chip design simultaneously projects RGB images that are coregistered; hence, these DLPs do not produce the sequential color artifacts.

While DLP displays are not used widely in visual psychophysics, they have been adapted for use in studies of color constancy (Brainard et al., 1997a), in vitro primate retina intracellular recordings (Packer et al., 2001), and functional magnetic resonance imaging (Engel, 1997).

7.3 STANDARD DISPLAY MODEL AND STIMULUS CHARACTERIZATION

7.3.1 OVERVIEW

Displays emit light in different ways. Nonetheless, it is possible to characterize a few general principles that describe the relationship between the electronic control signals and the display spectral radiance. These widely adopted principles are the basis of a standard display model (Brainard et al., 2002; Post, 1992). To calibrate a display effectively means establishing the parameters of the standard display model and using the model and calibration data to control the display spectral radiance.

A model is necessary because there are far too many images to calibrate individually (Brainard, 1989). For example, a static image on an 8-bit display has 2^{24} different (RGB) settings. A 1024×1024 (2^{20} pixels) display can render 2^{480} images. The standard display model defines a relatively small set of calibration measurements that can be used to calculate the expected spectral radiance for many of these images.

Several key measurements are necessary to specify a model for any particular display. First, each subpixel type has a characteristic SPD (Figure 7.2). The model assumes that the SPD is the same for all subpixels of a given type and is invariant when normalized for intensity level. Thus, the normalized SPD can be measured using a spectroradiometer that averages the spectral radiance emitted from a region of the display surface.

Second, the absolute level (peak radiance) of the SPD is set by the frame buffer value. The relationship between the frame buffer value and the SPD level is referred to as the gamma curve. The gamma curve is assumed to be the same for all subpixels of a given type (shift invariant), independent of the image content, and monotone increasing.

Third, the standard display model describes the spatial distribution of light emitted by each type of subpixel, called the point spread function (PSF). The standard display model assumes that the PSF is the same for subpixels of a given type (shift invariant) and independent of the image content.

Finally, most displays refresh the image (frame) at a rate between 30 and 240 times per second. Within each frame, the subpixel intensity can rise and fall, and the frame repetitions and pixel dynamics influence the visibility of motion and flicker. The standard display model assumes that each subpixel has a simple time-invariant impulse response function that is independent of the image content. This assumption is frequently violated because of the extensive engineering to control the dynamics of displays (see previous sections on LCDs and CRTs). Characterizing the display dynamics is particularly important for experiments involving rapidly changing high-contrast targets (e.g., random dots).

The standard display model clarifies the measurements needed to calibrate a display. The first two are to measure (1) the normalized spectral radiance distributions for each of the display primaries and (2) the gamma curve that specifies the absolute level of the spectral radiance given a particular frame buffer value. It is less common for scientists to measure the subpixel PSFs. These can be measured using a macro lens and the linear output of a calibrated digital camera (Farrell et al., 2008), but in most cases the function is treated as a single point (impulse). Characterizing the PSF can be meaningful for measurements of fine spatial resolution (e.g., quality of fonts, vernier resolution) where there are significant effects of human optics on retinal image formation. In the next section, we offer specific advice about making these calibration measurements and combining them into a computational implementation of the standard display model.

7.3.2 SPECTRAL RADIANCE AND GAMMA CURVES

It is common to use a spectral radiometer to measure the spectral radiance emitted by each of the three types of primaries. The standard display model assumes that for each primary the SPD takes the form $I(F)P(\lambda)$, where $P(\lambda)$ is the SPD of the display when the frame buffer is set to its maximum value and $0 < I(F) < 1$ is the relative intensity for a frame buffer value of f.

To estimate $I(F)$ and $P(\lambda)$, we measure the spectral radiance for a series of different frame buffer levels. An important detail is this: in most displays there is some stray light present even when $F = 0$. This light is usually treated as a fixed offset, $B(\lambda)$, and subtracted from the calibration data (Brainard et al., 2002). Hence, the measured spectral radiance curves have the form $R(\lambda, F) = I(F)P(\lambda) + B(\lambda)$.

The term I is the relative intensity of the primary and F is the frame buffer value. When F is set to the maximum value, the value of I is equal to 1. If one subtracts the background SPD, then $I(0) = 0$ and the relative intensity is typically modeled as a simple power law (Poynton and Funt, 2013) that gives the curve its name:

$$I = \alpha F^{\gamma} \tag{7.1}$$

For most displays $B(\lambda)$ is difficult to measure because it is small and negligible compared to the experimental stimuli. In such cases, the radiance is modeled by including a small, wavelength-independent, offset in the gamma curve:

$$R(\lambda, F) = I(F)P(\lambda) \tag{7.2}$$

$$I = \alpha F^{\gamma} + B_0 \tag{7.3}$$

Historically, the value of γ in manufactured displays has been between 1.8 and 2.4, which is quite significant. If one changes the γ of a display from 1.8 to 2.4, the same frame buffer values will produce very different spectral radiance distributions. Pixels set to the same frame buffer (RGB) produce spectral radiances that differ by as much as 10 CIELAB ΔE units (median ~6 ΔE). In recent years, manufacturers have converged to a function that is linear at small values, close to $\gamma = 2.4$ at high values, and overall similar to $\gamma = 2.2$ sRGB.*

The analytical gamma function is an approximation to the true $I(F)$. In modern computers, this approximation can be avoided by building a lookup table that stores the nonlinear relationship between the digital control values and the display output, $I(F)$.

This nonlinearity will continue across technologies because programmers prefer that equal spacing of the digital frame buffer values correspond to equal perceptual spacing (Poynton, 1993; Poynton and Funt, 2013). To maintain this relationship, the display intensity must be nonlinearly related to the frame buffer value (Stevens, 1957; Wandell, 1995).

7.3.3 SUBPIXEL POINT SPREAD FUNCTIONS

The spatial distribution of light from each subpixel is described by a PSF, $P(x, y, \lambda)$. The spatial spread of the light from each subpixel can be measured using a high-resolution digital camera with a

---

* International Color Consortium sRGB profiles, http://color.org/srgbprofiles.xalter.

closeup lens (Figure 7.1, Farrell et al., 2008). Furthermore, the spectral and spatial parts of the PSF are separable:

$$P(x, y, \lambda) = s(x, y)w(\lambda) \tag{7.4}$$

The subpixel point spread is assumed to have the same form across display positions, that is, the subpixel PSF at pixel (u, v) is $s(x - u, y - v)w(\lambda)$. And finally, the shape scales with intensity $I s(x - u, y - v)w(\lambda)$.

The standard display model assumes that PSFs from adjacent pixels sum. This linearity is ideal—no display is precisely linear. But display designs generally aim to satisfy these principles and implementations are close enough so that these principles are a good basis for display characterization and simulation.

7.3.4 LINEARITY

Apart from the nonlinear gamma curve, the standard display model is a shift-invariant linear system. That is, given the intensity of each subpixel, we compute the expected display spectral radiance as the weighted sum of the subpixel PSFs. If the subpixel intensities for one image are I_1 with corresponding spectral radiance $R_1(x, y, \lambda)$ and a second image is I_2 with corresponding spectral radiance $R_2(x, y, \lambda)$, then the radiance when the image is $I_1 + I_2$ will be $R_1(x, y, \lambda) + R_2(x, y, \lambda)$.

The calibration process should test the additivity assumption. Simple tests include checking that the light emitted from the ith subpixel does not depend on the intensity of other subpixels (Farrell et al., 2008; Lyons and Farrell, 1989; Pelli, 1997).

7.3.5 MODEL SUMMARY

The standard display model for a steady-state image can be expressed as a simple formula that maps the frame buffer values, F, to the display spatial–spectral radiance $R(x, y, \lambda)$.

Suppose the gamma function, PSF, and SPD of the jth subpixel type are $I_j(v)$, $p_j(x, y)$, and $w_j(\lambda)$. Suppose the frame buffer values for the jth subpixel type are $F_j(u, v)$. Then, the display spectral radiance across space is predicted to be

$$R(x, y, \lambda) = \sum_{u,v} \sum_{j} I_j(F_j(u,v))s_j(x-u, y-v)w_j(\lambda) \tag{7.5}$$

7.4 DISPLAY CALIBRATION

If the standard display model describes the device under test, then calibration requires a very small set of display measurements—gamma, SPD, PSF, and temporal response—to fully describe the physical radiance of displayed stimuli. Display calibration can be conceived as (1) measuring how well the key model assumptions hold (spectral homogeneity, pixel independence, spatial homogeneity) and (2) using the measurements to estimate the model parameters.

7.4.1 PIXEL INDEPENDENCE

The radiance emitted by a subpixel should depend only on the digital frame buffer value controlling that subpixel. Equivalently, the radiance emitted by a collection of pixels must not change as the digital values of other pixels change. Displays often satisfy this pixel independence principle for a large range of stimuli (Cooper et al., 2013; Farrell et al., 2008), but there are displays and certain types of stimuli that fail this test (Lyons and Farrell, 1989; Tobias and Tanner, 2012).

For example, CRTs must sweep the intensity of the electron beam very rapidly across each row of pixels. There are limits to how rapidly the beam intensity can change (a maximum "slew rate"). If a very different intensity is required for a pair of adjacent row pixels, the beam may not be able to adjust in time and independence is violated, and the standard display model will not be useful for characterizing the spatial–spectral radiance of such stimuli (Lyons and Farrell, 1989; Naiman and Makous, 1992).

LCDs are limited by rate at which LCs can change their state in response to a change in voltage polarity, as well as the asymmetry in their response to the "on" or "off" states. LCDs typically combine sample and hold circuitry to switch between different LC states and a flickering backlight to minimize the visibility of both motion blur. LCDs with these features (sample and hold circuitry with flickering LED or fluorescent backlights) can be modeled as a linear system (Farrell et al., 2008). Departure from display linearity occurs, however, when LCD manufacturers introduce "overdrive" and "undershoot" circuitry to minimize the visibility of motion blur or when they locally dim LED backlight panels to increase dynamic range. These new features make it very difficult to control and calibrate visual stimuli, particularly for studies that require precise control of timing (Tobias and Tanner, 2012).

There are several ways to test pixel independence (Farrell et al., 2008; Lyons and Farrell, 1989; Pelli, 1997), but the general principle is simple. Separately measure the radiance from the middle of a large patch of pixels. Make the measurement with a few different digital values. Then, create spatial patterns that are made up with half the pixels at one digital value and half at the other. The radiance from these mixed patches should be the average of the radiance from the large patches, measured individually.

A key assessment is to evaluate how well independence is satisfied for the planned experimental stimuli. For example, CRTs often fail pixel independence for high spatial frequency stimuli because of the finite slew rate of the electron beam. Nonetheless, CRTs are very useful for visual experiments that use low frequency stimuli, such as studies of human color vision. The standard display model, like any useful model, will have some compliance range, and the practical question is whether the model can be used for a specific set of experimental stimuli.

OLEDs are excellent devices for vision research because they typically meet the requirements of the standard display model (Cooper et al., 2013). Display electronics control the rate at which the pixel intensities can be changed (the update rate), but OLED pixels can be rapidly turned on and off. Thus, while the update rate limits the motion velocities that can be represented, the higher refresh rates minimize the visibility of motion blur and flicker. And, unlike the LCDs that modulate the intensity of a backlight, OLED pixels can be turned off, creating a perfectly black background.

Given these benefits, and the fact that the cost of manufacturing OLED displays is decreasing, one might consider these displays to be ideal devices for vision research. There is, however, one potentially problematic aspect of OLED development for vision research. OLED display manufacturers are experimenting with different types of color pixel patterns and developing proprietary methods for rendering images on these new displays. Unless it is possible to turn off or at least control the proprietary display rendering, it may be difficult to know the spatial distribution of the spectral energy in displayed stimuli.

7.4.2 SPECTRAL HOMOGENEITY

The relative spectral radiance from a subpixel should be the same as its intensity is varied. Any change in the relative spectral radiance will be manifested as an unwanted color shift, and the display will be difficult to calibrate. Recall that the intensity of the light from an LCD depends on the rotation of the polarization angle caused by the birefringent LC. In some displays, the polarization effect is wavelength dependent and this violates the spectral homogeneity assumption (Wandell and Silverstein, 2003). This failure occurs because the LC polarization is not precisely the same for all wavelengths and also as a result of spectral variations in polarizer extinction.

A second deviation from the standard display model occurs when the display emission is angle dependent. In fact, the first generation of LCDs had a very large angle dependence so that even small changes in the viewing position had a large impact on the spectral radiance at the cornea. The reason for this strong dependence is that the path followed by a ray through the LC and the polarizers has an influence on the likelihood of transmission, and this function is wavelength dependent (Silverstein and Fiske, 1993). Manufacturers have reduced these viewing angle dependencies by placing retardation films in the optical path (Yakovlev et al., 2015).

For visual psychophysics experiments, it is typical to fix the subject's head position relative to the screen, typically by using a chin rest or a bite bar placed on-axis in facing the middle of the display. Instruments used for display calibration should be placed at this position. If the spectrophotometer and the eye are located at any other angle, the spectral radiance from the display may be different.

7.4.3 SPATIAL HOMOGENEITY (SHIFT INVARIANCE)

When a subject is close to the display surface, the angle dependence of the spectral radiance appears as a spatial inhomogeneity: the spectral radiance at the cornea differs between on-axis (center) and off-axis (edge) pixels. At further distances, say 1 m away, the angle between the center and edge is smaller and the spatial homogeneity is better.

A second source of spatial inhomogeneity arises from the fact that it is difficult to maintain perfect uniformity of the pixels across the relatively large display surfaces. Such nonuniformities are referred to as "mura," which is a Japanese word for "unevenness." For LCDs, there are several sources of mura, including nonuniformity in the TFT thickness, LC material density, color filter variations, backlight illumination, and variations in the optical filters. Additional possible sources are impurities in the LC material, nonuniform gap between substrates, and warped light guides.

On LCDs, mura appears as blemishes and dark spots; manufacturers attempt to eliminate these sources during the manufacturing process. For OLEDs, mura is mainly due to nonuniformity in the currents in spatially adjacent diodes that appear as black lines, blotches, dots, and faint stains that are more visible in the dark areas of an image. This can be mitigated during the manufacturing process by introducing feedback circuitry that adjusts the pixel transistor current during a calibration procedure (McCreary, 2014).

7.5 DISPLAY SIMULATIONS

The standard display model serves as a foundation for display simulation technology. The model is implemented in the open-source ISETBIO distribution.* In this section, we present two examples that couple simulation to standard color image metrics. The examples illustrate the use of display simulation to answer questions about the appropriate use of display technology in vision research.

7.5.1 COLOR DISCRIMINATIONS: THE IMPACT OF BIT DEPTH

First, we consider how the number of digital steps (frame buffer levels) limits the ability to make threshold color and luminance discrimination measurements. Using the simulator, we calculated the CIE XYZ values for each of 27 different RGB levels, and we then calculated the CIELAB ΔE value between each of these 27 points and all of its neighbors within 2 digital steps. We repeated this calculation simulation assuming a frame buffer with 10 bits (1024 levels), the actual display resolution, and a coarser step size of 8 bits (256 levels) but equivalent gamma.

The distributions of CIELAB ΔE differences for the 10-bit and 8-bit displays are shown in the upper and lower histograms of Figure 7.6, respectively. For a 10-bit display, the signals within two digital steps are below $\Delta E = 1$. In this case, the visual discriminability is small enough to measure a psychophysical discrimination curve. If the display has only 8 bits of intensity resolution, the two digital steps frequently exceed $\Delta E = 1$. This explains why threshold measurements are impractical on 8-bit displays. For commercial purposes, however, one step is about $\Delta E = 1$, which explains why 8 bits renders a reasonable reproduction.

7.5.2 SPATIAL–SPECTRAL DISCRIMINATIONS

Next, we analyzed the visual impact of changing the subpixel PSF (see Figure 7.1). In this example, we compared two displays with the same primaries and spatial resolution (96 dots per inch), but with different pixel PSFs. In one case, the PSF is the conventional set of three parallel stripes (Dell LCD Display Model 1905FP), while in the second case the point spread is three adjacent chevrons (Dell LCD Display Model 1907FPc). We used the standard display model to calculate the spatial–spectral radiance of the 52 upper- and lowercase letters on both displays. The spatial–spectral radiance image data are represented as 3D matrices or hypercubes where each plane in the hypercube contains the stimulus intensity for points sampled across the display (x,y) for each of the sampled wavelengths (λ). To visualize the data, we map the vector describing the spectral radiance for each pixel into CIE XYZ values and convert these into sRGB display values (see inset in Figure 7.7).

* https://github.com/isetbio/isetbio: Tools for modeling image systems engineering in the human visual system front end.

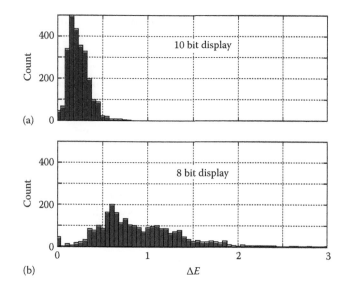

Figure 7.6 CIELAB ΔE differences between nearby values on a 10 bit and 8 bit display. Twenty-seven red, green, and blue points were selected, and the CIELAB ΔE values were calculated between the selected point and other points within two digital steps. The histograms shows the distribution of ΔE values for the 10 bit (a) and 8 bit (b) simulation. For the 10 bit display, two steps is below threshold, but for the 8 bit display one or two steps is at or above visual threshold. Hence, a 10 bit intensity resolution is necessary to measure psychophysical discrimination functions that require multiple near-threshold measures.

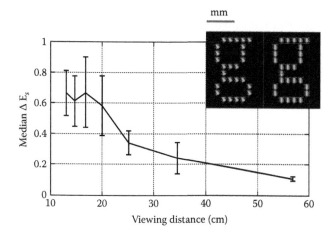

Figure 7.7 Visible difference between letters rendered on displays with different subpixel points spread functions but the same primary spectral power distributions (Figure 7.2b) and spatial resolution (96 dpi). The graph shows the median SCIELAB error, averaged cross 52 upper- and lowercase letters (+/− 1 s.d.) plotted as a function of viewing distances. The inset at the upper right is a magnified version of the letter "g" that illustrates the different subpixel point spread functions.

We used the spatial–spectral radiance data to calculate the spatial CIELAB (SCIELAB) ΔE difference (Zhang and Wandell, 1997, 1998) between each letter simulated on the two displays and viewed from different distances. Figure 7.7 plots the median SCIELAB ΔE value as function of viewing distance. The analysis predicts no visible differences between pairs of letters rendered on the two displays at any of the viewing distances. And indeed, we did not find significant differences between subject's judgments about the quality of letters rendered on the two different displays (Farrell et al., 2008).

7.6 SUMMARY

7.6.1 APPLICATIONS OF THE STANDARD DISPLAY MODEL

The standard display model guides both the calibration and simulation of visual stimuli. The model can be used to characterize visual stimuli so that others can replicate vision experiments. It can be used to simulate different types of displays and rendering algorithms and, in this way, makes it possible to evaluate the capabilities of displays during the engineering design process (Farrell et al., 2008). Finally, the standard display model supports the development of computational models for human vision by making it possible to calculate the irradiance incident at the eye (Farrell et al., 2014).

The standard display model assumes that the light generated by each subpixel is additive, independent, and shift invariant. These assumptions, referred to as spectral homogeneity, pixel independence, and spatial homogeneity, can be tested in the calibration process. A particular display may not meet these conditions for all stimuli, yet the model may still be used to predict the spatial–spectral radiance of a restricted class of visual stimuli. As an example, the standard display model does not predict the spectral radiance of high-frequency gratings presented on a CRT (Farrell et al., 2008; Lyons and Farrell, 1989; Pelli, 1997), but the model does predict the spectral–spectral radiance of large uniform color areas presented on a CRT (Brainard et al., 2002; Post, 1992). The standard display model can predict the steady-state spatial–spectral radiance of high-frequency gratings and text rendered on many LCDs (Farrell et al., 2008), particularly in the absence of complex circuitry to overdrive or undershoot pixel intensity (Lee et al., 2001, 2006) and locally dim LED backlights (Seetzen et al., 2004).

We present two examples that illustrate how to analyze display capabilities by coupling the standard model with color discrimination metrics. The first example shows why 10 bit intensity resolution is necessary to measure a psychophysical discrimination function. The second example analyzes the effect that different subpixel PSFs have on font discriminations. These examples illustrate how the standard display model can be used to analyze display capabilities in specific experimental conditions.

A further benefit of the standard model is to support reproducible research. Scientists can communicate about experimental stimuli by sharing the calibration parameters and a simulation of the standard display model. For the data and simulation, other scientists can reproduce and analyze experimental measurements by beginning with a complete spatial–spectral radiance of the experimental stimulus.

7.6.2 FUTURE DISPLAY TECHNOLOGIES

Advances in display technology transformed vision science over the past two decades. The growth in computing power continues to drive the development of display technologies that will influence visual psychophysics by broadening the scope of what we can control and study. Perhaps, the most exciting new developments are methods that expand the display from a passive device that emits a predetermined set of images to an interactive device that displays images that depend on continuous measurements of the viewer's head position.

A number of companies are developing head-mounted displays that are coupled with computer vision systems that sense the position and orientation of the head (Holliman et al, 2011; Kress and Starner, 2013). These systems comprise a pair of high quality displays, one for each eye, and a set of external cameras and algorithms that monitor the viewer's head position. The images presented to the two displays are approximations of what the viewer would see at each eye in a 3D environment. When the images are rapidly updated, and the computer graphics representation of the environment is detailed, the user has a compelling experience of being immersed within a virtual world. These systems—which include the displays, computer graphics programs, and head position–sensing systems—provide a "virtual reality" experience. A number of companies have developed products based on this technology and one hopes that these systems will be commercially viable products that can be controlled for scientific applications.

There are also new ideas about how to build displays that provide a relatively complete approximation of a full light field (Liu and Li, 2014). The goal of these "light field displays" is to control the intensity and color in each direction. The ability to control the rays in all directions generates a signal that is much closer to the physical reality. With light field displays, as one moves back and forth or side to side, the rays incident at the cornea change and match the experience of seeing through the window into a real 3D world, similar to looking at the scene through a window. Two viewers can stand next to one another and both see the same world, each from their own point of view. This type of display eliminates the need for head tracking and computationally intensive methods for rapidly updating the displayed image based on the viewer's head position. Such light field displays exist in early prototype form, and there is the hope that further engineering technology will produce viable commercial ventures.

To take advantage of these technologies in scientific applications will require further development of display calibration and simulation. The standard display model we explained here is woefully inadequate to characterize the stimuli delivered by head-mounted virtual reality systems or light field displays. The opportunities for using these systems for new scientific discovery are very great, and we are sure that scientists will develop principled approaches to calibration and simulation that will incorporate these new technologies into scientific practice and produce new insights about vision and the mind.

REFERENCES

Adelson, E. H. and Bergen, J. R. (1991) The plenoptic function and the elements of early vision, in *Computational Models of Image Processing*, (eds. M. S. Landy and J. A. Movshon). Cambridge, MA: MIT Press, pp. 3–20.

Bale, M., Carter, J. C., Creighton, C. J., Gregory, H. J., Lyon, P. H., Ng, P., Webb, L., and Wehrum, A. (2006) Ink-jet printing: The route to production of full-color P-OLED displays. *Journal of the Society for Information Display*, 14(5), 453–459.

Brainard, D. H. (1989) Calibration of a computer controlled color monitor, *Color Research & Application*, 14(1), 23–34.

Brainard, D. H., Brunt, W. A., and Speigle, J. M. (1997a) Color constancy in the nearly natural image. 1. Asymmetric matches, *Journal of the Optical Society of America A*, 14, 2091–2110.

Brainard, D. H., Pelli, D. G., and Robson, T. (2002) Display characterization, in *Encyclopedia of Imaging Science and Technology*, (ed. J. Hornak), Wiley, pp. 172–188.

Brainard, D. H., Wendy, A. B., and Speigle, J. M. 1997a Color constancy in the nearly natural image. 1. Asymmetric matchesm, *Journal of the Optical Society of America A*, 14(9), 2091.

Carter, J., Lyon, P., Creighton, C., Bale, M., and Gregory, H. (2005) Developing a scalable and adaptable ink jet printing process for OLED Displays, *SID Symposium Digest of Technical Papers*, Vol. 36, No. 1., pp. 523–525.

Castellano, J. A. (1992) *Handbook of Display Technology*. San Diego, CA: Academic Press.

Cooper, E. A., Jiang, H., Vildavski, V., Farrell, J. E., and Norcia, A. M. (2013) Assessment of OLED displays for vision research, *Journal of Vision*, 13(12), 16.

Engel, S. (1997) Retinotopic organization in human visual cortex and the spatial precision of functional MRI, *Cerebral Cortex*, 7(2), 181–192.

Farrell, J., Ng, G., Xiaowei, D., Larson, K., and Wandell, B. (2008) A display simulation toolbox for image quality evaluation, *Journal of Display Technology*, 4(2), 262–270.

Farrell, J. E. (1986) An analytical method for predicting perceived flicker, *Behavior and Information Technology*, 5(4), 349–358.

Farrell, J. E., Jiang, H., Winawer, J., Brainard, D. H., and Wandell, B. A. (2014) Modeling visible differences: The computational observer model, *SID Symposium Digest of Technical Papers*, Vol. 45, pp. 352–356.

Florence, J. M. and Yoder, L. A. (1996) Display system architectures for Ddigital micromirror device (DMD)-based projectors, *Proceedings of the SPIE* 2650, 193–208.

Gershun, A. (1939) "The Light Field" Translated by P. Moon and G. Timoshenko, *Journal of Mathematics and Physics*, 18, 51–151.

Holliman, N. S. et al. (2011) Three-dimensional displays: A review and applications analysis, *IEEE Transactions on Broadcasting*, 57(2), 362–371.

Kawamoto, H. (2002) The history of liquid-crystal displays, *Proceedings of the IEEE*, 90(4), 460–500.

Kelley, E. F., Lang, K., Silverstein, L. D., and Brill, M. H. (2009) Projector flux from color primaries SID, *Symposium Digest of Technical Papers*, 40(1), 224–227.

Kress, B. and Starner, T. (May 31, 2013) A review of head-mounted displays (HMD) technologies and applications for consumer electronics, *Proceedings of the SPIE, 8720, Photonic Applications for Aerospace, Commercial, and Harsh Environments IV*, Baltimore, MD, Vol. 87200A. doi:10.1117/12.2015654.

Law, H. B. (1976) The shadow mask color picture tube: How it began: An eyewitness account of its early history, *IEEE Transactions on Electron Devices*, 23(7), 752–759.

Lee, B.-W., Park, C., Kim, S., Jeon, M., Heo, J., Sagong, D., Kim, J., and Souk, J. (2001) Reducing gray-level response to one frame: Dynamic capacitance compensation, *SID Symposium Digest Technical Papers*, 32, 1260–1263.

Lee, S.-W., Kim, M., Souk, J. H., and Kim, S. S. (2006) Motion artifact elimination technology for liquid-crystal-display monitors: Advanced dynamic capacitance compensation method, *Journal of the Society for Information Display*, 14(4), 387–394.

Liu, X. and Li, H. (June 2014) The progress of light-field 3-D displays, *Information Display*, November–December, pp. 6–13.

Lyons, N. P. and Farrell, J. E. (1989) Linear systems analysis of CRT displays, *SID Digest*, 10, 220–223.

McCreary, J. L. (January 7, 2014). Correction of TFT Non-uniformity in AMOLED Display, Siliconfile Technologies Inc, assignee US Patent 8624805.

Naiman, A. C. and Makous, W. (1992) Spatial nonlinearities of gray-scale CRT pixels, *Proceedings of the SPIE*, 1666, 41–56.

Packer, O., Diller, L. C., Verweij, J., Lee, B. B., Pokorny, J., Williams, D. R., Dacey, D. M., and Brainard, D. H. (2001) Characterization and use of a digital light projector for vision research, *Vision Research*, 41(4), 427–439.

Pelli, D. G. (1997) Pixel independence: Measuring spatial interactions on a CRT display, *Spatial Vision*, 10(4), 443–446.

Post, D. L. (1992) Colorimetric measurement, calibration and characterization of self-luminous displays, in *Color in Electronic Displays*, (eds. H. Widdel and D. L. Post). New York: Plenum Press, pp. 299–312.

Poynton, C. and Funt, B. (2013) Perceptual uniformity in digital image representation and display, *Color Research & Application*, 39(1), 6–15.

Poynton, C. A. (1993) Gamma' and its disguises: The nonlinear mappings of intensity in perception, CRTs, film, and video, *SMPTE Motion Imaging Journal*, 102(12), 1099–1108.

Seetzen, H., Heidrich, W., Stuerzlinger, W., Ward, G., Whitehead, L., Trentacoste, M., Ghosh, A., and Vorozcovs, A. (2004) High dynamic range display systems, in *ACM SIGGRAPH 2004*. New York: ACM, pp. 760–768.

Silverstein, L. D. and Fiske, T. G. (1993) Colorimetric and photometric modeling of liquid crystal displays. *Color and Imaging Conference*. 1993(1). Society for Imaging Science and Technology.

Stevens, S. S. (1957) On the psychophysical law, *Psychological Review*, 64(3), 153–181.

Tang, C. W. and Vanslyke, S. A. (1987) Organic electroluminescent diodes. *Applied Physics Letters*, 51, 913.

Tobias, E. and Tanner, T. G. (2012) Temporal properties of liquid crystal displays: Implications for vision science experiments (ed. Bart Krekelberg). *PLoS One*, 7(9), E44048.

Tsujimura, T. (2012) *OLED Display: Fundamentals and Applications*. Hoboken, NJ: Wiley.

Wandell, B. A. (1995) Foundations of Vision, Sinauer Associates Inc., 476 pages.

Wandell, B. A. and Silverstein, L. D. (2003) *Digital Color Reproduction in The Science of Color*, 2nd edn., (ed. S. Shevell). Optical Society of America.

Watson, A. B., Ahumada, A. J. Jr., and J. E. Farrell. (1986) Window of visibility: A psychophysical theory of fidelity in time-sampled visual motion displays, *Journal of the Optical Society of America A*, 2(2), 300–307.

Wyszecki, G. and Stiles, W. S. (1967) *Color Science: Concepts and Methods, Quantitative Data and Formulas*. New York: Wiley, 1967 (1st ed.), 1982 (2nd ed.).

Yakovlev, D. A., Vladimir, G. C., and Kwok, H.-S. (2015) *Modeling and Optimization of LCD Optical Performance*. John Wiley & Sons.

Yang, D.-K. and Wu, S.T. (2006) *Fundamentals Of Liquid Crystal Devices*. Chichester, U.K.: John Wiley, 2006.

Younse, J.M. (1993) Mirrors on a chip, *IEEE Spectrum*, 30(11), 27–31.

Zhang, X. and Farrell, J. E. (2003) Sequential color breakup measured with induced saccades, *Proceedings of the SPIE*, Santa Clara, CA, Vol. 5007, pp. 210–217.

Zhang, X. and Wandell, B. A. (1997) A spatial extension of CIELAB for digital color-image reproduction, *Journal of the Society for Information Display*, 5(1), 61.

Zhang, X. and Wandell, B. A. (1998) Color image fidelity metrics evaluated using image distortion maps, *Signal Processing*, 70(3), 201–214.

Zhang, Y., Song, W., and Kees, K. (2008) A tradeoff between motion blur and flicker visibility of electronic display devices, *International Symposium on Photoelectronic Detection and Imaging 2007: Related Technologies and Applications*, (ed. L. Zhou), *Proceedings of SPIE* 6625, 662503.

8 Basic ophthalmic instruments

Walter D. Furlan

Contents

8.1 INTRODUCTION

This chapter describes the working principles and main characteristics of basic ophthalmic instruments used in a clinic by visual care professionals. It is divided in four sections: In Section 8.2, we describe the instruments used to explore and measure the anterior pole of the eye, such as keratometers, corneal topographers, slit-lamp biomicroscopes, and tonometers. In Section 8.3, we explain the working principles of instruments devoted to the observation of the retina: direct and indirect ophthalmoscopes and retinographers. In Section 8.4, we describe those instruments used in objective refraction: retinoscopes and refractometers; and finally, in Section 8.5, we report the set of instruments utilized in subjective refraction such as visual acuity (VA) charts and displays and phoropters.

8.2 INSTRUMENTS FOR THE EXAMINATION OF THE ANTERIOR SEGMENT

The examination of the anterior segment involves several instruments of different characteristics. On the one side, instruments such as keratometers and corneal topographers are intended for the optical characterization of the cornea, mainly its refractive power. On the other side, slit-lamp biomicroscopes are utilized for the observation of external ocular structures. The Goldmann applanation tonometer is included in this section as a slit-lamp accessory, and, for completeness, we also describe the noncontact tonometers and the ocular response analyzer (ORA).

8.2.1 KERATOMETERS

A keratometer is an instrument used for measuring the curvature of the central region of the anterior corneal surface. The invention of this instrument is usually attributed to H. von Helmholtz in 1851, although an earlier model was proposed in 1796 by J. Ramsden (Hirsch and Wick 1968). Clinical uses of keratometry are contact lens fitting and monitoring (Douthwaite 2009); measurement of corneal astigmatism, to obtain an approximated value of total ocular astigmatism by means of the Javal's rule (Remón et al. 2009); detection of irregular astigmatism and keratoconus; pre-refractive surgery evaluation; and intraocular lens (IOL) power determination (Yanoff and Duker 2009).

8.2.1.1 Principles of keratometry

The optical principle involved in keratometry is the relationship between the size of an object (illuminated mires) projected onto the cornea and the size of the virtual image of that object reflected from the cornea (the first Purkinje image) considered as a convex mirror. Assuming that the image is located near the focus of the *equivalent spherical convex mirror* (see Figure 8.1), the radius of the cornea surface (*r*) is obtained approximately by geometrical optics theory (Douthwaite 2009):

$$r = \frac{2dy}{h} \tag{8.1}$$

where

h is the object size
y is the image size (defined by the separation of the mires)
d is the distance between the object and the image, which is approximated to the distance between the object and the focal point (*F*)

According to Equation 8.1, the knowledge of the image size is necessary to obtain the radius of the cornea. However, this is a small image inside the eye and, therefore, a microscope has to be used in order to obtain a bigger real image outside the eye (*y'* in Figure 8.2a) to measure it.

One additional problem is that, due to involuntary eye movements, the image is constantly moving, so it is very difficult to measure it with a fixed graticule within the microscope. To overcome this problem, a prismatic system is employed to form a double image of the mires

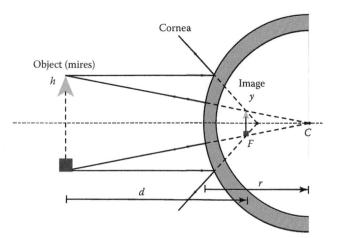

Figure 8.1 In keratometry, the cornea is considered as spherical convex mirror.

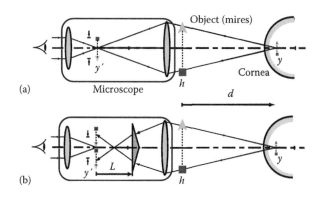

Figure 8.2 (a) A microscope is used to obtain a second real image of the mires outside the eye *y'*. (b) A biprism produces a double image of *y'*. When both images are just touching, their separation is equal to their size.

(see Figure 8.2b). This way, any movement of the eye will affect both images equally. The system is designed in such a way that the relative displacement between images can be modified by the practitioner until their separation equals the mire's image size: *y'*. This situation, which is achieved when the images are just touching (as represented in Figure 8.2b), can be reached in two different ways: (1) the separation of the images is adjusted by moving the prism along the optical axis on a given distance *L* from the image plane, or (2) the relative size of the images is varied by changing the size of the object *h* (separation of the mires). Therefore, by using Equation 8.1, the corneal radius curvature can be calculated directly if we know the distance *d* (which is very close to the distance between the eye and the mires), the prism power, and its axial position *L*.

To obtain the dioptric power of the cornea, keratometers assume that its surface is a single toroidal refractive surface. Therefore, it is necessary to measure the radius of curvature along both of its principal meridians at which the maximum and minimum magnifications of the mires are obtained. In fact, if the object's mire is a circle, the images produced by an astigmatic cornea is an ellipse with axes parallel to the corneal principal meridians.

The dioptric power (*P*) of the cornea is obtained approximately, assuming that its surface along the principal meridians can be approximated to a spherical diopter of radius *r*, using the following formula:

$$P = \frac{n-1}{r} \qquad (8.2)$$

where *n* is a "calibration" refractive index that takes different values depending on the manufacturer (thus, this assumed corneal index may cause problems when comparing keratometric power data from different instruments). The most common value of *n*, adopted by Bausch & Lomb, Haag-Streit (Javal-Schiötz), and many others, is 1.3375 that takes into account the negative power of the corneal back surface (Corbett et al. 1999).

8.2.1.2 One- and two-position keratometers

One- and *two-position* keratometers differ in the way that the system of doubling is used, as discussed in the previous section. In *one-position* keratometers, the mires have a fixed separation, and doubling is obtained independently in each principal meridian by varying the axial position of two prisms with mutually perpendicular bases. The instrument has a circular mire with plus and minus signs surrounding it. The combination of the objective and the prisms produce three images that are seen through the eyepiece, as shown in Figure 8.3.

The central image is formed by a Scheiner disc (see Section 8.4.2 and Figure 8.21); thus, it appears doubled if it is not correctly focused. This system, originally designed by Bausch & Lomb in 1932, allows for the curvature along both principal meridians of

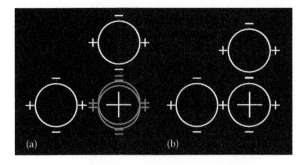

Figure 8.3 Images observed through a two-position type keratometers (Bausch & Lomb). (a) Out-of-focus image. (b) In-focus image with adjusted doubling along both principal meridians.

an astigmatic cornea to be obtained without repositioning the instrument between measurements. In this configuration, the fixed parameters in Equation 8.1 are *d* and *h*, whereas the variable is *y*. Consequently, the radius scale of the instrument is linear. On the other hand, in the *two-position* keratometers, based on a design proposed by Javal and Schiötz in 1881, a 90° rotation of the whole instrument is required between the measurements along each one of the two principal meridians. This is due to the fact that the doubling prism is fixed but the separation of the mires can be varied by moving them symmetrically around a circular path, approximately concentric with the cornea under test (Bennett and Rabbetts 1989). Figure 8.4 shows a typical pattern: the mires have to be set side by side to sequentially measure the flatter and steeper meridians when they are aligned. Due to their nearly complimentary colors, any overlap of their images is clearly seen in yellowish-white steps (see Figure 8.4c). The steps of the green mire's image correspond approximately to 1D. Thus, if an overlapping of the images is observed between measurements along the principal meridians, an approximated value of the corneal astigmatism can be deduced.

In Javal-Schiötz-type keratometers, the doubling prism and the separation of the images are both fixed, and the size of the images can be modified by altering the size of the object, that is, by changing the separation of the mires (see Figure 8.2). Therefore, in this case, the fixed parameters in Equation 8.1 are *d* and *y*, whereas the variable is *h*, and therefore, power scale of the instrument is linear.

Although the range of powers of most keratometers is 35D to 52D, it is possible to increase the range by placing lenses of low power in front of the microscope's objective. Typically, −1.00D and +1.25D lenses are used to extend the range to around 30D and 61D, respectively. Conversion tables are usually provided by manufacturers for these specified lenses.

8.2.1.3 Automated keratometry

Autokeratometry is a computer-determined measurement of the curvature of the cornea. The physical principle of automated keratometry is the same as for manual keratometers. As mires, most automated keratometers use light-emitting diodes (LEDs) arranged in different ways. A Placido disc (see Section 8.2.2) has also been incorporated in recent commercial models. In place of the eyepiece, automated keratometers use a digital camera to record the size of the mire's image and the instrument

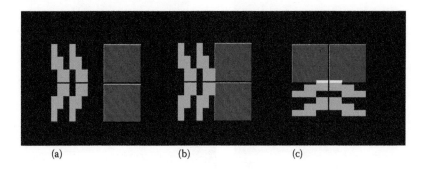

Figure 8.4 Sequence of images through a Javal-Schiötz keratometer. (a) Initial focused and aligned mires along one principal meridian (horizontal). (b) Image corresponding to the correct lecture of the power and radius along the horizontal meridian. (c) When the instrument's head is rotated 90°, a superposition of the mires is observed in the vertical meridian, indicating in this case a *with-the-rule* astigmatism of approximately 0.50D.

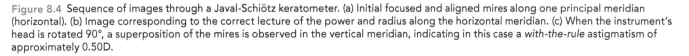

sofware calculates the cornea's radii. Several manufacturers have incorporated automated keratometry in autorefractometers (see Section 8.4).

8.2.2 CORNEAL TOPOGRAPHY

The area of the cornea used by keratometers to provide the corneal mire's image is an annular region of approximately 0.5 mm width and 3–4 mm diameter (depending on the dioptric power of the cornea). This means that the cornea's central and peripheral points outside this area are not be measured. The advantage of corneal topography is the ability to measure almost the entire area of the cornea and, then, to quantify irregular astigmatism that cannot be done with a keratometer. Applications of corneal topography include contact lens fitting, diagnosis of corneal irregularities (keratoconus, ectasias, dystrophies, and scars), and screening and management of refractive surgery patients. Most corneal topographers also use algorithms to evaluate several quantitative indices such as the *simulated keratometry*, which is an estimate of corneal curvature at the 3 mm zone; the *surface regularity index*, which measures the local regularity in the central 4.5 mm diameter; the *surface asymmetry index*, which is defined as the difference in corneal power between points situated 180° apart in each ring; and the *index of vertical symmetry* (Corbett et al. 1999).

8.2.2.1 Basic topographic principles

Corneal topographers can be classified into two main categories: Placido disc–based topographers (or videokeratoscopes) and triangulation systems (Horner et al. 2006). The underlying principle of videokeratoscopes is the same as for keratometers, but, instead of mires, they employ a Placido disc–based illumination system that consists of a series of dark and bright rings (see Figure 8.5a). The images of multiple reflected concentric circles projected on the corneal surface are captured by a CCD camera through a central aperture, and it is digitally analyzed to give different contour maps of the central cornea (of around 10–12 mm in diameter). Placido disc topography units do not have the ability to measure the central 1.8–2.0 mm of a cornea, and this information is currently extrapolated from the smallest reflected ring. On the other hand, peripheral rings are often limited by shadows from eyelashes and nose (see Figure 8.5b). Contour maps include *curvature*, *power*, *elevation*, and *difference*.

Curvature maps can be displayed in one of two formats: axial or tangential. For the axial map, the local radius of curvature is the distance from the corneal surface to the optical axis along a line perpendicular to the cornea. Thus, it is assumed that the rotation center of the best fit sphere lies on the optical axis. This map is a good estimate of overall corneal shape because it provides the average of curvature values from the optical axis to the reference corneal point (Swiegerling 2004). This is useful for monitoring global corneal changes, calculating IOL power, etc. Tangential maps (also called instantaneous curvature maps) represent localized changes and peripheral data better than axial maps because, in these maps, it is assumed that around each point on the corneal surface, there is a small spherical area with an independent radius of curvature. Thus, they are better indicators of corneal shape than axial

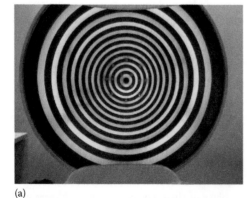

(a)

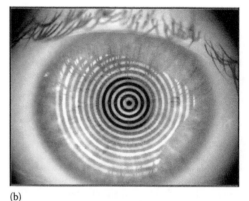

(b)

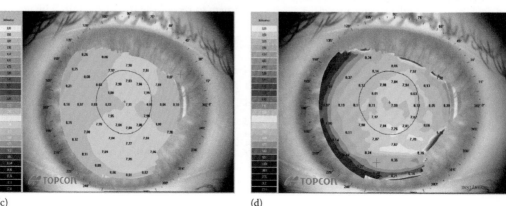

(c)　　　　　　　　　　　　　　　　　　　(d)

Figure 8.5 (a) Patient's view of a Placido disc in a videokeratoscope. (b) Projected rings of the Placido disc onto the patient's cornea. (c) Axial map. (d) Tangential map.

maps and are useful for evaluating surgically induced changes, ectasias, and contact lens fitting.

The curvature maps depict the underlying shape of the cornea by using standard colors to represent diopter changes (see Figure 8.5c and d). Warm colors depict the steeper areas and cool colors the flatter ones.

Topographers can display a refractive map by converting the radius of curvature data into dioptric power using Snell's law of refraction. This accounts for spherical aberration outside the central zone and provides information about the imaging power of the cornea.

An elevation map shows the measured height of the corneal curvature from a reference surface (iris plane, best fit, or reference sphere). Warm colors show points that are higher than the reference surface, and cool colors designate lower points. This map shows the three-dimensional shape of the cornea and is useful in measuring the amount of tissue removed by photorefractive surgery and assessing postoperative visual problems.

Triangulation systems use the optical triangulation principles to obtain elevation maps of the different surfaces of the anterior pole of the eye (not only the front surface of the cornea as with videokeratoscopes) by capturing the diffuse image of a projected target (a grid, a grating, or a single slit) on them. The Orbscan and Pentacam systems are probably the most popular instruments based on this principle. The information provided by these instruments is based on true elevation maps from which the other maps previously mentioned can be obtained mathematically (Swiegerling 2004).

8.2.3 SLIT LAMP

The slit-lamp biomicroscope is an instrument intended for use in eye examination, mainly of the anterior eye segment. It provides in detail a stereoscopic magnified view of the eye's structures. Several accessories can be adapted to a slit lamp to increase the range of applications, such as photo and video adapters, applanation tonometer (for measuring the intraocular pressure [IOP]), pachymeter (for measuring the thickness of the cornea and lens and the depth of the anterior chamber), and different types of lenses to perform gonyoscopy and fundus observation.

8.2.3.1 Fundamentals of slit-lamp biomicroscopy

The main components of the slit lamp are an illumination system consisting of a bright light slit source with variable width, height and orientation, and a binocular microscope providing variable magnifications. The main characteristic of the instrument is that both systems can be moved around a common center of rotation enabling a common point of focus that can also be uncoupled to employ different methods of inspection (see Figure 8.6).

The slit lamps use a Köeller illumination system, which is the same one used by slide projectors. A simplified scheme of the illumination system is shown in Figure 8.7. Instead of a slide, a variable aperture slit is used that is projected onto the eye by a positive lens of short focal distance. The light source (usually a halogen lamp or a white LED) is imaged by the condenser lens in the plane where the projection lens is located. This way, the projection lens captures all rays that illuminate

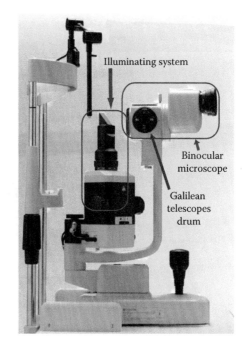

Figure 8.6 Slit lamp's main components. (Photograph supplied by Topcon España S.A., Barcelona, Spain.)

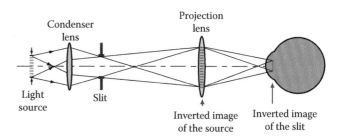

Figure 8.7 Köller illumination system in a slit-lamp biomicroscope.

the slit, and no further images of the filament are produced. The projection lens forms an image of the slit aperture on the anterior pole of the examined eye with maximum brightness. In most of the commercial slit lamps, the slit's width and length can vary continuously up to 15 mm. Different types of filters can be incorporated in the illumination system: a green (red-free) filter allows to view blood vessels more easily, a blue filter is used in combination with fluorescein to stain the tear film (mainly in contact lens practice), neutral density filters are applied to attenuate the intensity of the light, and diffuser filters are employed to obtain larger fields of illumination.

In general, the slit-lamp biomicroscopes allow magnifications between 5× and 40×. This range is covered by the combination of different objectives (usually 1× and 2×) and eyepieces (from 10× to 20×) and using a zoom system. As an alternative to the zoom, many commercial instruments incorporate a drum containing small Galilean telescopes located between the objective and the eyepiece to provide up to five different magnifications (see Figure 8.6). The diameter of the field of view varies (inversely proportional to the magnification) between approximately 5 and 35 mm. Like binoculars, slit-lamp microscopes have a pair of prisms (double Porro prisms) to obtain an erected image of the

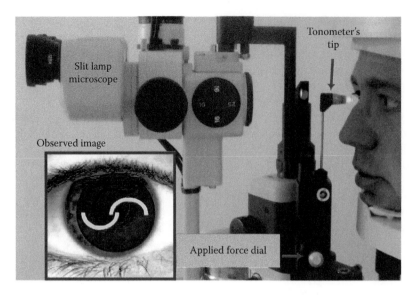

Figure 8.8 Goldmann applanation tonometer (inset: appearance of the tinted semirings of tear film trough the slit-lamp biomicroscope).

illuminated zone. Modern models can also incorporate accessories such as digital cameras and even modular Optical Coherence Tomography (OCT) units.

8.2.3.2 Slit-lamp accessories

High-power positive (60D, 78D, and 90D: aspheric Volk lenses) or negative (−55D: Hruby lens) lenses can be inserted in the observation system to visualize the fundus. In the first case, a real inverted image of the lens is formed in front of the microscope's objective; it is, thus, an alternative form of indirect ophthalmoscopy. Volk lenses produce an image with a field of view of around 70° with a magnification of 0.75×. The Hruby lens neutralizes the optical power of the eye and gives magnification of about 1.0× or less when used with emmetropic eyes: the field of view is limited by the pupil diameter and with a maximum of about 25°–30° when the lens is very close to the eye.

A versatile accessory is also the Goldmann three mirror contact lens. This is a −64D lens, mounted on a truncated-cone-like device that incorporates three mirrors to reflect the light from different parts of the eye (peripheral, mid-peripheral retina, and the iridocorneal angle) into the direction of the observer (Yanoff and Duker 2009).

8.2.3.3 Goldmann applanation tonometer

The Goldmann tonometer is an accessory to the slit lamp designed to measure the IOP. A precise assessment of the IOP is crucial for the diagnosis and decision-making regarding the treatment modalities in patients with glaucoma. Goldmann tonometry is considered the gold standard test. The physical basis of tonometry is related to Newton's third law of motion. It consists on exerting a force on the cornea with a specially designed tip (see Figure 8.8), assuming that the cornea reciprocally exerts a force equal in magnitude and opposite in direction on the tip. Contact tonometry assumes that the cornea is infinitely thin, perfectly elastic, and flexible. Although none of these assumptions are true, the surface tension force, which results from the tear film, counterbalance the rigidity of the cornea. The IOP is then defined as IOP = contact force/area of contact.

The area of contact between the tip and the cornea is a circle, and the tear film is moved to the borders of the applanating head of the tip forming a ring. During the examination, the tear film is tinted with a solution of sodium fluorescein. The examiner then uses a cobalt blue filter to view the applanated area. The head of the tip has two prisms inside that double the image and divide the field of view into two horizontally displaced yellowish-green semirings (see Figure 8.8, inset). The system is calibrated in such a way that, when the inner edges of the semirings are in contact, the applanated circle has a diameter of 3.06 mm. With this value, increments of 1 g of force exerted by the tip correspond to increments of 1 mmHg of pressure. The force applied to the tonometer's head is then adjusted to reach this applanated area using a dial connected to a variable tension spring (see Figure 8.8).

8.2.4 NONCONTACT TONOMETERS AND THE ORA

Noncontact tonometry uses an air pulse of a few microseconds (directed perpendicularly to the central cornea) to applanate the cornea; the corneal applanation is detected via an electrooptical system (Grolman 1972). As shown in Figure 8.9, a collimated beam of infrared (IR) light is directed onto the central cornea, forming a certain angle with the line of sight; the light reflected from the corneal surface is collected by an IR lens that focuses the beam in a point-like detector. IOP is computed by detecting two instants when the central cornea is nearly flat, that is, when the intensity of the reflected light reaches its maximum value (see Figure 8.9b). The first one is achieved between the initial convex form (see Figure 8.9a) and the concave form that it adopts when the deformation reaches its maximum value (see Figure 8.9c) due to the force of the air. Once the pulse ceased, the second one is achieved while the cornea is readopting its normal curvature, that is, when the sequence (a, b, c) in Figure 8.9 is reversed (see Figure 8.10).

IOP measurements are affected by their inherent uncertainty, in fact in normal eyes the IOP range is typically 10–20 mmHg, with diurnal variations up to 6 mmHg. As we have shown, contact and noncontact tonometers measure the IOP in fundamentally different ways. The resulting IOP values could vary between

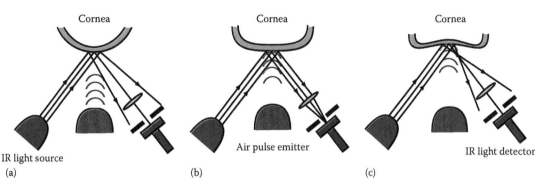

Figure 8.9 Sketch of the corneal deformation during a noncontact intraocular pressure measurement. (a) The air pulse is emitted by the tonometer. (b) The central cornea flattens, and the reflected IR detection reaches its maximum value. (c) The central cornea becomes concave, and the detected IR is minimum.

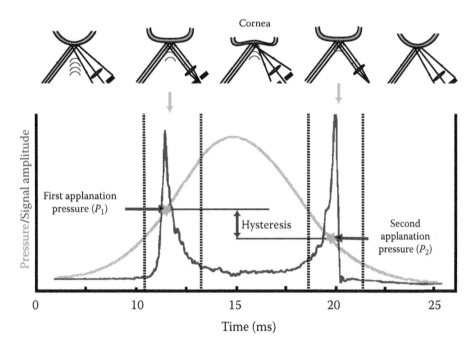

Figure 8.10 Sketch of the corneal deformation and the ocular response analyzer signals in a typical measurement. The signal amplitude peaks correspond to the corneal applanation moments.

these instruments, and these facts will need to be considered when comparing clinical measurements.

8.2.4.1 ORA

The ORA uses the same principle as the Grolman noncontact tonometer to provide additional information on the biomechanical properties of the cornea. In this instrument, the difference between the pressures at which the cornea flattens inward and outward is termed corneal hysteresis (CH), which is an indication of the biomechanical properties of the cornea. It is calculated as the difference in air pressures between force-in applanation (P_1) and force-out applanation (P_2) (see Figure 8.10). Thus CH also provides information about the contribution of corneal resistance to IOP measurements. *Corneal resistance factor* is another parameter provided by ORA and is derived as ($P_1 - kP_2$), where k is a constant associated with the central corneal thickness. Thus, in addition to providing information that aids in the diagnosis and management of glaucoma in refractive surgery, the ORA enables physicians

to better identify eyes at risk of keratoconus. The device also provides *corneal compensated IOP*; this parameter has been proven to be less influenced by corneal biomechanics than Goldmann or other methods of tonometry (Ehrlich et al. 2012).

8.3 RETINAL IMAGING INSTRUMENTS

Retinal imaging is useful to assess the health of a retina and helps to detect and manage several retinal pathologies such as glaucoma, diabetic retinopathy, retinal detachment, and macular degeneration. Several instruments have been developed for this purpose. In this section, we will describe the direct and indirect ophthalmoscopes and the retinal camera. More sophisticated instrumentation is discussed in other chapters of this handbook.

8.3.1 DIRECT OPHTHALMOSCOPE

The direct ophthalmoscope was introduced in the 1850s by Hermann von Helmholtz. It was the first proper device for examining the interior of the human eye by illuminating it

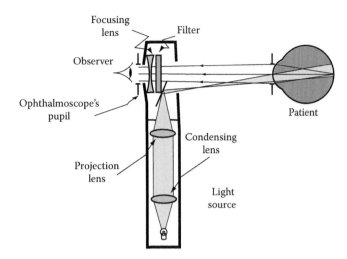

Figure 8.11 Direct ophthalmoscopy.

through the pupil (by means of a beam of light reflected by a semitransparent mirror) and observing in the same direction. Actually, at the time, Helmholtz used a naked candle as a light source (Keeler 2003). The operating principle of a direct ophthalmoscope is very simple: the practitioner uses the optical power of the examined eye to see its retina like a magnifier. In spite of its simplicity, the same technique is used today in modern ophthalmoscopes. As shown in Figure 8.11, the axes of the observation and illuminating beams are slightly displaced to remove the corneal reflex from the examiner field of view. To compensate for the ametropía of both the patient and the examiner, a rack of lenses of different powers is incorporated in the head of the instrument (see the Focusing lens in Figure 8.11). Additionally, to decrease glare and to enhance the visualization of retinal vessels, ophthalmoscopes have a range of filters and stops located between the condensing and the projection lens. The magnification of a direct ophthalmoscope (defined by convention as the ratio of the angle subtended by the object through the instrument, and the angle subtended by the object if it were viewed at 25 cm) depends on the distance from the practitioner to the patient (d), the eye's refraction (R), and the dioptric length of the

eye (R' being +60D for a reduced model eye). The magnification can then be approximately computed with the following expression (Bennett and Rabbetts 1989):

$$M = \frac{R'}{1 + Rd} \tag{8.3}$$

The field of view in direct ophthalmoscopy is directly proportional to the size of the pupil of the examined eye. For this reason, this technique is normally performed with mydriatics. If the examined eye has a pupil diameter ϕ, the field of view D (field of half illumination) can be expressed as (Bennett and Rabbetts 1989)

$$D = \phi \frac{R + d^{-1}}{R'} \tag{8.4}$$

Typical values of the magnification and fields of half illumination in direct ophthalmology are within the following ranges: $13\times \leq M \leq 18\times$ and $4° \leq D \leq 8°$.

8.3.2 INDIRECT AND BINOCULAR OPHTHALMOSCOPE

A few years after the invention of the ophthalmoscope, Christian Ruete introduced an indirect technique to observe a wider fundus area (Keeler 2003). In indirect ophthalmoscopy, the practitioner views a real inverted image of the retina, formed by a high-power positive lens in front of the eye. Several designs of instruments for this technique have been developed (Henson 1996). The most common is shown schematically in Figure 8.12. This corresponds to a binocular indirect ophthalmoscope. The power of the ophthalmoscope lens in front of the patient's eye normally ranges from +13D to +30D. This lens has two main functions: first, as shown in Figure 8.12, it forms a real inverted image of the patient's fundus at a plane between the ophthalmoscope's lens and the practitioner; second, it conjugates the entrance pupil of the viewing system and exit pupil of the illuminating system with the patient's pupil. As can be seen in Figure 8.13c, the observation and illumination systems do not overlap at the plane of the patient's pupil. This design completely removes the corneal reflex from the field of view. The paired

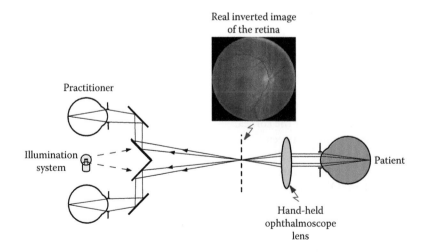

Figure 8.12 Scheme of an indirect binocular ophthalmoscope.

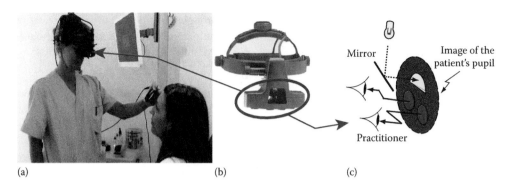

(a)   (b)   (c)

Figure 8.13 (a) Binocular indirect ophthalmoscope examination. (b) The illuminating and viewing systems are mounted on a helmet, and (c) the ophthalmoscope's lens images the patient's pupil (gray circle) at the same plane where both the exit pupil of the illuminating system (inner yellow semicircle) and the entrance pupil of the viewing system (red circles) are located.

mirrors mounted in the helmet before the practitioner's eyes (see Figure 8.13a and b) reduce the interpupillary distance, allowing a stereoscopic view of the fundus. Additionally, the ophthalmoscope's lens acts as a field lens. Therefore, its diameter is of fundamental importance to increase the observed retinal area. The magnification of an indirect ophthalmoscope depends on the distance from the lens to the patient (*a*), the lens power (*P*), and the dioptric length of the eye (*R'*). If we assume *R'* = 60D, it can be computed approximately with the following expression (Bennett and Rabbetts 1989):

$$M = 15(1 - Pa) \qquad (8.5)$$

As the product *Pa* in Equation 8.5 is bigger than the unity, the magnification in indirect ophthalmology is negative (inverted image). The field of view (*D*) in indirect ophthalmoscopy is directly proportional to the diameter of the ophthalmoscope's lens (ϕ) and inversely proportional to the distance between the lens and the patient's eye *a*. The field of view *D* (field of half illumination) can be expressed approximately as

$$D = \frac{\phi}{aR'} \qquad (8.6)$$

As the magnification and the field of view in indirect ophthalmology are dependent on the ophthalmoscope's lens, the range of values is wider than in direct ophthalmoscopy: −0.2× ≤ *M* ≤ −5× and 30° ≤ *D* ≤ 120°. Therefore, compared with direct ophthalmology, the indirect method allows a much larger (eventually stereo) field of view but a much lower magnification.

8.3.3 RETINOGRAPHY (FUNDUS CAMERAS)

Fundus cameras are indirect ophthalmoscopes in which the practitioner's eye is replaced by a CCD camera. In addition to the focusing light source (usually an IR source to avoid myosis), it includes an additional flash lamp to record the image (see Figure 8.14).

Most of the commercial fundus cameras do not require mydriatics and work with a pupil's diameter of 3–4 mm. The field of view of these instruments ranges between 30° and 45°, but external fixation targets allow to compose maps of larger regions. With fundus cameras, the technique of fluorescein angiography has become the gold standard of imaging of the ocular circulation

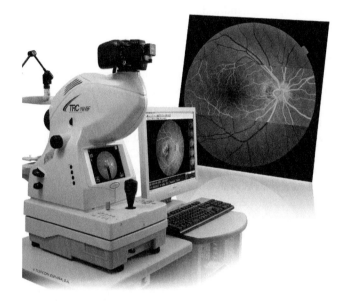

Figure 8.14 Retinal camera, the composed image on the right, shows the differences between normal view (top), a fluorescein angiography image (middle), and the same image (bottom) with a green filter. (Photograph supplied by Topcon España S.A., Barcelona, Spain.)

for the diagnosis of vascular disease. This technique requires two special filters. The first, a blue filter, is placed in the illumination system, limiting the wavelengths that enter into the eye to those that excite fluorescence. The second, a yellow filter, is placed in the viewing system, producing the same effect to the wavelengths that come from the fluorescein dye.

In modern retinographers, high-resolution digital IR cameras allow the addition of contrast studies using indocyanine green dye. The IR fluorescence of this dye highlights the circulation of the choroid, providing in real time an enhanced view of the deeper vascular structures. These images are complementary to those produced by fluorescein.

8.4 OBJECTIVE REFRACTION TESTS

The refractive power of the eye is assessed by locating the image on the far point of the patient's eyes. An objective refraction can be obtained by using different instruments without receiving any feedback from the patient.

One of the most common techniques used for objective refraction is to employ a handheld instrument called the retinoscope. With this instrument, the practitioner shines light into the patient's eye. The light travels through the front part of the eye to the retina. The diffusely reflected light on the retina bounces back to the retinoscope. It is then seen and analyzed by the retinoscopist who interposes a series of trial lenses in front of the eye to obtain an objective measurement of the eye's refractive error. Autorefractors, on the other hand, are computerized instruments that use a similar approach (but different principles) to obtain the same information. In this section, the optical principles of the retinoscope and refractors are explained.

8.4.1 RETINOSCOPE

A retinoscope is a handheld instrument with a very similar appearance to an ophthalmoscope. In fact, the first observations that led to clinical retinoscopy were made by Bowman in 1859 when examining an astigmatic eye with an ophthalmoscope. But it was not until 1873 that the use of retinoscopy to assess the refractive state of the eye was described by Cuignet (Millodot 1973). The principles of retinoscopy are essentially the same as those corresponding to Foucault knife-edge test; it consists of moving a knife-edge through the focus of a beam and observing the intensity pattern on a screen (Ojeda-Castañeda 2007).

As represented in Figure 8.15, in a streak retinoscope, the light produced by a light bulb with a fine linear filament within the base of the handle is projected by a positive lens and emanates from a mirror in the head of the instrument as a linear streak. The light bulb can be rotated 360° by means of a sleeve. By rotating the light source, the retinoscopist can align the orientation of the streak with the principal meridians of the patient's eye. The position of the lens in relation to the light filament can be altered by raising or lowering the sleeve. In this way, the vergence of the light emitted from the retinoscope can also be varied continuously. The streak of light passes through the patient's pupil and arrives at the retina. By observing the diffusely reflected light through the patient's pupil and the retinoscope's peephole, the practitioner can make deductions about the patient's refractive state (Michaels 1980, Safir 1982).

8.4.1.1 Illuminating systems

As mentioned before, the retinoscopist can control the vergence of the incident light. As shown in Figure 8.16, this is accomplished by changing the distance between the light filament **S** (normal to the plane of the figure) and the projection lens **L**. When the sleeve is up (plane mirror illumination), the streak is emitted as a divergent light beam that comes from a virtual image **S'** of the bulb located behind the retinoscope (see Figure 8.16a). As the sleeve is moved downward, this image moves to reach an end point when the image of the bulb is a real image located in front of the retinoscope (concave mirror illumination, see Figure 8.16b). The light entering the eye forms an out-of-focus patch of light upon the patient's retina **A**. During retinoscopy, the practitioner tilts the retinoscope sweeping a given meridian that is perpendicular to the axis of the streak. This means that if the streak is horizontal, the retinoscope should be tilted up and down in order to move the projected patch across the retina. When the retinoscope is tilted at an angle α in a given direction, the patch sweeps across the fundus in a direction that depends on the location of the image of the filament: if the image **S'** is located between the retinoscope and the eye, **A** moves in the opposite direction (see Figure 8.16d); otherwise (i.e., if the image is located behind the retinoscope, as shown in Figure 8.16c), or behind the examined eye), the patch on the patient's retina moves in the same direction. The patch on the retina becomes the object for the viewing system.

8.4.1.2 Viewing system

While moving the streak across the pupil, the examiner observes the relative movement of the fundus reflex at the pupil of the examined eye. What the retinoscopist sees depends on the vergence of the streak, the working distance and the refractive state of the eye. To describe how the reflex is seen by the practitioner, let us suppose that the illumination system is set in the plane mirror configuration (see Figure 8.16, left) and that the examined eye is myopic with its far

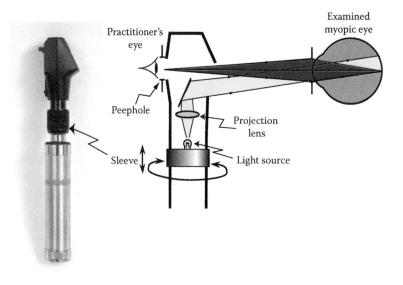

Figure 8.15 Cross section of a retinoscope. Illumination beam (yellow) and observation beam (red).

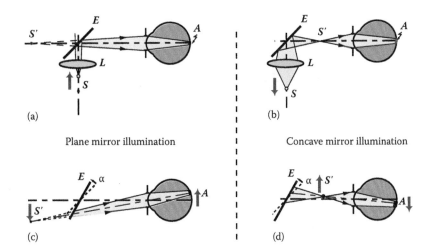

Figure 8.16 Illumination system: plane mirror (left) and concave mirror (right). Note the vergences of the incident rays: a concave mirror effect is produced when the light source is moved down; a plane mirror effect is produced when the bulb is moved up. The light source and the lens were removed in (c) and (d) for simplicity.

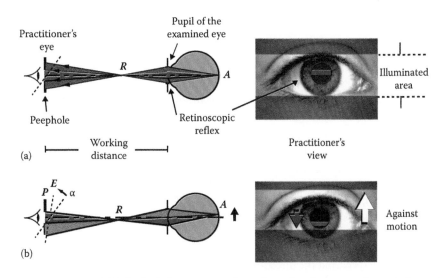

Figure 8.17 Viewing system in a myopic eye. (a) Left: the light emanating from the retina that passes through the eye's pupil converges to the eye's far point *R* and raises the peephole of the retinoscope. Part of the bundle of rays is blocked by the peephole. Therefore, the practitioner only sees the retinal reflex at the central zone of the pupil (right). (b) Left: sweeping the retinoscope upward an angle α produces an against motion of the observed reflex (right).

point (*R*) located between the restinoscope and the eye, as shown in Figure 8.17. We also assume that the illuminated patch at the retina *A* is so thin that it can be considered a line (normal to the plane of the figure). By looking through the peephole of the retinoscope at a given working distance, the practitioner can see at the patient's pupil some areas of the retinal reflex, while other portions are blocked by the peephole's borders (as in a Foucault-knife test) and then are perceived as a dark shadow in the subject's pupil. In this way, when the streak is centered on the optical axis, the retinoscopist sees a red streak reflex at the pupil, also centered on it as shown in Figure 8.17a. When the retinoscope is rotated upward at a small angle α, the patch on the retina *A* also moves upward. In this case, the peephole of the retinoscope blocks part of the light beam coming from the upper part of the patient's pupil (see Figure 8.17b) in such a way that the observer now sees that the red reflex moves in the opposite direction of the mirror, that is, an *against motion*. Following the same rationale, it is easy to deduce that if the far point of the eye (*R*) falls outside the interval

defined by the working distance, a *with motion* of the reflex will be seen by the practitioner. These results are summarized in Table 8.1.

A particular situation is achieved when the far point of the patient's eye lies just at the pupil's plane of the retinoscope (see Figure 8.18). In this case, the so-called "neutral point" is reached, which means that during the movement of the retinoscope, neither *with motion* nor *against motion* is observed.

Table 8.1 Retinoscopic reflex in different ametropias

	PLANE MIRROR	CONCAVE MIRROR
High miopía (far point in front of the retioscope)	Against motion	With motion
Low miopía (far point behind the retioscope) and hyperopia	With motion	Against motion

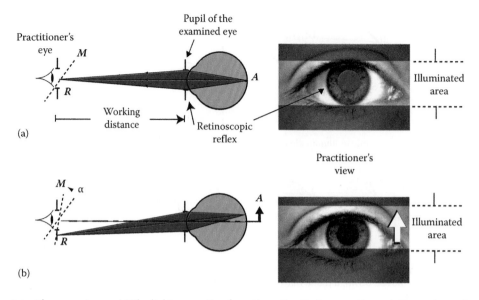

Figure 8.18 Neutral point with a myopic eye. (a) The light emanating from the retina that passes through the eye's pupil converges to the eye's far point R, where the peephole of the retinoscope is located. The reflex from the retina covers the whole patient's pupil. (b) When the retinoscope is tilted upward at a given angle, all rays coming from the retina are blocked by the peephole; therefore, the practitioner sees that the eye's pupil becomes suddenly dark.

Instead, the pupil appears uniformly illuminated or completely dark but no movement is seen.

8.4.1.3 Neutralization of the reflex: Relationship between refractive error and the working distance

The goal of retinoscopy is to bring the patient's far point to the plane where the peephole is located, that is, to obtain the neutralization of the reflex. This is usually achieved by means of trial lenses placed in front of the eye. If the far point is within the working distance (as in Figure 8.17), a negative lens is necessary; on the contrary, if the far point is out of the working distance interval, a positive lens would be needed. The power of the neutralization lens N is obtained by the practitioner using a trial and error approach, guided by the speed and the direction (see Table 8.1) of the reflex movements seen at the patient's pupil. In fact, when approaching the neutral point, the speed of the reflex increases (Bennett and Rabbetts 1989). For example, suppose we are searching for the compensation of a myopic eye as represented in Figure 8.19, following the trial and error approach, we can obtain the neutralization lens that in our case is a negative lens of power N (see Figure 8.19a). The power of this lens can be expressed as the sum of two powers: one is the lens that compensates or the eye ametropia C, that is, the lens that makes the patient's retina conjugate to infinity (see Figure 8.19b); the other one is a positive lens that makes the infinity conjugate to the retinoscope's pupil (see Figure 8.19c). The power of this positive lens, W, is just the inverse of the distance between the retinoscope and the eye (working distance). Thus, the power of the compensation lens is simply expressed as $C = N - W$. In other words, once the neutralization lens is obtained, the compensation lens is calculated by subtracting the inverse of the working distance, usually 66 cm (the length of an arm), and equivalent to 1.50D. Therefore, one source of error in retinoscopy is an inaccurate working distance that should result in a spherical error that becomes higher as the working distance is shorter; consequently, a minimum of 50 cm is recommended.

8.4.1.4 Retinoscopy of astigmatic eyes

In astigmatic eyes, the neutralization process should be performed sequentially along its two principal meridians in which, as represented in Figure 8.20, the reflex has different thickness, and only when the retinal reflex and the illuminating streak are aligned do they form an unbroken line. Otherwise, the reflex appears tilted and moves obliquely to the sweeping direction. Any irregularity in the width of the reflex indicates irregular astigmatism or moderate amount of aberrations. In fact, the retinoscopic principles can be applied to detect and measure optical aberrations (Caballero et al. 2006).

Another potential source of errors in retinoscopy is the oblique astigmatism induced by an inaccurate alignment between the observation line, and the line of sight of the examined eye. In this case, *an against*-the-rule astigmatism is induced that is usually not clinically important if this angle is lower than 5° but it becomes significant from 10° and increases nonlinearly with obliquity.

8.4.2 OPTOMETERS AND AUTOREFRACTORS

Optometers are instruments designed for the assessment of the refractive state of the eye. Historically, the first optometers developed were subjective instruments in which the sharpness of the image of a test object was used by the user to self-determine its refractive error. Most of these principles were adopted in modern autorefractors to give objective measurements.

8.4.2.1 The Badal optometer and the Scheiner disc

The Badal optometer consists of a fixed positive lens of power Φ placed at its focal distance in front of the examined eye and a movable target test (see Figure 8.21). As the test is moved a given distance z from its front focal point, the vergence of the light entering the eye changes, and it can be matched to the one corresponding to the eye's far point. Thus, the eye's refractive error R can be calculated by means of the Newton's lens equation.

$$zz' = -\Phi^{-2} \Rightarrow R = -\Phi^2 z \qquad (8.7)$$

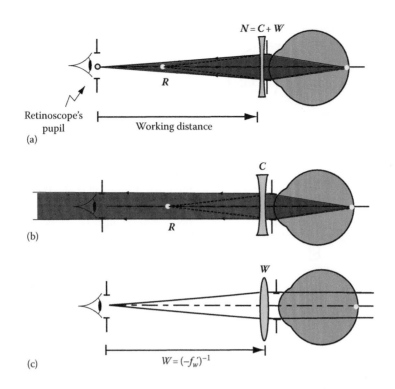

Figure 8.19 (a) The neutralization lens N brings the far point of the eye to the peephole's plane. This lens can be expressed as the sum of the compensation lens C that brings the far point to infinity (b), plus the working lens W. The working lens brings the infinite to the plane of the retinoscope, and thus, its power is the inverse of the working distance $-f'_W$ (c).

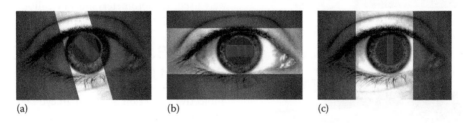

Figure 8.20 Restinoscopic reflex in an astigmatic eye. (a) Retinoscopic streak non-aligned with the principal meridians of the eye. (b) Retinoscopic streak aligned with one principal meridian (180°) of the eye. (c) Retinoscopic streak aligned with the other principal meridian (90°) of the eye.

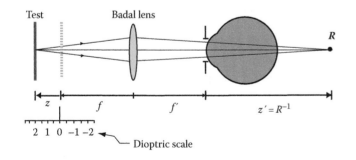

Figure 8.21 The Badal optometer. Example of the measurement of a hyperopic eye with its far point at R. The test image provided by the lens lies at the plane where R is located, and then the final image is focused at the patient's retina.

Note that as the eye's pupil lies near the image's focal point F' of the Badal lens, the eye–Badal lens' optical system is telecentric. As a result, the angular size subtended by the target and its image's size at the retina (the magnification) remain constant, irrespective of the distance between the target and the Badal lens.

When combined with an ophthalmoscope, a Badal optometer turns into an objective optometer. Thus, the image quality of the target lying at the patient's retina can be assessed externally by a practitioner by focusing the test (see Figure 8.22).

Although the first approach of this technique was proposed nearly 80 years ago (Fincham 1937), many modern autorefractometers use the same principle. In this case, the focusing and detection of the image at the retina is performed automatically by an opto-mechanical system.

8.4.2.2 Modern autorefractometers

Fundamentally, there are two types of autorefractors that derive objective refraction: image quality analysis and retinoscopy. In the first case, a Badal system is employed. Interestingly, to provide a more accurate focalization of the target upon the retina, most of the latest autorefractors employed today use a method proposed in the sixteenth century: the Scheiner disc (Salvesen and Kohler 1991). A Scheiner disc consists of a disc with two holes that is placed in front of the eye (see Figure 8.23). When viewing a distant object, the image perceived by the eye will

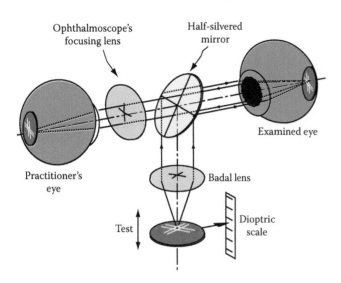

Figure 8.22 A simple objective optometer resulting as the combination of a Badal optometer and a direct ophthalmoscope. The practitioner focuses the test on the patient's eye.

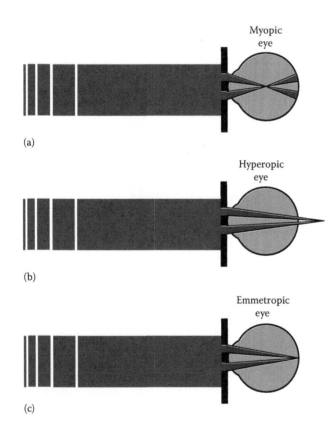

Figure 8.23 Scheiner principle to detect defocused images at the retina. (a, b) Double images are produced in myopic and hyperopic eyes. Note that in the myopic case, the images are crossed while in the hyperopic ones, they are not. (c) When the images are focused, they collapse in a single image.

be dependent on its refractive error. A myopic eye produces crossed diplopic images (see Figure 8.23a), whereas in hyperopia, the patient sees uncrossed images (see Figure 8.23b). Emmetropes or compensated eyes will see a single focused image (Figure 8.23c). In a modern autorefractor, the object test is illuminated by pairs of LEDs that are also imaged to the

pupillary plane. These effectively act as a modified Scheiner discs (Campbell et al. 2006).

In retinoscopy-based autorefractors, the retinoscopic streak sweeping is substituted by a rotating drum with alternate opaque and transparent apertures in order to create moving rectangular beams projected to the eye. The speed and direction of the movement of the reflex is detected by two photodetectors. The time difference from the slit image reaching each of the detectors allows the autorefractor to detect *with* or *against* motion, and a Badal system is used to neutralize motion along the explored meridian (Campbell et al. 2006).

Modern autorefractometers work with IR light mainly because at the near IR region (wavelengths around 900 nm), the reflectance from the eye's fundus is higher than in the visible. However, the optical stimulus for the patient is usually a visible fixation target that is introduced into the line of sight of the eye by means of a beam splitter. Fogging drives the fixation target in the hyperopic direction aiding the patient to relax the accommodation.

The IR image on the retina is reflected out of the eye and is imaged on an IR detector. The output signals are digitally analyzed and used to feedback the mechanical drivers to allow the best focus of the test at the retina. Several meridians are measured in a fraction of a second depending on the models.

To compensate for chromatic aberrations of the eye, resulting in a more hyperopic refraction than would be if visible light were used, IR optometers must be calibrated. Usually this task is performed simply with subjective measurements of refraction.

8.5 SUBJECTIVE REFRACTION EQUIPMENT

Subjective refraction is part of an eye examination where an optometrist or a physician determines the eye's refractive state. The term subjective means that this process necessarily does involve the patient's responses. Typically, the patient looks through trial lenses at a test chart. The eye care professional then changes lenses and other settings while asking the patient for subjective feedback on which settings gave the best vision. Several and well-established exam routines (Benjamin and Borish 2006) are performed with the equipment described in this section.

8.5.1 VISUAL ACUITY TESTING DEVICES, VISUAL ACUITY TEST CHARTS, AND OPTOTYPE PROJECTOR AND DISPLAYS

The measurement of VA is a psychophysical procedure that is a fundamental part of the clinical examination of the visual system. In fact, it is the standard parameter by which the outcome of most clinical trials are judged. VA is tested by requiring the patient to identify letters or symbols, called optotypes, on a printed or projected chart from a set viewing distance. Test charts are based on the measurement of the *minimum separable* that is the smallest visual angle at which two separate objects can be distinguished. Optotypes are normally represented at maximum contrast, that is, as black symbols against a white background, with a minimum luminance of around 80 cd/m² (Sheedy et al. 1984). The distance between the person's eyes and the testing chart is typically 5 or 6 m to approximate infinity in the

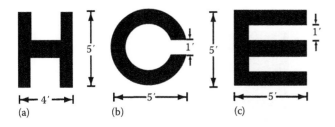

Figure 8.24 Optotypes construction. (a) Snellen letter. (b) Landolt C. (c) Tumbling E.

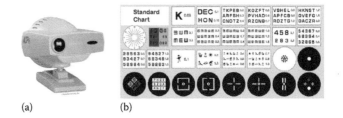

Figure 8.26 (a) Optotype projector and (b) different charts. (Photograph supplied by Topcon España S.A., Barcelona, Spain.)

measurement of *far* VA or at a defined reading distance (normally 30 or 40 cm) for the measurement of *near* VA.

VA is commonly specified as Snellen fractions in which the numerator is the testing distance and the denominator is the distance at which the minimum detail of the optotype subtends 1 min of arc. If distances are expressed in meters, an accepted *normal* VA is 6/6 (equivalent to 20/20 if distances are expressed in ft), although VA in most people is generally higher.

Most of the currently produced test charts in the Western world use non-serif Latin letters designed in a 5 × 4 box (see Figure 8.24a). Others utilize either a series of Cs or Es inscribed in a 5 × 5 box (see Figure 8.24b and c) arranged such that the gaps in the C, or the stems of the E, point in different directions.

Stating VA as a decimal number is the standard in European countries, as required by the norm EN ISO 8596. It is computed as VA= 1/MAR (MAR: minimum angle of resolution in minutes of arc). It means that, for example, a person that has 0.5 VA possesses half the resolution (and, therefore, needs twice the size to discern the optotype) than the lower limit of *normal* vision.

Many VA charts contain several lines of optotypes in which the letter sizes follow a geometrical progression, that is, the letters on each line are *t* times larger than those of the previous line that corresponds to a higher VA (see Figure 8.25a). The spacing between the letters and lines is commonly selected to overcome the so-called crowding effect that reduces the recorded VA (Ricci et al. 1998). A more homogeneous sampling of VA can be obtained when the size of the letters and the spacing between lines in a chart are ordered following the logarithm to the base

10 of decimal VA. Thus, in the Log MAR (from Logarithm of the MAR) VA scale, the 6/6 (or 20/20) line is LogMAR 0.00, and the 6/60 (20/200) line is LogMAR 1.0 (see Figure 8.25b). It means that the LogMAR scale converts the geometric sequence of a traditional chart to a linear scale in which increasing values in the LogMAR scale indicate vision loss, while decreasing values denote better VA. Although this scale is rarely used clinically, it has been recommended for research reports (Grosvenor 2007).

VA is affected by several factors such as the size of the pupil. Optical aberrations of the eye decrease VA as the pupil's diameter increases that occurs in scotopic conditions. On the other hand, for small pupils (lower than 2 mm), image sharpness may be limited by the diffraction of light by the pupil. The pupil's diameter that is generally best for VA in normal, healthy eyes is around 3–4 mm. For this reason, the ambient luminance should be controlled. Moreover, in order to obtain reliable and reproducible measurements of VA, care must be taken that viewing conditions correspond to the standard (Ricci et al. 1998).

Nowadays, printed charts (externally or internally illuminated) are still commercially available, but they are being gradually substituted by either projected charts or computer-controlled displays. Chart projectors are in fact slide projectors (see Figure 8.26a) that use the Köler illumination system already described in Section 8.1, in which, instead of a slit, they project a slide that contains several lines of optotypes (see Figure 8.26b). These systems can incorporate many charts and tests (up to 30 in many models) in a single compact unit. Optotype lines can be masked individually and can also be used with red-green overlay. In this case, the illuminance of the chart is dependent on both the intensity of the projector and the reflectivity of the screen. Metallic (aluminum) diffuser panels are often employed as screens in order to retain the polarization of the reflected light that is needed in some clinical tests (for instance, in those that need binocular dissociation).

Modern optotype units use computer-controlled LCD or LED displays instead of a slide projectors (see Figure 8.27), allowing the use of more than a hundred tests, including R/V tests, polarized tests, contrast tests, color vision, and low vision tests. These units are self-contained nearly flat-panel screens that can hang on the wall or sit on a small shelf or cabinet at the end of the refracting lane.

8.5.2 PHOROPTERS AND TRIAL CASE LENSES

Phoropters, also known as refractive units, are instruments designed to provide the eye care professional a comfortable way of placing trial lenses (and other optical elements described in the following text) in front of the patient's eye during the

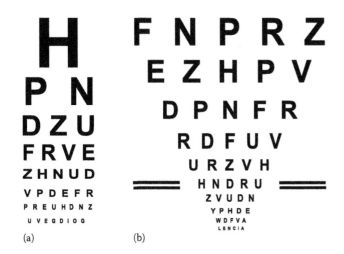

Figure 8.25 Visual acuity charts: (a) Snellen chart. (b) Bailey-Lovie (logMAR chart).

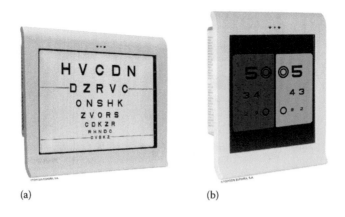

(a)   (b)

Figure 8.27 (a, b) Computer-controlled LCD displays. (Photograph supplied by Topcon España S.A., Barcelona, Spain.)

subjective refraction. By changing these lenses, the examiner is able to determine the spherical power, cylindrical power, and cylindrical axis necessary to correct the patient's refractive error.

The spherical and cylindrical lenses within a phoropter are mounted on several discs allowing for the resulting power in front of the eye to be changed in 0.25D or 0.12D steps, typically in the range of −19D to +16.5D for spherical lenses and from 0D to −6D for cylindrical lenses. Control of cylinder axis and power is usually performed by concentric control knobs within a double-calibration axis through the full 360° range. The axis scale reads in 5° increments (see Figure 8.28). The phoropters also include prismatic lenses (Risley prisms used to analyze binocular vision problems) and ±0.25D Jackson cross-cylinder that is automatically aligned with the cylinder axis. In some models for near vision test, the main optical axes of both lenses converge at 40 cm in front of the eyes by means of a tilt mechanism. For the binocular balance test, a rotary prism and polarizing filters are usually available. The auxiliary disc incorporates a retinoscopic lens of +1.50D power, Maddox rods, and a pinhole.

Automated phoropters are compact units with modern design that are remotely controlled by the practitioner using a personal computer or a *tablet* device. Therefore, they do not have knobs or manually operated parts in the phoropter's head, providing comfort for user and patient (see Figure 8.29). Computerized phoropters can include data links to an automated focimeter (lensmeter) and/or autorefractor.

A trial frame is an adjustable spectacle frame that includes cells into which all the various lenses (trial lenses), required to measure a patient's ametropia and accommodation, can be placed. A trial lens set provides around 270 lenses (with a standard 38 mm diameter lens size) packaged in a briefcase. Spherical and cylindrical lenses (both positive and negative) are included in pairs to cover powers in almost the same range and steps as by phoropters. Positive and negative lenses are differentiated by the color of the lens' rim: red for negative lenses and black (or green) for positive lenses. Auxiliary lenses are also included in the set.

In some refraction routines (e.g., with presbyopic patients), the trial frame is preferred for the final determination of the near addition, as the test can be performed at the patient's preferred working distance and position.

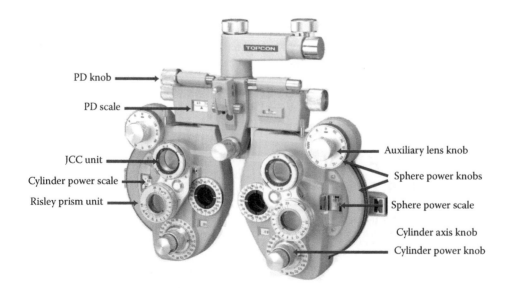

Figure 8.28 Manual phoropter parts. (Photograph supplied by Topcon España S.A., Barcelona, Spain.)

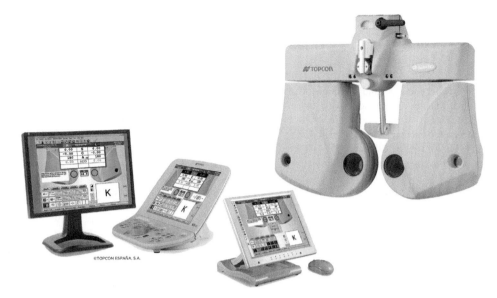

Figure 8.29 Automated phoropter parts. (Photograph supplied by Topcon España S.A., Barcelona, Spain.)

8.5.3 CONTRAST SENSITIVITY TESTS

The contrast sensitivity function (CSF) is a measure of the threshold contrast that the subject needs to see different stimuli with spatially varying frequencies. It is considered a more complete metric than VA. In fact, the CSF is nowadays considered a routine clinical tool in the optical quality assessment of the eye and in the diagnosis of several pathologies (Norton et al. 2002).

In clinical practice, the CSF is generally measured by means of optotypes of different contrasts, such as the Pelli-Robson (Pelli et al. 1988) chart (see Figure 8.30a) or by means of sinusoidal gratings of different spatial frequencies, orientation, and contrast (see Figure 8.30b and c; Ginsburg 1996). The main difference between these tests is that an optotype in the Pelli-Robson chart contains a wide range of spatial frequencies whose relative weights depend on the type of letter and its size, while a sinusoidal grating evaluates the response of the visual system to a single spatial frequency. These days, the most popular commercial tests for measuring CSF by means of sinusoidal gratings are Functional Acuity Contrast Test and the Vector Vision CSV-1000 (VectorVision, Greenville, OH). These tests commonly use nine patches for each spatial frequency, but they differ in the specific spatial frequencies evaluated, the step contrast sizes and ranges, and the psychophysical method to achieve the threshold.

Since tablets (mobile computers with display) appeared, new applications (apps) have been proposed for the measurement of both VA and CSF. The main advantage of these devices is their portability that could facilitate the standardization of rigorous visual screenings (Kollbaum et al. 2014) (Rodríguez-Vallejo et al. 2015).

REFERENCES

Benjamin, W. and Borish, I. Monocular and binocular subjective refraction. In *Borish's Clinical Refraction*, ed. W.J. Benjamin (St. Louis, MO: Elsevier Publications, 2006), pp. 790–898.

Bennett, A.G. and Rabbetts, R.B. *Clinical Visual Optics*, 2nd edn. (Oxford, U.K.: Butterworth-Heinemann, 1989).

Caballero, M.T., Furlan, W.D., Pons, A., Saavedra, G., and Martínez-Corral, M. Detection of wave aberrations in the human eye using a retinoscopy-like technique, *Opt. Commun.*, 260 (2006): 767–771.

Campbell, C.E., Benjamin, W.J., and Howland, H.C. Objective refraction: Retinoscopy, autorefraction, and photorefraction. In *Borish's Clinical Refraction*, ed. W.J. Benjamin (St. Louis, MO: Elsevier Publications, 2006), pp. 682–764.

Corbett, M.C., Rosen, E.S., and O'Brart, D.P.S. *Corneal Topography—Principles and Applications* (London, U.K.: BMJ Books, 1999).

Douthwaite, W. *Contact Lens Optics & Lens Design* (Philadelphia, PA: Elsevier, 2009).

Ehrlich, J., Radcliffe, N., and Shimmyo, M. Goldmann applanation tonometry compared with corneal-compensated intraocular pressure in the evaluation of primary open-angle Glaucoma. *BMC Ophthalmol.*, 12 (2012): 12–52.

Fincham, E.F. The coincidence optometer. *Proc. Phys. Soc.* 49 (1937): 456–468.

Ginsburg, A.P. Next generation contrast sensitivity testing. In *Functional Assessment of Low Vision*, eds. B. Rosenthal and R.E. Cole (St. Louis, MO: Mosby Year Book Inc., 1996), pp. 77–88.

Grolman, B. A new tonometer system. *Am. J. Optom. Arch. Am. Acad. Optom.*, 49 (1972): 646–666.

Grosvenor, T. *Primary Care Optometry* (St. Louis, MO: Elsevier, 2007).

Henson, D.B. *Optometric Instrumentation*, 2nd edn. (Oxford, U.K.: Butterworth-Heinemann, 1996).

Hirsch, M.J. and Wick, R.E. *Principles of Optometry Series*, Vol. 1 (Philadelphia, PA: Chilton, 1968).

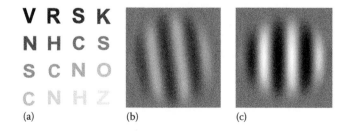

(a)   (b)   (c)

Figure 8.30 Contrast sensitivity function tests. (a) Pelli-Robson chart. (b, c) Patches of a single spatial frequency with different contrast and orientation.

Horner, D.G., Salmon, T.O., and Soni, P.S. Corneal topography. In *Borish's Clinical Refraction*, 2nd edn., ed. W.J. Benjamin (St. Louis, MO: Elsevier Publications, 2006), pp. 645–681.

Keeler, C.R. A brief history of the ophthalmoscope. *Optom. Pract.* 4 (2003): 137–145.

Kollbaum, P.S., Jansen, M.E., Kollbaum, E.J., and Bullimore, M.A. Validation of an iPad test of letter contrast sensitivity. *Optom. Vis. Sci.*, 91 (2014): 291–296.

Michaels, D.D. *Visual Optics and Refraction: A Clinical Approach*, 2nd edn. (St. Louis, MO: CV Mosby, 1980), pp. 357–376.

Millodot, M. A centenary of retinoscopy. *J. Am. Optom. Assoc.*, 44 (1973): 1057–1059.1

Norton, T.T., Corliss, D., and Bailey, J.E. *The Psychophysical Measurement of Visual Function* (Burlington, MA: Butterworth–Heinemann, 2002).

Ojeda-Castañeda, J. Foucault, wire, and phase modulation tests. In *Optical Shop Testing*, 3rd edn., ed. D. Malacara (Hoboken, NJ: John Wiley & Sons, 2007).

Pelli, D.G., Robson, J.G., and Wilkins, A.J. The design of a new letter chart for measuring contrast sensitivity. *Clin. Vis. Sci.* 2 (1988): 187–199.

Remón, L., Benlloch, J., and Furlan, W.D. Corneal and refractive astigmatism in adults: A power vectors analysis. *Optom. Vis. Sci.*, 86 (2009): 182–186.

Ricci, F., Cedrone, C., and Cerulli, L. Standardized measurement of visual acuity. *Ophthalmic. Epidemiol.* 5 (1998): 41–53.

Rodríguez-Vallejo, M., Remón, L., Monsoriu, J.A., and Furlan, W.D. Designing a new test for contrast sensitivity function measurement with iPad. *J. Optom.*, 8 (2015): 101–108.

Safir, A. Retinoscopy. In *Duane's Clinical Ophthalmology*, eds. W. Tasman and E.A. Jaeger (Philadelphia, PA: JB Lippincott, 1982).

Salvesen, S. and Kohler, M. Automated refraction: A comparative study of automated refraction with the Nidek AR-1000 autorefractor and retinoscopy. *Acta Ophthalmol.*, 69 (1991): 342–346.

Sheedy, J.E., Bailey, I.L., and Raasch, T.W. Visual acuity and chart luminance. *Am. J. Optom. Physiol. Opt.*, 61 (1984): 595–600.

Swiegerling, J. *Field Guide to Visual and Ophthalmic Optics* (Bellingham, WA: SPIE, 2004).

Yanoff, M. and Duker, J.S. *Ophthalmology*, 3rd edn. (Maryland Heights, MO: Mosby Elsevier, 2009).

9 Instrumentation for adaptive optics

Chris Dainty

Contents

9.1 INTRODUCTION AND CHAPTER OUTLINE

All optical systems—the eye is no exception—suffer from aberrations that degrade the image quality. In the case of the human eye, the aberrations have both a fixed and time-varying component. Individuals have different aberrations, and their left and right eyes also differ, and on top of the average aberration, there is a time variation. In the technique of adaptive optics (AO), the aberrations are partially compensated using an optical modulator such as a reflective deformable mirror (DM) or reflective or transmissive liquid crystal spatial light modulator (LCSLM). In the ideal case that the compensation is perfect, the optical performance is then limited only by diffraction, that is, the angular resolution according to the Rayleigh criterion is $1.22\lambda/D$, where λ is the mean wavelength and D is the diameter of the entrance pupil.

There are two reasons why one might wish to employ an AO system to improve the optics of the eye: (1) for imaging the retina and (2) for improved visual perception. Most instruments for retinal imaging, such as flood-illuminated fundus cameras, scanning laser ophthalmoscopes (SLOs), and optical coherence tomographers (OCTs), use a dilated pupil, but the aberrations are at their worst for this situation. By incorporating an AO system into the instrument—a nontrivial modification that adds significant practical complexity—one can achieve close to diffraction-limited imaging. As a consequence, cellular imaging becomes a possibility, not just the photoreceptors (cones and rods) but also other cells such as the retinal pigment epithelium and ganglion cells. With AO, imaging the retina becomes true microscopy, and many of the advanced techniques now employed in modern microscopy are applicable to retinal imaging.

A second application of AO in the eye is to provide improved vision. When AO was first demonstrated in the eye, there was much publicity surrounding the possibility of "super-vision"—visual acuity beyond the normal value of 6/6 (or 20/20)—and while

this is technically feasible in a very restricted field of view with a permanently dilated pupil, there is no practical value for most individuals with normal vision. There may be a benefit for those with below-average visual acuity in some circumstances. However, an important practical value of AO is for so-called visual simulation, either for experiments on visual perception or for demonstrating to patients the potential of various surgical procedures such as laser refractive surgery or of multifocal intraocular lenses (IOLs). Visual simulators not only allow us to compensate the eye's aberrations but also can add certain aberrations (such as spherical aberration) in a controlled way, so that we can investigate potential improvements to visual performance, for example, by a customized IOL.

AO was first proposed in astronomy, to enable ground-based telescopes to overcome the deleterious effects of atmospheric turbulence [1], and for many years the technique was restricted to large engineering teams in astronomy and military applications [2]. In fact, AO is rather straightforward in principle and is now increasingly used in vision science, microscopy, and other applications. Of the several books currently available, Hardy [3] is authoritative, Tyson [4] is comprehensive, and Porter et al. [5] cover applications in vision science.

One difficulty in building and operating an AO system lies in the fact that one is usually aiming for diffraction-limited performance (Strehl ratio > 0.8); building *any* diffraction-limited system at visible wavelengths is challenging, as root-mean-square wavefront errors smaller than approximately 40 nm have to be achieved. Good optical design and engineering are a necessity, not an optional extra. Another practical difficulty is that AO systems end up being quite cumbersome and complex, involving a large number of optical elements.

The three key components of a conventional AO system are the wavefront sensor, the wavefront corrector, and the control system. In some implementations—wavefront sensorless AO—there is no explicit wavefront sensing device, but rather the control signal is some measure of the image quality. In this chapter, we shall review the technology used in these three components for AO systems that are being used in vision science. First, we review the basic principles of AO.

9.2 PRINCIPLES OF ADAPTIVE OPTICS

Figure 9.1 shows the basic layout of a generic AO system designed for closed-loop retinal imaging: this is grossly simplified to illustrate the key components, and for visual simulation, the image plane (lower right) would be replaced by a visual display viewed by the subject. In Figure 9.1, it is assumed that light is emanating from a point in the retina; this point (or point-like) source is achieved by illuminating the eye by a narrow beam of light from a near-infrared laser or superluminescent diode, focused on the retina, at an eye-safe level, typically a few microwatts depending on the wavelength. It is possible to use visible light to generate this point-like source, although this is less comfortable for the subject. Light from this "laser guide star" exits the full (dilated) pupil of the eye, becoming aberrated in the process. The aberrations of this wavefront are compensated (cancelled), at least to some degree, by a wavefront corrector, which in Figure 9.1 is shown as a DM. This compensated wavefront can then form a high-quality image. Alternatively, the compensated wavefront enables a subject to view a display with close to diffraction-limited retinal image quality. The function of the wavefront sensor is to measure continuously the residual aberration, so that suitable signals can be sent to the wavefront corrector via a control system, enabling the wavefront compensator to continuously correct for time-varying changes in the eye's aberration. In this classical AO closed-loop architecture, there are three key components: (1) the wavefront sensor, (2) the wavefront corrector, and (3) the control system. Note in this system that the wavefront sensor acts as largely a null sensor, detecting only small departures from a flat,

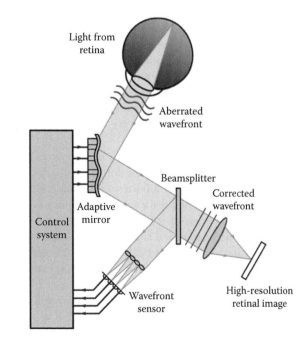

Figure 9.1 The principle of closed-loop adaptive optics in the eye. (Courtesy of Joseph Carroll; Devaney, N. et al., *Appl. Opt.*, 47, 6550, 2008.)

Light from retina

Aberrated wavefront

Beamsplitter

Corrected wavefront

Adaptive mirror

Control system

Wavefront sensor

High-resolution retinal image

unaberrated wavefront. This is quite different conceptually from a wavefront aberrometer, in which absolute measurements of large wavefront errors are required.

Before describing the three key components, we should ask: are they all absolutely necessary? In fact, considerable ingenuity has been devised by various researchers to eliminate one, or indeed all, of these components. For example, in the early days of astronomical AO, when control systems of sufficient bandwidth were difficult or expensive to implement, efforts were made to match the wavefront sensor and wavefront corrector so that the signals from the wavefront sensor could be directly amplified and fed into the wavefront corrector, with no further processing [6]: this elegant solution was based on linking a wavefront sensor that measured mirror modes directly to the mirror actuators. However, in practical terms, the control system, especially for AO in the eye, is very straightforward and low cost, so there is little advantage in eliminating it.

In contrast, using wavefront sensor can be problematical in some cases: is it possible to dispense with it and simply use image data to optimize the wavefront corrector? The answer is a qualified "yes," and some aspects of wavefront sensorless AO, sometimes called "wavefront control," are discussed in Section 9.3.

Finally, is it possible to correct for the eye's aberrations without using any AO hardware (sensor, corrector, control system)? By postprocessing image data taken using optical coherence tomography, some progress in aberration correction has been achieved [7–9], but it is limited to imaging, rather than for visual simulation.

9.3 WAVEFRONT SENSORS

The purpose of the wavefront sensor in an AO system is to provide a good estimate of the wavefront reflected from (or transmitted by) the wavefront corrector. It therefore has to be capable of estimating wavefronts emanating from the eye and, equally important, those imposed by the corrector. If the wavefront sensor fails to detect wavefront shapes (modes) that the corrector can produce, then inevitably these modes will build up, starting from system noise, and a poor image quality will result, even though the wavefront sensor records no errors. This is why is it crucial in any AO system to ensure the wavefront sensor is designed properly (detects all possible modes that can arise) and why the performance of an AO system should be measured by the attained system point spread function (or modulation transfer function), and not by the magnitude of the wavefront sensor residual error, which might provide a false indication if it is designed poorly. There is a considerable literature on wavefront sensor design (see, e.g., [10]).

The aberrations of the eye are frequently described by Zernike polynomials, with lower-order Zernike coefficients being associated with primary aberrations such as defocus, tilt, spherical aberration, coma, and astigmatism. The "pure" forms of these aberrations are better quantified by the Seidel coefficients (of the so-called aberration polynomial), but the lack of exact rotational symmetry in the eye leads to a preference for the Zernike description. A wavefront sensor must be able to estimate the magnitude of the Zernike coefficients in a wavefront and in addition be able to estimate the magnitude of terms contributed by the wavefront corrector. In the eye, unlike in astronomy, we are not (at least relatively speaking) short of light in the wavefront sensor: clearly light levels have to be eye-safe [11], but given the high sensitivity of modern detectors, there is rarely a shortage of signal in the wavefront sensor. For this reason, in vision science applications, wavefront sensors are usually overspecified and simply use a brute force design in which many more measurement points (or modes) than are strictly necessary are estimated.

The Shack–Hartmann device is by far the most widely used wavefront sensor in both retinal imaging and visual simulation; Figure 9.2a shows a schematic diagram, and Figure 9.3 shows a high-quality spot pattern (a) and wavefront estimate (b). It consists of a lenslet array placed one focal length in front of a pixelated detector, typically a CCD or CMOS detector. An incident plane wave produces an array of focused spots on the detector, spatially distributed in a regular array that mirrors the geometry of the lenslets, as shown by the dashed lines in Figure 9.2 and also Figure 9.3a. An aberrated wave forms a pattern of spots that deviate slightly from the reference positions

(solid lines in Figure 9.2), providing an estimate of the average wavefront slope over each lenslet.

There are a number of algorithms for reconstructing the aberrated wavefront from this array of average slope estimates [10]. However, in the AO control system, it is not necessary to estimate the wavefront; rather, the estimated average slopes are directly used in the matrix control algorithm (see Section 9.5). Displaying the residual compensated wavefront is a useful diagnostic tool, although as mentioned earlier, this estimated wavefront is only accurate if the wavefront sensor can measure all the modes present in the wavefront.

The Shack–Hartmann sensor is almost always the wavefront sensor used in vision science, but it certainly is not the only possible one. The sensitivity of the Shack–Hartmann sensor is fixed by its design (lenslet size, focal length), and it could be advantageous to have a sensor with a variable sensitivity. Both the pyramid sensor [12,13] (a wavefront slope or first-derivative sensor, like the Shack–Hartmann) and the curvature sensor [6] (a curvature or second-derivative sensor) offer the possibility of a variable sensitivity. However neither matches the practical simplicity of the

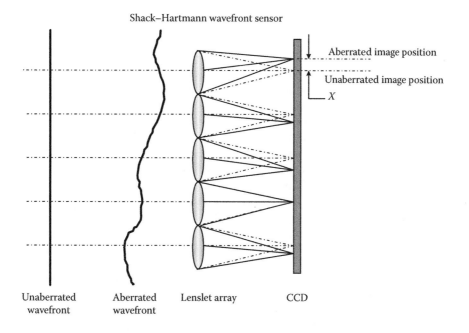

Figure 9.2 Schematic diagram of the Shack–Hartmann sensor.

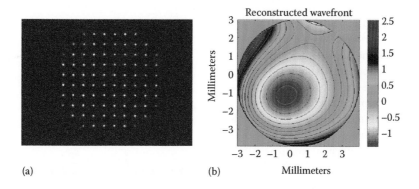

Figure 9.3 Outputs from the Shack–Hartmann sensor: (a) array of focused spots; (b) wavefront reconstructed using a modal method from the measured centroids of the spots.

Shack–Hartmann device. Other possible sensors include the phase diversity sensor [14], shearing interferometer [15], and holographic modal sensor [16].

The use of a wavefront sensor requires the presence of a known feature in the object, most commonly a point (or point-like) source. In astronomy, this is a (natural) star, or an artificially created point source, and such a guide star can be created by a laser illuminating the mesospheric sodium layer of Earth's atmosphere. In the case of the eye, a point source is created on the retina by focusing a laser, superluminescent diode or light-emitting diode, but this is not always desirable or satisfactory. Unless it has a very long wavelength (>900 nm approximately), it is visible to the subject, which is a problem for carrying out experiments of visual perception [17]. In that case, the sensing light has to be switched off just before the experiments are performed and switched on again afterward so that the level of wavefront compensation can be checked; in practice the compensation, without the wavefront sensing arm operating, will drift during the time of the psychophysical measurements. There is also a particular problem of wavefront sensing in rodents, whose retina-to-eye-length ratio is much larger than in humans, which causes the reference spot to be ill-defined, both laterally and in depth. A third potential problem with the use of a wavefront sensor is the fact that non-common-path errors (i.e., optical aberrations introduced after the beam splitter in Figure 9.1) are not measured, and hence not compensated.

For these reasons, there have been extensive studies of the use of image sharpening instead of wavefront sensing; this is referred to as "wavefront sensorless" AO. In essence, the wavefront modulator is adjusted until some metric of the image, such as sharpness, is maximized (or minimized). Clearly, as the number of degrees of freedom of the modulator, for example, the number of actuators on the DM, increases, this process becomes protracted. Although numerous standard maximization techniques have been suggested, including genetic algorithms and stochastic parallel gradients methods, the approach suggested by Booth [18] appears to offer the greatest hope for finding the global maximum in a guaranteed number of iterations. It should be noted that the wavefront sensorless approach is being widely adopted in AO microscopy and imaging through highly scattering media, where it is termed "wavefront control."

9.4 WAVEFRONT CORRECTORS: DEFORMABLE MIRRORS AND SPATIAL LIGHT MODULATORS

The choice of wavefront corrector is probably the most critical aspect of designing an AO system for imaging or visual simulation in the eye: it is typically also the single most expensive element. Since the role of the wavefront corrector (or wavefront modulator) is to compensate the time-varying aberrations of the eye, usually using signals derived from a wavefront sensor, it is crucial that it has the spatial and temporal resolution required for that task. Temporal issues are not a major problem in most AO systems for the eye, as the bandwidths encountered are relatively low, and most modulator technologies have the required bandwidth. However, the spatial requirements are more demanding.

Thibos et al. [19] measured the wavefront aberrations of 200 eyes in 100 normal young subjects and used these data to compile a computer model of the spatial wavefronts typical of the eyes. These data are invaluable for determining the required spatial performance of any wavefront corrector. The procedure for doing this is outlined in [20]: many samples of wavefronts are generated using the Thibos model, and each one is compensated, according to a criterion such as least squared residual error, by the chosen mirror technology using measured data for the influence functions of each actuator. The influence functions are measured by applying a small signal to each actuator, and measuring the resultant wavefront or phase produced, in this case using a laboratory interferometer. The set of influence functions for all actuators defines an "influence function matrix" **M**; this matrix is the crucial characteristic for any wavefront corrector and typically might be one column of Zernike coefficients for each actuator.

From a practical standpoint, commercial devices available for use as wavefront correctors are described by a set of specifications, and not by the influence function matrix or their ability to compensate the wavefront aberrations of the eye. For DMs, these specifications include the number and geometry of the actuators, the maximum stroke of each actuator, mirror diameter, and other information such as inter-actuator cross talk. For LCSLMs, the specifications include wavelength of operation, number of pixels, transmission and polarization properties, and phase stroke. Because of the differing specifications used for difference technologies, the only satisfactory way of choosing the best wavefront corrector is to carry out a simulation, along the lines of [20] for the type of aberrations one expects to compensate.

DM technologies used for AO have included
- Zonal continuous facesheet mirrors [21]
- Unimorph and bimorph mirrors [22]
- Membrane mirrors [23]
- MEMS mirrors [24]

Of these technologies, two have been favored in AO systems for the eye: magnetic actuator continuous facesheet mirrors and MEMS mirrors. The MEMS mirrors have the merit of small size and a large number of actuators, whereas the magnetic actuator mirrors have the advantage of a large stroke. All mirror technologies have the advantage that they are achromatic, introducing the same optical path difference for all wavelengths; this is clearly important in imaging or vision simulators that operate over a large spectral band.

Figure 9.4 illustrates the operation of a magnetically actuated continuous facesheet mirror (Alpao 97 actuator mirror). The main advantage of this technology is the very large stroke obtainable,

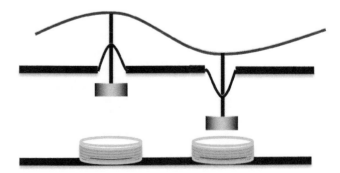

Figure 9.4 Schematic of the principle of a continuous facesheet magnetic actuator deformable mirror. (Courtesy of ALPAO sas.)

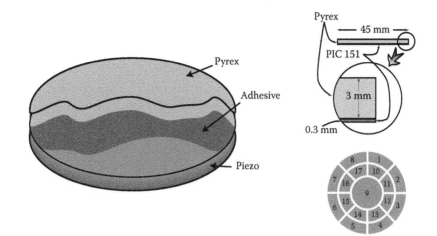

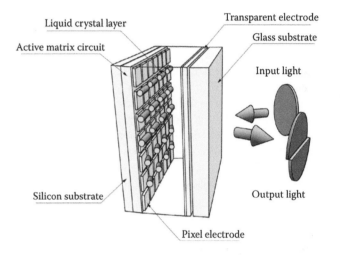

Figure 9.5 Schematic of the construction of a bimorph deformable mirror.

up to 50 μm. This is far more than is required for compensating the natural aberrations of the eye, but it can be used to add additional higher-order aberrations or low-order aberrations such as defocus and astigmatism. For many current research applications in vision science, magnetically actuated mirrors are the preferred DMs, but the cost is significant (€25K–€50K depending on specification).

Figure 9.5 shows the construction of a simple low-order bimorph mirror. In the bimorph design, actuators induce curvature in the mirror surface, and in some designs an additional actuator can introduce only focus as well. Bimorph mirrors have the potential to be lower cost than conventional push–pull continuous facesheet mirrors as they are easier to construct.

Figure 9.6 shows the principle of a MEMS mirror. MEMS mirrors have the advantage of small size (which enables compact AO systems to be built) but typically only have up to 5 μm stroke, which means they are not suitable for vision simulators.

The other major wavefront corrector technology is the LCSLM. This device introduces a voltage-dependent optical phase delay (see Figure 9.7) and can operate either in reflection (e.g., liquid crystal over silicon) or transmission [25]. These devices can contain large numbers of pixels (currently up to around 1 megapixel) and are always the modulator of choice for wavefront control in imaging through diffuse media, where a large number of degrees of freedom are required. Their main disadvantage is the low stroke (typically a phase slightly greater

Figure 9.7 Principle of operation of a reflective liquid crystal spatial light modulator. (Courtesy of Hamamatsu Corporation.)

than 2 pi), which means they have to be operated in "phase wrapping" mode and in principle with narrow bandwidth light. Since current usable devices are based on nematic liquid crystals, they operate quite slowly, although this is unlikely to be critical for AO in the eye.

9.5 CONTROL SYSTEMS

The control system involves both spatial and temporal aspects, and in general these are coupled and should be considered together. Here, for simplicity, we separate the two aspects.

Before discussing the mathematics of the control system, some remarks on the hardware required are relevant. The first controllers were custom-built circuits housed in several 19 in. racks. Nowadays, for a simple AO system using a single 100-actuator DM for the eye, a standard PC and instrument control software such as LabView is adequate for frame rates up to around 20–30 per second. Using low-cost graphics processing units or field-programmable gate arrays, frame rates up to 1000 per second can be achieved if desired provided the wavefront sensor data can be transferred fast enough [26]. In summary, control hardware is not a limiting issue for most control algorithms.

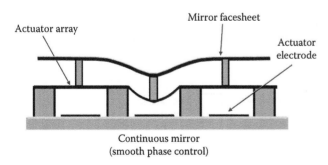

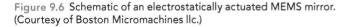

Figure 9.6 Schematic of an electrostatically actuated MEMS mirror. (Courtesy of Boston Micromachines llc.)

The operation of a wavefront sensor on an aberration described by the vector ϕ_0 (of, e.g., Zernike coefficients) can be expressed by the matrix equation:

$$\mathbf{s} = \mathbf{S}\phi_0$$

where

> $\mathbf{s}$ is the vector of sensor signals
> $\mathbf{S}$ is the wavefront sensor response matrix

For a Shack–Hartmann sensor, the components of $\mathbf{s}$ are the x- and y-slopes of the wavefront averaged over each lenslet (for N lenslets the vector has length 2N), and the elements $\mathbf{S}$ are the average x- and y-differentials of the Zernike polynomials averaged over each lenslet.

The actuator control signals form a vector $\mathbf{w}$ that is assumed to be a linear combination of the sensor signals:

$$\mathbf{w} = \mathbf{C}\,\mathbf{s}$$

where $\mathbf{C}$ is the "control matrix."

There are many approaches to finding $\mathbf{C}$ and the least squares method is outlined here. In a closed-loop AO system, the wavefront entering the sensor is the sum of the aberrated wave ϕ_0 and the mirror-induced phase $\phi = \mathbf{M}\,\mathbf{w}$, where M is the mirror influence function matrix introduced in Section 9.4. In the absence of noise, the wavefront sensor signal that results is given by

$$\mathbf{s} = \mathbf{S}\,(\phi_0 + \mathbf{M}\,\mathbf{w}) = \mathbf{s}_0 + \mathbf{B}\,\mathbf{w}$$

where $\mathbf{B} = \mathbf{S}\,\mathbf{M}$ is the "response matrix" for the mirror–sensor system. The matrix $\mathbf{B}$ plays an important role in least squares control. It is found by applying a fixed signal to each electrode of the mirror in turn and recording the sensor signals (e.g., the x- and y-slopes or simply the spot positions in a Shack–Hartmann sensor).

Since the wavefront sensor measures the net wavefront aberration after correction, we shall use this as a measure of the correction error. Using a standard least squares approach, it can be shown that the control matrix $\mathbf{C}$ that minimizes the wavefront sensor error is

$$\mathbf{C} = -[\mathbf{B}^{\mathrm{T}}\mathbf{B}]^{-1}\,\mathbf{B}^{\mathrm{T}}$$

where the quantity $[\mathbf{B}^{\mathrm{T}}\mathbf{B}]^{-1}\,\mathbf{B}^{\mathrm{T}}$ is known as the least squares inverse, or pseudoinverse, of $\mathbf{B}$.

The matrix $\mathbf{B}$ defines a mapping between the actuator signal vector ($\mathbf{w}$, length N_A) and sensor signal vector ($\mathbf{s}$, N_S), where N_A is the number of actuators and, for a Shack–Hartmann sensor, the number of lenslets is $N_S/2$. Any matrix of dimensions $N_S \times N_A$

can be written as a product of three matrices, the so-called singular value decomposition:

$$\mathbf{B} = \mathbf{U}\,\Lambda\,\mathbf{V}^{\mathrm{T}}$$

where

> $\mathbf{U}$ is an $N_S \times N_S$ orthogonal matrix
> $\mathbf{V}$ is an $N_A \times N_A$ orthogonal matrix
> Λ is an $N_S \times N_A$ diagonal matrix

We can write $\mathbf{U} = (\mathbf{u}_1, \mathbf{u}_2, \mathbf{u}_3, \ldots)$, $\mathbf{V} = (\mathbf{v}_1, \mathbf{v}_2, \mathbf{v}_3, \ldots)$, where $\mathbf{u}_i$ and $\mathbf{v}_i$ form complete sets of modes for the sensor signal and mirror control spaces, respectively, and Λ is a diagonal matrix of singular values λ_i of the matrix $\mathbf{B}$. Each nonzero value of λ_i relates the orthogonal basis component $\mathbf{v}_i$ in $\mathbf{w}$, the control signal space, to an orthogonal basis component $\mathbf{u}_i$ in $\mathbf{s}$, the sensor signal space.

We can now distinguish three possible situations:

1. *Correctable modes*: For these modes, $\lambda_i \neq 0$, the actuator control signal $\mathbf{w} = \mathbf{v}_i$ results in the sensor signal $\mathbf{s} = \lambda_i\,\mathbf{u}_i$. This mode can be corrected by applying the actuator model $\mathbf{w} = \lambda_i^{-1}\,\mathbf{v}_i$, the singular value λ_i being the sensitivity of the mode (clearly we not want λ_i to be too small).

2. *Unsensed mirror modes*: These are modes $\mathbf{v}_i$ for which there is no nonzero λ_i. Unsensed mirror modes would cause a big problem in an AO system and are to be avoided if at all possible by proper design of the wavefront sensor: some of the early zonal AO systems suffered from this defect, producing the so-called "waffle" modes in the mirror.

3. *Uncorrectable sensor modes*: These are modes $\mathbf{u}_i$ for which there is no nonzero λ_i. Nothing the mirror does affects these modes, and arguably there is no point in measuring them.

The control matrix $\mathbf{C}$ is now given by

$$\mathbf{C} = -\mathbf{B}^{-1*} = -\mathbf{V}\,\Lambda^{-1*}\,\mathbf{U}^{\mathrm{T}}$$

where Λ^{-1*} is the least-squares pseudoinverse of Λ formed by transposing Λ and replacing all nonzero diagonal elements by their reciprocals λ_i^{-1}, which now can be interpreted as the gains of the system modes. From a practical point of view, one discards modes that have a small value for λ_i as clearly they are susceptible to noise.

The least squares approach is simple and does not require much prior information about either the AO system or the statistics of the aberration to be corrected. It also has the advantage that the key matrix $\mathbf{B}$ can easily be measured. But prior knowledge, for example, about the aberration to be corrected or the noise properties of the sensor, could be an advantage in a well-designed control system. There are several other approaches to "optimal reconstructors," originating with the work of Wallner [27], and described in [28,29].

The temporal aspects of the control are determined by the time responses of the principal components of the AO system, as illustrated in Figure 9.8: the wavefront sensor detector (integrator), the wavefront reconstructor process (time lag), the DM

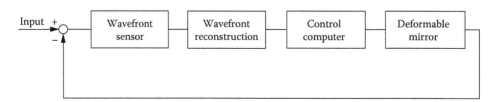

Figure 9.8 Elements of importance for temporal control.

(usually assumed to be fast compared to the others), and the control computer, which typically might implement a simple integrator. The approach to understand the temporal control aspects is through the transfer functions of each component [29].

The open-loop transfer function $G_{ol}(s)$ is the product of the transfer functions for each component:

$$G_{ol}(s) = G_{WS}(s) \times G_{WR}(s) \times G_{DM}(s) \times G_C(s)$$

And the closed-loop transfer function $G_{cl}(s)$ is given by

$$G_{cl}(s) = G_{ol}(s) \, [1 + G_{ol}(s) \, H(s)]^{-1}$$

where $H(s)$ is the feedback parameter.

The closed-loop bandwidth can be defined in a number of ways. One way is to define it as the frequency at which the closed-loop transfer function falls to the –3 dB level. However, a more informative measure is found by plotting the ratio of the input and output signals as a function of frequency, and defining the closed-loop bandwidth as the frequency at which this ratio is unity (i.e., above this frequency the system makes no correction). In general, this frequency is lower than the one given by the –3 dB definition.

9.6 OTHER CONSIDERATIONS

One of the most problematic issues in AO systems for imaging the retina are the elimination of back reflections that occur due to the illumination of retina: these back reflections are automatically damped in an optical coherence tomography (OCT) imager but do affect both AO-assisted fundus cameras and SLOs. In these systems, all-reflective optics is preferred, and if spherical mirrors are used, they have to be combined so as to minimize astigmatism [30].

A more fundamental problem of AO imaging is the lack of isoplanicity of the wavefront compensation. The isoplanatic patch is the region over which the point spread function (and wavefront aberration) is constant. The angle subtended by the isoplanatic patch varies considerably, but is in the range of 1°–5°, and consequently the region of good compensation by an AO system is in the same range. In principle, this can be increased using two DMs, conjugated to different plane of the eye (e.g., cornea and lens), but in practice this increases the system complexity and both the wavefront sensing and control become more complicated.

The process of designing an AO to assist imaging, or for vision simulation, is an iterative one and depends on the ultimate task to be implemented. Ideally, an end-to-end simulation of the system is desirable, but this is not always practical, although an approximate simulation is probably better than none at all. In practice, a frequent strategy is to select the wavefront corrector (probably the most expensive component) based on its ability to correct the aberrations to the required degree and on opto-mechanical issues such as the overall size of the optical system. Then the wavefront sensor is designed to sense all possible modes of the corrector, and finally the algorithm for the control loop is designed taking into account the required speed of correction.

If AO is to be used more widely, it has to be reduced to its simplest essentials and made as transparent as possible to the user. Current implementations of AO, using commercially available components, are cumbersome and require precise alignment. Integrated commercial AO retinal imaging [31] and vision simulation [32] systems are starting to become available.

REFERENCES

1. Babcock, H.W., The possibility of compensating astronomical seeing, *Publication of the Astronomical Society of Pacific*, **65**, 229–236 (1953).
2. Duffner, R.W., *The Adaptive Optics Revolution: A History*, University of New Mexico Press, Albuquerque, NM (2009).
3. Hardy, J.W., *Adaptive Optics for Astronomical Telescopes*, Oxford University Press, Oxford, U.K. (1998).
4. Tyson, R., *Principles of Adaptive Optics*, 3rd edition, CRC Press, Boca Raton, FL, ISBN 978-1439808580 (2010).
5. Porter, J. et al., *Adaptive Optics for Vision Science: Principles, Practices, Design and Applications*, Wiley-Interscience, Hoboken, NJ (2006).
6. Roddier, F., Curvature sensing and compensation: A new concept in adaptive optics, *Applied Optics*, **27**, 1223–1225 (1988).
7. Adie, S.G. et al., Computational adaptive optics for broadband optical interferometric tomography of biological tissue, *Proceedings of the National Academic Sciences*, **109**, 7175–7180 (2012).
8. Kumar, A. et al., Anisotropic aberration correction using region of interest based digital adaptive optics in Fourier domain OCT, *Biomedical Optics Express*, **6**, 1124–1134 (2015).
9. Shemonski, N.D. et al., A computational approach to high-resolution imaging of the living human retina without hardware adaptive optics, *Proceedings of SPIE*, **9307**, 930710 (2015).
10. Rousset, G., Wave-front sensors, in Part 2: The design of an adaptive optics system, *Adaptive Optics in Astronomy*, Ed. F. Roddier, Cambridge University Press, Cambridge, U.K., ISBN 978-0521553759 (1999).
11. IEC 60825-1:2014, Safety of laser products—Part 1: Equipment classification and requirements (2014) ICS 13.110; 31.260.
12. Ragazzoni, R., Pupil plane wavefront sensing with an oscillating prism, *Journal of Modern Optics*, **43**, 289–293 (1996).
13. Daly, E.M. and Dainty, C., Ophthalmic wavefront measurements using a versatile pyramid sensor, *Applied Optics,* **49**, G67–G77 (2010).
14. Gonsalves, R.A., Phase retrieval and diversity in adaptive optics, *Optical Engineering*, **21**, 215829 (1982). doi 10:1117/12.7972989.
15. Wyant, J.C., White light extended source shearing interferometer, *Applied Optics*, **13**, 200–202 (1974).
16. Neil, M.A.A., Booth, M.J., and Wilson, T., New modal wavefront sensor: A theoretical analysis, *Journal of Optical Society of America A*, **17**, 1098–1107 (2000).
17. Hofer, H.J., Blaschke, J., Patolia, J., and Keonig, D.E., Fixation light hue bias revisited: implications for using adaptive optics to study color vision, *Vision Research*, **56**, 49–56 (2012).
18. Booth, M.J., Wavefront sensorless adaptive optics for large aberrations, *Optics Letters*, **32**, 5–7 (2007).
19. Thibos, L.N., Bradley, A., and Hong, X., A statistical model of the aberration structure of normal well-corrected eyes, *Ophthalmic and Physiological Optics*, **22**, 427–433 (2002).
20. Devaney, N. et al., The correction of ocular and atmospheric wavefronts: A comparison of the performance of various deformable mirrors, *Applied Optics*, **47**, 6550–6562 (2008).
21. See, for example, Flexible Optical http://www.okotech.com/pdm, Imagine Eyes http://www.imagine-eyes.com/product/mirao52e/, and ALPAO http://www.alpao.com/Products/Deformable_mirrors.htm (last accessed October 17, 2016).
22. See, for example, AKA Optics http://www.akaoptics.com/bdm.php (last accessed October 17, 2016).

23. See, for example, Flexible Optical http://www.okotech.com/mmdm/ and Adaptica http://www.adaptica.com/site/en/pages/deformable-mirrors (last accessed October 17, 2016).

24. See, for example, Boston Micromachines http://www.bostonmicromachines.com/mems.htm and IrisAO http://www.irisao.com/product.ptt111.html (last accessed October 17, 2016).

25. See, for example, Hamamatsu Photonics http://www.hamamatsu.com/eu/en/product/category/3200/4015/index.html and HoloEye Photonics AG http://holoeye.com (last accessed October 17, 2016).

26. Kepa, K., Coburn, D., Dainty, J.C. and Morgan, F., High speed wavefront sensing with low cost FPGAs, *Measurement Science Review*, **8**, 87–93 (2008).

27. Wallner, E.P., Optimal wave-front correction using slope measurements, *Journal of Optical Society of America*, **73**, 1771–1776 (1983).

28. Roggemann, M.C. and Welsh, B., *Imaging Through Turbulence*, CRC Press, ISBN 9780849337871 (1996).

29. Kulscár, C. et al., Minimum variance prediction and control for adaptive optics, *Automatica*, **48**, 1939–1954 (2012).

30. Gómez-Vieyra, A. et al., First-order design of off-axis reflective ophthalmic adaptive optics systems using afocal telescopes, *Optics Express*, **17**, 18906–18919 (2009).

31. See, for example, Imagine Eyes, http://www.imagine-eyes.com/product/rtx1/ and PSI (Physical Sciences Inc.), http://www.psicorp.com/products/laser-based-sensors/compact-adaptive-optics-retinalimager-caori (last accessed October 17, 2016).

32. See, for example, http://www.voptica.com/ (last accessed October 17, 2016).

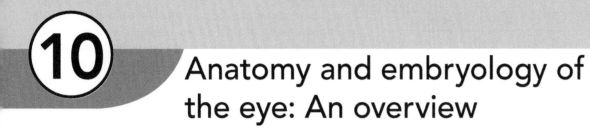

10 Anatomy and embryology of the eye: An overview

Vivian Choh and Jacob G. Sivak

CONTENTS

10.1 INTRODUCTION

The vertebrate eye can be described as a fluid-filled spherical structure designed to form and process images of the surrounding visual world. Anatomically, the eye consists essentially of three tissue layers: an outer fibrous layer made up of the sclera and cornea; an inner layer consisting largely of the retina, but including also parts of the ciliary body and iris; and an intermediate vascular layer made up of the choroid and portions of the ciliary body and iris (the uvea).

The human eye, approximately 24 mm in diameter, is a sophisticated neurosensory/optical instrument capable of detecting minute quantities of light and providing high-resolution ability. It is made up of a variety of cellular and noncellular components derived from ectodermal and mesodermal germinal sources. Its functional anatomy is best understood by understanding how it develops embryologically. Therefore, a major portion of this chapter will be devoted to eye development. Moreover, since this publication is a handbook of visual optics, the final section will focus on the anatomy and development of the optical components of the eye, namely, the cornea and the crystalline lens.

The section dealing with eye embryological development and the next concerning more specifically with the cornea and the lens have been written using a standard reference format in that each major statement and/or topic is given a primary source reference that is described in appropriate detail in the list of references provided at the end of the chapter. However, the section dealing with eye anatomy *per se* is not referenced from primary sources. Rather, the information presented in this section represents the accumulated body of knowledge provided by standard eye anatomy texts, including *Histology of the Human Eye* (Hogan et al., 1971), *Wolff's Anatomy of the Eye and Orbit* (8th edn.; Bron et al., 1997), and *The Human Eye: Structure and Function* (Oyster, 1999).

Clearly, a single chapter of many in a compendium dealing the many aspects of visual optics cannot be as detailed as a major volume dealing only with ocular anatomy. Thus, we have concentrated on a description of the ocular globe, with less emphasis on the ocular adnexa and little or no description of the visual pathways. However, we have made an effort to provide a comprehensive anatomical description of the eye that would be useful to those whose primary focus is the study of visual optics.

10.2 ANATOMY OF THE HUMAN EYE

10.2.1 OUTER TUNIC: CORNEA, SCLERA, AND LIMBUS

The sclera, comprising most of the outer layer (Figure 10.1a), consists of (1) a loose outer layer, the episclera, which is continuous with Tenon's capsule (or the fascia bulbi), a fibrous connective tissue layer that surrounds the eye and separates it from the orbital fat and the contents of the bony cavity in the skull in which the eye resides; (2) the sclera proper, a dense layer of mostly collagen fibers of varying diameters and with varying distances between them (Figure 10.1c), some elastic fibers and occasional cells, the keratocytes; and (3) and an inner layer, the

lamina fuscia, containing pigment cells (chromatophores) as well as connective tissue fibers. The sclera is one of the toughest tissues of the eye and acts as an anchor for the extraocular muscles responsible for moving the eye within the orbit.

The cornea, varying in thickness from about 0.6 to 1.1 mm, is also mainly fibrous in makeup. However, corneal anatomy is more regular than that of the sclera, a factor believed to be responsible for its transparency. The cornea consists of (1) an outer layer of stratified epithelium (typically 5 layers of cells); (2) a thin (6–9 μm) underlying collagenous layer, called the Bowman's membrane or layer; (3) a deeper stromal layer, also referred to as substantia propria, which consists largely of collagen fibers (and occasional keratocytes) arranged in about 200 lamellae

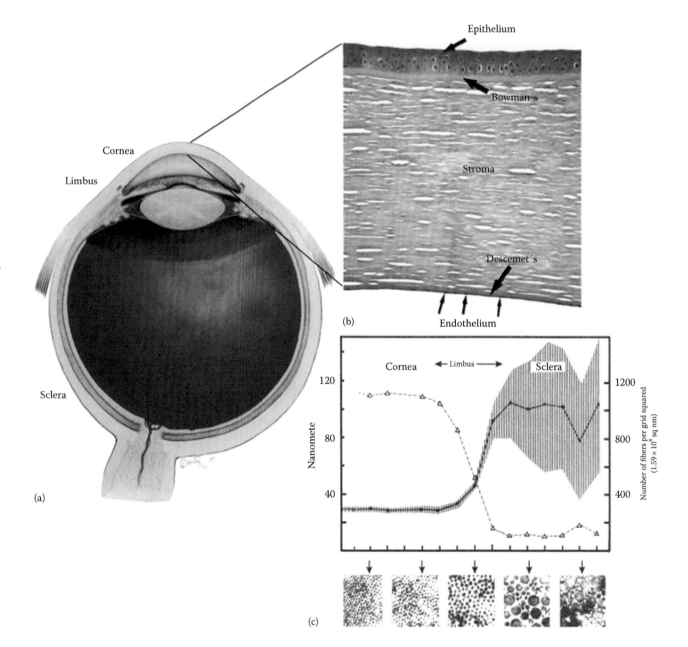

Figure 10.1 The outer tunic of the eye (a) consists of the five-layered cornea (b), the limbus, and the sclera. Collagen fibers are more organized in the cornea than in the sclera (c). ([a]: Adapted from National Eye Institute, National Institutes of Health (NEI Image Ref #NEA04); [b]: Reprinted by permission from Macmillan Publishers Ltd. *Eye*, Meeney, A. and Mudhar, H.S., Histopathological reporting of corneal pathology by a biomedical scientist: The Sheffield Experience, 27, 272–2766, Copyright 2013; [c]: Reprinted from *Exp. Eye Res.*, 21, Borcherding, M.S. et al., Proteoglycans and collagen fiber organization in human corneoscleral tissue, 59–70, Copyright (1975), with permission from Elsevier.)

parallel to the corneal surface and comprising about 90% of corneal thickness; (4) Descemet's membrane a thin (6 µm) collagenous layer that thickens with age; and (5) the endothelium, a single layer of flattened epithelial cells (Figure 10.1b). Recently it has been suggested that a "novel" sixth corneal layer, called Dua's layer, exists, named after the investigator who isolated this "layer" from the surrounding stroma. However, this "layer" is neither a separate structure, being, in reality, modified stroma, nor novel. A similar modified or alternate stroma, called the autochthonous layer, was described as early as 1876 by Rudolf Leuckart for fish, and in humans, a region of altered stroma, commonly referred to as pre-Descemet's stroma based on its location in the cornea, has already been identified (Binder et al., 1991). A distinct novel designation for this "layer" is therefore unnecessary and unwarranted.

The cells of the corneal epithelium are in a cycle of constant regeneration, with newer cells found at the basal columnar layer and older squamous shaped cells sloughing off from the surface. The cells of the corneal epithelium are bound tightly to each other by specialized cell structures such as zonula occludens in the case of the superficial squamous cells, and desmosomes and hemidesmosomes in the case of the basal columnar cells. Thus, the epithelium is a barrier to the movement of water into the cornea from the external tear layer. Small (less than a micron) protrusions and folds, microvilli and microplicae, are present on the surface of the superficial squamous cells. These are believed to help ensure the adherence of the tear film produced by the glands of the eyelids (see in the following text) to the surface of the eye.

The endothelium is made up of a single layer of metabolically active cells of uniform hexagonal shape. The cells contain numerous mitochondria and are bound to one another, but not as tightly as in the case of the epithelium. The metabolic activity of these cells is related to the maintenance of normal water content in the stroma, an important function in maintaining corneal transparency. In the adult eye, these cells do not exhibit a measurable regenerative cycle.

The rest of the cornea (Bowman's membrane, stroma, and Descemet's membrane) is made up of collagen fibers. The few cells that exist are mainly fibroblasts. In the stroma, the collagen fibers are of uniform thickness and arranged parallel to the corneal surface, with each lamella roughly orthogonal to the underlying and overlying lamellae immediately surrounding it (Figure 10.1c). Within each lamella, the collagen fibers within it are of equal diameters and are equally spaced from one another. The characteristic regularity of interfiber spacing is believed to be necessary to produce destructive interference, a prerequisite for corneal transparency. The corneal stroma has a very slow rate of turnover, if any (Rodrigues et al., 1982).

The limbus denotes the zone of transition between the cornea and sclera (Figure 10.2a). The outer layers of the limbus contain retes or ridges of fibrovascular tissue, called the "palisades of Vogt," in which limbal stem cells reside (Figure 10.2b). These stem cells are part of the natural cell cycle of proliferation, differentiation, migration, and exfoliation of the epithelial cells at the surface of the cornea. The limbus contains a circumferential vessel, the canal of Schlemm, which is involved in the drainage of aqueous humor, produced by the ciliary body. Ultimately, the humor drains into the anterior ciliary veins. The portion of

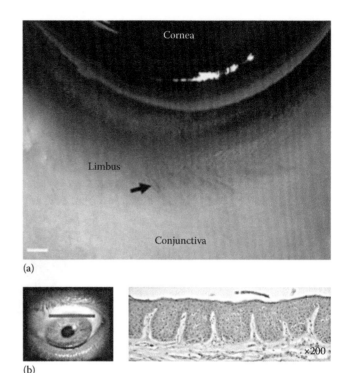

(a)

(b)

Figure 10.2 The limbus is a transition zone between the cornea and sclera (a). Stem cells reside within the palisades of Vogt (a, arrow and b). ([a]: Adapted by permission from Macmillan Publishers Ltd. *Cell Res.*, Li, W., Hayashida, Y., Chen, Y.-T., and Tseng, S.C.G., Niche regulation of corneal epithelial stem cells at the limbus, 17, 26–36, Copyright 2007; [b]: Chen, Z., De Paiva, C.S., Luo, L., Kretzer, F.L., Pflugfelder, S.C., and Li, D.-Q.: Characterization of putative stem cell phenotype in human limbal epithelia. *Stem Cells*. 2004. 22. 355–366. Copyright Wiley-VCH Verlag GmbH & Co. KGaA. Reproduced with permission.)

the limbus that is internal to the canal of Schlemm consists of a multilayered porous structure known as the trabecular meshwork. The meshwork, a narrow triangle of tissue made up of collagen and endothelial cells, is about 1.0 mm long. It extends from the peripheral cornea to the root of the iris and the forward extension of the ciliary body.

10.2.2 UVEA: CHOROID, CILIARY BODY, AND IRIS

The uvea, the middle vascular tunic of the eye, consists of a continuous bilayer of epithelium that forms the choroid and parts of the ciliary body and iris (Figure 10.3a).

The choroid (about 0.3 mm thick) is commonly described as consisting of four layers (Figure 10.3b). An outer epichoroid (suprachoroidal space) is located between the main body of the choroid and the sclera. This space provides access to the eye by the long and short posterior ciliary arteries and by the long and short ciliary nerves. The vascular layer, or stroma of the choroid, lies inward from the epichoroid and is made up of arteries and veins from the short and long posterior ciliary arteries and the anterior ciliary arteries, as well as the vortex and anterior ciliary veins. Continuing in an inward direction, the next layer is the choriocapillaris, a unique bed of capillaries, somewhat larger in width than typical of capillaries located elsewhere. The choriocapillaris is responsible for nourishing the outer (or proximal) layers of the retina. Below the choriocapillaris lies a thin (1–4 µm) intermediate layer between the choroid and retina known as Bruch's

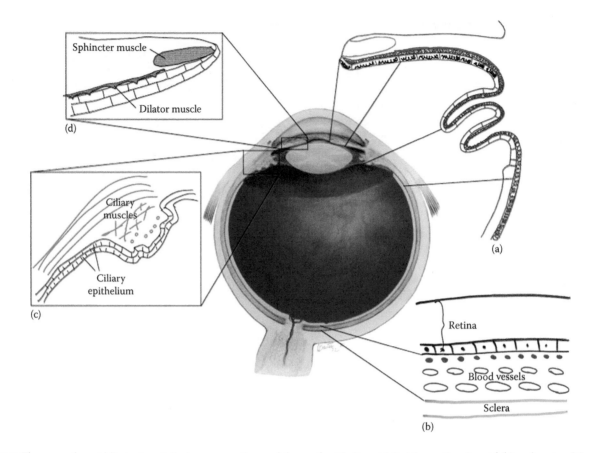

Figure 10.3 The uvea, the middle tunic, originates as a continuous bilayer of epithelium (a) that forms the choroid (b) and parts of the ciliary body (c) and iris (d). (Central image adapted from National Eye Institute, National Institutes of Health (NEI Image Ref #NEA04).)

membrane, which consists of the basement membranes of both the retina and choroid, along with other acellular components.

The ciliary body is both a vascular and muscular structure anterior to the choroid and intermediate between the choroid and the iris (Figure 10.3c). In sagittal section, the ciliary body appears as a shallow elongated triangle roughly 5.5–6.5 mm in length and about 1.0 mm in height. The innermost portion consists of two layers of epithelial cells, which are continuous with the retina. The innermost layer, adjacent to the ocular lens, is unpigmented while the external epithelial layer is pigmented. The blood vessels of the ciliary body originate from the same vessels that supply and drain the choroid and these vessels are the source of the aqueous humor of the eye, the fluid filling the anterior portions of the globe. The ciliary muscle is a smooth muscle innervated by the autonomic nervous system. It is made up of fibers with radial, oblique, and longitudinal orientations (Figure 10.3b). The muscle is responsible for the mechanism of accommodation in that its contraction releases tension on the suspensory ligaments that suspend the crystalline lens in the eye. The apical or anterior and inner portion of the ciliary body consists of ridges, approximately 70 in number, which are called the ciliary processes. The flatter posterior two-thirds of the ciliary body is referred to as the *pars plana*, while the ridged anterior third is known as the *pars plicata*. The suspensory ligaments, the *zonules*, that articulate between the lens and ciliary body travel toward the lens along the valleys between the processes.

The iris is the anterior-most portion of the uvea (Figure 10.3d). The primary function of the iris is to control the amount of light getting to the retina by constricting and dilating the pupil from diameters as small as 1.5 mm to as large as 8.0 mm, a change in area of about 30 times. The iris is described as consisting of five layers of tissue, the anterior most (closest to the cornea) of which is the endothelium, or anterior border layer, a layer consisting mainly of pigment cells and keratocytes (or fibroblasts), which is also described as being included in the iris stroma. The next layer, the vessel layer, forms the bulk of the human iris and is also referred to as the iris stroma. This layer contributes, in part, to the idiosyncratic appearance of the iris because of the large individual variation in vessel distribution. Variation in the number of pigment cells of the stroma is responsible for variation in the color of the iris. The iris stroma develops as two layers with the deeper layer extending further toward the pupil than the anterior layer. The edge of the anterior border layer overlying the stroma is referred to as the collarette. With the pupil constricted, the collarette is located approximately one-third of the distance from the pupil margin to the root of the iris. The collarette is more obvious to an observer when the stroma is less pigmented.

The arteries and veins are part of the same uveal circulation that originates from the ciliary arteries and veins referred to earlier. The arteries radiate from a circumferential vessel located at the root of the iris, which is called the major arterial circle of the iris and which represents an anastomosis of the long posterior ciliary arteries and the anterior ciliary arteries. In addition to vessels, the stroma contains a variety of cells, including pigment cells (melanocytes), clump cells (large round pigmented cells believed to be modified macrophages that stem from the pigmented epithelium of the iris), and mast cells containing heparin and histamine. Accumulations of melanocytes are referred to as iris freckles. A circumferential smooth muscle, the sphincter muscle, responsible for pupil constriction, is located in the stroma near the margin the pupil.

The deepest two layers of the iris consist of two layers of epithelial cells that are developmentally continuous with the two epithelial layers of the ciliary body. In the iris they are pigmented. However, the anterior-most layer is described as myoepithelial because these cells contain contractile elements (the iris dilator) that are responsible for dilating the pupil. The dilator muscle is located closer to the base of the iris than the sphincter. Posterior to the sphincter muscle, the two epithelial layers are simply pigmented epithelium. The pigmented epithelium of the iris curves around the pupil margin to form a thin pigmented and scalloped pupillary border (0.3–0.6 mm) as seen from the front, the iris frill.

The iris muscles are innervated by the sympathetic (via the superior cervical ganglion of the sympathetic spinal chain) and parasympathetic (via the ciliary ganglion and the short ciliary nerves) components of the autonomic nervous system.

10.2.3 RETINA

The neural retina forms the interior-most layer of the retina (Figure 10.4a). It is developmentally continuous with the epithelial components of the ciliary body and pigmented epithelium of the iris. The retina originates from both layers of the optic cup, which is itself an outpouching of the embryonic forebrain. Thus, the retina is a part of the central nervous system, as is the optic nerve. The external (or proximal) layer of the epithelium becomes pigmented during development but remains a single layer of cells and is referred to as the "pigmented epithelium of the retina" (RPE). The distal (or inner) layer of cells develops into a complex neural structure consisting essentially of three layers of cells and is commonly referred to as the neural retina. Historically, the importance of the retina was not recognized (e.g., it was often misidentified as a vascular tissue) until scholars such as Kepler (1571–1630) established that the eye is an optical instrument and the retina is its image plane.

Retinal thickness varies. It is thickest around the rim of the "fovea," a retinal depression demarking a small central retinal zone of high cone density. The thickness decreases to about 0.25 mm at the center of the foveal depression. At the anterior edge of the retina, the ora serrata, where it meets the epithelial layers of the ciliary body, the retina is about 0.1 mm thick.

From paraffin sections and light microscopy, using a standard stain for cell nuclei, the neural retina appears to consist of three layers of cells (Figure 10.4b). The outer or proximal layers of cells of the neural retina are the photoreceptors, the *rods* and *cones*, names that reflect the shape of their outer segments. Each of these segments is composed of a stack of double membrane discs that contain the visual pigment responsible for transducing light energy into neural signals. Fingerlike extensions from the cells of the RPE are found around the proximal ends of the outer segments. The cells of the RPE are involved in membrane disc breakdown and renewal, and stacked discs can be seen in the RPE cells under high magnification.

The rods and cones are oriented perpendicularly to the retinal surface. The outer segments containing the membrane discs with

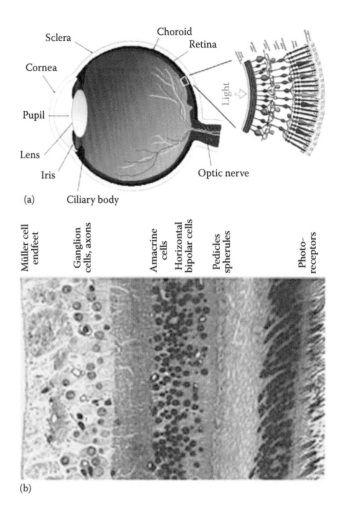

(a)

(b)

Figure 10.4 The retina, the innermost tunic of the eye, lines the posterior globe (a). It consists of 3 layers of cells, separated by two synaptic regions (b) Adapted from Kolb, H., Simple retinal anatomy, Webvision: The Organization of the Retina and Visual System, University of Utah Health Sciences Center, Salt Lake City, UT, January 31, 2012, Web. (September 21, 2014), http://webvision.med.utah.edu/book/part-i-foundations/simple-anatomy-of-the-retina.)

visual pigment are connected to the inner segments by way of a thin process containing the fibrils characteristic of a cilium, a fact that is indicative of the ciliated epithelium embryological origin of the rods and cones. The inner segment is densely populated with mitochondria and the nucleus. The nuclei of the photoreceptors localize to the same plane within the retina; this nuclei-containing layer is referred to as the "outer nuclear layer." The nucleus of each photoreceptor cells lies between the outer segment and a dendritic extension (the *myoid*) and a synaptic terminal ending (*pedicle* for cones and *spherule* for rods) that extends toward the inner globe. Thus, the vertebrate retina is inverted, such that light must penetrate the other layers of the retina before reaching the rods and cones.

The rods and cones synapse with the cells located in the "bipolar cell" (middle) layer of the retina. In addition to the nuclei of the bipolar cells, this *inner nuclear layer* also contains nuclei of the "horizontal cells" and the "amacrine cells." As is the case in the central nervous system, the retina (and the optic nerve) contains non-neural cells, the glial cells, the most prominent of which, in the retina, are *Müller cells*. Müller cells are oriented

Fundamentals

perpendicularly in the neural retina and extend its full thickness. The Müller cell nuclei are also found in the bipolar cell layer of the retina.

The bipolar cells extend both proximately (toward the receptors) and distally (toward the ganglion cells) as their name suggests. They are the primary neural link between the rods and cones and the third layer of retinal cells, the *ganglion cells*. It is the axons of the ganglion cells that form the bulk of the optic nerve fibers that travel from the eye to the brain.

While the rods and cones and the bipolar and ganglion cells are responsible for transmission of visual excitation along an axis perpendicular to the retinal surface, the horizontal cells and most of the amacrine cells provide horizontal or transverse retinal neural transmission, ultimately acting as modulators of the transmission of the visual signal to the visual cortex. The horizontal cells synapse with the rods and cones and the bipolar cells, while the amacrine cells synapse with the ganglion cells and the bipolar cells. While rods and cones represent the photoreceptors involved in vision, within the retina is a third set of nonvisual photoreceptors, known as the *intrinsically photosensitive retinal ganglion cells* (ipRGCs). The discovery of ipRGCs arose from observations that photoentrainment was still possible in individuals who were blind from degenerated rods and cones (Czeisler et al., 1995). Intrinsically photosensitive RGCs contain a different pigment, melanopsin, and their nuclei are localized to the ganglion cell layer, although some of these cells can be displaced to the inner nuclear layer (Provencio et al., 2000). At present, 5 subtypes of ipRGCs, with different morphologies and dendritic fields, are known to exist (Schmidt et al., 2011). Although the roles for ipRGCs are still being clarified, ipRGCs are believed to be involved in photoentrainment, the pupillary light response, and may even play a role in low visual acuity functions.

As mentioned earlier, nuclear staining of the neural retina shows up as three layers of nuclei. The outer layer contains the nuclei of the rods and cones, and the middle (the inner nuclear layer) contains the nuclei of the bipolar cells, the horizontal cells and the amacrine cells (as well as Müller cell nuclei), while the innermost layer (known as the ganglion cell layer) contains ganglion cell nuclei and some displaced amacrine cells. The two nonnucleated layers between the three layers of nuclei are the *outer and inner plexiform layers*, where retinal synapsing takes place.

There are a number of synapse morphologies in both the outer and inner plexiform layers. Ribbon synapses predominate in both layers. The name ribbon synapse arises from the appearance of the presynaptic ultrastructure, which includes electron-dense ribbonlike structures oriented perpendicularly to the synaptic surface, with presynaptic vesicles lined up on each side (Figure 10.5). The ribbon synapses always involve multiple cell articulations: rods or cones to bipolar and horizontal cells for the outer plexiform, and bipolar cells to ganglion and/or amacrine cells for the inner plexiform layer. In the outer layer, the synapses are found in indentations in the rod and cone spherule and pedicle, respectively.

Retinal synapse complexity, with respect to how direct the connections are between the receptor cells, the bipolar cells and the ganglion cells, is related to the degree of visual processing carried out by the retina (in comparison to the processing carried out by higher visual pathway centers), with more retinal processing taking place in eyes of lower vertebrates.

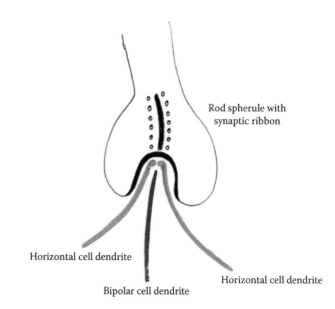

Figure 10.5 Rod spherule containing a ribbon synapse.

With respect to morphological diversity, it should be noted that each of the retinal cells mentioned here can be further subdivided into different types. For example, the largest bipolar cells are the rod bipolars, which, as the name indicates, synapse with rods. Midget bipolars are cells that synapse with a cone and a midget ganglion cell, while diffuse bipolars synapse with several cones and other ganglion cells.

In summary the retina is often described as consisting of ten layers: (1) RPE, (2) layer of rods and cones (outer and inner segments of the cells, but not the nucleus), (3) outer or external limiting membrane (consisting of a series of electron-dense adhering junctions at the inner segment of the photoreceptors), (4) outer nuclear layer (rod and cone nuclei), (5) outer plexiform layer, (6) inner nuclear layer (nuclei of bipolars, horizontal, and amacrine cells), (7) inner plexiform layer, (8) ganglion cell layer (cell bodies with nuclei), (9) nerve fiber layer or layer of optic nerve fibers (ganglion cell axons radiating toward the head of the optic nerve), and (10) inner or internal limiting membrane (hyaloid membrane).

The axons of the ganglion cells leave the globe as the principal fibers of the optic nerve. They travel to the optic chiasma, where the fibers from the nasal halves of the retinas cross to the opposite side of the brain and continue on as the optic tracts to synapse in the lateral geniculate body (LGB). Axons from the LGB travel to the visual cortices of the cerebral hemispheres via the visual pathways.

10.2.4 LENS

The human lens is a biconvex structure about 9.0 mm in diameter and 4.0–5.0 mm thick (anterior–posterior), depending on the accommodative state of the eye. It is located in the eye behind the iris and anterior to the vitreous chamber located posteriorly. The anterior surface is flatter than the posterior one in the unaccommodated state.

The lens is completely cellular, and like the cornea, its morphology is very regular, presumably in order to maximize transparency (Figure 10.6). It is not symmetrical in that its

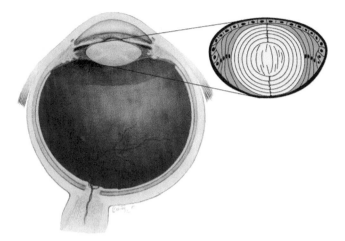

Figure 10.6 An anterior epithelium and primary and secondary lens fibers comprise the crystalline lens. The lens is surrounded by an acellular capsule. (Image on left adapted from National Eye Institute, National Institutes of Health (NEI Image Ref #NEA04).)

anterior surface is covered by a single layer of epithelial cells while the rest of the lens is made up of elongated cells (fibers) that interdigitate closely with one another by way of ball-and-socket articulations and form layers on top of each other. Fibers located deeper than the superficial layers lack cell organelles, including nuclei. The lens continues to develop through life. Continued mitotic cell development takes place at the lens equator, and the curved appearance in relation to the location of the nuclei of the superficial fiber layers is called the lens bow. The elongated lens cells (or fibers) extend almost from anterior pole to posterior pole. However, in the case of the human lens, they are not long enough to reach from pole to pole. Instead, the fibers articulate with each other a distance away from the poles, producing articulation lines referred to as the lens sutures. Suture architecture is simpler in lenses from young eyes (Y shaped). They become more complex with age to produce a star pattern.

The lens is surrounded by an acellular capsule believed to be derived from the basement membrane that surrounds the lens during early development of the embryonic lens vesicle. The capsule varies in thickness and this variation may play a role in determining the shape of the lens during accommodation. It articulates with the ciliary body by way of suspensory ligaments that merge with the lens capsule.

10.2.5 VITREOUS

The vitreous humor represents about 80% of the bulk of the eye. It is essentially a gel-like substance made up of water collagen and hyaluronic acid. The vitreous develops in stages. The *primary vitreous* develops in association with the hyaloid artery that nourishes the developing lens. The *secondary* vitreous forms most of vitreous of the adult human eye and it is believed to be derived from the retina. The zonular (elastic) ligaments that support the lens are derived from the *ciliary epithelium*.

10.2.6 OCULAR ADNEXA

In addition to the globe itself, a number of structures around the eye are important to its function. This review deals directly with the ocular globe itself. However, the adnexa can be simply

listed as the six extraocular muscles (superior, inferior, medial and lateral rectus muscles, and the superior and inferior obliques) responsible for eye movement, the eyelids containing the levator (upper lid) and orbicularis muscles, as well as the various glands of the lids (meibomian glands, goblet cells, and serous glands), and the lacrimal apparatus (lacrimal gland, tear ducts, etc.)

10.3 EMBRYONIC DEVELOPMENT OF THE EYE

10.3.1 EARLY EMBRYOGENESIS

A zygote is the single cell resulting from the fertilization of an ovum by a sperm. The first week of prenatal human development is devoted to *cleavage* (rapid proliferation). Zygotic cleavage results in the development of a hollow spherical structure, the blastocyst. The blastocyst differentiates at one pole into an *inner cell mass* from which the embryo will form. During the second week of embryonic development, the inner cell mass organizes into the primary germ layers: the ectoderm, mesoderm, and endoderm. Further differentiation of the primary germ layers will produce the major tissue types. In the development of the eye, the ectoderm (surface, neural, and neural crest cells) and the mesoderm predominate in eye development (Bron et al., 1997; Ozanics and Jakobiec, 1982).

As noted earlier, the optic nerve and the retina are extensions of the central nervous system (Mann, 1969). The central nervous system differentiates from the ectoderm on the dorsal surface of the embryo. The ectoderm thickens along the mid-dorsal region of the embryo to form the neural plate. A process of accelerated cell division along the two sides of the dorsal midline results in the formation of two folds of the neural plate to form a groove that becomes a tube once the two folds converge toward each other and fuse. The signals for the induction process are believed to emanate from the underlying notochord. The neural tube ultimately separates from the surface.

The cavity of the neural tube will become the ventricular system of the brain, while the epithelial cells that line the walls of the tube will produce the neurons and neuroglial cells of the central nervous system. Proliferation of the cells at the rostral end of the tube leads to the formation of the earliest components of the brain, the forebrain (prosencephalon), midbrain (mesencephalon), and hindbrain (rhombencephalon). In addition, neuroepithelial cells also give rise to a special group of cells, the neural crest cells, from the dorsal region of the neural tube soon after the neural folds have fused. Because there are no mesodermal somites in the head region, the cranial neural crest cells contribute to the formation of the connective tissues of the eye and its adnexal structures, in addition to the peripheral nervous system structures associated with the head (Noden, 1982).

10.3.2 FORMATION OF THE OPTIC CUP

The neural ectodermal components of the eye develop from the bilateral outgrowths of the forebrain (Duke-Elder, 1958; Mann, 1969). Even before the neural tube is completely closed, two lateral depressions (the optic pits) are visible in the developing forebrain (Figure 10.7). With further rapid growth, these pits become vesicles (pouches). Later, with still more growth, the outer

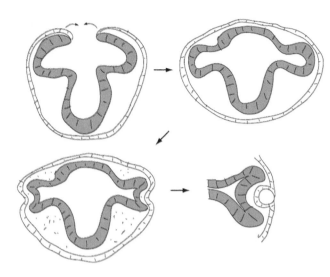

Figure 10.7 Development of the optic cup from the forebrain neural ectoderm, showing the neural tube with laterally placed optic vesicles, the optic cup. and the formation of the lens vesicle from the surface ectoderm.

wall of each vesicle collapses inward, or invaginates, to form the optic cup. The cup is the first indication of the spherical shape of the eye. The embryonic eye is, at this stage, connected to the forebrain by way of a hollow tube, the optic stalk. The invagination of the optic vesicle to form the cup results in the formation of two layers, one on top of the other. A ventral slit in the cup, the embryonic fissure, permits vascular contact with the inner structures of the cup; later, the fissure disappears.

The two layers of the cup eventually form all of the retina, parts of the ciliary body (excluding the musculature system that serves to control the shape or position of the lens) and the muscles and pigmented epithelium of the iris. The outer layer of the optic cup (the layer closest to the optic stalk) remains a relatively simple structure consisting of a single layer of epithelial cells, the retinal pigmented epithelium. As its name suggests, this layer is pigmented; it helps to form the dark chamber needed for image formation. The retinal pigmented epithelium also plays a role in the turnover of visual pigment of the specialized photoreceptor cells of the retina, the rods and cones, as indicated earlier.

The anterior region of the optic cup, the region adjacent to the lip of the cup, develops further to form the sphincter and dilator muscles of the iris and the two layers of the pigmented epithelium on the inner surface of the iris. The pigmented epithelium of the iris, is, in part, responsible for the ability of the iris to control the amount of light entering the eye.

The region of the optic cup between the developing retina and the iris eventually forms the inner lining of the ciliary body. Each of these two layers of the cup remain as a single layer of epithelial cells. The innermost layer remains unpigmented, whereas the layer below it (the layer that is continuous with the RPE) becomes pigmented.

Thus, the neural ectoderm produces all of the retina and the optic nerve and also parts of the ciliary body and iris. The rest of the ciliary body and iris develop from mesenchyme. This mesenchyme is derived mostly from cranial neural crest cells. It should be noted that all ocular structures develop simultaneously. Descriptions that specify the development of one or other

germinal tissues at one time do so for convenience only. Thus, while the optic cup is developing from the forebrain, a similar change is taking place in the surface ectoderm opposite of the open lips of the developing optic cups; the structures derived from the surface ectoderm include the lens and a portion of the cornea.

10.3.3 MESENCHYME

The mesenchymal structures of the eye are formed mainly around the developing optic cup. First, a vascular and pigmented layer of tissue is laid down to produce a continuous vascular region, the uvea. The uveal tract consists of the choroid, the vascular and muscular part of the ciliary body, and the anterior vascular region of the iris. A second mesenchymal layer of development overlies these vascular zones to form the fibrous outer tunic of the eye, the sclera and the majority of the cornea. Eventually, mesenchymal development is responsible for the production of other structures of the orbit, such as the extraocular muscles, the orbital bones, and the bulk of the eyelids.

The blood supply to the early developing eye is provided by the primitive dorsal and ventral ophthalmic arteries, which are in turn supplied by the internal carotid artery. A temporary embryonic vascular network, the hyaloid circulation, enters the optic cup from the mesenchyme around it through the embryonic fissure. The hyaloid artery stems from the primitive dorsal ophthalmic artery. This circulation supplies the early developing lens, forming a vascular network around the developing lens known as the *tunica lentis vasculosa*. Ultimately, the hyaloid vessels are resorbed, often leaving behind debris. With new growth peripherally, it becomes the definitive retinal circulation. As the embryonic fissure closes, in the distal to proximal direction, the hyaloid vessels are pushed to the region of the optic stalk. Growing retinal ganglion cell axons, during the formation of the optic nerve, surround the major artery and vein, which become known as the central retinal vessels.

At a later point, the ventral primitive ophthalmic artery loses its connection to the internal carotid artery and articulates instead with the primitive dorsal ophthalmic artery, which is now the definitive ophthalmic artery. The distal end of the ventral primitive ophthalmic artery becomes the nasal long posterior ciliary artery, while the distal end of the definitive ophthalmic artery, formerly dorsal primitive ophthalmic artery, will become the temporal long posterior artery. The short posterior ciliary arteries develop directly from the ophthalmic artery, while anterior ciliary arteries develop from the muscular branches of the ophthalmic artery in the four rectus muscles after the formation of the extraocular muscles.

10.3.4 RETINAL DEVELOPMENT

The inner part of the optic cup becomes an elaborate neural structure known as the neural retina (Bron et al., 1997; Jakobiec and Ozanics, 1982) (Figure 10.8). As described earlier, this layer differentiates extensively before birth into a region of three main cell layers: a layer of rods and cones (the layer closest to the pigmented epithelium); a middle layer, the inner nuclear layer; and the innermost layer, the ganglion cell layer. These three neural layers form a complex network that is, at least to some extent, involved in processing visual information. It is the nerve axons of the ganglion cells that form the optic nerve, the structure that carries visual information from the retina to the brain.

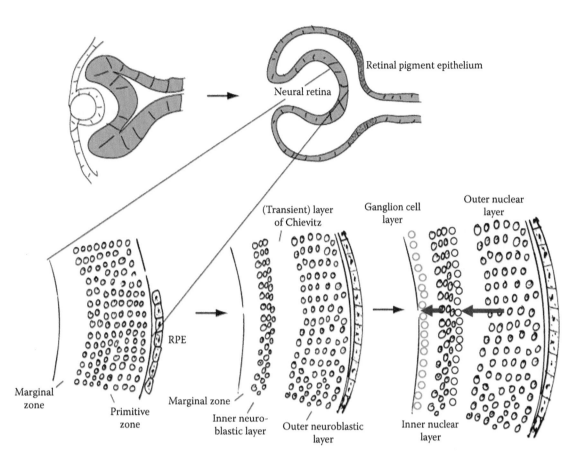

Figure 10.8 Formation of the retina starting with the retinal pigment epithelium, primitive and marginal zones, the inner and outer neuroblastic layers, and the 3 layers of cells of the developed retina.

The primitive neural retina differentiates into two zones: a cellular primitive zone (lying outermost, in close proximity to the outer layer of the optic cup) and an acellular marginal zone (lying innermost to the vitreous). The sequence of retinal maturation begins at the posterior pole and proceeds to the periphery. Cells in the primitive zone proliferate and migrate apart to form inner and outer neuroblastic layers. Between them is a narrow acellular layer, the transient fiber layer of Chievitz. Ganglion cells develop from the innermost part of the inner neuroblastic layer and migrate into the marginal zone of the nervous retina, forming a third layer of cells. Cells of the inner neuroblastic layer will also generate amacrine and Müller cells. Cells in the outer neuroblastic layer differentiate into photoreceptors, bipolar cells, and horizontal cell types. The bipolar and horizontal cells migrate toward the inner neuroblastic layer and take a position near the Müller and amacrine cells, thereby eliminating the space between the two layers of cells. The transient fiber layer of Chievitz disappears with the migration of bipolar and horizontal cells.

At this point in development, the cellular layers of the neural retina are in position for connections to be made between the layers. The ganglion cells send out fibers (axons) that grow toward the optic stalk and toward the LGB in the thalamus of the brain. Further modification with respect to the movement of ganglion and bipolar cells continues after birth to allow for the direct stimulation of the photoreceptors in the area of the macula and fovea.

At the fovea, a central retinal depression that coincides with the portion of the retina responsible for high-resolution vision and color vision is produced first by a migration of cones into the region, followed by the lateral (i.e., horizontal) diversion of the ganglion and bipolar cell layers so that light is incident on the receptors (cones) directly.

The RPE develops from the outer part of the optic cup. Pigmentation occurs earlier in the RPE than in the developing neural retina, iris, ciliary body, or choroid. The RPE remains one cell layer thick throughout its differentiation.

10.3.5 LENS DEVELOPMENT

The lens of the eye (also known as the crystalline lens) develops from the surface ectoderm. Optic cup development induces the cells of the surface ectoderm to thicken, forming a structure known as the lens placode (Figure 10.9). Rapid cell division in this region results in the invagination of the surface cells to form a lens vesicle, which ultimately separates from the surface to form a hollow spherical vesicle. The walls of this vesicle consist of a single layer of epithelial cells. Because of the inward invagination, the basement membrane of these cells is found around the outside of the vesicle. The lumen of the lens vesicle is slowly filled by elongation of cells from the posterior hemisphere of the original vesicle. These cells are identified as the primary lens fibers and form the embryonic nucleus of the lens. The single layer of cells remaining at the front (anterior) surface is known as the epithelium of the lens. Subsequent growth takes place at a mitotic

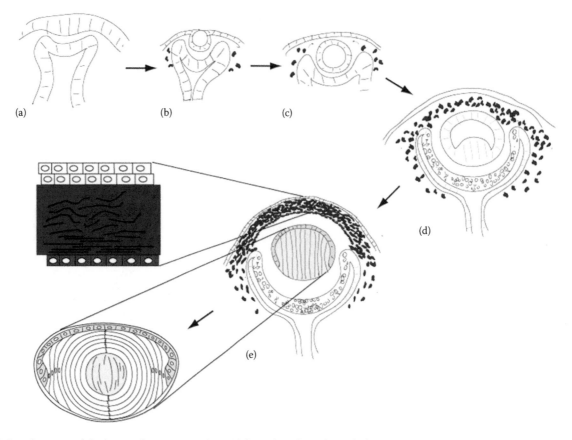

Figure 10.9 Development of the lens and cornea, starting with lens placode and vesicle formation (a). Ocular mesenchymal tissue migrates between the surface ectoderm and the developing lens to form the mesenchymal parts of the cornea (b–e).

zone around the equator of the embryonic nucleus. New cells grow over the old fibers and under the epithelium to form the secondary lens fibers. In the human lens, the fibers do not meet at a single point at the anterior and posterior poles, but rather form lines of articulation, the lens sutures, as noted earlier. The basement membrane around the outside of the developing lens thickens and becomes the lens capsule. This structure is, at least until later years, an acellular elastic body that plays an important role in determining the shape of the human lens during accommodation. Thus, the entire lens originates from the surface ectoderm.

10.3.6 CORNEAL DEVELOPMENT

Another significant ocular development involves further rapid cell division and invagination of the surface ectoderm (Figure 10.9). This event results in the formation of the epithelium of the cornea (which fuses to the mesenchymal portions), the limbus, and the inner and outer eyelids. The epithelia of these structures are continuous with each other, much like the two layers of epithelium in the uvea that are continuous across the iris, ciliary body, and retina.

The embryonic cornea starts out as a layer of cells of the surface ectoderm that reforms over the developing optic cup and lens vesicle when the lens vesicle detaches from the surface ectoderm. These cells are joined by migrating mesenchymal material, which develops in concert with the condensation of mesenchymal tissue around the developing optic cup. The most posterior of this tissue, Descemet's mesothelium, will lead to the formation of the endothelium of the cornea and Descemet's membrane. As the

surface ectodermal cells become stratified, a second mesenchymal cell layer develops below them that will produce the future stroma and form Bowman's layer of the cornea. Collagen fibers are laid down in a posterior to anterior direction by fibroblasts within the stroma.

10.3.7 DEVELOPMENT OF THE SCLERA, CHOROID, CILIARY BODY, AND IRIS

As stated earlier, the development of the sclera, choroid, ciliary body, and iris are intimately related to the condensation of mesenchyme cells in layers around the developing optic cup (Bron et al., 1997; Oyster, 1999; Ozanics and Jakobiec, 1982; Tripathi et al., 1989). The first condensation of mesenchyme around the optic cup will differentiate into the choroidal stroma. Future choroidal differentiation is directed to the development of the choroidal vasculature. The sclera forms from mesenchymal cells of neural crest origin that condense around the optic cup, beginning in the area of the future limbus and continuing in a posterior direction, thereby surrounding the future choroid.

In contrast to the choroid and sclera, where mainly mesenchymal cells are involved in embryological development, the ciliary body and iris contain neuroectodermal elements from the developing optic cup. The future ciliary epithelium differentiates from the advancing margins of the two layers of neuroectoderm of the optic cup. Further layering of mesenchyme of neural crest origin will contribute to the formation of the ciliary stroma and muscle, while the vascular portions of the ciliary body are mesodermal in origin. As pointed out earlier, the differentiation of the neuroectodermal component of the iris will form the future sphincter and the dilator muscles and the cells of the pigmented epithelium of the iris. The vascular framework of the iris, derived from mesoderm, develops along with the iris stroma, itself derived from mesenchyme of neural crest origin. The development of the iris stroma is closely associated with changes in the anterior portion of the tunica lentis vasculosa (primitive circulatory supply to the developing lens). The remnants of the atrophying tunica lentis vasculosa on the anterior surface of the lens will fuse with mesenchyme condensing along the anterior part of the optic cup to form the pupillary membrane. This layer of mesenchyme is pulled forward by the anterior growth of the edges of the optic cup. Subsequent central atrophy of the pupillary membrane leads to the formation of the collarette between the ciliary and pupillary portions of the iris.

10.4 EYE DEVELOPMENT AND OPTICAL FUNCTION

10.4.1 CORNEA

In the human (terrestrial) eye, both the cornea and lens are responsible for focusing an image on to the retina. Both have the same basic problem: how to maintain adequate physiological conditions for living tissue, while at the same time providing the image quality of a good optical device. Thus, both structures are anatomically and physiologically designed to function avascularly in order to maintain transparency yet also provide refractive power. Anatomically, both the lens and the cornea exhibit adaptations designed to minimize light scatter. Nevertheless, despite these and other similarities, differences in development (and location) have far-reaching consequences.

In humans, the cornea is an external ocular structure that represents the optical interface between the eye and the external environment. The external and internal optical surfaces are roughly parallel, so the refractive function of the cornea is a result of the difference in refractive index between the medium in front (air or water) and the medium behind (aqueous). Since the refractive index of aqueous is 1.335 and that of water is 1.333, corneal refractive power is virtually nil when the eye is in water (Sivak, 1980, 1990). This observation is true despite the fact that the overall refractive index of the cornea (1.376) is appreciably greater than that of water.

In view of the aquatic habitat of early vertebrates, it is obvious that the corneas of early species were in large measure transparent windows. This notion is reflected in the fact the lenses of fishes and aquatic mammals, the only refractive elements of these eyes, are spherical in shape and very high in overall refractive index (1.65; Walls, 1942). When the eye is in air, the refractive contribution of the cornea is considerable, due to the difference in refractive indices of air and aqueous humor. Thus, the refractive power of the human cornea (in air) is about twice that of the lens (Bennett and Francis, 1962).

The external location of the cornea is also important because the only part to undergo appreciable regeneration, the epithelium, can be lost and is not retained. Disruptions to the corneal cell cycle, where basal columnar cells reach the surface as squamous cells and are sloughed off to the external environment, can have unwanted optical consequences.

Although the stroma constitutes 80% of the thickness of the cornea, the orthogonal arrangement of the lamellae of the stroma and the highly regular arrangement, or lattice, of the collagen fibers and minimal interfiber spaces within each lamellae stroma contribute to the minimization of light scatter; as mentioned previously, these conditions promote mutually destructive interference. Highly regular arrangements of collagen are not limited to the stroma. The fine fibrils of collagen with Descemet's membrane are also arranged in a very regular lattice.

As described earlier, the cornea is essentially a collagen sandwich with epithelial tissue on both sides of a collagen core (the stroma) (Duke-Elder, 1958; Hogan et al., 1971). That the corneal epithelial cells adhere tightly to one another by means of desmosomes and other junctional devices is an important physiological factor that also minimizes light scatter. Unlike the epidermis, the superficial and deep epithelial surfaces are relatively smooth and parallel to one another. The endothelium at the posterior surface of the cornea contributes physiologically to the maintenance of transparency, housing the ionic pumps that reduce the water content in the stroma.

10.4.2 LENS

Like the cornea, lens morphology is also characterized by regularity of structure (Hogan et al., 1971), but the lens is entirely cellular. As already noted, a single layer of epithelium covers its anterior surface, and very elongated cells (fibers) fill the interior, extending to the posterior surface. The fibers are arranged in concentric shells of increasing diameter (Worgul, 1982). Their size and shape determine the ultimate size and shape of the lens (Sivak et al., 1985).

The lens fibers contain relatively few organelles and those in the central core lack organelles (nuclei, mitochondria, and endoplasmic reticula) entirely. These adaptations are important for the maintenance of transparency. The fibers are interconnected by means of numerous ball-and-socket articulations, which help produce a highly ordered geometric pattern. The tight articulations between fibers reduce intercellular space and minimize light scatter.

In many species, particularly in mammals, the secondary lens fibers do not taper sufficiently to meet at points at the anterior and posterior poles (Walls, 1942). Rather, the fiber ends form lines or sutures of varying complexity, extending in depth into the lens, depending on the species or age of the mammal. For example, the sutures of the early post-embryo human lens are Y shaped, erect anteriorly, and inverted posteriorly. The adult human lens can exhibit a more complex nine branched suture arrangement. Paradoxically, the sutures are located along the optical axis of the lens, where they can most affect the optical quality of the eye, as shown in a series of correlative morphological and optical studies (Kuszak et al., 1991, 1994; Sivak et al., 1994). This problem is believed to be mitigated by a slight tilt of the lens.

The fact that the lens is located within the eye and that it is made up of cells of surface origin that continue to multiply through life creates a unique set of circumstances; because the lens is surrounded by the humors of the eye, it must have a refractive index that is substantially greater than that of water in order to focus light. In fact, the equivalent refractive index of the human lens is 1.41 (Bennett and Francis, 1962), while in fishes, where the lens is the only refractive element of the eye, the index is as high as 1.65 (Walls, 1942). Thus, the cells of the lens have the highest protein concentration of any tissue of the body, an adaptation to the need for an elevated refractive index.

A second point related to the location of the lens, as well as to the fact that continued growth and development takes place peripherally, is that old lens cells are retained. In fact, the growth ring pattern of lens development results in the concentration of older tissue toward the center (Figure 10.9) and the formation of a gradient refractive index, the index at the center being higher than that at the periphery. This gradient plays an important optical role in reducing the optical aberrations of the eye, particularly spherical aberration (Bennett and Francis, 1962; Sivak and Bobier, 1990).

A final point to be mentioned is that in many species the lens provides the eye with a variable focus mechanism, accommodation. (In at least some birds the cornea is also involved in accommodation.) In humans, accommodative change in lens shape takes place in response to neural directives given to the ciliary muscle/zonular apparatus of the eye. Presbyopia, the loss of accommodation with age, is a consequence, at least in part, of continued lens development.

ACKNOWLEDGMENTS

We acknowledge that a number of the drawings were adapted from original art by Dr. Barbara Sivak. The authors thank Dr. Kevin van Doorn for tracking down the original paper by Leuckart. The authors also acknowledge the support of the Natural Sciences and Engineering Council of Canada.

REFERENCES

Bennett, A. G. and Francis, J. L. (1962). The eye as an optical system. In *Visual Optics and the Optical Space Sense*. Vol. 4: The Eye, (ed. Davson, H.), pp. 101–131. New York: Academic Press.

Binder, P. S., Rock, M. E., Schmidt, K. C., and Anderson, J. A. (1991). High-voltage electron microscopy of normal human cornea. *Invest Ophthalmol Vis Sci* 32, 2234–2243.

Borcherding, M. S., Blacik, L. J., Sittig, R. A., Bizzell, J. W., Breen, M., and Weinstein, H. G. (1975). Proteoglycans and collagen fibre organization in human corneoscleral tissue. *Exp Eye Res* 21, 59–70.

Bron, A. J., Tripathi, R. C., and Tripathi, B. J. (1997). *Wolff's Anatomy of the Eye and Orbit*, 8th edn. London, U.K.: Chapman & Hall, Ltd.

Chen, Z., De Paiva, C. S., Luo, L., Kretzer, F. L., Pflugfelder, S. C., and Li, D.-Q. (2004). Characterization of putative stem cell phenotype in human limbal epithelia. *Stem Cells* 22, 355–366.

Czeisler, C. A., Shanahan, T. L., Klerman, E. B., Martens, H., Brotman, D. J., Emens, J. S., Klein, T., and Rizzo, J. F. (1995). Suppression of melatonin secretion in some blind patients by exposure to bright light. *New Engl J Med* 332, 6–11.

Duke-Elder, S. (1958). *System of Ophthalmology*. Vol. 2: The Anatomy of the Visual System. London, U.K.: Henry Kimpton.

Hogan, M. J., Alvarado, J. A., and Weddell, J. E. (1971). *Histology of the Human Eye*. Philadelphia, PA: WB Saunders.

Jakobiec, F. A. and Ozanics, V. (1982). General topographic anatomy of the eye. In *Ocular Anatomy, Embryology and Teratology* (ed. Jakobiec, F. A.), pp. 1–9. Philadelphia, PA: Harper & Row.

Kolb, H. (2012). Simple retinal anatomy. Webvision: The Organization of the Retina and Visual System, University of Utah Health Sciences Center, Salt Lake City, UT, January 31, 2012. Web. (September 21, 2014). http://webvision.med.utah.edu/book/part-i-foundations/simple-anatomy-of-the-retina/.

Kuszak, J. R., Peterson, K. L., Sivak, J. G., and Herbert, K. L. (1994). The interrelationship of lens anatomy and optical quality. II. Primate lenses. *Exp Eye Res* 59, 521–535.

Kuszak, J. R., Sivak, J. G., and Weerheim, J. A. (1991). Lens optical quality is a direct function of lens sutural architecture. *Invest Ophthalmol Vis Sci* 32, 2119–2129.

Li, W., Hayashida, Y., Chen, Y.-T., and Tseng, S. C. G. (2007). Niche regulation of corneal epithelial stem cells at the limbus. *Cell Res* 17, 26–36.

Mann, I. (1969). *The Development of the Human Eye*. New York: Grune & Stratton.

Meeney, A. and Mudhar, H. S. (2013). Histopathological reporting of corneal pathology by a biomedical scientist: The Sheffield Experience. *Eye* 27, 272–276.

Noden, D. M. (1982). Periocular mesenchyme: Neural crest and mesodermal interactions. In *Ocular Anatomy, Embryology and Teratology* (ed. Jakobiec, F. A.), pp. 97–119. Philadelphia, PA: Harper & Row.

Oyster, C. W. (1999). *The Human Eye: Structure and Function*. Sunderland, MA: Sinauer Associates, Inc.

Ozanics, V. and Jakobiec, F. A. (1982). Prenatal development of the eye and its adnexa. In *Ocular Anatomy, Embryology and Teratology* (ed. Jakobiec, F. A.), pp. 11–96. Philadelphia, Pa: Harper & Row.

Provencio, I., Rodriguez, I. R., Jiang, G., Hayes, W. P., Moreira, E. F., and Rollag, M. D. (2000). A novel human opsin in the inner retina. *J Neurosci* 20, 600–605.

Rodrigues, M. M., Warring, G. O. I., Hackett, J., and Donohoo, P. (1982). Cornea. In *Ocular Anatomy, Embryology and Teratology* (ed. Jakobiec, F. A.), pp. 153–165. Philadelphia, PA: Harper & Row.

Schmidt, T. M., Do, M. T. H., Dacey, D., Lucas, R., Hattar, S., and Matynia, A. (2011). Melanopsin-positive intrinsically photosensitive retinal ganglion cells: From form to function. *J Neurosc* 31, 16094–16101.

Sivak, J. G. (1980). Accommodation in vertebrates: A contemporary survey. In *Current Topics in Eye Research*, Vol. 3 (eds. Davson, H. and Zadunaisky, J.), pp. 281–330. New York: Academic Press.

Sivak, J. G. (1990). Optical variability of the fish lens. In *The Visual System of Fish* (eds. Douglas, R. H. and Djamgoz, M. B. A.), pp. 63–80. London, U.K.: Chapman & Hall, Ltd.

Sivak, J. G. and Bobier, W. R. (1990). Optical components of the eye. In *Principles and Practice of Pediatric Optometry* (eds. Rosenbloom, A. A. and Morgan, M. W.), pp. 31–45. Philadelphia, PA: J B Lippincott.

Sivak, J. G., Herbert, K. L., Peterson, K. L., and Kuszak, J. R. (1994). The interrelationship of lens anatomy and optical quality. I. Non-primate lenses. *Exp Eye Res* 59, 505–520.

Sivak, J. G., Levy, B., Weber, A. P., and Glover, R. F. (1985). Environmental influence on shape of the crystalline lens: The amphibian example. *Exp Biol* 44, 29–40.

Tripathi, B. J., Tripathi, R. C., and Wisdom, J. (1989). Embryology of the anterior segment of the human eye. In *The Glaucomas* (eds. Ritch, R. M., Shields, M. B., and Krupin, T.), pp. 3–40. St. Louis, MO: Mosby.

Walls, G. L. (1942). *The Vertebrate Eye and Its Adaptive Radiation*. Bloomfield Hills, MI: Cranbrook Institute of Science.

Worgul, B. V. (1982). The lens. In *Ocular Anatomy, Embryology and Teratology*, Vol. 1 (ed. Jakobiec, F. A.), pp. 355–389. Philadelphia, PA: Harper & Row.

Michael A. Freed

Contents

11.1 INTRODUCTION

The retina is a thin (100–300 µm) transparent sheet of tissue at the back of the eye. The retina contains a variety of neurons: *photoreceptors*, which *transduce* light into electrical signals; secondary neurons, including *horizontal*, *bipolar*, and *amacrine* cells, which process these signals; and tertiary neurons, the *ganglion cells*, which stream the processed signals to the brain. This chapter describes how transduction and signal processing is accomplished. The chapter focuses on design principles, learned from the study of animals, which help an understanding of human retina.

11.2 PHOTORECEPTORS TRANSDUCE LIGHT INTO ELECTRICAL SIGNALS

11.2.1 PHOTORECEPTOR STRUCTURE

Photoreceptors have all the machinery necessary for the transduction of light into electrical signals and the transmission of these signals to other neurons. This machinery is divided into four compartments (Figure 11.1). (1) An *outer segment* contains stacks of membranous *disks* that are studded with proteins necessary for transduction. The outer membrane of the outer segment is fenestrated by channels through which ions flow to create electrical currents. (2) An *inner segment* has ion pumps that establish electrical gradients across the membrane, which are necessary for the flow of ions. The cytoplasm of the inner segment is filled with mitochondria that supply energy for ion pumps and filled with rough endoplasmic reticula that synthesize transduction proteins. (3) A *cell body* contains the nucleus with DNA that codes for transduction proteins. (4) An *axon*, whose termination is filled with ribbon synapses, transmits signals to other retinal neurons.

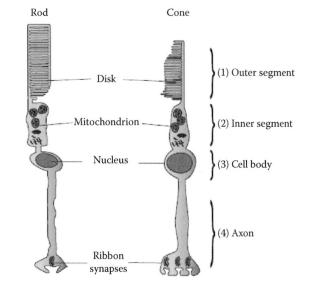

Figure 11.1 Rod and cones are divided into four compartments, listed at the right of the figure. (Adapted from Rodieck, R.W., *The First Steps in Seeing*, Sinauer Associates, Sunderland, MA, 1998.)

11.2.2 LIGHT IS TRANSDUCED INTO VOLTAGE

When light strikes a vertebrate retinal photoreceptor, its internal electrical potential, when referenced to the space outside of it, becomes more negative—it *hyperpolarizes*. When a photoreceptor hyperpolarizes, it signals the occurrence of light to secondary neurons. The vertebrate photoreceptor is unusual because most neurons signal by *depolarizing*—their internal potentials become more positive. Indeed, hyperpolarization to light sets the vertebrate photoreceptor apart from photoreceptors of invertebrates, from photoreceptors in the "3rd" (parietal) eye of some vertebrates, and from photosensitive ganglion cells in vertebrate retinas, all of which depolarize to light.

11.2.3 TRANSDUCTION IN DETAIL

Transduction is a sequence of molecular interactions that starts with light striking a molecule called "retinal" and ends with an electrical signal (for more detail, see Figure 11.2). Retinal is a ring structure attached to a bent chain and is held in the cleft of a protein molecule called "opsin." When retinal absorbs light, it untwists, which stretches the opsin into its meta form (R*). The meta opsin binds to *transducin*, which breaks apart into subunits. Transducin's alpha subunit binds to a phosphodiesterase (PDE), increasing the rate at which PDE hydrolyses cyclic guanosine monophosphate (cG) to its linear form, GMP.

To grasp how transduction hyperpolarizes the photoreceptor, it is necessary to understand that in the dark, before transduction had started, the photoreceptor was in balance. Positive ions exiting the inner segment were balanced by positive ions entering the outer segment in a cyclic movement of charges called the "dark current" (Figure 11.3a). About 2%–5% of ionic channels in the outer segment were open because they had bound cG (Korenbrot 2012). The creation of cG from GMP by a guanylate cyclase (GC) was balanced by the destruction by PDE. Transduction disturbs the balance: it activates PDE, reducing the concentration of cG, which reduces the number of open channels in the outer segment. As a result, the positive ions coming out of the inner segment are no longer balanced by ions coming into the outer segment. The disturbed balance causes a net outward electrical current called the "photocurrent." The photocurrent across the resistance of the membrane increases the transmembrane voltage difference, which hyperpolarizes the photoreceptor (Figure 11.3b).

The photocurrent and resulting hyperpolarization subside even when light is left on because reversing mechanisms engage with some delay after the onset of light (Figure 11.4). *G-protein-activating proteins* (GAPs), aka "regulators of G-protein signaling," bind to transducin's alpha subunit, which unbinds from PDE, which stops PDE from hydrolyzing cG. A two-step process impedes opsin's binding to transducin: first, *rhodopsin kinase* (RK), aka "G-protein-coupled receptor kinase," phosphorylates opsin; second, *arrestin* gloms onto the opsin (Figure 11.2, steps 5–10).

Reversing mechanisms also use calcium ions, which enter the outer segment through the channels that are opened by cG. When light closes these ionic channels, fewer calcium molecules enter, and thus the calcium concentration inside the photoreceptor falls. Consequently, calcium unbinds from GC-activating protein, which

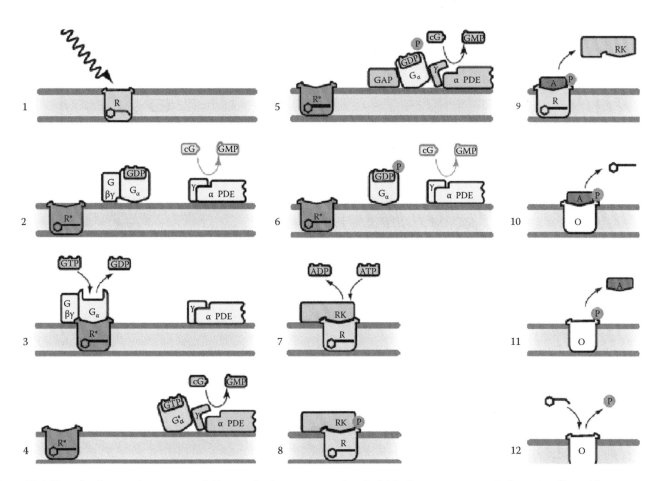

Figure 11.2 Transduction requires a sequential interaction between proteins. A disk in the outer segment is shown in yellow with gray membranes. (1) Light strikes an opsin molecule (R) that holds a molecule of retinal. (2) Retinal untwists, stretching opsin into its meta form (R*). (3) Meta opsin binds to transducin (subunits Gα, Gβ, and Gγ), which exchanges a guanosine diphosphate (GDP) for a guanosine triphosphate (GTP). (4) Gα splits away from Gβ/Gγ and binds to a phosphodiesterase's γ-subunit (PDEγ), uncovering catalytic sites on its α-subunit (PDEα), increasing the rate at which PDE_α hydrolyses cyclic guanosine monophosphate (cG) to its linear form, GMP. (5) G-protein-activating protein (GAP) binds to Gα, which gains a phosphate group (P) from GTP. (6) Gα unbinds from PDEγ, which deactivates PDEα. (7) Rhodopsin kinase (RK) gains a phosphate group from ATP. (8) RK phosphorylates R*, decreasing R*'s affinity for transducin. (9) Arrestin (A) replaces RK. (10) Retinal is removed from rhodopsin, leaving a naked opsin (O). (11) Arrestin unbinds from the opsin. (12) A phosphatase (not shown) removes the opsin's phosphate group, allowing a twisted retinal molecule to enter, ready to begin the transduction cycle again (back to step 1). (Adapted from Rodieck, R.W., *The First Steps in Seeing*, Sinauer Associates, Sunderland, MA, 1998.)

activates GC, which increases its production of cG. Falling calcium levels also dissociate RK from the protein *recoverin*, freeing RK to phosphorylate the opsin (Figure 11.2, step 8).

11.2.4 TRANSDUCTION FOLLOWS A 2nd MESSENGER SCHEME

A *2nd messenger scheme* is a sequence of molecular interactions: a *1st messenger* binds to a G-protein-coupled receptor (GPCR). The GPCR activates a G-protein, so called because it dephosphorylates guanosine triphosphate to GMP. The G-protein modulates the levels of a *2nd messenger*, which can modulate many other cellular processes. Transduction follows this scheme: retinal is the 1st messenger, opsin is a GPCR, transducin is a G-protein, and cG is a 2nd messenger. Yet opsin is different from most GPCRs, which must wait for their ligands to float toward them, often across the space that separates one cell from another. Opsin holds its ligand captive in a cleft. Captivity is temporary, however, because once retinal absorbs light and becomes a 1st messenger, it is released by the opsin. Retinal must then be transported to other

cells (pigmented epithelial cells, Müller cells) where it returns to the twisted state. Once twisted, retinal returns to the photoreceptor and eventually to the opsin (Figure 11.2).

11.2.5 OPSINS DETERMINE WAVELENGTH SENSITIVITY

There are many different opsins, distributed among disparate animal and plant phyla, all of which incorporate some form of retinal (Fernald 2006). The molecular structure of an opsin modifies retinal's wavelength selectivity, meaning that it absorbs some wavelengths best and other wavelengths to a lesser degree. Thus, for example, the human retina has 4 opsins in four different photoreceptors. *Rhodopsin in rods* absorbs green light centered on a wavelength of 498 nm. Three *photopsins* in *S, M, and L cones* absorb blue, green, and yellowish light, respectively (centered on 420–440, 534–545, and 564–580 nm).

Some ganglion cells in human and other mammals have an opsin called "melanopsin" (488 nm). Melanopsin is dissimilar to the photoreceptor opsins found in the very same retinas, but

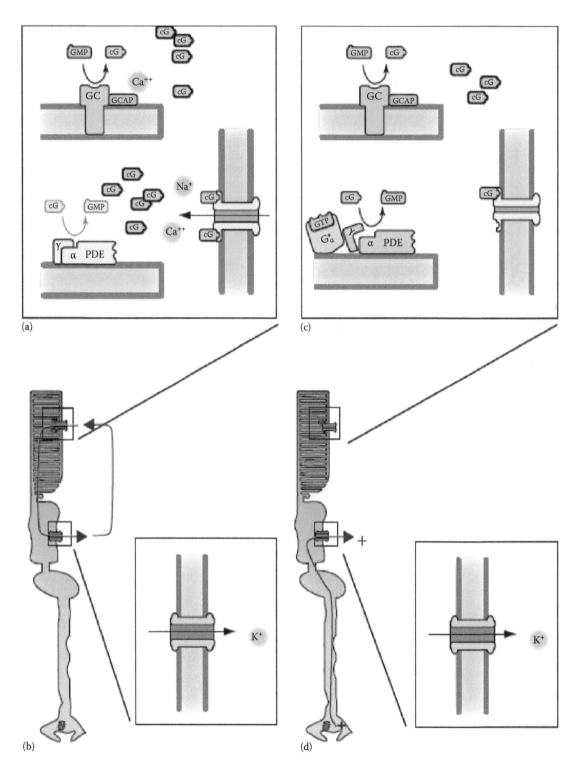

Figure 11.3 Transduction controls a cyclic flow of ions. (a) In the dark, guanylate cyclase (GC) produces cyclic guanylate monophosphate (cG) at a rate that is balanced by the rate at which phosphodiesterase (PDE) destroys cG. cG opens channels in the outer segment that let in sodium (Na$^+$) and calcium (Ca^{++}) ions. (b) There is just enough cG to ensure that the rate at which Na$^+$ and Ca^{++} enter the outer segment is balanced by the rate at which potassium ions (K$^+$) exit the inner segment. (c) When light strikes a disk in the outer segment, Gα binds to the γ-subunit of PDE, which uncovers catalytic sites on the PDE's α-subunit, increasing the rate at which it hydrolyses cG to GMP. PDE's ability to destroy cG overwhelms GC's ability to produce it, and the concentration of cG declines. cG unbinds from channels in the outer segment, causing them to close. (d) The channels in the inner segment continue to extrude K$^+$, but this outward current is no longer sourced from the outer segment. The source becomes positive charges inside the photoreceptor, which causes the photoreceptor to hyperpolarize. (Adapted from Rodieck, R.W., *The First Steps in Seeing*, Sinauer Associates, Sunderland, MA, 1998.)

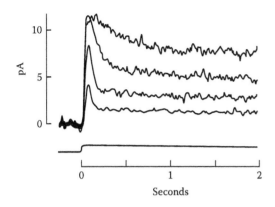

Figure 11.4 Photocurrents from a green-sensitive cone from monkey retina (macaque). Currents become larger as light intensity increases. Even though light remains on, currents decline over time due to reisomerization of R* and deactivation of the transduction cascade. (From Schnapf, J.L. et al., *J. Physiol. Lond.*, 427(AUG), 681, 1990.)

is homologous to photoreceptor opsins found in invertebrate retinas (Provencio et al. 1998). Moreover, melanopsin ganglion cells and invertebrate photoreceptors have similar 2nd messenger cascades that end by opening a transient receptor potential (TRP) ion channel and causing a depolarization (Schmidt et al. 2011). (A member of the TRP family of ion channels was first discovered to produce a transient receptor potential in mutant flies—hence the family name—but normally produces a sustained potential.) Therefore, the mammalian retina contains both photoreceptors and melanopsin ganglion cells, two different kinds of photosensitive cell, with two different transduction cascades. This situation may have arisen from an extinct vertebrate ancestor with two kinds of eyes, each with its own photoreceptor: one photoreceptor evolved into modern photoreceptors and the other into tertiary neurons such as ganglion cells (Lamb 2013). Indeed, an extant fishlike animal, whose phylum diverged long before a common vertebrate ancestor, has multiple eyes and photoreceptors of both kinds (*Amphioxus*).

11.3 IT TAKES BOTH RODS AND CONE PHOTORECEPTORS

Rods are more sensitive than cones, but cones respond to higher light levels. This essential difference is necessary for a *duplex retina*, which contains both rods and cones and which can respond to a wide range of light intensities.

11.3.1 RODS ARE MORE SENSITIVE TO LIGHT THAN CONES

Rods are more sensitive to light because they amplify a single photon into a larger photocurrent. In both rods and cones, amplification is achieved by having one molecule activate many molecules. For example, in a rod, when a photon untwists the retinal inside an opsin, it activates about 20 transducin molecules. Each of these transducin molecules activates one cG hydrolysis site on a PDE molecule, and each site hydrolyzes about 1000 cG molecules (mouse) (Arshavsky and Burns 2014). It takes 3 cGs to open a channel, which produces an accelerated

relationship between the drop in cG concentration and channel closings. A rod amplifies more than a cone because its opsin activates more transducin molecules and its transducin activates more PDE molecules (Kawamura and Tachibanaki 2008; Korenbrot 2012). Another reason that a rod is more sensitive is that it has weaker reversal mechanisms to tamp down transduction: the rod's RK and arrestin deactivate opsin slower and the rod expresses lower levels of GAP proteins that deactivate transducin (Vogalis et al. 2011; Tachibanaki et al. 2012). The weaker reversal mechanisms allow the photocurrent to last longer. Finally, a rod's sensitivity is increased because it has a higher stack of disks than a cone does; a photon entering the bottom of the stack will have greater probability to be absorbed before it exits the top of the stack.

11.3.2 RODS ARE MORE RELIABLE REPORTERS OF SINGLE PHOTONS THAN CONES ARE

The signal carried by the photocurrent and voltage is contaminated with noise. The culprit is heat, which bends and stretches molecules, and sends them skittering about the disk. As a result, crucial steps in the transduction cascade can occur spontaneously in the absence of light, causing fluctuations in the photocurrent. Retinal can untwist spontaneously, converting its opsin to the meta form (Barlow 1957). A spontaneous untwisting will cause a photocurrent indistinguishable from that which is generated by light. Therefore, spontaneous untwisting can cause retinal circuitry to give a false readout of photon rate. Here, rods and S cones have the advantage over L and M cones, because at a given temperature, their opsins allow spontaneous untwisting less often. A second source of noise is spontaneous activation of PDE, causing the single-photon current to vary in amplitude. Here again, rods have the advantage because their PDE molecules activate less readily (Rieke and Baylor 1996). A third cause of noise is the cG ion channel, which spontaneously flickers open and closed, adding smaller fluctuations to the photocurrent. Again, rods have less channel noise than cones do (Angueyra and Rieke 2013). For these reasons, rods provide a single-photon photocurrent that is relatively noiseless and that is therefore a reliable reporter of photon absorptions.

11.3.3 CONES RESPOND TO HIGHER INTENSITIES THAN RODS DO

Although a rod is more sensitive, a cone can transduce at higher photon rates. Consider that a bright light might untwist so many retinal molecules, activate so many PDEs, and destroy so many cGs that all the cG-gated ion channels would close. In this case, the photocurrent would be saturated because brighter light would no longer produce a larger photocurrent. Therefore, to avoid saturation, retinal must be retwisted faster than it is untwisted by light. A cone saturates at a higher photon rate than a rod because it has a 60,000 times higher rate of retinal untwisting (Kenkre et al. 2005). Also, a cone saturates at a higher photon rate because it has more powerful transduction reversing mechanisms than a rod does: the cone's transducin and PDE deactivate about 20 times faster (Kawamura and Tachibanaki 2008; Tachibanaki et al. 2012).

11.4 STRUCTURE OF THE RETINA AND ITS BASIC COMPONENTS

11.4.1 GENERAL ORGANIZATION OF THE RETINA

Light that enters the vertebrate eye must pass through the cornea, the lens, and secondary and tertiary neurons before it is absorbed by the outer segments. Because the retina is at the back of the eye, and at the focal point of a lens, an image spreads across the retina. Therefore, to capture every part of the image, there are photoreceptors at every location on the retina, with one exception. There are no photoreceptors or other neurons at the optic disk, where a bundle of ganglion cell axons exit the eye, forming the optic nerve. In some vertebrates (mouse, guinea pig), cones with short wavelength opsins are more dense where the sky is imaged, at the ventral (bottom) half of the retina (Rohlich et al. 1994).

In all vertebrates, high-resolution processing is not spread across the entire retina but instead is concentrated in smaller areas, usually near the center of the retina. High-resolution vision requires a patch of tightly packed cones and a concentration of neurons to process signals and send them to the brain. In humans and other simian primates, some reptiles, and some birds, secondary and tertiary neurons are displaced to the perimeter of the patch to allow better penetration of light: this causes a local thinning of the retina that is called a fovea (Latin for "pit"). Predators tend to have one or more high-resolution patches per retina and move their eyes and/or head to bring the image of the prey onto the fovea. Prey tends to have an elongated concentration of neurons that extends along where the image of the horizon normally lies. Prey do not move their eyes about to scan the world, but tend to hold them steady, in effect staring in all directions at once, on the lookout for predators.

11.4.2 LAYERING OF THE RETINA

The retina is strictly divided into *nuclear layers* for cell bodies and *plexiform layers* for making connections between neurons (Figure 11.5). This strict division is unusual: neurons and neuropile are more intermixed in most of the brain. The three nuclear layers are like the dough in a layered cake and two plexiform layers the filling between them. The dough layers are the outer nuclear layer with photoreceptor cell bodies, the inner nuclear layer with amacrine and bipolar cell bodies, and the ganglion cell layer with ganglion and amacrine cell bodies. The filling layers are the outer plexiform layer (OPL), where photoreceptors connect with bipolar cells, and the inner plexiform layer (IPL), where bipolar cells connect to ganglion cells.

Apparently, the vertebrate retina's 3-layer structure and its multiplicity of neuronal types are adaptations to image processing. This seems likely because there exist organs that merely sense light, for example, the parietal eye of nonmammalian vertebrates and the paired bilateral eyes of hagfish (slime eel). These simple visual organs have only two cell body layers: one layer for photoreceptors and another layer for neurons that stream photoreceptor signals to the brain. The intermediate 3rd layer for processing is missing (Pu and Dowling 1981).

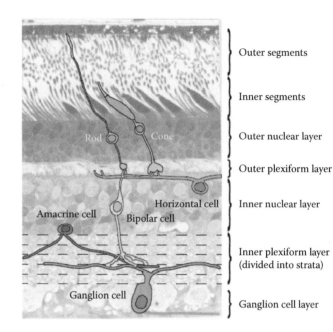

Figure 11.5 Layering of the retina and the position of neurons (*Macaca fascicularis*, courtesy of Noga Vardi). From top to bottom: the outer and inner segments belong to photoreceptors; next is an alteration of plexiform layers with nuclear layers; and the inner plexiform layer is divided into 5 strata.

11.4.3 STRATIFICATION OF THE INNER PLEXIFORM LAYER

In vertebrates, the IPL has a fine layering of its own and is divided into strata (Figure 11.5). These strata are like slots for expansion boards in a computer: each slot provides the space for different connections. Each stratum is where different types of bipolar cell send their axons; each type offers signals pertaining to color, temporal frequency, shadow, or light. Each stratum is where different types of ganglion or amacrine cell send their dendrites, there to receive a different blend of signals. Each ganglion cell type sends its signals to different brain nuclei. Thus, IPL stratification organizes connections between bipolar, amacrine, and ganglion cells and insures that each brain nucleus receives the signals needed for its particular kind of visual processing.

11.4.3.1 How neurons send signals from one to another

Neurons send signals to one another via *synapses*, places where two different neurons touch. A synapse sends a signal by changing the voltage across the neuronal membrane. Some ion channels can sense this voltage and modulate the flow of ions into the cell. In this way, internal ion concentrations are controlled, which is an effective way of controlling many cellular processes. There are three sorts of synapses: *chemical*, *gap junctions*, and *ephaptic*, which alter the postsynaptic voltage by different means.

11.4.4 CHEMICAL SYNAPSE

At a chemical synapse, a *presynaptic* neuron releases neurotransmitter onto a *postsynaptic* neuron (Figure 11.6a). Neurotransmitter is stored in tiny spheres of membrane called "synaptic vesicles." Vesicles are inside the presynaptic neuron and cluster near the membrane. To transmit a signal, vesicles fuse with the membrane and release their contents outside the neuron.

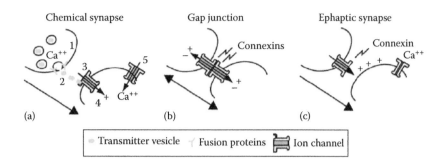

Figure 11.6 Different kinds of synapses. Large arrows show the direction of signal transmission. (a) Sign-conserving chemical synapse. Calcium (Ca⁺⁺) binds to specialized proteins in the presynaptic cell (1), causing a synaptic vesicle to fuse with the outer membrane and release an excitatory neurotransmitter (2). The neurotransmitter binds with a receptor on the postsynaptic cell, opening an ion channel (3). Positive ions enter the postsynaptic cell, depolarizing it (4). A channel senses this voltage change and opens, allowing calcium to enter (5). A sign-inverting chemical synapse is similar except the transmitter causes the net inward flow of negative ions, which hyperpolarizes the cell. (b) Gap junction. Positive or negative ions flow directly from one cell to another. (c) Ephaptic synapse. Positive or negative ions flow into a space outside the postsynaptic cell, altering the voltage across its membrane. A channel senses this voltage change and modulates the flow of calcium.

The neurotransmitter diffuses a short distance and binds to receptors on the membrane of the postsynaptic neuron, modulating current flow through ionic channels and thereby altering the postsynaptic cell's internal electrical potential. The packaging of neurotransmitter into vesicles significantly transforms the signal being transmitted: the signal is not continuous, but is discretized. Vesicles release their packets somewhat at random, which can add considerable noise to the transmitted signal (Freed and Liang 2014) (although see DeVries et al. 2006).

11.4.5 GAP JUNCTION

A gap junction is so called because it appears as a slightly widened gap between two neurons. Here, ions flow directly from one neuron to another across tubes constructed of proteins called "connexins" (Figure 11.6b). A gap junction is two-way, sending signals from cell A to cell B or vice versa. Because the signal is not divided into packets, as it would be by a chemical synapse, a gap junction provides a continuous and relatively noiseless transmission.

11.4.6 EPHAPTIC SYNAPSE

At an *ephaptic synapse*, currents flow from the presynaptic neuron into the space outside the postsynaptic neuron and alter the electrical potential of this space (Vroman et al. 2013) (Figure 11.6c). Altering the potential outside a neuron is an effective means of controlling ionic channels because ionic channels sense membrane voltage, which is the potential *difference* between inside and outside. Therefore, altering the external potential sends signals as effectively as altering the internal potential—the mode by which chemical and electric synapses transmit. An ephaptic synapse transmits a continuous and relatively noiseless signal.

11.4.7 INHIBITORY AND EXCITATORY SYNAPSES

A chemical synapse is *excitatory* if its transmitter decreases voltage across the postsynaptic neuron's membrane—it depolarizes (Figure 11.6a). Because depolarizing a neuron causes it to release transmitter, the release of transmitter engenders the release of transmitter and such a synapse is sign conserving. Contrariwise, at an inhibitory chemical synapse, transmitter releases the voltage across the postsynaptic membrane—it hyperpolarizes. Because

hyperpolarizing a neuron causes it to release less transmitter, the release of a transmitter *reduces* the release of transmitter, and such a synapse is *sign inverting*. A gap junction is sign conserving because a depolarization of the presynaptic cell engenders a depolarization of the postsynaptic cell. An ephaptic synapse is sign inverting because the depolarization of the presynaptic cell increases the voltage difference across the postsynaptic cell's membrane, which has the same effect on voltage-sensitive ion channels as hyperpolarizing the postsynaptic cell. Sign-inverting and sign-conserving signals antagonize one another and can cancel each other completely. For example, when a depolarization meets a hyperpolarization of equal but opposite amplitude, this can result in zero net voltage change.

11.4.8 DYADIC SYNAPSE

A dyadic synapse is a type of sign-conserving chemical synapse where a single transmitter packet can be sensed by two postsynaptic neurons (Figure 11.8). The bipolar cell has a dyadic synapse, where the postsynaptic neurons are some combination of amacrine and ganglion cells (Dowling and Boycott 1966; Kolb 1979) (Figure 11.8). Photoreceptors have higher-order synapses where a single transmitter packet can be sensed by as many as 11 different neurons (DeVries et al. 2006; Sterling 2013). All dyadic and higher-order synapses have an elongated or disk-shaped structure in the presynaptic cell, called the ribbon. Synaptic vesicles line up along the ribbon and are thought to release their contents sequentially; therefore, the ribbon may act as a conveyer belt (Parsons and Sterling 2003).

11.4.9 DIGITAL AND ANALOG SIGNALING

When synapses depolarize a neuron sufficiently, this initiates an *action potential*. During an action potential, a neuron generates a pulse of positive voltage: it suddenly reverses its internal potential from negative to positive and then almost as suddenly becomes negative again. Action potentials are due to currents through specialized sodium and potassium channels that open when they sense a depolarization and then close with a delay. Because action potentials are of uniform size, sending information by their absence or presence, they are a digital mode of signaling.

A chemical synapse discretizes signals into transmitter packets, which can transmit digitally—for example, when packets are triggered by action potentials. Yet in the retina, many synapses produce packets at such high frequency that their effects pile up within the postsynaptic neuron, caused gradated changes in voltage, thereby transmitting signals in an analog mode.

11.5 ORGANIZATION OF THE RETINA INTO CIRCUITS

11.5.1 NEURON TYPES

The terms bipolar, horizontal, amacrine, and ganglion all define *classes* of neurons with similar function. Within each class, there are multiple *types*: each type has a different operating characteristic and is connected to a different selection of neurons (Figure 11.7). Neuronal classes are analogous to the components used to build computers, for example, central processors, graphic processors, and coprocessors form a class because of their similar construction and/or purpose, but each type of processor performs a different function. Retinal neurons are wired into a modular circuit that is repeated across the retina. Each module is composed of all of the retinal neuronal classes—photoreceptors, horizontal, bipolar, amacrine, and ganglion (Figure 11.8). The basic module repeats across the retina but is modified for the high-resolution areas (fovea) and for areas of high sensitivity (just outside the fovea).

11.5.2 VERTICAL AND LATERAL PATHWAYS

Although the retinal circuit is complex, for simplicity's sake, the retina is described as having two orthogonal pathways (Figure 11.8). Vertical pathways transmit signals from outer to inner plexiform layers. Signals flowing along vertical pathways start at the photoreceptor, enter the bipolar cell, and end in the ganglion cell.

A *lateral pathway* spreads signals along a plexiform layer. In the OPL, signals spread through gap junctions between photoreceptors and through gap junctions between horizontal cells. In the IPL, signals spread through gap junctions and chemical synapses between amacrine cells. Signals in lateral pathways antagonize the signals in vertical pathways. In general, vertical signals are larger and faster than lateral ones and are easier to demonstrate, for example, by flashing a light (100 ms). Lateral signals carried by horizontal cells require lengthier patterned stimuli, like annuli or enlarging spots, to show antagonism of vertical signals.

11.5.3 NEGATIVE FEEDBACK LOOPS

The reason that lateral pathways are antagonistic to vertical ones is that their constituent neurons make feedback synapses (Figure 11.8). The horizontal cell receives a sign-conserving synapse from the photoreceptor axon terminal and returns a sign-inverting ephaptic synapse back onto the axon terminal. When a signal follows this loop, it starts in the cone outer segment, passes through the cone axon terminal to the horizontal

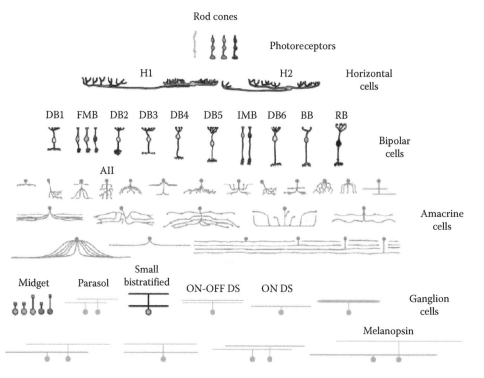

Figure 11.7 Cell types of primate retina (adapted from Masland 2001). Primate bipolar cells include ON and OFF midget bipolar cells (IMB, FMB), larger bipolar cells (DB1 through DB6), the blue bipolar cell (BB), and the rod bipolar cell (RB) (Boycott and Wässle 1999). The amacrine cells shown here are adopted from rabbit to provide an impression of the variety of forms (MacNeil and Masland 1998; MacNeil et al. 1999). Primate ganglion cells include types whose operating characteristics are inferred by homology to types from other mammals (ON–OFF DS, ON DS). (From Hannibal, J. et al., *J. Comp. Neurol.*, 522, 2231, 2014; Dacey, D.M. and Packer, O.S., *Curr. Opin. Neurobiol.*, 13, 421, 2003; Dacey, D.M. et al., *Neuron*, 37, 15, 2003.)

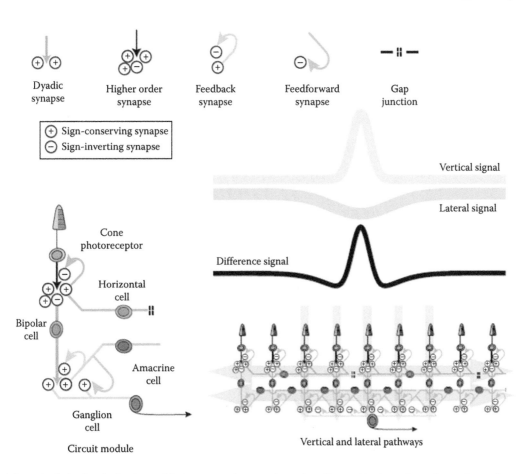

Figure 11.8 Modular retinal circuits. A circuit module is repeated across the retina. Cones, bipolar cells, and ganglion cells establish a vertical path that has a narrow distribution of signals representing the local image. Horizontal and amacrine cells establish two lateral paths with broader but inverted distributions that represent the global average. The lateral path makes inverting feedback and feedforward synapses onto the vertical path, creating a difference signal that emphasizes local changes in intensity differing from the global average.

cell, and then is inverted before it returns to the cone axon terminal: thus, the signal encounters an inverted version of itself (Figure 11.8). The amacrine cell also participates in a negative feedback loop: it receives a sign-conserving dyadic synapse from the bipolar cell axon and returns a sign-inverting chemical synapse back onto the axon (Figure 11.8). An amacrine cell's chemical synapse is termed a "conventional synapse" because it is not dyadic. Conventional synapses in the retina release the inhibitory amino acids glycine or GABA, the amino acid–derived dopamine, or acetylcholine, which is the same transmitter as the neuromuscular junction releases. Sometimes, conventional synapses release combinations of neurotransmitters (e.g., acetylcholine and GABA, dopamine and GABA) (Omalley et al. 1992; Hirasawa et al. 2012).

11.5.4 ANALOG-TO-DIGITAL CONVERSION

A ganglion cell receives transmitter packets and transmits action potentials to the brain. Because transmitter packets are an analog signal, and because action potentials are a binary signal, the ganglion cell accomplishes an analog-to-digital conversion. Some amacrine cells have multiple axons that transmit action potentials laterally across the retina over long distances (millimeters). All other neuronal classes, photoreceptor, horizontal, and bipolar cell, can transmit signals without action potentials in an analog mode (although see Protti et al. 2000, Baden et al. 2013).

11.6 HOW THE RETINA IS DESIGNED TO PROCESS INFORMATION

11.6.1 RETINA PROCESSING IN THE LIGHT IS COMPRESSIVE

To illustrate how the retina processes information, let us suppose that you, the reader, look at a picture on an old-fashioned green computer screen (luminance, 100 cd/m²; wavelength, 556 nm) (Figure 11.9). The reader scans the picture, looking at whatever is of interest. Scanning is accomplished by focusing different parts of the image on that region of the retina where processing power is concentrated, the fovea.

Photons fall on the fovea, their rate fluctuates, but their rate averaged over many seconds is about a thousand million photons per second (1×10^9 photons s^{-1}). About 40% of these photons will be absorbed, creating 4 hundred million meta opsins (R*) per second (4×10^8 R*/s).

This is an immense rate at which to absorb photons. Indeed, photons come too frequently for the brain to be informed about each and every one. If every R* were to be signaled by sending an action potential to the brain—a brute force method of processing—this would require each ganglion cell to fire action potentials at a rate of 20,000 per second, whereas normal rates evoked by scanning an image top out at 10 per second (Koch et al. 2006). Instead of brute force,

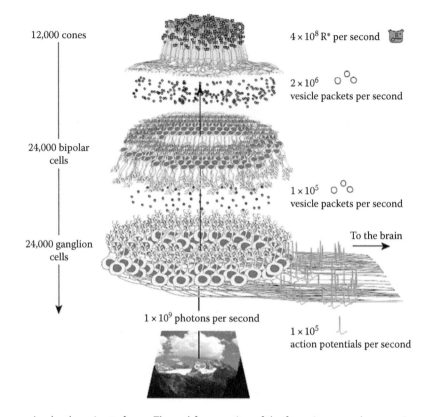

12,000 cones

4×10^8 R* per second

2×10^6 vesicle packets per second

24,000 bipolar cells

1×10^5 vesicle packets per second

24,000 ganglion cells

To the brain

1×10^9 photons per second

1×10^5 action potentials per second

Figure 11.9 Information processing by the primate fovea. The rod-free portion of the fovea is scanned over an image (at bottom). Photons from the image pass up through the vitreal (inner) surface of the retina, through the foveal pit where there are no ganglion or bipolar cells, and are absorbed by the inner segments of the cones. Information is transmitted down from the cones to the ganglion cells and then along ganglion cell axons to the brain. To the left are the numbers of neurons that transmit information. To the right are the rates of events that transmit this information. These rates fluctuate up and down as the image is scanned, so average rates are given.

the retina has a more efficient method. As signals travel through the retina, they are compressed into less and less frequent informative events, namely, vesicle packets and action potentials. Accordingly, the layer of cones generates 4×10^8 R* per second but release neurotransmitter packets much less frequently: about 2×10^6 packets per second (Choi et al. 2005). These transmitter packets are sensed by a layer of bipolar cells; this layer responds by releasing even fewer packets: 1×10^5 per second (Calkins et al. 1994; Freed and Liang 2010). These packets are sensed by a layer of ganglion cells that responds by firing 1×10^5 action potentials per second (Koch et al. 2006). Therefore, event rate, from R*s to action potentials, is compressed by about 1000 times (Figure 11.9).

The reader may have noticed, in this example of the fovea, that at the last stage of processing, there is an action potential for every vesicle, and thus no compression. The fovea is able to eschew compression because it has a massive number of ganglion cells: about one for every bipolar cell (Martin and Grünert 1992). Outside of the fovea, there are fewer ganglion cells for every bipolar cell, and compression is greater. Accordingly, in a mammalian retina that lacks a fovea, there is about 1 action potential for every 20 vesicles (Freed 2005; Freed and Liang 2010).

11.6.2 RETINAL COMPRESSION IS SIMILAR TO VIDEO COMPRESSION

Video compression and retinal compression have some similarities. Transmitting videos costs money to pay for bandwidth. Similarly, sending images through the retina and then to the brain costs metabolic energy in units of ATP. Indeed, every transmitter packet costs about 50,000 ATP and every action potential about 20,000,000 ATP (Koch et al. 2004; Alle et al. 2009). ATP is expended by ion pumps that maintain voltage across the membrane. Like video compression, retinal compression economizes by minimizing *redundant* events: events that convey the same information and that appear as spatial and temporal correlations between different neurons. Like video compression in its commonly used form, retinal compression loses information. Information is lost in the retina because it is obscured by noise generated by synapses and, to a lesser extent, by ion channels (Borghuis et al. 2009). Also, information is lost because retinal circuits act as filters that pass only certain kinds of information and block other kinds. The consequence of this loss is that all the action potentials fired by the layer of ganglion cells transmit less information than all the vesicles released from the overlying layer of cones (Borghuis et al. 2009).

Sparse coding Using infrequent events to transmit information is called "sparse coding." Sparse coding increases information per event because empty intervals between events (in space or time) convey information. Empty intervals are like zeros in binary words—they convey as much information as the 1's do. For example, an action potential is a binary unit worth only 1 bit by itself. Yet an action potential typically conveys 2 bits when surrounded by empty intervals (Koch et al. 2004). Sparse coding is economical because the empty intervals between action potentials cost very little, merely the metabolic energy to maintain the neuron in that interval.

Sparse coding makes a specific prediction about the information content of neural events. As event rate declines through the layers of the retina, and empty intervals increase, the information content of an event should also increase. As predicted, the cone layer releases transmitter packets at a high rate, and accordingly each packet conveys relatively little information: about 0.002 bits (Sterling and Freed 2007). The bipolar cell layer releases transmitter packets at a moderate rate, and each packet conveys a moderate amount of information: 0.1–0.4 bits (Freed 2005; Freed and Liang 2010). Ganglion cells outside the fovea fire action potentials relatively infrequently, and accordingly each action potential is worth a whopping 2 bits of information (Koch et al. 2004).

11.6.3 RETINAL PROCESSING IN THE DARK IS EXPANSIVE

If the reader were to turn off the computer monitor and sit in a very dark room (luminance: 1×10^{-6} cd/m^2), then the fovea would become useless because it contains cones and virtually no rods. To detect very dim objects in the room, the reader would need to look slightly to one side, so as to focus objects on the most sensitive part of the retina, just outside the fovea, where the rods are densest. The room is so dark that a single rod captures a single photon very infrequently: every 18 h. This is an astonishingly low rate, but signals converge so that at each stage of processing, a neuron receives photon signals more frequently. About 22 rods converge on a rod bipolar, which therefore collects a photon signal every 49 min (Grunert and Martin 1991). About 100 rod bipolar cells converge on a midget ganglion cell, through indirect pathways, which therefore collects a photon signal every 30 s (Dacey and Petersen 1992; Dacey 1993). Each photon signal causes a ganglion cell to fire 2–3 action potentials (Barlow et al. 1971; Mastronarde 1983). Because a photon signal is received by more than 20 ganglion cells, a single photon can trigger more than $2 \times 20 = 40$ spikes. Therefore, in the dark, the retina expands each photon into many spikes, an expansion very different from the compression that occurs in the light.

11.7 ON AND OFF PATHWAYS

ON cells depolarize when light turns on; OFF cells depolarize when light turns off (Figure 11.10). ON and OFF cells form vertical pathways that have some crosstalk between them, but which are separate for the most part. In this way, the brain receives separate signals from areas of the image that are brightening (ON) and from areas that are darkening (OFF).

The division into ON and OFF occurs at the very first synapse, the one that cones make onto bipolar cells. At this synapse, ON and OFF bipolar cells can sense the very same transmitter packet, but the packet hyperpolarizes the ON bipolar cell and depolarizes the OFF bipolar cell. Therefore, in effect, the cone synapse is both sign conserving and sign reversing. The trick here is that ON and OFF bipolar cells have different mechanisms to sense the neurotransmitter released by cones—which is the amino acid glutamate. For an ON bipolar cell, glutamate initiates a 2nd messenger cascade: it binds to a GPCR, which activates the

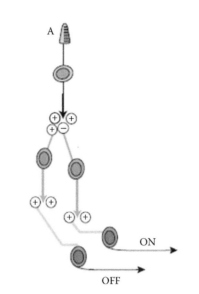

Figure 11.10 Circuits for lighting and darkening. The symbols for different synapses are those introduced in figure 8. ON and OFF bipolar cells receive sign-inverting and sign-conserving inputs from the same cone; ON and OFF signals are transmitted to ON and OFF ganglion cells.

G-protein G_O (Vardi 1998), which closes a TRP channel, and which hyperpolarizes the bipolar cell (Koike et al. 2010; Hughes et al. 2012). The GPCR is called mGluR6 and is classified as a "metabotropic synaptic receptor" because it initiates a 2nd messenger cascade (Slaughter and Miller 1981; Masu et al. 1991; Morigiwa and Vardi 1999). The 2nd messenger in this cascade is not known, but apparently it is *not* cG as it is for photoreceptors (Snellman et al. 2008).

For an OFF bipolar cell, glutamate binds to a receptor that is classified as *ionic* because it is also a channel—no 2nd messenger cascade required. When glutamate binds to this combined receptor–channel, the channel opens, allowing in positive ions, depolarizing the OFF bipolar cell. Note that light has an effect opposite to glutamate: when light turns on, the ON bipolar cell depolarizes; when light turns OFF, the OFF bipolar cell hyperpolarizes.

Due to this ability of the cone synapse to be both sign conserving and sign inverting, all retinal neuron classes are divided into ON and OFF. Horizontal cells receive sign-conserving signals from photoreceptors and are therefore OFF. Bipolar, ganglion, and amacrine cells are ON or OFF because they receive signals from ON or OFF bipolar cells. Some amacrine and ganglion cells are both ON and OFF, responding to both lightening and darkening, because they receive both ON and OFF signals.

The utility of dividing light from dark signals may be to conserve ATP energy (Liang and Freed 2010). Natural scenes have a disproportionate number of gray areas, intermediate between the brightest and darkest regions. If there were, hypothetically, only ON neurons to signal all areas of the image, these neurons would need to signal dark with a zero rate and bright with a high rate, leaving an intermediate rate to signal gray (if there were only OFF neurons, they too would need to signal gray with an intermediate rate). These hypothetical neurons would therefore spend a disproportionate amount of time at intermediate rate. The advantage of ON and OFF cells may be that they respond to gray areas with a low rate of transmitter packets or spikes, therefore conserving ATP.

11.8 ROD AND CONE PATHWAYS

Rod signals flow through neurons that are part of the cone circuit, and this, if effect, the rods "borrow" components of the cone circuit. Presumably, rods borrowed from cones in the course of vertebrate evolution, as rods evolved from cones (Okano et al. 1992; Lamb 2013). Borrowing occurs in mammals as in other vertebrates, but mammals have a specialized ON bipolar cell that only transmits rod signals and only at low light levels. Mammals also have a specialized ON amacrine cell that injects rod signals into the cone pathways, called type AII (Figure 11.11).

At the highest light levels, cone signals go through both vertical and lateral pathways and are processed for spatial contrast. At intermediate light levels, cone signals are depreciated, but rod signals flow through essentially these same vertical and lateral pathways as cone signals did and are processed for spatial contrast. In low light, rod signals flow exclusively through vertical pathways, and therefore spatial contrast is depreciated. The reason for this is that photons are absorbed infrequently by a few widely spread rods, yet all rods continuously generate noise. Lateral pathways would spread noise through gap junctions, drowning the photon signal, or add noise from inhibitory chemical synapses. Therefore, in low light, lateral pathways are shut down by closing gap junctions between photoreceptors and between AII amacrine cells (Smith et al. 1986; Bloomfield et al. 1997; DeVries et al. 2002).

11.8.1 DIFFERENT PATHWAYS FOR DIFFERENT LIGHT LEVELS

Figure 11.11 shows in detail the pathways for processing rod and cone signals. In bright light, there are separate ON and OFF vertical pathways for cone signals. Through the ON pathway, signals flow from cones to ON bipolar cells and then to ON ganglion cells (Figure 11.11, pathway 1 ON). Through the OFF pathway, signals flow from cones to OFF bipolar cells and then to OFF ganglion cells (Figure 11.11, pathway 1 OFF). In intermediate light, gap junctions between rods and cones open, allowing rod signals to enter the cone axon terminal, and thereafter enter both ON and OFF vertical cone pathways (Figure 11.11, pathways 2 ON and 2 OFF). At low light levels, signals travel directly from the rod axon terminal through its synapse onto the specialized rod bipolar cell. The rod bipolar cell, unlike all other types of ON bipolar cell, makes very few synapses onto ganglion cells (Freed and Sterling 1988). Instead, the rod bipolar cell transmits signals to the AII amacrine cell, which then transmits to ON and OFF cone bipolar cells (Figure 11.11, pathways 3 ON and 3 OFF).

The low-light circuit seems designed to add little extra noise: gap junctions between the AII amacrine and the ON bipolar cell are relatively noiseless. Also to reduce noise, the ON cone bipolar cell is hyperpolarized by transmitter release from the cone, which reduces the random, noisy thrum of transmitter packets onto the ON ganglion cell. Possibly—this is not yet known—the OFF bipolar cell is hyperpolarized by transmitter release from the AII amacrine cell, to reduce packet noise in the OFF ganglion cell.

The AII amacrine cell is a component of two circuits: in the bright-light circuit, it transmits signals from ON bipolar cells to OFF ganglion cells; in the the low-light circuit, it transmits signals from rod bipolar cells to ON and OFF cone bipolar cells. ON and OFF cone bipolar cells are also components of bright-light and low-light circuits. Consequently, different circuits can share component neurons, and therefore, a circuit should not be defined as an assemblage of neurons, but rather as a route traced by signals.

11.9 RETINAL CIRCUITS FOR FILTERING SIGNALS

Retinal circuits act as filters, selecting information about some features of the retinal image and discarding information about others. Each filter circuit ends in a distinct ganglion cell type that responds to certain stimuli and ignores others. Because each ganglion cell type sends its axon to its own combination of brain nuclei, filtering ensures that each part of the brain gets the

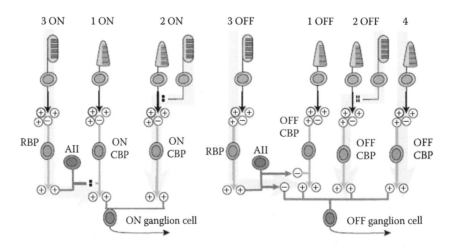

Figure 11.11 Vertical pathways for rod and cone signals (after Wässle 2004). Rods synapse on the rod bipolar cell (RBP); cones synapse on the cone bipolar cells (CBP). Rod pathways (highlighted in yellow) and cone pathways are numbered by decreasing light levels: from high (1 ON and 1 OFF) to intermediate (2 ON and 2 OFF) to low (3 ON and 3 OFF). Pathway 4 has been found in mouse. (From Tsukamoto, Y. et al., *J. Neurosci.*, 21, 8616, 2001.)

information required for its special visual processing. Filtering also helps in the compression of many photons into fewer events by eliminating information not necessary for further visual processing. The remainder of this chapter will enumerate these filters, and trace out the circuits that implement them.

11.9.1 TEMPORAL CHANGES

Almost all retinal neurons respond to changes more robustly than they respond to constancy. The photoreceptor is an example: if a light is turned on and then left on, cG levels are regenerated and ionic channels reopened, thus curtailing the photocurrent. Consequently, the photocurrent is large during the first 200 ms or so, but subsequently declines (Figure 11.4). Similarly, secondary and tertiary neurons respond most robustly to momentary changes in contrast. Depending on the neuron and the stimulus, this ability to signal change and to de-emphasize constancy works at different time spans and is variously called gain control, background adaptation, contrast adaptation, and light/dark adaptation, ranked from milliseconds to minutes. Change in an animal's environment provides greater danger and opportunity than constancy does, so responding to change is advantageous. Responding to change also contributes to sparse coding by saving transmitter packets and spikes for intermittent but important events.

11.9.2 LUMINANCE

Melanopsin ganglion cells do not change as most retinal neurons do but instead signal luminance. Their opsin and 2nd messenger cascade mediate enduring responses to constant light that last for hours and perhaps even longer (Wong 2012). A melanopsin ganglion cell's ability to respond steadily is put to use by brain nuclei that do not analyze images but regulate activity over the course of the day. The M1 brnb3$^-$ type of melanopsin ganglion cell sends signals to the hypothalamus that entrains the circadian clock (Schmidt et al. 2011); the M1 brnb3$^+$ type projects to the pretectum, which controls pupil size, moderating the rate at which photons strike the retina.

11.9.3 SPATIAL CONTRAST

Most retinal neurons signal local differences in intensity across the retina. This is accomplished by combining signals from vertical and lateral pathways (Figure 11.8). In the vertical pathway, signals spread over short distances, and the distribution of signal amplitude approximates a narrow Gaussian distribution. In the lateral pathway, signals spread over longer distances, and the distribution approximates a larger Gaussian. The lateral pathway receives inverted signals from the vertical pathway. Also, the lateral pathway makes feedforward and feedback sign-inverting synapses onto the vertical pathway. The result is that the lateral pathway averages signals across the retina, and this average is subtracted from the vertical local signal. Therefore, the retina responds to local changes that are different from the global mean.

11.9.4 MOTION

Bipolar cells contribute to motion sensitivity by transmitting depolarizations but not hyperpolarizations to ganglion cells, a form of *rectification*. Rectification occurs at the dyadic synapse between bipolar and ganglion cells. At this synapse, there is a threshold voltage for transmitter release: normally the bipolar cell is hyperpolarized below this threshold. Only when the bipolar cell depolarizes beyond this threshold does it release transmitter.

To understand the necessity for rectification, consider what would happen without it. A spot of light moves across a row of bipolar cells, causing each bipolar cell to respond in sequence (Figure 11.12a). A bipolar cell responds to the spot moving toward it with a depolarization and to the spot moving away with a hyperpolarization. If these signals were fed unaltered to the ganglion cell, then a depolarization from a bipolar cell underneath the spot would be canceled by a hyperpolarization from a bipolar cell that was underneath a moment ago. Therefore, only the first bipolar cell would get its signal through, and a moving spot would be indistinguishable from a spot over the first bipolar cell. With rectification, the depolarizations arrive at the ganglion cell uncanceled, producing a more robust response to movement (Figure 11.12b).

The ganglion cell receiving these motion signals have large dendritic trees: for example, ON and OFF parasol cells in primate and ON and OFF alpha cells in cat, mouse, and other mammals (Crook et al. 2008). Sometimes, separate motion-sensitive circuits converge on the same ganglion cell. An example is the ON alpha cell: one circuit is remote from its dendritic tree, and the other is just above it (Demb et al. 1999). Remote signals cancel local signals, so the ON alpha cell responds to local motion of an object against a background, but not global motion caused by eye movement (Olveczky et al. 2003).

11.9.5 DIRECTION OF MOTION

In the previous section, we specified a motion-detecting circuit with rectifying synapses. This circuit will respond the same whether the object goes left or right or—generalizing to the 2D retinal image—in any other direction (Figure 11.12a and b). The circuit lacks directional selectivity because it has the same structure whichever direction it is measured in and is therefore *isotropic*. Directionally selective (DS) ganglion cells fire spikes for objects moving in one direction, and not in other directions, which implicates a directional circuit with *anisotropic* structure.

The directional circuit relies on an intrinsic property of dendritic arbors: synapses activated in sequence, starting near the cell body and ending at the tip-most dendritic branches, will produce a larger voltage change at the dendritic tip than synapses activated in the opposite sequence (Euler et al. 2002; Tukker et al. 2004) (Figure 11.12c). This property causes a larger depolarization to objects moving radially away the soma than to objects moving inward toward the soma and results in a *centrifugal* selectivity.

The starburst is an amacrine cell whose strong centrifugal selectivity is critical for directional selectivity (Yoshida et al. 2001). The starburst is so called because its dendrites splay out from its cell body like an exploding starburst firework (Famiglietti 1983) (Figure 11.12c). The starburst receives signals from bipolar cells throughout its dendritic tree but transmits signals through synapses that are confined to tree branches distal to its cell body—where the depolarization from centrifugal currents is largest (Euler et al. 2002).

The DS ganglion cell fires action potentials when objects move in a line across its dendritic tree: a linear selectivity different from the centrifugal selectivity of the starburst

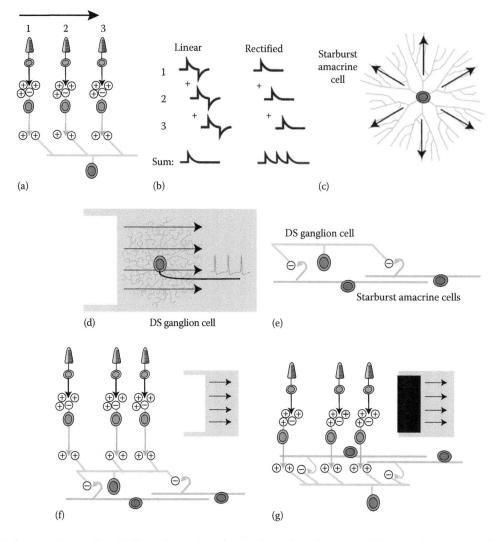

Figure 11.12 Circuits for detecting motion. (a) Circuit for motion. The circuit consists of an array of bipolar cells and a ganglion cell that sums their signals. An object moves from left to right (arrow). (b) Bipolar cells response to the moving object. If the bipolar cells were to have linear responses (left column), then their responses would tend to cancel one another. Instead, the responses are rectified (right column) and the sum is more robust. (c) The starburst amacrine cell is depolarized by movement away from the soma (centrifugal selectivity). (d) The directionally selective (DS) ganglion cell is depolarized by linear movement from left to right, which causes it to fire action potentials. Here, the moving object is a bright edge. (e) Anisotropic directional circuit. Starburst dendrites to the left of the starburst's soma make sign-inverting synapses onto the ganglion cell. (f) Motion and directional circuits converge on a DS ganglion cell. Motion circuit depolarizes ganglion cell for either direction, but directional circuit hyperpolarizes only for movement to the left. Therefore, only rightward movement causes depolarization unopposed by hyperpolarization and triggers spikes. Note this circuit includes ON bipolar and ON starbursts and confers directional sensitivity to bright edge a as in (d). (g) OFF DS circuit includes OFF starburst and OFF bipolar cells and confers directional selectivity to a dark edge moving rightward.

(Figure 11.12d). The directional circuit transforms the starburst's centrifugal selectivity into the ganglion cell's linear one. This transformation is accomplished by anisotropic wiring (Briggman et al. 2011). The DS ganglion cell receives many sign-inverting synapses from those starburst dendrites that select for the direction that is opposite to its preferred direction; the DS ganglion cell receives fewer synapses from those starburst dendrites that select for the same direction. Diagrammatically, a starburst dendrite to the left of the starburst cell body will synapse upon a ganglion cell that fires action potentials to rightward movement (Figure 11.12e).

An isotropic motion-sensitive array of bipolar cells also synapses on the dendritic tree of the DS cell, but with sign-conserving synapses (Figure 11.12f). The result is that leftward movement evokes depolarizing signals from bipolar cells that are antagonized by

hyperpolarizing signals from starburst cells. Rightward movement evokes depolarizing signals from bipolar cells that are unopposed, triggering action potentials. Predictably from this wiring, starbursts whose cell bodies are offset (to the right) from the DS ganglion cell provide sign-inverting signals (Fried et al. 2002).

The DS ganglion cell's dendritic tree has a centrifugal selectivity of its own. A centrifugal sequence of synaptic inputs initiates small spikes in the dendrites, which triggers larger spikes near the cell body (Oesch et al. 2005). The DS ganglion cell's centrifugal selectivity can either help or hurt the linear selectivity established by retinal circuitry (Schachter et al. 2010; Trenholm et al. 2011). For a DS ganglion cell that is selective for rightward movement, centrifugal and linear selectivity will coincide on the right side of its dendritic tree but will mutually interfere on the left side of its dendritic tree. This interference causes a "nondiscriminating zone," where the DS ganglion cell responds equally well to oppo-site directions of movement (Barlow and Levick 1965). Some DS ganglion cells avoid this nondiscriminating zone by the simple expedient of lacking dendrites where the nondiscriminating zone would have been (Trenholm et al. 2011).

Mice and rabbits have DS ganglion cells divided into 3 groups, which are further divided into more than 8 types. Whether primate retina has such a diverse collection of DS cells is not yet known. The first group, ON DS cells, may help keep the eyes focused on an object even as the head moves (vestibulo-ocular reflex). The reader can try this: focus on a stationary object while moving one's head. The eyes remain locked on the object, a natural ability that requires an accurate coordination of head and eye movements. There are several lines of circumstantial evidence that the ON DS cell helps coordinate such head and eye movements. Head movement traverses the entire image over the retina, a global movement that the ON DS cells can sense. In rabbits, there is a correspondence between the 3 types of ON DS cell and the 3 semicircular canals of the vestibular system: head movement that stimulates a single semicircular canal will be sensed by a single type of ON DS cell (Oyster and Barlow 1967; Oyster et al. 1972). ON DS cells send axons to the medial termi-nal nucleus of the accessory optic system (AOS), which controls eye movements (Simpson et al. 1988; Yonehara et al. 2009).

A second group, ON–OFF DS ganglion cells, may help track moving objects. The reader can try this: focus on a moving object and hold the head still. The eyes lock on the object, another remarkable natural ability. Yet tracking is inevitably disrupted by errors in eye movement and by unpredictable changes in object velocity. Therefore, an error signal is required to steer the eyes when an image moves over the retina. There are multiple lines of evidence that ON–OFF DS cells could generate this error signal. Tracking error causes local movements of the object's image over the retina that the ON–OFF DS cells can sense. In rab-bits, there is a correspondence between 4 types of ON–OFF DS cells and 4 extraocular muscles: each type of ON–OFF DS cell could provide an error signal for each muscle (Oyster and Barlow 1967; Rodieck 1998). ON–OFF DS cells receive inputs from congruent ON and OFF circuits and, thus, generalize movement direction to objects that brighten or darken. ON–OFF DS cells send signals to the superior colliculus, important for locating and targeting objects, and through the thalamus to the visual cortex for conscious perception of movement (Cruz-Martin et al. 2014). Some ON–OFF DS cells that sense vertical movement send axons to the AOS and might participate in eye movements (Kay et al. 2011; Rivlin-Etzion et al. 2011).

A third group of DS ganglion cell, the OFF DS ganglion cells, signals upward motion of objects—by sensing downward movement of images on the retina and projecting to the superior colliculus, less to the superior colliculus, but not at all to the AOS (Kim et al. 2008).

11.9.6 COLOR

Human color perception is the combination of two basic algorithms. First, there is *trichromacy*: the ability to distinguish colors. Second, there is *color opponency*: red against green and blue against yellow.

Trichromacy is based on different opsins that respond to short-, middle-, or long-wavelength light, and which are expressed in *S-, M-, and L-type cones*, respectively. Nature provides a natural experiment to show the contribution of cones to trichromacy, because different people have different numbers of cones. *Monochromats* are rare people who have only one type of cone. For these people, any number of visible wavelengths of light, if their intensities are adjusted correctly, will produce exactly the same cone response. Therefore, the monochromat, having but one cone type, confounds wavelengths with intensities and cannot tell colors apart. More numerous people are *dichromats* and more numerous still are *trichromats*, who have two or three types of cones, respectively. Increasing the number of cone types reduces the number of wavelength–intensity combinations that can pro-duce the same response and increases the number of distinguish-able colors. *Tetrachromats* are rare people, almost always female, who chimerically express two L opsins, resulting in four differ-ent cone types (Neitz and Neitz 2011). For tetrachromats, there should be exceedingly few wavelength–intensity combinations that produce the same responses in all four cones. Indeed, some tetrachromats seem able to take advantage of this neural disam-biguation and can tell apart colors than normal people cannot (Jordan and Mollon 1993).

Color opponency is based on retinal circuits that subtract the responses of different cone types. A well-researched example is a retinal circuit for blue–yellow opponency. The circuit imple-ments an algorithm that can be described mathematically in terms of S, M, and L signals and is divided into ON and OFF subcircuits. An ON subcircuit uses a vertical pathway to gener-ate an S signal and a lateral pathway to subtract an M+L+s signal, producing S−(M+L+s) (the small "s" denotes a small amount of S signal) (Figure 11.13a). An OFF subcircuit uses a vertical pathway to generate an M+L+s signal and a lateral pathway to subtract an L+M signal, producing (M+L+s) − (L+M) (Figure 11.13b). ON and OFF subcircuits converge on the small bistratified ganglion cell. ON and OFF signals subtract and this, the ganglion cell performs the operation [S−(M+L+s)] − [(M+L+s) − (L+M)], which results in S−(L+M) (Figure 11.13c).

How red–green opponency arises is a matter for discussion. Possibly red–green circuits are constructed from midget bipolar and midget ganglion cells (these cells are so called because they are so small in the fovea). For example, to construct an L–M circuit outside the fovea, an ON midget bipolar cell collects from an L cone that collects inverted M signals from H2 horizontal cells. A midget ganglion cell collects from midget bipolar cells with L–M signals and transmits these signals to the thalamus. Yet there is one important feature of the midget circuit that is not well understood: to what extent L and M signals mix. Mixing would diminish the selectivity of a midget ganglion cell for red or green. To avoid mixing requires that the bipolar, ganglion, and horizontal cells all pick signals according to opsin. Yet there is evidence for such choosiness and also evidence against it (Martin et al. 2001; Diller et al. 2004; Field et al. 2010).

Some scientists have suggested that the midget system is better suited for transmitting fine details of images than it is for color opponency (Calkins and Sterling 1999). Midget ganglion cells are tightly packed, and each transmits signals vertically from only one cone, thus rendering images in fine detail. Color vision, on the contrary, is blurred, suggesting a larger ganglion cell that col-lects from many cones. A large opponent ganglion cell may help

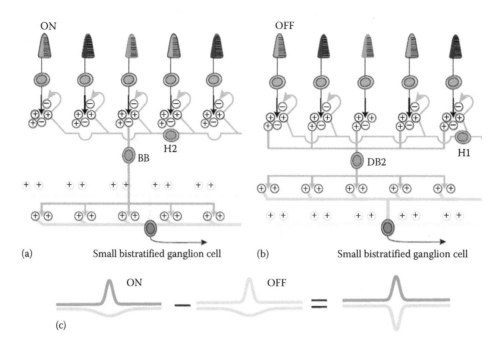

Figure 11.13 Circuit for blue/yellow opponency. (a) ON subcircuit. The H2 horizontal cell collects from L and M cones (red and green), but few S cones (blue), constructing a broad yellowish signal. The S cone receives this yellowish signal through an inverting feedback synapse, adding a narrow blue signal. The blue bipolar (BB) collects both blue and yellow signals from the S cone. (b) OFF subcircuit. The H1 horizontal cell collects from L and M cones, constructing a broad yellowish signal. M and L cones receive this yellow signal through an inverting feedback synapse, adding a narrow yellowish signal. The DB2 bipolar cell collect broad and narrow yellowish signals from a few S cones and more M and L cones. (c) ON and OFF signals are collected by the small bistratified ganglion cell. The broad yellowish signals from the two subcircuits cancel out, leaving narrow blue and yellowish signals. (After Dacey, D.M. et al., *Visual Neurosci.*, 1, 2013.)

explain why color circuits are incompletely described: relatively few large ganglion cells are required to cover the retina, suggesting that opponent cells are rare, and this may be why they have not been located often enough to establish their wiring.

What is known about color circuits in the retina cannot account fully for the human perception of color, in part because the correspondence between colors and cone signals is not simple. For example, the L cone is sometimes called "red" but is actually most sensitive to wavelengths that look yellowish. If people are asked to pick a red that is neither green nor blue nor yellow, they will pick a red that cannot be specified with L and M signals alone but requires some S signals too: (S+L)–M (Neitz and Neitz 2008). Asked to pick other unique hues, people will consistently choose a yellow, a blue, a green, and a red, all of which would require mixing of opponent signals from the retina or thalamus (Wuerger et al. 2005). Apparently then, color opponent signals do not constitute fully formed visual percepts but are reprocessed and remixed in the brain. If true, this would suggest that color opponent circuits in the retina do what other retinal circuits do: collect signals from one group of neurons and then subtract signals from another group, to reduce the redundancy of signals transmitted to the brain.

REFERENCES

Alle, H., Roth, A., and Geiger, J.R., 2009. Energy-efficient action potentials in hippocampal mossy fibers. *Science*, 325(5946), 1405–1408.

Angueyra, J.M. and Rieke, F., 2013. Origin and effect of phototransduction noise in primate cone photoreceptors. *Natural Neuroscience*, 16(11), 1692–1700.

Arshavsky, V.Y. and Burns, M.E., 2014. Current understanding of signal amplification in phototransduction. *Cellular Logistics*, 4, e29390.

Baden, T. et al., 2013. Spikes and ribbon synapses in early vision. *Trends in Neurosciences*, 36(8), 480–488.

Barlow, H.B., 1957. Purkinje shift and retinal noise. *Nature*, 179(4553), 255–256.

Barlow, H.B. and Levick, W.R., 1965. The mechanism of directionally selective units in rabbit's retina. *Journal of Physiology London*, 178(3), 477–504.

Barlow, H.B., Levick, W.R., and Yoon, M., 1971. Responses to single quanta of light in retinal ganglion cells of the cat. *Vision Research Supplements*, 3, 87–101.

Bloomfield, S.A., Xin, D.Y., and Osborne, T., 1997. Light-induced modulation of coupling between Aii amacrine cells in the rabbit retina. *Visual Neuroscience*, 0014, 565–576.

Borghuis, B.G., Sterling, P., and Smith, R.G., 2009. Loss of sensitivity in an analog neural circuit. *Journal of Neuroscience*, 29(10), 3045–3058.

Boycott, B. and Wässle, H., 1999. Parallel processing in the mammalian retina. *Investigative Ophthalmology and Visual Science*, 40, 1313–1327.

Briggman, K.L., Helmstaedter, M., and Denk, W., 2011. Wiring specificity in the direction-selectivity circuit of the retina. *Nature*, 471(7337), 183–188.

Calkins, D.J. and Sterling, P., 1999. Evidence that circuits for spatial and color vision segregate at the first retinal synapse. *Neuron*, 24(2), 313–321.

Calkins, D.J. et al., 1994. M and L cones in macaque fovea connect to midget ganglion cells by different numbers of excitatory synapses. *Nature*, 371(6492), 70–72.

Choi, S.Y. et al., 2005. Encoding light intensity by the cone photoreceptor synapse. *Neuron*, 48(4), 555–562.

Crook, J.D. et al., 2008. Y-cell receptive field and collicular projection of parasol ganglion cells in macaque monkey retina. *Journal of Neuroscience*, 28(44), 11277–11291.

Cruz-Martín, A. et al., 2014. A dedicated circuit links direction-selective retinal ganglion cells to the primary visual cortex. *Nature*, 507(7492), 358–361.

Dacey, D.M., 1993. The mosaic of midget ganglion cells in the human retina. *Journal of Neuroscience*, 13(12), 5334–5355.

Dacey, D.M. and Packer, O.S., 2003. Colour coding in the primate retina: Diverse cell types and cone-specific circuitry. *Current Opinion in Neurobiology*, 13(4), 421–427.

Dacey, D.M. and Petersen, M.R., 1992. Dendritic field size and morphology of midget and parasol ganglion cells of the human retina. *Proceedings of the National Academy of Sciences of the United States of America*, 89(20), 9666–9670.

Dacey, D.M. et al., 2003. Fireworks in the primate retina: In vitro photodynamics reveals diverse LGN-projecting ganglion cell types. *Neuron*, 37(1), 15–27.

Dacey, D.M., Crook, J.D., and Packer, O.S., 2013. Distinct synaptic mechanisms create parallel S-ON and S-OFF color opponent pathways in the primate retina. *Visual Neuroscience*, 31(2), 1–13.

Demb, J.B. et al., 1999. Functional circuitry of the retinal ganglion cell's nonlinear receptive field. *Journal of Neuroscience*, 19(22), 9756–9767.

DeVries, S.H. et al., 2002. Electrical coupling between mammalian cones. *Current Biology*, 12(22), 1900–1907.

DeVries, S.H., Li, W., and Saszik, S., 2006. Parallel processing in two transmitter microenvironments at the cone photoreceptor synapse. *Neuron*, 50(5), 735–748.

Diller, L. et al., 2004. L and M cone contributions to the midget and parasol ganglion cell receptive fields of macaque monkey retina. *Journal of Neuroscience*, 24(5), 1079–1088.

Dowling, J.E. and Boycott, B.B., 1966. Organization of the primate retina: Electron microscopy. *Proceedings of the Royal Society London* [Biology], 166, 80–111.

Euler, T., Detwiler, P.B., and Denk, W., 2002. Directionally selective calcium signals in dendrites of starburst amacrine cells. *Nature*, 418(6900), 845–852.

Famiglietti, E.V.J., 1983. "Starburst" amacrine cells and cholinergic neurons: Mirror-symmetric on and off amacrine cells of rabbit retina. *Brain Research*, 261, 138–144.

Fernald, R.D., 2006. Casting a genetic light on the evolution of eyes. *Science*, 313(5795), 1914–1918.

Field, G.D. et al., 2010. Functional connectivity in the retina at the resolution of photoreceptors. *Nature*, 467(7316), 673–677.

Freed, M.A., 2005. Quantal encoding of information in a retinal ganglion cell. *Journal of Neurophysiology*, 94(2), 1048–1056.

Freed, M.A. and Liang, Z., 2010. Reliability and frequency response of excitatory signals transmitted to different types of retinal ganglion cell. *Journal of Neurophysiology*, 103(3), 1508–1517.

Freed, M.A. and Liang, Z., 2014. Synaptic noise is an information bottleneck in the inner retina during dynamic visual stimulation. *Journal of Physiology*, 592(Pt 4), 635–651.

Freed, M.A. and Sterling, P., 1988. The ON-alpha ganglion cell of the cat retina and its presynaptic cell types. *Journal of Neuroscience*, 8, 2303–2320.

Fried, S.I., Munch, T.A., and Werblin, F.S., 2002. Mechanisms and circuitry underlying directional selectivity in the retina. *Nature*, 420(6914), 411–414.

Grunert, U. and Martin, P.R., 1991. Rod bipolar cells in the macaque monkey retina—Immunoreactivity and connectivity. *Journal of Neuroscience*, 11(9), 2742–2758.

Hannibal, J. et al., 2014. Central projections of intrinsically photosensitive retinal ganglion cells in the macaque monkey. *The Journal of Comparative Neurology*, 522(10), 2231–2248.

Hirasawa, H., Betensky, R.A., and Raviola, E., 2012. Corelease of dopamine and GABA by a retinal dopaminergic neuron. *Journal of Neuroscience*, 32(38), 13281–13291.

Hughes, S. et al., 2012. Profound defects in pupillary responses to light in TRPM-channel null mice: A role for TRPM channels in non-image-forming photoreception. *European Journal of Neuroscience*, 35(1), 34–43.

Jordan, G. and Mollon, J.D., 1993. A study of women heterozygous for colour deficiencies. *Vision Research*, 33(11), 1495–1508.

Kawamura, S. and Tachibanaki, S., 2008. Rod and cone photoreceptors: Molecular basis of the difference in their physiology. *Comparative Biochemistry and Physiology Part A: Molecular and Integrative Physiology*, 150(4), 369–377.

Kay, J.N. et al., 2011. Retinal ganglion cells with distinct directional preferences differ in molecular identity, structure, and central projections. *Journal of Neuroscience*, 31(21), 7753–7762.

Kenkre, J.S. et al., 2005. Extremely rapid recovery of human cone circulating current at the extinction of bleaching exposures. *Journal of Physiology London*, 567(Pt 1), 95–112.

Kim, I.J. et al., 2008. Molecular identification of a retinal cell type that responds to upward motion. *Nature*, 452(7186), 478–482.

Koch, K. et al., 2004. Efficiency of information transmission by retinal ganglion cells. *Current Biology*, 14(17), 1523–1530.

Koch, K. et al., 2006. How much the eye tells the brain. *Current Biology*, 16(14), 1428–1434.

Koike, C. et al., 2010. TRPM1 is a component of the retinal ON bipolar cell transduction channel in the mGluR6 cascade. *Proceedings of the National Academy of Sciences of the United States of America*, 107(1), 332–337.

Kolb, H., 1979. The inner plexiform layer in the retina of the cat: Electron microscopic observations. *Journal of Neurocytology*, 8, 295–329.

Korenbrot, J.I., 2012. Speed, sensitivity, and stability of the light response in rod and cone photoreceptors: Facts and models. *Progress in Retina and Eye Research*, 31(5), 442–466.

Lamb, T.D., 2013. Evolution of phototransduction, vertebrate photoreceptors and retina. *Progress in Retina and Eye Research*, 36, 52–119.

Liang, J. and Freed, M.A., 2010. The On pathway rectifies the Off pathway of the mammalian retina. *Journal of Neuroscience*, 30(16), 5533–5543.

MacNeil, M.A. and Masland, R.H., 1998. Extreme diversity among amacrine cells: Implications for function. *Neuron*, 20(5), 971–982.

MacNeil, M.A. et al., 1999. The shapes and numbers of amacrine cells: Matching of photofilled with Golgi-stained cells in the rabbit retina and comparison with other mammalian species. *Journal of Comparative Neurology*, 413(2), 305–326.

Martin, P.R. and Grünert, U., 1992. Spatial density and immunoreactivity of bipolar cells in the macaque monkey retina. *The Journal of Comparative Neurology*, 323(2), 269–287.

Martin, P.R. et al., 2001. Chromatic sensitivity of ganglion cells in the peripheral primate retina. *Nature*, 410(6831), 933–936.

Masland, R.H., 2001. The fundamental plan of the retina. *Nature Neuroscience*, 4(9), 877–886.

Mastronarde, D.N., 1983. Correlated firing of cat retinal ganglion cells. I. Spontaneously active inputs to X- and Y-cells. *Journal of Neurophysiology*, 49, 303–324.

Masu, M. et al., 1991. Sequence and expression of a metabotropic glutamate receptor. *Nature*, 349(6312), 760–765.

Morigiwa, K. and Vardi, N., 1999. Differential expression of ionotropic glutamate receptor subunits in the outer retina. *Journal of Comparative Neurology*, 405(2), 173–184.

Neitz, J. and Neitz, M., 2008. Colour vision: The wonder of hue. *Current Biology*, 18(16), R700–R702.

Neitz, J. and Neitz, M., 2011. The genetics of normal and defective color vision. *Vision Research*, 51(7), 633–651.

Fundamentals

Oesch, N., Euler, T., and Taylor, W.R., 2005. Direction-selective dendritic action potentials in rabbit retina. *Neuron*, 47(5), 739–750.

Okano, T. et al., 1992. Primary structures of chicken cone visual pigments - vertebrate rhodopsins have evolved out of cone visual pigments. *Proceedings of the National Academy of Sciences of the United States of America*, 89(13), 5932–5936.

Olveczky, B.P., Baccus, S.A., and Meister, M., 2003. Segregation of object and background motion in the retina. *Nature*, 423(6938), 401–408.

Omalley, D.M., Sandell, J.H., and Masland, R.H., 1992. Co-release of acetylcholine and GABA by the starburst amacrine cells. *Journal of Neuroscience*, 12(4), 1394–1408.

Oyster, C.W. and Barlow, H.B., 1967. Direction-selective units in rabbit retina: Distribution of preferred directions. *Science*, 155(3764), 841–842.

Oyster, C.W., Takahashi, E., and Collewijn, H., 1972. Direction-selective retinal ganglion cells and control of optokinetic nystagmus in the rabbit. *Vision Research*, 12(2), 183–193.

Parsons, T.D. and Sterling, P., 2003. Synaptic ribbon. Conveyor belt or safety belt? *Neuron*, 37(3), 379–382.

Protti, D.A., Flores-Herr, N., and Gersdorff, von, H., 2000. Light evokes Ca2+ spikes in the axon terminal of a retinal bipolar cell. *Neuron*, 25(1), 215–227.

Provencio, I. et al., 1998. Melanopsin: An opsin in melanophores, brain, and eye. *Proceedings of the National Academy of Sciences of the United States of America*, 95(1), 340–345.

Pu, G.A. and Dowling, J.E., 1981. Anatomical and physiological characteristics of pineal photoreceptor cell in the larval lamprey, Petromyzon marinus. *Journal of Neurophysiology*, 46(5), 1018–1038.

Rieke, F. and Baylor, D.A., 1996. Molecular origin of continuous dark noise in rod photoreceptors. *Biophysical Journal*, 71(5), pp.2553–2572.

Rivlin-Etzion, M. et al., 2011. Transgenic mice reveal unexpected diversity of on-off direction-selective retinal ganglion cell subtypes and brain structures involved in motion processing. *Journal of Neuroscience*, 31(24), 8760–8769.

Rodieck, R.W., 1998. *The First Steps in Seeing*, Sunderland, MA: Sinauer Associates.

Rohlich, P., van Veen, T., and Szel, A., 1994. Two different visual pigments in one retinal cone cell. *Neuron*, 13, 1159–1166.

Schachter, M.J. et al., 2010. Dendritic spikes amplify the synaptic signal to enhance detection of motion in a simulation of the direction-selective ganglion cell. *PLoS Computing Biology*, 6(8), e1000899.

Schmidt, T.M., Chen, S.-K., and Hattar, S., 2011. Intrinsically photosensitive retinal ganglion cells: Many subtypes, diverse functions. *Trends in Neurosciences*, 34(11), 572–580.

Schnapf, J.L. et al., 1990. Visual transduction in cones of the Monkey Macaca-Fascicularis. *Journal of Physiology—London*, 427(AUG), 681–713.

Simpson, J.I., Leonard, C.S., and Soodak, R.E., 1988. The accessory optic system. Analyzer of self-motion. *Annals of the New York Academy of Sciences*, 545, 70–179.

Slaughter, M.M. and Miller, R.F., 1981. 2-amino-4-phosphonobutyric acid: A new pharmacological tool for retina research. *Science*, 211, 182–185.

Smith, R.G., Freed, M.A., and Sterling, P., 1986. Microcircuitry of the dark-adapted cat retina: functional architecture of the rod-cone network. *Journal of Neuroscience*, 6, 3505–3517.

Snellman, J. et al., 2008. Regulation of ON bipolar cell activity. *Progress in Retinal and Eye Research*, 27(4), 450–463.

Sterling, P., 2013. Some principles of retinal design: The Proctor lecture. *Investigative Ophthalmology and Visual Science*, 54(3), 2267–2275.

Sterling, P. and Freed, M., 2007. How robust is a neural circuit? *Visual Neuroscience*, 24(4), 563–571.

Tachibanaki, S. et al., 2012. Low activation and fast inactivation of transducin in carp cones. *Journal of Biological Chemistry*, 287(49), 41186–41194.

Trenholm, S. et al., 2011. Parallel mechanisms encode direction in the retina. *Neuron*, 71(4), 683–694.

Tsukamoto, Y. et al., 2001. Microcircuits for night vision in mouse retina. *Journal of Neuroscience*, 21(21), 8616–8623.

Tukker, J.J., Taylor, W.R., and Smith, R.G., 2004. Direction selectivity in a model of the starburst amacrine cell. *Visual Neuroscience*, 21(4), 611–625.

Vardi, N., 1998. Alpha subunit of G(0) localizes in the dendritic tips of ON bipolar cells. *Journal of Comparative Neurology*, 395(1), 43–52.

Vogalis, F. et al., 2011. Ectopic expression of cone-specific G-protein-coupled receptor kinase GRK7 in zebrafish rods leads to lower photosensitivity and altered responses. *Journal of Physiology*, 589(Pt 9), 2321–2348.

Vroman, R., Klaassen, L.J., and Kamermans, M., 2013. Ephaptic communication in the vertebrate retina. *Frontiers in Human Neuroscience*, 7, 612.

Wässle, H., 2004. Parallel processing in the mammalian retina. *Natural Review Neuroscience*, 5(10), 747–757.

Wong, K.Y., 2012. A retinal ganglion cell that can signal irradiance continuously for 10 hours. *Journal of Neuroscience*, 32(33), 11478–11485.

Wuerger, S.M., Atkinson, P., and Cropper, S., 2005. The cone inputs to the unique-hue mechanisms. *Vision Research*, 45(25–26), 3210–3223.

Yonehara, K. et al., 2009. Identification of retinal ganglion cells and their projections involved in central transmission of information about upward and downward image motion. *PLoS One*, 4(1), e4320.

Yoshida, K. et al., 2001. A key role of starburst amacrine cells in originating retinal directional selectivity and optokinetic eye movement. *Neuron*, 30(3), 771–780.

12 Visual system architecture

Jonathan Winawer and Hiroshi Horiguchi

Contents

12.1 INTRODUCTION

Vision is the dominant sense in primates and gives rise to an enormous diversity of behavior. Vision is used to guide locomotion, coordinate hand movements, recognize objects and scenes, direct attention, and entrain circadian rhythms. To accomplish these many functions, the nervous system includes a sophisticated network of visual pathways and structures. The eye itself contains a great number of cell types specialized for extracting particular kinds of information from the light array impinging the retina. A large portion of the brain is used to analyze the signals from the eye, including many subcortical nuclei and about 25% of the cerebral cortex (Van Essen 2004). This chapter summarizes the visual system architecture in two parts: first, the pathways by which visual information from the eye is carried to the brain, and second, how different visual functions are distributed in cortex.

12.2 VISUAL INFORMATION FLOW FROM RETINA

12.2.1 HEMIDECUSSATION

Because of the complexity of visual system architecture, it is useful to identify a few organizing principles. One such principle is the segregation and recombination of signals at multiple stages of processing. Consider the visual image itself. A single visual scene gives rise to two distinct images, one in each eye. These two images are encoded by the two retinas and then combined into binocular representations in the brain. By first separating and then recombining the image, the visual system can extract useful information about the environment. In the following text, we describe the pathways involved in this process.

In humans and many other primates, the nerve fibers exit the eye in a bundle; the fiber bundle then splits at the optic chiasm, with about half the fibers going to each of the two hemispheres. As a result, each of the two retinal images is divided in the brain: signals from the temporal hemiretina are routed to the ipsilateral hemisphere of the brain, and signals from the nasal hemiretina are routed to the contralateral hemisphere. This splitting, called *hemidecussation*, was proposed by Isaac Newton (Figure 12.1a) (Brewster 1860). The splitting of the image presumably imposes a cost, as neighboring parts of the retinal image along the vertical midline are represented by distant structures in the brain. There is also a benefit: combining the inputs from corresponding points in the two retinas supports stereopsis and hence depth perception. In species with more lateralized eyes, there is less binocular vision, and most fibers

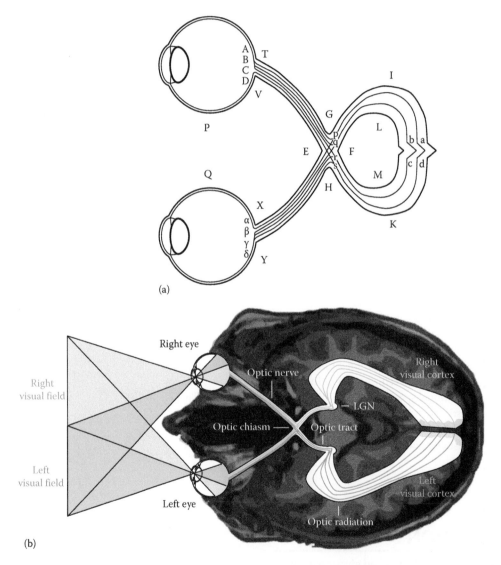

(a)

(b)

Figure 12.1 Representation of the visual fields. A key component of visual system architecture is the splitting of the visual field and the integration of the inputs from the two eyes at the optic chiasm (hemidecussation). (a) The first known diagram of hemidecussation at the optic chiasm, by Isaac Newton (Brewster 1860). Newton's early diagram of the visual pathways contains several accurate observations. First, the fibers leave the eye as a bundle, for example, ABCD leaving one eye and αβγδ leaving the other eye. Second, these fibers split at the optic chiasm, such that fibers originating from the temporal retina (AB and γδ) travel *ipsilaterally*, and fibers originating from the nasal retina (CD and αβ) travel *contralaterally*. Third, inputs originating from corresponding points of the two retinas are combined further downstream in the visual pathways (abcd). (b) A schematic of the visual pathways on an axial MRI slice shows a more modern version of Newton's drawing. The inputs from the right visual field (magenta) are routed to the left hemisphere, and the inputs from the left visual field (blue) are routed to the right hemisphere. The axons from the retinal ganglion cells hemidecussate at the optic chiasm and terminate in the LGN, where signals are then relayed to visual cortex.

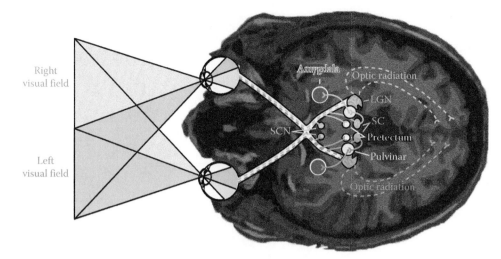

Figure 12.2 Subcortical targets of the optic nerve. Visual system architecture involves a complex network of pathways originating from the optic nerve. The largest targets of the optic nerve (~90% of fibers) are the two lateral geniculate nuclei (LGN) of the thalamus (purple), each representing one half of the visual field. The outputs of the LGN form the optic radiation, comprising the major pathway to primary visual cortex (dashed lines). Several additional nuclei are targeted by fibers branching from the optic nerve. These include the suprachiasmatic nucleus, just superior to the optic chiasm; the superior colliculus (SC), part of the midbrain; and the pretectum, just anterior to the superior colliculus. The superior colliculus projects anteriorly to the pulvinar, the largest thalamic nucleus, which in turn targets many other areas, including the amygdala. Each of the nuclei have numerous other inputs and outputs that are not depicted. The multiple pathways contribute to many functions, including perception, eye movements, pupil constriction, and regulation of circadian rhythms.

project to the contralateral hemisphere, for example, 97% in mice (Drager and Olsen 1980).

Because of hemidecussation, a lesion to the visual pathways has very different effects depending on where it occurs. A lesion peripheral to the optic chiasm (retina or optic nerve) disrupts vision through one eye. In contrast, a lesion central to the optic chiasm disrupts vision in one visual hemifield but through both eyes. Assessing whether a deficit is restricted to one eye or to one hemifield is an important tool for localizing visual disorders.

12.2.2 MAJOR SUBCORTICAL TARGETS OF THE OPTIC NERVE

All neural signals exit the eye in the optic nerve. This nerve is made up of the axons from retinal ganglion cells. There are several targets of the optic nerve (Figure 12.2), which we review in the following.

12.2.2.1 Lateral geniculate nucleus of the thalamus

About ninety percent of the axons exiting the eye terminate in the two lateral geniculate nuclei ("LGN") of the thalamus, as measured in macaque (Kandel et al. 2000, 528, Perry et al. 1984). The LGN in turn relays signals to the primary visual cortex via the optic radiation. This pathway is called the geniculostriate pathway, "geniculo" for LGN and "striate" for primary visual cortex.* The geniculostriate pathway is the dominant visual pathway in primates and supports many aspects of vision through cortical processing. In many other vertebrate species, including reptiles, birds, and rodents, the geniculostriate pathway is less dominant. For example, in mice, after lesions to this pathway, much (but not all) functional

vision remains (Prusky and Douglas 2004). In humans and other primates, damage to this pathway leads to blindness in a portion of the visual field, called a *scotoma*.

The geniculostriate pathway illustrates several important principles of visual system architecture. Two of these principles—information transfer via parallel pathways and the preservation of retinotopic maps—are discussed in the following.

12.2.2.1.1 Eye-of-origin parallel pathways

The LGN is bilaterally symmetric, one in the left hemisphere and one in the right hemisphere. Each LGN comprises 6 prominent layers as visualized by histological sections in postmortem tissue. Three of the layers are ipsilateral (left eye projections to left LGN) and three are contralateral (right eye projections to left LGN), so that information from the two eyes remains segregated in the LGN and is conveyed to the brain in parallel pathways where the inputs are combined (Figure 12.3).

12.2.2.1.2 Cell-type parallel pathways

The LGN layers are also separated by retinal ganglion cell type (Figure 12.3). The primate retina contains at least 17 types of ganglion cells (Field and Chichilnisky 2007). Pathways for a few of these cell types are known. The first two layers of the LGN (ventral) receive inputs from the large, parasol ganglion cells of the retina. The large cells in these layers of the LGN are referred to as "magnocellular." The four dorsal layers of the LGN receive inputs from the smaller midget ganglion cells of the retina; the small cells in these layers of the LGN are called "parvocellular." The two sets of layers—parvocellular (layers 3–6) and magnocellular (layers 1–2)—can be distinguished in the human brain using functional MRI (Denison et al. 2014), although resolution is not yet good enough to visualize the 6 individual layers. Interdigitating between these 6 prominent layers of the LGN are the more recently discovered koniocellular layers of the

---

* Primary visual cortex is called "striate cortex" after the stria of Gennari, a region of myelinated axons in primary visual cortex visible to the naked eye. Cortical visual areas outside of V1 are referred to as "extra-striate."

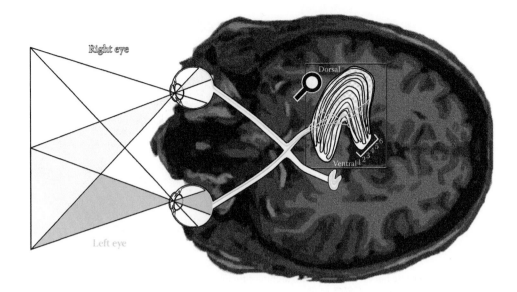

Figure 12.3 Parallel visual pathways via the lateral geniculate nucleus. A common feature of visual system architecture is the segregation and integration of neural signals transmitted in multiple parallel pathways. Signals from the left visual field, represented in the right half of the retina, are routed to the right LGN. The right LGN is shown in magnified view (red box). The magnified view is rotated relative to the underlay, exposing the dorsal–ventral axis, in order to highlight the layered structure of the LGN. The LGN layers segregate inputs by eye of origin and cell type. Layers 1, 4, and 6 (green) receive fibers from the contralateral eye, whereas layers 2, 3, and 5 receive fibers from the ipsilateral eye (yellow). Layers 1 and 2, called "magnocellular," are the targets of the parasol ganglion cells of the retina (large dotted textures), whereas layers 3–6, called "parvocellular," are the targets of the midget ganglion cells (small dotted texture). Outputs of the small bistratified retinal ganglion cells target the LGN between these layers, whose outputs comprise the koniocellular pathway (not shown). The inputs from the two eyes and the multiple cell types are combined in later stages of processing in the cortex.

LGN, which receive inputs from the small bistratified retinal ganglion cells.

The parasol, midget, and small bistratified cells comprise only a small subset of retinal ganglion cell classes. Pathways for the remaining cell classes are likely to be discovered in future work. But while these are only a few of the retinal ganglion cell classes, they comprise the vast majority of the retinal ganglion cells. The midget cells alone account for most of the ganglion cells in the human retina—about 45% in the periphery and 95% in the central retina (Dacey 1993).

12.2.2.1.3 Retinotopic maps

Just as there is organization between layers of the LGN (segregation by cell type and eye of origin), there is also organization within each layer. Organization within the layers embodies a second principle of visual system architecture, namely, preservation of the *retinotopic map*. Although the retinotopic map is not perfectly preserved within the optic nerve (Fitzgibbon and Taylor 1996, Horton et al. 1979), it is reconstituted in the LGN. Neighboring cells within a single layer of the LGN receive inputs from nearby retinal ganglion cells. While the retinal topology is mostly preserved in the LGN, it is not perfectly preserved, and the scale is significantly changed. For example, the LGN map splits the retinal image in two parts due to hemidecussation. The left LGN represents only the left half of the two retinas (right visual field), and the right LGN only the right half of the two retinas (left visual field). The LGN map also differs from the retina in that the foveal representation is greatly exaggerated in the LGN. For example, in macaque, the central 2.5° in the retina (less than 0.1% of the retinal image) projects to 10% of the LGN neurons (Connolly and Van Essen 1984, Schneider et al.

2004). Hence, an image from the retina visualized on the LGN will appear distorted in several ways. It will be split into two and larger in the center. Nonetheless, it is an image, as neighboring points in the LGN come from neighboring points in the retina. As we shall see later, the retinotopic map is preserved in many additional structures throughout the visual pathways.

12.2.2.1.4 Function of the LGN

The LGN is often described as a relay station because it receives inputs from the retina and sends outputs to the cortex. The functional properties of neurons in the LGN are not known to differ in a dramatic way from those in the retina. For example, in cat and macaque, the receptive fields tend to have center-surround organization, either with an excitatory center and inhibitory surround or vice versa, similar to receptive fields of retinal ganglion cells (Kandel et al. 2000). Rather than transforming the functional signals in a qualitative way, the LGN may serve other purposes. For example, it has been hypothesized to serve as a gating mechanism, allowing some signals to pass through to cortex while inhibiting other signals. This hypothesized gating mechanism is plausible because the LGN receives a large feedback projection from primary visual cortex, and this feedback could be used to control the responsivity of the LGN (Sherman and Koch 1986). Cognitive and task effects on the LGN have been measured in human neuroimaging experiments (Kastner et al. 2006), and in fact, it is estimated that 90% of the synapses in primate LGN are feedback from cortex, and only 10% feedforward from the retina. Nonetheless it is important to note that the feedforward signals, though smaller in number, drive the cells of the LGN very powerfully. Moreover, despite the lack of a known qualitative change in receptive field properties between retinal ganglion cells and LGN cells, there are quantitative changes. For example,

LGN cells have more suppressive surrounds and a different pattern of temporal responses; these changes may reflect cortical feedback or some degree of important visual analysis computed in the LGN (Sillito and Jones 2002).

12.2.3 SECONDARY PATHWAYS

While most of the projections from the eye project to the LGN of the thalamus, there are several secondary pathways that are also important.

12.2.3.1 Superior colliculus

After the LGN, the superior colliculus, sometimes called the "optic tectum," is the largest target of retinal ganglion cells, with about 10% of retinal ganglion cells terminating in the superior colliculi. Although the superior colliculus receives a small number of optic nerve fibers compared to the LGN, it is nonetheless a large number: about 100,000 nerve fibers, or 3 times more than the number of fibers in the auditory nerve. The superior colliculus is a paired structure, with one on each midbrain surface. This midbrain nucleus is a multilayered structure, with alternating layers of cell bodies and fibers. The superficial layers receive inputs from the eye and from primary visual cortex. The cells in these layers preserve a map of the retina, with an enhanced representation of the fovea. Although the superior colliculus is small (less than a cm long in each dimension), in recent years researchers have been able to visualize the retinotopic maps in living human brains using high-resolution fMRI (Katyal et al. 2010, Schneider and Kastner 2005). Each map represents the contralateral visual field, similar to the LGN and V1. Deeper layers of the superior colliculus receive inputs from extrastriate cortex, as well as other parts of the cerebral cortex including auditory and somatosensory areas. The representation from these sensory modalities is organized into a spatial map. Interestingly, the maps from these different sense modalities and from vision are in register. This organization is well suited for the superior colliculus to play a role in sensorimotor integration, coordinating inputs from various sense modalities and outputs to control eye movements (Stein et al. 2002). Outputs from the superior colliculus project both to midbrain nuclei involved in controlling eye movements and cortical areas such as the frontal eye fields.

12.2.3.2 Pulvinar

In primate, the pulvinar is the largest thalamic nucleus, with bidirectional pathways to all cortical lobes as well as subcortical regions including superior colliculus and amygdala. It receives input from the eye indirectly, via the superior colliculus, and comprises part of the extrageniculostriate pathway. Because the pulvinar is much smaller in carnivores and almost nonexistent in rodents, knowledge of the pulvinar's role in vision comes from primate research, including functional measurements and lesion studies in humans. Lesions to the pulvinar in human patients result in deficits of visual attention and awareness, consistent with evidence from nonhuman primate work implicating the pulvinar in visual attention and visual suppression during saccades (Snow et al. 2009). It is likely that the major projections to the various regions of cortex indicate a modulatory role, but the details are not understood. This pathway alone does not support high acuity, conscious vision because lesions to the primary visual pathway (geniculostriate) result in scotomas

or blindness. Nonetheless, there is increasing research in the role of this important secondary pathway. The pulvinar is sometimes contrasted with the LGN: the LGN is a primary relay, transmitting information from the retina to the cortex, whereas the pulvinar is a higher-order relay, transmitting information between different parts of the cortex (Sherman 2007).

12.2.3.3 Suprachiasmatic nucleus

The visual pathways have some important functions in addition to seeing. One such function is the regulation of circadian rhythms. Human behavior and physiology fluctuate on a cycle that is approximately 24 h. The most obvious example is the sleep–wake cycle. But there are many other functions such as body temperature that follow a daily rhythm. Circadian rhythms are only approximately 24 h; signals from the environment, called "zeitgebers," are needed to keep our rhythms synchronized to the 24 h light–dark cycle. The most important *zeitgeber* is light. A major nucleus involved in coordinating circadian rhythms is the suprachiasmatic nucleus (SCN). The SCN is a paired structure located just above the optic chiasm in the hypothalamus. It contains biological pacemakers and coordinates control of bodily rhythms in conjunction with other brain areas including the pineal gland. The SCN receives direct inputs from the retina. Some of the inputs to the SCN are intrinsically photosensitive ganglion cells, a recently discovered cell type containing the pigment melanopsin (Berson et al. 2002, Provencio et al. 2000). Patients who are blind due to photoreceptor loss may still have circadian rhythms entrained to the daily light cycle, supported by melanopsin in retinal ganglion cells, which transmit signals to the SCN; in contrast, patients who are blind due to enucleation or to loss of the retinal ganglion cell layer do not entrain to the daily light cycle (Flynn-Evans et al. 2014).

12.2.3.4 Pretectum

The pretectum is a midbrain nucleus just anterior to the superior colliculus and just posterior to the thalamus. It is a paired nucleus, one per hemisphere, and it receives inputs directly from the eye. Like the SCN, the pretectum receives some of its inputs from retinal ganglion cells containing the photopigment melanopsin. The pretectum is a complex structure containing 7 subnuclei. It is involved in several basic functions, including the control of pupil size, the optokinetic reflex, and entrainment of circadian rhythms.

12.3 VISUAL CORTEX

12.3.1 GENICULOSTRIATE PATHWAY

The main route by which visual information arrives in the cerebral cortex is the optic radiation. This pathway is greatly expanded in primates. In birds and rodents, for example, the pathway through the superior colliculus—or optic tectum—is the dominant visual pathway, and not the geniculostriate pathway (Hofbauer and Drager 1985). The optic radiation develops early and is one of the first major white matter pathways in the human brain to become densely myelinated (Dayan et al. 2013, Kinney et al. 1988), changing little in macromolecular tissue properties over the lifetime (Figure 12.1 in Yeatman et al. 2014).

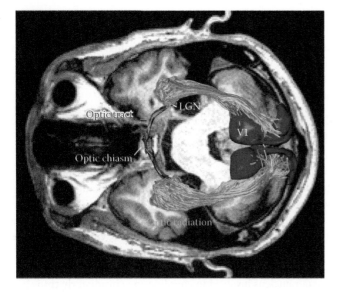

Figure 12.4 Geniculostriate pathway. The primary pathway by which the eye sends signals to the cortex is the geniculostriate pathway, consisting of the optic nerve/optic tract, the lateral geniculate nucleus, and the optic radiation. The image shows a computerized rendering of the optic tract (purple) and optic radiation (yellow), identified by fiber tractography and diffusion imaging of a living human brain. The underlay is an axial slice acquired from a T1-weighted magnetic resonance image. Image by Shumpei Ogawa.

Unlike the optic nerve, which is unidirectional, the optic radiation carries bidirectional signals between the LGN and visual cortex. The two ends of the optic radiation, the LGN and V1, are very different in sizes. V1 is about 50 times larger than the LGN (Andrews et al. 1997); each hemisphere's LGN contains about 1 million neurons, whereas each hemisphere's V1 contains about 150 million neurons, spanning 18 cm^2 (Wandell 1995). The optic radiation largely preserves the retinotopic map, so that damage to a restricted portion of the tract results in blindness in a restricted portion of the visual field. For example, one part of the optic radiation, called Meyer's loop, traverses around the occipital horn of the lateral ventricle and carries signals representing the upper visual field; a lesion to Meyer's loop results in upper field quadrantanopia (Figure 12.4).

12.3.2 V1 AND MAPS

12.3.2.1 Ocular dominance columns and parallel pathways

Primary visual cortex (V1) is located in posterior occipital cortex and distributes visual information to the rest of the brain. It receives its main input from the optic radiation, carrying signals from the LGN. As discussed in the prior section, inputs from the two eyes (but a single hemifield) are segregated into separate layers in the LGN. In V1, the inputs from these parallel pathways target interdigitated regions, called ocular dominance stripes or columns. Within each ocular dominance stripe, cells at the input layers are primarily driven by inputs originating from only one eye. These stripes can be visualized as a pattern of cytochrome oxidase activity in cortical layer IV (input layers) on the surface of a postmortem human brain of a patient who had one eye removed in adulthood (Horton and Hedley-Whyte 1984). (In the more superficial layers, the signals from the two eyes mix, and so the stripes are not clearly segregated.) In these patients, ocular dominance stripes from both

eyes exist, but stripes that would receive inputs from the enucleated eyes are inactive due to lack of input (Figure 12.5).

A very different result occurs if inputs from one eye are missing or abnormal in early childhood rather than adulthood. During early childhood, signals from the two eyes compete

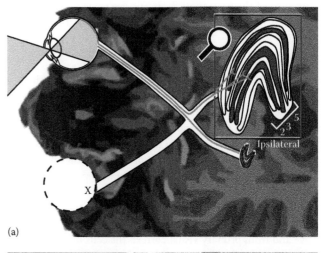

(a)

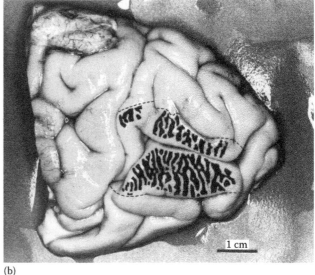

(b)

Figure 12.5 Eye-specific pathways. Eye-specific pathways in the human visual system can be studied by examining the postmortem brains of patients who had monocular enucleation earlier in life. (a) Schematic indicating eye-specific pathway in the lateral geniculate nucleus (LGN) in a patient whose left eye was enucleated (indicated by the dashed circle and "X"). Fibers from the right (intact) eye project to the ipsilateral layers in the LGN, layers 2, 3, and 5 (filled patterns in LGN). The red box is a magnified depiction of the right LGN. (b) A photograph of the right occipital lobe of a postmortem brain from a patient whose left eye was enucleated 23 years prior to his death. The dotted lines indicate the outline of primary visual cortex. The black painted regions show the locations of the right ocular dominance columns. These columns (sometimes called stripes) were identified by a method of cytochrome oxidase staining, which reveals regions of high metabolic activity. The stripes were identified on a flattened cortex and then rendered on a photograph taken of the intact occipital lobe prior to the staining and flattening procedure (Horton and Hedley-Whyte 1984). The regions between the black stripes receive inputs from a pathway originating at the enucleated eye; hence, they show lower metabolic activity. The eye-specific stripes are interdigitated throughout primary visual cortex.

for representation in V1. If signals from one eye are missing or abnormal from early infancy, then the usual pattern of ocular dominance is altered, as first demonstrated in animal experiments (Hubel et al. 1977) and later in postmortem human visual cortex (Adams et al. 2007). Even with correction of the abnormal inputs later in life, the person usually does not achieve high-quality vision through the eye with previously degraded images, likely in part due to the fact that the representation did not develop properly in V1 (Kiorpes and McKee 1999). The monocular representations in V1 are combined so that both other cells in V1 and all cells in visual areas beyond V1 are binocular. Hence, primary visual cortex is the last site in the visual processing stream in which the representations of the eyes are segregated.

12.3.3.2 Retinotopic map

The location of V1 was identified in the nineteenth century by neurologists and anatomists (Henschen 1893, Flechsig 1901). In the early twentieth century, the Japanese neuro-ophthalmologist Tatsuji Inouye first described the detailed retinotopic map in V1 (Inouye 2000). He took advantage of focal brain lesions in occipital cortex caused by bullet wounds from high velocity rifles in the Russo-Japanese war, comparing the location of the lesion in the brain to the visual field loss measured behaviorally. By combining data across many patients, each with a loss of visual field and a lesion site, he was able to derive a mapping of the visual field locations onto the visual cortex (Figure 12.6a). There were several main findings that have since been largely confirmed and described in

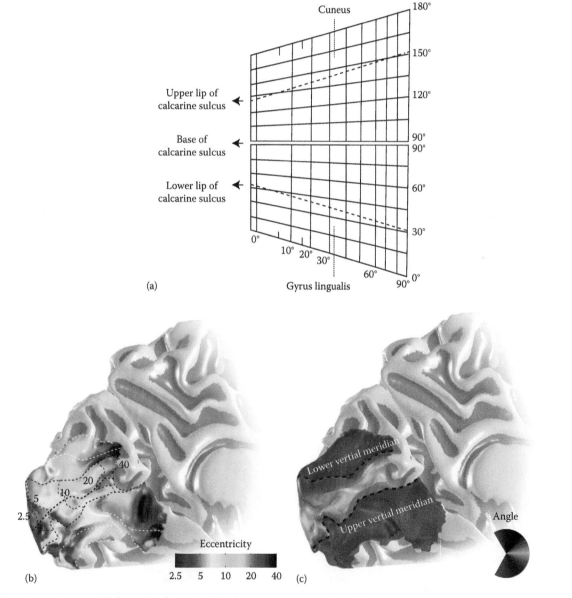

Figure 12.6 V1 retinotopic map. (a) Schematic of the visual field representation produced by Tatsuji Inouye in 1911 based on a comparison of visual field loss (scotomas) to lesions produced from bullet wounds in war (Inouye 2000). One visual hemifield is rendered in polar coordinates, with the distance from fixation (eccentricity) on the x-axis and angle representation on the y-axis. Anatomical labels indicate that the lower vertical meridian (180°) lies superior to the upper lip of the calcarine sulcus, and the upper meridian (0°) lies inferior to the lower lip of the calcarine. The horizontal representation (90°) is at the base of the calcarine sulcus. (b) Eccentricity representation in an individual observer derived from population receptive field mapping (Dumoulin and Wandell 2008) using functional MRI. (Adapted from (Wandell, B.A. and Winawer, J., *Vis. Res.*, 51(7), 718, 2011.).) Data were collected and analyzed by the authors (HH and JW). (c) Same as panel b, but showing the angle representation.

more detail. First, Inouye formulated the law of retinotopy: "neighboring points in one half of the retina are also next to one another in the corresponding principal visual area." Second, he observed that signals originating from the fovea are represented near the occipital pole and that the periphery is represented more anteriorly along the calcarine sulcus. This observation clarified views from the nineteenth century when there was uncertainty as to whether the fovea or the periphery was represented at the back of the occipital cortex (Henschen 1893). A third main finding is that the lower visual field is represented on the upper bank of the calcarine sulcus (the cuneus) and the upper visual field is represented on the lower bank of the calcarine sulcus (the lingual gyrus). Fourth, Inouye noted that the central visual field representation is greatly expanded in the cortex relative to the peripheral visual field representation, a phenomenon now called "cortical magnification."

Over the next hundred years, descriptions of the organization of primary visual cortex have become far more accurate and detailed, making it perhaps the best-studied region in the human cerebral cortex. About 10 years after Inouye's book, the V1 map was confirmed and described in more detail by Gordon Holmes and colleagues, using similar methods with patients in World War I (Holmes 1918, Holmes and Lister 1916). Decades later, lesions from strokes were identified more accurately using structural MRI and also compared to the location of visual scotomas (Horton and Hoyt 1991b) (Figure 12.6b). Currently, a detailed map of primary visual cortex can be obtained from a healthy, living individual human with less than 15 min of measurement using functional magnetic resonance imaging (Figure 12.6c). Maps of V1 derived from fMRI methods and from lesion studies using anatomical MRI show the same basic features described by Inouye and Holmes; the foveal-to-peripheral representation runs posterior-to-anterior; the lower vertical meridian is on the upper bank of the calcarine, and the upper meridian on the lower bank; and the further from the foveal representation, the less cortical territory allotted to a retinal region, so that the central 15° of the visual field occupies about half of visual cortex (*cortical magnification,* Figure 12.7). While the V1 map is generally continuous, so that adjacent locations in the retina are represented in adjacent locations in cortex (*retinotopic*), it is not perfectly continuous. Like the LGN map, there is a large discontinuity at the vertical midline, because the representation of either side of this midline projects to opposite hemispheres in the brain (*hemidecussation;* Figure 12.1).

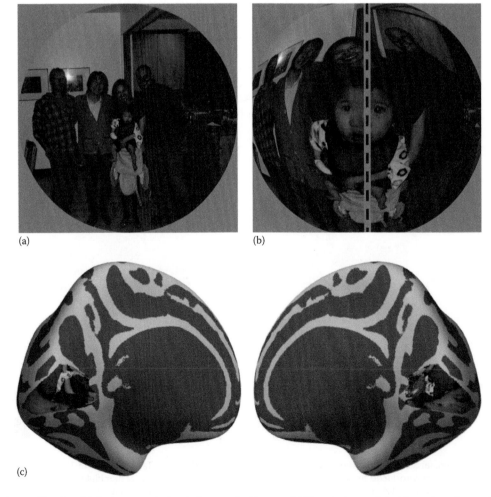

(a)

(b)

(c)

Figure 12.7 Cortical magnification. (a) A photograph simulating a retinal image. (b) The photograph in (a) was transformed to represent the approximate distortion in primary visual cortex due to cortical magnification. The center of the image is greatly expanded. The dashed line indicates that the two halves of the image are represented in the two hemispheres. (c) A more realistic rendering of the visual map in primary cortex was made by projecting the image onto a standard atlas of V1 (Benson et al. 2012, 2014), made by Noah Benson.

12.3.2.3 Measurement of retinotopic maps

A retinotopic map in visual cortex can now be measured quickly and reliably using functional MRI in healthy, awake, human subjects. In one method, a traveling wave is induced in visual cortex by presenting visual stimuli that slowly sweep across the visual field (Engel et al. 1994, Sereno et al. 1995). The method is called "traveling wave retinotopy" or "phase-encoded retinotopy" (Figure 12.8). During one measurement, a high-contrast ring slowly expands from the center of gaze (fovea) to the periphery. As the ring expands, a wave of cortical activity spreads from the posterior end of V1 to the anterior end. A single eccentricity value can be assigned to each measurement

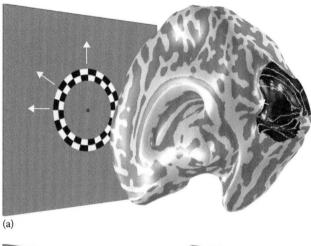

(a)

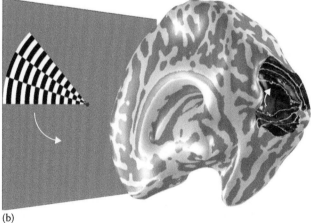

(b)

Figure 12.8 Traveling wave retinotopy. One method to obtain a retinotopic map using functional MRI is the traveling wave paradigm. (a) In one set of measurements, a ring aperture containing a high-contrast pattern cycles across eccentricities (e.g., expanding). This results in a wave of activity across the visual cortex, depicted in the rendering of an inflated right hemisphere. The pseudocolor map reveals the peak of the traveling wave at a point in time (shifted by 4 s to account for the delay in the hemodynamic signal). For each point on the cortex, the ring eccentricity that gives rise to the largest BOLD response is taken as the measurement of eccentricity for that voxel. (b) In another set of measurements, a wedge aperture containing a contrast pattern rotates around the fixation point. This results in a wave of activity across visual cortex shown on the same right hemisphere. For each point on the cortex, the wedge angle that gives rise to the largest BOLD response is taken as the measurement of polar angle for that voxel. The two coordinates, eccentricity and angle, comprise the position in visual space that correspond to a location in visual cortex.

point (voxel) in the fMRI experiment by identifying the stimulus eccentricity that results in the largest fMRI response. During a separate measurement, a high-contrast wedge rotates around a circle, eliciting a wave of cortical activity that follows the angle representation in visual cortex. A single angle value is assigned to each voxel by identifying the wedge position that results in the largest fMRI signal. By combining data across the ring and wedge measurements, each voxel is assigned two coordinates, an eccentricity and angle, creating a map. Using these methods, highly detailed maps of primary visual cortex have been measured in many fMRI studies, revealing both regularity in the maps across observers (Benson et al. 2012, Schira et al. 2009) and considerable individual differences in the size (3× range, [Dougherty et al. 2003]) and precise position of the map (Dumoulin et al. 2003, Stensaas et al. 1974).

12.3.3 MULTIPLE VISUAL FIELD MAPS: V1/V2/V3

The discovery that the cortex contains a map of the retina was a major discovery in the history of visual neuroscience. Perhaps more surprising and controversial was the set of discoveries documenting not just one, but many retinotopic maps. A second (V2), third (V3), and many more visual maps (Felleman and Van Essen 1991) have been reported in the animal literature over a period of decades. Using anatomical MRI (Horton and Hoyt 1991a) and then functional MRI (Sereno et al. 1995, Engel, Glover, and Wandell 1997), multiple visual field maps have also been shown in human, and it is now a well-accepted fact that visual cortex contains a large array of maps (Sereno and Tootell 2005, Wandell 1995, Wandell and Winawer 2011, Wandell et al. 2007, Zeki 1993) such that approximately 25% of the human cerebral cortex is estimated to be predominantly visual (Van Essen 2004). Because there are multiple visual field maps, each point in the image is represented in multiple cortical locations. In this sense, cortex is similar to the retina: just as the image is encoded in photoreceptors, bipolar cells, and multiple distinct classes of ganglion cells, each tiling the visual field, the image is also encoded in multiple cortical maps each tiling the visual field.

Using fMRI data and appropriate visualization, it is now possible to identify at least a dozen visual field maps in a single human observer from an experiment that lasts just 10–20 min. The locations of 12 such maps are shown as an example of surface rendering in the right hemisphere surface rendering (Figure 12.9). In addition to the 12 maps depicted, a number of additional maps have also been discovered, including several in parietal cortex (intraparietal sulcus [Swisher et al. 2007]) and ventral temporal cortex (parahippocampal cortex [Arcaro et al. 2009]) and anterior occipital cortex (V6 [Cardin et al. 2012, Pitzalis et al. 2006, Stenbacka and Vanni 2007]) as well as frontal cortex (Jerde and Curtis 2013, Jerde et al. 2012). With the exception of frontal cortex, the maps abut one another so that a large, contiguous region of cortex is tiled with maps. Identification of the multiple maps is generally accomplished by the same methods used to study V1, especially the traveling wave method for fMRI experiments (Engel et al. 1994, Sereno et al. 1995, Engel, Glover, and Wandell 1997). Using these methods, it is common to visualize maps on the cortical surface of the two polar coordinates of the stimulus location—angle and eccentricity—that most effectively drives the BOLD response in each voxel.

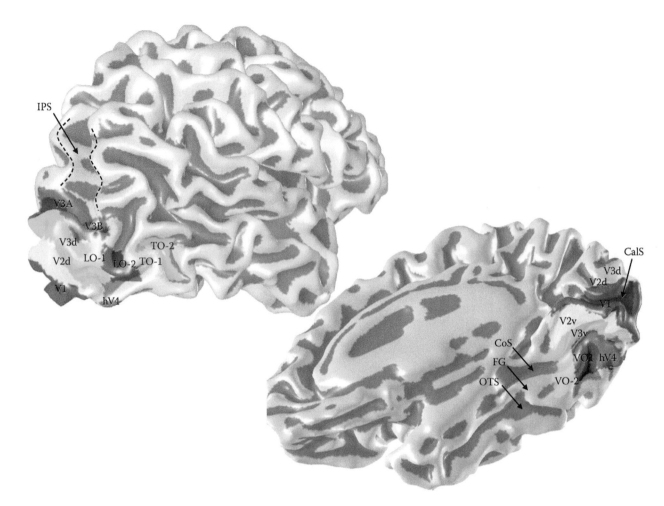

Figure 12.9 Locations of visual field maps. The locations of 12 visual field maps are indicated on the surface rendering of the right hemisphere. Two maps, V2 and V3, are each split into ventral and dorsal parts (V2v/V2d, V3v/V3d). The maps were derived from population receptive field modeling using fMRI. Several major sulci and gyri are labeled: intraparietal sulcus (IPS), calcarine sulcus (CalS), collateral sulcus (CoS), fusiform gyrus (FG), and occipitotemporal sulcus (OTS). (Reproduced from Wandell, B.A. and Winawer, J., *Vis. Res.*, 51, 718, 2011.)

A close-up view of the occipital cortex shows how these two color maps are used to identify multiple visual maps (Figure 12.10). We consider the V1, V2, and V3 maps first. The V1 map is surrounded by V2, which has an approximately horseshoe shape, and similarly, V2 is surrounded by V3. One reason these are considered separate maps is that each one—V1, V2, and V3—contains a complete representation of the contralateral visual hemifield. For this reason, borders between the maps are drawn at reversals in one of the polar coordinates. For V1, V2, and V3, the borders are reversals in the polar angle representation. One consequence of this arrangement is that the V2 and V3 maps contain a split hemifield representation, with a dorsal arm representing the lower visual quadrant and a ventral arm representing the upper visual quadrant. In contrast to the angle representation, which contains reversals at the V1/V2 and V2/V3 boundaries, the eccentricity representation is in register across several maps. This can be visualized as the large orange region at the occipital pole (foveal representation), surrounded by bands of increasing eccentricity (yellow, green, cyan, blue). Because the eccentricity map is in register across several visual field maps, and because the angle map boundaries are challenging to delineate near the foveal representation, the region at the occipital pole is sometimes referred to as the confluent fovea. The maps beyond V1, V2, and V3 are described in the following.

12.3.4 MAP ORGANIZATION

As the number of visual field maps described in the literature has grown well beyond V1, V2, and V3, to 20 or more maps, questions of large-scale map architecture become increasingly important: Are there organizing principles to the many maps? What is the scale over which visual functions are computed? How close is the homology between the human visual system and animal models? Several ideas have been advanced to explain why there are so many maps and how they are organized. These ideas are not necessarily in conflict, but rather each may highlight one or a few aspects of the cortical architecture.

12.3.4.1 Dorsal/ventral streams

One important proposal is that the maps can be separated into two processing streams, one more ventral and one more dorsal. According to this proposal, the two streams are, approximately, functionally and anatomically distinct (Ungerleider and Haxby 1994, Ungerleider and Mishkin 1982). The dorsal pathway includes a number of maps that support the representation of motion, spatial

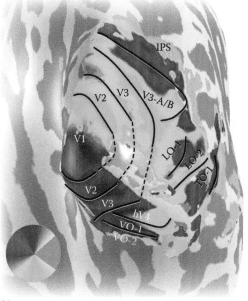
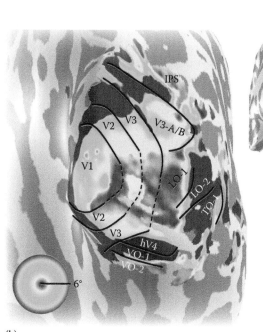

(a)   (b)

Figure 12.10 Retinotopic maps at occipital pole. The small inset on the right shows a rendering of a subject's right hemisphere, smoothed to show the sulci (dark grays) and gyri (light grays). The view is from behind the occipital pole. The dashed rectangle is a region magnified in the two main images. These images, both showing the same anatomical underlay, are overlaid with pseudocolor maps to show (a) the most effective angle or (b) the most effective eccentricity of visual stimulation for each point on the cortical surface. Data were modeled as population receptive fields in each voxel; locations are uncolored where the variance explained by the model is low. The images show a number of retinotopic maps tiling much of occipital cortex. Solid lines indicate boundaries between maps as revealed by the retinotopy data. Dashed lines interpolate over regions in which the measurements are poor, including a large region at the occipital pole where functional MRI data are obscured by a large vein (Winawer et al. 2010). (After Figure 1 in Wandell, B.A. et al., Computational modeling of responses in human visual cortex, in *Brain Mapping: An Encyclopedic Reference*, ed. A. Toga, 2015.)

location, and action. The ventral pathway includes maps involved in seeing color and form and recognizing objects and scenes. This notion is consistent with the idea of parallel pathways in the transmission of visual information, though it is important to note that the two streams are not entirely distinct. A major fiber pathway, the vertical occipital fasciculus (Yeatman et al. 2013), connects the two pathways and likely supports functions that require integrating object information (form, color, and so forth), which are action and location information. Reading, for example, requires recognizing shape and controlling eye movements.

One complication with the two-stream proposal is that the topography of human visual cortex is more consistent with three branches of visual field maps rather than two, each extending anteriorly from the early visual field maps at the occipital pole. The three streams are ventral (Arcaro et al. 2009), lateral/temporal (Amano et al. 2009), and dorsal/parietal (Swisher et al. 2007). It is not clear how the lateral/temporal maps would fit into the two-stream hypothesis.

12.3.4.2 Hierarchies and areas

A more detailed proposal is that the dorsal and ventral streams are organized as two branches in a visual hierarchy. The proposal that the multiplicity of areas forms a hierarchy is perhaps the most influential view of visual cortical architecture. The idea is an extension of the clear fact that the early visual system is hierarchical. For example, retinal ganglion cells can be said to lie above

the photoreceptors and below the thalamus in a visual hierarchy, as neural encoding of light first takes place in the photoreceptors, and signals are then transmitted primarily in one direction. However, once primary visual cortex distributes visual signals to many other parts of the brain, signals from any area can, in principle, affect responses anywhere else. Hence, whether there is a hierarchy of visual areas within the cortex is an empirical question. The notion of a hierarchy of visual areas is sometimes matched to the idea the visual perception unfolds in a series of discrete steps (Riesenhuber and Poggio 1999).

By analyzing many published data sets, Felleman and Van Essen (1991) showed that the pattern of connections between visual areas in macaque monkey was mostly, though not completely, consistent with a hierarchy of several levels, where one or more visual areas could be assigned to each level. The analysis was supported by the laminar pattern of connections. Similar studies have not been carried out in human, as tracer studies are generally not available.

In Felleman and Van Essen's hierarchy, the fundamental unit is the visual area. Visual areas were identified by several criteria, including connection patterns between areas, cytoarchitecture, and retinotopy. Because retinotopy is only one of several criteria, an area as treated by Felleman and Van Essen does not always have a one-to-one relationship with a retinotopic map. Some areas may contain a partial map, some may contain multiple maps, and some may not clearly contain a map at all.

12.3.4.3 Clusters

A different hypothesis is that the visual field maps are organized as several clusters (Wandell et al. 2005). According to this hypothesis, a cluster consists of several maps, arranged semicircularly around a common foveal representation. The eccentricity bands are in register across the maps within a cluster, and the angle bands are approximately radial, though the two types of measurements (angle and eccentricity) need not be precisely orthogonal. This proposal emphasizes visual field maps as the organizing unit, rather than cytoarchitecture, cell type, or connectivity. This is important because within a single map, there may be systematic variation in cytoarchitecture, such as the cytochrome oxidase "stripes" within V2. A hypothesis is that the maps within a cluster share computational resources such as circuitry for short-term memory and timing of neural signals. Furthermore, perceptual specializations may be organized in part at the cluster level, such as motion computations with the hMT+ cluster. Evidence for cluster organization has been found in both human and macaque using fMRI (Kolster et al. 2009, Wandell et al. 2005). The cluster proposal, like the hierarchy proposal, is also consistent with the theory that distinguishes between ventral and dorsal maps.

12.3.4.4 Diffusion imaging and tractography

A central question in map organization is how the maps communicate with one another. Historically, connections between the maps have been most extensively studied with postmortem anatomical tracer studies in animal models, especially the macaque. However, it is increasingly clear that beyond V1–V3, homology between the human visual system and the macaque visual system is at best highly uncertain (Sereno and Tootell 2005). Certain maps in human, like hV4 (human V4), differ in location and topology from macaque V4 (Brewer et al. 2005, Witthoft et al. 2014, Winawer et al 2010). Other maps, such as V3B and LO-2, exist in human but might not exist in macaque at all. Hence, it is critical to measure both map organization and connectivity between maps in the living human brain. In recent years, diffusion MRI combined with computational tract tracing has enabled the study of fiber pathways in living human brains. These tools are especially good at measuring large fiber bundles, tracing their pathways, and assessing tissue properties such as macromolecular tissue volume within the tracts (Mezer et al. 2013). Currently the method is less sensitive to small, local pathways. Being able to identify pathways in the living human brain has a number of applications. For example, once identified, quantitative MRI can be used to assess tissue properties, and these can be compared across development or between subject populations (healthy/disease). Moreover, tracts can then be followed over time within an individual, such as a patient with multiple sclerosis, to evaluate disease progression. Another important application of diffusion imaging and tractography is the identification of circuits for uniquely human behavior, such as reading. The major visual pathways involved in reading have been identified with diffusion imaging and studied extensively, in the context of both development and reading disorders (Ben-Shachar et al. 2007a) (Figure 12.11).

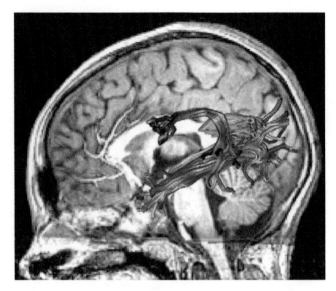

Figure 12.11 Reading circuitry. Diffusion magnetic resonance images were acquired in living human brains. The diffusion images, combined with fiber tractography, were used to identify several of the fiber bundles comprising the circuitry for reading: arcuate fasciculus (blue), inferior longitudinal fasciculus (orange), and temporal callosal projections (green) for a representative subject. The underlay is a sagittal slice from a T1-weighted image of the same subject. For the software and data used to create this image, see http://github.com/jyeatman/reading_circuits. (Reproduced from Wandell, B.A. and Yeatman, J.D., *Curr. Opin. Neurobiol.*, 23, 261, 2013.)

12.3.5 FUNCTIONAL MEASUREMENTS WITHIN MAPS AND VISUAL AREAS

A great deal of study has been dedicated to the response properties of individual neurons in primary visual cortex of cat and macaque, most famously by David Hubel and Torsten Wiesel (Hubel and Wiesel 1977). These studies, and the much more computational models that have followed, have characterized responses in visual cortex in terms of basic stimulus properties such as wavelength, contrast, orientation, and binocular disparity and in terms of computational principles such as rectification and normalization. There are many excellent reviews of the single-unit electrophysiology results from animal models, for both V1 and other visual areas (Carandini et al. 2005, Heeger et al. 1996, Maunsell and Newsome 1987, Shapley and Hawken 2002). We do not review this literature here, but instead focus primarily on studies of cortical response properties in the human visual system.

12.3.5.1 Population receptive fields

Given the large number of visual field maps in the human visual system, a natural question to ask is how the representation of the visual image differs between the maps. In the last decade, significant progress has been made in quantitative models of the fMRI signal in visual cortex. This modeling framework is often referred to as population receptive field (pRF) modeling. The word "population" indicates that the model measures responses from a population of neurons, not an individual neuron. The phrase "receptive field" is used by analogy to receptive field mapping of individual neurons, in which the brain response is expressed not in units of brain activity but in terms of stimulus properties such as contrast, position, and orientation. pRF modeling can also be fit to data from instruments other than MRI, such as the local field potential recorded from microelectrodes in animals (Victor et al. 1994) or intracranial electrodes in humans (Harvey et al. 2012, Winawer et al. 2013, Yoshor et al. 2007).

The pRF model predicts the entire time series of a voxel in an fMRI experiment. It does so by taking images as input and predicting the BOLD response as output (or ECoG or other measurement modalities). The first generation of pRF models assumed a linear pooling of spatial contrast and built predictions based on stimulus location and not the specific pattern comprising the stimulus (Dumoulin and Wandell 2008). The model identified the position and spatial extent of the region of visual space in which stimulus contrast results in a response at the measured site (e.g., an MRI voxel). As such, the pRF model goes beyond the travelling wave approach by quantifying the extent of spatial pooling, rather than a single point. The results of pRF modeling have revealed several patterns across visual cortex (Figure 12.12). First, the scale of spatial pooling (pRF size) increases as the voxel's pRF center location is more remote from the fovea (greater eccentricity). Second, pRF size differs across visual areas. For example, the pRF size in V1 is about half that of V3, which in turn is about half that of hV4. These general patterns are consistent with decades of single-unit recording in animal models, and they provide one example of how visual representations differ across the visual field maps.

To capture further aspects of visual representation, pRF models have expanded to account for a greater number of stimulus parameters. To do so, the models have necessarily become more complex (Figure 12.13). Newer models begin with the pixels in the image rather than the spatial locations (apertures) (Kay et al. 2008), and they incorporate a number of calculations

to summarize the relationship between stimulus and output, including spatial filtering, divisive normalization, sensitivity to second-order contrast, and nonlinear spatial summation (Kay et al. 2013b). While the model details are beyond the scope of the chapter, a few trends can be observed in the way in which model parameters differ between maps. For example, spatial summation is closest to linear in V1 and increasingly compressive (nonlinear) in visual field maps beyond V1. This pattern corresponds to the fact that in extrastriate maps like VO-1/2 or TO-1/2, even a small stimulus anywhere in the voxel receptive field produces a nearly maximal response; in V1, the response grows substantially as the spatial overlap between the stimulus and pRF gets larger. Another trend is that extrastriate maps show greater sensitivity to second-order contrast (variation in contrast level across the image) than V1, consistent with the idea that stimulus tuning becomes more complex in downstream visual areas.

PRF modeling in the human brain provides an opportunity to compare measurements made with different instruments. By using models, the comparisons can be made in model parameters that often correspond to stimulus features, like pRF location or size. This is an advantage because measurements made with different instruments such as BOLD and local field potential cannot easily be directly compared, as they have different units, different time scales, and many other differences. Comparison of pRF models made with different instruments have been informative. For example, one component of the electrical signal, an asynchronous, spectrally broadband component, is well matched to the fMRI measurement in terms of pRF position, size, and degree of spatial compression (Winawer et al. 2013). Other components of the ECoG signal, such as the amplitude of the visually evoked potential

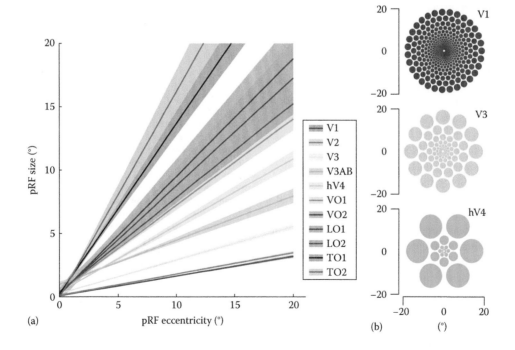

(a)

(b)

Figure 12.12 Population receptive field (pRF) size in human visual cortex. (a) pRF size is measured as a function of eccentricity and visual field map. In all visual areas, the pRF size increases with eccentricity. The pRF size is smallest in V1 (red) and increases with the visual hierarchy. The pRF size is one standard deviation of the spatial Gaussian describing the response to a point stimulus. (b) The pRF sizes of three areas are illustrated in visual space The circles are 1 standard deviation of the spatial Gaussian for a given eccentricity for each area. The images show the two trends in pRF size: larger pRFs with greater eccentricity and different pRF sizes in different visual areas. ([a]: Reproduced from Kay, K.N. et al., *J. Neurophysiol.*, 110, 481, 2013a; [b]: After Figure 1 from Freeman, J. and Simoncelli, E.P., *Nat. Neurosci.*, 14, 1195, 2011.)

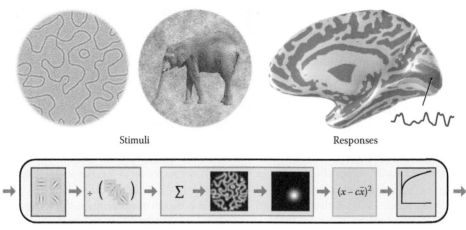

Stimuli   Responses

Models

Figure 12.13 Population receptive field (PRF) modeling from pixels to BOLD. A more complete pRF model of fMRI responses takes arbitrary visual stimuli as inputs (upper left) and predicts responses in a particular location in cortex (upper right). An example of a recent model that computes predicted responses for a variety of different spatial and texture patterns is shown in the lower panel (Kay, K.N. et al. 2013b). There are several computational stages in the model, including, from left to right, filtering and rectification (red), divisive normalization (beige), spatial pooling (yellow), second-order contrast nonlinearity (green), and a spatial nonlinearity (blue). (Figure provided by Kendrick Kay, http://kendrickkay.net/socmodel/.)

or the amplitude of narrowband gamma oscillations, show different patterns, indicating that neural responses at any one cortical location contain multiple signals. In future work, models that are integrated across measurement modalities will capture more and more aspects of the circuitry and visual representation.

12.3.5.2 Functional specialization within ventral maps and visual areas

In parallel with the study of visual field maps over the last several decades, another approach has been taken to study visual cortex. This approach focuses on tasks and behaviors, as well as stimulus properties other than spatial location, in order to characterize the functional architecture of visual cortex. Several principal findings have emerged. First, the ventral stream—ventral occipital and ventral temporal cortices—contains many regions that are highly responsive to particular stimulus features. These regions include specializations for color vision, faces, words, and scenes (Figure 12.14).

Much of the knowledge about functional specialization in the ventral visual pathways is derived from a century of lesion studies by neurologists (Zeki 1993). Patients with relatively focal cortical lesions to this pathway sometimes show deficits for specific visual functions, like face or color recognition. Much as Inouye's patients with lesions to primary visual cortex were blind in certain portions of the visual field (a scotoma), patients with ventral stream lesions can became blind to certain aspects of stimuli, such as color (achromatopsia) or facial identity (prosopagnosia).

Reports of such specific blindnesses were met with skepticism by many in the field (Zeki 1990). One reason for skepticism is that the observed deficits were rarely pure; a patient with cerebral achromatopsia might also have a visual field scotoma, for example. A second reason is that it just seemed unlikely that a person could be blind to a type of stimulus rather than a portion of the visual field. Nonetheless, over time many case studies have been reported, and some of the subjects have been studied in great detail in order to carefully characterize the perceptual deficits

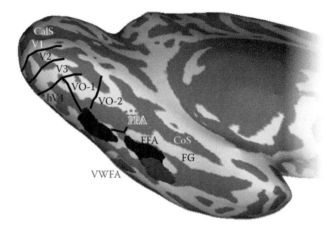

Figure 12.14 Stimulus-specific visual areas on the ventral surface of temporal cortex. A smoothed rendering of a left hemisphere, seen from below and medially, shows the locations of visual areas in ventral temporal cortex that are highly responsive to specific types of visual images. The visual word form area ("VWFA," blue) and fusiform face area ("FFA," red) were identified by functional MRI localizer experiments in which the subject viewed a variety of different stimuli; the blue and red regions were most strongly activated by images containing words or faces, respectively. Both the face- and word-selective responses comprise multiple patches (more posterior regions are filled with lighter shades), consistent with recent observations about category selective regions in ventral visual cortex (Grill-Spector and Weiner 2014, Weiner and Grill-Spector 2012). The approximate location of the parahippocampal place area ("PPA," green) is identified by comparison to previous papers. Several visual maps are shown in black outline. The PPA overlaps known visual field maps. (Adapted from a figure provided by Kendrick N. Kay, personal communication.)

(e.g., Duchaine et al. 2006). A century later, it appears that the neurologists reports were largely correct, and there are indeed locations on the ventral surface of the human visual system that show some degree of specialization for particular kinds of visual functions (Kanwisher 2010).

In the last quarter-century, functional specialization of the ventral stream has been characterized in much greater detail through the use of functional imaging with healthy subjects (PET and fMRI) and ECoG in patient volunteers (Allison et al. 1994, Grill-Spector and Weiner 2014, Kanwisher 2010). Across many studies, some patterns have become clear. First, there is a large-scale organization on the ventral temporal surface from medial to lateral (Figure 12.14). Medially, on the collateral sulcus, is a region highly responsive to scenes, sometimes called the "parahippocampal place area" (PPA) (Epstein and Kanwisher 1998). Lateral to the PPA are regions on the fusiform gyrus specialized for representing faces (Kanwisher et al. 1997). This region is sometimes called the fusiform face area (FFA), but because there may be several distinct clusters of face sensitive regions, it is also referred to by a series of anatomical names: pFus (posterior fusiform), mFus (middle fusiform), and IOG-faces (inferior occipital gyrus) (Weiner and Grill-Spector 2010). Quite close to the face area, typically slightly more laterally, is a region that is involved in seeing words, called the visual word form area (VWFA) (Ben-Shachar et al. 2007b, Cohen et al. 2000). This region is more prominent in the left hemisphere but can be identified bilaterally. More lateral are regions responsive to objects, called lateral occipital (LO) cortex (see Section 12.3.4.6) (Malach et al. 1995).

The names of these regions do not imply that the regions are *only* responsive to one kind of stimuli. The VWFA, for example, is also responsive to line drawings and many other kinds of stimuli. Nonetheless, there is enough specialization in each of these areas such that (1) they can be identified routinely in simple fMRI localizer experiments in nearly every subject based on the stimulus class that they are most responsive to and (2) cortical lesions to these locations result in perceptual deficits that are most severe in the expected domain (e.g., lesions to face areas impact face recognition more than other kinds of object recognition).

12.3.5.3 Functional specialization within lateral maps and visual areas

Several visual areas have been identified on the lateral surface of the temporal lobe. These include the areas known as "hMT+" (human middle temporal) and "LO" (lateral occipital) (Figures 12.9 and 12.10). The area known as "MT" (middle temporal) was first discovered in monkey and is a highly studied region in the visual pathways (Zeki 2004). It is a clear example of a cortical area with specialized function, in that it is highly sensitive to visual motion and binocular disparity. A neighboring region, known as MST (middle superior temporal), is also motion sensitive and appears to represent more complex forms of motion such as optic flow. The human homolog of these maps is known as hMT+ (DeYoe et al. 1994, Zeki et al. 1991). The "+" is included in the name because there are multiple visual field maps arranged in a cluster that are motion sensitive and because there is uncertainty about the exact homology between these multiple maps in human and the corresponding multiple maps in macaque. Two of these maps have been called TO-1/2 (temporal occipital 1 and 2), and they are likely homologs to MT and MST in the macaque (Amano et al. 2009, Huk et al. 2002).

Because of the very strong sensitivity to visual motion, this area can be easily identified in any human subject in a short period of fMRI scanning by contrasting the response to moving stimuli with the response to stationary stimuli. Stimulation of this region in patients with implanted electrodes causes motion illusions (Rauschecker et al. 2011), and lesions to this portion of cortex causes deficits in the ability to see visual motion ("akinetopsia") (Zeki 1991, Zihl et al. 1983). These areas are strongly modulated by top-down processes such as visual attention (Beauchamp et al. 1997).

Subsequent to the discovery of the MT maps, researchers identified an additional large visual area in the human brain, part of which was sandwiched between the early visual field maps (V1–V3) and MT and part of which extends down to the ventral surface. This area was first identified because of the functional specialization in object recognition and is generally known as LOC, the "lateral occipital complex" (Malach et al. 1995). More recently, the area has been subdivided into a more lateral and a more ventral portion (Sayres and Grill-Spector 2008). In functional MRI experiments, the LOC can be defined by having a greater response to images with intact objects compared to the same images that have been spatially scrambled in small parts. Lesions to this area cause deficits in the ability to recognize objects (object agnosia) (James et al. 2003). It was later discovered that there are two retinotopic maps between the early visual field maps (V1–V3) and MT, called LO-1 and LO-2 (Figures 12.9 and 12.10) (Larsson and Heeger 2006). The LOC overlaps LO-2.

The stimulus-selective regions of ventral and lateral occipital cortex should not be thought of as distinct from the retinotopic maps. For example, LOC, an object-selective region, overlaps the visual field map LO-2. Similarly, the PPA, a place-selective region, overlaps multiple visual field maps, including VO-2 and PHC-1/2 (Arcaro et al. 2009), and the "extrastriate body area," a region that responds to images of the human body (Epstein and Kanwisher 1998), overlaps multiple visual field maps on the dorsal surface of occipitotemporal cortex (Weiner and Grill-Spector 2011). In other cases, although maps have not yet been unambiguously identified, nonetheless there is evidence from fMRI studies that all visual regions tested to date have some sensitivity to stimulus position (Schwarzlose et al. 2008). If the position sensitivity turns out to be organized in such a way that neighboring cortical sites represent neighboring regions of visual space, then we would call the region a map. If, however, the position sensitivity does not have a clear spatial topography, then this would indicate a dissociation between spatial tuning and visual field maps. Future experiments will resolve this question in many more cortical areas such as the FFA and VWFA.

12.3.5.4 Functional specialization within dorsal maps and visual areas

Superior to the posterior maps (V1–V3), there is an ascending limb of the visual pathways including many visual field maps. Some of these maps, V3A/B, are highly responsive to visual motion and binocular disparity (Backus et al. 2001, Tootell et al. 1997). Other visual maps just superior to these, located in the intraparietal sulcus, are strongly modulated by attention (Silver et al. 2005). These maps were originally identified in memory-guided saccade tasks, in which subjects planned eye movements to locations in the visual field during fMRI scanning. For this reason, and because of the deficits associated with lesions to these maps, they are thought to play a role in coordinating visual representations and motor movements, including eye movements. These maps are strongly modulated by visual attention.

12.3.6 CORTICAL PLASTICITY AND STABILITY

One of the most important questions about visual system architecture is how it arises in development and how it changes when there is damage or unusual inputs to the visual system. Broadly speaking, the visual system faces two competing challenges: it must maintain a stable representation of the external world, so that other cortical areas can reliably interpret the outputs, and it must have some flexibility to learn from the environment. A general finding across decades of study in humans and animal models is that very early in life, the visual system has considerable flexibility to adapt to unusual inputs and circumstances. This ability to change and adapt is called plasticity. A complementary finding is that in later development and adulthood, there is a much greater tendency toward stability, meaning that there is less plasticity. These two competing demands on the visual system, and the apparent solution of favoring plasticity in early life and stability in later life, means that congenital disorders and disorders acquired later in life result in very different outcomes (Wandell and Smirnakis 2009).

12.3.6.1 Plasticity and stability in early development

12.3.6.1.1 Bilateral visual field maps in a patient with only one hemisphere

A striking example of a congenital deficit is the case of a girl born with a missing right cerebral hemisphere (Muckli et al. 2009). Such a case poses significant challenges for the development of visual field maps. In the normal developmental trajectory, the hemidecussation in the optic nerve results in each half of visual space being represented in the contralateral hemisphere. For this reason, each of the two LGNs and V1s represent only one half of space. In this case, the visual system developed very differently. Both halves of each retina send fibers to the left (intact) hemisphere. This hemisphere, unlike in most visual systems, represents the full visual field, not just the contralateral field. In her primary visual cortex, the left and right half of visual space are represented in two maps that overlap one another. In this case study, retinal ganglion cells were rerouted from the missing (right) LGN to the intact (left) LGN, demonstrating considerable flexibility of the early developing visual system to self-organize and alter the visual system architecture.

It is interesting to contrast this result with a different case study, in which one cerebral hemisphere was lost at the age of three for treatment of chronic encephalitis and epilepsy (Haak et al. 2014). In this case, unlike the congenital case, fMRI measurements showed that the early visual maps in the intact hemisphere retained the normal organization, representing only the contralateral hemisphere. The difference between these two studies suggests that by the age of three, many parts of the visual system are much more stable (and less plastic) than earlier in development.

12.3.6.1.2 Achiasma

In the vast majority of people, the optic nerves partially cross at the optic chiasm (hemidecussation); however, there is a rare congenital disorder in which people are born with no optic

chiasm. For these individuals, each eye sends retinal ganglion cell axons only to the ipsilateral LGN (no hemidecussation). Therefore, each LGN has a complete representation of the visual field from one eye, rather than a representation of one half of the visual field from two eyes. V1 inherits the unusual map from the LGN, so that each V1 represents the full visual field from one eye (Hoffmann et al. 2012, Victor et al. 2000). These individuals have an unusual visual system architecture. Similar to the case of the girl born with only one cerebral hemisphere, there appears to be a rather interesting retinotopic mapping solution for people born with no optic chiasm. They have a retinotopic map of each half of visual space folded over one other. This means that one small location in cortex represents not one contiguous region of space, but two regions that are mirror symmetric across the vertical meridian. Other than a loss of stereovision, these patients have relatively normal vision. The brain seems to learn which signals to combine and which signals to segregate; for example, even though the left and right visual field maps overlap, the subjects do not confuse inputs from the left and right half of space. This fact demonstrates that downstream visual areas, which rely on inputs from V1, have the flexibility to learn from experience how to read out an abnormal map in support of normal vision.

12.3.6.1.3 Rod monochromacy

In a rare congenital photoreceptor disorder, people are born with no functioning cones. Because the only working photoreceptors these individuals have are rods, they have no color vision and are called "rod monochromats." This poses an interesting problem for cortex and the development of the visual system architecture. Normally, the visual cortex devotes a large area to analyzing signals from the fovea; however, there are no rods in the fovea. Hence, if the cortex of rod monochromats developed like healthy controls, then there would be a large region near the occipital pole unresponsive to visual stimuli. This is not what is observed. Instead, in rod monochromats, the region of cortex that normally receives inputs from the fovea is instead responsive to rod inputs from the parafovea and periphery, reflecting plasticity in the early developing visual system (Baseler et al. 2002). It is important to note that this cortical response does not provide any vision at the fovea; after all, if there is no light absorption there can be no vision. What the plasticity means is that a certain portion of cortex functions differently in these individuals, contributing to visual analysis of signals originating outside the fovea.

12.3.6.2 Plasticity and stability in late development and adulthood

12.3.6.2.1 Retinal disorders (AMD/JMD, RP)

The effect of retinal disorders on brain function differs when the disorder happens later in life compared to when it is congenital, as in rod monochromacy. In several disorders, such as juvenile and age-related macular degeneration and retinitis pigmentosa, photoreceptors degenerate or become nonfunctional in a portion of the retina, creating a scotoma in the visual field, and a corresponding "lesion projection zone" in the cortex (Figure 12.15). In macular degeneration, the scotoma is in the central retina, and in retinitis pigmentosa, the scotoma is in the periphery. Because these disorders emerge well after the critical period in development, the brain shows little plasticity despite the rather

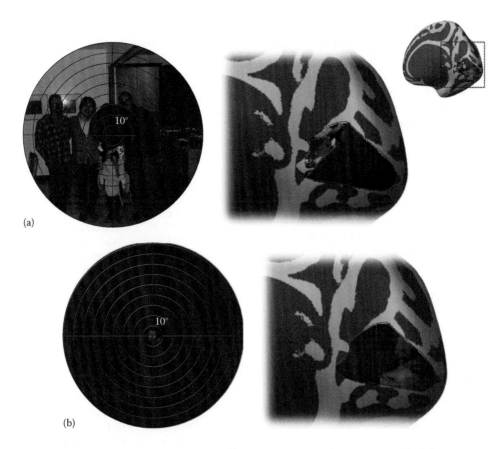

Figure 12.15 Retinal lesions. Various retinal diseases cause loss of function in regions of the retina. (a) The left panel shows an example of visual field loss from juvenile macular degeneration (JMD). The visual field loss is colored blue and is superimposed on an image (maximum visual angle of 90°). The cortical mesh on the right shows the projection of the image onto primary visual cortex. The lesion projection zone in blue. (b) An example of retinitis pigmentosa (RP), in which only a small central portion of the visual field is intact. All of the periphery indicates visual field loss (blue). The cortical mesh again shows the projection of the image onto primary visual cortex, with the lesion projection zone in blue. Note that the visual field loss is much larger in the case of the RP patient, but the cortical lesion projection zone is much larger for the JMD patient.

large changes in inputs associated with the diseases. As a result, large regions of visual cortex become relatively unresponsive to visual inputs (Baseler et al. 2011, Wandell and Smirnakis 2009). Because these visual areas were once functional, they can still be activated by feedback signals from higher brain regions, such as during memory or attention (Baker et al. 2005, Masuda et al. 2008, 2010).

12.3.6.2.2 Image restoration late in life

A question popular among philosophers is whether a person who has lived his or her life blind could learn to interpret images late in life if sight were restored. Such a situation has occasionally been borne out in the clinic. One case is the patient Mike May, who was blinded in a chemical accident at age 3 and remained blind until a corneal transplant surgery at age 46. Normally, the visual system is not fully mature by age 3. This means that when Mike lost his sight, his visual system was still developing. The development presumably stopped at this point so that when his retinal image was surgically repaired as an adult, his visual system was not fully mature. A key question was whether his visual system could mature at this late age, rendering a normal cortical architecture and visual perception. The answer, it seems, is no. Despite a clear image at the retina, even 10 years after surgery, his vision was still significantly impaired, and measurements of

his retinotopic maps, pRFs, and major fiber tracts all showed substantial differences from healthy controls (Fine et al. 2003, Levin et al. 2010). This case study indicates the limits of plasticity later in life.

12.4 SUMMARY

The visual system architecture reflects the wide array of functions the system performs, ranging from sensory to motor to circadian. All visual function originates with signals in the retina, processed by numerous cell types, including multiple classes of photoreceptors and ganglion cells. Signals leaving the eye are routed through a single nerve bundle, the optic nerve, which branches and targets numerous subcortical nuclei. Several important features of the visual system are exemplified in the branching pattern: (1) In hemidecussation, signals are separated by visual field, projecting to the contralateral hemisphere. (2) In parallel pathways, signals are segregated by cell type and eye of origin. The outputs of parallel pathways are then recombined to form new pathways. (3) In retinotopy, signals are represented by nearby points in the visual field project to nearby end points in the brain.

The visual pathways support many functions, ranging from circadian to motor to perception. The largest pathway is the geniculostriate pathway, in which signals are conveyed from retina to thalamus to cortex. Primary visual cortex, the largest recipient of visual fibers from the thalamus, distributes signals to many other areas of visual cortex. Most of these areas preserve the visual field map and are arranged in clusters of two or more maps. The different clusters emphasize different aspects of the stimulus and behavior. The more dorsal clusters contribute to spatial representations, spatial attention, and eye movements. The more ventral and lateral clusters contribute to recognition and appearance. Tools for studying these areas in the human brain have improved dramatically in the last few decades, allowing researchers to measure and quantitatively model the responses in many portions of the visual system as well as to measure the tissue properties of gray and white matter and fiber pathways connecting the visual areas. Studies of plasticity show that when inputs or pathways are abnormal early in development, there can be large changes to how the visual system develops.

ACKNOWLEDGMENTS

We thank Jennifer M Yoon, Kevin Weiner, and Brian Wandell for reading this chapter and providing many helpful suggestions.

REFERENCES

Adams, D. L., L. C. Sincich, and J. C. Horton. 2007. Complete pattern of ocular dominance columns in human primary visual cortex. *J Neurosci* 27 (39):10391–10403. doi: 10.1523/JNEUROSCI.2923-07.2007.

Allison, T., G. McCarthy, A. Nobre, A. Puce, and A. Belger. 1994. Human extrastriate visual cortex and the perception of faces, words, numbers, and colors. *Cereb Cortex* 4 (5):544–554.

Amano, K., B. A. Wandell, and S. O. Dumoulin. 2009. Visual field maps, population receptive field sizes, and visual field coverage in the human MT+ complex. *J Neurophysiol* 102 (5):2704–2718. doi: 10.1152/jn.00102.2009.

Andrews, T. J., S. D. Halpern, and D. Purves. 1997. Correlated size variations in human visual cortex, lateral geniculate nucleus, and optic tract. *J Neurosci* 17 (8):2859–2868.

Arcaro, M. J., S. A. McMains, B. D. Singer, and S. Kastner. 2009. Retinotopic organization of human ventral visual cortex. *J Neurosci* 29 (34):10638–10652.

Backus, B. T., D. J. Fleet, A. J. Parker, and D. J. Heeger. 2001. Human cortical activity correlates with stereoscopic depth perception. *J Neurophysiol* 86 (4):2054–2068.

Baker, C. I., E. Peli, N. Knouf, and N. G. Kanwisher. 2005. Reorganization of visual processing in macular degeneration. *J Neurosci* 25 (3):614–618. doi: 10.1523/JNEUROSCI.3476-04.2005.

Baseler, H. A., A. A. Brewer, L. T. Sharpe, A. B. Morland, H. Jagle, and B. A. Wandell. 2002. Reorganization of human cortical maps caused by inherited photoreceptor abnormalities. *Nat Neurosci* 5 (4):364–370. doi: 10.1038/nn817.

Baseler, H. A., A. Gouws, K. V. Haak, C. Racey, M. D. Crossland, A. Tufail, G. S. Rubin, F. W. Cornelissen, and A. B. Morland. 2011. Large-scale remapping of visual cortex is absent in adult humans with macular degeneration. *Nat Neurosci* 14 (5):649–655. doi: 10.1038/nn.2793.

Beauchamp, M. S., R. W. Cox, and E. A. DeYoe. 1997. Graded effects of spatial and featural attention on human area MT and associated motion processing areas. *J Neurophysiol* 78 (1):516–520.

Ben-Shachar, M., R. F. Dougherty, and B. A. Wandell. 2007a. White matter pathways in reading. *Curr Opin Neurobiol* 17 (2):258–270.

Ben-Shachar, M., R. F. Dougherty, G. K. Deutsch, and B. A. Wandell. 2007b. Differential sensitivity to words and shapes in ventral occipito-temporal cortex. *Cereb Cortex* 17 (7):1604–1611.

Benson, N. C., O. H. Butt, D. H. Brainard, and G. K. Aguirre. 2014. Correction of distortion in flattened representations of the cortical surface allows prediction of V1-V3 functional organization from anatomy. *PLoS Comput Biol* 10 (3):e1003538. doi: 10.1371/journal.pcbi.1003538.

Benson, N. C., O. H. Butt, R. Datta, P. D. Radoeva, D. H. Brainard, and G. K. Aguirre. 2012. The retinotopic organization of striate cortex is well predicted by surface topology. *Curr Biol* 22 (21):2081–2085. doi: 10.1016/j.cub.2012.09.014.

Berson, D. M., F. A. Dunn, and M. Takao. 2002. Phototransduction by retinal ganglion cells that set the circadian clock. *Science* 295 (5557):1070–1073. doi: 10.1126/science.1067262.

Brewer, A. A., J. Liu, A. R. Wade, and B. A. Wandell. 2005. Visual field maps and stimulus selectivity in human ventral occipital cortex. *Nat Neurosci* 8 (8):1102–1109.

Brewster, D. 1860. *Memoirs of the Life, Writings, and Discoveries of Sir Isaac Newton: 2*, Vol. 2. Edmonston and Douglas, London, U.K.

Carandini, M., J. B. Demb, V. Mante, D. J. Tolhurst, Y. Dan, B. A. Olshausen, J. L. Gallant, and N. C. Rust. 2005. Do we know what the early visual system does? *J Neurosci* 25 (46):10577–10597. doi: 10.1523/JNEUROSCI.3726-05.2005.

Cardin, V., R. Sherrington, L. Hemsworth, and A. T. Smith. 2012. Human V6: Functional characterisation and localisation. *PLoS One* 7 (10):e47685. doi: 10.1371/journal.pone.0047685.

Cohen, L., S. Dehaene, L. Naccache, S. Lehericy, G. Dehaene-Lambertz, M. A. Henaff, and F. Michel. 2000. The visual word form area: Spatial and temporal characterization of an initial stage of reading in normal subjects and posterior split-brain patients. *Brain* 123 (Pt 2):291–307.

Connolly, M. and D. Van Essen. 1984. The representation of the visual field in parvicellular and magnocellular layers of the lateral geniculate nucleus in the macaque monkey. *J Comp Neurol* 226 (4):544–564. doi: 10.1002/cne.902260408.

Dacey, D. M. 1993. The mosaic of midget ganglion cells in the human retina. *J Neurosci* 13 (12):5334–5355.

Dayan, M., M. Munoz, S. Jentschke, M. J. Chadwick, J. M. Cooper, K. Riney, F. Vargha-Khadem, and C. A. Clark. 2013. Optic radiation structure and anatomy in the normally developing brain determined using diffusion MRI and tractography. *Brain Struct Funct*. doi: 10.1007/s00429-013-0655-y.

Denison, R. N., A. T. Vu, E. Yacoub, D. A. Feinberg, and M. A. Silver. 2014. Functional mapping of the magnocellular and parvocellular subdivisions of human LGN. *Neuroimage* 102P2:358–369. doi: 10.1016/j.neuroimage.2014.07.019.

DeYoe, E. A., P. Bandettini, J. Neitz, D. Miller, and P. Winans. 1994. Functional magnetic resonance imaging (FMRI) of the human brain. *J Neurosci Methods* 54 (2):171–187.

Dougherty, R. F., V. M. Koch, A. A. Brewer, B. Fischer, J. Modersitzki, and B. A. Wandell. 2003. Visual field representations and locations of visual areas V1/2/3 in human visual cortex. *J Vis* 3 (10):586–598. doi: 10:1167/3.10.1.

Drager, U. C. and J. F. Olsen. 1980. Origins of crossed and uncrossed retinal projections in pigmented and albino mice. *J Comp Neurol* 191 (3):383–412. doi: 10.1002/cne.901910306.

Duchaine, B. C., G. Yovel, E. J. Butterworth, and K. Nakayama. 2006. Prosopagnosia as an impairment to face-specific mechanisms: Elimination of the alternative hypotheses in a developmental case. *Cogn Neuropsychol* 23 (5):714–747. doi: 10.1080/02643290500441296.

Dumoulin, S. O., R. D. Hoge, C. L. Baker, Jr., R. F. Hess, R. L. Achtman, and A. C. Evans. 2003. Automatic volumetric segmentation of human visual retinotopic cortex. *Neuroimage* 18 (3):576–587.

Dumoulin, S. O. and B. A. Wandell. 2008. Population receptive field estimates in human visual cortex. *Neuroimage* 39 (2):647–660. doi: 10.1016/j.neuroimage.2007.09.034.

Engel, S. A., G. H. Glover, and B. A. Wandell. 1997. Retinotopic organization in human visual cortex and the spatial precision of functional MRI. *Cerebral Cortex* 7 (2):181–192.

Engel, S. A., D. E. Rumelhart, B. A. Wandell, A. T. Lee, G. H. Glover, E. J. Chichilnisky, and M. N. Shadlen. 1994. fMRI of human visual cortex. *Nature* 369 (6481):525. doi: 10.1038/369525a0.

Epstein, R. and N. Kanwisher. 1998. A cortical representation of the local visual environment. *Nature* 392 (6676):598–601.

Felleman, D. J. and D. C. Van Essen. 1991. Distributed hierarchical processing in the primate cerebral cortex. *Cereb Cortex* 1 (1):1–47.

Field, G. D. and E. J. Chichilnisky. 2007. Information processing in the primate retina: Circuitry and coding. *Annu Rev Neurosci* 30:1–30. doi: 10.1146/annurev.neuro.30.051606.094252.

Fine, I., A. R. Wade, A. A. Brewer, M. G. May, D. F. Goodman, G. M. Boynton, B. A. Wandell, and D. I. MacLeod. 2003. Long-term deprivation affects visual perception and cortex. *Nat Neurosci* 6 (9):915–916.

Fitzgibbon, T. and S. F. Taylor. 1996. Retinotopy of the human retinal nerve fibre layer and optic nerve head. *J Comp Neurol* 375 (2):238–251. doi: 10.1002/(SICI)1096-9861(19961111)375:2<238::AID-CNE5>3.0.CO;2-3.

Flechsig, P. 1901. Developmental (myelogenetic) localisation of the cerebral cortex in the human subject. *Lancet* 158 (4077):1027–1030.

Flynn-Evans, E. E., H. Tabandeh, D. J. Skene, and S. W. Lockley. 2014. Circadian rhythm disorders and melatonin production in 127 blind women with and without light perception. *J Biol Rhythms* 29 (3):215–224. doi: 10.1177/0748730414536852.

Freeman, J. and E. P. Simoncelli. 2011. Metamers of the ventral stream. *Nat Neurosci* 14 (9):1195–1201. doi: 10.1038/nn.2889.

Grill-Spector, K., T. Kushnir, T. Hendler, S. Edelman, Y. Itzchak, and R. Malach. 1998. A sequence of object-processing stages revealed by fMRI in the human occipital lobe. *Hum Brain Mapp* 6 (4):316–328.

Grill-Spector, K. and K. S. Weiner. 2014. The functional architecture of the ventral temporal cortex and its role in categorization. *Nat Rev Neurosci* 15 (8):536–548. doi: 10.1038/nrn3747.

Haak, K. V., D. R. Langers, R. Renken, P. van Dijk, J. Borgstein, and F. W. Cornelissen. 2014. Abnormal visual field maps in human cortex: A mini-review and a case report. *Cortex* 56:14–25. doi: 10.1016/j.cortex.2012.12.005.

Harvey, B. M., M. J. Vansteensel, C. H. Ferrier, N. Petridou, W. Zuiderbaan, E. J. Aarnoutse, M. G. Bleichner et al. 2012. Frequency specific spatial interactions in human electrocorticography: V1 alpha oscillations reflect surround suppression. *NeuroImage*. doi: 10.1016/j.neuroimage.2012.10.020.

Heeger, D. J., E. P. Simoncelli, and J. A. Movshon. 1996. Computational models of cortical visual processing. *Proc Natl Acad Sci U S A* 93 (2):623–627.

Henschen, S.E. 1893. On the visual path and centre. *Brain* 16 (1–2):170–180.

Hofbauer, A. and U. C. Drager. 1985. Depth segregation of retinal ganglion cells projecting to mouse superior colliculus. *J Comp Neurol* 234 (4):465–474. doi: 10.1002/cne.902340405.

Hoffmann, M. B., F. R. Kaule, N. Levin, Y. Masuda, A. Kumar, I. Gottlob, H. Horiguchi, R. F. Dougherty et al. 2012. Plasticity and stability of the visual system in human achiasma. *Neuron* 75 (3):393–401. doi: 10.1016/j.neuron.2012.05.026.

Holmes, G. 1918. Disturbances of vision by cerebral lesions. *Br J Ophthalmol* 2 (7):353–384.

Holmes, G. and W. T. Lister. 1916. Disturbances of vision from cerebral lesions, with special reference to the cortical representation of the macula. *Brain* 39 (1–2):34–73.

Horton, J. C., M. M. Greenwood, and D. H. Hubel. 1979. Non-retinotopic arrangement of fibres in cat optic nerve. *Nature* 282 (5740):720–722.

Horton, J. C. and E. T. Hedley-Whyte. 1984. Mapping of cytochrome oxidase patches and ocular dominance columns in human visual cortex. *Philos Trans R Soc Lond B Biol Sci* 304 (1119):255–272.

Horton, J. C. and W. F. Hoyt. 1991a. Quadrantic visual field defects. A hallmark of lesions in extrastriate (V2/V3) cortex. *Brain* 114 (Pt 4):1703–1718.

Horton, J. C. and W. F. Hoyt. 1991b. The representation of the visual field in human striate cortex. A revision of the classic Holmes map. *Arch Ophthalmol* 109 (6):816–824.

Hubel, D. H. and T. N. Wiesel. 1977. Ferrier lecture. Functional architecture of macaque monkey visual cortex. *Proc R Soc Lond B Biol Sci* 198 (1130):1–59.

Hubel, D. H., T. N. Wiesel, and S. LeVay. 1977. Plasticity of ocular dominance columns in monkey striate cortex. *Philos Trans R Soc Lond B Biol Sci* 278 (961):377–409.

Huk, A. C., R. F. Dougherty, and D. J. Heeger. 2002. Retinotopy and functional subdivision of human areas MT and MST. *J Neurosc* 22 (16):7195–7205.

Inouye, T. 2000. Eye disturbances after gunshot injuries to the cortical visual pathways. Translated from the German by Glickstein M, and Fahle M. Oxford University Press, Oxford, U.K.

James, T. W., J. Culham, G. K. Humphrey, A. D. Milner, and M. A. Goodale. 2003. Ventral occipital lesions impair object recognition but not object-directed grasping: An fMRI study. *Brain* 126 (Pt 11):2463–2475. doi: 10.1093/brain/awg248.

Jerde, T. A. and C. E. Curtis. 2013. Maps of space in human fronto-parietal cortex. *J Physiol Paris* 107 (6):510–516. doi: 10.1016/j.jphysparis.2013.04.002.

Jerde, T. A., E. P. Merriam, A. C. Riggall, J. H. Hedges, and C. E. Curtis. 2012. Prioritized maps of space in human frontoparietal cortex. *J Neurosci* 32 (48):17382–17390. doi: 10.1523/JNEUROSCI.3810-12.2012.

Kandel, E.R., James, H. S., and Thomas, M. J. 2000. *Principles of Neural Science*, Vol. 4. McGraw-Hill, New York.

Kanwisher, N. 2010. Functional specificity in the human brain: A window into the functional architecture of the mind. *Proc Natl Acad Sci U S A* 107 (25):11163–11170. doi: 10.1073/pnas.1005062107.

Kanwisher, N., J. McDermott, and M. M. Chun. 1997. The fusiform face area: A module in human extrastriate cortex specialized for face perception. *J Neurosci* 17 (11):4302–4311.

Kastner, S., K. A. Schneider, and K. Wunderlich. 2006. Beyond a relay nucleus: Neuroimaging views on the human LGN. *Prog Brain Res* 155:125–143. doi: 10.1016/S0079-6123(06)55008-3.

Katyal, S., S. Zughni, C. Greene, and D. Ress. 2010. Topography of covert visual attention in human superior colliculus. *J Neurophysiol* 104 (6):3074–3083. doi: 10.1152/jn.00283.2010.

Kay, K. N., T. Naselaris, R. J. Prenger, and J. L. Gallant. 2008. Identifying natural images from human brain activity. *Nature* 452 (7185):352–355. doi: 10.1038/nature06713.

Kay, K. N., J. Winawer, A. Mezer, and B. A. Wandell. 2013a. Compressive spatial summation in human visual cortex. *J Neurophysiol* 110 (2):481–494. doi: 10.1152/jn.00105.2013.

Kay, K. N., J. Winawer, A. Rokem, A. Mezer, and B. A. Wandell. 2013b. A two-stage cascade model of BOLD responses in human visual cortex. *PLoS Comput Biol* 9 (5):e1003079. doi: 10.1371/journal.pcbi.1003079.

Kinney, H. C., B. A. Brody, A. S. Kloman, and F. H. Gilles. 1988. Sequence of central nervous system myelination in human infancy. II. Patterns of myelination in autopsied infants. *J Neuropathol Exp Neurol* 47 (3):217–234.

Kiorpes, L. and S. P. McKee. 1999. Neural mechanisms underlying amblyopia. *Curr Opin Neurobiol* 9 (4):480–486.

Kolster, H., J. B. Mandeville, J. T. Arsenault, L. B. Ekstrom, L. L. Wald, and W. Vanduffel. 2009. Visual field map clusters in macaque extrastriate visual cortex. *J Neurosci* 29 (21):7031–7039. doi: 10.1523/JNEUROSCI.0518-09.2009.

Larsson, J. and D. J. Heeger. 2006. Two retinotopic visual areas in human lateral occipital cortex. *J Neurosci* 26 (51):13128–13142.

Levin, N., S. O. Dumoulin, J. Winawer, R. F. Dougherty, and B. A. Wandell. 2010. Cortical maps and white matter tracts following long period of visual deprivation and retinal image restoration. *Neuron* 65 (1):21–31. doi: 10.1016/j.neuron.2009.12.006.

Malach, R., J. B. Reppas, R. R. Benson, K. K. Kwong, H. Jiang, W. A. Kennedy, P. J. Ledden, T. J. Brady, B. R. Rosen, and R. B. Tootell. 1995. Object-related activity revealed by functional magnetic resonance imaging in human occipital cortex. *Proc Natl Acad Sci U S A* 92 (18):8135–8139.

Masuda, Y., S. O. Dumoulin, S. Nakadomari, and B. A. Wandell. 2008. V1 projection zone signals in human macular degeneration depend on task, not stimulus. *Cereb Cortex* 18 (11):2483–2493.

Masuda, Y., H. Horiguchi, S. O. Dumoulin, A. Furuta, S. Miyauchi, S. Nakadomari, and B. A. Wandell. 2010. Task-dependent V1 responses in human retinitis pigmentosa. *Invest Ophthalmol Vis Sci* 51 (10):5356–5364. doi: 10.1167/iovs.09-4775.

Maunsell, J. H. and W. T. Newsome. 1987. Visual processing in monkey extrastriate cortex. *Ann Rev Neurosc* 10:363–401.

Mezer, A., J. D. Yeatman, N. Stikov, K. N. Kay, N. J. Cho, R. F. Dougherty, M. L. Perry et al. 2013. Quantifying the local tissue volume and composition in individual brains with magnetic resonance imaging. *Nat Med* 19 (12):1667–1672. doi: 10.1038/nm.3390.

Muckli, L., M. J. Naumer, and W. Singer. 2009. Bilateral visual field maps in a patient with only one hemisphere. *Proc Natl Acad Sci U S A* 106 (31):13034–13039. doi: 10.1073/pnas.0809688106.

Perry, V. H., R. Oehler, and A. Cowey. 1984. Retinal ganglion cells that project to the dorsal lateral geniculate nucleus in the macaque monkey. *Neuroscience* 12 (4):1101–1123.

Pitzalis, S., C. Galletti, R. S. Huang, F. Patria, G. Committeri, G. Galati, P. Fattori, and M. I. Sereno. 2006. Wide-field retinotopy defines human cortical visual area v6. *J Neurosci* 26 (30):7962–7973. doi: 10.1523/JNEUROSCI.0178-06.2006.

Provencio, I., I. R. Rodriguez, G. Jiang, W. P. Hayes, E. F. Moreira, and M. D. Rollag. 2000. A novel human opsin in the inner retina. *J Neurosci* 20 (2):600–605.

Prusky, G. T. and R. M. Douglas. 2004. Characterization of mouse cortical spatial vision. *Vision Res* 44 (28):3411–3418. doi: 10.1016/j.visres.2004.09.001.

Rauschecker, A. M., M. Dastjerdi, K. S. Weiner, N. Witthoft, J. Chen, A. Selimbeyoglu, and J. Parvizi. 2011. Illusions of visual motion elicited by electrical stimulation of human MT complex. *PLoS One* 6 (7):e21798. doi: 10.1371/journal.pone.0021798.

Riesenhuber, M. and T. Poggio. 1999. Hierarchical models of object recognition in cortex. *Nat Neurosc* 2 (11):1019–1025.

Sayres, R. and K. Grill-Spector. 2008. Relating retinotopic and object-selective responses in human lateral occipital cortex. *J Neurophysiol* 100 (1):249–267.

Schira, M. M., C. W. Tyler, M. Breakspear, and B. Spehar. 2009. The foveal confluence in human visual cortex. *J Neurosci* 29 (28):9050–9058.

Schneider, K. A. and S. Kastner. 2005. Visual responses of the human superior colliculus: A high-resolution functional magnetic resonance imaging study. *J Neurophysiol* 94 (4):2491–2503. doi: 10.1152/jn.00288.2005.

Schneider, K. A., M. C. Richter, and S. Kastner. 2004. Retinotopic organization and functional subdivisions of the human lateral geniculate nucleus: A high-resolution functional magnetic resonance imaging study. *J Neurosci* 24 (41):8975–8985. doi: 10.1523/JNEUROSCI.2413-04.2004.

Schwarzlose, R. F., J. D. Swisher, S. Dang, and N. Kanwisher. 2008. The distribution of category and location information across object-selective regions in human visual cortex. *Proc Natl Acad Sci U S A* 105 (11):4447–4452. doi: 10.1073/pnas.0800431105.

Sereno, M. I., A. M. Dale, J. B. Reppas, K. K. Kwong, J. W. Belliveau, T. J. Brady, B. R. Rosen, and R. B. Tootell. 1995. Borders of multiple visual areas in humans revealed by functional magnetic resonance imaging. *Science* 268 (5212):889–893.

Sereno, M. I. and R. B. Tootell. 2005. From monkeys to humans: What do we now know about brain homologies? *Curr Opin Neurobiol* 15 (2):135–144. doi: 10.1016/j.conb.2005.03.014.

Shapley, R. and M. Hawken. 2002. Neural mechanisms for color perception in the primary visual cortex. *Curr Opin Neurobiol* 12 (4):426–432.

Sherman, S. M. 2007. The thalamus is more than just a relay. *Curr Opin Neurobiol* 17 (4):417–422. doi: 10.1016/j.conb.2007.07.003.

Sherman, S. M. and C. Koch. 1986. The control of retinogeniculate transmission in the mammalian lateral geniculate nucleus. *Exp Brain Res* 63 (1):1–20.

Sillito, A. M. and H. E. Jones. 2002. Corticothalamic interactions in the transfer of visual information. *Philos Trans R Soc Lond B Biol Sci* 357 (1428):1739–1752. doi: 10.1098/rstb.2002.1170.

Silver, M. A., D. Ress, and D. J. Heeger. 2005. Topographic maps of visual spatial attention in human parietal cortex. *J Neurophysiol* 94 (2):1358–1371. doi: 10.1152/jn.01316.2004.

Snow, J. C., H. A. Allen, R. D. Rafal, and G. W. Humphreys. 2009. Impaired attentional selection following lesions to human pulvinar: Evidence for homology between human and monkey. *Proc Natl Acad Sci U S A* 106 (10):4054–4059. doi: 10.1073/pnas.0810086106.

Stein, B. E., M. W. Wallace, T. R. Stanford, and W. Jiang. 2002. Cortex governs multisensory integration in the midbrain. *Neuroscientist* 8 (4):306–314.

Stenbacka, L. and S. Vanni. 2007. fMRI of peripheral visual field representation. *Clin Neurophysiol* 118 (6):1303–1314. doi: 10.1016/j.clinph.2007.01.023.

Stensaas, S. S., D. K. Eddington, and W. H. Dobelle. 1974. The topography and variability of the primary visual cortex in man. *J Neurosurg* 40 (6):747–755. doi: 10.3171/jns.1974.40.6.0747.

Swisher, J. D., M. A. Halko, L. B. Merabet, S. A. McMains, and D. C. Somers. 2007. Visual topography of human intraparietal sulcus. *J Neurosci* 27 (20):5326–5337. doi: 10.1523/JNEUROSCI.0991-07.2007.

Tootell, R. B., J. D. Mendola, N. K. Hadjikhani, P. J. Ledden, A. K. Liu, J. B. Reppas, M. I. Sereno, and A. M. Dale. 1997. Functional analysis of V3A and related areas in human visual cortex. *J Neurosci* 17 (18):7060–7078.

Ungerleider, L. G. and J. V. Haxby. 1994. 'What' and 'where' in the human brain. *Curr Opin Neurobiol* 4 (2):157–165.

Ungerleider, L. G. and M. Mishkin. 1982. Two cortical visual systems. In *Analysis of Visual Behavior*, eds. D. Ingle, M. A. Goodale, and R. J. W. Mansfield, pp. 549–587. MIT Press, Cambridge, MA.

Van Essen, D. C. 2004. Organization of visual areas in macaque and human cerebral cortex. In *The Visual Neurosciences*, eds. L. M. Chalupa and J. S. Werner, pp. 507–521. MIT Press, Cambridge, MA.

Victor, J. D., P. Apkarian, J. Hirsch, M. M. Conte, M. Packard, N. R. Relkin, K. H. Kim, and R. M. Shapley. 2000. Visual function and brain organization in non-decussating retinal-fugal fibre syndrome. *Cereb Cortex* 10 (1):2–22.

Victor, J. D., K. Purpura, E. Katz, and B. Mao. 1994. Population encoding of spatial frequency, orientation, and color in macaque V1. *J Neurophysiol* 72 (5):2151–2166.

Wandell, B. A., A. A. Brewer, and R. F. Dougherty. 2005. Visual field map clusters in human cortex. *Philos Trans R Soc Lond B Biol Sci* 360 (1456):693–707.

Wandell, B. A., S. O. Dumoulin, and A. A. Brewer. 2007. Visual field maps in human cortex. *Neuron* 56 (2):366–383.

Wandell, B. A. and S. M. Smirnakis. 2009. Plasticity and stability of visual field maps in adult primary visual cortex. *Nat Rev Neurosci* 10 (12):873–884. doi: 10.1038/nrn2741.

Wandell, B. A. and J. Winawer. 2011. Imaging retinotopic maps in the human brain. *Vis Res* 51 (7):718–737. doi: 10.1016/j.visres.2010.08.004.

Wandell, B. A., J. Winawer, and K. N. Kay. 2015. Computational modeling of responses in human visual cortex. In *Brain Mapping: An Encyclopedic Reference*, ed. A. Toga.

Wandell, B. A. and J. D. Yeatman. 2013. Biological development of reading circuits. *Curr Opin Neurobiol* 23 (2):261–268. doi: 10.1016/j.conb.2012.12.005.

Wandell, B. A. 1995. *Foundations of Vision*. Sinauer Associates, Sunderland, MA.

Weiner, K. S. and K. Grill-Spector. 2010. Sparsely-distributed organization of face and limb activations in human ventral temporal cortex. *Neuroimage* 52 (4):1559–1573. doi: 10.1016/j.neuroimage.2010.04.262.

Weiner, K. S. and K. Grill-Spector. 2011. Not one extrastriate body area: Using anatomical landmarks, hMT+, and visual field maps to parcellate limb-selective activations in human lateral occipito-temporal cortex. *Neuroimage* 56 (4):2183–2199. doi: 10.1016/j.neuroimage.2011.03.041.

Weiner, K. S. and K. Grill-Spector. 2012. The improbable simplicity of the fusiform face area. *Trends Cogn Sci* 16 (5):251–254. doi: 10.1016/j.tics.2012.03.003.

Winawer, J., H. Horiguchi, R. A. Sayres, K. Amano, and B. A. Wandell. 2010. Mapping hV4 and ventral occipital cortex: The venous eclipse. *J Vis* 10 (5):1. doi: 10.1167/10.5.1.

Winawer, J., K. N. Kay, B. L. Foster, A. M. Rauschecker, J. Parvizi, and B. A. Wandell. 2013. Asynchronous broadband signals are the principal source of the BOLD response in human visual cortex. *Curr Biol* 23 (13):1145–1153. doi: 10.1016/j.cub.2013.05.001.

Witthoft, N., M. L. Nguyen, G. Golarai, K. F. LaRocque, A. Liberman, M. E. Smith, and K. Grill-Spector, 2014. Where is human V4? Predicting the location of hV4 and VO1 from cortical folding. *Cereb Cortex* 24 (9):2401–2408. doi:10.1093/cercor/bht092.

Yeatman, J. D., A. M. Rauschecker, and B. A. Wandell. 2013. Anatomy of the visual word form area: Adjacent cortical circuits and long-range white matter connections. *Brain Lang* 125 (2):146–155. doi: 10.1016/j.bandl.2012.04.010.

Yeatman, J. D., B. A. Wandell, and A. A. Mezer. 2014. Lifespan maturation and degeneration of human brain white matter. *Nat Commun* 5:4932. doi: 10.1038/ncomms5932.

Yoshor, D., G. M. Ghose, W. H. Bosking, P. Sun, and J. H. Maunsell. 2007. Spatial attention does not strongly modulate neuronal responses in early human visual cortex. *J Neurosc: Off J Soc Neurosc* 27 (48):13205–13209. doi: 10.1523/JNEUROSCI.2944-07.2007.

Zeki, S. 1990. A century of cerebral achromatopsia. *Brain* 113 (Pt 6): 1721–1777.

Zeki, S. 1991. Cerebral akinetopsia (visual motion blindness). A review. *Brain* 114 (Pt 2):811–824.

Zeki, S. 1993. *A Vision of the Brain*. Blackwell Scientific Publications, Oxford, U.K.

Zeki, S. 2004. Thirty years of a very special visual area, Area V5. *J Physiol* 557 (Pt 1):1–2. doi: 10.1113/jphysiol.2004.063040.

Zeki, S., J. D. Watson, C. J. Lueck, K. J. Friston, C. Kennard, and R. S. Frackowiak. 1991. A direct demonstration of functional specialization in human visual cortex. *J Neurosci* 11 (3):641–649.

Zihl, J., D. von Cramon, and N. Mai. 1983. Selective disturbance of movement vision after bilateral brain damage. *Brain* 106 (Pt 2):313–340.

Visual psychophysical methods

Denis G. Pelli and Joshua A. Solomon

Contents

13.1 INTRODUCTION

The optics of the eye limit what people can see, but they are not the only limit. Changes in visual optics do not necessarily alter vision. To measure vision itself, psychophysical methods are required. Psychophysics (Fechner 1860/1966) is a relatively young discipline, and its theory and practice continue to develop, affording ever better metrics of performance and appearance. This chapter is a practical guide for optical experts. It contains instructions for measuring contrast threshold and acuity. These measurements are especially relevant for the study of visual optics. More comprehensive discussions of the methods and history of psychophysics can be found elsewhere (e.g., Kingdom and Prins, 2010; Lu and Dosher, 2013; Pelli and Bex, 2013; Pelli and Farell, 2010; Solomon 2013).

13.2 THRESHOLD

Vision occurs in two steps, one optical and one neural. Visual optics are responsible for imaging the viewed object (or *stimulus*) onto the retina. Neurons in the retina and brain are responsible for transforming the retinal image into a percept. Stimulus and retinal image are related by a simple linear transformation, the optical transfer function (OTF), at each position in the retina. The relationship between retinal image and internal percept is more complex. Do not despair. Even without a complete model for this relationship, we can still usefully characterize visual ability by measuring "thresholds."

In any psychophysical experiment, the observer is assigned a task. For example, she might be asked to identify a black letter on a white background. Each combination of stimulus (in this case the specific letter) and response, which can be scored as right or wrong, is called a *trial*. After several such trials, we can estimate the probability of a correct response. If the observer were guessing randomly, this ("chance") probability would be $1/26 \approx 0.04$. If the letter were easy to see, the probability could be as high as 1.

Typically, we are interested in a physical parameter, such as contrast, which affects visual performance. The observer's *threshold* is the value of that parameter required for the observer to attain a criterion proportion correct. That criterion is arbitrary, but is typically roughly halfway between chance and perfect performance.*

Those interested in visual optics will usually want to describe visual performance in terms of threshold. It has three virtues.

---

* This chapter is concerned solely with *performance* thresholds, which should not be confused with *sensory* thresholds. The *performance* threshold is defined operationally and widely used, whereas the *sensory* threshold is a discredited theoretical assumption. Consider why an observer sometimes makes errors with faint stimuli. One possibility is that the observer hallucinated a stimulus on the blank interval, and that the hallucination was more intense than the veridical response to the actual stimulus. This seemingly psychedelic theory is in fact implied by the inescapable fact that the retina, like any physical sensor, must have internal random variations (noise) by which any response to a real stimulus must also occur with some nonzero probability in response to a blank. Another possibility is that the observer, having seen nothing in either interval, was forced to guess, and guessed wrong. In other words, there is a minimum perceptible intensity, and the stimulus did not exceed this *sensory threshold*. Attempts to confirm the existence of sensory thresholds have been largely unsuccessful (e.g., Green and Swets, 1966; Solomon, 2007).

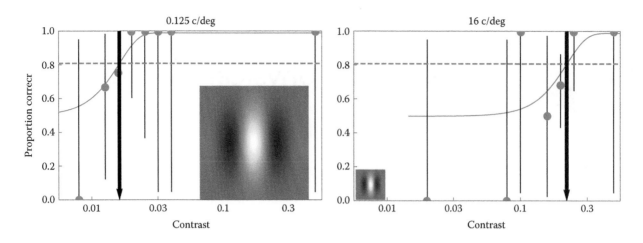

Figure 13.1 Two psychometric functions, differing only in threshold. The task was to detect a grating: 0.125 c/deg (a) or 16 c/deg (b). (The gratings shown are not to scale; the size ratio of the actual stimuli was 16:0.125 = 256:1.) The response on each two-alternative forced-choice trial was scored right or wrong. Each dot is the observer's proportion correct for however many trials that were tested at that contrast. Each dot has an error bar showing its 95% confidence interval. The smooth curve is a standard psychometric function fit to the data (Equation 13.1). The horizontal dashed line shows the arbitrary threshold criterion of 81% correct.

First, threshold can be precisely measured in just a few trials (20–40, see Figure 13.1). Second, threshold (unlike proportion correct) is a physical quantity (e.g., contrast) that is well understood in the context of visual optics. Third, a wide range of contrast thresholds (spanning more than 2 log units, 100:1) can be measured with high precision, whereas in limited time, proportion correct can be measured with useful precision over only a much smaller range of contrast, perhaps half a log unit, 3:1.

13.3 DETECTING A GRATING

13.3.1 FORCED-CHOICE METHOD: 2AFC

Imagine staring at a blank computer screen and hearing two beeps, one after the other. Each beep designates a time. A very faint picture (the stimulus) was displayed during one of the two beeps, randomly. You have to decide at which time the picture was shown. All of this constitutes one trial of a *two-alternative forced-choice* (*2AFC*) detection experiment. It is called 2AFC because you have been presented with 2 intervals and you have to choose one. This task is called "detection" because the other interval was blank. If you had your eyes closed, your expected response accuracy, given the two possible choices, would be 1/2 or 0.5. Response accuracy could be as high as 1, but you might still occasionally press the wrong button, perhaps because you were distracted or you blinked, so, in practice, maximum accuracy is typically closer to 0.99.

13.3.2 PSYCHOMETRIC FUNCTIONS

Figure 13.1 illustrates data from two 2AFC detection experiments. The left panel shows response accuracy when the stimulus was a sinusoidal luminance grating having 0.125 cycles per degree (c/deg) of visual angle. The right panel shows data for a 16 c/deg grating. As we shall see, gratings like these can be used to characterize the observer's contrast sensitivity function (CSF), of which the optical modulation transfer function (MTF) is one factor (Equation 13.2). The other factor is neural.

Figure 13.1 shows that the frequency of correct response tends to increase with stimulus contrast. Each data point in the figure conveys one such proportion correct. For example, the leftmost data point indicates that the observer responded correctly on zero (of one) trials at a contrast of 0.008, and the next data point indicates that he responded correctly on two of the three trials at a contrast of 0.013. Ultimately, however, we are interested in probabilities, not frequencies. We want to know, for example, what is the probability that the observer will respond correctly at a contrast of 0.013. Since 2 of the 3 responses collected were correct, 0.67 is the best guess for this probability, but a wide range of actual probabilities could give rise to the measured count of 2 out of 3. There are statistical techniques for determining the likelihood of any underlying (or true) probability given any observed frequency (i.e., both the number correct and the total number; Brown et al., 2001). Using one of these techniques, we can find the probability below which the chances of producing 2 out of 3 correct responses are less than 1/40. Similarly, we can find the probability above which the chances of producing 2 out of 3 correct responses are less than 1/40. The probabilities between these two comprise the 95% confidence interval.

Each error bar in Figure 13.1 contains a 95% confidence interval. Some confidence intervals are smaller than others, because more trials were used to determine those accuracies. You may wonder why. The answer is because this experiment employed an *adaptive staircase* that selected an informative stimulus contrast for each trial based on the previous trials.[*] Adaptive staircases, like QUEST (Watson and Pelli, 1983), are computer algorithms that estimate threshold quickly, with a minimum number of trials.

The psychometric functions in Figure 13.1 are Weibull functions with the form

$$p(c) = (1 - \lambda) - \left(\frac{1}{2} - \lambda \right) \exp\left[-(c/\alpha)^\beta \right], \tag{13.1}$$

---

[*] We used the *Psychophysica* (Watson & Solomon, 1997) implementation of Quest, using default values for all options and 32 trials per experiment.

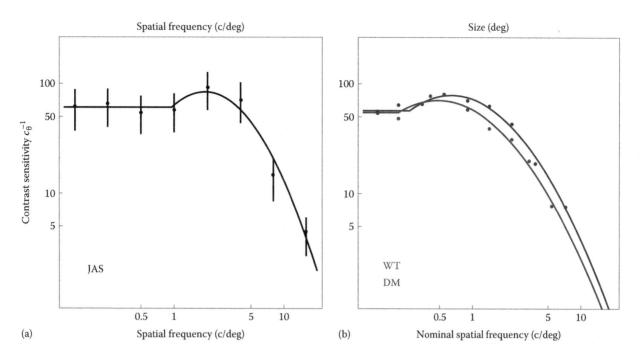

Figure 13.2 Contrast sensitivity, c_θ for (a) gratings, as a function of spatial frequency, and (b) letters (Bookman font), as a function of size. Smooth curves illustrate best-fitting, four-parameter summaries of these CSFs (Lesmes, et al., 2010). Letter thresholds from Pelli and Farell (1999). Nominal spatial frequency as specified by Majaj et al. (2002).

where c denotes stimulus contrast and $p(c)$ is the probability correct. Although other forms have been used, this particular form and the parameter values for psychometric slope ($\beta = 3$) and lapse rate ($\lambda = 0.01$) both have a long history in summarizing psychometric functions for 2AFC contrast detection (May and Solomon, 2013). The remaining parameter, α, establishes the psychometric function's horizontal position along the log contrast scale.

Any computational optimization routine can be used to find the best-fitting value of this parameter, but software specifically tailored to the needs of the psychophysicist are catalogued in Strasburger's (2015) regularly updated website *Software for visual psychophysics*. Current packages run in four programming environments: C++, MATLAB®, Mathematica, and R (e.g., Linares and López-Moliner, 2015; Watson and Pelli, 1983, Watson and Solomon 1997). The smooth curves in Figure 13.1 are maximum-likelihood fits.

13.3.3 CONTRAST SENSITIVITY FUNCTION

When plotted as a function of log contrast, as in Figure 13.1, the shape of the psychometric function is highly conserved. In most detection experiments, only its horizontal position varies (Foley and Legge, 1981). Thus, most studies plot only threshold, which is proportional to α (in Equation 13.1). In Figure 13.1, threshold contrast is more than ten times higher for the smaller stimulus. As noted before, threshold, for example, c_θ, is the stimulus strength at which performance reaches a given criterion level θ of accuracy. The criterion of 81% correct is convenient because $p^{-1}(\alpha) = 0.81$, however, any value can be used (preferably well below 100% and well above chance).

It is traditional to define sensitivity as the reciprocal of threshold (Campbell and Robson, 1968). When sensitivity is plotted against stimulus frequency, the result is known as a contrast sensitivity function (CSF). Figure 13.2a illustrates the CSF

estimated from the contrast threshold at eight spatial frequencies from 0.0125 to 16 c/deg. The precision of these measurements is illustrated by error bars (95% confidence intervals).*

The smooth curves drawn through the CSFs in Figure 13.2 have the form[†] described by Lesmes et al. (2010). They show that the CSF can be measured quickly by assuming a form and directly estimating its parameters from the trial data.

13.4 IDENTIFYING A LETTER

Most clinical vision tests are based on letter identification, not grating detection. Like 2AFC detection, letter identification is an objective, forced-choice task. There are several advantages to using letters (Pelli and Robson, 1991). First, letter identification is a familiar task, which clinical patients can perform with minimal instruction, which saves time in a busy clinic. Second, if there are many possible letters (at least five), then the guessing rate is low (at most 20%), which makes each trial more informative, which yields a good threshold estimate in fewer trials. Third, the ability to identify letters is important in many real-life tasks.

The spatial frequencies that mediate letter identification scale as the –0.7 power of letter size (Majaj et al., 2002). Each point in Figure 13.2b is the contrast threshold for 64% identification accuracy.

13.4.1 EYE CHARTS

Precise threshold measurement usually requires computer presentation of a custom stimulus for each trial. However, clinical

---

* These intervals were calculated from the distribution of thresholds produced by a simulated observer whose psychometric functions are identical to those that best fit the human observer's. This statistical technique is a form of parametric *bootstrapping* (Efron, 2012).
† There is a typo in Eq. 2 of Lesmes et al. (2010). The last γ_{max} should be $\log_{10}(\gamma_{max})$.

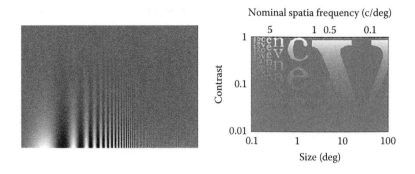

Figure 13.3 Eye charts for demonstration. Contrast sensitivity or threshold for gratings and letters (Campbell and Robson, 1964; Ohzawa, 2005; Pelli and Farell, 1999). Nominal spatial frequency as specified by Majaj et al. (2002).

assessment is mostly done using eye charts, because they are inexpensive and quick. They are also a good way of getting pilot data for many research projects. For a quick demonstration of effects to an audience, it can be useful to test contrast sensitivity at all scales at once. Figure 13.3 allows observers to visualize their contrast sensitivity as a function of spatial frequency of a grating or size of a letter.

13.4.2 ACUITY

Previously, we discussed threshold *contrast*. Threshold *size* is called "acuity," and it has been routinely measured in eye clinics since Herman Snellen (1862) introduced the acuity chart. Modern acuity charts (e.g., Figure 13.4a) are very similar to Snellen's original, differing mainly in using a fixed number of letters per line and using the same size reduction for each new line.

Like an acuity chart, the Pelli–Robson contrast sensitivity chart (Figure 13.4b) begins with large black letters, but, going down the chart, instead of shrinking, the letters remain the same large size and fade to lighter and lighter shades of gray, and eventually disappear.

Reading is an extraordinarily important visual task and easy to test, for example, by the MNREAD chart (Mansfield et al.,

1996). Reading speed is an important part of assessment of low vision, which may include optical and neural impairments.

13.5 CLASSIFYING TILT

13.5.1 SIGNAL-DETECTION THEORY

Psychophysical techniques extend far beyond the measurement of contrast thresholds. Indeed, contrast thresholds turn out to be poor predictors of human performance in many visual tasks. One such task is orientation discrimination, which is thought to be limited by poor memory for tilt (Pasternak and Greenlee, 2005). This limit can be quantified within the framework of signal-detection theory (SDT) (Green and Swets, 1966), the cornerstone of all contemporary psychophysical analysis.

One reason for SDT's dominance is its successful prediction of results with one task from results with another. We have already introduced the popular task of 2AFC detection (see Section 13.3.1). Tomassini et al. (2010) used 2AFC to measure memory for tilt. Two small luminance gratings were shown consecutively, and observers were asked to decide which of the two gratings was tilted clockwise with respect to the other. The points in Figure 13.5 illustrate one observer's accuracies in this task.

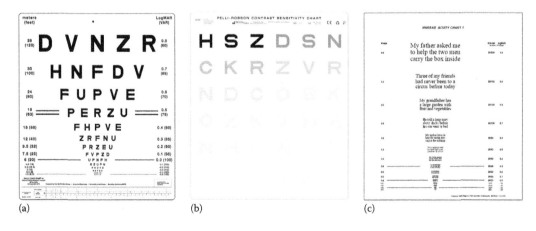

Figure 13.4 Eye charts for clinical use. (a) Acuity for letter identification can be quickly measured using the Revised ETDRS LogMAR chart. (b) Contrast sensitivity for letter identification can be measured quickly using the Pelli–Robson chart. (c) Acuity for reading can be quickly measured using the MNREAD chart. ([a]: Reprinted from Bailey, I.L. and Lovie, J.E., *Am. J. Optom. Physiol. Opt.*, 53, 740, 1976. With permission of Precision Vision; [b]: Reprinted from Pelli, D.G. et al., *Clin. Vis. Sci.*, 2, 187, 1988. With permission of Precision Vision; [c]: Reprinted from Mansfield, J.S. et al., A new reading acuity chart for normal and low vision, in: *Ophthalmic and Visual Optics/Noninvasive Assessment of the Visual System Technical Digest*, Vol. 3, Optical Society of America, Washington, DC, 1993, pp. 232–235; Mansfield, J.S. et al., *Invest. Ophthalmol. Vis. Sci.*, 37, 1492, 1996. With permission of Gordon Legge.)

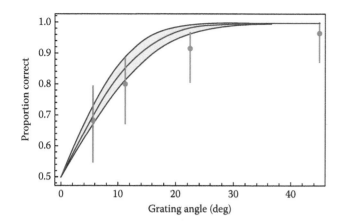

Figure 13.5 Two-alternative forced-choice tilt-classification accuracy (points) and a prediction (red curve) derived from a separate, method-of-adjustment experiment. Error bars and blue region contain 95% confidence intervals.

13.5.2 THE METHOD OF ADJUSTMENT

In a separate, but similar, experiment Tomassini et al. (2010) asked the same observer to rotate the second grating until its orientation best matched the observer's memory of the first. This task is called the method of adjustment. Often the matches were not veridical. Sometimes, the second grating was rotated too far clockwise; sometimes, it was rotated too far counterclockwise.

In SDT, all sensory quantities are corrupted by normally distributed noise. This includes memory of the first grating's orientation. Consequently, the 2AFC psychometric function should adhere to this normal distribution, and matching errors from the method of adjustment should be a sample from it.*

We used the standard deviation of matching errors as an estimate of memory noise. This standard deviation can be considered the threshold angle (corresponding to 84% correct in 2AFC) for discriminating clockwise from counterclockwise tilts. The red curve in Figure 13.5 is a normal distribution with this standard deviation. Note that the 2AFC data conform fairly well to the red curve. Thus, we can be reasonably confident that visual performances in both tasks were subject to the same limitation.

13.6 THE RELATIONSHIP BETWEEN VISUAL OPTICS, CONTRAST SENSITIVITY, AND ACUITY

There are several good ways to physically characterize the quality of the retinal image, but they all measure light reflected from the retina, whereas vision is mediated by neural response to photoisomerizations. For relevance to vision, the optical characterization with physical methods should be compared with optical characterization with psychophysics.

The linearity of optical imaging ensures that a sinusoidal luminance stimulus will produce a sinusoidal luminance image on the

retina. The amplitude (i.e., contrast) of that image is determined by the eye's optical MTF (or point spread):

$$MTF = \frac{c_{retina}}{c_\theta}, \tag{13.2}$$

where is the threshold contrast at the retina. Any change in the eye's optics therefore will produce a change in grating contrast at the retina, and this consequently will result in changed sensitivity to that spatial frequency. Thus,

$$c_{retina} = \frac{MTF}{CSF}, \tag{13.3}$$

at the spatial frequency of the grating.

The tight link between thresholds for grating detection and thresholds for letter identification (see Section 13.4) means that changes in letter acuity often reflect changes in the eye's optical MTF. Indeed, in central vision, low-to-normal acuity can be predicted from knowledge of the OTF (Watson and Ahumada, 2008). On the other hand, there is no correlation between acuity and optics among observers with excellent acuity (Villegas et al., 2008).

Visual optics affect visual sensitivity at all eccentricities, but optical factors account for very little of the normal worsening of vision with eccentricity. People often describe their peripheral vision as "blurry," but the poorness of normal human peripheral vision is largely a result of neural, not optical, factors. The drop of visual acuity with eccentricity is more closely related to retinal ganglion cell density that broadening of the optical point spread function (Lennie, 1977). Indeed, Lundström et al. (2007) found that resolution acuity in the periphery cannot be improved with optical correction.

Acuity for individual letters drops off with retinal eccentricity, but acuity for objects in cluttered scenes drops off much faster. This is called "crowding": the minimum spacing between target and clutter, for target recognition, increases linearly with eccentricity (for review see Pelli and Tillman, 2008). Our visual world is typically cluttered, and the resulting crowding typically prevents recognition of peripherally seen objects.

Crowding severely limits the speed of search and reading (Motter and Simoni, 2007; Pelli et al., 2007). Consequently, the identification of individual letters thus has very little relevance to patients unable to use the central portion of their visual fields. For this reason, Mansfield and colleagues (1993, 1996) developed the MNREAD chart (Figure 13.4c), which can more accurately characterize the everyday vision of these patients.

Declaration of commercial interest: Denis Pelli receives a royalty from sales of the Pelli–Robson contrast-sensitivity chart.

REFERENCES

Bailey, I. L. and Lovie, J. E. (1976). New design principles for visual acuity letter charts. *American Journal of Optometry and Physiological Optics*, 53, 740–745.

Brown, L. D., Cai, T. T., and DasGupta, A. (2001). Interval estimation for a binomial proportion. *Statistical Science*, 16, 101–133.

* In both paradigms, the second grating remained visible until the observer's response. Consequently, the memory noise for its orientation can be considered negligible.

Campbell, F. W. and Robson, J. G. (1964). Application of Fourier analysis to the modulation response of the eye. *Journal of the Optical Society of America* 54, 581A.

Campbell, F. W. and Robson, J. G. (1968). Application of Fourier analysis to the visibility of gratings. *The Journal of Physiology*, 197(3), 551–566.

Efron, B. (2012). Bayesian inference and the parametric bootstrap. *The Annals of Applied Statistics*, 6, 1971–1997.

Fechner, G. T. (1860/1966). *Elements of Psychophysics*, Vol. 1. Translated by H.E. Adler. New York: Holt, Rinehart & Winston.

Foley, J. M. and Legge, G. E. (1981). Contrast detection and near-threshold discrimination in human vision. *Vision Research*, 21, 1041–1053.

Green, D. M. and Swets, J. A. (1966). *Signal Detection Theory and Psychophysics*. New York: Wiley.

Kingdom, F. A. A. and Prins, N. (2010). *Psychophysics: A Practical Introduction*. London, U.K.: Academic Press.

Lennie, P. (1977). Neuroanatomy of visual acuity. *Nature*, 266, 496.

Lesmes, L. A., Lu, Z.-L., Baek, J., and Albright, T. D. (2010). Bayesian adaptive estimation of the contrast sensitivity function: The quick CSF method. *Journal of Vision* 10(3), 17, 1–21.

Lu, Z.-L. and Dosher, B. A. (2013). *Visual Psychophysics*. Cambridge, MA: MIT Press.

Linares, D. and López-Moliner, J. (2015). Quickpsy R package to quickly fit and plot psychometric functions for multiple conditions. http://dlinares.org/introquickpsy.html (accessed September 30, 2016).

Lundström, L., Manzanera, S., Prieto, P. M., Ayala, D. B., Gorceix, N., Gustafsson, J., Unsbo, P., and Artal, P. (2007). Effect of optical correction and remaining aberrations on peripheral resolution acuity in the human eye. *Optics Express*, 15(20), 12654–12661.

Majaj, N. J., Pelli, D. G., Kurshan, P., and Palomares, M. (2002). The role of spatial frequency channels in letter identification. *Vision Research*, 42, 1165–1184.

Mansfield, J. S., Ahn, S. J., Legge, G. E., and Luebker, A. (1993). A new reading acuity chart for normal and low vision. In: *Ophthalmic and Visual Optics/Noninvasive Assessment of the Visual System Technical Digest*, Vol. 3, pp. 232–235. Washington, DC: Optical Society of America.

Mansfield, J. S., Legge, G. E., and Bane, M. C. (1996). Psychophysics of reading. XV. Font effects in normal and low vision. *Investigative Ophthalmology and Visual Science*, 37, 1492–1501.

May, K. A. and Solomon, J. A. (2013). Four theorems on the psychometric function. *PLoS ONE*, 8, e74815. doi: 10.1371/journal.pone.0074815.

Motter, B. C. and Simoni, D. A. (2007). The roles of cortical image separation and size in active visual search performance. *Journal of Vision*, 7(2), 6.

Ohzawa, I. (2005). Make your own Campbell-Robson contrast sensitivity chart. http://www7.bpe.es.osaka-u.ac.jp/ohzawa-lab/izumi/CSF/A_JG_RobsonCSFchart.html (accessed September 30, 2016).

Pasternak, T. and Greenlee, M. W. (2005). Working memory in primate sensory systems. *Nature Reviews Neuroscience*, 6, 97–107.

Pelli, D. G. and Bex, P. (2013). Measuring contrast sensitivity. *Vision Research*, 90, 10–14. doi:10.1016/j.visres.2013.04.015.

Pelli, D. G. and Farell, B. (1999). Why use noise? *Journal of the Optical Society of America A*, 16, 647–653. http://psych.nyu.edu/pelli/pubs/pelli1999noise.pdf (accessed September 30, 2016).

Pelli, D. G. and Farell, B. (2010). Psychophysical methods. In M. Bass, C. DeCusatis, J. Enoch, V. Lakshminarayanan, G. Li, C. MacDonald, V. Mahajan, and E. V. Stryland (Eds.), *Handbook of Optics*, 3rd edn., Volume III: Vision and Vision Optics, pp. 3.1–3.12. New York: McGraw-Hill.

Pelli, D. G. and Robson, J. G. (1991). Are letters better than gratings? *Clinical Vision Sciences*, 6, 409–411. http://psych.nyu.edu/pelli/pubs/pelli1991letters.pdf (accessed September 30, 2016).

Pelli, D. G., Robson, J. G., and Wilkins, A. J. (1988). The design of a new letter chart for measuring contrast sensitivity. *Clinical Vision Sciences*, 2, 187–199. http://psych.nyu.edu/pelli/pubs/pelli1988chart.pdf (accessed September 30, 2016).

Pelli, D. G. and Tillman, K. A. (2008). The uncrowded window of object recognition. *Nature Neuroscience*, 11(10):1129–1135. http://www.nature.com/neuro/journal/v11/n10/index.html#pe (accessed September 30, 2016).

Pelli, D. G., Tillman, K. A., Freeman, J., Su, M., Berger, T. D., and Majaj, N. J. (2007). Crowding and eccentricity determine reading rate. *Journal of Vision*, 7(2), 20.

Snellen, H. (1862). *Scala Tipografica per Mesurare il Visus*. Utrecht, the Netherlands: P. W. Van de Weijer. http://books.google.com/books/about/Scala_tipografica_per_mesurare_il_visus.html?id=hq0GnQEACAAJ (accessed September 30, 2016).

Solomon, J. A. (2007). Contrast discrimination: Second responses reveal the relationship between the mean and variance of visual signals. *Vision Research*, 47, 3247–3258.

Solomon, J. A. (2013). Visual psychophysics. In D. S. Dunn (Ed.) *Oxford Bibliographies in Psychology*. New York: Oxford University Press. doi: 10.1093/OBO/9780199828340-0128.

Strasburger, H. (2015). Software for visual psychophysics: An overview. http://www.hans.strasburger.de/psy_soft.html (accessed September 30, 2016).

Tomassini, A., Morgan, M. J., and Solomon, J. A. (2010). Orientation uncertainty reduces perceived obliquity. *Vision Research*, 50, 541–547.

Villegas, E. A., Alcón, E., and Artal, P. (2008). Optical quality of the eye in subjects with normal and excellent visual acuity. *Investigative Ophthalmology & Visual Science*, 49(10), 4688–4696.

Watson, A. B. and Ahumada, A. J., Jr. (2008). Predicting visual acuity from wavefront aberrations. *Journal of Vision*, 8(4), 1–19, http://journalofvision.org/8/4/17/ (accessed September 30, 2016).

Watson, A. B. and Pelli, D. G. (1983). QUEST: A Bayesian adaptive psychometric method. *Perception & Psychophysics*, 33, 113–120. MATLAB code: https://github.com/kleinerm/Psychtoolbox-3/blob/master/Psychtoolbox/Quest/QuestDemo.m (accessed September 30, 2016).

Watson, A. B. and Solomon, J. A. (1997). Psychophysica: Mathematica notebooks for psychophysical experiments. *Spatial Vision*, 10, 447–466.

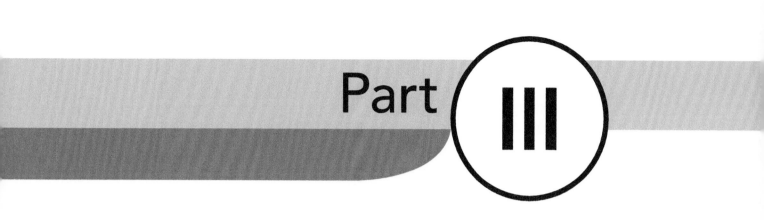

Part III

Optical properties of the eye

The cornea

Michael Collins, Stephen Vincent, and Scott Read

Contents

14.1 INTRODUCTION

The cornea is a living lens. Since its front surface is highly curved and has an interface to air (via the tears), the refractive power of the anterior cornea is high (~48 D), and it contributes most of the total optical power of the eye (~70%). While the posterior cornea is also highly curved, the similarity in refractive indices between the posterior cornea and the aqueous leads to the optical contribution from this surface being diminished and of opposite sign to the anterior cornea (~–7 D).

In some animal species (such as birds and reptiles), the cornea can change its focal power (accommodate) by steepening (Land and Nilsson 2012). However, in the human eye, it was Thomas Young's classic experiment involving submersing the eye in a solution (neutralizing most of the cornea's refractive power), which first showed that the crystalline lens and not the cornea was the source of accommodation (Young 1801).

14.1.1 CORNEAL ANATOMY

The cornea is now considered to have six anatomical layers: the epithelium, Bowman's layer, the stroma, Dua's layer (also referred to as pre-Descemet's stroma), Descemet's layer, and the endothelium. The total thickness of the central cornea is about 535 μm, becoming thicker in the periphery (Figure 14.1) (Doughty and Zaman 2000). The outermost epithelial layer is about 52 μm thick in the center of the cornea and like the overall cornea shows thickening toward the periphery (Li et al. 2012). The majority of the corneal thickness comprises the stroma, composed mostly of collagen fibrils suspended in a highly hydrated extracellular matrix. Bowman's layer separates the epithelium and stroma, while Dua's and Descemet's layers lie between the stroma and the endothelium. The endothelium is a single layer of cells interfacing to the anterior chamber of the eye.

14.1.2 CORNEAL SHAPE AND DIMENSIONS

The most common description of the shape of the normal anterior cornea is that of a prolate ellipse, where the radius of curvature of the cornea becomes progressively flatter away from center. The central radius of curvature is about 7.7–7.9 mm, and the Q value that effectively defines the rate of peripheral flattening is around –0.18 to –0.26 (with more negative values indicating greater peripheral flattening) for the central 6 mm diameter (Douthwaite et al. 1999, Eghbali et al. 1995, Guillon et al. 1986, Kiely et al. 1982, Read et al. 2006). The posterior corneal surface

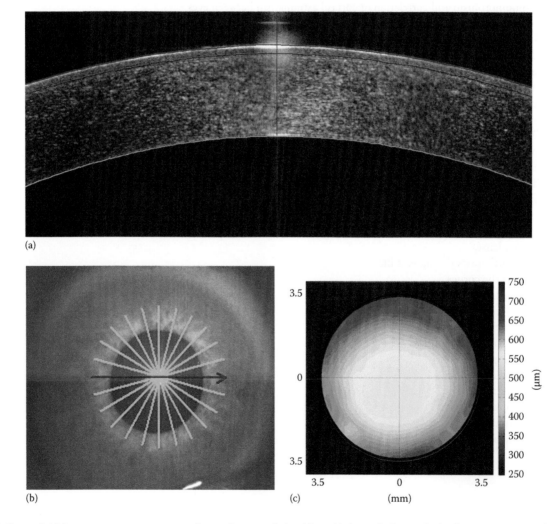

(a)

(b)

(c)

(mm)

Figure 14.1 Corneal thickness measurements over a 7 mm diameter derived from high-resolution optical coherence tomography: (a) segmented corneal cross section (red; anterior epithelium, blue; Bowman's layer, green; endothelium), (b) 12 radial scan protocol centered on the pupil, and (c) en face total corneal thickness.

is less well defined; however, Dubbelman et al. (2006) report a posterior central radius of curvature of about 6.5 mm and a prolate ellipse with Q of around –0.18, with some variability associated with age and axis. The cornea, when viewed from the front, is generally slightly wider in the horizontal meridian than the vertical, with the average horizontal and vertical visible iris diameters reported as 11.46 and 10.63 mm, respectively (Khng and Osher 2008).

14.1.3 REFRACTIVE INDICES OF THE CORNEA

Most studies agree that the refractive index of the corneal epithelium is slightly higher than that of the underlying stroma, with typical values in the range from 1.401 to 1.433 for the epithelium and 1.357 to 1.380 for the stroma (Lai and Tang 2014, Patel et al. 1995). The stromal refractive index has been shown to slightly increase with age (Patel et al. 2011), and the epithelial refractive index may be slightly lower in the peripheral cornea compared with the central cornea (Vasudevan et al. 2008). There may also be a slight gradient refractive index in the stroma, being higher in the anterior than posterior stroma (Patel and Alio 2012, Patel et al. 1995). Barbero (2006) showed that when the corneal stroma is assumed to have an axial gradient refractive index, this has the effect of slightly increasing the positive spherical aberration of the cornea for far object vergences.

14.1.4 TRANSMISSION PROPERTIES OF THE CORNEA

The relatively high transparency of the cornea to visible light (>95%) arises from its lack of blood vessels and myelinated nerve fibers and its underlying anatomy and composition. The stroma comprises stacked lamellae arranged approximately parallel to the corneal surface, and within the lamellae are collagen fibrils surrounded by extrafibrillar material, with a similar refractive index (Leonard and Meek 1997). The uniform size, spacing, and arrangement of the collagen fibrils produce transparency through the destructive interference of scattered light, and this transparency reduces when the cornea swells (Maurice 1957, Meek et al. 2003) or following injury (Torricelli and Wilson 2014). The anatomical spatial organization of the lamellae and fibrils is also thought to be primarily responsible for the birefringence properties of the cornea (Meek and Boote 2009). The spectral transmittance function of the cornea shows some absorption at the blue end of the visible spectrum, but little evidence of changes with age (Boettner et al. 1962, van den Berg and Tan 1994).

14.1.5 REFERENCE AXES FOR THE CORNEA

There are a wide variety of optical and clinical axes used to define the visual optics of the cornea. An understanding of these axes is important for the measurement of the optical properties of the cornea and for the spatial alignment of the various forms of optical correction of the eye. The classical landmarks of the surfaces of the eye are the Purkinje images, where the image of an external light source is seen reflected from the anterior and posterior corneal surfaces (1st and 2nd Purkinje images) and anterior and posterior crystalline lens surfaces (3rd and 4th Purkinje images).

There are a number of factors that complicate the measurement of the reference axes of the cornea. The center of the cornea is difficult to define because the limbus does not have a distinct boundary and the cornea's en face dimensions are elliptical (shorter in the vertical meridian) and not circular. The entrance pupil (i.e., the image of the pupil seen through the cornea) does represent the aperture of the eye; however, the relative location of the center of the pupil changes slightly as it varies in size through the influence of factors such as light level, accommodation, and the effect of cycloplegia (Mathur et al. 2014, Walsh 1988, Wilson et al. 1992, Yang et al. 2002). The center of the cornea and pupil are more accurately defined by centroids rather than a simple geometric center. As a further complication, when the eye looks up or down, there is also a small cyclotorsion of the eye by up to a few degrees (typically excyclotorsion in downward gaze and incyclotorsion in upward gaze), which changes the axis of the cornea with respect to the primary gaze eye position (Enright 1980).

The axes of the cornea that can be easily anatomically defined include the line of sight, pupil axis, and the "corneal vertex normal" (Mandell et al. 1995), also called the "subject fixated coaxially sighted corneal light reflex" (Chang and Waring 2014). The line of sight is typically used in calculating the optical aberrations of the cornea and can be defined by a line joining the fixation point to the center of the entrance pupil, with the assumption that the chief ray travels then to intersect the fovea (Figure 14.2). The point of intersection of this ray with the anterior corneal surface is called the "corneal sighting center" (Mandell 1995). However, both objective (Burns et al. 1995) and subjective (Applegate and Lakshminarayanan 1993) measurements of photoreceptor alignment at the fovea suggest some slight mismatch between the foveal photoreceptor's axis and the center of the exit pupil.

The "corneal vertex normal" is defined by the location of the first Purkinje image, when viewed coaxially with the fixation light source (or by the center of the ring pattern in a Placido disk videokeratoscope pattern). This vertex normal location is slightly down and in, with respect to the entrance pupil centroid or

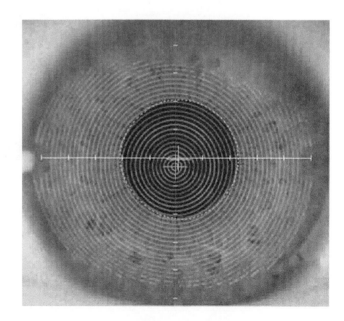

Figure 14.2 Placido ring pattern from a videokeratoscope showing relative locations of the geometric center of the cornea (white cross), entrance pupil center (yellow cross), and vertex normal at the center of the Placido ring pattern (blue cross).

corneal sighting center (0.38 mm) (Mandell et al. 1995). This axis provides some information about the shape of the corneal surface, but again does not coincide with the axis of the chief ray to the fovea (i.e., the line of sight). The geometric center of the cornea on average is located about 0.21 mm temporally to the line of sight (Pande and Hillman 1993). The pupil axis is defined by the line perpendicular to the cornea that passes through the center of the entrance pupil, but this has little value in terms of visual optics since the chief ray along this axis does not necessarily coincide with the fovea.

14.2 ASTIGMATISM OF THE CORNEA

14.2.1 CORNEAL ASTIGMATISM

Although the cornea exhibits a variety of optical imperfections, astigmatism is typically the most prominent corneal optical aberration in normal eyes. Since the cornea is the eye's most powerful optical component, it follows that corneal astigmatism is also often the largest contributor to the eye's total optical aberrations, with the potential to significantly impact visual performance as well as influencing ocular and visual development. Corneal astigmatism occurs due to differences in the cornea's curvature along its two perpendicular principal meridians, which results in light refracted by the cornea coming to a focus at two separate, orthogonal focal planes within the eye.

Although the reported prevalence of corneal astigmatism varies substantially across different studies, depending upon the demographics of the population examined and the criteria used to define astigmatism, small degrees of corneal astigmatism occur very commonly (Fledelius and Stubgaard 1986, Hoffmann and Hutz 2010, Huynh et al. 2006, Leung et al. 2012, Liu et al. 2011). A large population study of more than 15,000 European adults reported that around three quarters (74%) of this population exhibited 0.50 D or more corneal astigmatism (Hoffmann and Hutz 2010). Higher amounts of corneal astigmatism however are less common, with less than 3% of this population reported to have more than 3 D of corneal astigmatism.

Corneal astigmatism is most commonly classified according to the orientation of the cornea's principal meridians (Figure 14.3). When the steepest meridian is oriented close to the vertical (i.e., negative cylinder axis between 30° and 160°), it is referred to as "with-the-rule" (WTR) corneal astigmatism, and when the cornea's steep meridian is oriented close to the horizontal, it is referred to as "against-the-rule" (ATR) corneal astigmatism

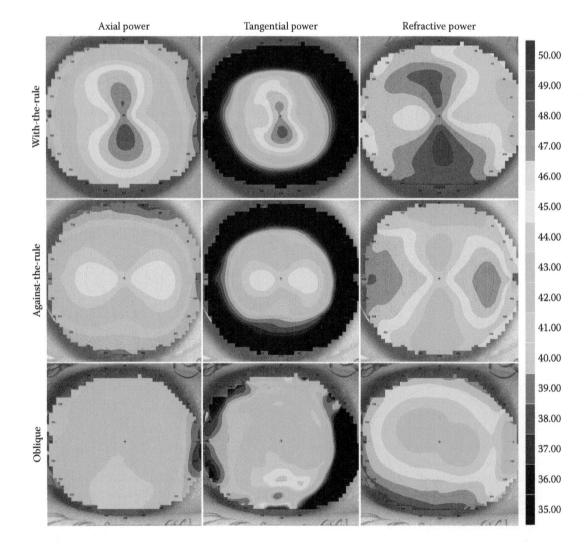

Figure 14.3 Axial, tangential, and refractive power topography maps (presented on the same dioptric scale) demonstrating "with-the-rule," "against-the-rule," and oblique corneal astigmatism.

(negative cylinder axis between 120 and 60). When the steep meridian is neither vertical nor horizontal (i.e., oriented between 159°and 129°, or 31° and 59°) it is referred to as oblique astigmatism. Most studies report a predominance of either WTR astigmatism (particularly in younger adult populations) or ATR corneal astigmatism (particularly in older adult populations), with oblique corneal astigmatism being relatively uncommon (Fledelius and Stubgaard 1986, Hoffmann and Hutz 2010, Huynh et al. 2006, Leung et al. 2012, Liu et al. 2011). In a study analyzing changes in corneal shape across both the central and peripheral cornea, Read et al. (2006) further classified corneal astigmatism according to the changes in astigmatism occurring in the peripheral cornea as being either stable (i.e., astigmatism magnitude similar in the central and peripheral cornea), increasing (i.e., astigmatism magnitude greater in the periphery compared to the center), or decreasing (i.e., astigmatism less in the periphery compared to the center). In 100 young adult subjects, astigmatism was found to most commonly reduce or remain stable from center to periphery (on average, corneal astigmatism reduced by ~0.2 D from the central 2 mm of the cornea to the peripheral cornea).

14.2.2 ANTERIOR AND POSTERIOR CORNEAL ASTIGMATISM

Both the anterior and posterior surfaces of the cornea exhibit astigmatism. Although the mean curvature and astigmatism of the anterior and posterior surfaces are reported to be highly correlated (Mas et al. 2009), when considered in terms of curvature, the posterior cornea is typically steeper and exhibits a higher degree of toricity compared to the anterior cornea (Dubbelman et al. 2006). However, the similarity in refractive index between the posterior cornea and aqueous means that the optical effect of posterior corneal astigmatism is typically small compared to the anterior cornea. Studies measuring the topography of the posterior corneal surface in normal adults report the posterior cornea on average contributes astigmatic refractive power of 0.30 D (ranging from a minimum of 0.01 D to a maximum of 1.10 D, with around 90% of eyes exhibiting 0.50 D or less of posterior corneal astigmatism) with the astigmatism from the posterior surface being predominantly ATR across a wide range of ages (Ho et al. 2009, Koch et al. 2012).

14.2.3 ENANTIOMORPHISM AND CORRELATION OF CORNEAL ASTIGMATISM BETWEEN EYES

It is generally accepted that the magnitude and axis of corneal astigmatism in the two eyes of the same individual are similar, with strong correlations reported in the corneal astigmatism of subjects' right and left eyes (Li and Bao 2014), even in subjects with relatively large differences in spherical refractive error between the two eyes (Vincent et al. 2011). These strong correlations suggest a high degree of symmetry exists in the corneal astigmatism of the two eyes. Although it is reported that most eyes show either mirror symmetry (e.g., right eye axis 10°, left eye axis 170°) or direct symmetry (e.g., right eye axis 10°, left eye axis 10°) in the axis of their corneal astigmatism, there are some conflicting reports upon which type of symmetry occurs more commonly (Dunne et al. 1994, Guggenheim et al. 2008, McKendrick and Brennan 1997). However, the largest

study examining this issue to date (including more than 50,000 subjects) reported that ocular refractive astigmatism (presumably of corneal origin in the majority of cases) showed mirror symmetry significantly more commonly than direct symmetry, suggesting that in the majority of the population, mirror symmetry (so-called enantiomorphism) of corneal astigmatic axis between fellow eyes is likely to be the most common pattern.

14.2.4 IMPACT OF CORNEAL ASTIGMATISM ON VISION

Although there is evidence that corneal astigmatism is compensated to some extent by the eye's internal optics (Artal et al. 2001, Kelly et al. 2004), ocular and corneal astigmatism are often highly correlated, and total ocular refractive astigmatism can be predicted with reasonable accuracy based upon corneal astigmatism (Grosvenor et al. 1988, Keller et al. 1996). The typically linear relationship between corneal and ocular astigmatism is utilized by a clinical "rule of thumb" (which relies on the fact that internal astigmatism is relatively consistent across the population and approximately 0.50 D and ATR in axis), first put forward in the 1800s by Javal and more recently simplified based upon empirical data by Grosvenor and colleagues (1988) that states

$$\text{Total ocular astigmatism} = \text{Corneal astigmatism} - 0.50 \times 90$$

Given the strong contribution of corneal astigmatism to the eye's total refractive astigmatism, it follows that corneal astigmatism has the potential to influence vision. The presence of astigmatic blur can lead to reductions in a range of clinical and functional vision measures including distance and near visual acuity, contrast sensitivity, stereo-acuity, and reading performance (Chen et al. 2005, Guo and Atchison 2010, Wolffsohn et al. 2011). Both distance and near visual acuity exhibit roughly linear reductions with increasing levels of astigmatic blur, with approximately a 1–2 line decrease in visual acuity reported with each 1 D increase in astigmatic blur (Wolffsohn et al. 2011). Under controlled laboratory conditions, Guo and Atchison (2010) demonstrated that on average 0.28 D of astigmatic blur was required for subjects to first notice a decrease in the clarity of their vision. However, in a group of near-emmetropic adults with small levels of astigmatism, Villegas et al. (2014) demonstrated that the correction of less than 0.50 D of astigmatism did not consistently result in measurable improvements in visual acuity. These findings suggest that for most clinical applications, correction of astigmatic refractive errors of greater than 0.50 D will be visually beneficial.

14.2.5 METHODS OF ASTIGMATISM CORRECTION

Since corneal astigmatism can cause decrements in clinical and functional visual performance, there is a need to correct corneal astigmatism (typically levels leading to 0.50 D or greater ocular astigmatism) in order to optimize vision. A range of clinical correction approaches are currently available including spectacles, contact lenses, and surgical options. The successful clinical correction of astigmatism relies upon neutralizing the magnitude of the eye's astigmatism along the astigmatic axes through optical and/or surgical means and then maintaining alignment of the correction over time.

Optical properties of the eye

The most commonly used clinical approach to correct astigmatism is spectacle lenses, which are capable of providing precise (typically prescribed to the nearest 0.25 D) and rotationally stable correction of large amounts of astigmatism. However, the use of spectacles can also induce spatial visual distortions due to meridional differences in image magnification (particularly for large magnitude astigmatic corrections). An alternative optical correction for astigmatism is the use of contact lenses (both soft and rigid lens designs), which typically do not result in the same spatial optical distortions associated with spectacle lenses. Soft contact lenses wrap to the corneal surface, and astigmatic lens designs typically employ mechanical stabilization zones (regions of increased lens thickness outside the optical zone of the lens) in order to align the astigmatic axis of the contact lens with that of the eye. The rotational stability of these lenses relies upon the interaction between these stabilization zones and the eyelids during blinking. Since spherical rigid contact lenses do not wrap to the eye, a tear fluid lens forms between the back surface of the lens and the anterior corneal surface that typically neutralizes/masks most of the anterior corneal astigmatism, without needing to rotationally align the lens. However, for high levels of corneal astigmatism and for patients with large amounts of internal ocular astigmatism, toric rigid lens designs are often required to optimize the lens fit and/or vision. The correction of relatively low magnitudes of astigmatism (up to ~1.5 D) with orthokeratology contact lenses (where a specially designed rigid contact lens is worn to reduce anterior corneal astigmatism) has also recently been reported (Chen et al. 2012).

Commonly used surgical approaches for astigmatism correction include limbal relaxing incisions (that are often used at the time of cataract surgery and involve incisions placed along the steep corneal meridian in order to flatten this meridian and reduce the corneal astigmatism), laser refractive surgery (where the corneal surface is ablated in a toric pattern in order to reshape the surface and correct astigmatism), and artificial lens implantation (where an artificial toric lens is implanted in the eye to correct astigmatism either in place of, or in addition to, the eye's crystalline lens).

14.3 HIGHER-ORDER ABERRATIONS OF THE CORNEA

14.3.1 ZERNIKE POLYNOMIALS

Aspheric models of the corneal surface have been used to characterize the spherical aberration of the cornea (Lotmar 1971). However, the most commonly used mathematical method today for describing the full optical properties of the cornea is to use the set of Zernike polynomials (Thibos et al. 2002). This approach begins with the height data describing the corneal surface and then uses the Zernike polynomial fit to characterize the surface shape (Schwiegerling and Greivenkamp 1997, Schwiegerling et al. 1995) or to derive the wavefront generated by the surface (Guirao and Artal 2000). While other mathematical models have been investigated (Howland et al. 1992, Iskander et al. 2002, Trevino et al. 2013), the Zernike polynomial approach remains popular because the basic components of the polynomial (e.g., lower-order terms of defocus and astigmatism and higher-order

terms such as coma, spherical aberration, and trefoil) are closely related to the traditional clinical measures of refractive errors of the eye (sphere and cylinder) and provide a close representation of the other optical errors (spherical aberration, coma, and trefoil) that are typically seen in the cornea. For normal corneas, up to the 4th order of the Zernike polynomial will generally provide an adequate fit with minimal residual error (Iskander et al. 2001).

14.3.2 SPHERICAL ABERRATION

The higher-order optical aberrations contributed by the anterior corneal surface make a major contribution to the total aberrations of the eye (Artal et al. 2001, He et al. 2003). The spherical aberration of the anterior corneal surface is primarily defined by the asphericity of the surface and for normal corneas with a typical prolate shape, the anterior corneal surface contributes positive spherical aberration to the eye (Kiely et al. 1982, Westheimer 1963). This positive spherical aberration from the anterior cornea is partially offset by the posterior corneal surface (Sicam et al. 2006) and the crystalline lens (el-Hage and Berny 1973). As the eye accommodates, the crystalline lens contributes increasing negative spherical aberration to the total optics (Atchison et al. 1995), and the spherical aberration of the total eye approaches zero at approximately intermediate focal distances (Atchison et al. 1995, Cheng et al. 2004). To eliminate the positive spherical aberration of the anterior cornea would require a significantly more prolate ($Q = -0.53$) shape, than the natural level of asphericity ($Q = -0.26$) (Kiely et al. 1982). On the other hand, if the corneal surface were spherical ($Q = 0$), it would contribute substantial amounts of positive spherical aberration to the eye.

14.3.3 COMA

The anterior corneal surface typically has small amounts of coma (Guirao et al. 2000, Llorente et al. 2004), while the posterior corneal surface contributes very little coma (Dubbelman et al. 2007). Since the anterior cornea is aspheric and the apex is slightly decentered from the corneal sighting center, some coma might be expected to result from this mismatch in axes (Navarro et al. 2006). If the corneal aberrations are calculated based on the vertex normal (the axis of the videokeratoscope) and not the line of sight (centered on the entrance pupil), this will produce errors in the calculation of the comatic and other asymmetric aberration terms of the cornea (Lu et al. 2008, Salmon and Thibos 2002).

The interaction between the coma produced by the anterior corneal surface and the coma of the total optics has been an issue of some interest. While the relationship between spherical aberration of the anterior cornea and internal optics follows consistent trends, the coma of the cornea and internal eye seems to be less well defined. Lateral coma is typically lower in magnitude in the total wavefront than the anterior corneal wavefront, while vertical coma shows no obvious trends (Figure 14.4). One of the first studies to conduct a detailed investigation of this relationship between corneal and internal coma (Kelly et al. 2004) suggested the potential for either an active mechanism (similar to emmetropization) or a passive mechanism associated with the relative offset between the line of sight and the entrance pupil center, altering the relative alignment and tilt of the cornea to the crystalline lens, such that the lateral coma of the crystalline lens largely compensates the anterior

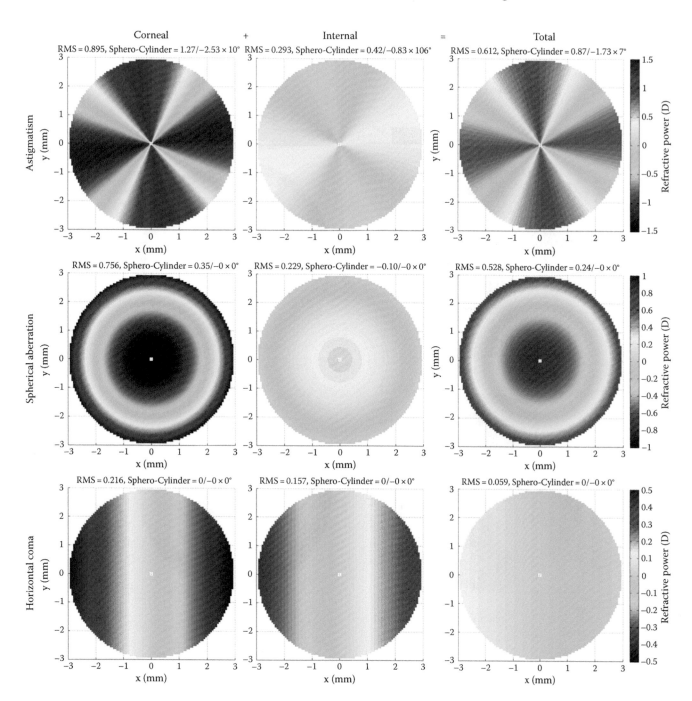

Figure 14.4 Example of partial compensation of corneal astigmatism, spherical aberration, and horizontal coma by the internal optics of the eye for a normal healthy subject with a moderate degree of WTR astigmatism. Refractive power maps generated from the anterior corneal wavefront (derived from videokeratoscope height data translated to the line of sight) and the total wavefront (obtained from a Hartmann–Shack wavefront sensor). The internal optics (contributions from the posterior corneal surface and the crystalline lens) are calculated by subtracting the corneal aberrations from the total ocular aberrations.

cornea's lateral coma. By measuring lateral coma in the cornea, internal optics, and angle kappa, Artal et al. (2006) were able to show that this compensation by the internal optics can be partly explained by the interaction between these factors. This was confirmed by Marcos and colleagues (2008) who also showed that in eyes with intraocular lenses designed to neutralize corneal spherical aberration, the lateral coma of the anterior cornea was largely compensated (passively) by the intraocular lens.

14.3.4 CORRELATION BETWEEN EYES

Analysis of the total wavefront of the eye has shown that many aberration terms are correlated between eyes (Porter et al. 2001). For the cornea, the right and left eyes of an individual also show correlation in some higher-order aberrations, as they do with corneal astigmatism (Figure 14.5). These correlations between right and left eyes are apparent for most of the higher-order terms including spherical aberration

Optical properties of the eye

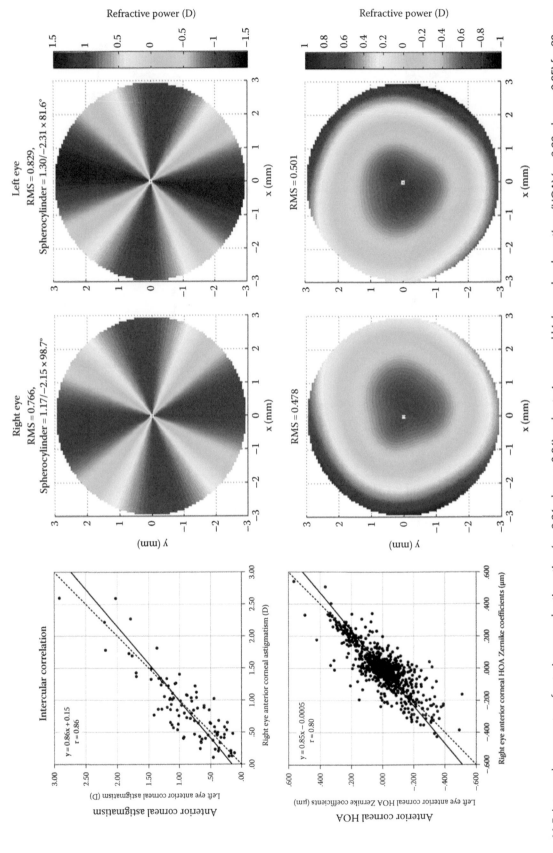

Figure 14.5 Interocular symmetry of anterior corneal astigmatism (r = 0.86, slope = 0.86) and anterior corneal higher-order aberrations (HOA) (r = 0.80, slope = 0.85) for 98 young isometropes. Refractive power maps generated from the corneal wavefront illustrate the right–left mirror symmetry (enantiomorphism).

(Wang et al. 2003), lateral coma (Lu et al. 2008, Wang et al. 2003), vertical coma (Wang et al. 2003), and the total higher-order root mean square (RMS) error (Lombardo et al. 2006, Wang et al. 2003).

14.3.5 HIGHER-ORDER ABERRATIONS IN ABNORMAL CORNEAS

The natural shape, and therefore the optical properties of the cornea, can be altered by conditions such as diseases affecting the cornea (e.g., keratoconus and pellucid marginal degeneration), through trauma or through surgical and refractive procedures (Figure 14.6). Because of the regional nature of many of these changes in corneal shape, the type and magnitude of changes in the corneal aberrations are highly dependent on the entrance pupil size used for analysis. Keratoconus causes a significant increase in the total higher-order aberrations of the cornea and eye, and the magnitude of higher-order aberrations increase if the cone is centered within the entrance pupil and also as the condition progresses and the cornea continues to thin and distort. The major higher-order aberrations produced by keratoconus are primary and secondary coma, mostly in the vertical direction (Buhren et al. 2007, Gobbe and Guillon 2005, Maeda et al. 2002), although increased trefoil (Gobbe and Guillon 2005, Kosaki et al. 2007) and spherical aberration are also present (Gobbe and Guillon 2005, Nakagawa et al. 2009). Pellucid marginal degeneration also affects the optical integrity of the cornea in a similar manner to keratoconus, increasing the total higher-order aberrations of the corneal surface (Gruenauer-Kloevekorn et al. 2006, Oie et al. 2008).

Surgical procedures to change the refractive properties of the cornea, such as laser-assisted in situ keratomileusis (LASIK), will also typically change both the lower-order and higher-order aberrations of the cornea. The total higher-order aberrations of the cornea may increase after myopic LASIK (Kohnen et al. 2005, Marcos et al. 2001, Serrao et al. 2011), with increased positive spherical aberration (Kohnen et al. 2005, Serrao et al. 2011). In hyperopic LASIK, the spherical aberration of the cornea may reduce, while the remaining total higher aberrations often increase (Kohnen et al. 2005, Wang and Koch 2003). Keratoplasty procedures such as penetrating keratoplasty (PK) and deep anterior lamellar keratoplasty (DALK) cause varying increases in the total higher-order aberrations of the cornea in comparison to control eyes (Koh et al. 2012), while procedures primarily affecting the posterior cornea, such as Descemet membrane endothelial keratoplasty (DMEK), appear to cause relatively minor changes in higher-order aberrations of the cornea (Rudolph et al. 2012, van Dijk et al. 2014).

The contact lens refractive procedure of orthokeratology changes the shape of the anterior corneal surface to alter the refractive power of the eye. Myopic orthokeratology flattens the central cornea (reducing myopia), but also affects the higher-order components of the corneal wavefront, producing substantial increases in the cornea's positive spherical aberration (Berntsen et al. 2005, Joslin et al. 2003) and in the total higher-order aberrations of the cornea (Berntsen et al. 2005, Hiraoka et al. 2007, Joslin et al. 2003).

14.3.6 IMPACT ON VISION AND METHODS OF CORRECTION

The effect of higher-order aberrations on visual performance has been shown to be dependent on factors such as the type of aberrations and their magnitude (Applegate et al. 2003). When the level of RMS error exceeds 0.05 μm (for a 3 mm pupil), all Zernike modes cause measureable losses of visual acuity, and modes closer to the center of the Zernike pyramid (e.g., spherical aberration) produce larger proportional losses of visual acuity than those modes near the edge of the Zernike pyramid (e.g., quadrafoil) (Applegate et al. 2003). Interestingly, the effects on visual acuity are also influenced by the interaction between Zernike modes, with some combinations of modes leading to improved vision (Applegate et al. 2003). In real eyes with a range of corneal conditions from normal to highly aberrated, there is a clear correlation between aspects of visual performance (high and low contrast visual acuity and contrast sensitivity) and the total RMS of the corneal surface (Applegate et al. 2000), and this relationship is dependent on pupil size, with larger pupils leading to higher corneal RMS.

The correction of corneal higher-order aberrations is often considered in conjunction with the correction of the total higher-order aberrations of the eye, since any change to the corneal aberrations will impact on the total aberrations. The total wavefront of the eye is typically measured with a wavefront sensor and may be used in combination with the subjective spherical or cylinder component of refraction to define the total optics of the eye (Thibos et al. 2004). The anterior corneal wavefront can be derived from topographical measurements of the surface height or curvature (Guirao and Artal 2000). The difference between the total wavefront and the anterior corneal wavefront is called the internal wavefront and comprises contributions from both the posterior corneal surface and the crystalline lens, both of which are difficult to measure.

Correcting the higher-order aberrations of the cornea or the anterior cornea or total eye can be undertaken with adaptive optics or wavefront-based customized corrections with contact lenses, refractive surgery, or spectacles. In an experimental setting, there are visual benefits in correcting the total wavefront of the eye with adaptive optics (Liang et al. 1997, Sawides et al. 2010, Yoon and Williams 2002), and the degree of advantage increases as the level of aberrations of the eye become greater, such as in keratoconus (Williams et al. 2000). However, the correction of higher-order aberrations is complicated by the apparent capacity of the visual system to partially adapt to the existing aberrations, so that full optical correction of the higher-order aberrations does not immediately produce optimal vision outcomes (Artal et al. 2004).

Achieving a clinical higher-order aberration correction of the cornea or total eye with contact lenses, spectacle, or refractive surgery has proven to be challenging. But as the level of higher-order aberrations of the eye increase, there are examples of visual benefit through the use of customized correction of higher-order aberrations by soft contact lenses (Marsack et al. 2007, Sabesan et al. 2007) and rigid contact lenses (Marsack et al. 2007, 2014, Sabesan et al. 2013). Wavefront-optimized and wavefront-guided ablation patterns used in refractive surgery have shown relatively

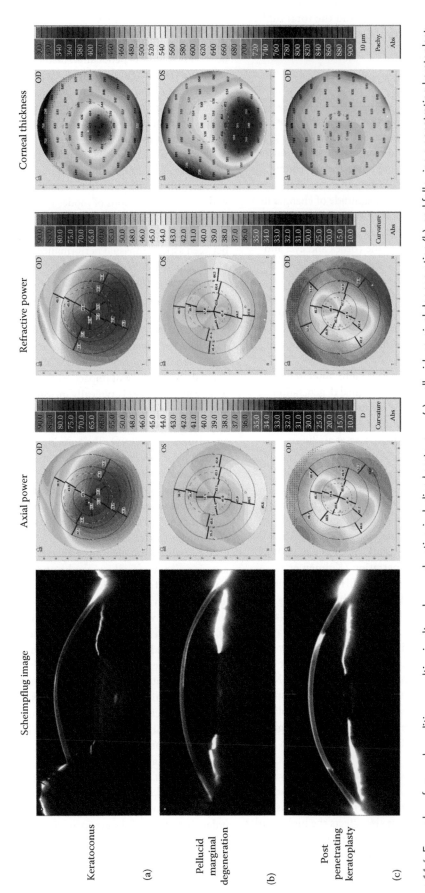

Figure 14.6 Examples of corneal conditions resulting in altered corneal optics, including keratoconus (a), pellucid marginal degeneration (b), and following a penetrating keratoplasty procedure (corneal graft) (c). For each condition, a cross-sectional Scheimpflug image, an anterior corneal axial power, refractive power map, and a corneal thickness map are shown to illustrate the structural and optical corneal changes occurring.

small (Mrochen et al. 2000) or little (Netto et al. 2006) visual advantage for eyes with normal levels of higher-order aberrations. Each of these correction modalities presents challenging and specific clinical demands, but they all require highly precise wavefront measurements, sophisticated calculation of the required optical correction, and accurate and consistent alignment of the higher-order correction with the center of the entrance pupil.

14.4 SHORT-TERM CHANGES IN THE OPTICS OF THE CORNEA

A range of intrinsic and external factors have been shown to induce short-term, transient changes in the optical properties of the cornea. While such changes or physiological fluctuations are typically small in magnitude and short in duration, they may potentially impact upon the accurate measurement of the cornea's true optical properties and also influence the quality of vision.

14.4.1 DIURNAL CHANGES

The cornea exhibits significant diurnal variation in both thickness and curvature, undergoing a small amount of thinning (~15 μm) and anterior steepening (~0.03 mm) in the first few waking hours (Read and Collins 2009) (Figure 14.7). The largest changes occur during the immediate recovery from physiological edema induced by overnight eyelid closure. Steepening of the anterior and posterior corneal surfaces (0.2 D and 0.03 D, respectively) are observed in the first 2 h after waking (Read and Collins 2009), with smaller fluctuations in anterior and posterior corneal astigmatism observed throughout the day (J_0 and J_{45} corneal power vectors <0.02 D) (Read and Collins 2009, Read et al. 2005). Two anterior surface higher-order aberrations, primary vertical coma and trefoil along 30°, also display diurnal fluctuations. The combination of these two aberrations represents a "wavelike" corneal distortion increasing throughout the day most likely associated with eyelid pressure (Read et al. 2005).

14.4.2 EYELID PRESSURE

Several studies have further examined near work–induced regional changes in corneal optics, which appear to be related to eyelid forces exerted on the cornea as a result of the narrowed palpebral aperture during downward gaze (Buehren et al. 2003, 2005, Collins et al. 2005, Shaw et al. 2008, 2009, Vincent et al. 2013). A horizontal band of distortion in the superior cornea is typically observed, along with an increase in ATR corneal astigmatism (Buehren et al. 2005, Shaw et al. 2008) that may take up to 2 h to completely regress (Collins et al. 2005) (Figure 14.8). These temporary changes in corneal optics associated with eyelid pressure are thought to be superficial (i.e., epithelial cell redistribution).

14.4.3 ACCOMMODATION AND CONVERGENCE

The forces exerted upon the cornea during ciliary muscle contraction were initially thought to potentially alter corneal astigmatism (Fairmaid 1959, Mandell and Helen 1968). However, recent studies accurately measuring the changes in corneal topography during accommodation confirmed Thomas Young's observation in the 1800s that the cornea remains stable with increasing levels of ciliary muscle contraction (Bayramlar et al. 2013, Buehren et al. 2003, Read et al. 2007). Small variations in corneal optics during accommodation appear to be a result of cyclotorsional eye movements rather than a true change in corneal curvature (i.e., when ocular rotation during accommodation is corrected for, the accommodative-related changes in corneal parameters are negligible).

It is also conceivable, given the proximity of the medial rectus muscle point of insertion relative to the limbus (~5–6 mm) and its location (aligned with the horizontal meridian of the cornea), that extraocular muscle forces transmitted during convergence could induce changes in corneal optics (Lopping and Weale 1965). Small but significant increases in ATR astigmatism and vertical coma have been observed following convergence

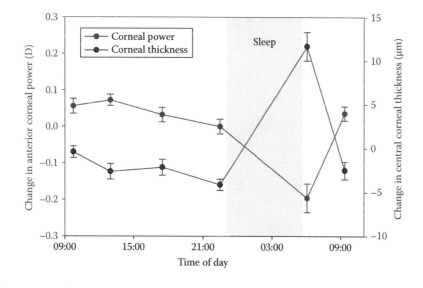

Figure 14.7 Illustration of the average diurnal changes observed in the central anterior corneal power (red) and central corneal thickness (blue) over a 24 h period in young subjects with healthy corneas (n = 15). Note the largest changes in corneal thickness (a swelling of the central cornea) and power (a flattening of the anterior corneal surface) occur immediately after waking (Read and Collins 2009). Error bars represent the standard error of the mean.

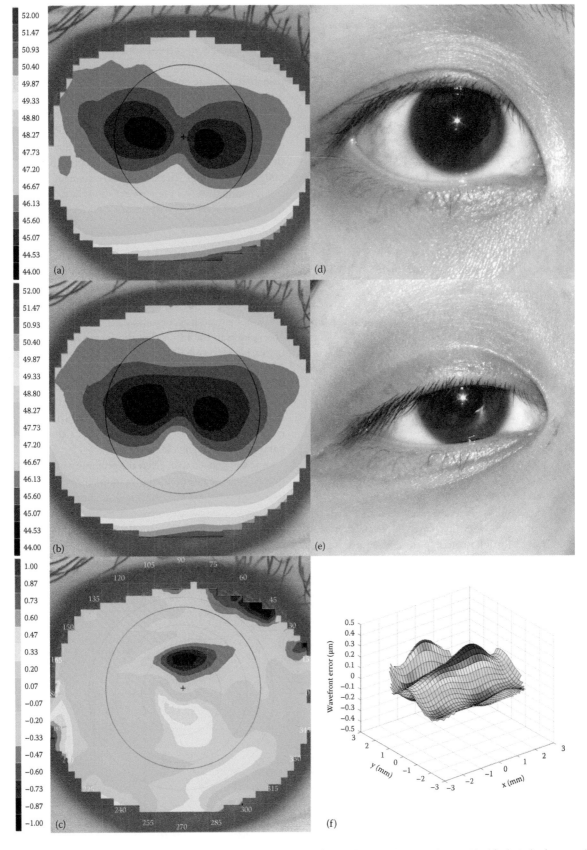

Figure 14.8 Corneal refractive power maps before (a) and after (b) a 10 min reading task at 25° downward gaze. The black circle denotes the pupil outline detected by the videokeratoscope. The refractive power difference map (post minus pre-reading) (c) displays a horizontal band of corneal flattening within the superior pupil and a region of corneal steepening within the inferior pupil (a vertical coma-like effect). The morphology of the palpebral aperture for the same participant during primary gaze (d) and 25° downward gaze (e) highlights the change in upper eyelid position during the reading task and corresponds with the regional topographical change in (c). The change in the anterior corneal surface wavefront (based on higher-order Zernike terms) over a 6 mm pupil is illustrated in (f) and displays a "wavelike" distortion in the corneal optics.

(without accommodation); however, these optical changes appeared to be a result of the change in eyelid position (relative to the cornea) during convergence (i.e., a nasal rotation of the globe resulted in a relative narrowing of the palpebral aperture) (Read et al. 2010).

14.4.4 EXTERNAL FORCES

External forces applied to the cornea through digital pressure (Mandell and Helen 1968), eye rubbing (Mansour and Haddad 2002), or a modified applanation tonometer (Carney and Clark 1972) have been shown to significantly alter its shape and optics in the short term (for up to several minutes). Transient changes in corneal optics following applied external forces are thought to be a result of corneal surface deformation and changes in the tear film. The repeated application of external forces to the cornea has been linked with the development of permanent changes in corneal structure and optics (e.g., the development of keratoconus) (Gunes et al. 2014, Kandarakis et al. 2011, Sugar and Macsai 2012).

14.4.5 CONTACT LENSES

Forces applied during contact lens wear may alter the shape and, subsequently, the optics of the cornea. In modern orthokeratology, reverse geometry rigid gas-permeable contact lenses are worn overnight to deliberately flatten the central cornea to temporarily reduce myopia and/or astigmatism. The observed changes in corneal shape are thought to occur as a result of epithelial tissue redistribution (from the central to midperipheral cornea) (Nichols et al. 2000, Wang et al. 2003) and a slight thickening (Alharbi and Swarbrick 2003) and bending (Owens et al. 2004) of the corneal stroma, with the refractive effects as a result of the superficial (epithelial) changes (Reinstein et al. 2009).

Other forms of contact lens correction may unintentionally alter corneal optics following lens wear. The short-term wear of scleral contact lenses that vault the cornea entirely and rest upon the sclera have been shown to induce a transient increase in corneal astigmatism (up to ~0.7 D) that dissipated within 3 h of lens removal (Vincent et al. 2014). This may be a result of fluid forces within the post-lens tear film, gradual compression of the tissues adjacent to the cornea during lens settling (conjunctiva and sclera), or superior eyelid pressure upon the lens. Similarly, the fitting approach adopted for keratoconic corneas (e.g., apical bearing or three-point touch) may also influence the magnitude of change in corneal optics following lens wear (Romero-Jimenez et al. 2014). Cosmetic tinted soft contact lenses have also been shown to substantially alter corneal topography and subsequently degrade vision following lens removal. As the eyelid contacts the pigmented annular region of such lenses (up to 30 μm in pigment thickness), a differential pressure may be applied to the cornea with each blink resulting in a ring-shaped impression in the epithelium (Voetz et al. 2004).

14.5 CHANGES IN CORNEAL OPTICS THROUGHOUT LIFE

14.5.1 SHAPE OF THE CORNEA

The shape of the cornea varies with age (Figure 14.9). The corneal curvature of newborns and infants is initially steep (Friling et al. 2004) and flattens rapidly during childhood, before stabilizing and reaching adult levels by 4 years (Asbell et al. 1990). Measures of corneal curvature appear to be relatively stable during adulthood (~20–60 years) (Dubbelman et al. 2006, Fledelius and Stubgaard 1986, Wei et al. 2014). However, in older age groups (up to 80 years), a steepening of the horizontal corneal meridian

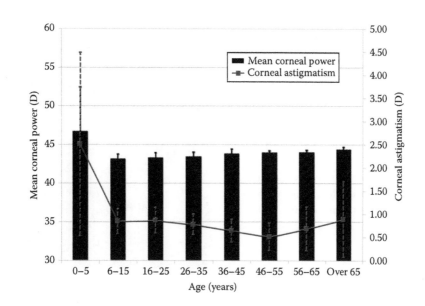

Figure 14.9 Mean corneal power and anterior corneal astigmatism as a function of age (based on data from various cross-sectional studies of corneal power and astigmatism including Anera et al. 2009, Anstice 1971, Baldwin et al. 1981, Fledelius and Stubgaard 1986, Friling et al. 2004, Ho et al. 2010, Isenberg et al. 2004, Koch et al. 2012, Lam et al. 1999, Leung et al. 2012, Liu et al. 2011, Mallen et al. 2005, Mutti et al. 2004, O'Donoghue et al, 2011, Sawada et al. 2008, Shankar and Bobier. 2004, Wickremasinghe et al. 2004, Zhang et al. 2000). Error bars represent the standard deviation. Mean corneal power and anterior corneal astigmatism decreases during youth, remains relatively stable throughout middle age, and increases during older age.

is typically observed (Goto et al. 2001, Hayashi et al. 1995, Ninn-Pedersen 1996, Pardhan and Beesley 1999), along with potential hormonal-related variations between genders: a greater corneal steepening in females and a concomitant flattening of the vertical meridian in males (Goto et al. 2001). Corneal asphericity also changes over a lifetime, typically a small increase in anterior surface asphericity and a decrease in posterior surface asphericity (Dubbelman et al. 2006, Pardhan and Beesley 1999), along with a greater degree of irregularity and asymmetry is found to occur in older age (Goto et al. 2001).

14.5.2 ASTIGMATISM

In early life, infants often display steep corneas with a wide range of corneal astigmatism of significant magnitude (Friling et al. 2004, Isenberg et al. 2004), typically either WTR (Ehrlich et al. 1997, Isenberg et al. 2004) or ATR in orientation (Friling et al. 2004, Gwiazda et al. 1984). During childhood, corneal astigmatism diminishes with corneal flattening, but persistent astigmatism has been associated with both meridional amblyopia and the development of refractive errors (discussed later under "Emmetropization"). During adulthood (up to 40 years), a large proportion of the population exhibit low levels of WTR corneal astigmatism (Fledelius and Stubgaard 1986). Changes in corneal curvature during older adult life (a steepening of the horizontal meridian) result in a shift toward ATR corneal astigmatism (Anstice 1971, Baldwin and Mills 1981), potentially due to age-related changes in eyelid tension (Read et al. 2007, 2014).

14.5.3 HIGHER-ORDER ABERRATIONS

While total ocular higher-order aberrations increase dramatically with older age, this is a result of an increase in internal higher-order aberrations due to changes in crystalline lens geometry (with minimal contribution from the posterior cornea) rather than changes in anterior corneal aberrations (Berrio et al. 2010). Corneal spherical aberration is typically stable throughout life or becomes very slightly more positive with age (Amano et al. 2004, Artal et al. 2002, Berrio et al. 2010, Fujikado et al. 2004, Lyall et al. 2013, Oshika et al. 1999, Wang et al. 2003, Wei et al. 2014). Another consistent finding is an increase in anterior corneal coma (horizontal coma, the sum of horizontal and vertical coma, or the RMS of coma terms) with age, in both younger and older populations (Amano et al. 2004, Lyall et al. 2013, Oshika et al. 1999, Wang et al. 2003, Wei et al. 2014,, Zhang et al. 2011,) which suggests that the cornea becomes less symmetric with age. Anterior surface comatic RMS increases from ~0.1 to 0.3 μm between 20 and 40 years of age to ~0.3 to 0.5 μm between 70 and 80 years (Amano et al. 2004, Oshika et al. 1999, Wei et al. 2014).

14.5.4 TRANSPARENCY

Since corneal transparency is dependent upon the uniform size and spacing of collagen fibrils within the stroma (Maurice 1957), any increase in fibril size or spacing results in a decrease in transparency (i.e., corneal haze or opacification). An increase in the number of collagen molecules, collagen cross-linking, and resultant growth of collagen fibrils leads to reduced collagen

fiber spacing (Kanai and Kaufman 1973) and consequently a less elastic (Elsheikh et al. 2007) and less transparent cornea with age (Daxer and Fratzl 1997).

In a large sample of healthy participants, Ni Dhubhghaill and colleagues (2014) reported that corneal densitometry (a measure of back scattered light or corneal transparency) increases significantly with age, but only for the peripheral cornea (6–12 mm) (Figure 14.10). A 46% increase in corneal densitometry values were observed from age 20 to 80 averaged over the entire cornea.

The density of stromal keratocytes, corneal fibroblast cells that produce and remodel the stroma during wound healing, may also be related to age-related changes in corneal transparency, with an approximate 3%–5% reduction in density per decade of life (based on in vitro corneal donor tissue analysis and in vivo confocal microscopy) (Moller-Pedersen 1997, Patel et al. 2001). Exposure to ultraviolet (UV) radiation over a lifetime may also contribute to a reduction in corneal transparency. UV radiation in the mouse model appears to downregulate enzymatic activity within the corneal epithelium that results in protein aggregation (Manzer et al. 2003).

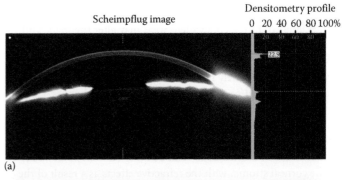

Figure 14.10 Scheimpflug images and corresponding optical densitometry profiles (expressed in percentage): (a) a 10-year-old male (22.9%), (b) a 75-year-old male (26.2%), and (c) a 36-year-old male with advanced keratoconus and associated central corneal scarring (92.4%). Corneal transparency decreases with age, most noticeably in the peripheral cornea, and as a result of edema, corneal disease or injury, and resultant scarring.

14.6 REFRACTIVE ERROR AND CORNEAL OPTICS

14.6.1 EMMETROPIZATION

Astigmatic emmetropization (the reduction of neonatal refractive errors) refers to the significant reduction in the magnitude of corneal and therefore total ocular astigmatism during childhood (Atkinson et al. 1980, Gwiazda et al. 1984) possibly associated with increased eyelid pressure with age (Grosvenor 1976). A failure of this mechanism during infancy, particularly for oblique astigmatism, increases the risk of the development of meridional amblyopia (Abrahamsson and Sjostrand 2003). Specifically, during the first 2 years of life, the cornea undergoes a significant reduction in refractive power (over 1 D) (Mutti et al. 2005), in conjunction with crystalline lens thinning and flattening and axial elongation as part of the emmetropization process. The magnitude of corneal astigmatism decreases with age and changes from WTR to ATR in orientation, approaching sphericity, with the majority of these corneal changes occurring within the first 9 months (Mutti et al. 2004). Oblique astigmatism in human infants is relatively rare (Gwiazda et al. 1984, Mayer et al. 2001, Mohindra et al. 1978) and longitudinal studies of corneal astigmatism suggest that vision-dependent emmetropization of spherical refractive errors is largely unaffected by low magnitude corneal astigmatism (Mutti et al. 2004).

Higher levels of corneal astigmatism could potentially influence eye growth and result in the development of refractive errors by degrading retinal image quality (Green 1871). However, longitudinal studies of children have reported equivocal findings regarding the relationship between the magnitude and orientation of total astigmatism and the development of myopia (Fulton et al. 1982, Goss and Grosvenor 1990, Gwiazda et al. 2000, Parssinen 1991). Greater levels of corneal higher-order aberrations could also potentially alter emmetropization through a vision-mediated mechanism. While this does not appear to be the case for surgically induced nonrotationally symmetric corneal aberrations in chicks (de la Cera et al. 2007), modelling suggests that such aberrations in combination with (rotationally symmetric) spherical aberration could provide optical cues with the potential to promote the development of myopia and astigmatism (Buehren et al. 2007).

14.6.2 SPHERICAL AMETROPIA

A number of studies have reported an association between the presence and magnitude of astigmatism and both myopic (Farbrother et al. 2004, Fulton et al.1982, Heidary et al. 2005, Kaye and Patterson 1997) and hyperopic (Farbrother et al. 2004, Guggenheim and Farbrother 2004, Kronfeld and Devney 1930) refractive errors. While the majority of these studies refer to the total ocular astigmatism, it is reasonable to assume the astigmatism in question was primarily corneal in origin. With respect to astigmatic orientation, high myopia and hyperopia tend to be associated with WTR astigmatism, whereas emmetropic eyes typically exhibit small amounts of ATR astigmatism (Farbrother et al. 2004). While a positive correlation has consistently been observed between the presence and magnitude of astigmatism and the degree of spherical ametropia (in particular, WTR astigmatism and myopia), it remains unknown if corneal astigmatism, or uncorrected ocular astigmatism, is a stimulus for refractive error development or a by-product of anomalous eye growth.

Corneal higher-order aberrations also vary with refractive error type; however, the relationship is less clear than the observed association between astigmatism and spherical ametropia. Myopes typically display greater levels of corneal spherical aberration in comparison to emmetropes (Vasudevan et al. 2007), while adult hyperopes manifest greater corneal spherical aberration compared to myopes (Llorente et al. 2004). In a younger population (~17 years) with a lower magnitude of hyperopia, corneal aberrations were not significantly different between hyperopes, myopes, and emmetropes (Philip et al. 2012), suggesting that differences in corneal aberrations between refractive error groups may be dependent upon both the magnitude of refractive error and age. Young adults with spherical ametropia and total astigmatism >1.00 D display greater levels of ocular total higher-order RMS compared to non-astigmats (≤1.00 D) (Cheng et al. 2003), most likely due to greater levels of corneal higher-order aberrations.

Given the association between near work and myopia, the relationship between reading, corneal aberrations, and refractive error is of particular interest since eyelid pressure during downward gaze has been shown to induce transient changes in corneal topography and increases in higher-order aberrations (Buehren et al. 2003). Buehren et al. (2005) noted that progressing myopes displayed significantly greater changes in corneal optics compared to emmetropes following a sustained 2 h reading task, due to a reduced palpebral aperture size. However, these trends are less apparent for interrupted reading tasks (Vasudevan et al.2007). Conversely, Zhu et al. (2013) observed similar levels of corneal higher-order aberrations over the course of a day in young emmetropes and stable and progressing myopes, but significant between group differences for total ocular aberrations (more negative spherical aberration in myopes), suggesting that variations of the internal optics may play a role in myopia development or progression rather than the anterior corneal surface.

14.6.3 ANISOMETROPIA

Anisometropic eyes exhibit a high degree of interocular symmetry for measures of corneal higher-order aberrations (Vincent et al. 2011) and central corneal power (i.e., keratometry) (Huynh et al. 2006, Kim et al. 2013, Logan et al. 2004, Tong et al. 2004); however, the mean refractive corneal power (the average of the steep and flat corneal meridians) has been shown to be significantly greater (steeper) in the more myopic eye of adult anisometropes when assessed using videokeratoscopy (Vincent et al. 2011). In addition, following a short-duration reading task, the more myopic eye of anisometropes display a small but significantly greater increase in corneal ATR astigmatism compared to the fellow eye along with a greater reduction in image quality (Vincent et al. 2013). This finding lends some support to the notion of a corneal-induced astigmatic image-mediated mechanism associated with the development of anisomyopia.

14.6.4 PRESBYOPIA

During the onset and progression of presbyopia (~45 years), the cornea's optics are relatively stable, with minimal change in curvature, spherical higher-order aberrations, astigmatism or transparency, or evidence of compensatory effects for an age-related reduction in accommodation. In fact, during middle age, the internal optics of the eye begins to display an additive effect rather than partial compensation for corneal aberrations (Artal et al. 2002). As a result, in recent years, a range of corneal procedures have been used in an attempt to correct presbyopia in the nondominant eye by increasing the corneal refractive power through corneal steepening (e.g., intrastromal incisions [Holzer et al. 2012], corneal reshaping inlays [Porter et al. 2012], and hyperopic orthokeratology [Gifford and Swarbrick 2013]), inducing corneal multifocality (e.g., excimer laser wavefront-guided ablation profiles [Telandro 2009], and refractive corneal inlays [Limnopoulou et al. 2013]), or extending the eye's depth of focus (e.g., intracorneal small aperture [pinhole] inlays [Tomita et al. 2012]). While such interventions have been shown to be relatively safe (and, in the case of orthokeratology, reversible), their functionality and degree of patient satisfaction is highly influenced by lighting conditions and pupil size.

14.6.5 AMBLYOPIA

The optical properties of the cornea associated with amblyopia have been examined by a number of studies. The fellow eyes of pediatric amblyopes display similar levels of corneal power and astigmatism (Cass and Tromans 2008, Debert et al. 2011, Wang and Taranath 2012). However, older pediatric and adult cohorts consistently display a greater amount of corneal astigmatism in the amblyopic eye compared to the fellow non-amblyopic eye, irrespective of the spherical refractive error or amblyogenic factor (Patel et al. 2010, Plech et al. 2010, Vincent et al. 2012). This increased level of astigmatism may be a result of abnormal ocular development since optic disk anomalies have been linked with corneal astigmatism (Jonas et al. 1997) and altered optic nerve head morphology (hypoplasia) is often observed in amblyopic eyes (Lempert and Porter 1998). However, visual experience and the development of spherical refractive error may also play a role, since corneal astigmatism is relatively symmetrical in pediatric amblyopes.

Relatively few studies have examined the higher-order corneal aberrations of amblyopic eyes. Plech et al. (2010) found similar levels of corneal higher-order aberrations between the fellow eyes of isometropic and anisometropic unilateral amblyopes. Conversely, Vincent et al. (2012) observed interocular asymmetries in corneal aberrations of monocular amblyopes dependent upon the amblyogenic factor. Strabismic amblyopes displayed greater levels of corneal trefoil in the turned eye (after excluding potential artifacts of eccentric fixation, horizontal pupil offset, and alterations in extraocular muscle tension post strabismus surgery), while anisometropic amblyopes displayed significant interocular differences in fourth-order terms C(4,2) secondary astigmatism, C(4,−2) secondary astigmatism along 45 degrees, and C(4,0) spherical aberration. These between-eye differences are potentially indicative of some form of feedback between the visual experience of the amblyopic eye and corneal optics.

14.6.6 IMPOSED ASTIGMATISM IN ANIMALS

Imposed astigmatic defocus in various animal models alters emmetropization (irrespective of axis orientation) (Kee et al. 2004); however, the resultant end point of eye growth is less predictable than the trends observed for imposed spherical defocus. During imposed astigmatic defocus, animal eyes may grow toward the circle of least confusion (Irving et al. 1995), the less hyperopic or more myopic meridian (Schmid and Wildsoet 1997), or a bimodal shift toward each principal meridian (Kee et al. 2004, McLean and Wallman 2003).

Young animals reared with imposed astigmatic defocus develop not only axial myopic and hyperopic refractive errors but also significant corneal astigmatism (Kee et al. 2003, 2004). This change in corneal optics is also consistent between experimental paradigms of form deprivation or lens-induced spherical ametropias in primates (Kee et al. 2005) and also for imposed astigmatic defocus in chicks (Irving et al. 1995, Schmid and Wildsoet 1997). However, since the axis of the resultant astigmatism (typically oblique corneal astigmatism) is not appropriately oriented to counteract the imposed astigmatic defocus, the corneal changes that occur as a result of altered visual experience appear to be a consequence of altered eye growth rather than a corneal emmetropization mechanism.

14.7 CONCLUSIONS

The cornea makes an important contribution to the optics of the human eye. Its transparency is derived from a delicate balance of collagen lamellae and fibril spacing and controlled hydration. Its shape provides a balance between the various optical requirements of the eye, with spherical power to focus light at the retinal plane both at the fovea and peripherally. The prolate elliptical shape defines the spherical aberration of the anterior corneal surface, so that this surface balances the internal spherical aberration of the eye across a range of object vergences. Its astigmatism is also partly offset by the astigmatism of the internal components, and it is one of the major aberrations of the eye to show consistent systematic changes throughout life. Given the ability of the eye to self-regulate its growth with respect to focus (emmetropization), it seems possible that these changes in astigmatism could have functional importance for the growth and/or vision of the eye at various stages of life.

The cornea is the primary source of astigmatism and higher-order aberrations for the total eye, and the quality of the retinal image can be improved through their correction. While the correction of astigmatism has been undertaken for almost 200 years, the correction of higher-order aberrations of the cornea and eye are still in their infancy. As these techniques develop, there are important opportunities to provide better vision correction outcomes, particularly for patients with significant levels of corneal and ocular higher-order aberrations.

As the most powerful and accessible refracting surface of the eye, various clinical methods to reshape the cornea have been developed and refined over recent years. Orthokeratology and laser refractive surgery have proven successful in correcting myopia, hyperopia, astigmatism, and presbyopia. However, the physiological mechanisms that underpin the natural shape of the cornea can often counteract these optical changes, and this provides another challenge for researchers to understand how to predict and compensate for these corneal responses.

The cornea has evolved to form an optically balanced combination with the crystalline lens to create a high-quality retinal image across a range of object vergences. An understanding of the visual optics of the cornea needs to consider not only the cornea in isolation but also its optical role in combination with the crystalline lens.

REFERENCES

Abrahamsson, M. and J. Sjostrand, Astigmatic axis and amblyopia in childhood, *Acta Ophthalmol Scand* 81 (2003): 33–37.

Alharbi, A. and H. A. Swarbrick, The effects of overnight orthokeratology lens wear on corneal thickness, *Invest Ophthalmol Vis Sci* 44 (2003): 2518–2523.

Amano, S., Y. Amano, S. Yamagami et al., Age-related changes in corneal and ocular higher-order wavefront aberrations, *Am J Ophthalmol* 137 (2004): 988–992.

Anera, R. G., M. Soler, J. de la Cruz Cardona, C. Salas, and C. Ortiz, Prevalence of refractive errors in school-age children in Morocco, *Clin Exp Ophthalmol* 37 (2009): 191–196.

Anstice, J., Astigmatism—Its components and their changes with age, *Am J Optom Arch Am Acad Optom* 48 (1971): 1001–1006.

Applegate, R. A., C. Ballentine, H. Gross, E. J. Sarver, and C.A. Sarver, Visual acuity as a function of Zernike mode and level of root mean square error, *Optom Vis Sci* 80 (2003): 97–105.

Applegate, R. A., G. Hilmantel, H. C. Howland, E. Y. Tu, T. Starck, and E. J. Zayac, Corneal first surface optical aberrations and visual performance, *J Refract Surg* 16 (2000): 507–514.

Applegate, R. A. and V. Lakshminarayanan, Parametric representation of Stiles-Crawford functions: Normal variation of peak location and directionality, *J Opt Soc Am A* 10 (1993): 1611–1623.

Artal, P., A. Benito, and J. Tabernero, The human eye is an example of robust optical design, *J Vis* 6 (2006): 1–7.

Artal, P., E. Berrio, A. Guirao, and P. Piers, Contribution of the cornea and internal surfaces to the change of ocular aberrations with age, *J Opt Soc Am A Opt Image Sci Vis* 19 (2002): 137–143.

Artal, P., L. Chen, E. J. Fernandez, B. Singer, S. Manzanera, and D. R. Williams, Neural compensation for the eye's optical aberrations, *J Vis* 4 (2004): 281–287.

Artal, P., A. Guirao, E. Berrio, and D. R. Williams, Compensation of corneal aberrations by the internal optics in the human eye, *J Vis* 1 (2001): 1–8.

Asbell, P. A., B. Chiang, M. E. Somers, and K. S. Morgan, Keratometry in children, *CLAO J* 16 (1990): 99–102.

Atchison, D. A., M. J. Collins, C. F. Wildsoet, J. Christensen, and M. D. Waterworth, Measurement of monochromatic ocular aberrations of human eyes as a function of accommodation by the Howland aberroscope technique, *Vision Res* 35 (1995): 313–323.

Atkinson, J., O. Braddick, and J. French, Infant astigmatism: Its disappearance with age, *Vision Res* 20 (1980): 891–893.

Auger, P., Confirmation of the Simplified Javal's Rule, *Am J Optom Physiol Opt* 65 (1988): 915.

Baldwin, W. R. and D. Mills, A longitudinal study of corneal astigmatism and total astigmatism, *Am J Optom Physiol Opt* 58 (1981): 206–211.

Barbero, S., Refractive power of a multilayer rotationally symmetric model of the human cornea and tear film, *J Opt Soc Am A Opt Image Sci Vis* 23 (2006): 1578–1585.

Bayramlar, H., F. Sadigov, and A. Yildirim, Effect of accommodation on corneal topography, *Cornea* 32 (2013): 1251–1254.

Berntsen, D. A., J. T. Barr, and G. L. Mitchell, The effect of overnight contact lens corneal reshaping on higher-order aberrations and best-corrected visual acuity, *Optom Vis Sci* 82 (2005): 490–497.

Berrio, E., J. Tabernero, and P. Artal, Optical aberrations and alignment of the eye with age, *J Vis* 10 (2010).

Boettner, E. A. and J. R. Wolter, Transmission of the ocular media, *Invest Ophthalmol Vis Sci* 1 (1962): 776–783.

Buehren, T., M. J. Collins, and L. Carney, Corneal aberrations and reading, *Optom Vis Sci* 80 (2003): 159–166.

Buehren, T., M. J. Collins, and L. G. Carney, Near work induced wavefront aberrations in myopia, *Vis Res* 45 (2005): 1297–1312.

Buehren, T., M. J. Collins, J. Loughridge, L. G. Carney, and D. R. Iskander, Corneal topography and accommodation, *Cornea* 22 (2003): 311–316.

Buehren, T., D. R. Iskander, M. J. Collins, and B. Davis, Potential higher-order aberration cues for sphero-cylindrical refractive error development, *Optom Vis Sci* 84 (2007): 163–174.

Buhren, J., C. Kuhne, and T. Kohnen, Defining subclinical keratoconus using corneal first-surface higher-order aberrations, *Am J Ophthalmol* 143 (2007): 381–389.

Burns, S. A., S. Wu, F. Delori, and A. E. Elsner, Direct measurement of human-cone-photoreceptor alignment, *J Opt Soc Am A Opt Image Sci Vis* 12 (1995): 2329–2338.

Carney, L. G. and B. A. Clark, Experimental deformation of the in vivo cornea, *Am J Optom Arch Am Acad Optom* 49 (1972): 28–34.

Cass, K. and C. Tromans, A biometric investigation of ocular components in amblyopia, *Ophthalmic Physiol Opt* 28 (2008): 429–440.

Chang, D. H. and G. O. Waring, The subject-fixated coaxially sighted corneal light reflex: A clinical marker for centration of refractive treatments and devices, *Am J Ophthalmol* 158 (2014): 863–874.

Chen, C. C., S. W. Cheung, and P. Cho, Toric orthokeratology for highly astigmatic children, *Optom Vis Sci* 89 (2012): 849–855.

Chen, S. I., M. Hove, C. L. McCloskey, and S. B. Kaye, The effect of monocularly and binocularly induced astigmatic blur on depth discrimination is orientation dependent, *Optom Vis Sci* 82 (2005): 101–113.

Cheng, H., J. K. Barnett, A. S. Vilupuru et al., A population study on changes in wave aberrations with accommodation, *J Vis* 4 (2004): 272–280.

Cheng, X., A. Bradley, X. Hong, and L. N. Thibos, Relationship between refractive error and monochromatic aberrations of the eye, *Optom Vis Sci* 80 (2003): 43–49.

Collins, M. J., K. Kloevekorn-Norgall, T. Buehren, S. C. Voetz, and B. Lingelbach, Regression of lid-induced corneal topography changes after reading, *Optom Vis Sci* 82 (2005): 843–849.

Daxer, A. and P. Fratzl, Collagen fibril orientation in the human corneal stroma and its implication in keratoconus, *Invest Ophthalmol Vis Sci* 38 (1997): 121–129.

de la Cera, E. G., G. Rodriguez, A. de Castro, J. Merayo, and S. Marcos, Emmetropization and optical aberrations in a myopic corneal refractive surgery chick model, *Vis Res* 47 (2007): 2465–2472.

Debert, I., L. M. de Alencar, M. Polati, M. B. Souza, and M. R. Alves, Oculometric parameters of hyperopia in children with esotropic amblyopia, *Ophthalmic Physiol Opt* 31 (2011): 389–397.

Doughty, M. J. and M. L. Zaman, Human corneal thickness and its impact on intraocular pressure measures: A review and meta-analysis approach, *Surv Ophthalmol* 44 (2000): 367–408.

Douthwaite, W. A., T. Hough, K. Edwards, and H. Notay, The EyeSys videokeratoscopic assessment of apical radius and p-value in the normal human cornea, *Ophthalmic Physiol Opt* 19 (1999): 467–474.

Dubbelman, M., V. A. Sicam, and G. L. Van der Heijde, The shape of the anterior and posterior surface of the aging human cornea, *Vis Res* 46 (2006): 993–1001.

Dubbelman, M., V. A. Sicam, and R. G. van der Heijde, The contribution of the posterior surface to the coma aberration of the human cornea, *J Vis* 7 (2007): 10.1–10.8.

Dunne, M. C., M. E. Elawad, and D. A. Barnes, A study of the axis of orientation of residual astigmatism, *Acta Ophthalmol (Copenh)* 72 (1994): 483–489.

Eghbali, F., K. K. Yeung, and R. K. Maloney, Topographic determination of corneal asphericity and its lack of effect on the refractive outcome of radial keratotomy, *Am J Ophthalmol* 119 (1995): 275–280.

Ehrlich, D. L., O. J. Braddick, J. Atkinson et al., Infant emmetropization: Longitudinal changes in refraction components from nine to twenty months of age, *Optom Vis Sci* 74 (1997): 822–843.

el-Hage, S. G. and F. Berny, Contribution of the crystalline lens to the spherical aberration of the eye, *J Opt Soc Am* 63 (1973): 205–211.

Elsheikh, A., D. Wang, M. Brown, P. Rama, M. Campanelli, and D. Pye, Assessment of corneal biomechanical properties and their variation with age, *Curr Eye Res* 32 (2007): 11–19.

Enright, J. T., Ocular translation and cyclotorsion due to changes in fixation distance, *Vis Res* 20 (1980): 595–601.

Fairmaid, J. A., The constancy of corneal curvature: An examination of corneal response to changes in accommodation and convergence, *Br J Physiol Opt* 16 (1959): 2–23.

Farbrother, J. E., J. W. Welsby, and J. A. Guggenheim, Astigmatic axis is related to the level of spherical ametropia, *Optom Vis Sci* 81 (2004): 18–26.

Fledelius, H. C. and M. Stubgaard, Changes in refraction and corneal curvature during growth and adult life. A cross-sectional study, *Acta Ophthalmol (Copenh)* 64 (1986): 487–491.

Friling, R., D. Weinberger, I. Kremer, R. Avisar, L. Sirota, and M. Snir, Keratometry measurements in preterm and full term newborn infants, *Br J Ophthalmol* 88 (2004): 8–10.

Fujikado, T., T. Kuroda, S. Ninomiya et al., Age-related changes in ocular and corneal aberrations, *Am J Ophthalmol* 138 (2004): 143–146.

Fulton, A. B., R. M. Hansen, and R. A. Petersen, The relation of myopia and astigmatism in developing eyes, *Ophthalmology* 89 (1982): 298–302.

Gifford, P. and H. A. Swarbrick, Refractive changes from hyperopic orthokeratology monovision in presbyopes, *Optom Vis Sci* 90 (2013): 306–313.

Gobbe, M. and M. Guillon, Corneal wavefront aberration measurements to detect keratoconus patients, *Cont Lens Anterior Eye* 28 (2005): 57–66.

Goss, D. A. and T. Grosvenor, Rates of childhood myopia progression with bifocals as a function of nearpoint phoria: Consistency of three studies, *Optom Vis Sci* 67 (1990): 637–640.

Goto, T., S. D. Klyce, X. Zheng, N. Maeda, T. Kuroda, and C. Ide, Gender- and age-related differences in corneal topography, *Cornea* 20 (2001): 270–276.

Green, J., On astigmatism as an active cause of myopia, *Trans Am Ophthalmol Soc* 1 (1871): 105–107.

Grosvenor, T., What causes astigmatism? *J Am Optom Assoc* 47 (1976): 926–932.

Grosvenor, T., S. Quintero, and D. M. Perrigin, Predicting refractive astigmatism: A suggested simplification of Javal's rule, *Am J Optom Physiol Opt* 65 (1988): 292–297.

Gruenauer-Kloevekorn, C., U. Fischer, K. Kloevekorn-Norgall, and G. I. Duncker, Pellucid marginal corneal degeneration: Evaluation of the corneal surface and contact lens fitting, *Br J Ophthalmol* 90 (2006): 318–323.

Guggenheim, J. A. and J. E. Farbrother, The association between spherical and cylindrical component powers, *Optom Vis Sci* 81 (2004): 62–63.

Guggenheim, J. A., T. Zayats, A. Prashar, and C. H. To, Axes of astigmatism in fellow eyes show mirror rather than direct symmetry, *Ophthalmic Physiol Opt* 28 (2008): 327–333.

Guillon, M., D. P. Lydon, and C. Wilson, Corneal topography: A clinical model, *Ophthalmic Physiol Opt* 6 (1986): 47–56.

Guirao, A. and P. Artal, Corneal wave aberration from videokeratography: Accuracy and limitations of the procedure, *J Opt Soc Am A Opt Image Sci Vis* 17 (2000): 955–965.

Guirao, A., M. Redondo, and P. Artal, Optical aberrations of the human cornea as a function of age, *J Opt Soc Am A Opt Image Sci Vis* 17 (2000): 1697–1702.

Gunes, A., L. Tok, O. Tok, and L. Seyrek, The youngest patient with bilateral Keratoconus secondary to chronic persistent eye rubbing, *Semin Ophthalmol* 30 (2015): 454–456.

Guo, H. and D. A. Atchison, Subjective blur limits for cylinder, *Optom Vis Sci* 87 (2010): E549–E559.

Gwiazda, J., K. Grice, R. Held, J. McLellan, and F. Thorn, Astigmatism and the development of myopia in children, *Vis Res* 40 (2000): 1019–1026.

Gwiazda, J., M. Scheiman, I. Mohindra, and R. Held, Astigmatism in children: Changes in axis and amount from birth to six years, *Invest Ophthalmol Vis Sci* 25 (1984): 88–92.

Hayashi, K., H. Hayashi, and F. Hayashi, Topographic analysis of the changes in corneal shape due to aging, *Cornea* 14 (1995): 527–532.

He, J. C., J. Gwiazda, F. Thorn, and R. Held, Wave-front aberrations in the anterior corneal surface and the whole eye, *J Opt Soc Am A Opt Image Sci Vis* 20 (2003): 1155–1163.

Heidary, G., G. S. Ying, M. G. Maguire, and T. L. Young, The association of astigmatism and spherical refractive error in a high myopia cohort, *Optom Vis Sci* 82 (2005): 244–247.

Hiraoka, T., C. Okamoto, Y. Ishii, T. Kakia, and T.Oshika, Contrast sensitivity function and ocular higher-order aberrations following overnight orthokeratology, *Invest Ophthalmol Vis Sci* 48 (2007): 550–556.

Ho, J. D., S. W. Liou, R. J. Tsai, and C. Y. Tsai, Effects of aging on anterior and posterior corneal astigmatism, *Cornea* 29 (2010): 632–637.

Ho, J. D., C. Y. Tsai, and S. W. Liou, Accuracy of corneal astigmatism estimation by neglecting the posterior corneal surface measurement, *Am J Ophthalmol* 147 (2009): 788–795, 795. e1–e2.

Hoffmann, P. C. and W. W. Hutz, Analysis of biometry and prevalence data for corneal astigmatism in 23,239 eyes, *J Cataract Refract Surg* 36 (2010): 1479–1485.

Holzer, M. P., M. C. Knorz, M. Tomalla, T. M. Neuhann, and G. U. Auffarth, Intrastromal femtosecond laser presbyopia correction: 1-year results of a multicenter study, *J Refract Surg* 28 (2012): 182–188.

Howland, H. C., R. H. Rand, and S. R. Lubkin, A thin-shell model of the cornea and its application to corneal surgery, *Refract Corneal Surg* 8 (1992): 183–186.

Huynh, S. C., A. Kifley, K. A. Rose, I. Morgan, G.Z. Heller, and P. Mitchell, Astigmatism and its components in 6-year-old children, *Invest Ophthalmol Vis Sci* 47 (2006): 55–64.

Huynh, S. C., X. Y. Wang, J. Ip et al., Prevalence and associations of anisometropia and aniso-astigmatism in a population based sample of 6 year old children, *Br J Ophthalmol* 90 (2006): 597–601.

Irving, E. L., M. G. Callender, and J. G. Sivak, Inducing ametropias in hatchling chicks by defocus—Aperture effects and cylindrical lenses, *Vis Res* 35 (1995): 1165–1174.

Isenberg, S. J., M. Del Signore, A. Chen, J. Wei, and P. D. Christenson, Corneal topography of neonates and infants, *Arch Ophthalmol* 122 (2004): 1767–1771.

Iskander, D. R., M. J. Collins, and B. Davis, Optimal modeling of corneal surfaces with Zernike polynomials, *IEEE Trans Biomed Eng* 48 (2001): 87–95.

Iskander, D. R., M. R. Morelande, M. J. Collins, and B. Davis, Modeling of corneal surfaces with radial polynomials, *IEEE Trans Biomed Eng* 49 (2002): 320–328.

Jonas, J. B., F. Kling, and A. E. Grundler, Optic disc shape, corneal astigmatism, and amblyopia, *Ophthalmology* 104 (1997): 1934–1937.

Joslin, C. E., S. M. Wu, T. T. McMahon, and M. Shahidi, Higher-order wavefront aberrations in corneal refractive therapy, *Optom Vis Sci* 80 (2003): 805–811.

Kanai, A. and H. E. Kaufman, Electron microscopic studies of corneal stroma: Aging changes of collagen fibers, *Ann Ophthalmol* 5 (1973): 285–287.

Kandarakis, A., M. Karampelas, V. Soumplis et al., A case of bilateral self-induced keratoconus in a patient with tourette syndrome associated with compulsive eye rubbing: Case report, *BMC Ophthalmol* 11 (2011): 28.

Kaye, S. B. and A. Patterson, Association between total astigmatism and myopia, *J Cataract Refract Surg* 23 (1997): 1496–1502.

Kee, C. S., Astigmatism and its role in emmetropization, *Exp Eye Res* 114 (2013): 89–95.

Kee, C. S., L. F. Hung, Y. Qiao-Grider, R. Ramamirtham, and E. L. Smith, 3rd, Astigmatism in monkeys with experimentally induced myopia or hyperopia, *Optom Vis Sci* 82 (2005): 248–260.

Kee, C. S., L. F. Hung, Y. Qiao-Grider, A. Roorda, and E. L. Smith, 3rd, Effects of optically imposed astigmatism on emmetropization in infant monkeys, *Invest Ophthalmol Vis Sci* 45 (2004): 1647–1659.

Kee, C. S., L. F. Hung, Y. Qiao, and E. L. Smith, 3rd, Astigmatism in infant monkeys reared with cylindrical lenses, *Vis Res* 43 (2003): 2721–2739.

Keller, P. R., M. J. Collins, L. G. Carney, B. A. Davis, and P. P. van Saarloos, The relation between corneal and total astigmatism, *Optom Vis Sci* 73 (1996): 86–91.

Kelly, J. E., T. Mihashi, and H. C. Howland, Compensation of corneal horizontal/vertical astigmatism, lateral coma, and spherical aberration by internal optics of the eye, *J Vis* 4 (2004): 262–271.

Khng, C. and R. H. Osher, Evaluation of the relationship between corneal diameter and lens diameter, *J Cataract Refract Surg* 34 (2008): 475–479.

Kiely, P. M., L. G. Carney, and G. Smith, Diurnal variations of corneal topography and thickness, *Am J Optom Physiol Opt* 59 (1982): 976–982.

Kim, S. Y., S. Y. Cho, J. W. Yang, C. S. Kim, and Y. C. Lee, The correlation of differences in the ocular component values with the degree of myopic anisometropia, *Korean J Ophthalmol* 27 (2013): 44–47.

Koch, D. D., S. F. Ali, M. P. Weikert, M. Shirayama, R. Jenkins, and L. Wang, Contribution of posterior corneal astigmatism to total corneal astigmatism, *J Cataract Refract Surg* 38 (2012): 2080–2087.

Koh, S., N. Maeda, T. Nakagawa et al., Characteristic higher-order aberrations of the anterior and posterior corneal surfaces in 3 corneal transplantation techniques, *Am J Ophthalmol* 153 (2012): 284–290 e1.

Kohnen, T., K. Mahmoud, and J. Buhren, Comparison of corneal higher-order aberrations induced by myopic and hyperopic LASIK, *Ophthalmology* 112 (2005): 1692.

Kosaki, R., N. Maeda, K. Bessho et al., Magnitude and orientation of Zernike terms in patients with keratoconus, *Invest Ophthalmol Vis Sci* 48 (2007): 3062–3068.

Kronfeld, P.C. and C. Devney, The frequency of astigmatism, *Arch Ophthalmol* 4 (1930): 873–884.

Lai, T. and S. Tang, Cornea characterization using a combined multiphoton microscopy and optical coherence tomography system, *Biomed Opt Express* 5 (2014): 1494–511.

Lam, A. K., C. C. Chan, M. H. Lee, and K. M. Wong, The aging effect on corneal curvature and the validity of Javal's rule in Hong Kong Chinese, *Curr Eye Res* 18 (1999): 83–90.

Land, M.F. and D. Nilsson. 2012. *Animal Eyes, Oxford Animal Biology Series*. Oxford University Press, Oxford, U.K.

Lempert, P. and L. Porter, Dysversion of the optic disc and axial length measurements in a presumed amblyopic population, *J AAPOS* 2 (1998): 207–213.

Leonard, D. W. and K. M. Meek, Refractive indices of the collagen fibrils and extrafibrillar material of the corneal stroma, *Biophys J* 72 (1997): 1382–1387.

Leung, T. W., A. K. Lam, L. Deng, and C. S. Kee, Characteristics of astigmatism as a function of age in a Hong Kong clinical population, *Optom Vis Sci* 89 (2012): 984–992.

Li, Y. and F. J. Bao, Interocular symmetry analysis of bilateral eyes, *J Med Eng Technol* 38 (2014): 179–187.

Li, Y., O. Tan, R. Brass, J. L. Weiss, and D. Huang, Corneal epithelial thickness mapping by Fourier-domain optical coherence tomography in normal and keratoconic eyes, *Ophthalmology* 119 (2012): 2425–2433.

Liang, J., D. R. Williams, and D. T. Miller, Supernormal vision and high-resolution retinal imaging through adaptive optics, *J Opt Soc Am A Opt Image Sci Vis* 14 (1997): 2884–2892.

Limnopoulou, A. N., D. I. Bouzoukis, G. D. Kymionis et al., Visual outcomes and safety of a refractive corneal inlay for presbyopia using femtosecond laser, *J Refract Surg* 29 (2013): 12–18.

Liu, Y. C., P. Chou, R. Wojciechowski et al., Power vector analysis of refractive, corneal, and internal astigmatism in an elderly Chinese population: The Shihpai Eye Study, *Invest Ophthalmol Vis Sci* 52 (2011): 9651–9657.

Llorente, L., S. Barbero, D. Cano, C. Dorronsoro, and S. Marcos, Myopic versus hyperopic eyes: Axial length, corneal shape and optical aberrations, *J Vis* 4 (2004): 288–298.

Logan, N. S., B. Gilmartin, C. F. Wildsoet, and M. C. Dunne, Posterior retinal contour in adult human anisomyopia, *Invest Ophthalmol Vis Sci* 45 (2004): 2152–2162.

Lombardo, M., G. Lombardo, and S. Serrao, Interocular high-order corneal wavefront aberration symmetry, *J Opt Soc Am A Opt Image Sci Vis* 23 (2006): 777–787.

Lopping, B. and R. A. Weale, Changes in corneal curvature following ocular convergence, *Vis Res* 5 (1965): 207–215.

Lotmar, W., Theoretical eye model with asphorics, *J Opt Soc Am* 61 (1971): 1522–1529.

Lu, F., J. Wu, J. Qu et al., Association between offset of the pupil center from the corneal vertex and wavefront aberration, *J Optom* 1 (2008): 8–13.

Lu, F., J. Wu, Y. Shen et al., On the compensation of horizontal coma aberrations in young human eyes, *Ophthalmic Physiol Opt* 28 (2008): 277–282.

Lyall, D. A., S. Srinivasan, and L. S. Gray, Changes in ocular monochromatic higher-order aberrations in the aging eye, *Optom Vis Sci* 90 (2013): 996–1003.

Maeda, N., T. Fujikado, T. Kuroda et al., Wavefront aberrations measured with Hartmann-Shack sensor in patients with keratoconus, *Ophthalmology* 109 (2002): 1996–2003.

Mallen, E. A., Y. Gammoh, M. Al-Bdour, and F. N. Sayegh, Refractive error and ocular biometry in Jordanian adults, *Ophthalmic Physiol Opt* 25 (2005): 302–309.

Mandell, R. B., Locating the corneal sighting center from videokeratography, *J Refract Surg* 11 (1995): 253–259.

Mandell, R. B., C. S. Chiang, and S. A. Klein, Location of the major corneal reference points, *Optom Vis Sci* 72 (1995): 776–784.

Mandell, R. B. and R. S. Helen, Stability of the corneal contour, *Am J Optom Arch Am Acad Optom* 45 (1968): 797–806.

Mansour, A. M. and R. S. Haddad, Corneal topography after ocular rubbing, *Cornea* 21 (2002): 756–758.

Manzer, R., A. Pappa, T. Estey, N. Sladek, J. F. Carpenter, and V. Vasiliou, Ultraviolet radiation decreases expression and induces aggregation of corneal ALDH3A1, *Chem Biol Interact* 143–144 (2003): 45–53.

Marcos, S., S. Barbero, L. Llorente, and J. Merayo-Lloves, Optical response to LASIK surgery for myopia from total and corneal aberration measurements, *Invest Ophthalmol Vis Sci* 42 (2001): 3349–3356.

Marcos, S., P. Rosales, L. Llorente, S. Barbero, and I. Jimenez-Alfaro, Balance of corneal horizontal coma by internal optics in eyes with intraocular artificial lenses: Evidence of a passive mechanism, *Vis Res* 48 (2008): 70–79.

Marsack, J. D., K. E. Parker, Y. Niu, K. Pesudovs, and R. A. Applegate, On-eye performance of custom wavefront-guided soft contact lenses in a habitual soft lens-wearing keratoconic patient, *J Refract Surg* 23 (2007): 960–964.

Marsack, J. D., A. Ravikumar, C. Nguyen et al., Wavefront-guided scleral lens correction in Keratoconus, *Optom Vis Sci* 91 (2014): 1221–1230.

Mas, D., J. Espinosa, B. Domenech, J. Perez, H. Kasprzak, and C. Illueca, Correlation between the dioptric power, astigmatism and surface shape of the anterior and posterior corneal surfaces, *Ophthalmic Physiol Opt* 29 (2009): 219–226.

Mathur, A., J. Gehrmann, and D. A. Atchison, Influences of luminance and accommodation stimuli on pupil size and pupil center location, *Invest Ophthalmol Vis Sci* 55 (2014): 2166–2172.

Maurice, D. M., The structure and transparency of the cornea, *J Physiol* 136 (1957): 263–286.

Mayer, D. L., R. M. Hansen, B. D. Moore, S. Kim, and A. B. Fulton, Cycloplegic refractions in healthy children aged 1 through 48 months, *Arch Ophthalmol* 119 (2001): 1625–1628.

McKendrick, A. M. and N. A. Brennan, The axis of astigmatism in right and left eye pairs, *Optom Vis Sci* 74 (1997): 668–675.

McLean, R. C. and J. Wallman, Severe astigmatic blur does not interfere with spectacle lens compensation, *Invest Ophthalmol Vis Sci* 44 (2003): 449–457.

Meek, K. M. and C. Boote, The use of x-ray scattering techniques to quantify the orientation and distribution of collagen in the corneal stroma, *Prog Retin Eye Res* 28 (2009): 369–392.

Meek, K. M., D. W. Leonard, C. J. Connon, S. Dennis, and S. Khan, Transparency, swelling and scarring in the corneal stroma, *Eye (Lond)* 17 (2003): 927–936.

Mohindra, I., R. Held, J. Gwiazda, and J. Brill, Astigmatism in infants, *Science* 202 (1978): 329–331.

Moller-Pedersen, T., A comparative study of human corneal keratocyte and endothelial cell density during aging, *Cornea* 16 (1997): 333–338.

Mrochen, M., M. Jankov, M. Bueeler, and T. Seiler, Correlation between corneal and total wavefront aberrations in myopic eyes, *J Refract Surg* 19 (2003): 104–112.

Mrochen, M., M. Kaemmerer, and T. Seiler, Wavefront-guided laser in situ keratomileusis: Early results in three eyes, *J Refract Surg* 16 (2000): 116–121.

Mutti, D. O., G. L. Mitchell, L. A. Jones et al., Refractive astigmatism and the toricity of ocular components in human infants, *Optom Vis Sci* 81 (2004): 753–761.

Mutti, D. O., G. L. Mitchell, L. A. Jones et al., Axial growth and changes in lenticular and corneal power during emmetropization in infants, *Invest Ophthalmol Vis Sci* 46 (2005): 3074–3080.

Nakagawa, T., N. Maeda, R. Kosaki et al., Higher-order aberrations due to the posterior corneal surface in patients with keratoconus, *Invest Ophthalmol Vis Sci* 50 (2009): 2660–2665.

Navarro, R., L. Gonzalez, and J. L. Hernandez, Optics of the average normal cornea from general and canonical representations of its surface topography, *J Opt Soc Am A Opt Image Sci Vis* 23 (2006): 219–232.

Netto, M. V., W. Dupps, Jr., and S. E. Wilson, Wavefront-guided ablation: Evidence for efficacy compared to traditional ablation, *Am J Ophthalmol* 141 (2006): 360–368.

Ni Dhubhghaill, S., J. J. Rozema, S. Jongenelen, I. Ruiz Hidalgo, N. Zakaria, and M. J. Tassignon, Normative values for corneal densitometry analysis by Scheimpflug optical assessment, *Invest Ophthalmol Vis Sci* 55 (2014): 162–168.

Nichols, J. J., M. M. Marsich, M. Nguyen, J. T. Barr, and M. A. Bullimore, Overnight orthokeratology, *Optom Vis Sci* 77 (2000): 252–259.

Ninn-Pedersen, K., Relationships between preoperative astigmatism and corneal optical power, axial length, intraocular pressure, gender, and patient age, *J Refract Surg* 12 (1996): 472–482.

O'Donoghue, L., A. R. Rudnicka, J. F. McClelland, N. S. Logan, C. G. Owen, and K. J. Saunders, Refractive and corneal astigmatism in white school children in northern ireland, *Invest Ophthalmol Vis Sci* 52 (2011): 4048–4053.

Oie, Y., N. Maeda, R. Kosaki et al., Characteristics of ocular higher-order aberrations in patients with pellucid marginal corneal degeneration, *J Cataract Refract Surg* 34 (2008): 1928–1934.

Oshika, T., S. D. Klyce, R. A. Applegate, and H. C. Howland, Changes in corneal wavefront aberrations with aging, *Invest Ophthalmol Vis Sci* 40 (1999): 1351–1355.

Owens, H., L. F. Garner, J. P. Craig, and G. Gamble, Posterior corneal changes with orthokeratology, *Optom Vis Sci* 81 (2004): 421–426.

Pande, M. and J. S. Hillman, Optical zone centration in keratorefractive surgery. Entrance pupil center, visual axis, coaxially sighted corneal reflex, or geometric corneal center?, *Ophthalmology* 100 (1993): 1230–1237.

Pardhan, S. and J. Beesley, Measurement of corneal curvature in young and older normal subjects, *J Refract Surg* 15 (1999): 469–474.

Parssinen, O., Astigmatism and school myopia, *Acta Ophthalmol (Copenh)* 69 (1991): 786–790.

Patel, S. and J. L. Alio, Corneal refractive index-hydration relationship by objective refractometry, *Optom Vis Sci* 89 (2012): 1641–1646.

Patel, S., J. L. Alio, F. Amparo, and J. L. Rodriguez-Prats, The influence of age on the refractive index of the human corneal stroma resected using a mechanical microkeratome, *Cornea* 30 (2011): 1353–1357.

Patel, S., J. Marshall, and F. W. Fitzke, 3rd, Refractive index of the human corneal epithelium and stroma, *J Refract Surg* 11 (1995): 100–105.

Patel, S., J. McLaren, D. Hodge, and W. Bourne, Normal human keratocyte density and corneal thickness measurement by using confocal microscopy in vivo, *Invest Ophthalmol Vis Sci* 42 (2001): 333–339.

Patel, V. S., J. W. Simon, and R. L. Schultze, Anisometropic amblyopia: Axial length versus corneal curvature in children with severe refractive imbalance, *J AAPOS* 14 (2010): 396–398.

Philip, K., A. Martinez, A. Ho et al., Total ocular, anterior corneal and lenticular higher order aberrations in hyperopic, myopic and emmetropic eyes, *Vis Res* 52 (2012): 31–37.

Plech, A. R., D. P. Pinero, C. Laria, A. Aleson, and J. L. Alio, Corneal higher-order aberrations in amblyopia, *Eur J Ophthalmol* 20 (2010): 12–20.

Porter, J., A. Guirao, I. G. Cox, and D. R. Williams, Monochromatic aberrations of the human eye in a large population, *J Opt Soc Am A Opt Image Sci Vis* 18 (2001): 1793–1803.

Porter, T., A. Lang, K. Holliday et al., Clinical performance of a hydrogel corneal inlay in hyperopic presbyopes, *ARVO* 53 (2012): abstract 4056.

Read, S. A., T. Buehren, and M. J. Collins, Influence of accommodation on the anterior and posterior cornea, *J Cataract Refract Surg* 33 (2007): 1877–1885.

Read, S. A. and M. J. Collins, Diurnal variation of corneal shape and thickness, *Optom Vis Sci* 86 (2009): 170–180.

Read, S. A., M. J. Collins, and L. G. Carney, The diurnal variation of corneal topography and aberrations, *Cornea* 24 (2005): 678–687.

Read, S. A., M. J. Collins, and L. G. Carney, A review of astigmatism and its possible genesis, *Clin Exp Optom* 90 (2007): 5–19.

Read, S. A., M. J. Collins, L. G. Carney, and R. J. Franklin, The topography of the central and peripheral cornea, *Invest Ophthalmol Vis Sci* 47 (2006): 1404–1415.

Read, S. A., M. J. Collins, S. H. Cheong, and E. C. Woodman, Sustained convergence, axial length, and corneal topography, *Optom Vis Sci* 87 (2010): E45–E52.

Read, S. A., S. J. Vincent, and M. J. Collins, The visual and functional impacts of astigmatism and its clinical management, *Ophthalmic Physiol Opt* 34 (2014): 267–294.

Reinstein, D. Z., M. Gobbe, T. J. Archer, D. Couch, and B. Bloom, Epithelial, stromal, and corneal pachymetry changes during orthokeratology, *Optom Vis Sci* 86 (2009): E1006–E1014.

Romero-Jimenez, M., J. Santodomingo-Rubido, P. Flores-Rodriguez, and J. M. Gonzalez-Meijome, Short-term corneal changes with gas-permeable contact lens wear in keratoconus subjects: A comparison of two fitting approaches, *J Optom* 8 (2014): 48–55.

Rudolph, M., K. Laaser, B. O. Bachmann, C. Cursiefen, D. Epstein, and F. E. Kruse, Corneal higher-order aberrations after Descemet's membrane endothelial keratoplasty, *Ophthalmology* 119 (2012): 528–535.

Sabesan, R., T. M. Jeong, L. Carvalho, I. G. Cox, D. R. Williams, and G. Yoon, Vision improvement by correcting higher-order aberrations with customized soft contact lenses in keratoconic eyes, *Opt Lett* 32 (2007): 1000–1002.

Sabesan, R., L. Johns, O. Tomashevskaya, D. S. Jacobs, P. Rosenthal, and G. Yoon, Wavefront-guided scleral lens prosthetic device for keratoconus, *Optom Vis Sci* 90 (2013): 314–323.

Salmon, T. O. and L. N. Thibos, Videokeratoscope-line-of-sight misalignment and its effect on measurements of corneal and internal ocular aberrations, *J Opt Soc Am A Opt Image Sci Vis* 19 (2002): 657–669.

Sawada, A., A. Tomidokoro, M. Araie, A. Iwase, T. Yamamoto, and Group Tajimi Study, Refractive errors in an elderly Japanese population: The Tajimi study, *Ophthalmology* 115 (2008): 363–370 e3.

Sawides, L., E. Gambra, D. Pascual, C. Dorronsoro, and S. Marcos, Visual performance with real-life tasks under adaptive-optics ocular aberration correction, *J Vis* 10 (2010): 19.

Schmid, K. and C. F. Wildsoet, Natural and imposed astigmatism and their relation to emmetropization in the chick, *Exp Eye Res* 64 (1997): 837–847.

Schwiegerling, J. and J. E. Greivenkamp, Using corneal height maps and polynomial decomposition to determine corneal aberrations, *Optom Vis Sci* 74 (1997): 906–916.

Schwiegerling, J., J. E. Greivenkamp, and J. M. Miller, Representation of videokeratoscopic height data with Zernike polynomials, *J Opt Soc Am A Opt Image Sci Vis* 12 (1995): 2105–2113.

Serrao, S., G. Lombardo, P. Ducoli, and M. Lombardo, Optical performance of the cornea six years following photorefractive keratectomy for myopia, *Invest Ophthalmol Vis Sci* 52 (2011): 846–857.

Shankar, S. and W. R. Bobier, Corneal and lenticular components of total astigmatism in a preschool sample, *Optom Vis Sci* 81 (2004): 536–542.

Shaw, A. J., M. J. Collins, B. A. Davis, and L. G. Carney, Corneal refractive changes due to short-term eyelid pressure in downward gaze, *J Cataract Refract Surg* 34 (2008): 1546–1553.

Shaw, A. J., M. J. Collins, B. A. Davis, and L. G. Carney, Eyelid pressure: Inferences from corneal topographic changes, *Cornea* 28 (2009): 181–188.

Sicam, V. A., M. Dubbelman, and R. G. van der Heijde, Spherical aberration of the anterior and posterior surfaces of the human cornea, *J Opt Soc Am A Opt Image Sci Vis* 23 (2006): 544–549.

Sugar, J. and M. S. Macsai, What causes keratoconus? *Cornea* 31 (2012): 716–719.

Telandro, A., The pseudoaccommodative cornea multifocal ablation with a center-distance pattern: A review, *J Refract Surg* 25 (2009): S156–S159.

Thibos, L. N., R. A. Applegate, J. T. Schwiegerling, R. Webb, VSIA Standards Taskforce Members, Vision science and its applications, Standards for reporting the optical aberrations of eyes, *J Refract Surg* 18 (2002): S652–S660.

Thibos, L. N., X. Hong, A. Bradley, and R. A. Applegate, Accuracy and precision of objective refraction from wavefront aberrations, *J Vis* 4 (2004): 329–351.

Tomita, M., T. Kanamori, G. O. 4th Waring et al., Simultaneous corneal inlay implantation and laser in situ keratomileusis for presbyopia in patients with hyperopia, myopia, or emmetropia: Six-month results, *J Cataract Refract Surg* 38 (2012): 495–506.

Tong, L., S. M. Saw, K. S. Chia, and D. Tan, Anisometropia in Singapore school children, *Am J Ophthalmol* 137 (2004): 474–479.

Torricelli, A. A. and S. E. Wilson, Cellular and extracellular matrix modulation of corneal stromal opacity, *Exp Eye Res* 129 (2014): 151–160.

Trevino, J. P., J. E. Gomez-Correa, D. R. Iskander, and S. Chavez-Cerda, Zernike vs. Bessel circular functions in visual optics, *Ophthalmic Physiol Opt* 33 (2013): 394–402.

van den Berg, T. J. and K. E. Tan, Light transmittance of the human cornea from 320 to 700 nm for different ages, *Vis Res* 34 (1994): 1453–1456.

van Dijk, K., K. Droutsas, J. Hou, S. Sangsari, V. S. Liarakos, and G. R. Melles, Optical quality of the cornea after Descemet membrane endothelial keratoplasty, *Am J Ophthalmol* 158 (2014): 71–79 e1.

Vasudevan, B., K. J. Ciuffreda, and B. Wang, Nearwork-induced changes in topography, aberrations, and thickness of the human cornea after interrupted reading, *Cornea* 26 (2007): 917–923.

Vasudevan, B., T. L. Simpson, and J. G. Sivak, Regional variation in the refractive-index of the bovine and human cornea, *Optom Vis Sci* 85 (2008): 977–981.

Villegas, E. A., E. Alcon, and P. Artal, Minimum amount of astigmatism that should be corrected, *J Cataract Refract Surg* 40 (2014): 13–19.

Vincent, S. J., D. Alonso-Caneiro, and M. J. Collins, Corneal changes following short-term minisceral contact lens wear, *Cont Lens Anterior Eye* 37 (2014): 461–468.

Vincent, S. J., M. J. Collins, S. A. Read, and L. G. Carney, Monocular amblyopia and higher order aberrations, *Vis Res* 66 (2012): 39–48.

Vincent, S. J., M. J. Collins, S. A. Read, L. G. Carney, and M. K. Yap, Interocular symmetry in myopic anisometropia, *Optom Vis Sci* 88 (2011): 1454–1462.

Vincent, S. J., M. J. Collins, S. A. Read, L. G. Carney, and M. K. Yap, Corneal changes following near work in myopic anisometropia, *Ophthalmic Physiol Opt* 33 (2013): 15–25.

Voetz, S. C., M. J. Collins, and B. Lingelbach, Recovery of corneal topography and vision following opaque-tinted contact lens wear, *Eye Contact Lens* 30 (2004): 111–1117.

Walsh, G., The effect of mydriasis on the pupillary centration of the human eye, *Ophthalmic Physiol Opt* 8 (1988): 178–182.

Wang, B. Z. and D. Taranath, A comparison between the amblyopic eye and normal fellow eye ocular architecture in children with hyperopic anisometropic amblyopia, *J AAPOS* 16 (2012): 428–430.

Wang, J., D. Fonn, T. L. Simpson, L. Sorbara, R. Kort, and L. Jones, Topographical thickness of the epithelium and total cornea after overnight wear of reverse-geometry rigid contact lenses for myopia reduction, *Invest Ophthalmol Vis Sci* 44 (2003): 4742–4746.

Wang, L., E. Dai, D. D. Koch, and A. Nathoo, Optical aberrations of the human anterior cornea, *J Cataract Refract Surg* 29 (2003): 1514–1521.

Wang, L. and D. D. Koch, Anterior corneal optical aberrations induced by laser in situ keratomileusis for hyperopia, *J Cataract Refract Surg* 29 (2003): 1702–1708.

Wei, S., H. Song, and X. Tang, Correlation of anterior corneal higher-order aberrations with age: A comprehensive investigation, *Cornea* 33 (2014): 490–496.

Westheimer, G., Optical and motor factors in the formation of the retinal image, *J Opt Soc Am* 53 (1963): 86–93.

Wickremasinghe, S., P. J. Foster, D. Uranchimeg et al., Ocular biometry and refraction in Mongolian adults, *Invest Ophthalmol Vis Sci* 45 (2004): 776–783.

Williams, D., G. Y. Yoon, J. Porter, A. Guirao, H. Hofer, and I. Cox, Visual benefit of correcting higher order aberrations of the eye, *J Refract Surg* 16 (2000): S554–S559.

Wilson, M. A., M. C. Campbell, and P. Simonet, The Julius F. Neumueller Award in Optics, 1989: Change of pupil centration with change of illumination and pupil size, *Optom Vis Sci* 69 (1992): 129–136.

Wolffsohn, J. S., G. Bhogal, and S. Shah, Effect of uncorrected astigmatism on vision, *J Cataract Refract Surg* 37 (2011): 454–460.

Yang, Y., K. Thompson, and S. A. Burns, Pupil location under mesopic, photopic, and pharmacologically dilated conditions, *Invest Ophthalmol Vis Sci* 43 (2002): 2508–2512.

Yoon, G. Y. and D. R. Williams, Visual performance after correcting the monochromatic and chromatic aberrations of the eye, *J Opt Soc Am A Opt Image Sci Vis* 19 (2002): 266–275.

Young, T., On the mechanism of the eye, *Philos Trans R Soc Lond* 91 (1801): 23–88.

Zhan, M. Z., S. M. Saw, R. Z. Hong et al., Refractive errors in Singapore and Xiamen, China—A comparative study in school children aged 6 to 7 years, *Optom Vis Sci* 77 (2000): 302–308.

Zhang, F. J., Z. Zhou, F. L. Yu, Z. L. Lu, T. Li, and M. M. Wang, Comparison of age-related changes between corneal and ocular aberration in young and mid-age myopic patients, *Int J Ophthalmol* 4 (2011): 286–292.

Zhu, M., M. J. Collins, and A. C. Yeo, Stability of corneal topography and wavefront aberrations in young Singaporeans, *Clin Exp Optom* 96 (2013): 486–493.

The lens

Fabrice Manns, Arthur Ho, and Jean-Marie Parel

Contents

15.1 OVERVIEW

From a basic point of view, the crystalline lens of the eye (the "lens") can be described as a biconvex lens that provides approximately one-third (20D) of the total dioptric power of the eye. In reality, the lens is a much more complex optical system:

- The lens is aspheric and inhomogeneous, with a refractive index gradient (the "lens gradient") that plays a key role in determining its optical properties.
- The lens is a dynamic optical element that changes shape during accommodation. Changes in lens shape allow the eye to change focus.

- The lens continuously grows throughout life. Lens growth produces age-dependent changes in lens shape and refractive index gradient that affect the optical properties.
- The lens transmission progressively decreases with age. Increased scattering within the lens produces a loss of transparency that eventually leads to age-related cataract.

The primary optical function of the lens is to allow the eye to accommodate, but age-dependent changes in shape and gradient also play a key role in the optics of the eye over the entire lifespan. Lens growth in childhood is a key factor in the development of the refractive state of the eye. In adulthood, lens growth results in a progressive loss of balance between corneal and internal ocular aberrations that deteriorates retinal image quality.

Continued lens growth produces changes in the lens optical and mechanical properties that are ultimately also responsible for the progressive loss of the ability of the eye to accommodate, eventually leading to presbyopia.

A complete optical model of the lens that provides realistic predictions of the lens power and aberrations must to take into account the contribution of the gradient, it must be age-dependent, and it must be able to model accommodative changes. An additional challenge when studying the optics of the lens is that the location of the lens inside the eye behind the iris prevents direct measurements of the lens shape and optical properties.

This chapter describes the basic anatomy of the lens from an optical point of view, with a focus on lens shape and refractive index. The chapter concludes with a review of the lens optical properties focused on lens power and spherical aberration.

15.2 BASIC ANATOMY OF THE LENS

The following description is limited to aspects of the basic anatomy of the lens that are most relevant from a visual optics point of view. For more detailed descriptions, the reader is referred to textbooks of ocular anatomy and relevant review articles (Beebe et al., 2008, 2010).

15.2.1 LOCATION OF THE LENS INSIDE THE EYE

The crystalline lens is located in the posterior chamber of the eye, right behind the iris (see Figure 15.1).

The central zone of the iris is in contact with the anterior lens surface, and it glides on the anterior lens surface during pupil constriction or dilation. The lens is surrounded by aqueous humor. The posterior surface of the lens is separated from the anterior surface of the vitreous by a small space called Berger's space.

The lens is suspended in the eye by suspensory ligaments, or zonular fibers, which connect the lens to the ciliary body of the eye. Changes in tension of the zonular fibers during contraction and relaxation of the ciliary muscle induce the changes in lens shape that allow the eye to change focus during accommodation.

15.2.2 LENS ANATOMY

15.2.2.1 Lens geometry

The lens is shaped approximately like an oblate spheroid (a surface of revolution obtained by rotating an ellipse about its minor axis), with a minor axis located in the anterior to posterior direction (see Figure 15.2).

The anterior pole and posterior pole of the lens are, respectively, the most anterior point and the most posterior point on the lens surface. The lens equator is the contour formed by joining the most peripheral points of the lens surface. If we assume that the lens is rotationally symmetric, then the poles are the intersection points of the axis of rotational symmetry with the anterior and posterior surfaces, and the equator is the circle formed by the intersection of the lens surface with the plane perpendicular

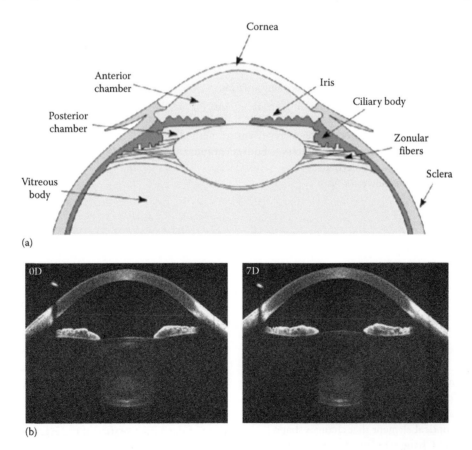

(a)

(b)

Figure 15.1 (a) Basic anatomy of the anterior segment showing the location of the lens and its relation to the accommodative structures. (b) Optical coherence tomography (OCT) images of the lens of a 24-year-old subject in the relaxed and accommodated (7D stimulus) state showing the relation between iris and anterior lens surface. (OCT images courtesy of Marco Ruggeri, PhD.)

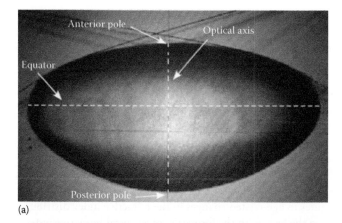

(a)

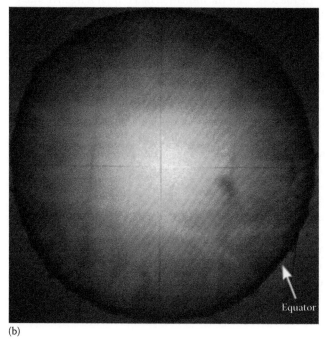

(b)

Figure 15.2 Sagittal (a) and coronal (b) shadow: Photographs of an isolated crystalline lens showing the general shape of the lens and the location of the equator, optical axis, anterior pole, and posterior pole. The equator is located slightly anterior to the midpoint between anterior and posterior poles.

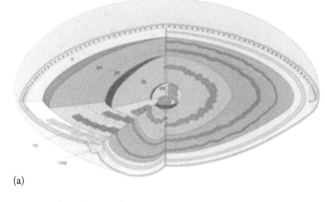

(a)

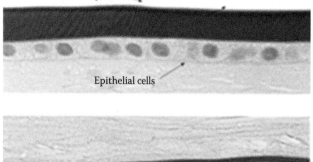

(b)

Figure 15.3 (a) Schematic representation of the lens structure, showing the lens capsule (cap), epithelium (ep), and lens fibers classified according to Vogt (c, cortex; an, adult nucleus; jn, juvenile nucleus; fn, fetal nucleus; and en, embryonic nucleus). The hexagonal shape of the lens fibers is shown in the cross-sectional representation on the left. (b) Histological images showing the anterior and posterior capsule near the poles, the single layer of epithelial cells beneath the anterior capsule, and the absence of epithelial cells in the posterior lens surface. (Sketch from Taylor, V.L. et al., *Invest. Ophthamol. Vis. Sci.*, 37(7), 1396, 1996; Histology image courtesy of Noel Ziebarth, PhD.)

to the axis of symmetry at the axial position where the lateral dimension of the lens is maximal. The lens equator is located slightly anterior to the midpoint between anterior and posterior poles (see Figure 15.2).

15.2.2.2 Lens structure

The lens is avascular and transparent. It consists of the lens capsule, lens epithelium, and lens fiber cells (see Figure 15.3).

The lens capsule is a thin acellular membrane that envelops the lens. The capsule plays an important role in the mechanics of accommodation by distributing the force produced by zonular tension across the lens surface. The capsule consists of extracellular matrix components, primarily type IV collagen. The human lens capsule thickness varies across the lens surface. The anterior lens capsule is thicker than the posterior lens capsule. The lens capsule is thickest at the anterior mid-periphery of the lens (Fincham, Barraquer). The thickness of the anterior human

lens capsule increases with age (Krag, Barraquer). The central thickness of the anterior capsule in the adult eye reported in recent studies ranges from approximately 8 to 25 μm on average (Krag, Ziebarth, Barraquer). The central thickness of the posterior human lens capsule is on the order of 5–10 μm (Krag, Ziebarth, Barraquer).

The lens epithelium is a single layer of cells located beneath the capsule in the anterior and equatorial regions of the lens. There are no epithelial cells in the posterior part of the lens. Epithelial cells located near the pole have a width of 10–20 μm and a thickness of 5–10 μm (Bron, Tasman). Epithelial cells become smaller toward the lens periphery. The most peripheral zone of epithelial cells, located just anterior to the lens equator, is called the germinative zone. Cells in the germinative zone slowly proliferate. New epithelial cells formed in the germinative zone migrate mostly posteriorly, and they differentiate into elongated lens fiber cells when they reach the lens equator (Bron et al., 2008).

Fiber cells elongate both in the anterior (apical) and posterior (basal) direction until they meet cells propagating from opposite sides of the lens. The junctions of lens fibers occur

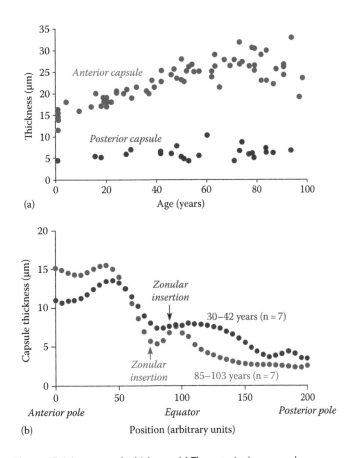

(a)

(b)

Figure 15.4 Lens capsule thickness. (a) The anterior lens capsule becomes thicker with age. The thickness of the posterior capsule is independent of age. (b) The thickness of the lens capsule varies across the lens, with a sharp decrease slightly anterior to the equator. The anterior capsule is thicker than the posterior capsule. (Graphs (a) From: Krag, S. and Andreassen, T.T., *Invest. Ophthalmol. Vis. Sci.*, 44(2), 691, 2003.; (b) From Barraquer, R.I. et al., *Invest. Ophthalmol. Vis. Sci.*, 47, 2053, 2006.)

along planes and form lines, called the lens sutures, which are visible upon examination of the lens (Bron et al., 2008) (see Figure 15.4).

Lens fiber cells are arranged in layers, with new fibers being formed on top of older fibers, beneath the epithelium in the anterior lens or beneath lens capsule in the posterior lens. Lens fiber cells have an hexagonal cross section with a thickness of approximately 0.5–2 μm and a width of 8–15 μm depending on their location within the lens (Bron et al., 1998; Augusteyn, 2010) (see Figure 15.4). As new fibers are formed in the superficial layer of the lens, there is compaction of older fibers located toward the center of the lens. There is evidence that fiber compaction contributes to age-related changes in the refractive index distribution inside the lens, with the formation of a central refractive index plateau (Augusteyn, 2010).

15.2.2.3 Cortex and nucleus

The internal region of the lens beneath the capsule and epithelial cells is generally classified into two distinct regions: the cortex and the nucleus (see Figure 15.5).

There are different definitions for these regions, but most commonly the nucleus is defined as the part of the lens that is present at birth (Augusteyn, 2010). With this definition, the cortex corresponds to the lens fibers that have been added to the lens after birth. The nucleus has a central thickness of 2–3 mm and an equatorial diameter on the order of 7 mm (Augusteyn, 2010). The cortex and nucleus of the lens can be further separated into subregions corresponding to zones of increased or decreased reflectivity (zones of discontinuity) that can be seen on cross-sectional optical images acquired for instance with a slit lamp or with optical coherence tomography (OCT) (Koretz et al., 1984; Dubbelman et al., 2003; Augusteyn, 2010; Ruggeri et al., 2012).

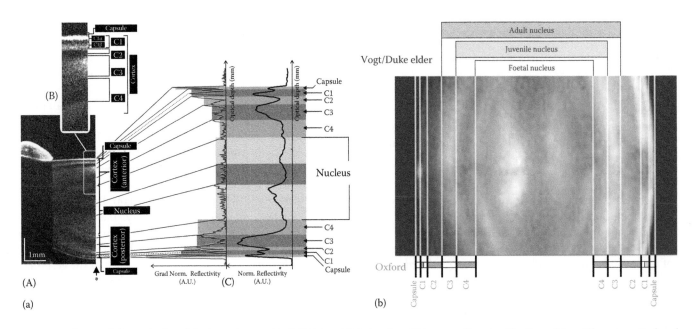

(a)

(b)

Figure 15.5 Zones of discontinuity of the lens observed in OCT (a) and Scheimpflug (b) images. The classification of the different cortical and nuclear zones is shown, following the Oxford or Vogt nomenclatures. (a): (A), OCT image of the lens; (B), Magnified view; and (C), OCT intensity profiles along the central A-line. (OCT image courtesy of Marco Ruggeri; Scheimpflug image from Augusteyn, 2010.)

15.3 LENS SHAPE AND DIMENSIONS

In optical models of the eye, the crystalline lens of the eye is generally represented as a thick spherical or aspheric lens characterized by the radius of curvature and asphericity of its anterior and posterior surfaces, its central thickness, and its diameter. The following section briefly reviews the techniques that have been employed to measure the lens shape and summarizes values obtained from key studies, including age dependence and changes with accommodation.

15.3.1 IN VIVO MEASUREMENT METHODS

In vivo measurements of the shape or dimensions of the lens have been acquired using optical techniques, ultrasound, and magnetic resonance imaging (MRI) (see Figure 15.6).

Optical techniques generally provide measurements with higher resolution and precision than ultrasound or MRI. However, due to strong absorption of optical radiation by the iris, optical techniques can only measure the central zone of the lens that is not hidden by the iris. In addition, optical measurements

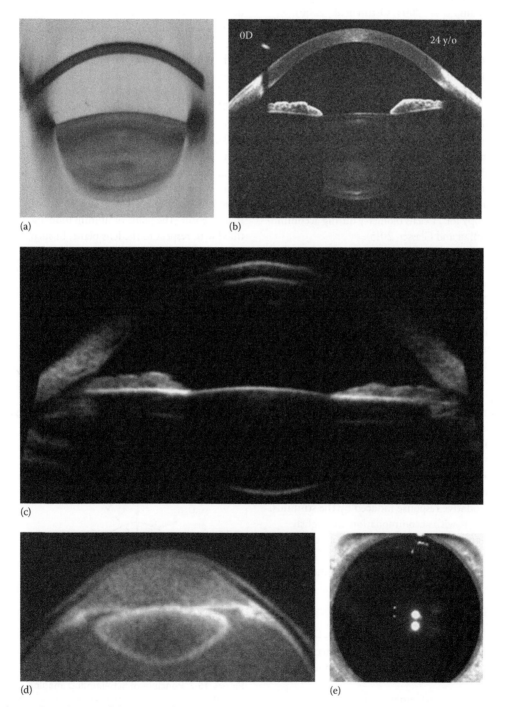

(a)

(b)

(c)

(d)

(e)

Figure 15.6 *In vivo* human lens shape and dimensions have been measured using several techniques: Scheimpflug imaging (a), OCT (b), ultrasound biomicroscopy (c), MRI (d), and phakometry (e). (Ultrasound image from Ramasubramanian, V. and Glasser, R., *J. Cataract Refract. Surg.*, 41, 511, 2015; Phakometry image from Rosales, P. et al., *J. Vis.*, 6 (10), 105710, 2006; Scheimpflug image from Koretz, J.F. et al., *J. Opt. Soc. Am. A, Opt. Image Sci. Vis.*, 19(1), 144, 2002; MRI image from Strenk, S.A. et al., *Invest. Ophthalmol. Vis. Sci.*, 40, 1162, 1999; OCT image from Ruggeri, M. et al., *Biomed. Opt. Express*, 3, 1506, 2012.)

of lens curvatures must be corrected for distortions due to refraction by the cornea for the anterior lens surface and by the cornea, anterior lens and lens structure for the posterior lens. In particular, there is some uncertainty in optical measurements of the posterior lens surface due to refractive distortions caused by the lens refractive index gradient. Optical techniques for lens biometry are discussed in more detail in the next section.

Ultrasound biometry (A-mode ultrasound) has been used to primarily measure lens thickness and its changes with accommodation (Shum et al., 1993; Beers and Van der Heijde, 1994, 1996; Garner and Yap, 1997; Kirschkamp et al., 2004; Ostrin et al., 2006). Since ultrasound energy penetrates through the iris, it is possible to image both the central and peripheral lens shape in two or three dimensions (B-mode or ultrasound biomicroscopy). However, it is difficult to obtain accurate measurements of lens curvature with ultrasound because current ultrasound imaging systems do not provide simultaneously the axial and lateral resolution, field of view, penetration, and depth of focus required for precise measurements of 2D or 3D lens shape. In addition, special care must be taken to ensure precise positioning and alignment of the acoustic transducer during image acquisition and the images must be corrected to take into account refraction of the ultrasound beam (Mateo et al., 2014). However, recent studies have shown the potential of high-frequency ultrasound biomicroscopy for 2D lens biometry during accommodation (Ramasubramanian and Glasser, 2015).

MRI is currently the only technique that is capable of producing distortion-free images of the entire lens. MRI images have been used to measure lens thickness, lens diameter, and radius of curvature (Strenk et al., 1999; Hermans et al., 2009; Kasthurirangan et al., 2011; Sheppard, 2011). Unfortunately, the spatial resolution of MRI is currently not sufficient to provide precise measurements of the lens radii of curvature. High cost, limited accessibility, and relatively long imaging times limit the applicability of MRI as a technique for large scale routine lens biometry studies.

One of the specific challenges faced when measuring the lens shape *in vivo* is that reliable measurement can only be obtained if the accommodation state of the eye is controlled. Measurements of the relaxed lens should ideally be performed under cycloplegia to ensure that accommodation is fully relaxed. Measurements of accommodation-induced changes in lens shape require a controlled accommodation stimulus and ideally a device to quantify the objective accommodative response induced by the stimulus. The changes in lens shape with accommodation are slightly overestimated if they are expressed in terms of the accommodation stimulus instead of the accommodative response, since the accommodative response is expected to be less than the accommodative stimulus (accommodation lag) (Hermans et al., 2008).

15.3.2 OPTICAL TECHNIQUES FOR *IN VIVO* LENS BIOMETRY

Optical techniques that have been used for lens biometry can be divided into three broad classes: phakometry, Scheimpflug imaging, and OCT.

In phakometry, the anterior and posterior lens radii of curvature are calculated from measurements of the relative magnification of virtual images of luminous objects formed by reflection from the anterior corneal surface, anterior lens surface, and posterior lens surface (Purkinje images I, III, and IV) (Rabbetts, 2007, pp. 411–414).

The first practical instrument that employed this principle to measure lens curvatures was the ophthalmometer developed by Helmholtz (1855). Helmholtz based his theory of accommodation on measurements of lens curvature in the relaxed and accommodated state using his ophthalmometer. Gullstrand adopted Helmholtz' values of the lens radius of curvature in his classic eye model (Helmholtz, 2005). Phakometry is a relatively simple method, but it calculates the lens radii of curvature with the assumption that the corneal and lens surfaces are spherical. Phakometry does not readily provide information on peripheral lens shape or asymmetry. Despite this limitation, due to its relative simplicity and low cost, phakometry has remained the most commonly used method to measure lens curvatures *in vivo* since the early measurements of Helmholtz (Tron, 1929; Mutti et al., 1992, 1998; Ooi, 1995; Garner, 1997; Goss et al., 1997; Van Veen and Goss, 1998; Kirschkamp et al., 2004; Garner et al., 2006; Rosales et al., 2006; Atchison, 2008; Richdale, 2013).

The first quantitative measurements of cross-sectional lens shape were obtained by Brown (1972, 1973; Koretz et al., 1984) using a slit lamp biomicroscope modified to satisfy the Scheimpflug condition. In the Scheimpflug imaging system, the anterior segment is illuminated at approximately normal incidence with a narrow slit beam of light. The diffuse reflection from the illuminated ocular structures is collected with a lens that produces images onto an image sensor. The plane of illumination and the image plane are tilted with respect to the lens plane, in such a way that the two tilted planes are perfectly conjugate with each other (Scheimpflug condition) (Scheimpflug, 1906) (see Figure 15.7).

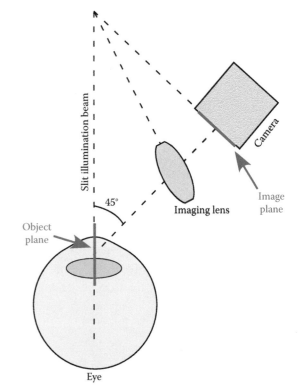

Figure 15.7 Principle of Scheimpflug imaging of the crystalline lens. A meridional plane of the crystalline lens is illuminated with a narrow slit-shaped beam. The diffuse reflection from the crystalline lens is collected with an objective lens and imaged on a camera. The planes of the crystalline lens (object plane) and camera (image plane) are tilted with respect to the plane of the objective. The three planes intersect. (From Dubbelman, M. et al., *Vis. Res.*, 41(14), 1867, 2001.)

This configuration ensures that the entire anterior segment, from the anterior corneal surface to the posterior lens surface is imaged sharply on the image sensor. However, this configuration results in a depth dependent magnification. Scheimpflug images must therefore be corrected for both keystone distortion and for refractive distortions to obtain accurate measurements of corneal and lens cross-sectional shape and thickness (Kampfer et al., 1989; Cook and Koretz, 1991, 1994a,b, 1998; Dubbelman et al., 2001, 2003, 2005; Dubbelman and Van der Heijde, 2001; Koretz, 2001, 2002; Hermans et al., 2007).

OCT is an optical imaging modality that relies on the principles of low-coherence interferometry to produce 2D or 3D images of tissue microstructure, with a resolution that typically ranges from 3 to 10 μm (Huang et al., 1991; Wojkowski, 2012; Drexler et al., 2014). Low-coherence interferometry can also be used to measure ocular distances along a single A-line (optical biometry), as an alternative to A-mode ultrasound biometry. Optical biometry based on low-coherence interferometry has been used to measure lens thickness and its changes with age and accommodation (Drexler et al., 1997; Boltz, 2007; Tsorbatzoglou et al., 2007; Alderson et al. 2012).

There are several commercially available OCT systems designed for anterior segment imaging. However, these devices currently have an imaging depth limited that is insufficient for the acquisition of anterior segment images that include the cornea and entire lens. In addition, the acquisition of 3D images of the anterior segment requires high speed to minimize motion artifacts. Several approaches have been implemented recently to produce customized OCT systems that provide sufficient imaging depth and speed to allow acquisition of complete anterior segment images including the cornea and lens (Grulkowski et al., 2009, 2012; Jungwirth et al., 2009; Zhou et al., 2009; Furukawa et al., 2010; Shen et al., 2010; Ruggeri et al., 2012; Satoh, 2012). Recent studies have demonstrated the use of OCT for the acquisition of 3D lens topography maps and for real-time *in vivo* imaging of changes in lens shape during accommodation (Ortiz, 2012; Gambra et al., 2013; Satoh et al., 2013). To produce accurate measurements, OCT images must be corrected for distortions caused by the scan geometry, distortions caused by refraction through the ocular surfaces, and distortions caused by eye motion (Ortiz, 2012). Given these advances and the expectation of further advances in imaging speed, imaging depth, and image analysis, it is very likely that OCT will become the gold standard for optical biometry of the lens.

15.3.3 *IN VITRO* MEASUREMENTS OF LENS SHAPE

Measurements of the lens shape and dimensions in two or three dimensions can also be acquired on lenses that have been extracted from cadaver eyes (see Figure 15.8).

Several methods have been used for such measurements, including photography techniques (Parker, 1972; Howcroft and Parker, 1977; Glasser and Campbell, 1999; Augusteyn et al., 2006; Rosen, 2006; Borja, 2008; Urs et al., 2009), corneal topography systems (Manns, 2004; Schachar, 2004), OCT (De Castro et al., 2010, 2011a,b; Maceo et al., 2011; Birkenfeld et al., 2013,

2014, 2015; Sun et al., 2014, Marussich et al., 2015), and MRI (Jones et al., 2005).

Since the isolated lens is free of external forces, its shape is expected to correspond to the shape of the *in vivo* lens when zonular tension is completely released (i.e., at maximal possible accommodation). *In vitro* measurements of accommodative changes in human lens shape and power have been obtained using mechanical systems that simulate accommodation by applying radial stretching forces to the lens (Pau, 1951; Fisher, 1977; Pierscionek, 1993, 1995; Glasser, 1998; Manns et al., 2007; Ehrmann et al., 2008; Michael et al., 2012; Pinilla Cortes et al., 2015). For these experiments, the lens is isolated from the eye together with the surrounding accommodative structures (zonules, ciliary body, and sometimes sclera).

Extraction of the lens from the eye eliminates the need to correct for distortions introduced by the cornea. It also allows measurements of the entire lens shape, including the peripheral regions that would otherwise be hidden by the iris during *in vivo* measurements using optical techniques. However, special care must be taken when using lenses from cadaver eyes to ensure that the shape of the lens is unaltered. Lenses must be kept immersed in an aqueous medium to avoid dehydration. They are subject osmotic swelling, particularly if they have been stored for extended periods of time before use or if the lens capsule has been damaged during dissection (Augusteyn et al., 2006). Dehydration and swelling can significantly alter the lens shape and optical properties. Some studies have relied on flash freezing to circumvent these issues (Parker, 1972; Howcroft and Parker, 1977). However, freezing may also alter the lens shape. Due to the difficulty in preserving the lens shape, data from studies using cadaver lenses are variable and sometimes of uncertain reliability.

15.3.4 BIOMETRIC DATA

15.3.4.1 Equatorial lens diameter (Tables 15.1 and 15.2)

Since the iris prevents measurements of lens diameter using optical techniques, there are limited *in vivo* data on lens diameter, all from studies using MRI. Most of the human lens diameter measurements published to date have been acquired on cadaver lenses and therefore corresponds to the diameter of the lens free of external forces (maximal accommodation). The lens equatorial diameter is age-dependent. It increases at a high rate during childhood and adolescence and remains approximately constant, with a much slower rate of increase, in adult eyes after the age of 30–40 years (see Figures 15.9 and 15.10).

15.3.4.2 Lens axial thickness (Tables 15.3 and 15.4)

Measurements of lens thickness and its changes with age and accommodation, both *in vitro* and *in vivo*, suggest that lens thickness decreases with age until adolescence and that it slowly increases with age in the adult eye (see Figure 15.9). Lens thickness is found to increases linearly with accommodation, at approximately 0.04 mm per diopter of accommodation, independent of age (see Figure 15.11). In general, there is good agreement between data from *in vivo* and *in vitro* studies. If we consider the lens cortex to include the lens fibers added after birth, then most of the change in lens thickness

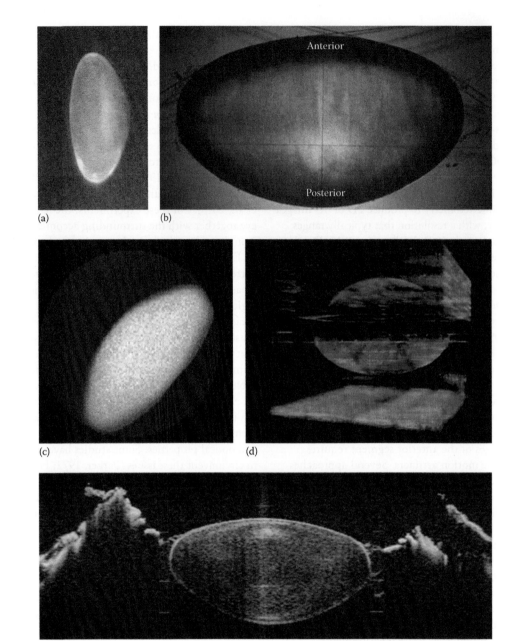

Figure 15.8 *In vitro* lens shape has been measured on isolated lenses or lenses mounted in lens stretching devices that simulate accommodation by applying radial stretching forces. Projection photography (a), shadow photography (b), T2-weigthed MRI (c), 3D OCT (d), and 2D OCT of a lens mounted in a lens stretching system (e). (Photography from Glasser, A. and Campbell, M.C.W., *Vis. Res.*, 39, 1991, 1999; Shadow photography from Urs, R. et al., *Vis. Res.*, 49(1), 74, 2009; MRI from Jones, C.E. et al., *Vis. Res.*, 45(18), 2352, 2005a; Jones, C.E., et al., *Vis. Res.*, 46, 2317, 2005b; 3D OCT from de Castro, A. et al., *Opt. Express*, 18(21), 21905, 2010.)

Table 15.1 **Results of selected studies providing values of *in vivo* human lens diameter change with accommodation**

STUDY	AGE RANGE (YEARS) AND # OF EYES	CHANGE WITH ACCOMMODATION
Strenk et al. (1999)	22–83 ($n = 25$)	Age-dependent (see Figure 15.10)
Jones et al. (2007)	18–33 ($n = 26$)	–0.067 ± 0.030 mm/D stimulus on average (range: –0.01 to –0.12 mm/D)
Kasthuriangan et al. (2008, 2011)	19–29 ($n = 15$)	–0.35 mm on average for a near stimulus ranging from 4.8D to 6.9D
Hermans et al. (2009)	18–35 ($n = 5$)	–0.074 ± 0.008 mm/D demand
Sheppard et al. (2011)	19–30 ($n = 19$)	–0.14 ± 0.17 mm/D on average for 4D demand (variable)
Richdale et al. (2013)	30–50 ($n = 26$)	–0.075 mm/D response

All results were obtained using MRI.

Table 15.2 **Results of selected studies providing the age dependence of the lens thickness in the adult human lens**

STUDY	METHOD	AGE RANGE (YEARS) AND # OF EYES	AGE DEPENDENCE OF LENS THICKNESS (mm)
Dubbelman et al. (2001)	Scheimpflug imaging, *in vivo*	16–65 (n = 90)	2.93 + 0.024 age
Atchison (2008)	A-mode ultrasound, *in vivo*	18–69 (n = 106)	3.13 + 0.0235 age
Rosen et al. (2006)	Shadow photography, *in vitro*	20–99 (n = 37)	3.97 + 0.0123 age
Birkenfeld et al. (2014)	OCT, *in vitro*	19–71 (n = 35)	3.52 + 0.0196 age

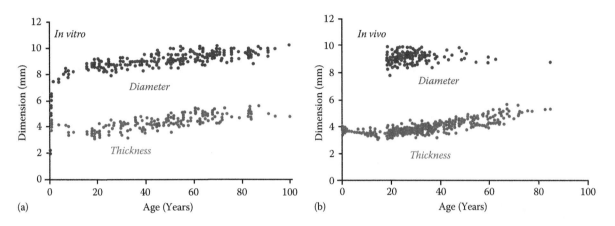

Figure 15.9 Lens thickness (blue circles) and lens diameter (orange circles) vary with age. Lens diameter increases rapidly in childhood and at a much slower rate in adulthood. Lens thickness decreases with age until adolescence and slowly increases with age in adulthood. The two plots show data compiled from multiple studies where lens dimensions were measured *in vitro* (a) and *in vivo* (b). (From Augusteyn, R.C., *Exp. Eye Res.*, 90, 643, 2010.)

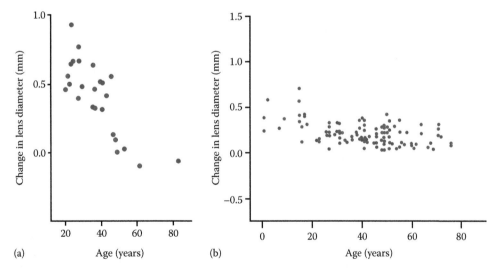

Figure 15.10 Changes in lens diameter during accommodation, plotted as a function of age. (a) The graph shows data acquired *in vivo* in response to an 8D stimulus using MRI. (b) The graph shows the change in diameter measured in a lens stretching system. ([a]: From Strenk, S.A. et al., *Invest. Ophthalmol. Vis. Sci.*, 40, 1162, 1999; [b]: From Augusteyn et al., *Exp. Eye Res.*, 2011.)

Table 15.3 **Results of selected studies providing the change in human lens thickness with accommodation**

STUDY	METHOD	AGE RANGE (YEARS) AND # OF EYES	CHANGE WITH ACCOMMODATION
Dubbelman et al. (2005)	Scheimpflug imaging	16–51 (n = 65)	0.045 mm/D stimulus
Ostrin et al. (2006)	A-mode ultrasound	21–30 (n = 22)	0.067 mm/D response
Richdale et al. (2013)	OCT	30–50 (n = 26)	0.064 mm/D response
Ramasubramanian and Glasser (2015)	MRI	21–36 (n = 26)	0.065 mm/D stimulus

Note: Results are from *in vivo* studies.

Table 15.4 Results of selected studies providing the change in the anterior and posterior lens radii of curvature with age

STUDY	METHOD	AGE RANGE (YEARS) AND # OF EYES	RADIUS OF CURVATURE (mm)
Koretz et al. (2001)	Scheimpflug imaging, *in vivo*	18–70 ($n = 100$)	Anterior: $11{\cdot}155 - 0{\cdot}02 * age$
Dubbelman et al. (2001)	Scheimpflug imaging, *in vivo*	16–65 ($n = 102$)	Anterior: $12.9 - 0.057\ age$ Posterior: $6.5 - 0.017\ age$ ($n = 42$)
Atchison et al. (2008)	Phakometry, *in vivo*	18–69 ($n = 106$)	Anterior: $12.3 - 0.044\ age$ Posterior: -6.86 ± 0.85 (no age dependence)
Richdale et al. (2013)	Phakometry, *in vivo*	30–50 ($n = 26$)	Anterior: $11.82 - 0.11\ age$ Posterior: 7.43 ± 0.46 (no age dependence)
Borja et al. (2008)	Shadow photography, *in vitro*	6–57 ($n = 34$)	Anterior: $4.46 + 0.14\ age$ (from 6 to 50 years) Posterior: $-3.47 - 0.06\ age$
Birkenfeld et al. (2014)	OCT, *in vitro*	19–71 ($n = 35$)	Anterior: $-0.00006\ age^3 + 0.0082\ age^2 - 0.2203\ age + 7.4769$ Posterior: $-0.0006\ age^2 + 0.073\ age + 3.5139$

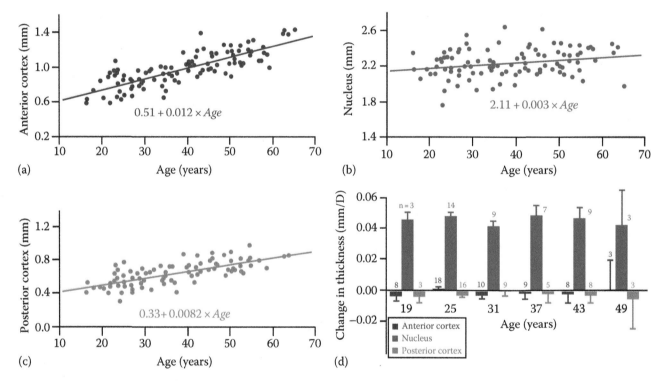

Figure 15.11 Changes in cortex and nucleus thickness with age (a) and with accommodation (b), measured with Scheimpflug imaging. The data show that in the adult eye, the increase in lens thickness with age is due to changes in increase in the thickness of the lens cortex. On the other hand, the increase in lens thickness with accommodation is due primarily to an increase in the thickness of the lens nucleus. The thickness of the lens cortex remains approximately constant with accommodation. (From Dubbelman, M. et al., *Vis. Res.*, 43(22), 2363, 2003.)

with age corresponds to a change in the thickness of the cortex (see Figure 15.11). On the other hand, most of the change in lens thickness during accommodation is caused by changes in the thickness of the nucleus. The thickness of the lens cortex remains approximately constant with accommodation (see Figure 15.11).

15.3.4.3 Lens curvatures (Tables 15.5 and 15.6)

In adulthood, at 0 diopter of accommodation (relaxed lens), both the anterior and posterior lens surfaces progressively become steeper with age. On the other hand, the anterior and posterior surfaces of the *in vivo* accommodated lens are found to progressively flatten with age (see Figure 15.12).

The *in vivo* finding is consistent with measurements on isolated lenses, which show that the anterior and posterior surfaces become flatter with age. There are limited data available for young lenses during childhood or adolescence, particularly *in vivo*.

The anterior lens curvature is found to increase approximately linearly with accommodation. The posterior lens curvature is found to undergo minimal changes with accommodation (see Figure 15.13).

In general, measurements of the posterior radius of curvature obtained *in vivo* using optical techniques are more uncertain, due to refractive distortions at the anterior lens surfaces and within the refractive index gradient of the lens.

Table 15.5 **Results of selected studies providing the change in the anterior and posterior lens radii of curvature and/or curvature with accommodation**

STUDY	METHOD	AGE RANGE (YEARS) AND # OF EYES	CHANGE WITH ACCOMMODATION (IN mm/D FOR RADIUS AND D/D FOR CURVATURE) A = ACCOMMODATION STIMULUS IN D
Koretz et al. (2002)	Scheimpflug imaging	18–70 ($n = 100$)	Anterior radius: $-0.4736 + 0.0047$ age in mm/D Anterior curvature: $2.903 - 0.03976$ age in D/D Posterior radius: $0.2778 - 0.004375$ age in mm/D Posterior curvature: $-2.426 + 0.0372$ age D/D
Dubbelman et al. (2005)	Scheimpflug imaging	16–51 ($n = 65$)	Anterior curvature: 6.7D/D stimulus Posterior curvature: 3.7D/D stimulus
Ramasubramanian and Glasser (2015)	MRI	21–36 ($n = 26$)	Anterior radius: $0.1058 A^2 - 1.4953 A + 0.8607$ Posterior radius: $0.0157 A^2 - 0.2971 A + 0.1694$

Note: Results are from *in vivo* studies.

Table 15.6 **Results of selected studies providing values of the anterior and posterior lens asphericity**

STUDY	METHOD	AGE RANGE (YEAR) AND # OF EYES	SHAPE FACTOR VALUE ($p = 1 + Q$)
Dubbelman et al. (2001)	Scheimpflug, *in vivo*	16–65 ($n = 90$)	Anterior: $-5.4 + 0.03$ age Posterior: $-5.0 + 0.07$ age ($n = 41$)
Manns et al. (2004)	Stereophotogrammetry	46–93 ($n = 26$)	Anterior: 4.27 ± 2.01 Posterior: -0.64 ± 1.85
Birkenfeld et al. (2014)	OCT, *in vitro*	19–71 ($n = 35$)	Anterior: $-14.8 + 0.23$ age Posterior: $-1.48 + 0.0275$ age

Note: The value provided is for the shape factor of a conic or conicoid fit of the lens surfaces. (Shape factor, $p = 1 + Q$, where Q is the asphericity.)

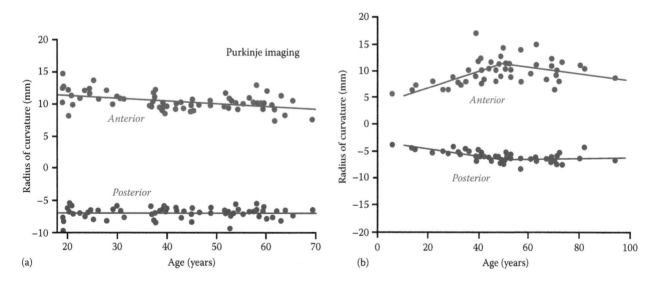

Figure 15.12 Age dependence of the human lens radii of curvature measured *in vivo* in the relaxed state (a) and *in vitro* (b). In the adult eye, *in vivo*, the anterior lens surface progressively becomes steeper with age. The radius of the posterior surface is independent of age. *In vitro*, both surfaces progressively become flatter with age, until around the age of 50, after which there is no clear age dependence. The isolated lens free of zonular tension corresponds to the maximal possible accommodative state. (*In vivo* data from Atchison, D.A. et al., *J. Vis.*, 8(4), 29.1, 2008. *In vitro* data from Borja, D. et al., *Invest. Ophthalmol. Vis. Sci.*, 49(6), 2541, 2008.)

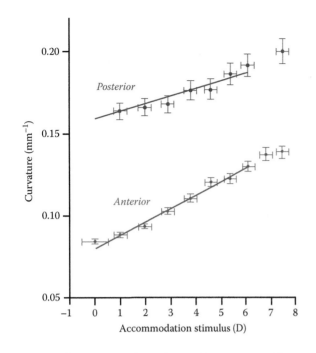

Figure 15.13 Change in anterior and posterior curvature with accommodation stimulus for a 29-year-old subject measured using Scheimpflug imaging. (From Dubbelman, M. et al., *Vis. Res.*, 45(1), 117, 2005.)

15.3.4.4 Lens asphericity

It is difficult to obtain precise and reliable measurements of lens asphericity *in vivo* because the peripheral regions are hidden by the iris and refractive distortions by the cornea and/or anterior lens produce uncertainty in the measurements. On the other hand, there is also a large variability in the published data acquired *in vitro*, probably due to issues with preservation of lens shape or due to difficulties in acquiring measurements of lens shape with sufficient resolution and precision. Generally, the anterior lens surface is found to be elliptical and the posterior lens surface is found to be hyperbolic, but there is a large variability within studies and between studies. There are limited and conflicting data available on changes of lens asphericity with age or accommodation.

15.4 LENS REFRACTIVE INDEX GRADIENT

The crystalline lens of the eye is an inhomogeneous optical element with a refractive index that increases from the lens surface (outer cortex) to the lens center (nucleus) (see Figure 15.14).

The contours (in two dimensions) or surfaces (in three dimensions) that join the points of equal refractive index are called the iso-indicial contours or surfaces (see Figure 15.14). The refractive index gradient is steepest in the cortex. In the nucleus, the refractive index is approximately uniform. This section briefly summarizes the techniques that have been used to characterize the refractive index gradient and the values that have been obtained. A comprehensive discussion of the refractive index gradient and its implications for the optics of the crystalline lens can be found in the review article by Piersconiek and Regini (2012).

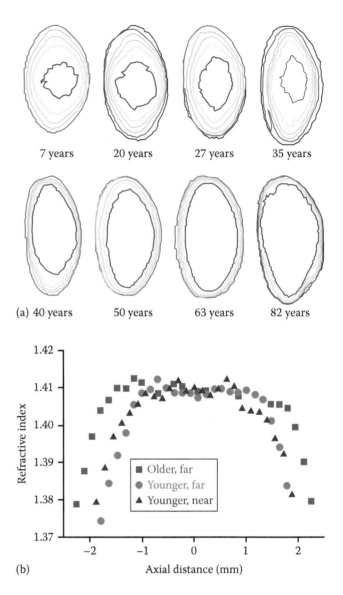

Figure 15.14 (a) Iso-indicial contours of isolated human lenses of different age reconstructed from T2-weighted MRI images. The contours correspond to the iso-indicial shells with refractive index ranging from 1.36 to 1.43 in 0.01 increments. (b) Refractive index profile along the optical axis reconstructed from an *in vivo* MRI image. Squares: Older subject far viewing; Triangles: Younger subject near viewing; Circles: Younger subject far viewing. ([a]: From Jones C.E. et al., *Vis. Res.*, 45(18), 2352, 2005a; Jones, C.E., et al., *Vis. Res.*, 46, 2317, 2005b; [b]: From Kasthurirangan, S. et al., *Invest. Ophthalmol. Vis. Sci.*, 49, 2531, 2008.)

15.4.1 MEASUREMENT OF THE LENS REFRACTIVE INDEX

The first reliable measurements of the lens refractive index were obtained using Abbe refractometers (Huggert, 1948; Nakao et al., 1969). In his classic eye model, Gullstrand used values that were published in 1907 by Freytag, who measured the refractive index of cadaver lenses using a refractometer and found values of 1.386 for the outer cortex and 1.406 for the lens center (Hemlholtz, 1924b). The Abbe refractometer provides a direct measurement of the refractive index, but it is a destructive technique that requires dissection and sometimes frozen sectioning of the lens. Dissection and freezing may alter the tissue properties.

The refractive index distribution within the lens can also be estimated from measurements of the spatial variation of the relative protein or water content within the lens (Philipson, 1969; van Heyningen, 1972; Fisher and Pettet, 1973; Fagerholm et al., 1981; McEwan and Farnsworth, 1987; Deussen and Pau, 1989; Siebinga et al., 1991; De Korte et al., 1994). The concentration of protein in the human lens increases from approximately 15%–20% (0.15–0.20 g/cm³) in the outer cortex, close to the lens surfaces, to approximately 35%–40% (0.35–0.40 g/cm³) in the nucleus (Fagerholm et al., 1981; Siebinga, 1991). The refractive index, n, can be estimated from the protein concentration, C (in g/cm³), by assuming a linear relationship:

$$n = n_w + \alpha \times C \qquad (15.1)$$

where

n_w is the refractive index of water

α is the refractive index increment, approximately equal to 0.19–0.20 g/cm³ for lens proteins (Pierscionek et al., 1987; Zhao et al., 2011; Pierscionek and Regini, 2012)

The refractive index increment is equal to the difference in refractive index between protein and water. Using a value of $n_w = 1.335$ (refractive index of saline, Pierscionek et al., 1987), this formula gives a range of refractive index of 1.364–1.385 for the outer cortex (15%–25% protein concentration) and 1.402–1.415 for the lens nucleus (35%–40% protein concentration).

Recently, several nondestructive methods have been implemented to derive values of the lens refractive index in different species, including laser ray tracing (LRT) (Campbell and Hughes, 1981; Campbell, 1984; Axelrod et al., 1988; Chan et al., 1988; Pierscionek et al., 1988, 2005; Pierscionek, 1989; Perscionek and Chan, 1989; Acosta et al., 2005; Vazquez et al., 2006), fiber-optic reflectometry (Pierscionek, 1993, 1994, 1995), MRI (Moffat et al., 2002; Jones et al., 2004, 2005; Kasthurirangan et al., 2008), OCT (Verma et al., 2007; De Castro et al., 2010, 2011a,b, 2013; Siedlecki et al., 2012; Birkenfeld et al., 2013, 2014), and Talbot x-ray imaging (Hoshino et al., 2011; Bahrami et al., 2014, 2015; Pierscionek et al., 2015). These methods do not provide a direct measurement of the refractive index. They rely on measurements of parameters that are combined with calibration curves or optical models to derive values of the refractive index.

LRT (Campbell and Hughes, 1981; Campbell, 1984; Axelrod et al., 1988; Chan et al., 1988; Pierscionek et al., 1988, 2005; Pierscionek, 1989; Perscionek and Chan, 1989; Acosta et al., 2005; Vazquez et al., 2006) relies on the measurement of the exit slopes of narrow laser beams that are transmitted through the lens. The measured slopes are compared with theoretical predictions calculated assuming a model of the refractive index distribution using approximations of the ray path. The parameters of the refractive index gradient model are found by minimizing the error between the theoretical predictions and measurements.

Fiber optic reflectometry (Pierscionek, 1993, 1994, 1995) uses a small fiber optic probe that is placed in contact with the lens surface or inserted into the lens. The power returning from the probe is measured. The system is calibrated to provide a value

of the reflectance at the fiber-lens interface. The lens refractive index is calculated from the reflectance measurement by using the Fresnel equations.

The application of MRI to produce refractive index maps relies on a calibration curve that relates refractive index to the spin–spin relaxation time (Moffat et al., 2002; Jones et al., 2004, 2005) (see Figure 15.15).

The calibration curve can be produced by measuring the relaxation time and refractive index of homogenized protein solutions of known concentration. So far, MRI is the only technique that has been used to reconstruct the refractive index gradient *in vivo* (Kasthurirangan et al., 2008) (see Figure 15.15).

Reconstruction of the gradient from OCT images is based on inverse methods similar to those used with LRT. OCT images are maps of the optical path length of rays propagating through the lens. The optical path lengths measured along each A-line are compared with theoretical predictions calculated assuming a

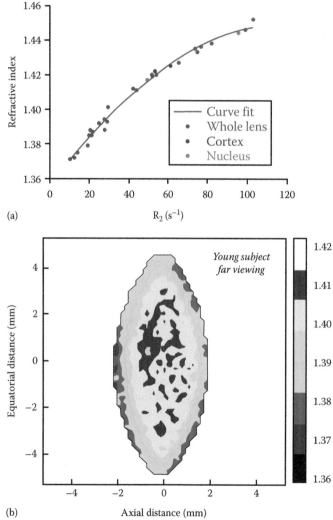

Figure 15.15 Calibration curve of refractive index as a function of the T2 (spin–spin) relaxation time constant (a) and example of 2D refractive index maps acquired *in vivo* (b). ([a]: From Jones C.E. et al., *Vis. Res.*, 45(18), 2352, 2005a; Jones, C.E., et al., *Vis. Res.*, 46, 2317, 2005b; [b]: Kasthurirangan, S. et al., *Invest. Ophthalmol. Vis. Sci.*, 49, 2531, 2008.)

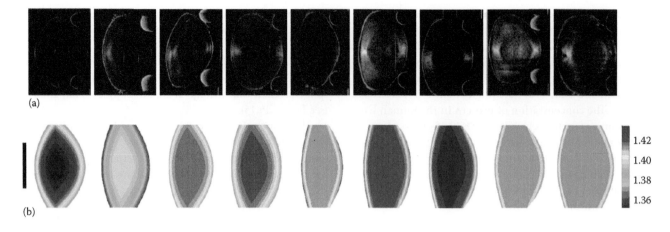

(a)

(b)

Figure 15.16 Refractive index maps of isolated lenses (a) reconstructed from OCT images (b). (From de Castro et al., 2009.)

model of the refractive index distribution. The parameters of the refractive index gradient model are found by minimizing the difference between the optical path lengths obtained experimentally from the OCT images and the values derived from computational ray traces (Verma et al., 2007; De Castro et al., 2010, 2011a,b, 2013; Siedlecki et al., 2012; Birkenfeld et al., 2013, 2014) (see Figure 15.16).

X-ray Talbot interferometry is a special microtomography technique that provides tomographic differential phase shift images. Each pixel of the image corresponds to a phase shift. The phase shift is converted into a value of protein volume fraction by using a calibration equation. A second calibration equation is then used to convert protein concentration into refractive index (Hoshino et al., 2011; Bahrami et al., 2014, 2015; Piersioneck et al., 2015).

Overall, all of the techniques that have been used to measure the refractive index all suffer from a relatively large variability and/or uncertainty. An accuracy of ±0.005 in the values of the index corresponds to a measurement accuracy better than 0.4%, which is very difficult to achieve experimentally. The estimates of cortex and nucleus index obtained from measurements of the concentration and specific index of lens proteins may be the most reliable values available to date.

15.4.2 SUMMARY OF VALUES: EFFECTS OF AGE AND ACCOMMODATION

Along the optical and equatorial axes of the lens, the variation of the refractive index from the lens surface to the lens center is commonly modeled with a power function of the form (JMO; Smith et al., 1991, 1992; De Castro et al., 2001; Jones et al., 2005; Kasthurirangan et al., 2008; Birkenfeld et al., 2013; Bahrami et al., 2014):

$$n_0(r) = n_N - (n_N - n_C) \times \left(\frac{r}{t}\right)^p \qquad (15.2)$$

where

- r is the position along the axis ($r = 0$ at the lens center)
- t is the length of the semiaxis from the lens center to the lens surface
- n_C is the refractive index of the outer cortex (at $r = t$)
- n_N is the refractive index at the center of the lens (at $r = 0$)

The coefficient p determines the steepness of the gradient. Typical distributions produced with this model are shown in Figure 15.17.

A value of $p = 2$ corresponds to a parabolic profile. Larger values of p correspond to flatter central profiles.

For human lenses, published values of the refractive index values range from $n_C = 1.36$ to 1.38 for the outer cortex and from $n_N = 1.40$ to 1.43 for the nucleus. The difference between nucleus and cortex index ranges from 0.04 to 0.05. The refractive index of the cortex and nucleus are generally found to be constant with age and with accommodation.

Values of the power coefficient, p, are variable, reflecting the challenge in obtaining precise measurements of the refractive index profile. The power coefficient is generally found to range from 2 to 4 in young lenses along the optical axis. There is evidence that the values of the power coefficient along the

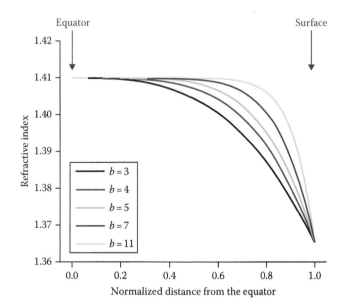

Figure 15.17 Axial refractive index profile corresponding to the formula of Equation 15.2. In this calculation, the equatorial index is $n_N = 1.41$ and the outer cortical index is $n_C = 1.365$. The value of the power coefficient b is varied from 3 to 11. As the value of the power coefficient increases, the refractive index becomes more uniform in the center of the lens (central refractive index plateau).

axial and equatorial directions increase with age (Siebinga et al., 1991; Jones et al., 2005; Kasthurirangan et al., 2008; de Castro et al., 2011a,b) that corresponds to the progressive formation of a refractive index plateau in the central region of the lens (Augusteyn, 2008). This finding is consistent with the compaction theory of lens fibers. The changes in the shape of the gradient with age are believed to play a key role in the age dependence of lens power and aberrations. However, recent studies *in vitro* study using OCT or Talbot x-ray interferometry for gradient reconstruction found a constant axial power coefficient with age (Bahrami et al., 2014; Birkenfeld et al., 2014). Further studies are needed to better characterize the refractive index profile and its age dependence.

Studies *in vivo* using MRI (Kasthurirangan et al., 2008) suggest that there are only small changes in the power coefficient of the axial or equatorial refractive index profile during accommodation.

15.5 LENS POWER

Lens power is one of the ocular components of refraction, together with corneal power, axial eye length, and anterior chamber depth. The paraxial lens power is determined by the anterior and posterior lens curvatures, lens thickness, and refractive index gradient.

The power of the adult lens in its relaxed state is generally considered to be around 20D, close to the value used by Gullstrand in his eye model, or approximately one-third of the total power of the eye. More recent studies suggest that the lens power progressively decreases with age throughout life, from an average of approximately 40D in full-term infants to 24D in young adults and 22D in older adults (Jones et al., 2005; Mutti et al., 2005; Olsen et al., 2007; Atchison et al., 2008; Irribaren et al., 2014; Irribaren, 2015; Jongenelen et al., 2015; Li et al., 2015) (see Figure 15.18).

15.5.1 MEASUREMENT OF LENS POWER *IN VIVO*

Given that the lens is located inside the eye behind the cornea, it is not possible to directly measure its power *in vivo*. The *in vivo* lens power can be calculated either from measurements of ocular distances, corneal power, and refractive error or from measurements of lens curvature and thickness.

15.5.1.1 Calculation from ocular biometry (Bennett method)

To date, most of the *in vivo* data on lens power available in the literature have been acquired using an indirect method that relies on measurement of corneal power, ocular distances, and refractive error. Using a paraxial model of the eye, it is relatively straightforward to find the value of lens power that, combined with the measured ocular parameters, places the conjugate of the retina at the far point of the eye. Assuming that the refractive indices of aqueous and vitreous are equal to 1.336, the lens power is given by

$$L = \frac{1.336}{s'} - \frac{1.336}{s} \tag{15.3}$$

where

  s' is the distance from the image principal plane of the lens to the retina

  s is the distance from the object principal plane of the lens to the primary image of the far point formed by the cornea

  (s, s' are both positive and in units of meters)

In its simplest form, the method assumes that the eye is a system of two thin lenses (cornea and lens) separated by a distance equal to the anterior chamber depth (Senström method, Rozema et al., 2011).

This method has been refined by Bennett (1988). The method of Bennett relies on measurement of corneal power, anterior chamber

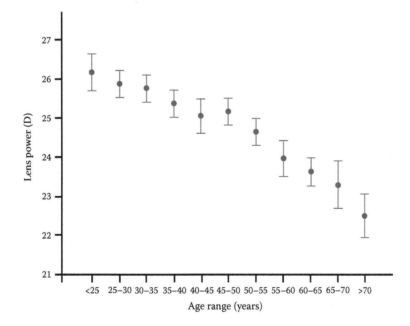

Figure 15.18 Age dependence of lens power in a study on 1069 adults of age ranging from 20 to 85 years showing a progressive decrease in lens power with age, from an average of around 26D in the age group of 20–25 years to an average of around 22.5D in the age group of 70–85 years. (From Jongenelen, S. et al., *Invest. Ophthalmol. Vis. Sci.*, 56, 7026, 2015.)

depth, lens thickness, axial eye length, and the subject's refraction. The calculation of lens power relies on an estimate of the position of the principal planes of the lens relative to the lens vertices. The location of the principal planes of the lens is determined by assuming that the ratio of anterior to posterior lens radius is a constant, equal to the value used in the Emsley–Gullstrand model. Once the location of the principal planes is found, it is straightforward to calculate the object and image distances for the lens (s and s') if refractive error, corneal power, anterior chamber depth, lens thickness, and axial eye length are known. The accuracy of the Bennett method in estimating the lens power is on the order of ±1D (Bennett, 1988; Dunne et al., 1989; Royston et al., 1989; Rozema et al., 2011). The Bennett method relies on two constants that provide the position of the anterior and posterior principal planes in terms of lens thickness. Modified versions of the Bennett method have been developed using different values of the constants to increase prediction accuracy or a thin lens approximation to allow lens power calculations when the lens thickness is not available (Rozema et al., 2011; Hernandez et al., 2015).

To date, the Bennett method has been the most common method to calculate lens power *in vivo*, particularly in large scale studies, because it relies on measurements that can be acquired using standard clinical devices (keratometry, ultrasound or optical biometry, refraction). However, caution must be exercised when comparing lens power data from different studies. Differences in the approach used to calculate corneal power can produce large differences in the value of lens power. The original Bennett formula uses the keratometric index (1.3375) to estimate corneal power from the radius of the anterior corneal surface. Some studies used an index of 1.332 or 1.3315. For a cornea with an anterior radius of curvature of 7.7 mm and a posterior radius of 6.3 mm, corresponding to a power of 42.5D, the corneal power predicted from keratometry is 43.8D if the index is 1.3375 and 43.1D if the index is 1.332. In an average emmetropic eye, the Bennett method will underestimate the lens power by approximately 2D when the index is 1.3375 and 1D when the index is 1.332. With modern anterior segment biometry devices, it is now possible to take into account the posterior corneal surface in the calculation of corneal power, which will help produce more accurate estimates of lens power (Hernandez et al., 2015). Another potential source of error is the difference in the position of the posterior boundary that is used as a reference for the measurement of axial eye length. Measurements acquired using ultrasound or commercial optical biometry systems use the inner limiting membrane as the reference. Axial eye length measured with these devices must be corrected by adding the retinal thickness (approximately 0.2 mm) to avoid overestimating the lens power (Li et al., 2015).

15.5.1.2 Calculation from lens biometry—Equivalent index

The *in vivo* lens power can be calculated from measurements of lens curvature and thickness obtained using phakometry or anterior segment imaging. In these calculations, the lens is generally assumed to be homogeneous, with an "equivalent index." The effective power of the equivalent lens is then calculated using the thick lens formula:

$$L = L_1 + L_2 - \frac{t}{n_{eq}} \times L_1 \times L_2 \qquad (15.4)$$

where L_1 and L_2 are the anterior surface power and posterior surface power, respectively. If we assume a refractive index of 1.336 for aqueous and vitreous, the anterior and posterior surface powers are given by

$$L_1 = \frac{n_{eq} - 1.336}{R_1} \qquad L_2 = \frac{1.336 - n_{eq}}{R_2} \qquad (15.5)$$

where R_1 and R_2 are the anterior and posterior radii, respectively ($R_1 > 0$ and $R_2 < 0$).

The equivalent index is an artificial value calculated so that the equivalent homogeneous lens has the same shape and power as the crystalline lens. The equivalent index can be calculated if the lens radii, thickness, and power are known. The equivalent index can be calculated *in vivo* by combining measurements of lens shape with calculations of lens power using the Bennett method (Mutti et al., 1995; Dubbleman et al., 2001; Jones et al., 2005; Mutti et al., 2005; Atchison et al., 2008). The equivalent index can also be calculated from measurements of lens shape and power acquired on isolated lenses (Borja et al., 2008; Birkenfeld et al., 2014). The equivalent index is generally found to be higher than the peak refractive index within the lens. Most studies also find that the equivalent index decreases with age in adult eyes (see Figure 15.19), with average values ranging from approximately 1.425 to 1.445 at 20 years of age and from 1.410 to 1.430 at 60 years of age, depending on the studies (Dubbelman et al., 2001; Atchison et al., 2008; Borja et al., 2008), even though a recent study found an average value of 1.415 with no age dependence (Birkenfeld et al., 2014).

There are less data available on age dependence of the equivalent index in childhood. The mean value in the age range from 6 to 18 years is around 1.425 (Mutti et al., 1995, 1998; Jones et al., 2005; Garner et al., 2006), with one study finding an increase with age (Garner et al., 2006), while another study found a decrease (Jones et al., 2005). In infants between the age of 3 and 18 months, the equivalent refractive index was found to range from around 1.450 to 1.460 on average (Wood et al., 1996; Mutti et al., 2005).

In his eye model, Gullstrand assumed that the equivalent index increased with accommodation. However, recent studies suggest that the equivalent index remains constant with accommodation (Garner, 1997; Hermans et al., 2008). The constancy of refractive index with accommodation implies that values the equivalent index obtained from isolated lenses, which are expected to be in a state of maximal accommodation, and from *in vivo* measurements, are expected to be equivalent.

Special caution must be exercised when comparing the values of lens power obtained in different studies using phakometry. Many studies use a fixed value of the equivalent index, generally equal to the value of the Gullstrand eye model (1.416) (Mutti et al., 1995; Ooi and Grosvenor, 1995; Jones et al., 2005). When this value of the equivalent index is used, the lens power is underestimated. The same caution must be exercised when comparing values of the equivalent index across studies. As discussed in the previous section, the lens power calculated using the Bennett method is dependent on the value of the keratometric index that is used for calculation of corneal power. When the standard keratometric index of 1.3375 is used, both the lens power and equivalent refractive index are underestimated.

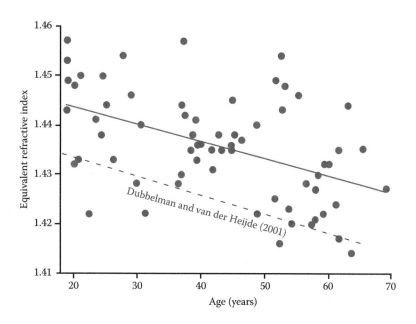

Figure 15.19 Age dependence of the lens equivalent index in adult eyes measured using phakometry. The equivalent lens index is generally found to progressively decrease with age. The dashed line shows values obtained by Dubbelman et al. using Scheimpflug imaging. (From Atchison, D.A. et al., *J. Vis.*, 8(4), 29.1, 2008.)

15.5.1.3 Comparison of methods

The Bennett method and phakometry provide comparable values of lens power (Dunne et al., 1987; Rozema et al., 2011). The Bennett method is used more commonly, particularly in large-scale studies, because there are currently no standard commercial devices that provide measurements of lens curvatures. In addition, the lens power calculated from lens curvature using Equations 15.4 and 15.5 is highly sensitive to errors in the value of the equivalent refractive index. A precision of ±0.5D in lens power requires a precision of ±0.002 in the value of the equivalent refractive index, which is difficult to achieve in practice (Hernandez et al., 2015). Due to this uncertainty, the Bennett method is generally thought to provide more precise measurements of lens power.

15.5.2 MEASUREMENT OF LENS POWER *IN VITRO*

Lens power or focal length can be measured directly on isolated postmortem lenses by adapting standard optical metrology techniques. Most *in vitro* studies of lens power rely on an adaptation of the LRT technique discussed in Section 15.4 (Glasser and Campbell, 1999; Jones et al., 2005; Borja et al., 2008; Birkenfeld et al., 2014) (see Figure 15.20).

The advantage of *in vitro* measurements is that they allow direct measurements of the lens power, unaffected by uncertainty in the measurements of lens dimensions or in the value of the refractive index. However, as with all other measurements on *in vitro* lenses, special care must be taken to avoid alteration of the lens properties, for instance due to swelling. In addition, the *in vitro* lens corresponds to the maximally accommodated state, free of all zonular tension. *In vitro* measurements can therefore not be compared with measurements obtained *in vivo* on relaxed lenses, expect in lenses from older donors who are of presbyopic age.

The power of the *in vitro* lens decreases with age at a rate that closely matches the loss of accommodative power (see Figure 15.21).

This observation is consistent with the fact that the lens taken out of the eye is in its most accommodated state.

15.5.3 LENS PARADOX

The concept of the "lens paradox" was introduced by Nicholas Brown, following his observations that the lens surfaces become steeper with age in adults (Koretz and Handelman, 1988). The steepening of the lens surfaces is expected to produce an increase in lens power that should produce a myopic shift of the adult eye. Yet, the refractive state of the adult eye is found to shift toward hyperopia, and the lens power is found to remain approximately constant with age. Brown called this discrepancy in the relation between surface curvature and lens power the "lens paradox."

Pierscionek hypothesized that the lens paradox could be explained by age-related changes in the refractive index gradient profile of the lens (Pierscionek, 1990, 1993). This hypothesis is consistent with the finding that the equivalent index of the lens decreases with age (Ooi and Grosvenor, 1995; Dubbelman et al., 2001; Atchison et al., 2008; Borja et al., 2008) and the observation of a progressive age-related formation of a refractive index plateau within the lens (Augusteyn, 2008).

15.6 LENS ABERRATIONS

The aberrations of the lens are determined by the shape of the lens surfaces and the refractive index gradient. Since the lens shape and refractive index gradient change as the lens continuously grow with age, it is reasonable to expect that the lens aberrations are age-dependent. Age-related changes in lens aberrations are the primary cause for the observed change in balance between corneal and internal aberration with age (Artal et al., 2002). Similarly, since the lens shape changes with accommodation, lens aberrations are also expected to be accommodation-dependent.

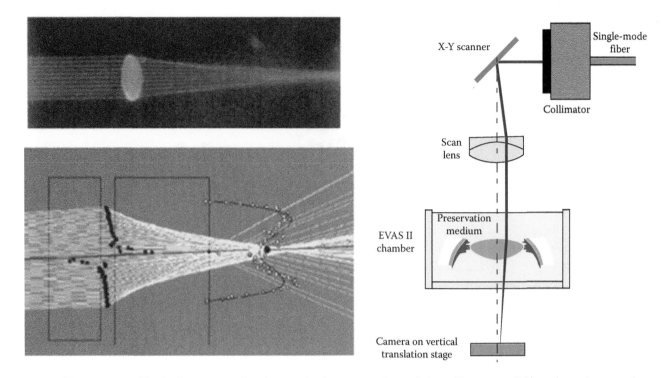

Figure 15.20 Measurement of *in vitro* lens power using the scanning laser ray tracing technique. Narrow parallel laser beam is scanned across the aperture of the lens. In the implementation on the left, the path of rays leaving the lens is recorded by taking photographs. The intersection of the rays with the optical axis is calculated in terms of the input ray height to provide the paraxial focal length and longitudinal spherical aberration. In the implementation on the right, the ray path or the position of best focus is found by recording spot diagrams at different axial positions by using a camera mounted on a translation stage beneath the lens. (Image on the left from Glasser and Campbell, 1999; Image on the right courtesy of Bianca Maceo.)

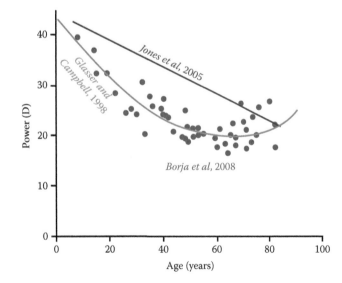

Figure 15.21 Power of the isolated human lens from three different studies (Borja et al., 2008; Jones et al., 2005; Glasser and Campbell, 1998). The isolated lens, which is free of zonular tension, is in the maximal possible accommodative state. The isolated lens power therefore decreases with age at a rate that approximates the loss of accommodative power. (From Borja, D. et al., *Invest. Ophthalmol. Vis. Sci.*, 49(6), 2541, 2008.)

Due to the position of the lens inside the eye behind the iris, it is not possible to obtain direct measurements of the lens aberrations *in vivo*. In principle, lens aberrations could be calculated from *in vivo* measurements of the lens curvature, asphericity, and thickness (Smith and Atchison, 2001) or from 3D lens topography (Ortiz et al., 2012), but these calculations would require a knowledge of the refractive index gradient within the lens. In addition, the location of the lens inside the eye behind the iris makes it challenging to obtain accurate measurements of the asphericity or peripheral shape of the anterior and posterior lens surfaces.

The aberrations of the lens inside the eye are generally estimated by subtracting corneal aberrations from the total ocular aberrations (El Hage and Berny, 1973; Tomlinson et al., 1993; Artal and Guirao, 1998; Artal et al., 2001, 2002; Smith et al., 2001; Kelly et al., 2004) or by measuring the ocular aberrations after cancelling the refractive power of the cornea by immersing the corneal surface in aqueous media (Millodot and Sivak, 1979; Artal and Guirao, 1998; Artal et al., 2001). These methods yield the internal aberrations of the eye, which also include the posterior corneal surface aberrations. Data from *in vivo* studies show that the spherical aberration of the lens is negative (Tomlison et al., 1993; Smith et al., 2001), and that it has a compensatory effect for the positive spherical aberration of the anterior corneal surface (El Hage and Berny, 1973; Artal et al., 2001, 2002; Smith et al., 2001; Kelly et al., 2004). The lens spherical aberration is found to become less negative with age (Smith et al., 2001).

To date, direct measurements of lens aberrations have only been obtained on isolated lenses, primarily using LRT (Sivak and Kreuzer, 1983; Glasser and Campbell, 1998, 1999; Roorda and Glasser, 2004; Birkenfeld et al., 2013, 2014, 2015; Maceo Heilman et al., 2015), but also using point-diffraction interferometry (Acosta et al., 2009, 2010; Gargallo et al., 2013).

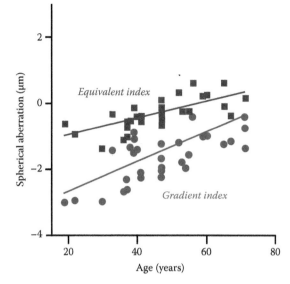

Figure 15.22 Spherical aberration (Zernike coefficient Z_{40}) of isolated lenses calculated using a ray trace through a model of the lens reconstructed from OCT images. The squares are values obtained using a homogeneous model of the lens with a refractive index equal to the equivalent index. The circles are values obtained using the refractive index gradient reconstructed from the OCT images. The lens spherical aberration is found to be negative in young adult lenses. It becomes less negative with age. Since the data were acquired on isolated lenses free of zonular tension, the results are expected to correspond to the *in vivo* lens in its accommodated state. (From Birkenfeld, J. et al., *Invest. Ophthalmol. Vis. Sci.*, 55, 2599, 2014.)

Results form *in vitro* studies are generally consistent with the *in vivo* findings. The lens is found to have negative spherical aberration that becomes less negative with age (see Figure 15.22).

Experiments on human and monkey lenses also show that the lens aberrations become more negative with accommodation (De Castro et al., 2013; Maceo Heilman et al., 2015). Recent studies on isolated lenses also suggest that the refractive index gradient makes a significant contribution to the aberrations of the lens and its changes with age (Birkenfeld et al., 2013, 2014, 2015; De Castro et al., 2013). If the gradient is replaced with a homogeneous index, the spherical aberration becomes positive or significantly less negative (see Figure 15.22).

REFERENCES

Acosta E, Bueno JM, Schwarz C, Artal P. Relationship between wave aberrations and histological features in ex vivo porcine crystalline lenses. *Journal of Biomedical Optics* 2010; 15 (5): 055001.

Acosta E, Vázquez D, Castillo LR. Analysis of the optical properties of crystalline lenses by point-diffraction interferometry. *Ophthalmic and Physiological Optics* 2009; 29: 235–246.

Acosta E, Vazquez D, Garner L, Smith G. Tomographic method for measurement of the gradient refractive index of the crystalline lens. I. The spherical fish lens. *Journal of the Optical Society of America A* 2005; 22 (3): 424–433.

Alderson A, Davies LN, Mallen EAH, Sheppard AL. A method for profiling biometric changes during disaccommodation. *Optometry and Vision Science* 2012; 89 (5): E738–E748.

Artal P, Berrio E, Guirao A, Piers P. Contribution of the cornea and internal surfaces to the change of ocular aberrations with age. *Journal of the Optical Society of America A* 2002; 19: 137–143.

Artal P, Guirao A. Contributions of the cornea and lens to the aberrations of the human eye. *Optics Letters* 1998; 23: 1713–1715.

Artal P, Guirao A, Berrio E, Williams DR. Compensation of corneal aberrations by the internal optics of the human eye. *Journal of Vision* 2001; 1 (1): 1–8.

Atchison DA, Markwell EL, Kasthurirangan S, Pope JM, Smith G, Swann PG. Age-related changes in optical and biometric characteristics of emmetropic eyes. *Journal of Vision* 2008; 8 (4): 29.1–29.20.

Augusteyn RC. Growth of the lens: In vitro observations. *Clinical and Experimental Optometry* 2008; 91: 226–239.

Augusteyn RC. On the growth and internal structure of the lens. *Experimental Eye Research* 2010; 90: 643–654.

Augusteyn RC, Rosen AM, Borja D, Ziebarth NM, Parel JM. Biometry of primate lenses during immersion in preservation media. *Molecular Vision* 2006; 12 (7): 740–747.

Axelrod D, Lerner D, Sands PJ. Refractive index within the lens of a goldfish eye determined from the paths of thin laser beams. *Vision Research* 1988; 28: 57–65.

Bahrami M, Hoshino M, Pierscionek B, Yagi N, Regini J, Uesugi K. Optical properties of the lens: An explanation for the zones of discontinuity. *Experimental Eye Research* 2014; 124: 93–99.

Bahrami M, Hoshino M, Pierscionek B, Yagi N, Regini J, Uesugi K. Refractive index degeneration in older lenses: A potential functional correlate to structural changes that underlie cataract formation. *Experimental Eye Research* 2015; 140: 19–27.

Barraquer RI, Michael R, Abreu R, Lamarca J, Tresserra F. Human lens capsule thickness as a function of age and location along the sagittal lens perimeter. *Investigative Ophthalmology and Visual Science* 2006; 47: 2053–2060.

Beebe DC. The lens. In *Adler's Physiology of the Eye*, 10th edn. Kaufman PL and Alm A, eds. 2003, Chapter 5, pp. 117–153.

Beers APA, Van der Heijde GL. In vivo determination of the biomechanical properties of the component elements of the accommodation mechanism. *Vision Research* 1994; 34 (21): 2897–2905.

Beers APA, Van der Heijde GL. Age-related changes in the accommodation mechanism. *Optometry and Vision Science* 1996; 73 (4): 235–242.

Bennett AG. A method of determining the equivalent powers of the eye and its crystalline lens without resort to phakometry. *Ophthalmic and Physiological Optics* 1988; 8: 53–59.

Birkenfeld J, de Castro A, Marcos S. Contribution of shape and gradient refractive index to the spherical aberration of isolated human lenses. *Investigative Ophthalmology and Visual Science* 2014; 55: 2599–2607.

Birkenfeld J, de Castro A, Marcos S. Astigmatism of the ex vivo human lens: Surface and gradient refractive index age-dependent contributions. *Investigative Ophthalmology and Visual Science* 2015; 56: 5067–5073.

Birkenfeld J, de Castro A, Ortiz S, Pascual D, Marcos S. Contribution of the gradient refractive index and shape to the crystalline lens spherical aberration and astigmatism. *Vision Research* 2013; 86: 27–34.

Bolz M, Prinz A, Drexler W, Findl O. Linear relationship of refractive and biometric lenticular changes during accommodation in emmetropic and myopic eyes. *British Journal of Ophthalmology* 2007; 91: 360–365.

Borja D, Manns F, Ho A, Ziebarth N, Rosen AM, Jain R, Amelinckx A, Arrieta E, Augusteyn RC, Parel JM. Optical power of the isolated human crystalline lens. *Investigative Ophthalmology and Visual Science* 2008; 49 (6): 2541–2548.

Bron AJ, Tripathi RC, Tripathu BJ. The lens and zonules. In *Wolff's Anatomy of the Eye and Orbit*, 8th edn. Chapman & Hall, 1998, Chapter 12, pp. 411–435.

Brown N. Quantitative slit-image photography of the lens. *Transactions of the Ophthalmological Societies of the United Kingdom* 1972; 92: 303–307.

Brown N. The change in shape and internal form of the lens of the eye on accommodation. *Experimental Eye Research* 1973; 15: 441–459.

Campbell MCW, Hughes A. An analytic gradient index schematic lens and eye for the rat that predicts aberrations for finite pupil. *Vision Research* 1981; 21: 1129–1148.

Chan DYC, Ennis J, Pierscionek B, Smith G. Determination and modelling of the 3 D gradient refractive indices in crystalline lenses. *Applied Optics* 1988; 27: 926–931.

Clinical and Experimental Optometry 2013; 96: 486–493.

Cook CA, Koretz JF. Acquisition of the curves of the human crystalline lens from slit lamp images: An application of the Hough transform. *Applied Optics* 1991; 30 (16): 2088–2099.

Cook CA, Koretz JF. Methods to obtain quantitative parametric descriptions of the optical surfaces of the human crystalline lens from Scheimpflug slit-lamp images. I. Image processing methods. *Journal of the Optical Society of America A, Optics, Image, Science, and Vision* 1998; 15 (6): 1473–1485.

Cook CA, Koretz JF, Pfahnl A, Hyun J, Kaufman PL. Aging of the human crystalline lens and anterior segment. *Vision Research* 1994; 34 (22): 2945–2954.

de Castro A, Barbero S, Ortiz S, Marcos S. Accuracy of the reconstruction of the crystalline lens gradient index with optimization methods from ray tracing and Optical Coherence Tomography data. *Optics Express* 2011a; 19 (20): 19265–19279.

de Castro A, Birkenfeld J, Maceo B, Manns F, Arrieta E, Parel JM, Marcos S. Influence of shape and gradient refractive index in the accommodative changes of spherical aberration in nonhuman primate crystalline lenses. *Investigative Ophthalmology and Visual Science* 2013; 54 (9): 6197–6207.

de Castro A, Ortiz S, Gambra E, Siedlecki D, Marcos S. Three-dimensional reconstruction of the crystalline lens gradient index distribution from OCT imaging. *Optics Express* 2010; 18 (21): 21905–21917.

de Castro A, Siedlecki D, Borja D, Uhlhorn S, Parel JM, Manns F, Marcos S. Age-dependent variation of the Gradient Index profile in human crystalline lenses. *Journal of Modern Optics* 2011b; 58: 1781–1787.

de Korte CL, van der Steen AF, Thijssen JM, Duindam JJ, Otto C, Puppels JG. Relation between local acoustic parameters and protein distribution in human and porcine eye lenses. *Experimental Eye Research* 1994; 59: 617–627.

Deussen A, Pau H. Regional water content of clear and cataractous human lenses. *Ophthalmic Research* 1989; 21: 374–380.

Drexler W, Baumgartner A, Findl O, Hitzenberger CK, Fercher AF. Biometric investigation of changes in the anterior eye segment during accommodation. *Vision Research* 1997; 37 (19): 2789–2800.

Drexler W, Liu M, Kumar A, Kamali T, Unterhuber A, Leitgeb RA. Optical coherence tomography today: Speed, contrast, and multimodality. *Journal of Biomedical Optics* 2014; 19 (7): 071412.

Dubbelman M, Van der Heijde GL. The shape of the aging human lens: Curvature, equivalent refractive index and the lens paradox. *Vision Research* 2001; 41 (14): 1867–1877.

Dubbelman M, van der Heijde GL, Weeber HA. The thickness of the aging human lens obtained from corrected Scheimpflug images. *Optometry and Vision Science* 2001; 78 (6): 411–416.

Dubbelman M, Van der Heijde GL, Weeber HA. Change in shape of the aging human crystalline lens with accommodation. *Vision Research* 2005; 45 (1): 117–132.

Dubbelman M, Van der Heijde GL, Weeber HA, Vrensen GF. Changes in the internal structure of the human crystalline lens with age and accommodation. *Vision Research* 2003; 43 (22): 2363–2375.

Dunne MCM, Barnes DA, Royston JM. An evaluation of Bennett's method for determining the equivalent powers of the eye and its crystalline lens without resort to phakometry. *Ophthalmic and Physiological Optics* 1989; 9: 69–71.

Ehrmann K, Ho A, Parel J-M. Biomechanical analysis of the accommodative apparatus in primates. *Clinical and Experimental Optometry* 2008; 91: 302–312.

El Hage S, Berny F. Contribution of the crystalline lens to the spherical aberration of the eye. *Journal of the Optical Society of America* 1973; 63: 205–211.

Fincham EF. The mechanism of accommodation. *British Journal of Ophthalmology* 1937 (suppl. 8): 1–80.

Fisher RF, Pettet BE. Presbyopia and the water content of the human crystalline lens. *Journal of Physiology* 1973; 234: 443–447.

Fisher RF. The force of contraction of the human ciliary muscle during accommodation. *Journal of Physiology* 1977; 270: 51–74.

Fagerholm PP, Philipson BT, Lindstrom B. Normal human lens—The distribution of protein. *Experimental Eye Research* 1981; 33: 615–620.

Furukawa H, Hiro-Oka H, Satoh N, Yoshimura R, Choi D, Nakanishi M, Igarashi A, Ishikawa H, Ohbayashi K, Shimizu K. Full-range imaging of eye accommodation by high-speed long-depth range optical frequency domain imaging. *Biomedical Optics Express* 2010; 1 (5): 1491–1501.

Gambra E, Ortiz S, Perez-Merino P, Gora M, Wojtkowski M, Marcos S. Static and dynamic crystalline lens accommodation evaluated using quantitative 3-D OCT. *Biomedical Optics Express* 2013; 4: 1595–1609.

Gargallo A, Arines J, Acosta E. Lens aberrations and their relationship with lens sutures for species with Y-suture branches. *Journal of Biomedical Optics* 2013; 18 (2): 25003.

Garner LF. Calculation of the radii of curvature of the crystalline lens surfaces. *Ophthalmic and Physiological Optics* 1997; 17 (1): 75–80.

Garner LF, Smith G. Changes in equivalent and gradient refractive index of the crystalline lens with accommodation. *Optometry and Vision Science* 1997; 74 (2): 114–119.

Garner LF, Stewart AW, Owens H, Kinnear RF, Frith MJ. The Nepal longitudinal study: Biometric characteristics of developing eyes. *Optometry and Vision Science* 2006; 83: 274–280.

Garner LF, Yap MK. Changes in ocular dimensions and refraction with accommodation. *Ophthalmic and Physiological Optics* 1997; 17: 12–17.

Glasser A, Campbell MCW. Presbyopia and the optical changes in the human crystalline lens with age. *Vision Research* 1998; 38: 209–229.

Glasser A, Campbell MCW. Biometric, optical and physical changes in the isolated human crystalline lens with age in relation to presbyopia. *Vision Research* 1999; 39: 1991–2015.

Goss DA, Van Veen HG, Rainey BB, Feng B. Ocular components measured by keratometry, phakometry, and ultrasonography in emmetropic and myopic optometry students. *Optometry and Vision Science* 1997; 74 (7): 489–495.

Grulkowski I, Gora M, Szkulmowski M, Gorczynska I, Szlag D, Marcos S, Kowalczyk A, Wojtkowski M. Anterior segment imaging with Spectral OCT system using a high-speed CMOS camera. *Optics Express* 2009; 17 (6): 4842–4858.

Grulkowski I, Liu JJ, Potsaid B, Jayaraman V, Lu CD, Jiang J, Cable AE, Duker JS, Fujimoto JG. Retinal, anterior segment and full eye imaging using ultrahigh speed swept source OCT with vertical-cavity surface emitting lasers. *Biomedical Optics Express* 2012; 3 (11): 2733–2751.

Helmholtz H. Uber die Akkommodation des Auges. *Graefes Archiv fuer Ophthalmology* 1855; 1: 1–74.

Helmholtz H. *Treatise on Physiological Optics*, Vol. I, Southall JPC, ed. 1924. Dover Publications, Mineola, NY, 2005, pp. 337, 344.

Hermans E, Dubbelman M, van der Heijde R, Heethaar R. The shape of the human lens nucleus with accommodation. *Journal of Vision* 2007; 7(10): 16.1–16.10.

Hermans EA, Dubbelman M, Van der Heijde R, Heethaar RM. Equivalent refractive index of the human lens upon accommodative response. *Optometry and Vision Science* 2008; 85: 1179–1184.

Hermans EA, Pouwels PJW, Dubbelman M, Kuijer JPA, van der Heijde RGL, Heethaar RM. Constant volume of the human lens and decrease in surface area of the capsular bag during accommodation: An MRI and Scheimpflug study. *Investigative Ophthalmology & Visual Science* 2009; 50: 281–289.

Hernandez VM, Cabot F, Ruggeri M, De Freitas C, Ho A, Yoo S, Parel JM, Manns F. Calculation of crystalline lens power using a modification of the Bennett method. *Biomedical Optics Express* 2015; 6: 4501–4515.

Heyningen R. The human lens III: Some observations on the post-mortem lens. *Experimental Eye Research* 1972; 13: 155–160.

Hoshino M, Uesugi K, Yagi N, Mohri S, Regini J, Pierscionek B. Optical properties of in situ eye lenses measured with X-ray Talbot interferometry: A novel measure of growth processes. *PLoS One* 2011; 6 (9): e25140.

Howcroft MJ, Parker JA. Aspheric curvatures for the human lens. *Vision Research* 1977; 17: 1217–1223.

Huang D, Swanson EA, Lin CP et al. Optical coherence tomography. *Science* 1991; 254: 1178–1181.

Huggert A. Chapter IV: The correlation between zones of optical discontinuity and iso-indicial surfaces. *Acta Ophthalmologica* 1948; S30: 30–94.

Irribaren R. Crystalline lens and refractive development. *Progress in Retinal and Eye Research* 2015; 47: 86–106.

Irribaren R, Morgan IG, Hashemi H, Khabazkhoob M, Emamian MH, Shariati M, Fotouhi A. Lens power in a population-based cross-sectional sample of adults aged 40 to 64 years in the Shahroud Eye Study. *Investigative Ophthalmology and Visual Science* 2014; 55: 1031–1039.

Jones CE, Atchison DA, Pope JM. Changes in lens dimensions and refractive index with age and accommodation. *Optometry and Vision Science* 2007; 84: 990–995.

Jones CE, Atchison DA, Meder R, Pope JM. Refractive index distribution and optical properties of the isolated human lens measured using magnetic resonance imaging (MRI). *Vision Research* 2005a; 45 (18): 2352–2366.

Jones LA, Mitchell L, Mutti DO, Hayes JR, Moeschberger ML, Zadnik K. Comparison of ocular component growth curves among refractive error groups in children. *Investigative Ophthalmology and Visual Science* 2005b; 46: 2317–2327.

Jones CE, Pope JM. Measuring optical properties of an eye lens using magnetic resonance imaging. *Magnetic Resonance Imaging* 2004; 22: 211–220.

Jongenelen S, Rozema JJ, Tassignon MJ, on behalf of the EVICR.net and Project Gullstrand Study Group. Distribution of crystalline lens power in vivo as a function of age. *Investigative Ophthalmology and Visual Science* 2015; 56: 7026–7035.

Jungwirth J, Baumann B, Pircher M, Götzinger E, Hitzenberger CK. Extended in vivo anterior eye-segment imaging with full-range complex spectral domain optical coherence tomography. *Journal of Biomedical Optics* 2009; 14 (5): 050501.

Kampfer A, Wegener A, Dragomirescu V, Hockwin O. Improved biometry of the anterior segment. *Ophthlamic Research* 1989; 21: 239–248.

Kasthurirangan S, Markwell EL, Atchison DA, Pope JM. In vivo study of changes in refractive index distribution in the human crystalline lens with age and accommodation. *Investigative Ophthalmology and Visual Science* 2008; 49: 2531–2540.

Kasthurirangan S, Markwel EL, Atchison DA, Pope JM. MRI study of the changes in crystalline lens shape with accommodation and aging in humans. *Journal of Vision* 2011; 11 (3): 1–16.

Kelly JE, Mihashi T, Howland HC. Compensation of corneal horizontal/vertical astigmatism, lateral coma, and spherical aberration by internal optics of the eye. *Journal of Vision* 2004; 4 (4): 262–271.

Kirschkamp T, Dunne M, Barry JC. Phakometric measurement of ocular surface radii of curvature, axial separations and alignment in relaxed and accommodated human eyes. *Ophthalmic and Physiological Optics* 2004; 24: 65–73.

Kleiman NJ, Worgul BV. Lens. In *Duane's Foundations of Clinical Ophthalmology*, Vol. 1. Tasman W and Jaeger EA, eds. Lippincott Williams & Williams, 2001, Chapter 15.

Koretz JF, Cook CA. Aging of the optics of the human eye: Lens refraction models and principal plane locations. *Optometry and Vision Science* 2001; 78 (6): 396–404.

Koretz JF, Cook CA, Kaufman PL. Aging of the human lens: Changes in lens shape at zero-diopter accommodation. *Journal of the Optical Society of America, A, Optics, Image, Science and Vision* 2001; 18 (2): 265–272.

Koretz JF, Cook CA, Kaufman PL. Aging of the human lens: Changes in lens shape upon accommodation and with accommodative loss. *Journal of the Optical Society of America A, Optics, Image, Science and Vision* 2002; 19 (1): 144–151.

Koretz JF, Cook CA, Kuszak JR. The zones of discontinuity in the human lens: Development and distribution with age. *Vision Research* 1994; 34 (22): 2955–2962.

Koretz JF, Handelman GH. How the human eye focuses. *Scientific American* 1988; 259 (7): 92–97.

Koretz JF, Handelman GH, Brown NP. Analysis of human crystalline lens curvature as a function of accommodative state and age. *Vision Research* 1984; 24 (10): 1141–1151.

Krag S, Andreassen TT. Mechanical properties of the human posterior lens capsule. *Investigative Ophthalmology & Visual Science* 2003; 44 (2): 691–696.

Krag S, Olsen T, Andreassen TT. Biomechanical characteristics of the human anterior lens capsule in relation to age. *Investigative Ophthalmology and Visual Science* 1997; 38: 357–363.

Li SM, Wang N, Zhou Y et al. Paraxial schematic eye models for the 7- and 14- year-old Chinese children. *Investigative Ophthalmology and Visual Science* 2015; 56: 3577–3583.

Maceo BM, Manns F, Borja D, Nankivil D, Uhlhorn S, Arrieta E, Ho A, Augusteyn RC, Parel JM. Contribution of the crystalline lens gradient refractive index to the accommodation amplitude in non-human primates: In vitro studies. *Journal of Vision* 2011; 13 (11): 23.

Maceo Heilman B, Manns F, de Castro A, Durkee H, Arrieta E, Marcos S, Parel JM. Changes in monkey crystalline lens spherical aberration during simulated accommodation in a lens stretcher. *Investigative Ophthalmology and Visual Science* 2015; 56 (3): 1743–1750.

Manns F, Parel JM, Denham D et al. Optomechanical response of human and monkey lenses in a lens stretcher. *Investigative Ophthalmology and Visual Science* 2007; 48 (7): 3260–3268.

Marussich L, Manns F, Nankivil D, Maceo Heilman B, Yao Y, Arrieta-Quintero E, Ho A, Augusteyn R, Parel JM. Measurement of crystalline lens volume during accommodation in a lens stretcher. *Investigative Ophthalmology and Visual Science* 2015; 56: 4239–4248.

Mateo T, Chang A, Mofid Y, Pisella PJ, Ossant F. Axial ultrasound B-scans of the entire eye with a 20-MHz linear array: Correction of crystalline lens phase aberration by applying Fermat's principle. *IEEE Transactions on Medical Imaging* 2014; 33 (11): 2149–2166.

McEwan JR, Farnsworth PN. Regional resistivity variations in lens homogenates. *Experimental Eye Research* 1987; 44 (4): 567–576.

Michael R, Mikielewicz M, Gordillo C, Montenegro GA, Pinilla Cortés L, Barraquer RI. Elastic properties of human lens zonules as a function of age in presbyopes. *Investigative Ophthalmology and Visual Science* 2012; 53 (10): 6109–6114.

Millodot M, Sivak J. Contribution of the cornea and lens to the spherical aberration of the eye. *Vision Research* 1979; 19: 685–687.

Moffat BA, Atchison DA, Pope JM. Age-related changes in refractive index distribution and power of the human lens as measured by magnetic resonance micro-imaging in vitro. *Vision Research* 2002; 42: 1683–1693.

Mutti DO, Zadnik K, Adams AJ. A video technique for phakometry of the human crystalline lens. *Investigative Ophthalmology and Visual Science* 1992; 33 (5): 1771–1782.

Mutti DO, Zadnik K, Adams AJ. The equivalent refractive index of the crystalline lens in childhood. *Vision Research* 1995; 35: 1565–1573.

Mutti DO, Zadnik K, Fusaro RE, Friedman NE, Sholtz RI, Adams AJ. Optical and structural development of the crystalline lens in childhood. *Investigative Ophthalmology and Visual Science* 1998; 39: 120–133.

Mutti DO, Mitchell L, Jones LA, Friedman NE, Frane SI, Lin WK, Moeschberger ML, Zadnik K. Axial growth and changes in lenticular and corneal power during emmetropization in infants. *Investigative Ophthalmology and Visual Science* 2005; 46: 3074–3080.

Nakao S, Ono T, Nagata R, Iwata K. Model of refractive indices in the human crystalline lens. *Japanese Journal of Ophthalmology* 1969; 23: 903–906.

Ooi CS, Grosvenor T. Mechanisms of emmetropization in the aging eye. *Optometry and Vision Science* 1995; 72 (2): 60–66.

Olsen T, Arnarsson A, Sasaki H, Sasaki K, Jonasson F. On the ocular refractive components: The Reykjavik Eye Study. *Acta Ophthalmologica Scandinavica* 2007; 85: 361–365.

Ortiz S, Pérez-Merino P, Gambra E, de Castro A, Marcos S. In vivo human crystalline lens topography. *Biomedical Optics Express* 2012; 3 (10): 2471–2488.

Ostrin L, Kasthurirangan S, Win-Hall D, Glasser A. Simultaneous measurements of refraction and A-scan biometry during accommodation in humans. *Optometry and Vision Science* 2006; 83: 657–665.

Parker, JA. Aspheric optics of the human lens. *Canadian Journal of Ophthalmology* 1972; 7: 168–175.

Pau H. Dependence of the shape of the lens on physical factors. *Ophthalmologica* 1951; 122: 308–314.

Philipson B. Distribution of protein within the normal rat lens. *Investigative Ophthalmology* 1969; 8: 258–270.

Pierscionek B, Bahrami M, Hoshino M, Uesugi K, Regini J, Yagi N. The eye lens: Age-related trends and individual variations in refractive index and shape parameters. *Oncotarget* 2015; 6 (31): 30532–30544.

Pierscionek B, Smith G, Augusteyn RC. The refractive increments of bovine α-, β- and γ- crystallins. *Vision Research* 1987; 27 (9): 1539–1541.

Pierscionek BK. Growth and ageing effects on the refractive index in the equatorial plane of the bovine lens. *Vision Research* 1989; 29: 1759–1766.

Pierscionek BK. Presbyopia—Effect of refractive index. *Clinical and Experimental Optometry* 1990; 73: 23–30.

Pierscionek BK. What we know and understand about presbyopia. *Clinical and Experimental Optometry* 1993a; 76: 83–90.

Pierscionek BK. In vitro alteration of human lens curvatures by radial stretching. *Experimental Eye Research* 1993b; 57: 629–635.

Pierscionek BK. Surface refractive index of the eye lens determined with an optic fibre sensor. *Journal of the Optical Society of America A* 1993c; 10: 1867–1871.

Pierscionek BK. Refractive index of the human lens surface measured with an optic fibre sensor. *Ophthalmic Research* 1994; 26: 32–36.

Pierscionek BK. Age-related response of human lenses to stretching forces. *Experimental Eye Research* 1995a; 60: 325–332.

Pierscionek BK. The refractive index along the optic axis of the bovine lens. *Eye* 1995b; 9: 776–782.

Pierscionek BK, Belaidi A, Bruun HH. Refractive index gradient in the porcine lens for 532 and 633 nm light. *Eye* 2005; 19: 375–381.

Pierscionek BK, Chan DYC. The refractive index gradient of the human lens. *Optometry and Vision Science* 1989; 66: 822–829.

Pierscionek BK, Chan DYC, Ennis JP, Smith G, Augusteyn RC. A non-destructive method of constructing three-dimensional gradient index models for crystalline lenses: I. Theory and experiment. *American Journal of Optometry and Physiological Optics* 1988; 65: 481–491.

Pierscionek BK, Regini JW. The gradient index lens of the eye: An opto-biological synchrony. *Progress in Retinal and Eye Research* 2012; 31: 332–349.

Pinilla Cortés L, Burd HJ, Montenegro GA, D'Antin JC, Mikielewicz M, Barraquer RI, Michael R. Experimental protocols for ex vivo lens stretching tests to investigate the biomechanics of the human accommodation apparatus. *Investigative Ophthalmology and Visual Science* 2015; 56 (5): 2926–2932.

Rabbetts RB. *Clinical Visual Optics*, 4th edn. Elsevier, Philadelphia, PA, 2007, pp. 411–414.

Ramasubramanian V, Glasser A. Objective measurement of accommodative biometric changes using ultrasound biomicroscopy. *Journal of Cataract & Refractive Surgery* 2015; 41: 511–526.

Richdale K, Sinnott LT, Bullimore MA et al. Quantification of age-related and per diopter accommodative changes of the lens and ciliary muscle in the emmetropic human eye. *Investigative Ophthalmology and Visual Science* 2013; 54 (2): 1095–1105.

Roorda A, Glasser A. Wave aberrations of the isolated crystalline lens. *Journal of Vision* 2004; 4 (4): 250–261.

Rosales P, Dubbelman M, Marcos S, van der Heijde R. Crystalline lens radii of curvature from Purkinje and Scheimpflug imaging. *Journal of Vision* 2006; 6 (10): 105710–105767.

Rosen AM, Denham DB, Fernandez V, Borja D, Ho A, Manns F, Parel JM, Augusteyn RC. In vitro dimensions and curvatures of human lenses. *Vision Research* 2006; 46: 1002–1009.

Royston JM, Dunne MCM, Barnes DA. Calculation of crystalline lens radii without resort to phakometry. *Ophthalmic and Physiological Optics* 1989; 9: 412–414.

Rozema JJ, Atchison DA, Tassignon MJ. Comparing methods to estimate the human lens power. *Investigative Ophthalmology and Visual Science* 2011; 52: 7937–7942.

Ruggeri M, Uhlhorn S, De Freitas C, Ho A, Manns F, Parel JM. Imaging and full-length biometry of the eye during accommodation using spectral-domain OCT with an optical switch. *Biomedical Optics Express* 2012; 3: 1506–1520.

Satoh N, Shimizu K, Goto A, Igarashi A, Kamiya K, Ohbayashi K. Accommodative changes in human eye observed by Kitasato anterior segment optical coherence tomography. *Japanese Journal of Ophthalmology* 2013; 57: 113–119.

Schachar RA. Central surface curvatures of postmortem-extracted intact human crystalline lenses: Implications for understanding the mechanism of accommodation. *Ophthalmology* 2004; 111(9): 1699–16704.

Scheimpflug T. Der Photospektrograph und seine Anwendungen. Photographische Korrespondenz, 1906; 43: 516–531, 1906.

Shen M, Wang MR, Yuan Y, Chen F, Karp CL, Yoo SH, Wang J. SD-OCT with prolonged scan depth for imaging the anterior segment of the eye. *Ophthalmic Surgery, Lasers & Imaging* 2010; 41: S65–S69.

Sheppard AL, Evans CJ, Singh KD, Wolffsohn JS, Dunne MCM, Davies LN. Three-dimensional magnetic resonance imaging of the phakic crystalline lens during accommodation. *Investigative Ophthalmology and Visual Science* 2011; 52: 3689–3697.

Shum PJ, Ko LS, Ng CL, Lin SL. A biometric study of ocular changes during accommodation. *American Journal of Ophthalmology* 1993; 115 (1): 76–81.

Siebinga I, Vrensen GFJM, de Mul FFM, Greve J. Age-related changes in local water and protein content of human eye lenses measured by Raman microspectroscopy. *Experimental Eye Research* 1991; 53: 233–239.

Siedlecki D, de Castro A, Gambra E, Ortiz S, Borja D, Uhlhorn S, Manns F, Marcos S, Parel JM. Distortion correction of OCT images of the crystalline lens: Gradient index approach. *Optometry and Vision Science* 2012; 89 (5): E709–E718.

Sivak JG, Kreuzer RO. Spherical aberration of the crystalline lens. *Vision Research* 1983; 23: 59–70.

Smith G, Atchison DA, Pierscionek BK. Modeling the power of the aging human eye. *Journal of the Optical Society of America A* 1992; 9: 2111–2117.

Smith G, Atchison DA. The gradient index and spherical aberration of the lens of the human eye. *Ophthalmic and Physiological Optics* 2001; 21: 317–326.

Smith G, Cox MJ, Calver R, Garner LF. The spherical aberration of the crystalline lens of the human eye. *Vision Research* 2001; 41: 235–243.

Smith G, Pierscionek BK, Atchison DA. The optical modelling of the human lens. *Ophthalmic and Physiological Optics* 1991; 11: 359–369.

Strenk SA, Semmlow JL, Strenk LM, Munoz P, Gronlund-Jacob J, De Marco JK. Age-related changes in human ciliary muscle and lens: A magnetic resonance imaging study. *Investigative Ophthalmology and Visual Science* 1999; 40: 1162–1169.

Sun M, Birkenfeld J, de Castro A, Ortiz S, Marcos SOCT. 3-D surface topography of isolated human crystalline lenses. *Biomedical Optics Express* 2014; 5: 3547–3561.

Taylor VL, Al-Ghoul KJ, Lane CW, Davis VA, Kuszak JR, Costello MJ. Morphology of the normal human lens. *Investigative Ophthalmology and Visual Science* 1996; 37: 1396–1410.

Tomlinson A, Hemenger RP, Garriott R. Method for estimating the spheric aberration of the human crystalline lens in vivo. *Investigative Ophthalmology and Visual Science* 1993; 34: 621–629.

Tsorbatzoglou A, Nemeth G, Szell N, Biro Z, Berta A. Anterior segment changes with age and during accommodation measured with partial coherence interferometry. *Journal of Cataract & Refractive Surgery* 2007; 33: 1597–1601.

Tron E. Variationsstatistische Untersuehungen über Refraktion. *Graefe's Archive for Clinical and Experimental Ophthalmology* 1929; 122 (1): 1–33.

Urs R, Manns F, Ho A, Borja D, Amelinckx A, Smith J, Jain R, Augusteyn R, Parel JM. Shape of the isolated ex-vivo human crystalline lens. *Vision Research* 2009; 49 (1): 74–83.

Van Veen HG, Goss DA. Simplified system of Purkinje image photography for phakometry. *American Journal of Optometry and Physiological Optics* 1988; 65 (11): 905–908.

Vazquez D, Acosta E, Smith G, Garner L. Tomographic method for measurement of the gradient refractive index of the crystalline lens. II. The rotationally symmetrical lens. *Journal of the Optical Society of America A* 2006; 23 (10): 2551–2565.

Verma Y, Rao KD, Suresh MK, Patel HS, Gupta PK. Measurement of gradient refractive index profile of crystalline lens of fisheye in vivo using optical coherence tomography. *Applied Physics B* 2007; 87: 607–610.

Wood ICJ, Mutti DO, Zadnik K. Crystalline lens parameters in infancy. *Ophthalmic and Physiological Optics* 1996; 16: 310–317.

Wojtkowski M, Kaluzny B, Zawadzki RJ. New directions in ophthalmic optical coherence tomography. *Optometry and Vision Science* 2012; 89 (5): 524–542.

Zhao H, Brown PH, Magone T, Schuck P. The molecular refractive function of lens γ-crystallins. *Journal of Molecular Biology* 2011; 411: 680–699.

Zhou J, Wang J, Jiao S. Dual channel dual focus optical coherence tomography for imaging accommodation of the eye. *Optics Express* 2009; 17 (11): 8947–8955.

Ziebarth NM, Manns F, Uhlhorn SR, Venkatraman AS, Parel JM. Noncontact optical measurement of lens capsule thickness in human, monkey, and rabbit postmortem eyes. *Investigative Ophthalmology and Visual Science* 2005; 46: 1690–1697.

16 Schematic eyes

David A. Atchison

Contents

16.1 INTRODUCTION

Models that show optical-related biometry of the human eye are referred to as schematic eyes. These are usually based on average values of population values available to the authors and are thus subject to biases such as refraction range, age, gender, and race and being clinically based.

Schematic eyes can be developed at different levels of complexity. Simple models, involving as few as a single optical surface, simplify calculations but can only give limited approximations to the real eye's optical performance. Complex models involving multiple aspheric surfaces, gradients of refractive index, surface tilts and misalignments, etc., can give a much more complete description of the eye's normal levels of aberration and off-axis optical performance, but involve much more elaborate calculations.

One group of schematic eyes are the paraxial schematic eyes, for which the refractive surfaces are spherical and centered on a common optical axis. Usually the media of these eyes have uniform refractive indices, but they can have gradient refractive indices. The term "paraxial" means that these schematic eyes can only be considered accurate for describing optics of real eyes accurately within the paraxial region for which rays are close to the optical axis and subtend small angles to it. They do not predict aberrations and retinal image quality accurately for large pupils or for angles more than a few degrees from the optical axis. The paraxial region can be considered as that region for which replacing sines of angles by the angles in radians gives little error using geometrical optics. Practically, this means pupil sizes less than 0.5 mm and field angles less than 2°.

For predicting aberrations and the image quality due to such aberrations, more realistic schematic eyes are required. These are the finite, or wide-angle, schematic eyes. These may include one or more nonspherical refractive surfaces, a lack of surface alignment along a common axis, and media whose refractive indices vary with wavelength. Many of these are developments of paraxial schematic eyes and, by ignoring their refinements, can be evaluated in the same way as for the latter. While most finite eyes are based on population averages, they can be customized to more accurately describe individual eyes (e.g., Navarro et al. 2006).

A range of optical properties can be examined with paraxial schematic eyes including the cardinal point locations, retinal illumination, retinal size magnification, surface reflections (e.g., the Purkinje image sizes and positions), entrance and exit pupils, and effects of changes in biometry on refractive errors. These have practical applications such as calculations of safe light levels and power of intraocular lenses. Finite schematic eyes enable refinement of these properties and examination of ocular aberrations and image quality for both central and peripheral vision.

16.2 A BRIEF HISTORY OF SCHEMATIC EYES

Polyak (1957) gave a history of the understanding of the optical system of the human eye. Following Galen in about AD 200, the lens was believed to be the receptive element of the eye for 1300 years until Leonardo da Vinci proposed that the lens is only one element of the refractive system that forms a real image on the retina. In the early seventeenth century, Kepler and Scheiner showed that the retinal image was inverted relative to the object. Descartes gave an accurate description of the optical system in his "La Dioptrique" in 1637.

Thomas Young (1801) made measurements of his myopic left eye, including its anterior corneal radius of curvature and axial length. These were combined with available information on excised lenses (Petit 1730) and refractive indices. Young developed a three-refracting-surface schematic eye, in both unaccommodated and accommodated forms, although this was not formally given as such. Young was aware of the influence of the lens gradient index on power and his high lens equivalent refractive index of 1.436 made allowance for this. Young used his model for calculating image positions in peripheral vision (Atchison and Charman 2010).

In 1844, Moser constructed a schematic eye that was hypermetropic because of a low lens refractive index (Le Grand and El Hage 1980). In 1851, Listing described a three-refracting-surface schematic eye with a single surface cornea and an aperture stop placed 0.5 mm in front of a homogeneous lens. Helmholtz (1909) modified Listing's schematic eye by changing the positions of the lenticular surfaces. He also gave this model in a form accommodated by 7.69D (object distance 130.1 mm). Helmholtz (1909) described a simpler schematic eye of Listing, called a reduced eye, with only one refracting surface (the anterior cornea). Tscherning (1900) published a four-refracting-surface schematic eye with a posterior corneal surface.

Gullstrand (1909) developed a six-refracting-surface schematic eye. Based on modeling of the gradient index of the lens (see Atchison and Smith 2004), he used a two-shell structure for the lens with different refractive indices for the outer cortex and the core shells. This schematic eye is known as Gullstrand's number 1, or exact, eye. Following Helmholtz, Gullstrand presented this eye at two levels of accommodation. Gullstrand also developed a three-refracting-surface version with a lens of zero thickness; this Gullstrand's number 2 (simplified) eye also has two levels of accommodation.

In 1945, Le Grand modified the Tscherning four-refracting-surface eye schematic eye to become Le Grand's full theoretical eye (Le Grand and El Hage 1980). He presented refractive index dispersion data. Following Gullstrand, Le Grand also presented a simplified three-refracting-surface model with a single corneal surface and a lens of zero thickness.

The lack of lens thickness limits the usefulness of three-refracting-surface eyes. Emsley (1952) overcame this problem by developing the Gullstrand–Emsley eye, a modification of Gullstrand's simplified eye with a 3.6 mm thick lens. Bennett and Rabbetts (1988) and Rabbetts (2007) presented a modification of the Gullstrand–Emsley eye to take into account more recent biometric data.

Emsley (1952) also presented a reduced schematic eye with 60D power. The Gullstrand's number 1, the Le Grand full theoretical, the Gullstrand–Emsley, and Emsley reduced eyes are probably the most popular schematic eyes.

Possibly the first finite schematic eye was that of Lotmar (1971) and incorporated a conicoidal anterior cornea for the purpose of predicting off-axis astigmatism. This was followed by Drasdo and Fowler's (1974) eye for the purpose of predicting image sizes and Kooijman's (1983) eye for the purpose of predicting light levels on the retina. Other well-known finite schematic eyes are those of Navarro (Navarro et al. 1985; Escudero-Sanz and Navarro 1999) and Liou and Brennan (1997). Evaluations of the off-axis performance of paraxial and finite schematic eyes will not be given here, but are covered by Atchison and Smith (2000) and Bakaraju et al. (2008). One word of caution is that schematic eyes should not be judged harshly if their optical performances are poor at tasks for which they were not intended.

Blaker (1980) described an adaptive schematic eye that is a modified Gullstrand number 1 eye, in which the lens has been reduced to two surfaces but has a gradient refractive index. The lens gradient index, lens surface curvatures, lens thickness, and the anterior chamber depth vary as linear functions of accommodation. Blaker (1991) revised his model to include aging effects, with the lens curvatures, lens thickness, and anterior chamber depth altering in the unaccommodated state as a function of age. Several other adaptive schematic eyes have been developed for accommodation (e.g., Navarro et al. 1985, 2007a,b; Atchison and Smith 2000; Norrby 2005; Rabbetts 2007), refractive errors (Atchison 2006), and aging (Smith et al. 1992; Norrby 2005; Goncharov and Dainty 2007; Navarro et al. 2007a,b; Rabbetts 2007; Díaz et al. 2008; Atchison 2009). A recent sophisticated schematic eye includes age and accommodation effects, chromatic dispersion, and tilts and decentrations of surfaces (Navarro 2014).

Some of the schematic eyes mentioned earlier will be discussed in greater detail later.

16.3 GAUSSIAN PROPERTIES

Gaussian optics refers to the assumption that rays beyond the paraxial region behave as if they are paraxial rays, that is, they are aberration-free. One of the main applications of paraxial schematic eyes is predicting the Gaussian properties of real eyes. Of these, the most important are the equivalent power F, positions of the six cardinal points ($\mathbf{F}$, $\mathbf{F}'$, $\mathbf{P}$, $\mathbf{P}'$, $\mathbf{N}$, and $\mathbf{N}'$), and the positions and magnifications of the pupils. Gaussian properties are given for several schematic eyes in Tables 16.1 and 16.2.

Table 16.1 **Gaussian properties of paraxial schematic eyes (distances are in millimeters)**

	GULL. 1 RELAXED	GULL 1. ACCOMM.	LE GRAND RELAXED	LE GRAND ACCOMM.	GULL–EMS. RELAXED	GULL–EMS. ACCOMM.	B AND R RELAXED	EMSLEY REDUCED
Power (D)	58.636	70.576	59.940	67.677	60.483	69.721	60	60
Eye length **VR**′	24.385	24.385	24.197	24.197	23.896	23.896	24.086	22.222
VV′	7.2	7.2	7.6	7.7	7.2	7.2	7.3	0.0
OV	∞	92.000	∞	141.792	∞	116.298	∞	∞
Acccom. (D)	0	10.870	0	7.053	0	8.599	0	0
Cardinal point positions								
VF	−15.706	−12.397	−15.089	−12.957	−14.983	−12.561	−15.156	−16.667
VF′	24.385	21.016	24.197	21.932	23.896	21.252	24.086	22.222
VP	1.348	1.772	1.595	1.819	1.550	1.782	1.511	0.000
VP′	1.601	2.086	1.908	2.192	1.851	2.128	1.819	0.000
VN	7.078	6.533	7.200	6.784	7.062	6.562	7.111	5.556
VN′	7.331	6.847	7.513	7.156	7.363	6.909	7.419	5.556
PN = P′N′	5.730	4.761	5.606	4.965	5.511	4.781	5.600	5.556
FP = N′F′	17.054	14.169	16.683	14.776	16.534	14.343	16.667	16.667
PF′ = FN	22.785	18.930	22.289	19.741	22.045	19.124	22.267	22.222
N′R′	17.054	17.539	16.683	17.041	16.534	16.987	16.667	16.667
F′R′	0	3.370	0	2.264	0	2.644	0	0
Pupil properties								
VE	3.047	2.668	3.038	2.660	3.052	2.674	3.048	0.000
VE′	3.665	3.212	3.682	3.255	3.687	3.249	3.699	0.000
$\bar{M}_{EA}$	1.133	1.117	1.131	1.115	1.130	1.114	1.131	1.000
$\bar{M}_{E'A}$	1.031	1.051	1.041	1.055	1.036	1.049	1.036	1.000
E′R′	20.720	21.173	20.515	20.942	20.209	20.647	20.387	22.222

Note: **E, E**′, entrance and exit pupils; **F, F**′, front and back focal points; $\bar{M}_{EA}$ and $\bar{M}_{E'A}$, magnifications of entrance and exit pupils; **N, N**′, front and back nodal points; **O**, object point; **P, P**′, front and back principal points; **R**′, axial retinal point; **V, V**′, anterior vertex of cornea and vertex of last refracting surface.

16.3.1 CARDINAL POINTS AND EQUIVALENT POWER

A centered optical system has three pairs of cardinal points (Figure 16.1). For the eye, light leaving the front (also called the first or anterior) focal point **F** passes into the eye and would, without the back of the eye being in the way, be imaged at infinity. Light coming into the eye from infinity and parallel to the axis is imaged at the back (also second and posterior) focal point **F**′. The principal points are images of each other (they are said to be conjugate) with a transverse magnification between them of +1. If the equivalent power of the eye is known, for light travelling into the eye, the object position can be considered relative to the front principal point **P**, and refraction can be treated as if it occurred at the back principal point **P**′. The reverse is the case for light travelling from the retina out of the eye. The nodal points are conjugates of each other in which light travelling into the eye toward the front nodal point **N**′ appears to pass through the back nodal point **N** on the image side with the angle of inclination unchanged.

The equivalent power is a measure of the ability of an optical system to deviate rays of light. The equivalent power F is related to the distances between the focal and principal points by the equation

$$F = -\frac{n}{PF} = \frac{n'}{P'F'} \tag{16.1}$$

where
  n is the refractive index of object space (usually air for the eye)
  n′ is the refractive index of image space (the vitreous chamber for the eye)

For refraction at the principal points,

$$\frac{n'}{l'} - \frac{n}{l} = F \tag{16.2}$$

where
  l is the distance from the front principal point to the object point
  l′ is the distance from the back principal plane to the image point

Table 16.2 **Gaussian properties of finite schematic eyes (distances are in millimeters)**

	LOTMAR, KOOIJMAN[a]	NAVARRO RELAXED	NAVARRO ACCOMM. 10D	LIOU AND BRENNAN
Power (D)	59.940	60.416	70.145	60.343
Eye length **VR′**	24.197	24.004	24.004	23.950
VV′	7.6	7.6	7.2	7.73
OV	∞	∞	100	∞
Cardinal point positions				
VF	−15.089	−14.969	−12.051	−15.040
VF′	24.197	24.004	21.172	23.950
VP	1.595	1.583	2.005	1.532
VP′	1.908	1.890	2.393	1.810
VN	7.200	7.145	6.727	7.100
VN′	7.513	7.452	7.116	7.378
PN = P′N′	5.606	5.561	4.723	5.568
FP = N′F′	16.683	16.552	14.056	16.572
PF′ = FN	22.289	22.114	18.779	22.140
N′R′	16.683	16.552	16.888	16.572
F′R′	0	0	2.832	0
Pupil properties				
VE	3.038	3.042	2.928	3.098
VE′	3.682	3.682	3.551	3.720
$\overline{M}_{EA}$	1.131	1.133	1.128	1.133
$\overline{M}_{E′A}$	1.041	1.041	1.058	1.035
E′R′	20.515	20.322	20.453	20.230

Note: Symbols in the first column are the same of those for Table 16.1.
[a] Same as Le Grand full theoretical eye.

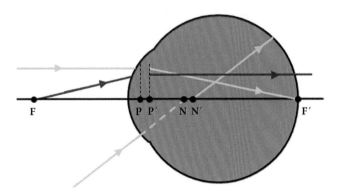

Figure 16.1 Cardinal points of the eye.

Useful equations connecting the cardinal points and the equivalent power are, without derivation,

$$\mathbf{PN} = \mathbf{P'N'} = \frac{n'-n}{F} \qquad (16.3)$$

$$\mathbf{FN} = \mathbf{P'F'} \qquad (16.4)$$

$$\mathbf{N'F'} = \mathbf{FP} \qquad (16.5)$$

16.3.2 APERTURE STOP AND THE ENTRANCE AND EXIT PUPILS

The aperture stop of an eye is its iris. In most schematic eyes, this is assumed to coincide with the vertex of the anterior lens surface. The basic reduced eyes do not have an iris, but an aperture stop can be placed in the plane of the cornea or at some other suitable position. The image of the aperture stop formed by the cornea is called the entrance pupil, and the image of the aperture formed by the lens is called the exit pupil. The entrance pupil is of more interest than the exit pupil because it is what is seen when an eye is viewed.

The pupil ray is the name that is given to the ray, from a particular off-axis position, passing through the center of the aperture stop and appearing to pass through the centers of the entrance and exit pupils. Figure 16.2 shows the anterior path of a paraxial pupil ray to locate the entrance pupil of a simplified schematic eye with a single surface cornea. Here l becomes the anterior chamber depth, l′ becomes the apparent anterior chamber depth, n becomes the refractive index of the aqueous, and n′ becomes the refractive index of air (=1.0). Using the standard sign convention with distances to the left of the refracting surface

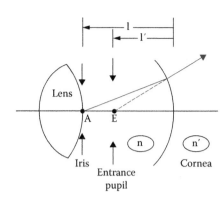

Figure 16.2 Formation of the entrance pupil for a schematic eye with a single surface cornea. **A** is the anterior vertex of the lens and **E** is the center of the entrance pupil.

being negative and distances to the right being positive, the paraxial refraction Equation 16.2 can be rearranged as

$$l' = \frac{n'l}{n + lF} \qquad (16.6)$$

where F here is the power of the cornea. The pupil magnification $\bar{M}_{EA}$, or the ratio of the entrance pupil diameter to that of the stop, is given by

$$\bar{M}_{EA} = \frac{nl'}{n'l} \qquad (16.7)$$

In this treatment, distances l and l' are negative, but the results are usually given in a positive form.

For a two-surface cornea, a slightly more involved treatment is needed to obtain the entrance pupil position and its magnification, in which refraction takes place at the back surface, the ray is transferred to the front surface, and refraction takes place at the front surface.

Properties of the paraxial exit pupil, including its magnification $\bar{M}_{E'A}$, can be obtained by ray tracing toward the back of the eye from the aperture stop.

16.3.3 EFFECT OF ACCOMMODATION ON CARDINAL POINTS

The cardinal points of the relaxed (zero accommodation) and accommodated versions of three schematic eyes are shown in Figures 16.3 through 16.5. With accommodation, the principal points move away from the cornea, the nodal points move toward the cornea, and the focal points move toward the cornea.

16.4 "EXACT" PARAXIAL SCHEMATIC EYES

In the "exact" paraxial schematic eyes, the optical structure of real eyes is modeled as closely as possible while using spherical surfaces. The exact eye has at least four refracting surfaces, two for the cornea and at least two for the lens. The cardinal properties of these and other paraxial schematic eyes are given in Table 16.1.

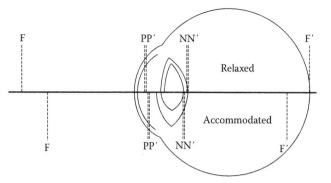

Figure 16.3 The Gullstrand number 1 schematic eye.

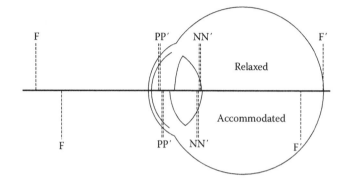

Figure 16.4 The Le Grand full theoretical schematic eye.

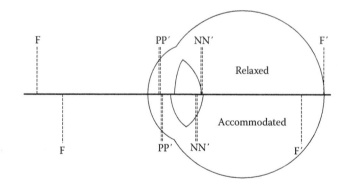

Figure 16.5 The Gullstrand–Emsley schematic eye.

16.4.1 GULLSTRAND NUMBER 1 (EXACT) EYE

This schematic eye accounts for refractive index variation within the lens (Figure 16.3, Tables 16.3 and 16.4). It has six refracting surfaces, two for the cornea and four for the lens. It is presented in both relaxed and accommodated (10.9D) versions. The lens contains a central nucleus (core) of high refractive index surrounded by a cortex of lower refractive index. Gullstrand had an axial length of 24.0 mm, thus placing the retina 0.39 mm short of the back focal point **F'** because he reasoned that positive spherical aberration would produce the best image plane slightly in front of the paraxial image. Here the length of the eye has been increased to 24.39 mm so that the retina coincides with **F'**.

Table 16.3 **Gullstrand exact or number 1 schematic eye—Relaxed form**

MEDIUM	REFRACTIVE INDEX	RADIUS OF CURVATURE (mm)	DISTANCE TO NEXT SURFACE (mm)	EQUIVALENT POWERS (D)		
				SURFACE	COMPONENT	WHOLE EYE
Air	1.000					
		7.700		48.831		
Cornea	1.376		0.500		43.053	
		6.800		−5.882		
Aqueous	1.336		3.100			
		10.000		5.000		58.636
Lens cortex	1.386		0.546			
		7.911		2.528		
Lens core	1.406		2.419		19.111	
		−5.760		3.472		
Lens cortex	1.386		0.635			
		−6.000		8.333		
Vitreous	1.336		17.1854			

Table 16.4 **Gullstrand exact or number 1 schematic eye—Accommodated form, object distance −92.00 mm (≈10.878D)**

MEDIUM	REFRACTIVE INDEX	RADIUS OF CURVATURE (mm)	DISTANCE TO NEXT SURFACE (mm)	EQUIVALENT POWERS (D)		
				SURFACE	COMPONENT	WHOLE EYE
Air	1.000					
		7.700		48.831		
Cornea	1.376		0.500		43.053	
		6.800		−5.882		
Aqueous	1.336		2.700			
		5.333˙		9.376		70.576
Lens cortex	1.386		0.6725			
		2.655		7.533		
Lens core	1.406		2.6550		33.057	
		−2.655		7.533		
Lens cortex	1.386		0.6725			
		−5.333˙		9.376		
Vitreous	1.336		17.1854			

16.4.2 LE GRAND FULL THEORETICAL EYE

The lens of this eye has a constant refractive index and thus has only two refracting surfaces (Figure 16.4 and Tables 16.5 and 16.6). It has both relaxed and accommodated (7.1D) forms.

16.5 SIMPLIFIED PARAXIAL SCHEMATIC EYES

The Gullstrand number 1 eye and the Le Grand full theoretical eye are more complex than is required for many paraxial optical calculations. Simpler eyes are often adequate because errors that

arise in using them are usually less than the expected variations between real eyes. The cornea of simplified schematic eyes is reduced to a single refracting surface and the lens has two surfaces and a uniform refractive index.

16.5.1 GULLSTRAND–EMSLEY EYE

Emsley (1952) modified Gullstrand's number 2 eye in order to simplify computations (Figure 16.5 and Tables 16.7 and 16.8). The aqueous and vitreous refractive indices became 4/3, the lens refractive index became 1.416, the lens was thickened, and the accommodated version had lens surface radii of curvature of ±5.00 mm.

Table 16.5 **Le Grand full theoretical eye—Relaxed form**

MEDIUM	REFRACTIVE INDEX	RADIUS OF CURVATURE (mm)	DISTANCE TO NEXT SURFACE (mm)	EQUIVALENT POWERS (D)		
				SURFACE	COMPONENT	WHOLE EYE
Air	1.000					
		7.8		48.346		
Cornea	1.3771		0.55		42.356	
		6.5		−6.108		
Aqueous	1.3374		3.05			59.940
		10.2		8.098		
Lens	1.4200		4.00		21.779	
		−6.0		14.000		
Vitreous	1.3360		16.5966			

Table 16.6 **Le Grand full theoretical eye—Accommodated form, object distance −141.793 mm (≈7.053D)**

MEDIUM	REFRACTIVE INDEX	RADIUS OF CURVATURE (mm)	DISTANCE TO NEXT SURFACE (mm)	EQUIVALENT POWERS (D)		
				SURFACE	COMPONENT	WHOLE EYE
Air	1.000					
		7.800		48.346		
Cornea	1.3771		0.550		42.356	
		6.500		−6.108		
Aqueous	1.3374		2.650			67.677
		6.000		14.933		
Lens	1.4200		4.500		30.700	
		−5.500		16.545		
Vitreous	1.3360		16.4965			

Table 16.7 **Gullstrand–Emsley eye—Relaxed form**

MEDIUM	REFRACTIVE INDEX	RADIUS OF CURVATURE (mm)	DISTANCE TO NEXT SURFACE (mm)	EQUIVALENT POWERS (D)		
				SURFACE	COMPONENT	WHOLE EYE
Air	1.000					
		7.800		42.735	42.735	
Cornea	4/3		3.6			60.483
		10.000		8.267		
Lens	1.416		3.6		21.755	
		−6.000		13.779		
Vitreous	4/3		16.6962			

16.5.2 BENNETT AND RABBETTS SIMPLIFIED EYE

Bennett and Rabbetts (1988) modified the relaxed version of the Gullstrand–Emsley eye (Table 16.9), and Rabbetts (2007) introduced versions for accommodation of 2.5D, 5.0D, 7.5D, and 10D. Rabbetts introduced an "elderly" version of the eye with a lower lens refractive index than other forms and with a refractive error of 1D hypermetropia to account for the usual shift in refraction beyond 30 years of age (Saunders 1981, 1986).

Table 16.8 **Gullstrand–Emsley eye—Accommodated form, object distance −116.298 mm (≈8.599D)**

MEDIUM	REFRACTIVE INDEX	RADIUS OF CURVATURE (mm)	DISTANCE TO NEXT SURFACE (mm)	EQUIVALENT POWERS (D)		
				SURFACE	COMPONENT	WHOLE EYE
Air	1.000					
		7.800		42.735	42.735	
Cornea	4/3		3.2			69.721
		5.000		16.533		
Lens	1.416		4.0		32.295	
		−5.000		16.533		
Vitreous	4/3		16.6962			

Table 16.9 **Bennett and Rabbetts eye—Relaxed form**

MEDIUM	REFRACTIVE INDEX	RADIUS OF CURVATURE (mm)	DISTANCE TO NEXT SURFACE (mm)	EQUIVALENT POWERS (D)		
				SURFACE	COMPONENT	WHOLE EYE
Air	1.000					
		7.800		43.077	43.077	
Cornea	1.336		3.6			60.000
		11.000		7.818		
Lens	1.422		3.6		20.828	
		−6.47515		13.280		
Vitreous	1.336		16.7863			

16.6 REDUCED PARAXIAL SCHEMATIC EYES

Further simplifications are possible that may give models accurate enough for some calculations such as estimates of retinal image size. Reduced eyes contain only an anterior cornea as a refracting surface. In the more sophisticated eyes, the two principal points and the two nodal points are separated, but in reduced eyes, the principal points **P** and **P′** must be at the corneal vertex **V**, and the nodal points **N** and **N′** must be at the corneal center of curvature. For the eye power to be similar to those of more sophisticated eyes, reduced eyes must have shorter axial lengths. As the cornea has absorbed the power of the lens, its radius of curvature is much smaller than those of more sophisticated eyes. As reduced eyes lack a lens, optical parameters associated with accommodation cannot be investigated.

16.6.1 EMSLEY REDUCED EYE (1952)

The eye has a power of 60D produced by a corneal radius of curvature of 50/9 mm and a refractive index of 4/3 (Figure 16.6 and Table 16.10).

16.7 FINITE SCHEMATIC EYES

As mentioned in Section 16.1, paraxial schematic eyes do not accurately predict the aberrations of real eyes. This section describes the modifications that are used by finite schematic eyes. Following this, some of the better known finite eyes are described. Gaussian properties of these eyes are listed in Table 16.2.

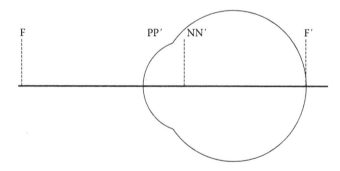

Figure 16.6 The Emsley reduced schematic eye.

16.7.1 MODELING SURFACE SHAPES

In finite schematic eyes, one or more spherical surfaces are replaced by aspheric surfaces. A simple aspheric surface is the conicoid, which may be described by the following equations:

$$(X^2 + Y^2) + (1 + Q)Z^2 - 2ZR = 0 \tag{16.8}$$

$$Z = \frac{(X^2 + Y^2)}{R + \sqrt{R^2 - (1 + Q)(X^2 + Y^2)}} \tag{16.9}$$

Here

Z is the optical axis to which the X- and Y-axes are perpendicular

R is the vertex radius of curvature

Q is the surface asphericity for which Q < 0 indicates that the surface is flattening away from the vertex, Q = 0 specifies a sphere, and Q > 0 indicates that the surface is steepening away from the vertex

Table 16.10 **Emsley reduced eye**

MEDIUM	REFRACTIVE INDEX	RADIUS OF CURVATURE (mm)	DISTANCE TO NEXT SURFACE (mm)	EQUIVALENT POWERS (D)		
				SURFACE	COMPONENT	WHOLE EYE
Air	1.000					
		5.555˙				60.000
Vitreous	4/3		22.222˙			

Table 16.11 **Coefficients of gradient index distribution of lenses of schematic eyes**

REFRACTIVE INDEX	GULLSTRAND UNACCOMM.[a]	GULLSTRAND ACCOMM.[a]	LIOU AND BRENNAN ANTERIOR PORTION	LIOU AND BRENNAN POSTERIOR PORTION
$N_{0,0}$	1.386000	1.386000	1.368	1.407
$N_{0,1}$	+0.0246370	+0.0349938	+0.049057	
$N_{0,2}$	-8.22384×10^{-3}	−0.0237423	−0.015427	-6.605×10^{-3}
$N_{0,3}$	$+3.83400 \times 10^{-4}$	$+7.49690 \times 10^{-3}$		
$N_{0,4}$		-9.37113×10^{-4}		
$N_{1,0}$	$+3.79333 \times 10^{-4}$	-3.28065×10^{-4}	-1.978×10^{-3}	-1.978×10^{-3}
$N_{1,1}$	-1.66853×10^{-3}	$+2.10850 \times 10^{-3}$		
$N_{1,2}$	$+2.78750 \times 10^{-4}$	-5.27125×10^{-4}		
$N_{2,0}$	-6.67170×10^{-5}	-3.49958×10^{-5}		

Note: The relevant distance unit is millimeter.
[a] Distributions upon which the shell indices and curvatures of the lens are based.

There are representations of the conicoid other than those given by the previous equations (Atchison and Smith 2000). Often other surface shapes can be reasonably approximated by conicoids. There are more complex aspheric shapes such as figured conicoids (Atchison and Smith 2000), nonrotationally symmetric conicoids (Burek and Douthwaite 1993), and surfaces containing Zernike polynomial terms (Navarro et al. 2006).

16.7.2 MODELING REFRACTIVE INDEX DISTRIBUTION OF THE LENS

The internal structure of the lens can be represented by a multiple layered shell structure, such as for the Gullstrand number 1 eye, and a continuously varying index (e.g., Gullstrand 1909; Blaker 1980, 1991; Smith et al. 1991; Al-Ahdali and El-Messiery 1995; Siedlicki et al. 2004; Goncharov and Dainty 2007; Navarro et al. 2007a; Díaz et al. 2008; Bahrami and Goncharov 2012; Bahrami et al. 2014). While ray tracing through a gradient index is more complex than for shell structures, routines for performing the computations are readily available.

A general form for representing refractive index distribution N(Y, Z) in the YZ section, assuming rotational symmetry about the optical (Z) axis, is

$$N(Y,Z) = N_0(Z) + N_1(Z)Y^2 + N_2(Z)Y^4 \ldots \quad (16.10)$$

with

$$N_0(Z) = N_{0,0} + N_{0,1}Z + N_{0,2}Z^2 + N_{0,3}Z^3 + N_{0,4}Z^4 \ldots$$

$$N_1(Z) = N_{1,0} + N_{1,1}Z + N_{1,2}Z^2 + N_{1,3}Z^3 + N_{2,4}Z^4 \ldots \quad (16.11)$$

$$N_2(Z) = N_{2,0} + N_{2,1}Z + N_{2,2}Z^2 + N_{2,3}Z^3 + N_{2,4}Z^4 \ldots$$

...

The refractive index distributions, upon which Gullstrand's number 1 eye relaxed and accommodated lenses are based (Gullstrand 1909), can be described in these terms, and the $N_{i,j}$ coefficients are given in Table 16.11 with iso-incidal contours shown in Figure 16.7a.

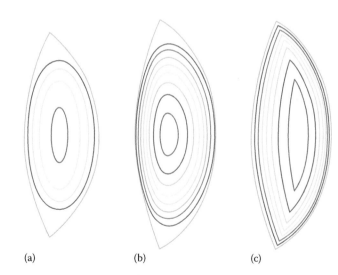

(a)   (b)   (c)

Figure 16.7 Lens shapes and iso-incidal contours of (a) the model on which the unaccommodated version of the Gullstrand number 1 eye is based, (b) the Liou and Brennan eye described in the text and Table 16.17, and (c) the Navarro et al. (2007a) distribution with the parameters of the Liou and Brennan eye lens except that p = 2 in Equation 16.16 and the highest axial index is 60% of the lens thickness from the anterior surface. Index contours are shown in 0.05 steps, with the outermost contour of each lens obtained from rounding the surface refractive index upward to the nearest 0.05 interval.

Smith et al. (1991) presented a gradient index model of the lens in which the index in any direction at relative distance r between the lens center and edge is described by

$$N(r) = c_0 + c_1 r^2 + c_2 r^4 + c_3 r^6 \qquad (16.12)$$

The refractive distribution was divided in two half ellipses (Figure 16.7b), with the front and back ellipses having semiaxes along the optical axis of a_1 and a_2, respectively, and both ellipses having a semiaxis b along the equatorial meridian. Smith et al. gave a set of $N_{i,j}$ coefficients in terms of the c coefficients and semiaxes. Some $N_{i,j}$ coefficients for the front half of the ellipse are given by

$$N_{0,0} = c_0 + c_1 + c_2 + c_3$$

$$N_{0,1} = \frac{(-2c_1 - 4c_2 - 6c_3)}{a_1}$$

$$N_{0,2} = \frac{(c_1 + 6c_2 + 15c_3)}{a_1^2} \qquad (16.13)$$

$$N_{1,0} = \frac{(c_1 + 2c_2 + 3c_3)}{b^2}$$

and some $N_{i,j}$ coefficients for the back half of the ellipse are given by

$$N_{0,0} = c_0$$

$$N_{0,1} = 0$$

$$N_{0,2} = \frac{c_1}{a_2^2} \qquad (16.14)$$

$$N_{1,0} = \frac{c_1}{b^2}$$

Another gradient index model has the index in any direction at relative distance r between the lens center and edge described by

$$N(r) = c_0 + c_1 r^P \qquad (16.15)$$

or

$$N(r) = c_0 + c_1 r^{2p} \qquad (16.16)$$

This has been employed in several eye models (e.g., Smith et al. 1992; Navarro et al. 2007a,b; Bahrami and Goncharov 2012).

As can be seen in Figure 16.7b, the distributions given by Equations 16.12, 16.13, and 16.14 mean that the interior iso-indical contours do not mimic the external shape of the lens. Navarro et al. (2007a) developed a lens model in which the axial point of highest index was manipulated to make this happen, but the equatorial plane becomes curved away from the lens center (Figure 16.7c). In this model, the anterior and posterior iso-indical surfaces meet at sharp angles. Bahrami and Goncharov (2012) developed a "geometric-invariant" refractive index structure, similar to that of Navarro et al., but with smoothing so that the anterior and posterior surfaces and iso-indical contours meet smoothly.

Several authors have modeled changes in the gradient index with accommodation and/or age (e.g., Smith et al. 1992; Goncharov and Dainty 2007; Navarro et al. 2007a,b; Díaz et al. 2008).

16.7.3 MODELING THE RETINA

Paraxial schematic eyes do not consider retinal surface shape. A curved surface must be used for analyzing the off-axis aberrations of the eye, in particular the second-order aberrations of defocus and astigmatism. Retinal shapes are given for three finite schematic eyes in Tables 16.12 through 16.14.

16.7.4 MODELING CHROMATIC DISPERSION

Equations for chromatic dispersion of ocular media are given in the following. In these equations, λ represents wavelength and the constants have different values for the different media.

The Cauchy dispersion equation is

$$n(\lambda) = A + \frac{B}{\lambda^2} + \frac{C}{\lambda^4} + \frac{D}{\lambda^6} \cdots \qquad (16.17)$$

Table 16.12 **Lotmar eye (1971)—Same as Le Grand except for surface asphericities Q and curved retina**

MEDIUM	REFRACTIVE INDEX	RADIUS OF CURVATURE (mm)	Q	DISTANCE TO NEXT SURFACE (mm)	EQUIVALENT POWERS (D) SURFACE	COMPONENT	WHOLE EYE
Air	1.000						
		7.8	−0.286[a]		48.346		
Cornea	1.3771			0.550		42.356	
		6.5	0		−6.108		
Aqueous	1.3374			3.050			59.940
		10.2	0		8.098		
Lens	1.4200			4.000		21.779	
		−6.0	−1.0		14.000		
Vitreous	1.3360			16.5966			
Retina		−12.3					

[a] Conicoid equivalent of asphericity given by Lotmar.

Table 16.13 **Kooijman eye (1983)—Same as Le Grand except for surface asphericities Q and curved retina**

MEDIUM	REFRACTIVE INDEX	RADIUS OF CURVATURE (mm)	Q	DISTANCE TO NEXT SURFACE (mm)	EQUIVALENT POWERS (D)		
					SURFACE	COMPONENT	WHOLE EYE
Air	1.000						
		7.8	−0.25		48.346		
Cornea	1.3771			0.55		42.356	
		6.5	−0.25		−6.108		
Aqueous	1.3374			3.05			59.940
		10.2	−3.06		8.098		
Lens	1.4200			4.00		21.779	
		−6.0	−1.0		14.000		
Vitreous	1.3360			16.5966			
Retina[a]							

[a] Two retinas included: one with a radius of curvature of −10.8 and Q 0 and the other one with a radius of curvature of −14.1 and Q 0.346.

Table 16.14 **Navarro et al. eye (1985)**

MEDIUM	REFRACTIVE INDEX	RADIUS OF CURVATURE (mm)	Q	DISTANCE TO NEXT SURFACE (mm)	EQUIVALENT POWERS (D)		
					SURFACE	COMPONENT	WHOLE EYE
Air	1.000						
		7.72	−0.26		48.705		
Cornea	1.3771			0.55		42.882	
		6.50	0		−5.983		
Aqueous	1.3374			3.05 d_2			60.416
		10.2 R_3	−3.1316 Q_3		8.098		
Lens	1.420 n_3			4.00 d_3		21.779	
		−6.0 R_4	−1.0 Q_4		14.000		
Vitreous	1.336			16.4040 d_4			
Retina		−12.0					

Note: The numbers are for the relaxed form. The model is set to accommodation level A with the symbols varying by $n_3 = 1.42 + 0.00009 *$ $\ln(10A + A^2)$; $R_3 = 10.2 − 1.75 * \ln(A + 1)$; $R_4 = − 6.0 + 0.2294 * \ln(A + 1)$; $Q_3 = − 3.1316 − 0.34 * \ln(A + 1)$; $Q_4 = − 1.0 − 0.125 *$ $\ln(A + 1)$; $d_2 = 3.05 − 0.05 * \ln(A + 1)$; $d_3 = 4.0 + 0.1 * \ln(A + 1)$; $d_4 = 16.40398 − 0.05 * \ln(A + 1)$.

Atchison and Smith (2006) found that use of four terms gives excellent fits to real eye data. Navarro (2014) used this equation in recent modeling.

The Cornu dispersion equation is

$$n(\lambda) = n_\infty + \frac{K}{\lambda - \lambda_0} \quad (16.18)$$

where constants n_∞, K, and λ_0 have different values for the different media. This equation was used in association with the Le Grand full theoretical eye (Le Grand 1967). Thibos et al. (1992) described a chromatic version of Emsley reduced eye with wavelength in nanometers and $n_\infty = 1.320535$, K = 4.685, and $\lambda_0 = 214.102$.

Bennett and Tucker (1975) used the form

$$n^2(\lambda) = a_1 + a_2\lambda^2 + \frac{a_3}{\lambda^2} + \frac{a_4}{\lambda^4} \quad (16.19)$$

to describe the refractive index of water, with λ in nanometers and $a_1 = 1.7632$, $a_2 = -1.38 \times 10^{-8}$, $a_3 = 6.12 \times 10^3$, and $a_4 = 1.41 \times 10^8$, but this underestimates the chromatic aberration of the eye.

Navarro et al. (1985) used the equation

$$n(\lambda) = a_1(\lambda)n^{**} + a_2(\lambda)n_F + a_3(\lambda)n_C + a_4(\lambda)n^* \quad (16.20)$$

where a further set of equations gives the $a_i(\lambda)$ functions:

$$a_i(\lambda) = A_0 + A_1^2 + \frac{P}{\left(\lambda^2 - \lambda_0^2\right)} + \frac{R}{\left(\lambda^2 - \lambda_0^2\right)^2} \quad (16.21)$$

Liou and Brennan (1997) used

$$n(\lambda) = n_{555} + 0.0512 - 0.01455\lambda + 0.0961\lambda^2 \quad (16.22)$$

16.7.5 LOTMAR EYE (1971)

Lotmar (1971) modified the Le Grand full schematic eye by aspherizing the anterior corneal and posterior lens surfaces (Table 16.12).

Atchison and Smith (2000) showed that the corneal asphericity as given was approximated well by conicoidal asphericity Q = -0.286, while the posterior lens surface was a paraboloid (Q = -1).

16.7.6 KOOIJMAN EYE (1983)

Again, this is a modification of the Le Grand full schematic eyes with asphericities for the anterior and posterior corneal surfaces (both Q -0.25), anterior lens surface (Q -3.06), and posterior lens surface (Q -1) (Table 16.13). Two forms of the retinal surface were provided, one being spherical with a radius of curvature of -10.8 mm and the other being elliptical with a radius of curvature of -14.1 mm and asphericity Q +0.346.

16.7.7 NAVARRO EYE (1985)

Navarro et al. (1985) developed a variable accommodation schematic eye based on the Le Grand full schematic eye

(Table 16.14). Escudero-Sanz and Navarro (1998) added a spherical retina with a radius of curvature of -12 mm to make the model suitable for determining off-axis aberrations. Lens refractive index n, lenticular radii of curvature R and asphericities Q, and intraocular distances d are altered as a function of accommodation A according to the equations given at the bottom of Table 16.4, with subscripts 2, 3, and 4 referring to surfaces.

The model incorporates refractive index dispersion according to Equations 16.20 and 16.21, and the constants are given in Tables 16.15 and 16.16.

16.7.8 LIOU AND BRENNAN EYE (1997)

This schematic eye was designed to model spherical aberration and has anatomical values based on 45-year eyes (Table 16.17). It includes four conicoidal surfaces and a gradient index lens.

Table 16.15 **Constants for the Navarro et al. (1985) eye, with chromatic dispersion coefficients according to Equation 16.20**

	n** (0.365 μM)	n_F (0.4861 μM)	n_C (0.6563 μM)	n* (1.014 μM)
Cornea	1.3975	1.3807	1.37405	1.3668
Aqueous	1.3593	1.3422	1.3354	1.3278
Lens	1.4492	1.42625	1.4175	1.4097
Vitreous	1.3565	1.3407	1.3341	1.3273

Table 16.16 **Further constants for the Navarro et al. (1985) eye, with chromatic dispersion coefficients according to Equation 16.21**

COEFFICIENT	A_0	A_1	P	R
a_1	0.66147196	-0.40352796	-0.28046790	0.03385979
a_2	-4.20146383	2.73508956	1.50543784	-0.11593235
a_3	6.29834237	-4.69409935	-1.57508650	0.10293038
a_4	-1.75835059	2.36253794	0.35011657	-0.02085782

Table 16.17 **Liou and Brennan eye (1997)**

MEDIUM	REFRACTIVE INDEX	RADIUS OF CURVATURE (mm)	Q	DISTANCE TO NEXT SURFACE (mm)	EQUIVALENT POWERS (D) SURFACE	EQUIVALENT POWERS (D) COMPONENT	EQUIVALENT POWERS (D) WHOLE EYE
Air	1.000						
		7.77	-0.18		48.391		
Cornea	1.376			0.55		42.262	
		6.40	-0.60		-6.250		
Aqueous	1.336			3.16			60.314
		12.40	-0.94		2.581		
Lens	Gradient anterior[a]			1.59	6.283	22.134	
		∞	—				
Lens	Gradient posterior[a]			2.43	9.586		
		-8.10	+0.96		3.950		
Vitreous[b]	1.3360			16.59655			

[a] Gradient index coefficients are in given in Table 16.11.
[b] No retinal radius of curvature provided. Stop is displaced 0.5 mm nasally from the optical axis.

The model includes media whose indices vary with wavelength according to Equation 16.22, but chromatic aberration occurs only at the front surface and is about half of that of real eyes. The aperture stop is displaced 0.5 mm nasally from the line of sight and the angle alpha between the line of sight and the optical axis is 5°. The model does not specify retinal shape, but a radius of curvature of −12 mm gives good off-axis performance (Atchison and Smith 2000).

The gradient index is based on Equation 16.12 and has a parabolic index in the equatorial direction (c_2, c_3 are zero in the equation) so that it fits Equation 16.16 with p = 1. $N_{i,j}$ coefficients for the front and back halves of the lens are determined by setting a_1, a_2, and b to 1.59, 2.43, and 4.4404 mm, respectively, in Equations 16.13 and 16.14; these coefficients are shown in Table 16.11 and iso-incidal contours are shown in Figure 16.7b.

REFERENCES

Al-Ahdali, I. H. and El-Messiery, M. A., Examination of the effect of the fibrous structure of a lens on the optical characteristics of the human eye: A computer-simulated model, *Applied Optics* 25 (1995): 5738–5745.

Atchison, D. A., Optical models for human myopic eyes, *Vision Research* 46 (2006): 2236–2250.

Atchison, D. A., Age-related paraxial schematic emmetropic eyes, *Ophthalmic and Physiological Optics* 29 (2009): 58–64.

Atchison, D. A. and Charman, W. N., Thomas Young's contribution to visual optics: The Bakerian lecture "On the mechanism of the eye", *Journal of Vision* 10 (2010): 6, 1–16.

Atchison, D. A. and Smith, G., *Optics of the Human Eye* (Oxford, U.K.: Butterworth-Heinemann, 2000).

Atchison, D. A. and Smith, G., Possible errors in determining axial length changes during accommodation with the IOLMaster, *Optometry and Vision Science* 81 (2004): 282–285.

Atchison, D. A. and Smith, G., Chromatic dispersions of the ocular media of human eyes, *Journal of the Optical Society of America A* 22 (2006): 29–37.

Bahrami, M. and Goncharov, A. V., Geometric-invariant gradient index lens: Analytical ray raytracing, *Journal of Biomedical Optics* 17 (2012): 055001-1–055001-9.

Bahrami, M., Goncharov, A. V., and Pierscionek, B. K., Adjustable internal structure for reconstructing gradient index profile of crystalline lens, *Optics Letters* 39 (2014): 1310–1313.

Bakaraju, R. C., Ehrmann, K., Papas, E., and Ho, A., Finite schematic eye models and their accuracy to in-vivo data, *Vision Research* 48 (2008): 1681–1694.

Bennett, A. G. and Rabbetts, R. B., Schematic eyes—Time for a change?, *Optician* 196(5169) (1988): 14–15.

Bennett, A. G. and Tucker, J., Correspondence: Chromatic aberration of the eye between 200 and 2000 nm, *British Journal of Physiological Optics* 30 (1975): 132–135.

Blaker, J. W., Toward an adaptive model of the human eye, *Journal of the Optical Society of America* 70 (1980): 220–223.

Blaker, J. W., A comprehensive model of the aging, accommodative adult eye, in *Technical Digest on Ophthalmic and Visual Optics*, vol. 2 (Washington, DC: Optical Society of America, 1991), pp. 28–31.

Burek, H. and Douthwaite, W. A., Mathematical models of the general corneal surface, *Ophthalmic and Physiological Optics* 13 (1993): 68–72.

Díaz, J. A., Pizarro, C., and Arasa, J., Single dispersive gradient-index profile for the aging human lens, *Journal of the Optical Society of America A* 25 (2008): 250–261.

Drasdo, N. and Fowler, C. W., Non-linear projection of the retinal image in a wide-angle schematic eye, *British Journal of Ophthalmology* 58 (1974): 709–714.

Emsley, H. H., *Visual Optics*, vol. 1, 5th edn. (London, U.K.: Butterworths, 1952), pp. 343–348.

Escudero-Sanz, I. and Navarro, R., Off-axis aberrations of a wide-angle schematic eye model, *Journal of the Optical Society of America A* 15 (1999): 1881–1891.

Goncharov, A. V. and Dainty, C., Wide-field schematic eye models with gradient-index lens, *Journal of the Optical Society of America A* 24 (2007): 2157–2174.

Gullstrand, A., Appendices II and IV, in *Helmholtz's Handbuch der Physiologischen Optik*, vol. 1, 3rd edn. 1909 (trans. edited by J. P. C. Southall, Optical Society of America, 1924).

Helmholtz, H. von, *Handbuch der Physiologischen Optik*, vol. 1, 3rd edn. (1909) (English trans. edited by J. P. C. Southall, Optical Society of America, 1924), pp. 94–96, 152.

Kooijman, A. C., Light distribution on the retina of a wide-angle theoretical eye, *Journal of the Optical Society of America* 73 (1983): 1544–1550.

Le Grand, Y., *Form and Space Vision* (Revised edition trans. M. Millodot and G. Heath, Indiana University Press, Bloomington, IN, 1967).

Le Grand, Y. and El Hage, S. G., *Physiological Optics*, trans. and update of Le Grand Y, La dioptrique de l'oeil et sa correction (1968). In *Optique Physiologique*, vol. 1. (Heidelberg, Germany: Springer-Verlag, 1980), pp. 57–69.

Liou, H.-L. and Brennan, N. A., Anatomically accurate, finite model eye for optical modeling, *Journal of the Optical Society of America A* 14 (1997): 1684–1695.

Lotmar, W., Theoretical eye model with aspheric surfaces, *Journal of the Optical Society of America* 61 (1971): 1522–1529.

Navarro, R., Adaptive model of the aging emmetropic eye and its changes with accommodation, *Journal of Vision* 14(13) (2014): 1–17, 21.

Navarro, R., González, L., and Hernández-Matamoros, J. L., On the prediction of optical aberrations by personalized eye models, *Optometry and Vision Science* 83 (2006): 371–381.

Navarro, R., Palos, F., and Gonzáles, L., Adaptive model of the gradient index of the human lens. I. Formulation and model of aging ex vivo lenses, *Journal of the Optical Society of America A* 24 (2007a): 2175–2185.

Navarro, R., Palos, F., and Gonzáles, L., Adaptive model of the gradient index of the human lens. II. Optics of the accommodating aging lens, *Journal of the Optical Society of America A* 24 (2007b): 2911–2920.

Navarro, R., Santamaría, J., and Bescós, J., Accommodation-dependent model of the human eye with aspherics, *Journal of the Optical Society of America A* 2 (1985): 1273–1281.

Norrby, S., The Dubbelman eye model analysed by ray tracing through aspheric surfaces, *Ophthalmic and Physiological Optics* 25 (2005): 153–161.

Petit, F. P., Mémoire sur le cristallin de l'oeil de l'homme, des animaux á quatre pieds, des oiseaux et des poisons, *Mémoires de l'Académie royale des sciences* (1730): 4–26.

Polyak, S. L., *The Vertebrate Visual System* (Chicago, IL: University of Chicago Press, 1957), Chapter 1.

Rabbetts, R. B., *Bennett and Rabbetts' Clinical Visual Optics*, 4th edn. (Oxford, U.K.: Butterworth-Heinemann, 2007), pp. 223–227.

Saunders, H., Age dependence of human refractive errors, *Ophthalmic and Physiological Optics* 1 (1981): 159–174.

Saunders, H., A longitudinal study of the age dependence of human ocular refraction, *Ophthalmic and Physiological Optics* 6 (1986): 39–46.

Siedlicki, D., Kasprzak, H., and Pierscionek, B. K., Schematic eye with a gradient-index lens and aspheric surfaces, *Optics Letters* 29 (2004): 1197–1199.

Smith, G., Atchison, D. A., and Pierscionek, B. K., Modelling the power of the aging human eye, *Journal of the Optical Society of America A* 10 (1992): 2111–2117.

Smith, G., Pierscionek, B. K., and Atchison, D. A., The optical modelling of the human lens, *Ophthalmic and Physiological Optics* 11 (1991): 359–369.

Thibos, L. N., Ye, M., Zhang, X., and Bradley, A., The chromatic eye: A new reduced-eye model of ocular chromatic aberration in humans, *Applied Optics* 31 (1992): 3594–3600.

Tscherning, M., *Physiologic Optics*, trans C. Weiland (Philadelphia, PA: The Keystone, 1900), pp. 26–32.

Young, T., On the mechanism of the eye, *Philosophical Transactions of the Royal Society of London* 92 (1801): 23–88.

17 Axes and angles of the eye

David A. Atchison

Contents

17.1 INTRODUCTION

Most man-made optical systems are rotationally symmetric about the optical axis. If the surfaces of the system are spherical, it is the line joining their centers of curvatures. If one or more surfaces are astigmatic or toroidal and have two planes of symmetry, the line of intersection of these two planes forms part of the optical axis.

Because of the lack of symmetry of the eye and because the fixation point and fovea are not along a best-fit axis of symmetry, in order to fully describe the optical properties of the eye a number of axes are required. The axes influence the retinal image quality and have clinical applications, including the diagnosis of binocular anomalies and determining ablation patterns in corneal refractive surgery. This chapter describes these axes, their importance, their applications, and how they may be determined.

The validity of some axes is dependent upon idealized properties of the eye. The visual axis requires the existence of nodal points, which would truly only exist if the eye is rotationally symmetric. The fixation axis requires the existence of a unique center of rotation of the eye. Some axes pass through the center of the entrance pupil (the image of the aperture stop by the cornea) and their directions depend upon pupil size because this affects the pupil center. Astigmatism of surfaces is a complicating factor that is disregarded in this chapter.

The directions of the axes are defined relative to each other, and the angles between them are often discussed in the following sections.

Many terms used in this chapter were introduced in previous chapters, for example, nodal points and entrance pupil.

17.2 DEFINITIONS AND IMPORTANCE OF AXES

Definitions of optical axis, line of sight, visual axis, pupillary axis, and fixation axis, and the angles between, are similar to those provided in dictionaries of visual science (Cline et al. 1989; Millodot 2009).

17.2.1 OPTICAL AXIS

This was defined in the previous section. As the eye is not a centered system, it does not contain a true optical axis, but the concept of optical axis can be applied by defining the optical axis as the line of "best fit" through the centers of curvature of the "best fit" spheres to each surface.

17.2.2 LINE OF SIGHT

This is the line joining the fixation point and the fovea through the center of the pupil. The fovea is usually to the temporal side of the optical axis, and hence the fixation point (see **T** in Figure 17.1) is usually on the nasal side of the optical axis. The position at which the line of sight intercepts the cornea is called the "corneal sighting center" (Mandell 1995) or "visual center of the cornea" (Cline et al. 1989).

The line of sight is the most important axis from the point of view of visual function because it defines the center of the beam of light entering the eye from the fixation point. As mentioned in Section 17.1, unfortunately, it is not fixed because the pupil center may move when pupil size changes.

17.2.3 VISUAL AXIS

This is the line joining the fixation point and the fovea through the nodal points. The visual axis is the line segments **TN** and **N'T'** shown in Figure 17.1. It is usually close to the line of sight at the cornea and entrance pupil. The position at which the visual axis intercepts the cornea is called the "ophthalmometric pole" (Le Grand and El Hage 1980).

The visual axis is a convenient reference point for visual functions, particularly as it does not depend on pupil size.

The *foveal achromatic axis* is closely related to the visual axis. It is the path from the fixation point to the fovea so that the ray does not suffer from transverse chromatic aberration. Ignoring the small change in nodal points that occurs with change in wavelength, this axis is identical to the visual axis and its definition can be used as a basis for locating the visual axis.

Rabbetts (2007) did not like the use of the term "visual axis" for the ray passing through the nodal points because he argued

that this ray is not a representative of the beam passing into the eye from a fixation target. He preferred to call it the "nodal axis" and reserved the term "visual axis" for the axis defined in Section 17.2.2 as the line of sight. His approach is contrary to the majority of literature.

17.2.4 PUPILLARY AXIS

This is the line passing through the center of the entrance pupil, and that is normal to the anterior corneal surface.

The pupillary axis is used to estimate eccentric fixation, the condition in which a retinal point other than the foveal center is used for fixation. Eccentric fixation is often an adaptation to heterotropia (squint or turned eye); this is discussed further in Section 17.4.2.

The pupillary axis would lie along the optical axis if the eye, including the pupil, was a centered system. However, the pupil is often not centered relative to the cornea and the cornea may not be in a regular shape. These factors cause the pupil axis to lie in some other direction, and usually it does not pass through the fixation point **T** (Figure 17.1).

17.2.5 FIXATION AXIS

This is the line passing through the fixation point and the center-of-rotation **C** of the eye (Figure 17.1). It is the reference for measuring eye movements. The idea of a fixation axis is an approximation because there is no unique center of rotation, and the estimates of it depend upon the direction of rotation of the eye (Alpern 1969).

17.2.6 KERATOMETRIC AXIS

This is the axis of instruments, such as keratometers and corneal topographers, for measuring corneal shape, and it contains the center of curvature **C_c** of the anterior cornea (Figure 17.1).

The keratometric axis intercepts the line of sight at the fixation target (Figure 17.1). The keratometric axis intercepts the cornea at the *vertex normal*, which is the center of corneal topographic maps. This position does not coincide with the *anterior corneal apex* (Mandell and St. Helen 1969), which is the point with

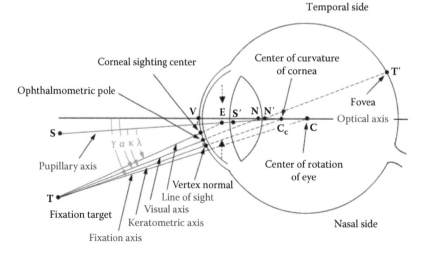

Figure 17.1 Most of the axes and angles of the eye. For clarity, the object has been shown very close to the eye and the angles are exaggerated. **E** is the center of the entrance pupil, and **N** and **N'** are the nodal points of the eye.

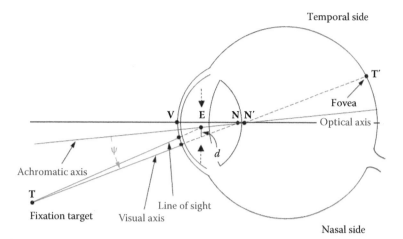

Figure 17.2 The relationship between the achromatic axis, optical axis, visual axis, and line of sight. The angle between the achromatic axis and the visual axis is ψ. The separation between the line of sight (or achromatic axis) and the visual axis at the entrance pupil is *d*.

the smallest radius of curvature. Variations in the keratometric axis between instruments occur because of differences in the distance to the fixation target. Applegate et al. (2009) discussed the implications of this distance, through its effects on tilt and decentration of the keratometric axis, on determining the corneal contribution to the aberrations of the eye. Mathur et al. (2012) presented equations to correct for the fixation distance when determining the contribution.

17.2.7 ACHROMATIC AXIS

Thibos et al. (1990) used this term to refer to the *pupil nodal ray*, which is the ray passing through the center of pupil and which has no transverse chromatic aberration. It is shown in Figure 17.2. It is similar to the optical axis but, unlike the optical axis, it is dependent on pupil position. The achromatic axis is distinct from the foveal achromatic axis mentioned in Section 17.2.3.

17.2.8 PUPILLARY CIRCULAR AXIS

Atchison et al. (2014) coined the term pupillary circular axis to refer to the visual field position at which the pupil appears most nearly circular. The word "circular" is included to distinguish this axis from the pupillary axis (Section 17.2.4). The pupil appears approximately circular when viewed along the line of sight, but when viewed at a considerable angle to the line of sight, it appears elliptical. The minor axis of the ellipse is nearly parallel to the meridian of the viewing angle and the size in the perpendicular direction remains approximately constant (Mathur et al. 2013). Changes in pupil size and shape are relevant for peripheral imagery.

17.3 LOCATING AXES

17.3.1 LINE OF SIGHT

Many ophthalmic instruments have a video display showing the anterior eye. To achieve coincidence of the line of sight with the instrument optical axis, a participant fixates a target and the instrument is moved vertically and horizontally until the pupil is correctly centered.

17.3.2 VISUAL AXIS

This can be measured with laboratory techniques. The visual axis intercept at the cornea (the ophthalmometric pole) can be determined by a participant viewing a vernier target, half of which is blue and half of which is red (Figure 17.3), through an artificial pupil of about 1 mm diameter. When the artificial pupil is not centered on the visual axis, there is a break in the alignment of the halves of the target, so it is moved until alignment is obtained.

Thibos et al. (1990) showed an adaption of this approach to estimate the separation *d* of the visual axis from the line of sight at the cornea (Figure 17.2). The small artificial pupil is scanned across the eye to find the eye pupil limits at which the vernier target disappears. The center of these positions corresponds to the corneal sighting center on the line of sight. This position is then compared with the ophthalmometric pole. Thibos et al. obtained

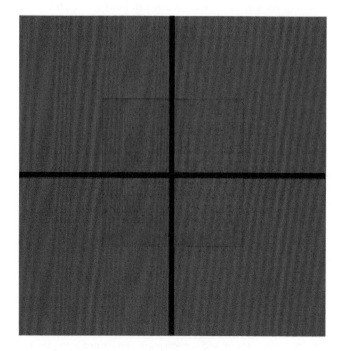

Figure 17.3 A target for locating the visual axis.

values of d between –0.1 and 0.4 mm along the horizontal meridian for five subjects with drug-induced pupil dilation, with a mean of +0.14 mm (the positive sign means that the visual axis was nasal to the line of sight in object space).

Simonet and Campbell (1990) made estimates of separation d with a different method. For natural (although generally large) pupils, the differences for eight eyes of five subjects were between –0.1 and +0.5 mm, with a mean of +0.11 mm.

17.3.3 KERATOMETRIC AXIS

Mandell et al. (1995) found a mean difference between the keratometric axis and the corneal sighting center of 0.4 ± 0.1 mm, with most of 20 participants having the keratometric axis nasal and inferior to the corneal sighting center. The corresponding mean angle between the keratometric axis and the line of sight was 2.7°. The mean difference between the keratometric axis and the corneal apex was 0.6 ± 0.2 mm, with the most participants having the keratometric axis above the corneal apex. Applegate et al. (2009) found the mean position of the keratometric axis to be 0.15 ± 0.14 mm nasal and 0.04 ± 0.12 mm inferior to the corneal sighting center. Note that the horizontal component of the difference between the keratometric axis and the corneal sighting center is affected by pupil size as the natural pupil moves temporally with dilation (Walsh 1988; Wilson et al. 1992; Yang et al. 2002; Wildenmann and Schaeffel 2013; Mathur et al. 2014).

17.3.4 PUPILLARY CIRCULAR AXIS

Sets of images of the pupil have been taken along the horizontal visual field (Spring and Stiles 1948; Sloan 1950; Jay 1961; Haines 1969; Mathur et al. 2013). Plotting the ratio of the dimensions of minor and major axes against visual field position, the pupillary circular axis corresponds to the peak of the fit (Figure 17.4). Atchison et al. (2014) developed a procedure by which the vertical position of the pupillary circular axis can be estimated from horizontal visual field data.

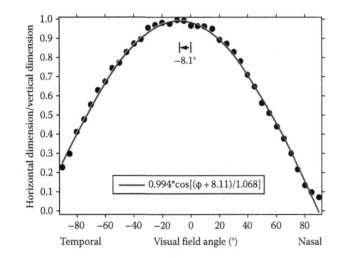

Figure 17.4 Determining the pupillary circular axis for one person from ratios of horizontal and vertical pupil dimensions along the horizontal visual field. The axis is (–)8.1° in the temporal visual field.

17.4 ANGLES BETWEEN AXES

17.4.1 VISUAL AXIS AND OPTICAL AXIS: ANGLE ALPHA (α)

Angle alpha is the angle between the optical and visual axes (Le Grand and El Hage 1980; Cline et al. 1989; Millodot 2007). As measuring this angle involves a participant fixating a target, perhaps the line of sight would be a better reference than the visual axis, but the distinction is of no practical importance.

One method of locating the optical axis of a centered optical system is to direct a point source into the system and to find the position of the beam for which there is alignment of reflections from the different surfaces. The eye has four reflecting surfaces, giving four main reflections (the Purkinje images). As the eye is not a centered system, all that can be done is to minimize the spread of these images, and the corresponding direction of the source estimates the best fit to an optical axis.

The angle α can be determined with the ophthalmophakometer (Figure 17.5). This instrument contains a graduated arc, with an observing telescope mounted centrally in the arc and the patient's eye at the center of curvature of the arc. The arc has a pair of light sources, one source slightly above and the other source slightly below the telescope. The sources give pairs of Purkinje images. A small fixation target **T** is moved along the arc until an observer looking through the telescope judges that the Purkinje images are in the best possible alignment. This best fit optical axis of the eye now corresponds with the instrument axis, and angle α is given by the scale reading at the position of the fixation target. If the instrument arc and sources can be rotated through 90°, angle α can be obtained in the vertical direction. Usually, the distance between the eye and the arc is large (typically 86 cm) so that a discrepancy between the front nodal point and the center of curvature of the arc is not critical.

The angle between the visual axis and the optical axis is taken as positive if the visual axis is on the nasal side of the optical axis in object space. The mean value of angle α is considered to be about +5° horizontally and is usually in the range +3° to +5°. The visual axis is usually superior to the optical axis by (+) 2°–3° (Tscherning 1900).

Peripheral refraction varies considerably with visual field position, and finding the visual field angle about which astigmatism

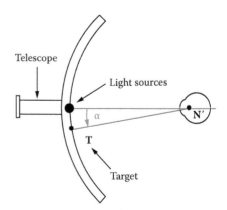

Figure 17.5 The ophthalmophakometer.

shows symmetry is in an indirect way of determining angle α. The astigmatism is plotted as a function of visual field position and angle α corresponds to where the astigmatism passes through a turning (stationary) point (Lotmar and Lotmar 1974; Dunne et al. 1993; Atchison et al. 2006; Atchison et al. 2014). Using this approach, Atchison et al. (2006) obtained estimates of +4.8° ± 5.1° horizontally and +1.5° ± 3.4° vertically (the means have been changed in sign from those given by the authors to be consistent with the convention used in the previous paragraph). These values are similar to those reported for the ophthalmophakometer, and the angles determined by the two methods are highly correlated for the horizontal field (Atchison et al. 2006). The angle determined by both methods varies with refraction in the horizontal field so that it becomes smaller as myopia increases (Atchison et al. 2006, 2014).

17.4.2 PUPILLARY AXIS AND LINE OF SIGHT: ANGLE LAMBDA (λ)

This angle has been called both λ and κ, with Lancaster (cited by Le Grand and El Hage 1980) using λ and Le Grand (Le Grand and El Hage 1980) and more recently Artal's group (e.g., Tabernero et al. 2007) using κ. Here, λ is used and the angle between the pupillary and visual axes is called κ, which is consistent with dictionaries of vision science (Cline et al. 1989; Millodot 2009).

This angle can be determined using the ophthalmophakometer just described, but it can be determined with even simpler equipment. A participant fixates a target **T**, and an observer watches the anterior corneal reflection of a small source of light **S** (Figure 17.1). The position of the light is changed, with the observer's eye maintained next to the light source, until the reflected image **S′** appears centered in the pupil **E**. Now the angle between the observation axis and the line of sight is λ because the observation axis coincides with the pupillary axis.

The pupillary axis is usually temporal to the line of sight in object space. Artal (2014) obtained horizontal and vertical components of +3.9° ± 2.2° and +0.2° ± 1.7°, with positive signs indicating temporal and superior displacements of the pupillary axis from the line of sight (Figure 17.6). The horizontal component decreases as eye length or the magnitude of myopia increases (Tabernero et al. 2007; Artal 2014).

Angle λ is used to diagnose eccentric fixation and heterotropia. To test for eccentric fixation, the angle is determined monocularly, that is, the fellow eye is occluded. A large angle indicates the likely presence of eccentric fixation. To test for direction and amount of heterotropia in the *Hirschberg test*, the angle is estimated binocularly, that is, both eyes are open. A large angle is observed in the presence of heterotropia because one eye rotates so that its fovea is not being used to align the fixation target. Either the fovea is being suppressed, another retinal point is being used for fixation (anomalous correspondence), or both are occurring. We are not strictly measuring angle λ in either the monocular or binocular cases because the patient has rotated the eye to align a retinal point, eccentric to the fovea, with the fixation target.

The usual clinical application of these tests involves a clinician shining a penlight, in front of his/her face, at the patient's eye or eyes. The patient is instructed to look at the penlight. This is not moved, but the clinician judges the position of the corneal reflection relative to the pupil center (Figure 17.7). Normally the reflection is about half a millimeter nasal to the center of the pupil (Figure 17.7a) and each millimeter change in this position

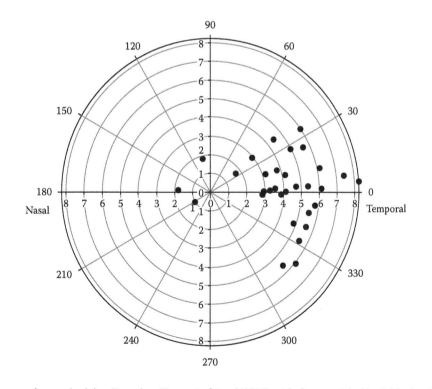

Figure 17.6 Angle α in a group of normal adults. (Based on Figure 6 of Artal (2014), with data provided by Pablo Artal and published with kind permission from the Optical Society of America.)

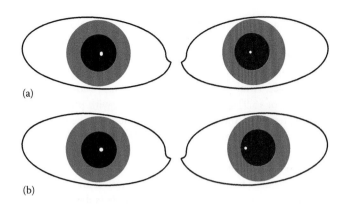

(a)

(b)

Figure 17.7 The Hirschberg test for measuring the angle of heterotropia. (a) Normal appearance of corneal reflexes; (b) appearance of reflexes for left exotropia (left eye turned out).

corresponds to approximately 13° of eye rotation (Grosvenor 1996). The appearances of the reflexes in left exotropia are shown in Figure 17.7b. As some patients may have normal fixation with unusual angles, results from the two eyes should be compared in both monocular and binocular tests.

17.4.3 PUPILLARY AXIS AND VISUAL AXIS: ANGLE KAPPA (κ)

In practical terms, this is the same as angle λ. It is shown in Figures 17.1.

17.4.4 VISUAL AXIS AND ACHROMATIC AXIS: ANGLE PSI (Ψ)

Thibos et al. (1990) estimated this angle from

$$\sin(\psi) = \frac{d}{\mathbf{EN}} \tag{17.1}$$

where
  d is the distance between the visual axis and line of sight at the entrance pupil
  $\mathbf{EN}$ is the estimate of the distance between the entrance pupil and the nodal point (Figure 17.2)

Based on the Gullstrand number 1 schematic eye (Chapter 16), $\mathbf{EN}$ is 4.0 mm. A sufficiently accurate method for determining d

was given in Section 17.3.2. For five participants, Thibos et al. determined a range of horizontal angles between –1.2° and +5.3°, with a mean of +2.1° (positive angles mean the visual axis is inclined nasally to the achromatic axis in object space).

17.4.5 FIXATION AXIS AND OPTICAL AXIS: ANGLE GAMMA (γ)

Figure 17.8 shows the relationship between angles γ and α. Here, y is the distance between the optical axis and fixation target $\mathbf{T}$, $\mathbf{N}$ is front nodal point, $\mathbf{C}$ is center of rotation of the eye, and w is the distance from the projection of $\mathbf{T}$, onto the optical axis, to the cornea at $\mathbf{V}$. Trigonometry gives

$$\tan(\alpha) = \frac{y}{w + \mathbf{VN}} \tag{17.2}$$

and

$$\tan(\gamma) = \frac{y}{w + \mathbf{VC}} \tag{17.3}$$

which combine to give

$$\tan(\gamma) = \tan(\alpha)\frac{w + \mathbf{VN}}{w + \mathbf{VC}} \tag{17.4}$$

Using $\mathbf{VN}$ for the Gullstrand number 1 schematic eye (7.1 mm in Table 16.1) and $\mathbf{VC}$ as 14 mm, the difference between angle γ and angle α is less than 1% for object distances greater than 70 cm.

17.4.6 LINE OF SIGHT AND PUPILLARY CIRCULAR AXIS

No name has been given to the angle between the line of sight and the pupillary axis; the angle is simply the visual field position at which the pupillary circular axis occurs. Mathur et al. (2013) and Atchison et al. (2014) assigned horizontal and vertical components with positive angles if the pupillary circular axis was in the nasal and superior visual fields, respectively. For a group of 30 adults, along the horizontal field, they obtained a range of positions of –1° to –9° with a mean and standard deviation of –5.3° ± 1.9°. Corresponding values for the vertical visual field were –7° to 0° and –3.2° ± 1.5°.

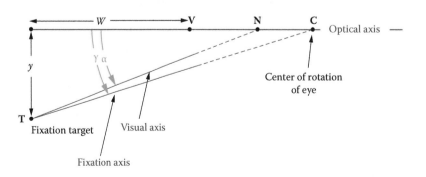

Figure 17.8 Determination of angle γ. Based on Figure 17.1, with extraneous detail removed.

REFERENCES

Alpern, M., Specification of the direction of regard, in *Muscular Mechanisms*, H. Davson, ed., Vol. 3, 2nd edn of *The Eye* (Academic Press, London, U.K., 1969): pp. 5–12.

Applegate, R. A., Thibos, L. N, Twa, M., and Sarver, E. J., Importance of fixation, pupil center, and reference axis in ocular wavefront sensing, videokeratography, and retinal image quality, *Journal of Cataract and Refractive Surgery* 35 (2009): 139–152.

Artal, P., Optics of the eye and its impact on vision: A tutorial, *Advances in Optics and Photonics* 6 (2014): 340–367.

Atchison, D. A., Mathur, A., Suheimat, M., and Charman, W. N., Visual field co-ordinates of pupillary and optical axes, *Optometry and Vision Science* 91 (2014): 582–587.

Atchison, D. A., Pritchard, N., and Schmid, K. L., Peripheral refraction along the horizontal and vertical visual fields in myopia, *Vision Research* 46 (2006): 1450–1458.

Cline, D., Hofstetter, H. W., and Griffin, J. R., *Dictionary of Visual Science*, 4th edn. (Chilton Trade Book Publishing, Radnor, PA, 1989).

Dunne, M. C., Misson, G. P., White, E. K., and Barnes, D. A., Peripheral astigmatic asymmetry and angle alpha, *Ophthalmic and Physiological Optics* 13 (1993): 303–305.

Grosvenor, T. P., *Primary Care Optometry*, 3rd edn, Chap. 6 (Butterworth-Heinemann, St. Louis, MO, 1996).

Haines, R. F., Dimensions of the apparent pupil when viewed at oblique angles, *American Journal of Ophthalmology* 68 (1969): 649–656.

Jay, B. S., The effective pupillary area at varying perimetric angles, *Vision Research* 1 (1962): 418–424.

Le Grand, Y. and El Hage, S. G., *Physiological Optics*, trans. and update of Le Grand, Y., *La dioptrique de l'oeil et sa correction* (1968). In *Optique Physiologique*, Vol. 1 (Springer-Verlag, Heidelberg, Germany, 1980): pp. 71–74.

Lotmar, W. and Lotmar, T., Peripheral astigmatism in the human eye: Experimental data and theoretical model predictions, *Journal of the Optical Society of America A* 64 (1974): 510–513.

Maloney, R. K., Corneal topography and optical zone location in photorefractive keratectomy, *Refractive and Corneal Surgery* 6 (1990): 363–371.

Mandell, R. B., Locating the corneal sighting center from videokeratography, *Journal of Cataract and Refractive Surgery* 11 (1995): 253–258.

Mandell, R. B., Chiang, C. S., and Klein, S. A., Location of the major corneal reference points, *Optometry and Vision Science* 72 (1995): 776–784.

Mandell, R. B. and St. Helen, R., Position and curvature of the corneal apex, *American Journal of Optometry and Archives of the American Academy of Optometry* 46 (1969): 25–29.

Mathur, A., Atchison, D. A., and Tabernero, J., Effect of age on components of peripheral ocular aberrations, *Optometry and Vision Science* 89 (2012): 967–976.

Mathur, A., Gehrmann, J., and Atchison, D. A., Pupil shape as viewed along the horizontal visual field, *Journal of Vision* 13 (2013), 6: 1–8.

Mathur, A., Gehrmann, J., and Atchison, D. A., Influences of luminance and accommodation stimuli on pupil size and pupil center location, *Investigative Ophthalmology and Vision Science* 55 (2014): 2166–2172.

Millodot, M., *Dictionary of Optometry*, 7th edn. (Butterworth-Heinemann, Oxford, U.K., 2009).

Rabbetts, R. B., *Bennett and Rabbetts' Clinical Visual Optics*, 4th edn. (Butterworth-Heinemann, Oxford, U.K., 2007): p. 235.

Simonet, P. and Campbell, M. C. W., The optical transverse chromatic aberration of the fovea of the human eye, *Vision Research* 30 (1990): 187–206.

Sloan, L. L., The threshold gradients of the rods and cones: In the dark-adapted and in the partially light-adapted eye, *American Journal of Ophthalmology* 33 (1950): 1077–1089.

Spring, K. H. and Stiles, W. S., Apparent shape and size of the pupil viewed obliquely, *British Journal of Ophthalmology* 32 (1948): 347–354.

Tabernero, J., Benito, A., Alcón, E., and Artal, P., Mechanism of compensation of aberrations in the human eye, *Journal of the Optical Society of America A* 24 (2007): 3274–3283.

Thibos, L. N., Bradley, A., Still, D. L. et al., Theory and measurement of ocular chromatic aberration, *Vision Research* 30 (1990): 33–49.

Tscherning, M., *Physiologic Optics*, trans C. Weiland (The Keystone, Philadelphia, PA, 1900), pp. 63–66.

Walsh, G., The effect of mydriasis on the pupillary centration of the human eye, *Ophthalmic and Physiological Optics* 8 (1988): 178–182.

Wildenmann, U. and Schaeffel, F., Variations of pupil centration and their effects on video eye tracking, *Ophthalmic and Physiological Optics* 33 (2013): 634–641.

Wilson, M. A., Campbell, M. C., and Simonet, P., Change of pupil centration with change of illumination and pupil size, *Optometry and Vision Science* 69 (1992): 129–136.

Yang, Y., Thompson, K., and Burns, S. A., Pupil location under mesopic, photopic, and pharmacologically dilated conditions, *Investigative Ophthalmology and Vision Science* 43 (2002): 2508–2512.

18 The retina and the Stiles–Crawford effects

Brian Vohnsen

Contents

18.1 RETINAL PHOTORECEPTOR CONES AND RODS

Refraction by the anterior eye is essential for proper focusing of light in the formation of images of the exterior world onto the retina. However, it is the absorption by pigments located within the photoreceptor cells that triggers the visual system. Understanding light–photoreceptor interactions is therefore necessary to unravel the complexities in the last optical step in the eye prior to subsequent neural responses. In the human retina, there are two kinds of photoreceptors that are responsible for vision, namely, the rods and the cones. The rods are responsible for dim light (scotopic) vision, whereas the cones are responsible for vision in normal and bright light (photopic) conditions. The transition from pure rod to cone-mediated vision is a combination of the two (mesoscopic) whereby the visual system has an astonishingly large dynamical range that spans about 12 log units, most of which is accomplished by the retina as changes in pupil size account for little more than 1 log unit.

The visual pigment of the rods is rhodopsin whose peak of absorption is at a wavelength of approximately 496 nm, whereas three different types of photopsin pigments are present in the cones categorized as S, M, and L (short, medium, and long wavelength) or blue, green, and red sensitive cones with broad absorption peaks centered at wavelengths of 419, 531, and 559 nm, respectively [1] as seen in Figure 18.1. The density of each type varies across the retina, and between individuals, but it is typically on the order of 5% S, 30% M, and 65% L. In unbleached conditions each photoreceptor can absorb up to 2/3 of the incident light for wavelengths near its absorption peak and 2/3 of the absorbed photons will result in photoisomerization (cis → trans) of the visual pigment and thus a maximum catch efficiency of approximately 45% [2]. The absorption of a quantum of light triggers a biochemical cascade of events that results in a modified electrical current and associated photovoltage across the membrane wall, which is communicated via bipolar cells to ganglion cells at the inner retina and further on via the optic nerve to the visual cortex.

The rod and cone photoreceptors are elongated cells with a length of close to 100 µm (shorter in the parafovea), and in the healthy eye, they are oriented with their central axis aligned toward a common point near the pupil center [3]. Incident light reaches each photoreceptor at its synaptic end followed by the cell nucleus, the outer limiting membrane, and the inner segment that contains high-refractive-index mitochondria at the far end ellipsoid [4,5], before entering the outer segment that contains approximately 1000 layers of pigment, and, if not absorbed or scattered, reaches the outer segment termination beyond which a monolayer of retinal pigment epithelium cells is located to absorb remnant light, nourish the outer segment, and provide phagocytic removal of cellular waste. The layered outer segments of rods consist of stacked lipid bilayer discs, whereas for cones they consist of stacked membrane invaginations each of which contains thousands of pigment molecules [6] that are organized into an approximately square grid on each [7]. The pigment-containing discs and membranes are continuously renewed at a daily rate of approximately 10% with the oldest pigments being shredded at

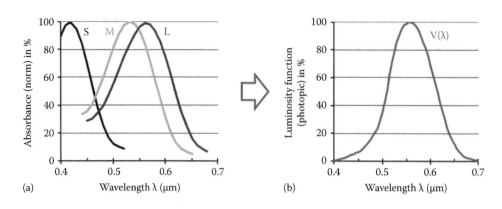

Figure 18.1 Normalized spectral sensitivities of the S-, M-, and L-cone pigments based on data from Reference 1 (a) and the associated CIE V-λ luminosity function with highest sensitivity at 555 nm wavelength (b).

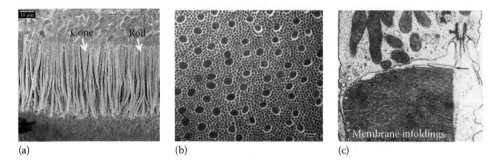

Figure 18.2 Histological images of the parafoveal rod–cone photoreceptor mosaic: side view (a) and front view (b). Also shown (c) is a zoom-in on dense outer-segment pigment layers above which large mitochondria cells can be seen. (a and b: Courtesy of Dr. Peter Munro, Institute of Ophthalmology, University College London, U.K.; From Steinberg, R.H., Fisher, S.K., and Anderson, D.H.: Disc morphogenesis in vertebrate photoreceptors. *J. Comp. Neurol.* 1980. 190. Figure 3. Copyright Wiley-VCH Verlag GmbH & Co. KGaA. Reproduced with permission.)

the outer segment terminations [8,9]. The fact that pigments are packed into the outer rather than the wider inner segments economizes parafoveal photoreceptor metabolism by making more efficient use of pigments where visual acuity is lowest.

The total number of rods in the retina of a human eye equals approximately 120 million, which is vastly superior to the approximately 6 million cones. Rods are absent at the fovea centralis but present in the spacing between cones at other eccentricities, whereas cones have their highest density in a hexagonal packing at the fovea reaching a peak value of approximately 160,000/mm² but falling off rapidly towards the parafovea with some 7,000 cones/mm² [10]. The outer segment length is largest for the foveal cones ~35 μm and drops to ~20 μm for the parafoveal cones. The inner segment diameter of cones increases from approximately 2.2 μm at the fovea to approximately 8.0 μm at the parafovea, whereas the outer segment is tapered down to approximately 2.0 μm at the outer segment termination. The diameter of rods is smallest being below 2.0 μm. Figure 18.2 shows histological images of rods and cones and a magnified view of outer segment membranes.

18.2 INTRODUCTION TO THE STILES–CRAWFORD EFFECTS AND RETINAL DIRECTIONALITY

The Stiles–Crawford effect refers commonly to the psychophysical observation reported by Walter. S. Stiles and Brian H. Crawford in 1933 that in photopic and mesoscopic conditions,

the eye is sensitive to the entrance point of light at the pupil [11]. When light enters the eye at an off-axis pupil point, it is less efficient in triggering a visual sensation due to its oblique incidence on the retina [12]. This phenomenon is termed the Stiles–Crawford effect of the first kind (SCE-I). The reduction in visibility for obliquely incident light is described by a relative visibility, η, which is the ratio of subjectively determined brightness for a ray of light that enters the eye at a given pupil point with respect to that of a parallel ray of light (for the unaccommodated emmetropic eye) that enters the eye at the pupil point of highest visibility located near the pupil center though typically displaced nasally by approximately 0.5 mm [13]. Thus, the SCE-I is described by the location of the pupil point of highest visibility and the corresponding distribution of the relative visibility across the pupil plane. This is normally expressed by a Gaussian function:

$$\eta(r) = 10^{-\rho r^2} \tag{18.1}$$

where ρ denotes the characteristic directionality parameter and r is the distance in the pupil plane from the point of highest visibility (if the distance is measured with respect to the geometrical pupil center, this equation is modified to $\eta(r) = 10^{-\rho|r-r_0|^2}$ where $r_0 = (x_0, y_0)$ denotes the location of the point of highest visibility and $r = (x,y)$ denotes any point across the pupil). Equation 18.1 was introduced by Stiles and Crawford using, respectively, a base-10 logarithm [14] and the natural base-e logarithm [15]. The value of the directionality parameter is lowest at the fovea (with a typical value of 0.05/mm²) and increases with retinal eccentricity (to approximately 0.10/mm²) although the accuracy of parafoveal SCE-I measurements is lower [16,17]. In general, the directionality might differ in the vertical and horizontal directions across the pupil (whereby the exponent in Equation 18.1 is replaced using $\rho r^2 \rightarrow \rho_x x^2 + \rho_y y^2$), but observed differences are in general small [13] and rotational symmetry $\rho \equiv \rho_x \equiv \rho_y$ will for the most be used in the remainder of this chapter. Although the SCE-I was first discovered using white light, it was soon repeated with narrow-bandwidth light, which led to the discovery of a subtle wavelength dependence of the directionality parameter, whereas the pupil location of the peak of visibility remains stable across the visible spectrum [12,14]. In scotopic conditions a psychophysical directionality is also present, though commonly neglected, as its impact is much smaller ($\rho \approx 0.01$/mm²) [18,19]. Measurement results for photopic and scotopic conditions of the SCE-I are shown in Figure 18.3.

With the incident light power adjusted to compensate for the reduction in brightness caused by the SCE-I, a minor change in color appearance was first observed by Stiles [14] and termed the Stiles–Crawford effect of the second kind (SCE-II) by Hansen [20]. This hue shift remains less understood than the SCE-I, but the two effects are believed to share a common origin with the SCE-II containing additional information about the availability of visual pigments and the role of saturation.

Light that is backscattered from the retina displays also a directionality that is even more pronounced and has a spectral dependence that differs from its psychophysical counterpart. This effect is known as the optical Stiles–Crawford effect (OSCE) [21–25].

The fact that the SCE-I, SCE-II, and OSCE share a common central pupil point with an associated directionality has led the present author to search for a unified model that will encapsulate the main characteristics of all three effects [26–28]. Measurement uncertainties, variations among subjects, and the fact that different methodologies have been used are the main reasons why even today the origins of the Stiles–Crawford effects have still not been fully comprehended although significant insight has been gained. This was marked with a symposium in 2008 at the Frontiers in Optics conference commemorating "The Stiles-Crawford effects of the first and second kinds, 75 years of scientific achievements," and related publications summarize both the original discoveries and significant progress made [29,30] in the now more than 80 years since the discovery of the SCE-I.

That photoreceptors have directional properties has actually been known since 1844 when Brücke reported that these elongated cylindrical cells are endowed with an elevated refractive index and that their axes are aligned and point toward the eye pupil [31,32]. The pointing of cones is adaptable showing a kind of phototropism that has been demonstrated following cataract surgery and eye patching after which the point of highest visibility moves gradually toward the new pupil center in the course of some 10 days [33,34]. Confocal microscopy imaging of frog retinas removed from the eye cup, but maintained in an isotonic liquid environment to maintain cellular activity (physiological environment), has confirmed a remarkable reorientation of photoreceptor cells that adjust toward the incident rays so as to capture most light [35] although the underlying biomechanics involved remain little understood.

Already in 1843, Hannover had reported on detailed hexagonal microscopic patterns in individual photoreceptors of frog and fish retinas observed shortly after their death [36]. In 1961, Enoch provided experimental evidence of modal waveguide radiation patterns observed in transmission from single photoreceptors in isolated retinal tissues from primates including humans [37,38] examples of which are shown in Figure 18.4. Packer et al. made

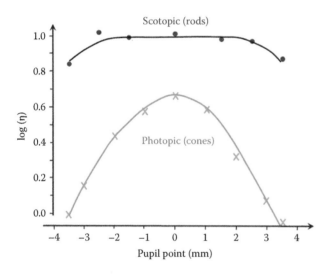

Figure 18.3 Photopic and scotopic Stiles–Crawford effect of the first kind visibility data (semilog plot) for one subject at 6° nasal obtained with orange-red light. For photopic conditions the directionality is $\rho \approx 0.045$/mm², whereas for scotopic conditions it is $\rho \approx 0.010$/mm² (but only half of that if the outermost points are excluded, in which case the distribution is almost flat). (Reprinted from *Vision Res.*, 15, Van Loo, J.A. and Enoch, J.M., The scotopic Stiles-Crawford effect, 1005–1009, Figure 5, Copyright 1975, with permission from Elsevier.)

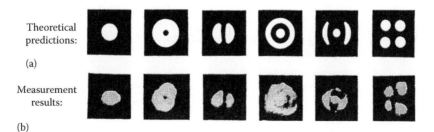

Figure 18.4 Waveguide intensity distributions measured in transmission (b) from outer segments in retinal tissues (rat, monkey, and human) compared with the theoretical expectations based on individual cylindrical waveguide modes (a). The associated *V*-number increases from left (single mode) to right (multimode). (Reproduced from Enoch, J.M., *J. Opt. Soc. Am.*, 53, 71, 1963, Figure 6. With permission of Optical Society of America.)

microscopic photopigment density imaging using primate retinas before and after bleaching and found also a difference in optical path length of the rod and cone outer segment terminations [39]. It is important to keep in mind that all such observations are made only after having removed the retina from the eye. This makes it prone to alterations at short time scales that may lead to modified optical properties and induce geometrical distortions when compared to those of the living retina in the intact eye. To overcome these limitations there is a great need for new methods that can accurately probe the photoreceptors in the living eye even at a subcellular level.

18.3 METHODS USED TO EXAMINE THE STILES–CRAWFORD EFFECTS

The SCE-I is typically analyzed using a dual-path projection setup in which a reference light enters the eye at the pupil point of highest visibility and a separate test light enters the eye at sequentially different positions across the pupil either along a single horizontal cross section or in both horizontal and vertical traverses. The displacement of the test light is realized by rotation of a beam splitter [11,14] or a half mirror mounted at the front focal plane of a lens used to generate a Maxwellian view [40] at its back focal plane [41,42] or by linear translation of the projected light source onto the pupil plane [13,43]. Computer control of intensity and wavelength with liquid crystal filters [41,42] can speed up measurements although most time is required for the subjective determination of effective visibility at each entrance point. A uniaxial flickering system can minimize system errors caused by slight differences in dual-path systems such as minor spectral differences in the two paths [44]. Dilation of the eye pupil is required to approach an 8 mm pupil and a bite bar is typically used to minimize head motion during measurements. Both the reference and the test lights are observed in a Maxwellian view whereby the light enters the eye through a small pupil area to minimize the impact of ocular aberrations and to make it appear uniform within the field of view [40]. The apparent brightness of the reference field is then compared with the test field and one of them is adjusted

in illumination power until a satisfactory match between the two has been accomplished, as judged by the subject, from which a measurement of the relative visibility can be made. Measurements are done monocularly (albeit binocular analysis has been attempted with one eye used for the test and the other eye used for the reference illumination [12]) either using a bipartite field where the test and reference fields are viewed simultaneously side by side extending typically 1–2 visual degrees or letting the two fields overlap in space but flicker sequentially in time so that at any instant only the reference or the test is visible [11]. Minor variations of these methods exist in which the test and reference fields have different sizes or systems in which the viewer sees the reference field continuously onto which the test field is projected for incremental visibility determination.

It is vital that the size as well as the uniform appearance of both the test and reference fields remain stable throughout the experiment to avoid systematic errors in the visibility determination. Aberrations will tend to displace the test field when the entrance point is shifted across the pupil, and therefore lower-order refractive errors are typically corrected using ophthalmic lenses [13], a Badal system [43], or current-driven tuneable lenses [42,44]. The two methods (bipartite and flicker) for SCE-I characterization are shown schematically in Figure 18.5. In either case, it is an enduring task to perform accurate measurements across the entire pupil without unintended movements of the eye for extended periods of time from minutes to hours, and thus only with highly collaborative subjects can accurate measurements be performed. To combat this, the characterization can be made objective by recording electroretinograms using different angles of incidence although the relation to the psychophysical SCE-I is nontrivial with signal contributions originating both from post-photoreceptor bipolar cells and from inner retinal neurons [45–47].

Rapid changes in the illumination direction can result in sudden brightness increases known as the transient Stiles–Crawford effect [48]. These can almost halve the apparent directionality when periodic flickering at 1 Hz is compared to that of 10 Hz (with the latter producing similar results to that of the conventional SCE-I) [44]. Following bleaching a change in directionality has also been observed once visual pigments recover from the intense illumination [49]. Even a sudden change in the direction of linear polarization of obliquely incident light can cause a

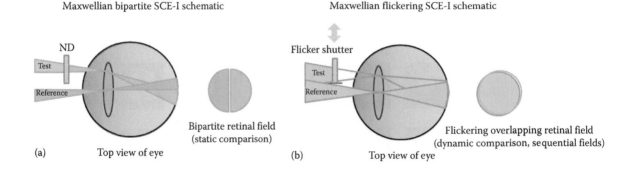

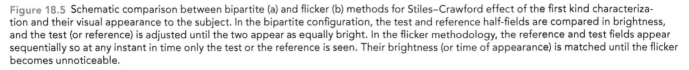

Figure 18.5 Schematic comparison between bipartite (a) and flicker (b) methods for Stiles–Crawford effect of the first kind characterization and their visual appearance to the subject. In the bipartite configuration, the test and reference half-fields are compared in brightness, and the test (or reference) is adjusted until the two appear as equally bright. In the flicker methodology, the reference and test fields appear sequentially so at any instant in time only the test or the reference is seen. Their brightness (or time of appearance) is matched until the flicker becomes unnoticeable.

change in apparent brightness by modified absorption [50]. These effects give insight into the temporal dynamics of the photoreceptor responses and the role of the spatial distribution of visual pigments.

How the SCE-I adds up across the pupil has been a central question already in the original work by Stiles and Crawford [11]. This question is pertinent to normal vision and pupil size using a Newtonian rather than a Maxwellian view since the angular spectrum of light at the retina differs in the two cases which may cause brightness [28] and color differences [51]. From integration of Equation 18.1 across a pupil diameter $d = 2R$, an effective pupil diameter d_{eff} ($<d$) can be derived as

$$d_{eff} = 2\sqrt{\frac{1-10^{-\rho R^2}}{\rho \ln(10)}} \tag{18.2}$$

which approaches $1.32/\sqrt{\rho}$ for large pupils (and thus $d_{eff} \sim 6$ mm for a directionality parameter $\rho = 0.05/\text{mm}^2$). Additivity has been found valid for normal pupil sizes using incoherent and partially coherent light, but for larger pupils differences appear due to blur even after correction of defocus [52–54]. Interestingly, this is not the case for highly coherent monochromatic laser light using annular pupils where light is symmetrically incident on the retina leading to complete cancellation of the SCE-I. In this case, there is no wave front slope across the resulting speckle pattern perceived by the photoreceptors as expressed by the Poynting vector of the light in the retinal plane [55–57]. The case of SCE-I characterization with coherent light is shown in Figure 18.6.

Characterization of the SCE-II is performed using the same kind of experimental setting as for the SCE-I with the added complexity of an adjustable wavelength for the test to allow for better color matching with the reference field [14,41] or, ideally, three separate adjustable sources with distinct wavelengths and brightness so that a complete color match can be obtained [58]. The measurement is realized by first making a standard visibility match to compensate for the SCE-I and only thereafter to determine the best color match to correct for the SCE-II.

Finally, characterization of the OSCE relies also on Maxwellian illumination to irradiate a bleached retinal area (which increases directionality and signal [59]) in order to capture an image of backscattered light with a camera that is located in a conjugate plane of the eye pupil [22,25] although a variant is possible where both the incident light and a closely spaced small collection aperture are scanned in tandem across the eye pupil with light being captured point by point using a photomultiplier tube instead of a camera [60].

Imaging of the retina can be used to characterize the Stiles–Crawford effect both in terms of an average directionality with scanning laser ophthalmoscopy (SLO) [61] and optical coherence tomography (OCT) [62] and, when resolution allows it, at the level of single photoreceptors [63,64]. In such measurements only the incoming light [63,64] or simultaneously both the incoming light and the collection aperture [61,62] are displaced across the eye pupil while capturing images with a brightness that is used for the analysis of directionality. It must be stressed that this objective method is not probing the visual response to oblique incidence but rather the directionality of light acceptance by and

backscattering from the photoreceptors. When both the incident and the collected light is oblique, the resulting effect may be viewed as a combination of the SCE-I and the OSCE with a correspondingly higher directionality [26]. Finally, the fact that the beam used in scanning retinal imaging technologies is being tightly focused onto the retina [61,62,64] may result in differences when compared to flood illumination of an extended retinal area [22,63] due to the match of illumination spot size to the photoreceptors being imaged [64].

18.4 EXPERIMENTAL RESULTS AND SUBJECTIVE VARIATIONS

Psychophysical analysis of SCE-I is a cumbersome process that requires not only accurate alignment and system calibration but also significant collaboration by the subject to minimize measurement errors. As a result, most studies have examined a very limited number of subjects although Applegate and Lakshminarayanan did report on monocular measurements with 670 nm red light and individual refractive corrections for 49 young subjects finding a directionality parameter in the range of $0.048/\text{mm}^2$–$0.053/\text{mm}^2$ (horizontal, vertical) [13]. Figure 18.7 shows SCE-I and SCE-II results obtained in the right eye of the author using a bipartite field at three distinct wavelengths. Each entrance point has been measured four times to determine average visibility and standard derivations for the SCE-I, whereas the SCE-II results are based on eight repeated measurements at each pupil entrance point.

Since photoreceptors point toward a common pupil point, eyes with a short axial length may have more inclined photoreceptors than eyes having large axial length as shown schematically in Figure 18.8. In developing eyes, growth mechanisms work toward maintaining emmetropization and in studies of myopia using chicks fitted with contact lenses altered eye growth can indeed be observed [65]. Relatedly, the SCE-I directionality and photoreceptor pointing might play a role for accommodation [66–68] although this remains somewhat an open question. In a study of highly myopes, a 16% reduction in effective directionality was found [69], which agrees qualitatively well with the previous argument. A simple geometrical analysis [42] suggests that directionality for the myopic eye, ρ_M, is reduced when compared to that of an emmetropic eye, ρ, as

$$\rho_M \approx (1 + DL)^2 \rho \tag{18.3}$$

where L is the effective axial length (length normalized by the its refractive index) of an emmetropic eye ($\sim$17 mm) and the refractive error ($D < 0$) is expressed in diopters. In turn, from Equation 18.3 it is expected that hyperopic eyes ($D > 0$) have a higher directionality although this remains to be studied.

As a potential clinical tool, the SCE-I, the SCE-II, and the OSCE have still not been fully explored. The retinal appearance of drusen, age-related macular degeneration [70], cone dystrophy [71], and retinitis pigmentosa [45,71] may all perturb photoreceptor alignment and thereby alter the Stiles–Crawford functions. Indeed, a reduced directionality and dislocated visibility peak locations in the pupil plane have been reported for patients using both psychophysical (SCE-I) and objective (OSCE) measures.

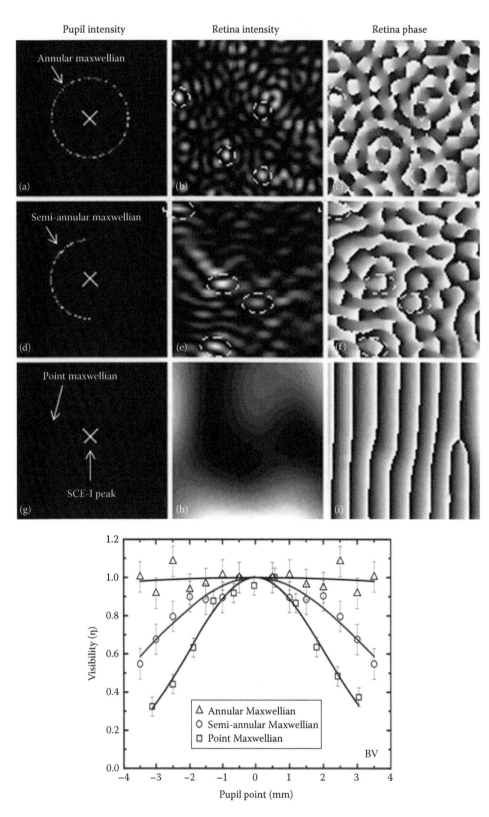

Figure 18.6 Calculated intensities at the pupil for a 5 mm annular (a, b, c), semi-annular (d, e, f), and point-like (g, h, i) Maxwellian illumination with randomized phase and corresponding retinal images (33 × 33 μm) showing the intensity and phase (2π-wrapped) in the retinal field. Yellow circles mark selected bright speckles and their corresponding phase maps of the wave front across them. Below, corresponding Stiles–Crawford effect of the first kind (SCE-I) measurement results for the author's right eye showing a partial (semi-annular) and complete (annular) SCE-I cancellation once the wave front tilt across the speckles is absent. The directionality parameter decreases from 0.051/mm² (Maxwellian point) to 0.019/mm² (semi-annular) and to 0.0008/mm² (annular). The simulations and the experiments used a 632.8 nm wavelength (HeNe laser) except for the Maxwellian point results that have been obtained in a bipartite setup using a tungsten–halogen source and a 620 nm wavelength filter (10 nm band pass). (Reproduced from Vohnsen, B. and Rativa, D., *J. Vis.*, 11, 19, Figures 3 and 7. Copyright 2011. With permission from Association for Research in Vision and Ophthalmology.)

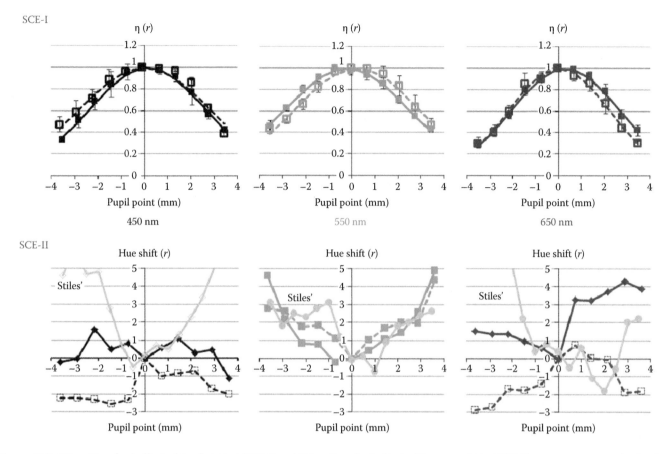

Figure 18.7 Stiles–Crawford effect of the first kind (SCE-I) and Stiles–Crawford effect of the second kind (SCE-II) measurement results for the author's right eye at three selected wavelengths. The directionality parameter is in the range of 0.029–0.045/mm² dependent on the wavelength and bandwidth (with defocus corrected the directionality increases to about 0.050–0.068/mm² [42]). In the SCE-II plots, results from Stiles (1937) have been included for comparison (at 457, 542, and 636 nm wavelengths: yellow curves) with the hue shift expressed in nm-units. The bandwidth of the color filters used increases with wavelength. Solid lines (solid square symbols) have been obtained with a wide-bandwidth setting, whereas dashed lines (open square symbols) have been obtained with a narrow-bandwidth setting of the illumination. (The SCE-I images have been reproduced from Lochocki, B. et al., *J. Mod. Opt.*, 58, 1817, 2011, Figure 4. With permission from Taylor & Francis.)

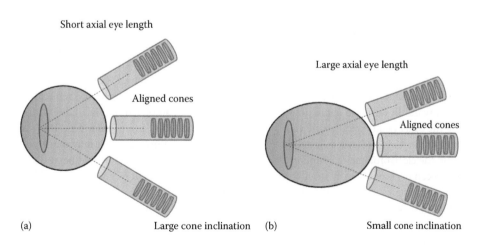

Figure 18.8 Schematic of expected cone pointing in eyes having (a) short or (b) large axial length (assuming that cones are packed equally dense).

18.5 OPTICAL MODELS OF THE PHOTORECEPTOR CONES

Different models of photoreceptor–light interactions have been proposed based on principles that range from geometrical optics to electromagnetic wave propagation. For the most, models include only light in the forward direction as reflections are weak because of the low contrast of refractive indices across the photoreceptors. Nonetheless, objective methods and retinal imaging relies on backscattered light and therefore some models have included reflections in the analysis.

18.5.1 MODELS OF THE SCE-I AND THE OSCE

It was clear from the outset that the SCE-I must originate in the retina as no other structure in the eye, or reflective differences at the cornea, could explain the significant reduction in visibility observed once an incident ray of light is displaced towards the pupil rim [11,12]. The first models of the SCE-I expanded on Brücke's description of the photoreceptor cell. Wright and Nelson followed by O'Brien argued that on account of total internal reflection light rays would be guided from the inner to the outer segment [12] and concentrated in the ellipsoid of each cone [72]. Multiple reflections in the ellipsoid could increase the angle of propagation and eventually lead to leakage. This would be most prominent for rays that enter the photoreceptor at an angle rather than along its cellular axis thereby mimicking the SCE-I before light reaches the pigments and stimulates vision by absorption in the outer segment. Rather than the angular dependence of light acceptance by single cones, Safir and Hyams [73] suggested a smearing out in the acceptance angle, and thus a reduced directionality, as a collective effect caused by photoreceptor disarray. However, subsequent work has found no significant cone disarray near the fovea of the healthy retina [63].

Toraldo di Francia argued that on account of the small size of photoreceptors, they should be considered as electromagnetic waveguides that would invalidate a purely geometrical approach [74]. Since then, waveguide models have taken sway not least due to the strong evidence of mode-like radiation patterns from photoreceptors as reported by Enoch [37,38]. Röhler and Fischer discussed waveguide modes and light absorption in photoreceptors [75], whereas Snyder and Pask introduced a complete cylindrical step-index waveguide analysis of a single nonabsorbing photoreceptor cone [76,77] that addressed directly the angular dependence of the SCE-I and has come to form the basis for much subsequent theoretical work. In doing so approximations have to be made regarding the exact geometrical shape and refractive index distribution across the photoreceptor and its surrounding medium, while the electromagnetic coupling between neighboring photoreceptors has typically been ignored despite of their tight cellular packing. The waveguide modes for a hexagonal arrangement of cylindrical waveguides representative of the cone photoreceptor mosaic has slow modes confined to each cylindrical waveguide (resembling the modes of an isolated waveguide) and faster modes confined to the space between the waveguides [78,79]. Accurate measurement of refractive indices in photoreceptors is challenging, and some variations exist in published values, but they generally find a higher refractive index

of the outer than the inner segment [80]. Thus, the ellipsoid may be understood as a matching element between the two [74,81]. Waveguide characteristics are best described in terms of the characteristic V-number defined by

$$V = \frac{\pi d_w}{\lambda}\sqrt{n_1^2 - n_2^2} \qquad (18.4)$$

where

- n_1 is the refractive index of the waveguide core (inner or outer photoreceptor segment with diameter d_w)
- n_2 is the lower refractive index of the surrounding medium

Reported refractive indices for the inner segment myoid are in the range 1.353–1.361 and in the outer segment 1.353–1.430 with the surrounding interstitial matrix having a refractive index of 1.340 (wavelength-dependent dispersion of these is typically omitted) [76,80]. When $V < 2.405$, the waveguide becomes single mode (i.e., two orthogonal linear polarization modes with identical propagation constants), which is believed to be the case for foveal cones at least toward the long-wavelength end of the visible spectrum. Scalar optics can be used for ballistic light because of the small angles of propagation from the pupil to the retina (excluding intraocular scattering), and thus across any plane in the retina, the forward-propagating electromagnetic field, ψ_r, can be expressed as a sum of a discrete set of M guided modes, ψ_m, and a continuous set of radiative (nonguided) modes, ψ_{ng}:

$$\psi_r(\mathbf{r}) = \sum_{m=1}^{M} c_m \psi_m(\mathbf{r}) + \int_{-\infty}^{\infty}\int_{-\infty}^{\infty} c_{ng}(\mathbf{k}_\parallel)\psi_{ng}(\mathbf{r},\mathbf{k}_\parallel)\,d\mathbf{k}_\parallel \quad (18.5)$$

where

- c_m and c_{ng} are scaling factors for each guided and nonguided mode, respectively
- $\mathbf{k}_\parallel$ is the in-plane wave-vector component (the integral limits will be effectively limited by the angular spectrum of the incident light)

Due to the low-refractive-index contrast of the photoreceptors and the surrounding matrix, linear-polarized modes (LP_{lm}) can be assumed which simplifies the calculations considerably [26]. It has been argued that once the light is guided the radiative components will have little impact on the absorption in the outer segments [82]. Absorption may be included in a waveguide model by use of a complex refractive index within the outer segment and possibly also in its surrounds. This results in leaking modes and for dim illumination it may modify the predictions considerably as more pigments would be available to absorb the light. Nonetheless, a reduced coupling for obliquely incident light is still seen that mimics the angular dependence of the SCE-I [83].

The fraction of incident light coupled to any waveguide can be calculated by the mode overlap integral across the entrance facet so that the total guided power (prior to absorption) in a waveguide is

$$P = \sum_{m=1}^{M} \left| \iint \psi_r \psi_m^* \, dx\, dy \right|^2 . \qquad (18.6)$$

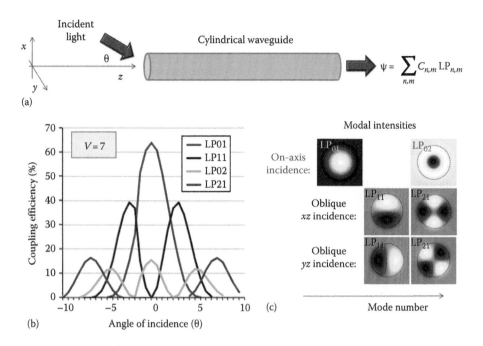

Figure 18.9 Angular dependence of (a) light coupling to different linear-polarized modes in (b) a cylindrical waveguide and (c) corresponding modal waveguide intensities. The chosen waveguide has $V = 7$ (representative of the inner segment of parafoveal cones). For on-axis incidence, light couples only to the fundamental mode LP_{01} and, when present, LP_{02}. The Gaussian angular dependence expressed by the Stiles–Crawford effect of the first kind function is approximated by the coupling efficiency for the fundamental mode (foveal cones with a small V) or by its sum for all modes (for large V).

The angular dependence of coupled light power for a plane incident wave to different LP modes of a cylindrical step-index waveguide is shown in Figure 18.9.

The fundamental mode LP_{01} is almost identical to a Gaussian mode that is the fundamental mode of waveguides having a parabolically graded index [26,84]. This is attractive from a mathematical point of view since it allows a number of analytical derivations to be made including the analysis of tilt on the SCE-I, the matching of a focused beam to the waveguide, and the role of an off-axis incidence on the coupling efficiency. These examples are shown in Figure 18.10. From the Gaussian mode an expression of

the directionality parameter for a single-mode waveguide with a mode radius w (approximately equal to the waveguide core radius) can be found as

$$\rho = 2\log(e)\left(\frac{\pi n_{eye} w}{\lambda f_{eye}}\right)^2 \qquad (18.7)$$

for a schematic eye having focal length $f_{eye} = 22.2$ mm and refractive index $n_{eye} = 1.33$. The factor of 2 in Equation 18.7 appears when illuminated with a plane wave, whereas if exposed

Coupling of a Gaussian beam $\exp(-r^2/w_r^2)$ to a Gaussian waveguide $\exp(-r^2/w_m^2)$

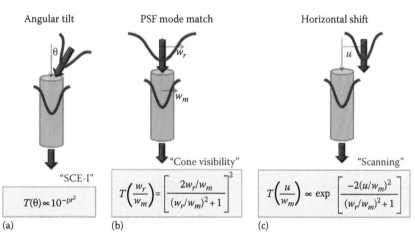

Figure 18.10 Three different cases of light coupling for an incident Gaussian beam to the Gaussian mode of a cylindrical waveguide. The transmitted power fraction is expressed by the function T. The directionality is $\rho = 2\log(e)\left(\frac{\pi n_{eye}}{\lambda f_{eye}}\right)^2\left(\frac{w_r^2 w_m^2}{w_r^2 + w_m^2}\right)$. (a) Angular tilt, (b) PSF mode match, and (c) Horizontal shift.

to a Gaussian beam of light with a waist that is matched to the waveguide mode, the factor of 2 cancels out. Equation 18.7 shows a $1/\lambda^2$ wavelength dependence that agrees well with the dependence found from the OSCE [24]. The diameter dependence is in fair correspondence with the fact that parafoveal directionality (where the inner segment diameter and thus the mode radius is larger) is higher than at the fovea [16,17]. Although Equation 18.7 has been derived for the incoming light, the same holds true in the reverse situation of light diffracted from the waveguide and propagated toward the pupil. A numerical example with $w = 1.0$ µm and $\lambda = 0.55$ µm results in $\rho \sim 0.10/\text{mm}^2$, which is larger than the common SCE-I value at the fovea but similar to what has been measured for the OSCE [22,23].

As either the wavelength or the size of the analyzed photoreceptor waveguide is changed, higher-order modes may appear that invalidate a simple Gaussian model. When the wavelength is modified, it produces sudden variations of the directionality parameter once the number of allowed modes becomes altered. These variations smoothen across a retinal patch containing many photoreceptors with slightly different waveguide parameters whereby it resembles the spectral distribution of the directionality parameter first observed by Stiles [14]. The two are shown for comparison in Figure 18.11. In the short-wavelength limit, age-dependent absorption in the crystalline lens contributes slightly to the effective directionality [85].

Detailed numerical electromagnetic finite-difference time-domain calculations in 2-D and 3-D for individual cylindrical waveguides representative of a single photoreceptor cone or rod have confirmed the angular dependence of light coupling [86] and also found a uniform spectral dependence [87] that gives added insight into the SCE-I and a plausible role for vision of the layered packing in the outer segments (although absorption was not included).

The drawbacks of any photoreceptor waveguide model are that the refractive indices are not fully known and that the shape of each cell (even the average cell) differs from that of the perfect cylindrical waveguide typically considered. Another concern is that their short length may prevent multiple reflections to occur when light is incident from normal pupil sizes being nearly parallel to the axis of each photoreceptor cell. This can severely limit the number of multiple reflections that may take place for the effective buildup of waveguide modes before possible absorption occurs. Moreover, for oblique incidence it is not clear what happens to the nonguided radiative modes that cannot contribute to vision as their inclusion would annul the directionality of the SCE-I. Thus, a waveguide model might oversimplify the actual interaction of light with visual pigments in normal nonbleaching illumination conditions unless both guided and radiative components of the light are fully accounted for. Enoch had stressed a similar concern prior to publishing his finding of transmission modes in bright light conditions of retinal tissue removed from the eye [88]. In the context of the OSCE, a single aperture diffraction model has been used to model waveguiding from photoreceptors in conjunction with a rough surface model of the retina [23]. Backscattering of light from the choroid has also been used to describe the wavelength dependence of the directionality parameter although this approach appears most appropriate in intense bleaching light where the likelihood of absorption is decreased [89].

The present author has taken a new approach to address the light-gathering capabilities of outer segment pigment layers using optical reciprocity and antenna principles to argue for identical radiative lopes and collective acceptance angles of the photoreceptors [28]. Thus, the light acceptance angle by a single layer of dense pigments in an outer segment should correspond to the diffraction angles from an aperture with equal dimensions.

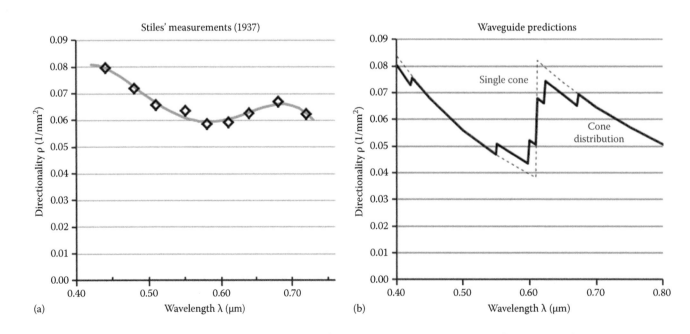

Figure 18.11 Spectral dependence of the directionality parameter from measurements by Stiles (a) and from waveguide predictions (the single cone and cone distribution plots, the latter with contributions from five different waveguide diameters) showing a similar though more pronounced reduction in directionality in the green-orange wavelength range (b). ([a]: Reproduced from Stiles, W.S., *Proc. R. Soc. Lond. B*, 123, 90, 1937, Figure 7. With permission by the Royal Society, U.K.; [b]: Reproduced from Vohnsen, B., *J. Mod. Opt.*, 56, 2261, Figure 2. Copyright 2009. With permission from Taylor & Francis.)

Subdividing each layer into distinct light-induced point dipoles (representative of visual pigment molecules) allows light scattering and acceptance calculations to be made in any direction when assuming that all other refractive indices beyond the outer segment can be taken as equal (this is for simplicity rather than a fundamental limitation). Pigment molecules are densely packed on each layer [6,7], which makes them collectively to have similar light-gathering capabilities as uniform layers. Thus, by stacking apertures to represent one or multiple outer segments, it allows for modelling of their collective light acceptance. This approach ignores multiple scattering between layers [28] notwithstanding that it may produce resonant effects at specific wavelengths [87]. Light scattering between adjacent photoreceptors can easily be included, and multiple scattering could potentially be included via a self-consistent set of equations that can account for coupled scattering between layers and neighboring outer segments [28,90]. Optical reciprocity implies that the outer segment receives light in the same way that it would emit had it been the source of the light and the propagation direction reversed. This is true for the waveguide model of the retinal receptor, and it is equally valid for a layered model of the outer segment. Therefore, such an approach is useful to determine the pupil field for light propagated from (or to) each layer in the outer segment.

This pupil field can be calculated using the standard Fraunhofer diffraction equation for single circular apertures each having the diameter d_w of the outer segment layer that it represents. For N parallel apertures spaced at a layer-to-layer distance of δ, the diffracted fields in the pupil plane can be added resulting in

$$\psi_{pupil}(r) = \sum_{n=1}^{N} \frac{A_n n_{eye}}{i\lambda z_n} \exp\left(ik\left(z_n + \frac{r^2}{2z_n}\right)\right)\left(\frac{2J_1(krd_w/2z_n)}{krd_w/2z_n}\right) \quad (18.8)$$

where J_1 is the first-order Bessel function of the first kind and the wave number is $k = 2\varphi n_{eye}/\lambda$. The $z_n = f_{eye} + h + (n-1)\delta$ is the axial distance from aperture n to the pupil of the schematic eye and h is the distance from the first outer segment layer to the inner limiting membrane. The total length of the outer segment is $L = N(\delta - 1)$. Since all the apertures at different depths in the outer segment are driven by the same incident light, this must be phase-locked at the pupil, which is accomplished by connecting the amplitude factors between the layers using $A_{n+1} = A_n \exp(-ik\delta)$.

The straightforward addition from each layer in Equation 18.8 is based on the assumption that all layers capture light independently. The electromagnetic wave is thus allowed to diffract beyond the cell boundaries of the photoreceptor and waveguiding is not enforced. The high refractive index of outer segments is due to the high density of pigments contained in the stacked membrane infoldings or discs. Dim light will be absorbed in a fraction of each outer segment, and only intense light will propagate all the way to the outer segment terminations. Absorption can be introduced ad hoc in the axial direction by use of Beer–Lambert's law in the amplitude factor A_n to exponentially dampen the impact of pigments located deep within the outer segment. Disarray of the layers in a single photoreceptor, or between adjacent photoreceptors, would still introduce a smearing out and reduction of the directionality. The layered diffraction model and resulting directionality curves (intensity at the pupil plane $\left|\psi_{pupil}\right|^2$) for a schematic eye for dim and bright light, respectively, is shown in Figure 18.12. In this model, the illumination brightness becomes a deciding factor as for dim light (absorbed in a fraction of the full outer segment length) directionality is low, but in bright light conditions more pigment is bleached and the directionality parameter increases (due to the contribution from deeply located layers) thus giving a direct physical meaning to the role of bleaching for the SCE-I [91]. Thus the model does not only give directionality predictions for the SCE-I but also for the OSCE usually analyzed in bleached conditions where contributions from the entire outer segment length will increase the effective directionality.

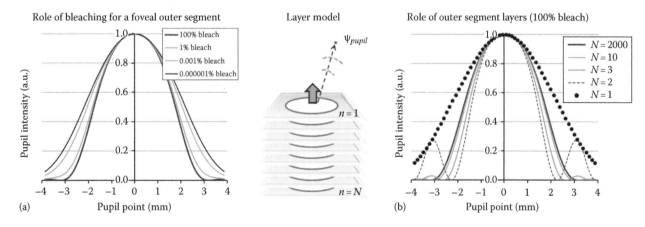

Figure 18.12 Layered scattering model showing calculated pupil intensity distributions at $\lambda = 550\,nm$ for a foveal outer segment in a schematic eye model. The chosen outer segment has dimensions $L = 40\,\mu m$, $d_w = 2\,\mu m$, and $N = 2000$. Beer–Lambert's law has been employed to show the role of bleaching (a) to modify the contribution from deeply located layers. This shows that for dim light the Stiles–Crawford effect of the first kind directionality is low ($\rho = 0.066/mm^2$: blue curve), whereas when fully bleached the directionality is high ($\rho = 0.113/mm^2$: red curve) as representative of the optical Stiles–Crawford effect. The directionality is lower for a shorter outer segment if its diameter is kept unaltered (e.g., for $L = 20\,\mu m$ the directionality becomes $\rho = 0.0585/mm^2$ and $\rho = 0.0759/mm^2$, respectively, for the same two cases of low or high bleach). The number of outer segment layers included in the calculation (b) plays no significant role once $N \geq 10$.

18.5.2 MODELS OF THE SCE-II

The SCE-II has most commonly been considered a result of self-screening by pigments and light leakage rather than waveguiding [11,92,93]. When oblique light traverses the retina, the distance through which it can interact with visual pigments is shortened as compared to on-axis incidence. This ray-optical approach seems somewhat at odds with the waveguide model. Independent of the exact mechanisms involved, it is worth noting that most studies have made use of a finite bandwidth illumination and thus differences in visibility across the spectral band should be expected. To examine this further one may express the spectral sensitivity curve (luminous function) for the eye as $\eta_{max}(\lambda)$, with the highest sensitivity equal to unity at a wavelength of 555 nm, whereby the SCE-I function from Equation 18.1 can be modified into a spectrally dependent "SCE-I color visibility function" [27]:

$$\eta(r,\lambda) = \eta_{max}(\lambda)10^{-\rho(\lambda)r^2} \qquad (18.9)$$

where the wavelength dependencies of the effective visibility and directionality parameter have been made explicit. The wavelength derivative of this function is

$$\frac{\partial \eta(r,\lambda)}{\partial \lambda} = \left[\frac{d\eta_{max}}{d\lambda} - \eta_{max}\frac{d\rho}{d\lambda}r^2\ln(10)\right]10^{-\rho(\lambda)r^2}. \qquad (18.10)$$

Any visibility difference across the spectral bandwidth of the illumination, $\Delta\lambda$, for a chosen central wavelength λ_0, would cause an incremental visibility difference (and thus a hue shift) equal to $\Delta\eta = \partial\eta/\partial\lambda|_{\lambda_0} \Delta\lambda$. Analysis of the SCE-II consists of a subjective color comparison between the test and reference fields that enter the eye pupil through pupil point $r = r_0$ and $r = 0$, respectively, and only after having compensated for the SCE-I by multiplying with $10^{\rho(\lambda_0)r_0^2}$ at the chosen wavelength. Equation 18.10 can be rewritten to consider this comparative case resulting in

$$\Delta f = \left(\left.\frac{\partial\eta(r,\lambda)}{\partial\lambda}\right|_{r=r_0} - \left.\frac{\partial\eta(r,\lambda)}{\partial\lambda}\right|_{r=0}\right)10^{\rho(\lambda_0)r_0^2}$$

$$= \frac{d\eta_{max}}{d\lambda}(10^{-(\rho(\lambda)-\rho(\lambda_0))r_0^2}-1) - \eta_{max}\left(\frac{d\rho}{d\lambda}r_0^2\ln(10)\right)10^{-(\rho(\lambda)-\rho(\lambda_0))r_0^2}$$

$$(18.11)$$

which can be interpreted as a "SCE-II hue-shift function" per unit wavelength caused by a finite bandwidth and the spectrally non-uniform sensitivity curve and directionality parameter of the eye [27]. The incremental visibility is found by multiplying Equation 18.11 with the spectral bandwidth $\Delta\lambda$. If the dispersion of the directionality parameter is partially ignored by setting $d\rho/d\lambda = 0$ (but kept in the exponent), Equation 18.11 can be simplified to

$$\Delta f \approx \frac{d\eta_{max}}{d\lambda}(10^{-(\rho(\lambda)-\rho(\lambda_0))r_0^2}-1) \approx \frac{d\eta_{max}}{d\lambda}\ln(10)(\rho(\lambda)-\rho(\lambda_0))r_0^2. \qquad (18.12)$$

In this approximation, no hue shift should be expected at the spectral sensitivity peak of the eye where $d\eta_{max}/d\lambda = 0$. In turn, the largest predicted hue shift (positive or negative) should occur at

wavelengths that coincide with the largest slope of the spectral sensitivity curve of the eye shown in Figure 18.1. Based on the CIE (international commission on illumination) data, the largest slope is at wavelengths of 510 nm and 605 nm, respectively, where the largest positive or negative hue shifts should therefore be expected.

Near to the pupil point of highest visibility ($r = 0$), the predicted hue shift has a parabolic dependence on the tested pupil point location r_0 centered at the peak of visibility, and it may cause either a negative or a positive hue shift. At short wavelengths (below 555 nm), $d\eta_{max}/d\lambda > 0$ for which the upper end of the spectral band dominates, whereas at long wavelengths (above 555 nm), $d\eta_{max}/d\lambda < 0$, and the lower end of the spectral band would dominate in the visual response. For any finite bandwidth, the pre-compensation of the SCE-I will therefore be incomplete when analyzing the SCE-II. The added complexity of the second term in Equation 18.11 containing the spectral derivative of the directionality parameter complicates the analysis and may modify somewhat the predictions. Assuming the same spectral dependence as for the OSCE, that is, $\rho \propto 1/\lambda^2$, then the derivative $d\rho/d\lambda < 0$ and the second term in Equation 18.11 will raise the sum of the terms. Equations 18.9 and 18.11 are both shown in Figure 18.13 that also highlights selected cross sections. As can be seen, Δf is in qualitatively good agreement with reported SCE-II distributions, both as a function of pupil point and as a function of wavelength [14,41]. It must be stressed that in this model entirely monochromatic (zero bandwidth) illumination produces no hue shift. Differences in the directionality parameter of different cone types could be included in the model (S-cones are slightly wider than M and L cones and may therefore have a slightly higher directionality) as well as partial bleaching of S-, M-, or L-cone pigments (modifying the initial 5%, 30%, and 65% distribution) allowing apparent color changes even for entirely monochromatic light.

18.6 FITTING FUNCTIONS FOR THE STILES–CRAWFORD EFFECTS AND THE DIRECTIONALITY PARAMETER

It is often convenient to plot the logarithm of Equation 18.1 rather than the Gaussian function of Stiles and Crawford [14,15] as this gives a parabolic dependence on pupil point, that is, $\log(\eta) = -\rho r^2$. The Gaussian SCE-I function (or its parabolic variant) is the most commonly used model when fitting visibility measurement data not least because of its elegant simplicity but also because it makes directly use of the directionality parameter that simplifies the comparison of results with other studies. It has been verified that when an inverse Gaussian SCE-I absorption filter is applied to the eye, it is possible to annul the directionality effect observed [94]. Rativa and Vohnsen introduced a modified super-Gaussian function that includes higher-order terms in the exponent to allow flattening of the visibility function and found that it can be adapted to fit experimental results (at short wavelengths where a multimodal waveguide dependence would be expected), but it also increases the complexity by requiring higher-order directionality parameters [95]. Moon and Spencer introduced a polynomial (conceptually similar to a super-Gaussian) and a nonpolynomial function for SCE-I fitting [96], and

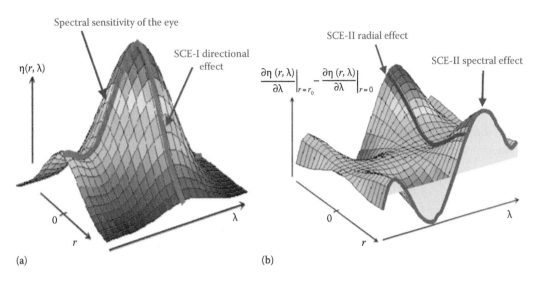

Figure 18.13 Stiles–Crawford effect of the first kind (SCE-I) color visibility function (a) and Stiles–Crawford effect of the second kind hue-shift function (b). Note that the latter has not been pre-compensated for the SCE-I, which, if included, will raise the function. (Reproduced from Vohnsen, B., *J. Mod. Opt.*, 56, 2261, Figure 8. Copyright 2009. With permission from Taylor & Francis.)

Enoch introduced a function that improved the quality of fitting for data collected near the pupil rim [52]. The author of this paper has recently introduced a scaled Airy disc distribution as a new fitting function for the SCE-I based on the aperture diffraction model for outer segment layers [28,97] and expressed by

$$\eta(r) = \left(\frac{2 J_1(\alpha r)}{\alpha r}\right)^2 \qquad (18.13)$$

where to a first approximation $\alpha \approx 2\sqrt{\rho \ln(10)} \approx 3.03\sqrt{\rho}$. The Airy disc function is compared to the Gaussian SCE-I function in Figure 18.14, and all the various fitting functions are summarized in Table 18.1.

Table 18.1 **Rotationally symmetric visibility functions used to represent the SCE-I as a function of pupil point ($r = 0$ at the point of highest visibility)**

AUTHOR(S)	SCE-I FUNCTIONS
Stiles [14]	$\eta(r) = 10^{-\rho r^2}$
Crawford [15]	$\eta(r) = e^{-\beta r^2}$
Moon and Spencer [96]	$\eta(r) = 1 - \beta r^2 + \gamma r^4$
Moon and Spencer [96]	$\eta(r) = (1 - a) + a\cos(br)$
Enoch [52]	$\eta(r) = A(1 + \cos Br)^2$
Rativa and Vohnsen [95]	$\eta(r) = a_1 10^{-\rho_1 r^2} + a_2 10^{-\rho_2 r^4} + a_3 10^{-\rho_3 r^6}$
Vohnsen [28]	$\eta(r) = \left(\dfrac{2 J_1(\alpha r)}{\alpha r}\right)^2$

18.7 VISUAL IMPLICATIONS OF THE STILES–CRAWFORD EFFECTS

In optical models of the eye, the retina is often considered as a screen onto which the anterior eye projects images of the outside world whereby a visual response is triggered. The SCE-I is typically included as a Gaussian pupil apodization notwithstanding that it is of retinal origin. Thus care needs to be taken as the directionality has been determined with Maxwellian illumination that may give rise to differences from that of normal vision. This can be seen in observed deviations of the integrated Stiles–Crawford effect from the mathematical integration of the SCE-I [52–54]. However, in the healthy eye with small aberrations, and photoreceptors oriented toward a common point, the approach is expected to be valid.

The common explanation of the SCE-I is that it is a mechanism of the eye that serves to dampen the role of intraocular scattering [98] as well as aberrations [99] that could otherwise degrade the effective retinal images captured by the photoreceptor cells.

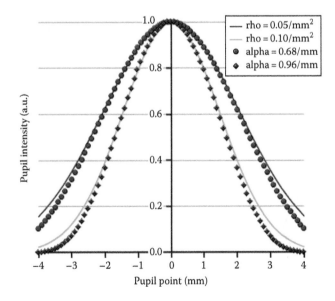

Figure 18.14 Comparison of the standard Gaussian Stiles–Crawford effect of the first kind function for two different directionality parameters (lines) with corresponding Airy disc functions (symbols) based on Equation 18.13. Differences become only apparent near the pupil rim.

Intensity point spread function (PSF) analysis and its associated modulation transfer function (MTF) are widespread measures used to determine the role of aberrations and SCE-I pupil apodization when analyzing the optics of the eye as a function of spatial frequency. For a schematic eye model, the PSF (excluding constants) can be expressed as the absolute square of the Fourier transform (FT) of the pupil function P_{eye} (equal to unity within the circular pupil and zero outside), the wave front aberrations of the eye Φ_{WA} propagated to the pupil plane, and a Gaussian amplitude pupil apodization $G_{SCE-I}(x,y) = 10^{-\rho(x^2+y^2)/2}$, that is,

$$I_{PSF}(u,v) = \left| FT_{\frac{n_{eye}u}{\lambda f_{eye}},\frac{n_{eye}v}{\lambda f_{eye}}} \left\{ P_{eye} G_{SCE-I} \exp\left(i\Phi_{WA}\right) \right\} \right|^2$$

$$= \left| FT_{\frac{n_{eye}u}{\lambda f_{eye}},\frac{n_{eye}v}{\lambda f_{eye}}} \left\{ P_{eye} \exp\left(i\Phi_{WA}\right) \right\} * FT_{\frac{n_{eye}u}{\lambda f_{eye}},\frac{n_{eye}v}{\lambda f_{eye}}} \left\{ G_{SCE-I} \right\} \right|^2 \quad (18.14)$$

where

$\dfrac{n_{eye}u}{\lambda f_{eye}}, \dfrac{n_{eye}v}{\lambda f_{eye}}$ are spatial frequencies for retinal coordinates (u,v)

$*$ denotes a convolution

The convolution produces a smoothening effect of the retinal field amplitude *and* phase when compared to the case of excluding the SCE-I. As the FT of a Gaussian function is itself a Gaussian, the pupil apodization corresponds exactly to the smearing of the PSF with a fundamental Gaussian waveguide mode of foveal cones [100], which explains the role of G_{SCE-I} when transferred to the retinal plane. Thus, if higher-order waveguide modes are present, or if there is photoreceptor disarray, the Gaussian pupil apodization would be invalidated. The MTF can be calculated from the absolute value of the FT of Equation 18.14 resulting in (excluding normalization)

$$MTF(u,v) = \left| P_{eye} G_{SCE-I} \exp(i\bar{\Phi}_{WA}) * P_{eye} G_{SCE-I} \exp(-i\Phi_{WA}) \right| \quad (18.15)$$

where the bar above the wave front aberration in the first term signifies a coordinate inversion ($x \to -x$ and $y \to -y$). From Equation 18.15 it can be seen that the SCE-I reduces high-frequency contents in the MTF by damping contributions from light near to the pupil rim [98,101]. Figure 18.15 shows the impact of the SCE-I on the calculated PSF and MTF without and with the presence of defocus.

Equation 18.14 shows that the resulting PSF is sensitive to the wave front slope at the retina present in $FT_{\frac{n_{eye}u}{\lambda f_{eye}},\frac{n_{eye}v}{\lambda f_{eye}}} \left\{ P_{eye} \exp(i\Phi_{WA}) \right\}$ whether the SCE-I is applied in the pupil plane or is included directly as a photoreceptor convolution in the retinal plane when determining the effective retinal image. Only if replacing $FT_{\frac{n_{eye}u}{\lambda f_{eye}},\frac{n_{eye}v}{\lambda f_{eye}}} \left\{ G_{SCE-I} \right\}$ by a delta function (to remove the role of the SCE-I) will the effective PSF be insensitive to any retinal wave front slope.

For a schematic eye with a circular pupil being uniformly illuminated by coherent light, the field incident on the retina (excluding the pupil SCE-I) can be calculated as

$$\psi_r(u,v) = \frac{4\sqrt{\pi}n_{eye}}{id\lambda f_{eye}} \int_0^{d/2} r \exp(i\Phi_{WA}) J_0\left(\frac{2\pi n_{eye}}{\lambda f_{eye}} r\rho \right) dr \quad (18.16)$$

where

$$\rho = \sqrt{u^2 + v^2}$$
$$r = \sqrt{x^2 + y^2}$$

J_0 is the zeroth-order Bessel function of the first kind

The wave front aberrations at each pupil point ($r\cos\theta$, $r\sin\theta$) may be expanded as a series of Zernike polynomials Z_q [102] normalized across the pupil radius $d/2$ whereby $\Phi_{WA}(r,\theta) = k \sum_{q=1}^{\infty} c_q Z_q(r,\theta)$ and the c_q's are scaling parameters. When the field with the aberrated wave front is propagated to the retinal plane, only Zernike modes of even radial order (defocus, astigmatism, spherical aberrations, …) will produce a wave front

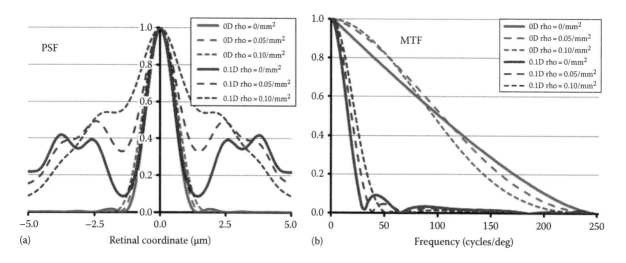

Figure 18.15 The role of the Stiles–Crawford effect of the first kind (SCE-I) on the point spread function (PSF) (a) and the modulation transfer function (MTF) (b) at 0.555μm wavelength for a schematic eye model with a large 8 mm pupil without (green lines) and with (red lines) 0.1 diopters defocus. The SCE-I ($\rho = 0.05/mm^2$ dashed line and $\rho = 0.10/mm^2$ dotted line) widens the PSF compared to the case of no SCE-I ($\rho = 0$ solid line) and dampens the off-axis intensity ringing which is particularly apparent once defocus is introduced. Likewise, the SCE-I dampens the MTF at high frequencies while increasing the impact of low frequencies.

slope for light incident on the photoreceptors, whereas for odd radial orders (coma, trefoil, …), the wave front will impinge directly along the axis of each photoreceptor, and the image will only be blurred in the field amplitude [100]. Rather than introducing the SCE-I in the pupil plane, a Gaussian waveguide mode, ψ_m, will be assumed in each photoreceptor whereby the effective retinal image can be written as $\left| FT_{\frac{n_{eye}u}{\lambda f_{eye}}, \frac{n_{eye}v}{\lambda f_{eye}}} \left\{ P_{eye} \exp\left(i\Phi_{WA}\right) \right\} * \psi_m \right|^2$.

The incident retinal field from Equation 18.16 may be expressed in terms of an amplitude and a phase factor as $\psi_r(u,v) = B(u,v)$ $\exp(i\phi_r(u,v))$. A series expansion of this field at and near the photoreceptor centered at (u_c, v_c) gives that $\psi_r(u,v) \approx B_0(u_c,v_c)$ $\exp[i(\phi_0 + \phi_u(u - u_c) + \phi_v(v - v_c))]$ where B_0 is the amplitude and ϕ_0 is the phase of the retinal field at the central point of the waveguide entrance facet and the phase derivatives across the waveguide aperture are $\phi_u = \partial\phi/\partial u\big|_{u_c,v_c}$ and $\phi_v = \partial\phi/\partial v\big|_{u_c,v_c}$. In this slowly varying field approximation, the power coupled to the fundamental Gaussian mode of the waveguide can be calculated as

$$P \approx 2\pi\tau w_m^2 B_0^2 \exp\left(-\frac{w_m^2}{2}\left(\varphi_u^2 + \varphi_v^2\right)\right) \tag{18.17}$$

where τ is a constant. Examples of using Equation 18.17 on the effective PSF are shown in Figure 18.16 in the presence of different Zernike aberrations.

The exact spatial distribution of visual pigments across the length of outer segments in the cone–rod photoreceptor mosaic makes it challenging to determine exactly where the light is likely be absorbed when triggering vision. The diameter of photoreceptors is approximately matched to the pupil size and the diffraction limit of resolution. Likewise, a geometrical extension of the conical outer segment of parafoveal cones to the pupil plane is approximately matched to the pupil size [28]. This intricate balance suggests that the entire outer segment shape may play an important role for vision when the retina is exposed to different lighting conditions [16]. With this in mind the layered light scattering and receiving model can also be used to predict light distributions and estimated visibility across the outer segments [28]. Figure 18.17 shows the outcome of such an analysis when illuminating a hexagonal packing of layered outer segments with a plane wave being incident on or off axis to simulate the SCE-I and when a focused beam of light is incident on layered outer segments and focused either on or between the photoreceptors to trigger vision. As this approach does not enforce waveguiding, light that is scattered between the receptors may stimulate a visual response even if not directly incident on the photoreceptor in question [103]. It should be noted that self-screening has only been included in the axial absorption in the model as the off-axis angles are small.

The fact that photoreceptors are oriented to capture a maximum of light also means that they are pointing so that wave front

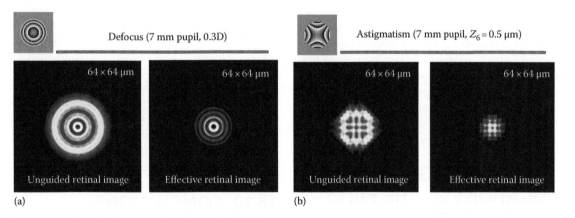

Figure 18.16 Comparison of monochromatic point spread function intensity images without (unguided) and with (effective) inclusion of the Stiles–Crawford effect of the first kind as a retinal Gaussian waveguide mode with $w_m = 1.5\,\mu m$ that dampens the role of Zernike modes having even radial order away from the geometrical image point as shown here for defocus (a) and astigmatism (b), respectively.

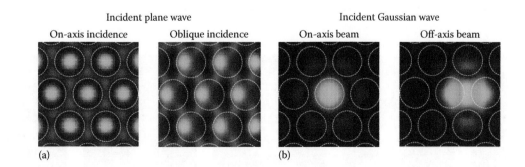

Figure 18.17 Simulated scattering of light in a close-packed hexagonal arrangement of outer segments each with a 2 μm diameter. Plane wave incidence on axis and 10° oblique (a) simulates Stiles–Crawford effect of the first kind characterization with increased scattering into the interstitial matrix for oblique incidence. Simulating single-cone vision of a focused Gaussian beam (b) incident on a single or between two outer segments shows that light may enter neighboring segments due to scattering. (Reproduced from Vohnsen, B., *Biomed. Opt. Express*, 5, 1569, 2014, Figures 8 and 10. With permission of Optical Society of America.)

tilt is small or entirely absent at their receiving aperture. This fact has been elegantly shown by Artal et al. who found that the eye is best adapted to its own aberrations [104]. Although a neural explanation is argued for, an entirely optical effect based on local photoreceptor pointing adapted to the incident wave front in the plane of the retina may encompass these findings and agrees qualitatively well with recent observations in retinal tissues [35].

Finally, the role of the SCE-II in a normal visual situation may also be considered by integration of Equations 18.10 through 18.12 across the pupil when considering a possible hue shift induced by a finite bandwidth of the illumination. However, its influence would be interwoven with the chromatic aberrations of the eye.

18.8 PHYSICAL MODELS OF THE RETINA

Rather than attempting to study the retina itself under the microscope or using mathematical models for light–tissue interactions, the construction of retinal simulators that would have similar characteristics and be useful to study the role of the Stiles–Crawford effects in a complementary physical setting has been carried out. Such approaches have also inspired biomimetic applications in solar cells [105] and imaging technologies [106]. Physical models of retinal waveguides have been made for use in the microwave range with enlarged waveguides of polystyrene foam to match the cm-range wavelengths employed [107,108] and smaller dielectric rods to match mm-range wavelengths [109] that have all confirmed a directionality similar to that of the SCE-I. Rativa and Vohnsen introduced a liquid-filled photonic crystal fiber as a retinal simulator with waveguide dimensions similar to those of the parafoveal retina and with temperature tuning that allow the exact waveguide parameters to be set from a single to multimode range [110]. This simulator has been used to model the angular dependence of the SCE-I as well as to examine the role of defocus and other aberration modes [111,112]. An example on the use of this simulator is shown in Figure 18.18.

18.9 RETINAL IMAGING IMPLICATIONS OF THE STILES–CRAWFORD EFFECTS

When light is obliquely incident on the retina, photoreceptors reflect less light back through the eye pupil, and thus the highest visibility of photoreceptors is obtained when the imaging light enters near the pupil center [63,113]. Adaptive optics wave front correction [114,115] and beam apodization [116,117] can be used in scanning retinal imaging applications to match the size of the focused incident beam to the width of the photoreceptors being imaged.

As discussed in Section 18.3, fundus photography, SLO, and OCT have all been used to probe photoreceptor directionality [61–64] showing the combined role of an oblique incidence on the individual photoreceptors as well as the directional dependence of the backscattered light. The pigments are commonly bleached or infrared light is used to ensure that the light traverses the full length of the photoreceptors. It must be stressed that the signal that facilitates the visualization of photoreceptors in high-resolution retinal images originates in scattering of light from refractive index differences within the photoreceptors themselves that may well have a different directionality than the psychophysical SCE-I. A somewhat open question is the origin of the multilayered retinal reflections and in particular the middle band at or near the ellipsoid where high-refractive-index mitochondria organelles are located, seen in high-resolution OCT images [5,118,119]. OCT has also revealed reflective gaps in certain outer segments possibly caused by irregular spacing of the membrane invaginations [120] and temporal variations in the photoreceptor brightness believed to originate in changes in the outer segment length as pigment layers detach from the photoreceptors [121].

The present author has made use of the layered scattering model [28] to also analyze fundus imaging in relation to oblique incidence of light [122]. In doing so it has been assumed that the main contributions to the retinal images are from a maximum of three layers spaced at distances that have been chosen to match the reflective layers seen in OCT images. Figure 18.19 shows examples of calculated fundus images and the same model has been found valid when analyzing image brightness in simulated retinal images as a function of the angle of incidence of the light to mimic the SCE-I showing good correspondence with parafoveal measurements of cone photoreceptor directionality.

18.10 CONCLUSIONS

Although questions remain on the exact retinal mechanisms involved in the Stiles–Crawford effects as well as their role for vision, imaging, and photoreceptor diagnostics, there can be little doubt that they play an active role for our visual system. Although most of this chapter

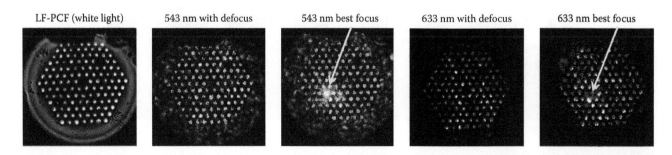

| LF-PCF (white light) | 543 nm with defocus | 543 nm best focus | 633 nm with defocus | 633 nm best focus |

Figure 18.18 Liquid-filled photonic crystal fiber. Castor oil and temperature tuning are used to ensure waveguiding in a single-to-multimode regime in the individual densely packed columns rather than in the core of the fiber whereby it resembles the parafoveal cone mosaic. Each column in the assembly has a 6.4 μm diameter. Images show transmitted light though 30 mm length of fiber with the incident being defocused and tightly focused, respectively, only a single column. HeNe lasers (543 and 633 nm wavelength) have been used for the illumination.

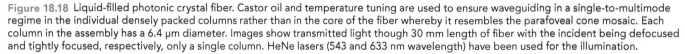

Cone mosaic (triple layer)   Simulated cone–rod mosaic

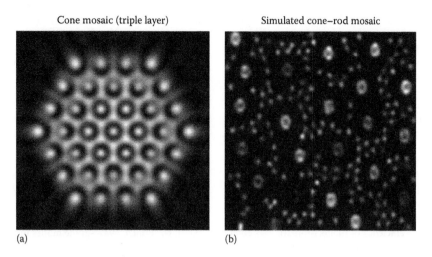

(a)   (b)

Figure 18.19 Calculated flood-illumination retinal images for a schematic eye with 8 mm pupil. The cone mosaic image of 82,000 cones/mm² (a) has been obtained from the backscattering of light for a hexagonal arrangement of 37 identical cones having 2.5 μm diameter and three internal reflective layers (at the ellipsoid, at the inner–outer segment junction, and at the posterior end of the outer segment). The image shows the interference between light from different depths whereby a dark ring from destructive interference is produced at each. The same model has also been used to model a parafoveal rod–cone mosaic (b) showing details that resemble multimodal waveguides but are scattering patterns from the larger cones. (From Vohnsen, B., *Front. Opt.*, 2014, Abstract FW5F.5.)

has described their role for the human eye, very similar cellular mechanisms are found in a wide range of animal eyes ranging from mammals to invertebrates with the latter having directional light-guiding mechanisms in elongated cylindrical rhabdomeres that are located behind the anterior corneal lens where they direct light across visual pigments toward posterior axons [123].

Apart from the photoreceptors in the human retina themselves, another light-guiding mechanism has been identified in the parafoveal Muller glial cells that, due to an increased refractive index, are believed to act as a supportive mechanism to aid the guiding of light across the neural retina toward the photoreceptors with a minimum of perturbations [124] and to contribute with a redistribution of the retinal image to better match the spectral properties of the receiving pigments of the cones and rods [125]. The refractive indices across the Muller cells and their surrounds suggest a multimode waveguide behavior that is less efficient to confine the light than the photoreceptors. Their directionality is expected to be low and therefore no SCE-I function has yet been assigned to the Muller cells.

In a significant part of this chapter, the possible absence or reduced role of waveguiding in the individual photoreceptors has been analyzed based on the fact that the directionality seems to be inherent in the proper organization of the visual pigments themselves and that the role of nonguided components (normally ignored in waveguide models) cannot simply be ignored as it accounts for a significant fraction of the total light transmitted across the retina. Further progress along this line will require improved data on the exact optical properties of the photoreceptors and their surrounds in the living eye and retina as well as more advanced modelling tools. It seems therefore fitting to end this chapter with the insightful statement of Prof. N. Bohr that "...owing to the very limits imposed by the properties of light, no instrument is imaginable which is more efficient for its purpose than the eye" [126], which makes perfect sense considering how the retinal cone and rod photoreceptors have developed to match the optics of our eye, the light in our environment, and to provide accurate vision in the healthy eye.

ACKNOWLEDGMENTS

I am indebted to my former and current students who have all contributed to my present understanding of the Stiles–Crawford effects. For the work presented here, I thank in particular Dr. Diego Rativa, Dr. Benjamin Lochocki, Dr. Sara Castillo, Ms. Denise Valente, and Mr. Fabio Rodrigo. I am also very grateful to Prof. Gerald Westheimer and Prof. Jay M. Enoch whose work and insight has inspired me enormously. This research has been made possible, thanks to the funding from Science Foundation Ireland (grants 07/SK/B1239a and 08/IN.1/B2053).

REFERENCES

1. J. K. Bowmaker and H. J. Dartnall, Visual pigments of rods and cones in human retina, *J. Physiol.* **298** (1980) 501–511.
2. R. W. Rodieck, *The First Steps in Seeing* (Sinauer Associates, Sunderland, MA, 1998).
3. A. M. Laties and J. M. Enoch, An analysis of retinal receptor orientation. I. Angular relationship of neighboring photoreceptors, *Invest. Ophthalmol.* **10** (1971) 69–77.
4. Q. V. Hoang, R. A. Linsenmeier, C. K. Chung, and C. A. Curcio, Photoreceptor inner segments in monkey and human retina: Mitochondrial density, optics, and regional variation, *Vis. Neurosci.* **19** (2002) 395–407.
5. D. C. Hood, X. Zhang, R. Ramachandran, C. L. Talamini, A. Raza, J. P. Greenberg, J. Sherman, S. H. Tsang, and D. G. Birch, The inner segment/outer segment border seen on optical coherence tomography is less intense in patients with diminished cone function, *Invest. Ophthalmol. Vis. Sci.* **52** (2011) 9703–9709.
6. J. J. Wolken, *Light Detectors, Photoreceptors and Imaging Systems in Nature* (Oxford University, New York, 1995).
7. D. Fotiadis, Y. Liang, S. Filipek, D. A. Saperstein, A. Engel, and K. Palczewski, Atomic-force microscopy: Rhodopsin dimers in native disc membranes, *Nature* **421** (2003) 127–128.
8. R. W. Young, The renewal of rod and cone outer segments in the rhesus monkey, *J. Cell Biol.* **49** (1971) 303–318.

9. J. Nguyen-Legros and D. Hicks, Renewal of photoreceptor outer segments and their phagocytosis by the retinal pigment epithelium, *Int. Rev. Cytol.* **196** (2000) 245–313.

10. C. A. Curcio, K. R. Sloan, R. E. Kalina, and A. E. Hendrickson, Human photoreceptor topography, *J. Comp. Neurol.* **292** (1990) 497–523.

11. W. S. Stiles and B. H. Crawford, The luminous efficiency of rays entering the eye pupil at different points, *Proc. R. Soc. Lond. B* **112** (1933) 428–450.

12. W. D. Wright and J. H. Nelson, The relation between the apparent intensity of a beam of light and the angle at which the beam strikes the retina, *Proc. Phys. Soc.* **48** (1936) 401–424.

13. R. A. Applegate and V. Lakshminarayanan, Parametric representation of Stiles–Crawford functions: Normal variation of peak location and directionality, *J. Opt. Soc. Am. A* **10** (1993) 1611–1623.

14. W. S. Stiles, The luminous efficiency of monochromatic rays entering the eye pupil at different points and a new colour effect, *Proc. R. Soc. Lond. B* **123** (1937) 90–118.

15. B. H. Crawford, The luminous efficiency of light rays entering the eye pupil at different points and its relation to brightness threshold measurements, *Proc. R. Soc. B* **124** (1937) 81–96.

16. G. Westheimer, Dependence of the magnitude of the Stiles-Crawford effect on retinal location, *J. Physiol.* **192** (1967) 309–315.

17. J. M. Enoch and G. M. Hope, Directional sensitivity of the foveal and parafoveal retina, *Invest. Ophthalmol. Vis. Sci.* **12** (1973) 497–503.

18. F. Flamant and W. S. Stiles, The directional and spectral sensitivities of the retinal rods to adapting fields of different wave-lengths, *J. Physiol.* **107** (1948) 187–202.

19. J. A. Van Loo, Jr. and J. M. Enoch, The scotopic Stiles-Crawford effect, *Vision Res.* **15** (1975) 1005–1009.

20. G. Hansen, Zur Kenntnis des physiologischen Apertur-Farbeffektes Stiles-Crawford-Effekt II. Art, *Naturwissenschaften* **31** (1943) 416–417.

21. G. Toraldo di Francia and L. Ronchi, Directional scattering of light by the human retina, *J. Opt. Soc. Am.* **42** (1952) 782–783.

22. S. A. Burns, S. Wu, F. Delori, A. E. Elsner, Direct measurement of human-cone-photoreceptor alignment, *J. Opt. Soc. Am. A Opt. Image Sci. Vis.* **12** (1995) 2329–2338.

23. S. Marcos and S. A. Burns, Cone spacing and waveguide properties from cone directionality measurements, *J. Opt. Soc. Am. A Opt. Image Sci. Vis.* **16** (1999) 995–1004.

24. N. P. A. Zagers, J. van de Kraats, T. T. J. M. Berendschot, and D. van Norren, Simultaneous measurement of foveal spectral reflectance and cone-photoreceptor directionality, *Appl. Opt.* **41** (2002) 4686–4696.

25. J.-M. Gorrand and M. Doly, Alignment parameters of foveal cones, *J. Opt. Soc. Am. A Opt. Image Sci. Vis.* **26** (2009) 1260–1267.

26. B. Vohnsen, I. Iglesias, and P. Artal, Guided light and diffraction model of human-eye photoreceptors, *J. Opt. Soc. Am. A Opt. Image Sci. Vis.* **22** (2005) 2318–2328.

27. B. Vohnsen, On the spectral relation between the first and second Stiles–Crawford effect, *J. Mod. Opt.* **56** (2009) 2261–2271.

28. B. Vohnsen, Directional sensitivity of the retina: A layered scattering model of outer-segment photoreceptor pigments, *Biomed. Opt. Express* **5** (2014) 1569–1587.

29. J. M. Enoch and V. Lakshminarayanan, 75th Anniversary of the Stiles-Crawford effect(s): A celebratory special symposium: October 23 2008, Rochester, New York, *J. Mod. Opt.* **56** (2009) 2164–2175.

30. G. Westheimer, Directional sensitivity of the retina: 75 years of Stiles-Crawford effect, *Proc. R. Soc. B* **275** (2008) 2777–2786.

31. E. W. Brücke, Ueber die physiologische Bedeutung der stab-förmigen Körper und der Zwillingszapfen in den Augen der Wirbelthiere, *Archiv für Anatomie, Physiologie und wissenschaftliche Medicin* (1844) 444–451.

32. H. von Helmholtz, *Treatise on Physiological Optics* (Dover Publications, New York, 1962).

33. H. S. Smallman, D. I. A. MacLeod, and P. Doyle, Vision: Realignment of cones after cataract removal, *Nature* **412** (2001) 604–605.

34. M. Kono, J. M. Enoch, E. Strada, P. Shih, R. Srinivasan, V. Lakshminarayanan, W. Susilasate, and A. Graham, Stiles-Crawford effect of the first kind: Assessment of photoreceptor alignments following dark patching, *Vision Res.* **41** (2001) 103–118.

35. B. Wang, Q. Zhang, R. Lu, Y. Zhi, and X. Yao, Functional optical coherence tomography reveals transient phototropic change of photoreceptor outer segments, *Opt. Lett.* **39** (2014) 6923–6926.

36. A. Hannover, Mikroskopiske undersøgelser af nervesystemet, *Kong. Danske Videns. Sels. Naturv. Math. Afh.* **X** (1843) 1–111.

37. J. M. Enoch, Wave-guide modes in retinal receptors, *Science* **133** (1961) 1353–1354.

38. J. M. Enoch, Optical properties of the retinal receptors, *J. Opt. Soc. Am.* **53** (1963) 71–85.

39. O. S. Packer, D. R. Williams, and D. G. Bensinger, Photopigment transmittance imaging of the primate photoreceptor mosaic, *J. Neurosci.* **16** (1996) 2251–2260.

40. G. Westheimer, The maxwellian view, *Vision Res.* **6** (1966) 669–682.

41. B. Lochocki, D. Rativa, and B. Vohnsen, Spatial and spectral characterisation of the first and second Stiles-Crawford effects using tuneable liquid-crystal filters, *J. Mod. Opt.* **58** (2011) 1817–1825.

42. B. Lochocki and B. Vohnsen, Defocus-corrected analysis of the foveal Stiles–Crawford effect of the first kind across the visible spectrum, *J. Opt.* **15** (2013) 125301.

43. N. Singh, D. A. Atchison, S. Kasthurirangan, and H. Guo, Influences of accommodation and myopia on the foveal Stiles–Crawford effect, *J. Mod. Opt.* **56** (2009) 2217–2230.

44. B. Lochocki and B. Vohnsen, The 1st Stiles-Crawford effect: A single beam frequency flickering system in *Proceedings VII European/I World Meeting in Visual and Physiological Optics*, R. Iskander and H. T. Kasprzak, Eds. (Wroclaw, Poland, 2014), pp. 183–185, ISBN: 978-83-7493-847-1.

45. D. G. Birch, M. A. Sandberg, and E. L. Berson, The Stiles-Crawford effect in retinitis pigmentosa, *Invest. Ophthalmol. Vis. Sci.* **22** (1982) 157–164.

46. D. L. McCulloch and V. Lakshminarayanan, The Stiles-Crawford effect of the first kind and the full-field electroretinogram (ERG), *J. Mod. Opt.* **56** (2009) 2176–2180.

47. C. S. Matsumoto, K. Shinoda, H. Matsumoto, S. Satofuka, A. Mizota, K. Nakatsuka, and Y. Miyake, Stiles–Crawford effect in focal macular ERGs from macaque monkey, *J. Vision* **12** (2012) 6.

48. W. L. Makous, A transient Stiles-Crawford effect, *Vision Res.* **8** (1968) 1271–1284.

49. P. L. Walraven, Recovery from the increase of the Stiles-Crawford effect after bleaching, *Nature* **210** (1966) 311–312.

50. P. J. de Groot, Transient threshold increase due to combined changes in direction of propagation and plane of polarization, *Vision Res.* **19** (1979) 1253–1259.

51. J. Gordon and I. Abramov, Color appearance: Maxwellian vs. Newtonian views, *Vision Res.* **48** (2008) 1879–1883.

52. J. M. Enoch, Summated response of the retina to light entering different parts of the pupil, *J. Opt. Soc. Am.* **48** (1958) 392–405.

53. B. Drum, Additivity of the Stiles–Crawford effect for a Fraunhofer image, *Vision Res.* **15** (1975) 291–298.

54. B. Vohnsen, B. Lochocki, and C. Vela, Integrated Stiles-Crawford effect in normal vision, in *Proceedings VII European/I World Meeting in Visual and Physiological Optics*, R. Iskander and H. T. Kasprzak, Eds. (2014), pp. 368–371, ISBN: 978-83-7493-847-1.

55. B. Vohnsen and D. Rativa, Absence of an integrated Stiles-Crawford function for coherent light, *J. Vis.* **11** (2011) 19.

56. S. Castillo and B. Vohnsen, Exploring the Stiles-Crawford effect of the first kind with coherent light and dual Maxwellian sources, *Appl. Opt.* **52** (2013) A1–A8.

57. G. Westheimer, Retinal light distributions, the Stiles–Crawford effect and apodization, *J. Opt. Soc. Am. A Opt. Image Sci. Vis.* **30** (2013) 1417–1421.

58. J. M. Enoch and W. S. Stiles, The colour change of monochromatic light with retinal angle of incidence, *Opt. Acta* **8** (1961) 329–358.

59. G. J. van Blockland and D. van Norren, Intensity and polarization of light scattered at small angles from the human fovea, *Vision Res.* **26** (1986) 485–494.

60. J.-M. Gorrand and F. Delori, A reflectometric technique for assessing photoreceptor alignment, *Vision Res.* **35** (1995) 999–1010.

61. P. J. Delint, T. J. M. Berendschot, and D. van Norren, Local photoreceptor alignment measured with a scanning laser ophthalmoscope, *Vision Res.* **37** (1997) 243–248.

62. W. Gao, B. Cense, Y. Zhang, R. S. Jonnal, and D. T. Miller, Measuring retinal contributions to the optical Stiles-Crawford effect with optical coherence tomography, *Opt. Express* **16** (2008) 6486–6501.

63. A. Roorda and D. R. Williams, Optical fiber properties of individual human cones, *J. Vis.* **2** (2002) 404–412.

64. D. Rativa and B. Vohnsen, Analysis of individual cone-photoreceptor directionality using scanning laser ophthalmoscopy, *Biomed. Opt. Express* **2** (2011) 1423–1431.

65. F. Schaeffel, A. Glasser, and H. C. Howland, Accommodation, refractive error and eye growth in chickens, *Vision Res.* **28** (1988) 639–657.

66. E. F. Fincham, The accommodation reflex and its stimulus, *Br. J. Ophthalmol.* **35** (1951) 381–393.

67. P. B. Kruger, N. Lopez-Gil, and L. R. Stark, Accommodation and the Stiles-Crawford effect: Theory and a case study, *Ophthalmic. Physiol. Opt.* **21** (2001) 339–351.

68. K. Blank, R. R. Provine, and J. M. Enoch, Shift in the peak of the photopic Stiles-Crawford function with marked accommodation, *Vision Res.* **15** (1975) 499–507.

69. S. S. Choi, L. F. Garner, and J. M. Enoch, The relationship between the Stiles-Crawford effect of the first kind (SCE-I) and myopia, *Ophthalmic. Physiol. Opt.* **23** (2003) 465–472.

70. V. C. Smith, J. Pokorny, and K. R. Diddie, Color matching and the Stiles–Crawford effect in observers with early age-related macular changes, *J. Opt. Soc. Am. A* **5** (1988) 2113–2121.

71. P. K. DeLint, T. T. Berendschot, and D. van Norren, A comparison of the optical Stiles-Crawford effect and retinal densitometry in a clinical setting, *Invest. Ophthalmol. Vis. Sci.* **39** (1998) 1519–1523.

72. B. O'Brien, A theory of the Stiles and Crawford effect, *J. Opt. Soc. Am.* **36** (1946) 506–509.

73. A. Safir and L. Hyams, Distribution of cone orientations as an explanation of the Stiles-Crawford effect, *J. Opt. Soc. Am.* **59** (1969) 757–765.

74. G. Toraldo di Francia, Retina cones as dielectric antennas, *J. Opt. Soc. Am.* **39** (1949) 324.

75. R. Röhler and W. Fischer, Influence of waveguide modes on the light absorption in photoreceptors, *Vision Res.* **11** (1971) 97–101.

76. A. W. Snyder and C. Pask, The Stiles-Crawford effect: Explanation and consequences, *Vision Res.* **13** (1973) 1115–1137.

77. A. W. Snyder and C. Pask, Waveguide modes and light absorption in photoreceptors, *Vision Res.* **13** (1973) 2605–2608.

78. W. Wijngaard, Some normal modes of an infinite hexagonal array of identical circular dielectric rods, *J. Opt. Soc. Am.* **64** (1974) 1136–1144.

79. L. Fischer, A. Zvyagin, T. Plakhotnik, and M. Vorobyev, Numerical modeling of light propagation in a hexagonal array of dielectric cylinders, *J. Opt. Soc. Am. A* **27** (2010) 865–872.

80. R. L. Sidman, The structure and concentration of solids in photoreceptor cells studied by refractometry and interference microscopy, *J. Biophys. Biochem. Cytol.* **3** (1957) 15–30.

81. C. Pask and A. W. Snyder, Theory of the Stiles–Crawford effect of the second kind, in *Photoreceptor Optics*, A. W. Snyder and R. Menzel, Eds. (Springer-Verlag, Berlin, Germany, 1975), pp. 152–156.

82. R. Sammut and A. W. Snyder, Contribution of unbound modes to light absorption in visual photoreceptors, *J. Opt. Soc. Am.* **64** (1974) 1711–1714.

83. W. Fischer and R. Röhler, The absorption of light in an idealized photoreceptor on the basis of waveguide theory—II: The semi-infinite cylinder, *Vision Res.* **14** (1974) 1115–1125.

84. M. P. Rowe, N. Engheta, S. S. Easter, Jr., and E. N. Pugh, Jr., Graded-index model of a fish double cone exhibits differential polarization sensitivity, *J. Opt. Soc. Am. A Opt. Image Sci. Vis.* **11** (1994) 55–70.

85. J. J. Vos and F. L. van Os, The effect of lens density on the Stiles-Crawford effect, *Vision Res.* **15** (1975) 749–751.

86. A. M. Pozo, F. Pérez-Ocón, and J. R. Jiménez, FDTD analysis of the light propagation in the cones of the human retina: An approach to the Stiles-Crawford effect of the first kind, *J. Opt. A: Pure Appl. Opt.* **7** (2005) 357–363.

87. M. J. Piket-May, A. Taflove, and J. B. Troy, Electrodynamics of visible-light interactions with the vertebrate retinal rod, *Opt. Lett.* **18** (1993) 568–570.

88. J. M. Enoch, Waveguide modes: Are they present, and what is their possible role in the visual mechanism? *J. Opt. Soc. Am.* **50** (1960) 1025–1026.

89. T. T. J. M. Berendschot, J. van de Kraats, and D. Van Norren, Wavelength dependence of the Stiles–Crawford effect explained by perception of backscattered light from the choroid, *J. Opt. Soc. Am. A Opt. Image Sci. Vis.* **18** (2001) 1445–1451.

90. B. Chen and W. Makous, Light capture by human cones, *J. Physiol.* **414** (1989) 89–109.

91. J. R. Coble and W. A. H. Rushton, Stiles-Crawford effect and the bleaching of cone pigments, *J. Physiol.* **217** (1971) 231–242.

92. P. L. Walraven and M. A. Bouman, Relation between directional sensitivity and spectral response curves in human cone vision, *J. Opt. Soc. Am.* **50** (1960) 780–784.

93. M. Alpern, The Stiles-Crawford effect of the second kind (SCII): A review, *Perception* **15** (1986) 785–799.

94. D. A. Atchison, D. H. Scott, N. C. Strang, and P. Artal, Influence of Stiles–Crawford apodization on visual acuity, *J. Opt. Soc. Am. A* **19** (2002) 1073–1083.

95. D. Rativa and B. Vohnsen, Single- and multimode characteristics of the foveal cones: The super-Gaussian function, *J. Mod. Opt.* **58** (2011) 1809–1816.

96. P. Moon and D. E. Spencer, On the Stiles-Crawford effect, *J. Opt. Soc. Am.* **34** (1944) 319–329.

97. B. Vohnsen, Multi-aperture model of photoreceptor outer segments, in *Proceedings VII European/I World Meeting in Visual and Physiological Optics*, R. Iskander and H. T. Kasprzak, Eds. (2014), pp. 372–375, ISBN: 978-83-7493-847-1.

98. H. Metcalf, Stiles-Crawford apodization, *J. Opt. Soc. Am.* **55** (1965) 72–74.

99. X. Zhang, M. Ye, A. Bradley, and L. Thibos, Apodization by the Stiles–Crawford effect moderates the visual impact of retinal image defocus, *J. Opt. Soc. Am. A Opt. Image Sci. Vis.* **16** (1999) 812–820.

100. B. Vohnsen, Photoreceptor waveguides and effective retinal image quality, *J. Opt. Soc. Am. A Opt. Image Sci. Vis.* **24** (2007) 597–607.

101. J. P. Carroll, Apodization model of the Stiles-Crawford effect, *J. Opt. Soc. Am.* **70** (1980) 1155–1156.

102. R. J. Noll, Zernike polynomials and atmospheric turbulence, *J. Opt. Soc. Am.* **66** (1976) 207–211.

103. W. M. Harmening, W. S. Tuten, A. Roorda, and L. C. Sincich, Mapping the perceptual grain of the human retina, *J. Neurosci.* **34** (2014) 5667–5677.

104. P. Artal, L. Chen, E. J. Fernández, B. Singer, S. Manzanera, and D. R. Williams, Neural compensation for the eye's optical aberrations, *J. Vision* **4** (2004) 281–287.

105. V. G. Kravets and A. N. Grigorenko, Retinal light trapping in textured photovoltaic cells, *Appl. Phys. Lett.* **97** (2010) 133701.

106. P. Kornreich and B. Farell, True three-dimensional camera, *J. Electron. Imaging* **22** (2013) 013028.

107. B. O'Brien, Vision and resolution in the central retina, *J. Opt. Soc. Am.* **41** (1951) 882–894.

108. J. M. Enoch and G. A. Fry, Characteristics of a model retinal receptor studied at microwave frequencies, *J. Opt. Soc. Am.* **48** (1958) 899–911.

109. P. J. de Groot and R. E. Terpstra, Millimeter-wave model of a foveal receptor, *J. Opt. Soc. Am.* **70** (1980) 1436–1452.

110. D. Rativa and B. Vohnsen, Simulating human photoreceptor optics using a liquid-filled photonic crystal fiber, *Biomed. Opt. Express* **2** (2011) 543–551.

111. B. Vohnsen, D. Rativa, C. Vela, B. Lochocki, and P. Kruger, The role of defocus on photoreceptor light coupling analyzed with a waveguide-based retinal simulator, *Investig. Ophthalmol. Vis. Sci.* **54** (2013), 3429. ARVO-abstract 3429-C0150.

112. D. Valente, D. Rativa, and B. Vohnsen, The role of defocus analyzed with a waveguide-based retinal simulator, in *Proceedings VII European/I World Meeting in Visual and Physiological Optics*, R. Iskander and H. T. Kasprzak, Eds. (Wroclaw, Poland, 2014), pp. 357–360, ISBN: 978-83-7493-847-1.

113. B. Vohnsen, I. Iglesias, and P. Artal, Directional imaging of the retinal cone mosaic, *Opt. Lett.* **29** (2004) 968–970.

114. A. Roorda, F. Romero-Borja, W. J. Donnelly III, H. Queener, T. J. Hebert, and M. C. W. Campbell, Adaptive optics scanning laser ophthalmoscopy, *Opt. Express* **10** (2002) 405–412.

115. A. Dubra, Y. Sulai, J. L. Norris, R. F. Cooper, A. M. Dubis, D. R. Williams, and J. Carroll, Noninvasive imaging of the human rod photoreceptor mosaic using a confocal adaptive optics scanning ophthalmoscope, *Biomed. Opt. Express* **2** (2011) 1864–1876.

116. B. Vohnsen and D. Rativa, Ultrasmall spot size scanning laser ophthalmoscopy, *Biomed. Opt. Express* **2** (2011) 1597–1609.

117. Y. N. Sulai and A. Dubra, Adaptive optics scanning ophthalmoscopy with annular pupils, *Biomed. Opt. Express* **3** (2012) 1647–1661.

118. R. F. Spaide and C. A. Curcio, Anatomical correlates to the bands seen in the outer retina by optical coherence tomography: Literature review and model, *Retina* **31** (2011) 1609–1619.

119. R. S. Jonnal, O. P. Kocaoglu, R. J. Zawadzki, S.-H. Lee, J. S. Werner, and D. T. Miller, The cellular origins of the outer retinal bands in optical coherence tomography images, *Invest. Ophthalmol. Vis. Sci.* **14** (2014) 14907.

120. M. Pircher, E. Götzinger, H. Sattmann, R. A. Leitgeb, and C. K. Hitzenberger, In vivo investigation of human cone photoreceptors with SLO/OCT in combination with 3D motion correction on a cellular level, *Biomed. Opt. Express* **18** (2010) 13935–13944.

121. R. S. Jonnal, J. R. Besecker, J. C. Derby, O. P. Kocaoglu, B. Cense, W. Gao, Q. Wang, and D. T. Miller, Imaging outer segment renewal in living human cone photoreceptors, *Opt. Express* **18** (2010) 5227–5270.

122. B. Vohnsen, Modeling photoreceptor mosaic imaging as backscattering of light from multilayered discs, *Front. Opt.* (2014). Abstract FW5F.5, https://www.osapublishing.org/abstract. cfm?uri=fio-2014-FW5F.5.

123. D. G. Stavenga, Waveguide modes and refractive index in photoreceptors of invertebrates, *Vision Res.* **15** (1975) 323–330.

124. K. Franze, J. Grosche, S. N. Skatchkov, S. Schinkinger, C. Foja, D. Schild, O. Uckerman, K. Travi, A. Reichenbac, and J. Guck, Müller cells are living optical fibers in the vertebrate retina, *Proc. Natl. Acad. Sci.* **104** (2007) 8287–8292.

125. A. M. Labin, S. K. Safuri, E. N. Ribak, and I. Perlman, Müller cells separate between wavelengths to improve day vision with minimal effect upon night vision, *Nat. Commun.* **5** (2014) 4319.

126. N. Bohr, Light and life, *Nature* **131** (1933) 457–459.

127. R. H. Steinberg, S. K. Fisher, and D. H. Anderson, Disc morphogenesis in vertebrate photoreceptors, *J. Comp. Neurol.* **190** (1980) 501–518.

19 Refractive errors

David A. Wilson

Contents

19.1 INTRODUCTION

The eye can be treated as an optical instrument. Indeed, it is often compared to a camera or, perhaps more accurately, a video camera. The lack of homogeneity of its media and the misalignment of its elements, however, set it apart from other instruments. Helmholtz once said of the eye (quoted in Fishman) that "it is not too much to say that if an optician wanted to sell me an instrument that had all these defects, I should think myself quite justified in blaming his carelessness in the strongest terms, and giving him back his instrument" (Fishman 2010). However, despite the lack of homogeneity of its media and the misalignment of its elements, which differentiates the eye from other optical instruments and which "offended" Helmholtz's sense of optical order, it is capable of creating a sharp image on the retina. Under normal conditions, the eye can provide more than adequate acuity and can adjust for various viewing distances. In practice, if not theory, it is a very effective optical instrument. Nevertheless there are several factors that either individually or in combination can contribute to a blurred image, with a resulting reduction in visual acuity. Unlike the theoretical errors referred to by Helmholtz, these are variations from the normal functioning of the eye that do affect vision. While the optics of the normal eye are fully covered elsewhere in this book, this chapter looks at the various forms of refractive errors that can affect vision and their modes of correction.

19.2 SIGNIFICANCE OF REFRACTIVE ERROR

Ametropia and presbyopia are experienced by up to one-third of the world's population. Objective presbyopia (loss of accommodation with binocular near vision <N8 and improving to ≥N8 with near addition lenses) alone affects almost all people over the age of 45. Thus, everyone over this age will need correction for near vision, distance vision, or both. If uncorrected, refractive errors can be responsible for anything from mild visual impairment to blindness. It has also been shown that poor visual acuity affects quality of life regardless of the cause (Brown et al. 2002). Consequently, reduced visual acuity as a result of uncorrected

refractive error has the same effect on quality of life as reduced acuity resulting from ocular pathology despite the ease of its correction. All refractive errors are easily corrected with spectacles, contact lenses, or refractive surgery in most cases; however, in the small proportion with high myopia, there is the potential for other, irreversible, blinding conditions such as retinal detachments and myopic macular degeneration (Foster and Jiang 2014, Wong et al. 2014). Since health has been defined as "a state of complete physical, mental and social well-being and not merely the absence of disease or infirmity" (WHO 1946), refractive errors represent a major health concern.

19.2.1 PREVALENCE OF AMETROPIA

There are no accurate global prevalence rates for ametropia; however, there have been many country-based or regional epidemiological studies. A meta-analysis of population-based studies found that the prevalences of distance refractive errors among adults over 40 years in the United States, Western Europe, and Australia were one-third, one-third, and one-fifth of their respective populations (The Eye Diseases Prevalence Working Group 2004). Myopia has the highest prevalence rates (of ametropia) globally, largely through its high rates in the most populous countries such as China, although the study cited earlier also found the levels of myopia to be significantly higher than hyperopia (well over twice in all three regions). We were unable to find any region with more hypermetropia than myopia; however, there are regions where the prevalences are closer. Asian countries, notably China, have particularly high rates of myopia and the incidence is increasing. "In 1988, over 80% of children in Grade 1 (age 6 years) and about 30% in Grade 12 (age 17 years) had normal unaided VA. By 2007 this dropped to only 60% in Grade 1 and about 10% in Grade 12" (Xiang and He 2013). The prevalence of myopia in children between 5 and 15 years in China has been found to be 42.4% (He et al. 2007). Increasing prevalence of myopia, however, is not restricted to Asia. The United States has experienced an increase in myopia in 12- to 54-year-olds between 1971–1972 and 1999–2004 from 25% to 41.6% (Vitale 2009).

19.2.2 PREVALENCE OF PRESBYOPIA

The optics of presbyopia are considered in Section 19.6. The prevalence of presbyopia, however, is dependent on the definition used, that is, whether we are considering functional presbyopia or objective presbyopia. Functional presbyopia is effectively a question of whether a person needs an optical correction to achieve a particular level of near visual acuity. A functional presbyope requires an optical correction to be added to the presenting distance refractive correction to achieve a given near visual acuity. Various prevalence studies have used slightly different criteria for the required level; for example, an improvement of greater than or equal to one line of acuity improvement (Burke et al. 2006) or a specific level of acuity such as Jaeger J1 print (Duarte et al. 2003) or N8 print (Ramke et al. 2007). Another definition of functional presbyopia is "a net positive value from the addition of distance and near addition" (Burke et al. 2006). Under this definition, most myopes, other than low myopes with advanced objective presbyopia, would not be considered as functional presbyopes since the aggregate of their distance correction and near addition

would be negative or zero. Myopes with a distance correction of –2.50 D, for example, would effectively have a +2.50 D near correction merely by removing their distance spectacles. Thus, while these myopes may have the same level of objective presbyopia as emmetropes or hyperopes of similar age, with a similar loss of amplitude of accommodation, they would not be considered as functional presbyopes since clear near vision can still be achieved without correction.

The broader concept of objective presbyopia is defined as needing an optical correction of greater than or equal to +1.00 D added to the best distance optical correction to improve near vision to a near visual acuity of N8 (Nirmalan et al. 2006). The prevalence of objective presbyopia will always exceed functional presbyopia since a large number of uncorrected myopes will fall into the former category. Functional presbyopia is therefore a subset of objective presbyopia that includes almost all people over the age of 45 years. The current estimate of the size of this group is just over two billion people in the United States (United States Census Bureau 2013).

In 2008, functional presbyopia was estimated at approximately 1.04 billion globally, of whom 517 million were uncorrected and 410 million have been prevented from carrying out their required near tasks (Holden et al. 2008). Functional presbyopia has also been recognized as having a significant effect on quality of life as people value near vision just as highly as their distance vision (Tahhan et al. 2013).

The numbers affected by presbyopia can also be expected to grow substantially with increasing life expectancy. Weighted global life expectancy has increased from 34.1 years in 1913 to 66.6 years in 2001 (Riley 2005). In the early twentieth century, relatively few people reached a presbyopic age. However, in the early twenty-first century, not only is the global population growing, but also most people, particularly those in developed countries, can expect to live well beyond the advent of presbyopia.

19.2.3 PREVALENCE OF UNCORRECTED REFRACTIVE ERROR

Uncorrected distance refractive error has been identified as the largest cause (48.3%) of visual impairment, affecting approximately 107.8 million people globally (Bourne et al. 2013). It is also the second largest cause of blindness at 20%, or about 20.9 million people (Bourne et al. 2013). Its share of blindness, however, is even greater if pathologic myopia, which alone has been estimated in population-based studies to be the first to third most common causes of blindness (Wong et al. 2014), is considered.

The enormity of the problem of uncorrected refractive error has become clearer since the World Health Organization changed the criterion for testing acuity in its ICD-10 categories of visual impairment and blindness from best corrected vision, to presenting vision (World Health Organization 2010). While the use of best-corrected vision effectively gives the prevalence of uncorrectable refractive error, it underestimates the size of the problem by implicitly assuming that best-corrected vision is achievable. Unfortunately, the necessary correction is not available in many parts of the world. The use of presenting vision, however, allows for people with no access to refractive services and people wearing inadequate correction to be included in the estimates. It also highlights the imbalance of the burden of

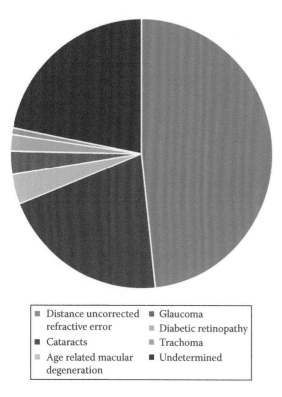

Figure 19.1 Causes of visual impairment.

uncorrected refractive error that falls disproportionately on those in under resourced, developing countries.

The economic burden, from lost productivity, of uncorrected refractive error has been estimated at $269 billion International dollars (Smith et al. 2009) or about US $202 billion (Fricke et al. 2012), while correcting the problems has been estimated to have a cost as low as US $S20 billion (Fricke et al. 2012). Indeed, correcting refractive error is much more cost effective than most other health interventions, up to 40 times more cost effective than cataract surgery, for example (Tahhan et al. 2013).

Although refractive error has now been recognized as a major global health problem, presbyopia is still largely ignored by governments, nongovernment organizations, and the World Health Organization. Neither the statistics in Figure 19.1 nor the estimate of economic burden includes uncorrected presbyopia that has been estimated to affect 517 million people (Holden et al. 2008). This is approximately five times as many as affected by distance refractive error.

19.3 EMMETROPIA

Emmetropia is the state of the eye in which parallel incident light is focused on the retina when the eye is at rest. That is, with accommodation relaxed, the focus conjugate with the retina, known as the far point or punctum remotum, is at infinity. An emmetropic eye is often referred to as a normal eye. The aggregate powers of the optical elements of the eye are appropriate for its axial length. Assuming no other problems, the emmetropic eye will produce a clear image of distant objects at any age. For the eye's pupil size, a distant object is taken to be greater than or equal to 6 m from the eye. This is the reason for the visual acuity

test distance being 6 m and consequently the denominator of the Snellen visual acuity fraction being 6 (or 20, if feet are used).

19.4 AMETROPIA

Ametropia is the state in which parallel incident light is not focused on the retina when the eye is at rest. As with emmetropia, ametropia applies to distance vision and is, thus, a distance refractive error. With accommodation relaxed, the focus conjugate with the retina, the far point, does not lie at infinity. There are three forms of ametropia: myopia, hypermetropia (often referred to as hyperopia), and astigmatism. All forms of ametropia are normally corrected by spectacles or contact lenses; however, some cases are also suited to refractive surgery. While the broader term, refractive error, is sometimes taken to include presbyopia, ametropia does not.

19.4.1 MYOPIA

Myopia is the state where parallel light incident to the eye at rest comes to a focus short of the retina, thereafter diverging to create a blur circle on the retina. The far point, the point conjugate with the retina, is no longer at infinity but moves closer to the eye by an amount that depends on the magnitude of myopia. Light diverging from the far point will be brought to a focus on the retina with the eye at rest. The negative power of the divergent light compensates for the excess strength of the eye. If the far point is rotated three dimensionally around the center of rotation of the eye, it creates the far point sphere. This represents the ideal image plane for any correcting lens. Effectively a myopic eye is too strong. This may be caused by the length of the eye (axial myopia) or the strength of the refractive media (refractive myopia) or a combination of both. This distinction will be discussed further in Section 19.4.4.

Because of its excess strength, the eye can provide clear near vision, closer than an emmetropic eye of the same age. And it is for this reason that myopia is often referred to as shortsightedness or nearsightedness. Distance vision will be blurred.

19.4.1.1 Effect of a pinhole

While the image plane for many instruments, such as cameras, is flat, the object will be at varying distances from the instrument. That is, the object will be three dimensional while the intended image plane is two dimensional. So for a fixed focus system, the foci from various parts of the three-dimensional object will fall at various distances from the screen.

In Figure 19.2, light from three points on an object (O′, O″, O‴) is refracted by a lens and passes through an aperture to form

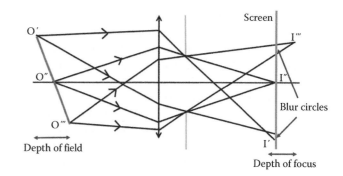

Figure 19.2 Depth of field and depth of focus.

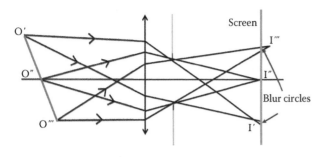

Figure 19.3 The effect of a smaller pupil.

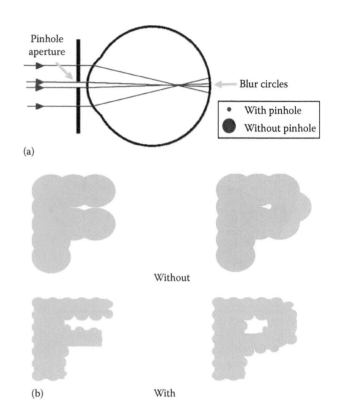

(a)

(b) Without / With

Figure 19.4 (a) Effect of a pinhole on the myopic eye. (b) Effect of a pinhole.

three foci (I′, I″, I‴). Only the image I″ falls on the screen with the other images on the screen being blur circles. In the case of human vision, these blur circles, if sufficiently small, will appear to be clear foci. The range of foci, measured along the axis, in which clarity is not compromised is referred to as the depth of focus. The corresponding object field distance is referred to as the depth of field.

The reduction of the aperture within the system will increase the depth of focus, until limited by the effects of diffraction. In Figure 19.3, light from three points on an object (O′, O″, O‴) is refracted by a lens and passes through a reduced aperture compared to the previous slide. Once again three foci (I′, I″, I‴) are formed and only the image I″ falls on the screen. In this case, however, the blur circles produced by the points O′ and O‴ are smaller than with the previous aperture size and more likely to be perceived as being in focus.

The use of a pinhole will, therefore, improve the depth of focus for a myopic person by reducing the diameter of the blur circle produced. It is this for reason that myopes squint to see without their correction. By reducing the aperture, squinting gives the effect of a pinhole. This principle can also be used to determine whether a refractive error is the cause of poor visual acuity. By increasing the depth of focus, a pinhole will improve the visual acuity for any ametrope. Therefore, if acuity does not improve with the use of a pinhole, the loss of visual acuity is not the result of refractive error.

The pinhole reduces the diameter of blur circles, as shown in Figure 19.4b. The limit is determined by diffraction. In the figure, the large blur circles evident without the pinhole overlap, making it difficult, if not impossible, to distinguish the letters. Reduction of the aperture diameter reduces the blur circles and allows the letters to be distinguished.

19.4.2 HYPERMETROPIA

Hypermetropia or hyperopia is the state where parallel light incident to the eye at rest converges toward a focus behind the retina, with the converging light creating a blur circle on the retina. The far point lies behind the retina and incident light converging toward the far point will be brought to a focus on the retina with the eye at rest. The positive power of the convergent light compensates for the weakness of the eye. If the far point is rotated three dimensionally around the center of rotation of the eye, it creates the far point sphere. As with myopia this represents the ideal image plane for any correcting lens. Effectively a hyperopic eye is optically too weak and, again, this may be axial or refractive in nature.

Because of its lack of strength, the eye can provide clear vision at distance in a young eye with sufficient reserves of accommodation. By exerting accommodation, young hyperopes can increase the power of their eyes and overcome the deficit caused by the hyperopia. For this reason, hyperopia is often referred to as far-sightedness. Near vision, however, will be blurred because there is insufficient accommodation to cover the lack of power and still focus on the near object. Clear vision is obtained through accommodation but near vision is blurred. Depending on the amount of hyperopia, aging reduces the amplitude of accommodation, resulting in blurred distance vision in addition to the blurred near vision.

Hyperopes cannot normally totally relax their accommodation. This leaves what is referred to as latent hyperopia, which cannot be detected through subjective refraction unless cycloplegics are used. The remaining detectable refractive error is referred to as manifest hyperopia. Younger people will still be able to overcome some of their manifest hyperopia by exerting accommodation. This is referred to as facultative hyperopia with the remaining refractive error being absolute hyperopia. The relationship between the types of hyperopia can be illustrated as

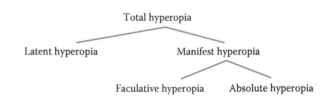

Total hyperopia

Latent hyperopia / Manifest hyperopia

Faculative hyperopia / Absolute hyperopia

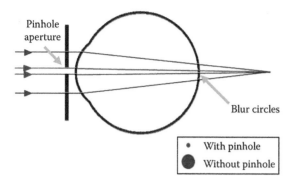

Figure 19.5 Effect of a pinhole on the hypermetropic eye.

As a person ages, the amplitude of accommodation diminishes and, as a result, manifest hyperopia will increase relative to latent hyperopia (Atchison and Smith 2000). Absolute hyperopia will also increase relative to facultative hyperopia and so an aging hyperope will require a stronger distance correction, in addition to a correction for presbyopia.

19.4.2.1 Effect of a pinhole

As with myopia, the use of a pinhole will improve the depth of focus for a hyperopic person by reducing the diameter of the blur circle produced. Hyperopes also squint to see without their correction (Figure 19.5).

19.4.3 ASTIGMATISM

Astigmatism is the state of an eye where parallel light incident to the eye at rest creates two line foci at varying distances and, in the case of regular astigmatism, perpendicular to each other. In the much rarer case of irregular astigmatism, the lines are not perpendicular. This form of astigmatism cannot be corrected by spectacle lenses but may be corrected by contact lenses that create a tear lens between the deformed cornea and the back surface of the lens with the spherical front surface of the contact lens effectively becoming the front surface of the lens/eye system.

In a normal eye, the front surface of the cornea is prolate ellipsoid in shape, rotationally symmetrical, somewhat like the end of a rugby ball or egg. Astigmatism is normally caused by the front surface of the cornea being toric in shape, not rotationally symmetrical, somewhat like the side of a rugby ball with one principal meridian having a different radius of curvature than the other principal meridian. It may also be the result of a tilted crystalline lens.

As with myopia and hyperopia, astigmatism can also be axial or refractive in nature (discussed in Section 19.4.4) combining the effects of myopia or hyperopia with the effect of the toric shaped cornea.

19.4.3.1 Sturm's conoid

Sturm's conoid, also known as the astigmatic pencil, is the shape of a pencil of light after refraction through a spherocylindrical lens, first described by German physician Johann Sturm (1635–1703). Two focal lines are formed, each perpendicular to the principal meridian that creates it. The larger the difference between the principal powers (the cylinder power), the greater the separation of the focal lines, known as the interval of Sturm. The lengths of the focal lines (h_1 and h_2) are affected by the lens diameter (d), the interval of Sturm ($\ell_2' - \ell_1'$), and the distance of the other focal line. That is,

$$h_1 = \frac{d(\ell_2' - \ell_1')}{\ell_2'} \quad h_2 = \frac{d(\ell_2' - \ell_1')}{\ell_1'}$$

The larger the cylinder power and larger the lens diameter, the longer will be the lines. The second focal line will always be larger than the first, as can be seen from the smaller denominator for h_2 in the previous formulae and in Figure 19.6.

Figure 19.6 shows an astigmatic pencil for a plus spherocylinder with a minus cylinder axis of 180. This would be used to correct with-the-rule compound hyperopic astigmatism (discussed in the next section). At a point between the focal lines, created by the average of the two principal powers, the cross section of the pencil is spherical. This is known as the circle of least confusion.

The circle of least confusion is also the basis of the concept of best vision sphere. That is, if an astigmat were to be corrected with a spherical lens, as may occur in remote service delivery where no processing facilities exist, then the power of the lens should be the power that coincides with the circle of least confusion.

Figure 19.7 shows the changing shape of the cross section of the pencil as it leaves the lens. As it first leaves the lens, it will be circular than a gradually narrowing ellipse that collapses to form the first focal line. After the first line, the pencil will become a gradually widening ellipse of the same orientation, eventually becoming a circle (the circle of least confusion). The pencil then becomes a narrowing ellipse of the opposite orientation, collapsing to form the second focal line.

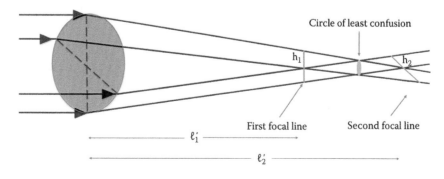

Figure 19.6 Sturm's conoid (minus cylinder correcting lens axis 180).

Optical properties of the eye

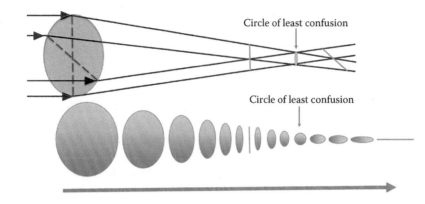

Figure 19.7 Changing shape of Sturm's conoid (minus cylinder correcting lens axis 180).

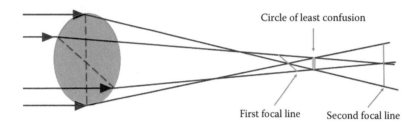

Figure 19.8 Sturm's conoid (minus cylinder correcting lens axis 90).

The orientation of the focal lines will, of course, depend on the orientation of the lens. Figure 19.8 shows the pencil for a plus spherocylinder lens with a minus cylinder axis of 90, which would correct against-the-rule astigmatism.

19.4.3.2 Types of regular astigmatism

There are five types of regular astigmatism (discussed in the following text), depending on the positions of the two focal lines relative to the retina. In addition, astigmatism can be classified according to the orientation of the strongest meridian of the eye. If the strongest meridian, that is the most myopic or least hyperopic meridian, is at or close to vertical, it is referred to as with-the-rule astigmatism. Vertical is considered to be between 70° and 110°. This is the most common form among adults, which is the reason for the term with-the-rule. The correcting lens will have a minus cylinder axis at or close to 180° (or plus cylinder axis 90°).

If the strongest meridian of the eye is at or close to horizontal, it is referred to as against-the-rule astigmatism. Horizontal is considered to be between 160° and 20°. The correcting lens will have a minus cylinder axis at, or close to, 90° (or plus cylinder close to 90°). If the axis is outside the vertical or horizontal zones, that is, between 20° and 70° or between 110° and 160°, it is referred to as oblique.

19.4.3.2.1 Simple astigmatism

Simple astigmatism is the state where parallel light incident to the eye at rest forms one line on the retina and another either in front or behind. If the off retina line focus forms in front of the retina, it is referred to as simple myopic astigmatism. The correcting lens will have no power in one meridian; for example, Plano/–2.00 × 180° (the principal powers here are 0.00 and –2.00 D). This prescription could also be written in plus cylinder form as

–2.00/+2.00 × 90°. Figure 19.9a shows an example of with-the-rule simple myopic astigmatism with the stronger vertical meridian of the eye forming the horizontal line focus in front of the retina and the weaker horizontal meridian of the eye forming the vertical image on the retina.

An astigmatic eye will have two far points, one for each of its principal powers. Figure 19.9b shows the far points for the case of simple myopic astigmatism shown in Figure 19.9a. The far point for the myopic vertical meridian lies in front of the eye. The far point for the horizontal meridian lies at infinity.

If the off retina line focus would form behind the retina, it is referred to as simple hyperopic astigmatism. Again, the correcting lens will have zero power in one meridian; for example, +2.00/–2.00 × 180° (the principal powers here are 0.00 and +2.00 D). This prescription could also be written in plus cylinder form as Plano/+2.00 × 90°. Figure 19.10a shows an example of with-the-rule simple hyperopic astigmatism with the stronger vertical meridian of the eye forming the horizontal line focus on the retina and the weaker horizontal meridian of the eye forming the vertical hyperopic image behind the retina.

Figure 19.10b shows the far points for the case of simple hyperopic astigmatism shown in Figure 19.10a. The far point for the stronger vertical meridian lies at infinity. The far point for the weaker horizontal meridian lies behind the retina.

19.4.3.2.2 Compound astigmatism

Compound astigmatism is the state where parallel light incident to the eye at rest forms both line foci on either side of the retina. If both line foci lie in front of the retina, it is referred to as compound myopic astigmatism. The correcting lens will have minus power in both meridians; for example, –3.00/–2.00 × 180° (the principal powers here are –3.00 and –5.00 D).

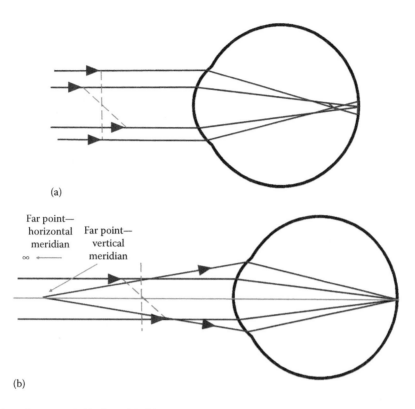

(a)

Far point—
horizontal
meridian

Far point—
vertical
meridian

∞

(b)

Figure 19.9 (a) Simple myopic astigmatism (with-the-rule). (b) Far points.

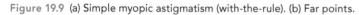

(a)

Far point—
vertical
meridian

Far point—
horizontal
meridian

∞

(b)

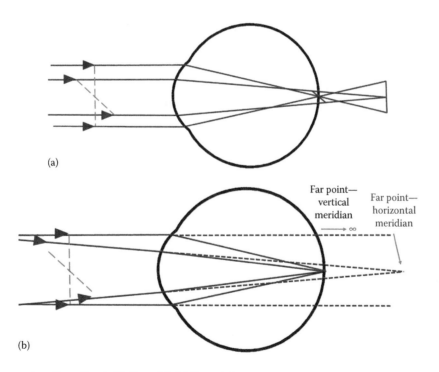

Figure 19.10 (a) Simple hyperopic astigmatism (with-the-rule). (b) Far points.

This prescription could also be written in plus cylinder form as −5.00/+2.00 × 90°. Figure 19.11a shows an example of with-the-rule compound myopic astigmatism with the stronger vertical meridian of the eye forming the horizontal line focus furthest from the retina and the weaker horizontal meridian of the eye forming the vertical image closer to the retina.

Figure 19.11b shows the far points for the case of compound myopic astigmatism shown in Figure 19.11a. Both lie in front of the eye but the far point for the weaker horizontal meridian is further away than that of the stronger horizontal meridian.

If both line foci would lie behind the retina, it is referred to as compound hyperopic astigmatism. The correcting

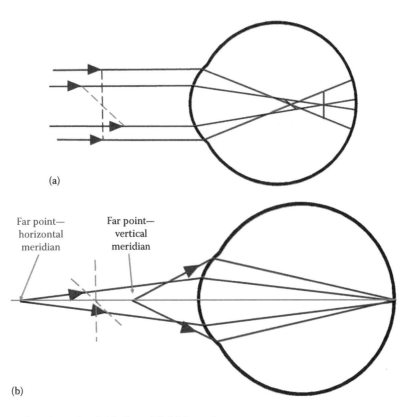

(a)

Far point—
horizontal
meridian

Far point—
vertical
meridian

(b)

Figure 19.11 (a) Compound myopic astigmatism (with-the-rule). (b) Far points.

lens will have plus power in both meridians; for example, +3.00/–2.00 × 180° (the principal powers here are +3.00 and +1.00 D). This prescription could also be written in plus cylinder form as +1.00/+2.00 × 90°. Figure 19.12a shows an example of with-the-rule compound hyperopic astigmatism with the stronger vertical meridian of the eye forming the horizontal line focus closest to the retina and the weaker

horizontal meridian of the eye forming the vertical image furthest from the retina.

Figure 19.12b shows the far points for the case of compound hyperopic astigmatism shown in Figure 19.12a. The far point for the stronger vertical meridian lies behind the retina. The far point for the weaker horizontal meridian also lies behind, but further from the retina.

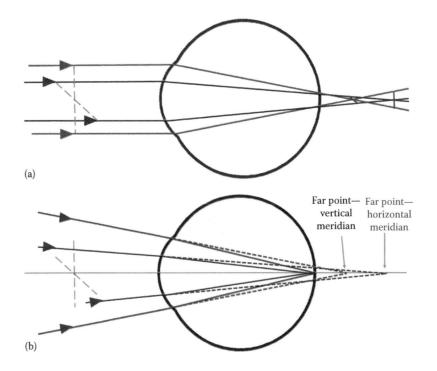

(a)

Far point—
vertical
meridian

Far point—
horizontal
meridian

(b)

Figure 19.12 (a) Compound hyperopic astigmatism (with-the-rule). (b) Far points.

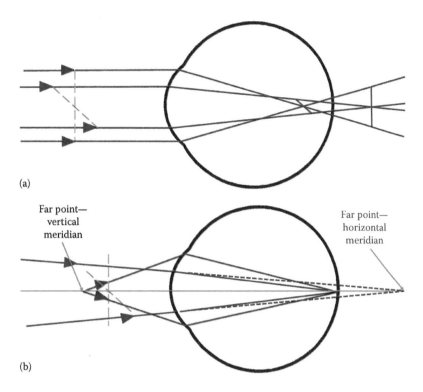

(a)

Far point—
vertical
meridian

Far point—
horizontal
meridian

(b)

Figure 19.13 (a) Mixed astigmatism (with-the-rule). (b) Far points.

19.4.3.2.3 Mixed astigmatism

Mixed astigmatism is the state where parallel light incident to the eye at rest forms one line foci on either side of the retina. The correcting lens will have minus power in one meridian and plus in the other meridian; for example, +1.00/−2.50 × 180° (the principal powers here are +1.00 and −1.50 D). This prescription could also be written in plus cylinder form as −1.50/+2.50 × 90°. Figure 19.13a shows an example of with-the-rule mixed astigmatism with the stronger vertical meridian of the eye forming the horizontal line focus in front of the retina and the weaker horizontal meridian of the eye forming the vertical image behind the retina.

In Figure 19.13a, the circle of least confusion lies in front of the retina, implying that the dioptric power of the vertical myopic meridian is greater than the dioptric power of the horizontal meridian. If the dioptric powers were equal in the absolute sense, for example, +1.00/−2.00 × 180° (vertical power of −1.00 D, horizontal power of +1.00 D), the circle of least confusion would lie on the retina. This prescription could also be written in plus cylinder form as −1.00/+2.00 × 90°. If we were applying the principle of best vision sphere in this case, then no lens would be given since the dioptric average of the principal powers is zero.

Figure 19.13b shows the far points for the case of mixed astigmatism shown in Figure 19.13a. The far point for the stronger vertical meridian lies in front of the eye. The far point for the weaker horizontal meridian lies behind the retina.

19.4.4 AXIAL VERSUS REFRACTIVE AMETROPIA

Ametropia may be caused by either the powers of the cornea and crystalline lens or the axial length of the eye or a combination of both. Using the reduced eye, Figure 19.14a shows a case of axial

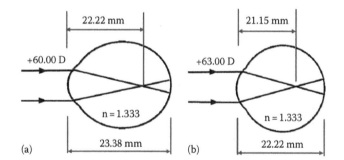

(a)   (b)

Figure 19.14 (a) Axial myopia. (b) Refractive myopia.

myopia. The emmetropic reduced eye is considered to have a power of +60.00 D and an axial length of 22.22 mm, coinciding with the reduced focal length (the medium in the reduced eye is not air) of 22.22 mm.

$$f' = \frac{1000n}{F}$$

$$f' = \frac{1000(1.333)}{+60.00}$$

$$f' = +22.22 \text{ mm}$$

In the case of axial myopia, the eye's power is normal but the eye is too long, causing the image to fall in front of the retina. The length of 23.38 mm gives a required dioptric power of +57.00 D. Since the power of the eye is the "normal" +60.00 D, this gives ocular myopia of −3.00 D.

$$F = \frac{1000n}{f'}$$

$$F = \frac{1000(1.333)}{+23.38}$$

$$F = +57.00 \text{ D}$$

Figure 19.14b shows a case of refractive myopia. For the refractive myope, the eye's power is too strong while the axial length is normal, causing the image to fall in front of the retina. The axial length of 22.22 mm would be correct for a power of +60.00 D; however, the shorter focus of 21.15 mm gives a power of +63.00 D or ocular myopia of –3.00 D.

$$F = \frac{1000n}{f'}$$

$$F = \frac{1000(1.333)}{+21.15}$$

$$F = +63.00 \text{ D}$$

Figure 19.15a shows a case of axial hyperopia. For the axial hyperope, the eye's power is normal, but the eye is too short, causing the image to fall behind the retina. The length of 21.15 mm gives a required dioptric power of +63.00 D. Since the power of the eye is the "normal" +60.00 D, this gives ocular hyperopia of +3.00 D.

$$F = \frac{1000n}{f'}$$

$$F = \frac{1000(1.333)}{+21.15}$$

$$F = +63.00 \text{ D}$$

Figure 19.15b shows a case of refractive hyperopia. For the refractive hyperope, the eye's power is too weak while the axial length is normal, causing the image to fall behind the retina. The axial length of 22.22 mm would be correct for a power of +60.00 D; however, the longer focus of 23.38 mm gives a power of +57.00 D or ocular hyperopia of –3.00 D.

$$F = \frac{1000n}{f'}$$

$$F = \frac{1000(1.333)}{+23.38}$$

$$F = +57.00 \text{ D}$$

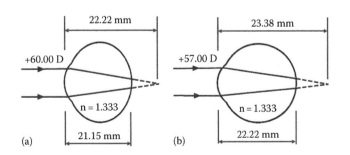

Figure 19.15 (a) Axial hypermetropia. (b) Refractive hypermetropia.

It is not possible for an eye care practitioner to determine whether a person is an axial ametrope of refractive ametrope without taking an axial measurement using ultrasound or making assumptions based on corneal curvature.

19.5 CORRECTION OF AMETROPIA

Regardless of whether the cause of the ametropia is axial or refractive, the method of correction is the same, though the cause does have implications for the type of correction chosen (as will be discussed further in Section 19.5.1).

In the case of myopia, the second principal focus (the focus formed by a lens with parallel light incident of the front surface) of the correcting lens should coincide with the far point (Figure 19.16). This requires a minus lens where the virtual second principal focus lies in front of the lens. The divergence created by the minus lens is neutralized by the excess strength of the eye.

In the case of hyperopia, the second principal focus of the correcting lens should coincide with the far point (Figure 19.17). This requires a plus lens where the real second principal focus lies behind the lens. The convergence created by the plus lens supplements the eye's weaker power.

For astigmatism each meridian requires a different power. Figure 19.18a shows an example of a case of compound myopic

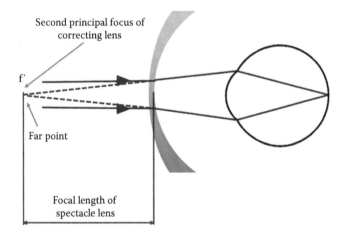

Figure 19.16 Correction of myopia.

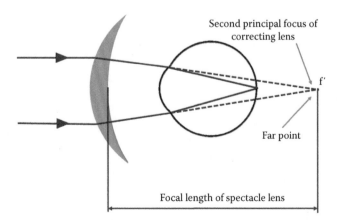

Figure 19.17 Correction of hypermetropia.

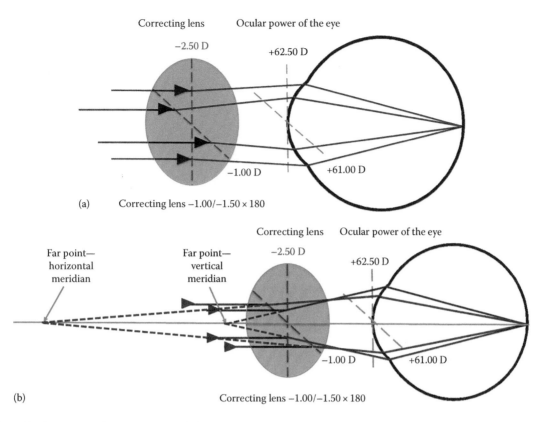

(a) Correcting lens −1.00/−1.50 × 180

(b) Correcting lens −1.00/−1.50 × 180

Figure 19.18 (a and b) Correction of astigmatism (with-the-rule example).

astigmatism (with-the-rule). The vertical meridian of the eye is 2.50 D too strong and the horizontal meridian 1.00 D too strong. The vertical power of the lens of −2.50 D neutralizes the extra 2.50 D of the eye, while the horizontal power of the lens of −1.00 D neutralizes the extra 1.00 D in the horizontal meridian. Thus, the lens together with the eye produces a point focus on the retina.

In the case of astigmatism, the second principal focus of each principal meridian of the correcting lens should coincide with its respective far point (Figure 19.18b). In this example of compound myopic astigmatism, this requires a minus spherocylindrical lens where the virtual second principal focus for each meridian lies at different distances in front of the lens.

19.5.1 RELATIVE SPECTACLE MAGNIFICATION

The ratio of the retinal image size in a corrected ametropic eye to the retinal image size in a standard emmetropic eye is known as relative spectacle magnification is. It can be shown that relative spectacle magnification is affected by the type of ametropia, that is, whether it is axial or refractive. If the second principal plane of a spectacle lens is placed at the first principal focus of an axial ametropic eye, then the image size will remain constant for any length of eye (see Figure 19.19). This relationship is known as Knapp's law and this implies that spectacle lenses are a better form of correction for axial ametropes, particularly in cases of anisometropia (defined in the following section). By contrast a contact lens will form a smaller image for a hyperope than an emmetrope and larger for a myope (Figure 19.19). In anisometropia this would (theoretically) lead to spectacle-induced aniseikonia (different image sizes) (Figure 19.20).

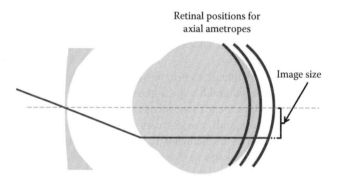

Figure 19.19 Axial ametropia corrected by a spectacle lens.

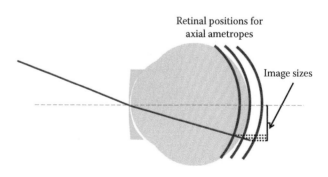

Figure 19.20 Axial ametropia corrected by a contact lens.

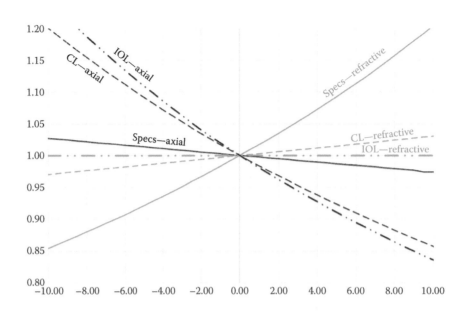

Figure 19.21 Relative spectacle magnification of axial and refractive ametropias.

For refractive ametropes, contact lens is a better option than spectacle lens. A spectacle lens positioned at or near the first principal focus of the eye would create a smaller image size for myopes and larger for hyperopes. A contact lens will create a larger image size.

Figure 19.21 shows the relationship between correcting lens power and magnification for different modes of correction. The figure also shows that intraocular lenses have a similar effect to contact lenses. For refractive ametropes, intraocular lenses have no magnification effect since the lenses are at (or very close to) the eye's entrance pupil.

In reality, it is difficult to determine whether the ametropia is axial or refractive, and it is quite likely that it may be a combination of both in many cases. It has also been shown that, while Knapp's law holds true in theory and in the design of some instruments such as the focimeter, it does not apply as predicted clinically. It has been argued that this may be due to confounding influences such as retinal stretching (Kramer et al. 1999, Romano and von Noorden 1999). As a result, while we might expect the image size to be the same for an axial anisometrope corrected by spectacles, because the retina is also stretched in the longer eye,

the image will cover a smaller number of receptors and would be perceived as being smaller.

19.5.2 ANISOMETROPIA

Anisometropia is the condition of unequal refractive state for the two eyes. That is, where the refractive error of each eye varies from the other by a significant amount. This is generally considered to be a difference of greater than or equal to 1.00 D. This can create visual discomfort for the wearer, notably in the form of differential prismatic effect as the lines of sight are directed away from the optical centers of the lenses. This rarely troubles young anisometropes since they can direct the lines of sight through any part of the single vision lenses and thus avoid areas where they are likely to encounter significant differential prismatic effect. Presbyopic anisometropes wearing progressive lenses or bifocals, however, are required to direct their lines of sight through the reading section of the lenses. Since the reading section of these lenses is significantly removed from the distance optical centers, this creates differential prismatic effect as they read. Figure 19.22 shows the increasing differential prismatic effect as the lines of sight drop into the reading area of a progressive lens.

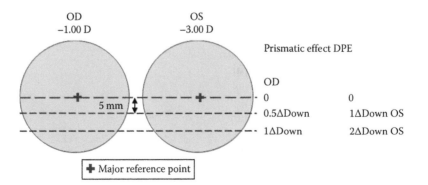

Figure 19.22 Differential prismatic effect at the reading point of progressive lenses in anisometropia.

Anisometropes are also potentially troubled by aniseikonia, or different image sizes as a result of different amounts of relative spectacle magnification, as discussed in the previous section. Theoretically, this is likely to affect axial anisometropes wearing contact lenses or refractive anisometropes wearing spectacles. As discussed earlier, however, this theoretical distinction does not necessarily follow in practice. Nevertheless, some anisometropes do experience aniseikonia, regardless of the nature of their ametropia (Kramer et al. 1999, Romano and von Noorden 1999).

19.6 PRESBYOPIA

From an optical point of view, presbyopia is the condition where the loss of accommodation occurring with aging is such that the person can no longer read at his or her habitual position. That is, the eye cannot add the extra strength necessary to overcome the divergence of the incident light from a near object; it no longer has sufficient reserves of accommodation.

The near point of accommodation (punctum proximum) is the closest point of clear vision conjugate with the retina when maximum accommodation is applied. As the eye ages, the near point gradually recedes and the amplitude of accommodation, the difference between the near point and far point in diopters, diminishes. The eye is considered to be presbyopic when the near point recedes beyond a comfortable reading position.

While objective presbyopia is independent of distance refractive error, functional presbyopia will be affected by myopia. A myope for whom the distance correction and near addition aggregate to zero is effectively corrected for near work by removing the distance spectacles. The extra power of the eye overcomes the divergent light and allows the image to focus on the retina. Effectively, the light is diverging from the eye's far point and focusing on the retina without the need for accommodation.

If we assume that the addition required to correct the presbyopia is +2.50 D and the ocular correction for distance is –2.50 D, then the eye is effectively 2.50 D too strong for distance with parallel incident light being brought to a focus in front of the retina. However, this extra 2.50 D of plus power will render the divergent light from the object at 250 mm parallel.

19.7 CORRECTION OF PRESBYOPIA

Presbyopia is corrected by adding plus power to the distance prescription. For a working distance is 0.4 m and the divergence of the light striking the lens has a power of –2.50 D (the reciprocal of –0.4 m). Thus, the plus power required to render the divergent light parallel is +2.50 D. For younger presbyopes with a reasonable amplitude of accommodation, the addition will not be the reciprocal of the working distance since the eye will still apply some accommodation to the system.

Presbyopia has been corrected for centuries with the use of spectacle lenses and, before that, handheld plus lenses. Spectacles were originally only single vision readers, until the invention of bifocals, often credited to the American polymath Benjamin Franklin who reported their invention in a letter to a friend in

1784 without naming an inventor. The next major breakthrough in the correction of presbyopia came with the invention of progressive lenses by French scientist, Bernard Maitenez, in 1959. Although the concept of progressive powered lenses dates back to the early twentieth century, the lenses invented by Maitenaz were the first commercially successful design. Unlike bifocals and trifocals, which have discrete zones, progressive addition lenses increase gradually in plus power toward the bottom of the lens, having no obvious dividing lines. In the case of bifocals, trifocals, and progressive lenses, presbyopia is corrected by the addition of the plus power to the distance prescription in the one lens. Progressive lenses have improved significantly over the past 60 years, and they are rapidly becoming the preferred mode of correction globally.

Figure 19.23 shows the effects of correcting low presbyopia with a bifocal. If we assume that the useable amplitude of accommodation is +2.00 D, then the person can see comfortably to a distance of 0.5 m (the near point). In order to read at 0.33 m, however, an addition of +1.00 D is required. This will allow the person to see to 1 m through the segment (the focal length of +1.00 D) and as close as 0.33 m. Thus, there is an overlap in the fields between distance and near and objects at an intermediate distance could be viewed through either the distance portion or the segment.

In more advanced presbyopia, however, there will be a gap between the field through the distance portion, due to the receding near point, and the shorter depth of field of the stronger segment. In Figure 19.24 we assume that the useable amplitude of accommodation is +1.00 allowing clear vision through the distance portion from infinity to 1 m. An addition of +2.00 D is now required to read at 0.33 m. This stronger addition, however, only allows clear vision out to its focal length of 0.5 m (the focal length of +2.00 D), leaving a gap between 1.0 m and 0.5 m.

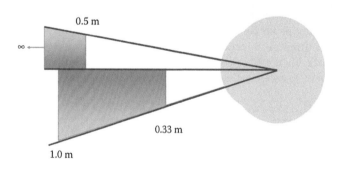

Figure 19.23 Effect of correcting low presbyopia with a bifocal lens.

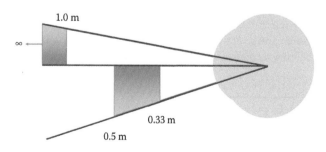

Figure 19.24 Effect of correcting low presbyopia with a bifocal lens in a subject with a residual amplitude of accommodation of 1 D.

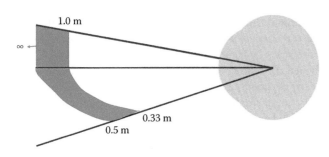

Figure 19.25 Effect of correcting low presbyopia with a progressive lens.

A progressive lens provides appropriate powers for any distance from infinity to the chosen working distance. In Figure 19.25, we again assume a useable amplitude of +1.00 D and an addition of +2.00 D. But, because the addition gradually increases inferiorly, there is a power to suit any working distance.

More recent developments include the use of adjustable liquid lenses together with a fixed distance cover lens. The adjustable rear lens has a fixed back surface and a changeable front surface activated by a slide on the frame that changes the profile of the liquid-filled membrane to increase or decrease convexity as required.

Another recent development involves the use of an electrical impulse to change the refractive index of the lens in a particular section of the lens or varying from center to periphery.

Until relatively recently, contact lenses were not an appropriate form of correction for presbyopia, given the inconvenience of inserting and then removing to see in the distance once more. However, bifocal contact lenses are now available and widely used. They come in various forms including a lens with diffraction rings, employing both geometric optics principles (refraction) and physical optical principles (diffraction) to obtain the two powers necessary for the correction of presbyopia. Monovision has also been used to correct presbyopia with one eye provided with a distance power contact lens and the other with the near correction. Surgical corrections are now also available with intraocular lenses. These will be covered more fully in *Handbook of Visual Optics: Instrumentation and Vision Correction, Volume Two*.

19.8 CONCLUDING COMMENTS

In the field of visual optics, refractive error is deserving of particular attention if only because it affects a significant proportion of the world's population. Very few people will not at some point be affected, even if they are fortunate enough in their youth to be free of distance refractive error. There is also ample evidence of substantial increases in the incidence of myopia around the world (He et al. 2007, Vitale 2009), as well as rapid improvements to the modes of correction and several advance in the control of myopic progression. Other sections of this book will examine these in more detail.

REFERENCES

Atchison, D. and G. Smith. 2000. *Optics of the Human Eye.* Oxford, U.K.: Butterworth Heinemann.

Bourne, R.R.A., G.A. Stevens, R.A. White, J.L. Smith, S.R. Flaxman, H. Price, J.B. Jonas et al. 2013. Causes of vision loss worldwide, 1990–2010: A systematic analysis. *The Lancet Global Health.* Accessed November 13, 2013. http://www.thelancet.com/journals/langlo/article/PIIS2214-109X(13)70113-X/abstract.

Brown, M.M., G.C. Brown, S. Sharma, J. Landy, and J. Bakal. 2002. Quality of life with visual acuity loss from diabetic retinopathy and age-related macular degeneration. *Archives of Ophthalmology* 120:481–484.

Burke, A.G., I. Patel, B. Munoz, A. Kayongoya, W. McHiwa, A.W. Schwarzwalder, and S.K. West. 2006. Population-based study of presbyopia in rural Tanzania. *American Journal of Ophthalmology* 113 (5):723–727.

Duarte, W.R., A.J.D. Barros, J.S. Dias-da-Costa, and J.M. Cattan. 2003. Prevalência de deficiência visual de perto e fatores associados: um estudo de base populacional. *Cadernos de Saúde Pública* 19:551–559.

Fishman, R.S. 2010. Darwin and Helmholtz on imperfections of the eye. *Archives of Ophthalmology* 128 (9):1209–1211.

Foster, P.J. and Y. Jiang. 2014. Epidemiology of myopia. *Eye (London)* 28 (2):202–208.

Fricke, T.R., B.A. Holden, D.A. Wilson, G. Schlenther, K.S. Naidoo, S. Resnikoff, and K.D. Frick. 2012. Global cost of correcting vision impairment from uncorrected refractive error. *Bulletin of the World Health Organization* 90:728–738.

He, M., W. Huang, Y. Zheng, L. Huang, and L.B. Ellwein. 2007. Refractive error and visual impairment in school children in rural southern China. *Ophthalmology* 114 (2):374.e1–382.e1.

Holden, B.A., T.R. Fricke, S.M. Ho, R. Wong, G. Schlenther, S. Cronje, A. Burnett, E. Papas, K.S. Naidoo, and K.D. Frick. 2008. Global vision impairment due to uncorrected presbyopia. *Archives of Ophthalmology* 126 (12):1731–1739.

Kramer, P., S. Shippman, G. Bennett, D. Meininger, and V. Lubkin. 1999. A study of aniseikonia and Knapp's law using a projection space eikonometer. *Binocular Vision & Strabismus Quarterly* 14 (3):197–201.

Nirmalan, P.K., S. Krishnaiah, B.R. Shamanna, G.N. Rao, and R. Thomas. 2006. A population-based assessment of presbyopia in the state of Andhra Pradesh, South India: The Andhra Pradesh eye disease study. *Investigative Ophthalmology and Visual Science* 47 (6):2324–2328.

Ramke, J., R. du Toit, A. Palagyi, G. Brian, and T. Naduvilath. 2007. Correction of refractive error and presbyopia in Timor-Leste. *British Journal of Ophthalmology* 91:860–866.

Riley, J.C. 2005. Estimates of regional and global life expectancy, 1800–2001. *Population and Development Review* 31 (3):537–543.

Romano, P.E. and G.K. von Noorden. 1999. Knapp's law and unilateral axial high myopia. 1970. *Binocular Vision & Strabismus Quarterly* 14 (3):215–222.

Smith, T.S.T., K.D. Frick, B.A. Holden, T.R. Fricke, and K.S. Naidoo. 2009. Potential lost productivity resulting from the global burden of uncorrected refractive error. *Bulletin of the World Health Organization* 87. Accessed April 9, 2009. http://www.who.int/bulletin/volumes/87/6/08-055673/en/.

Tahhan, N., E. Papas, T.R. Fricke, K.D. Frick, and B.A. Holden. 2013. Utility and uncorrected refractive error. *Ophthalmology* 120 (9):1736–1744

The Eye Diseases Prevalence Working Group. 2004. The prevalence of refractive errors among adults in the United States, Western Europe, and Australia. *Archives of Ophthalmology* 122:495–505.

United States Census Bureau. 2013. International data base. Accessed April 24, 2014. http://www.census.gov/population/international/data/idb/region.php?N=%20Results%20&T=10&A=aggregate&RT=0&Y=2014&R=1&C=.

Vitale, S., R.D. Sperduto, and F.L. Ferris III. 2009. Increased prevalence of myopia in the United States between 1971–1972 and 1999–2004. *Archives of Ophthalmology* 127 (12):1632–1639.

WHO. 1946. Preamble to the constitution of the World Health Organization as adopted by the International Health Conference, New York, 19–22 June, 1946. In (Official Records of the World Health Organization, no. 2, p. 100) and entered into force on 7 April 1948.

Wong, T.Y., A. Ferreira, R. Hughes, G. Carter, and P. Mitchell. 2014. Epidemiology and disease burden of pathologic myopia and myopic choroidal neovascularization: An evidence-based systematic review. *American Journal of Ophthalmology* 157 (1):9.e12–25.e12.

World Health Organization. 2010. International statistical classification of diseases and related health problems 10th revision (ICD-10) version for 2010. http://www.who.int/classifications/icd/en/.

Xiang, F. and M. He. 2013. Increases in the prevalence of reduced visual acuity and myopia in Chinese children in Guangzhou over the past 20 years. 27 (12):1353–1358.

20 Monochromatic aberrations

Susana Marcos, Pablo Pérez-Merino, and Carlos Dorronsoro

Contents

20.1 INTRODUCTION

The eye is an optical instrument that projects scenes of the visual world onto the retina. The eye is far from being a perfect optical system, in particular for large pupil diameters. Refractive errors (defocus or astigmatism) occur frequently in the eye. Apart from these low-order aberrations, the eye suffers also from high-order *aberrations* (HOA). The existence of significant amounts of HOA in the eye, higher than in conventional man-made optical instruments, was first acknowledged by Helmholtz (1986). Like defocus, optical aberrations blur the retinal image, reducing image contrast and limiting the range of spatial frequencies available to further stages of the visual processing. The contribution of aberrations to optical degradation is typically smaller than defocus or astigmatism. The blurring effect of aberrations becomes more noticeable for large pupils. For small pupil sizes, *diffraction* effects, associated to the limited aperture size, predominate over the aberrations. Along with diffraction and aberrations, *scattering* also contributes to degradation of retinal image quality. This chapter will describe basic concepts of ocular aberrometry; methods to measure the aberrations of the eye; the aberrations in the normal eye; population variability; changes

with accommodation, aging, and refractive errors; and aberrations in pathology and following treatment. Aberrometry has become a widespread technique in the ophthalmology clinic, as a guiding tool both to select surgical parameter in refractive surgery, in corneal procedures, and in contactology and to evaluate the outcomes of ocular treatment. With the availability of new tools to quantify ocular geometry and biometry, the sources of monochromatic aberrations in the eye can be identified, surpassing the simple description of the aberrations provided by aberrometry alone and allowing a deeper understanding of the factors leading to these aberrations and more accurate treatments.

20.2 OCULAR WAVE ABERRATION

20.2.1 BASIC CONCEPTS IN OCULAR ABERROMETRY

In a perfect optical system, rays entering through different parts of the pupil hit the image plane (the retina in case of the eye) at the same location. Imperfections of the optics cause departure of those rays from the ideal location. These angular deviations are called "transverse aberrations" (Malacara 1992, Born and Wolf 1993). The most common representation of the aberrations of the optical system is in terms of the "wave aberration," which is defined as the departure of the wave aberration from its ideal wave form. The wavefront is normal to the trajectories of the rays. This wavefront is a spherical wave for the perfect optical system and a distorted wave for an aberrated system. The difference of the aberrated wavefront from the ideal spherical wavefront is called wave aberration. The wave aberration is measured at the pupil plane and is represented as "topographical" map. For a perfect optical system, the wave aberration is flat across the pupil. A typical wave to describe the ocular wave aberration is in terms of a "Zernike polynomial expansion" (Malacara 1992, Mahajan 1994, Thibos et al. 2000). Zernike coefficients represent the weight of each of simpler polynomial in the wave aberration. The low-order terms correspond to conventional refractive errors: first-order terms represent prism and second-order terms represent defocus or astigmatism. High-order terms include other well-known monochromatic aberrations: that is, spherical aberration (representing changes in defocus with pupil size) or coma (a third order, nonrotationally symmetric aberration). The root-mean-square (RMS) wavefront error can be estimated from the wave aberration, and it is used as a global optical quality metric. Fourier optics allows direct estimation of the *point spread function* (PSF) and *modulation transfer function* (MTF) from the wave aberration (Goodman 1996). The phase of the pupil function is proportional to the wave aberration. The PSF is the squared modulus of the Fourier transform of the pupil function and the MTF is the modulus of the inverse Fourier Transform of the PSF. It should be noted that, unlike the MTF estimated from double-pass measurements, the MTF computed from the wave aberration does not contain the effects of scattering. However, while the MTF is easily obtained from the wave aberration (for any pupil size, and computationally, for every focus), the wave aberration or the PSF cannot be estimated directly from the double-pass MTF estimates. *Retinal image quality metrics* computed from the MTF and PSF have revealed good correlations with functional visual quality. These include *Strehl ratio* and more specifically visual Strehl ratio (computed from the MTF, weighted by the contrast sensitivity function).

20.2.2 MEASUREMENT OF MONOCHROMATIC ABERRATIONS

The first technique for measuring aberrations in the human eye was described by Tscherning (1894). In a psychophysical method, a grid superimposed on a 5-D spherical lens was projected on the retina and the subject could see a shadow image of the grid on the retina. The aberrations of the eye were inferred from the distortions of the grid reported by the subject. A modification of this technique (the crossed cylinder aberroscope) was proposed by Howland and Howland in 1968, in which the spherical lens was replaced by a crossed cylinder lens of 5 D power. An objective version (using light reflected off the retina) of the Tscherning aberroscope was proposed by Mierdel et al. (1997) and developed for clinical applications. Smirnov (1961) proposed another psychophysical technique (coined aberrometer for the first time) whereby a grid is viewed through the entire eye's pupil except a single central point, which is viewed through a small aperture that scans the entire pupil sequentially. The spatially resolved refractometer (Webb et al. 1992, He et al. 1998, Burns and Marcos 2000,

Moreno-Barriuso et al. 2001a), based on a similar concept, uses a efficient pupil scanning system, and a rapid feedback from the subject, allowing measurement of the wave aberration in much shorter time scales, despite the psychophysical nature of the technique. Objective techniques, relying on light reflected off the retina, include the laser ray tracing (LRT) (an objective version of the Smirnov's aberrometer) and the Shack–Hartmann wavefront sensor, initially developed by Shack, based on principles proposed by Hartmann in 1900, to improve the quality of satellite images, and first applied by Liang et al. (1994) to measure the aberrations of the eye in 1994.

Most current aberrometers measure the transverse aberration as a function of pupil position. The transverse aberrations are proportional to the local derivatives (slope) of the wave aberration, and therefore the wave aberration can be easily retrieved from the transverse aberration. Transverse aberration can be measured as the test beam goes into the eye (incoming aberrometry) or as the wavefront emerges from the eye (outgoing aberrometry). Figure 20.1 shows the basic principles of these two types of aberrometers, exemplified by the *Shack–Hartmann* wavefront sensor (S-H), which is an outgoing aberrometer, and the LRT, an incoming aberrometer. In the S-H (Liang et al. 1994), a narrow beam from a point light source is imaged by the eye onto the retina. The reflected wave travels through a lenslet array that focuses multiple spots (one per lenslet) onto a CCD camera. Each lenslet samples a small part of the wavefront corresponding to a certain pupil location. For a perfect optical system, the spots will be imaged at the focal point of each lenslet. The aberrations will cause local tilts of the wavefront, and therefore the spots will get deviated from the focal points. The transverse ray aberration associated to each lenslet can be determined from the departure of the centroid of its corresponding image with respect to the ideal position. In the LRT (Moreno-Barriuso et al. 2001a, Marcos et al. 2002), the pupil is sampled sequentially, as a laser

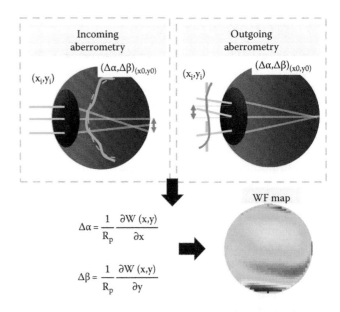

$$\Delta\alpha = \frac{1}{R_p}\frac{\partial W(x,y)}{\partial x}$$

$$\Delta\beta = \frac{1}{R_p}\frac{\partial W(x,y)}{\partial y}$$

Figure 20.1 Principles of incoming (i.e., laser ray tracing, upper left) and outgoing (i.e., Shack–Hartmann, upper right) aberrometers. In both cases, ray aberrations ($\Delta\alpha,\Delta\beta$) are measured for each pupil position (x_i,y_i), which are proportional to the derivatives of the wavefront aberration. Wavefront aberration is described by Zernike polynomial expansions.

beam is scanned across the dilated pupil and projects a spot onto the retina. A CCD in a plane conjugate to the retina captures the aerial images as a function of entry pupil. By the effect of aberrations, rays entering the eye through eccentric locations get deviated from the central ray. The local transverse aberration is measured as the angular distance between the centroid of each of the aerial images and the centroid of the image corresponding to a centered entry pupil. Again, the wave aberration is estimated from the set of local transverse aberrations. These two techniques have been used extensively in the laboratory to understand the optical properties of the normal eye, as well as in clinical applications.

The methods described earlier measure the aberrations of the entire optical system of the eye. Conventional *corneal topography* can be used to measure the aberrations of the cornea alone (Applegate et al. 2000, Guirao and Artal 2000, Barbero et al. 2002a,b). By performing virtual ray tracing on corneal elevation maps (obtained from Placido disk corneal topography or other techniques), the transverse aberration of the anterior corneal surface can be measured. From these data, the corneal wave aberration can be obtained as described earlier.

20.3 MONOCHROMATIC ABERRATIONS IN THE NORMAL HUMAN EYE

20.3.1 VARIATION OF ABERRATIONS IN THE POPULATION

The amount and distribution of aberrations vary greatly in the population. Figure 20.2a shows examples of the wave aberrations of a group of normal, young subjects. Several population studies show a wide distribution of the aberrations

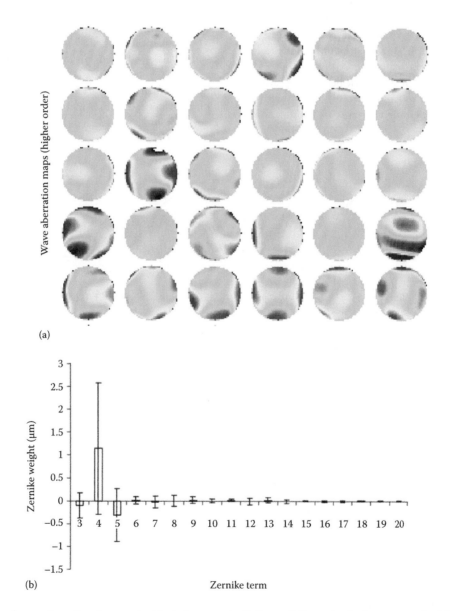

(a)

(b)

Figure 20.2 (a) Examples of wave aberrations in normal subjects (measured with the spatially resolved refractometer, for 6.5 mm pupils). (b) Average Zernike terms in a population of 108 eyes, for 5 mm pupils. (a: Adapted from Marcos, S. and Burns, S.A., *Vision Res.*, 40, 2437, 2000; b: Adapted from Castejon-Mochon, F.J. et al., *Vision Res.*, 42, 1611, 2002.)

Optical properties of the eye

in human eyes (Porter et al. 2001, Castejon-Mochon et al. 2002, Plainis and Pallikaris 2008). Castejon-Mochon et al. reported monochromatic ocular aberrations in 108 eyes of a normal young population studied using a near-infrared (IR) Shack–Hartmann wavefront sensor. For that population and a 5 mm pupil, more than 99% of the RMS wavefront error was contained in the first four orders of a Zernike expansion and about 91% corresponded only to the second order (Figure 20.2b). Plainis and Pallikaris studied the statistics of ocular HOA in an emmetropic population, using a commercial Shack–Hartmann aberrometer. Population average values of Zernike coefficients were almost zero, with the exception of primary spherical and oblique trefoil: mean higher-order RMS error was 0.26 μm, corresponding to an equivalent defocus of 0.20 D (for 6 mm pupils). Porter et al. (2001) studied wave aberrations in 109 eyes, using Hartmann–Shack aberrometry, and found that the means of almost all Zernike modes, except for spherical aberration, were approximately zero and had a large intersubject variability. Spherical aberration had a mean value of 0.138 ± 0.103 μm (for a 5.7 mm pupil) and was the only mode to have a mean significantly different from zero.

Aberrations tend to be mirror symmetric between left and right eyes, although similarly to anisometropia, subjects with very different left to right wave aberration patterns are not uncommon (Marcos and Burns 2000).

20.3.2 MONOCHROMATIC ABERRATIONS AND AGING

Aberration measurements have revealed that part of the decrease in retinal image quality with age is due to an increase in ocular aberrations. McLellan et al. (2001) using a spatially resolved refractometer aberrometer showed (in a group of 38 subjects, ranging from 23 to 65 years) that third- and higher-order aberrations increase with age (from 0.7 to 1 μm, for a 7.3 mm pupil, on average). Results from this study are shown in Figure 20.3. While no significant changes were found for third-order terms, the correlation of spherical aberration and fifth- and higher-order terms with age was highly statistically significant. Artal et al. (2002) measured corneal (from videokeratography) and total aberrations (using Shack–Hartmann aberrometry) in a group of 17 subjects (from 20 to 70 years) and showed that part of the increase of optical aberrations with age was due to the disruption of the compensation of corneal and internal aberrations, which was common in young subjects (Artal and Guirao 1998). This is not surprising for spherical aberration, since *ex vivo* measurements have shown that the spherical aberration of the crystalline lens shifted toward positive values with age (Glasser and Campbell 1998, Birkenfeld et al. 2013; see Section 20.4.3). Several studies reported a shift of ocular spherical aberration toward more positive values with age (Figure 20.3b). Amano et al. (2004) in a group of 75 eyes (from 18 to 69 years) showed a statistically significant increase of corneal and total coma RMS and of ocular spherical

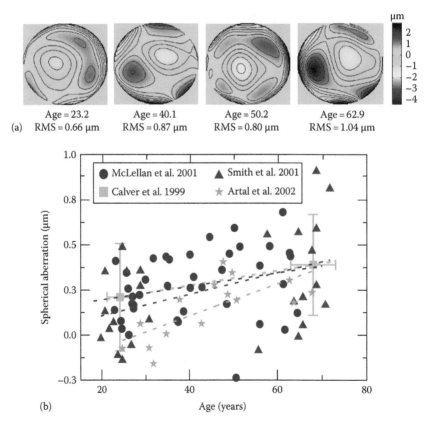

Figure 20.3 (a) Examples of wave aberrations in normal subjects of different ages (measured with the spatially resolved refractometer, for 6.5 mm pupils). (b) Spherical aberration of the eye as a function of age, from different studies. (a: Adapted from McLellan, J. et al., *Invest. Ophtalmol. Vis. Sci.*, 42, 1390, 2001; b: Adapted from Marcos, S. and Burns, S.A., *Vision Res.*, 40, 2437, 2000.)

aberration with age. Jahnke et al. (2006) measured corneal and total (using a commercial Tscherning aberrometer) in a group of 98 eyes (from 17 to 65 years) for 6 mm pupils and reported a decrease of the correlation of corneal and total aberrations and an increase in total coma, spherical aberration, and third- and fourth-order RMS with age. Lyall et al. (2013) measured corneal and total aberrations in a group of 300 eyes in a clinical setting and found significant changes with age in fourth-, fifth-, and sixth-order RMS.

20.3.3 MONOCHROMATIC ABERRATIONS AND ACCOMMODATION

The aberrations of the crystalline lens (and therefore the aberrations of the eye) change with accommodation as a result of shape changes in the lens (see examples of low-order aberrations and HOA changes with age in Figure 20.4a). He et al. (2000) reported aberration measures in a group of eyes for accommodation stimuli ranging from 0 D (infinity) to 6 D, using the spatially resolved refractometer. As shown in Figure 20.4b, optimal optical quality (in terms of RMS excluding defocus and astigmatism) was attained for about 2 D of accommodative demand,

while the RMS increased by about 1 mm on average for 6 D of accommodation. The most systematic changes were found for the spherical aberration term (decreasing in all subjects, and going from positive to negative in several subjects) and the higher-order aberrations. These changes are likely related to changes in the shape of the crystalline lens during the accommodation process. Hofer et al. (2001) measured dynamic changes of HOA during the accommodation process (from 0 to 2 D) and found temporal changes even when the accommodation was paralyzed by topical drugs.

Cheng et al. (2004a) measured total aberrations as a function of accommodation (up to 6 D) in a large population (91 young adults) and confirmed that spherical aberration shows the greatest change with accommodation, the change being always negative and proportional to the change in accommodative response. Coma and astigmatism also changed with accommodation, but the direction of the change was variable. Even though aberrations change with accommodation, the magnitude of the aberration change remains less than the magnitude of the uncorrected aberrations, even at high accommodative levels. Plainis et al. (2005) acknowledged that the estimation of the accommodative response from wave

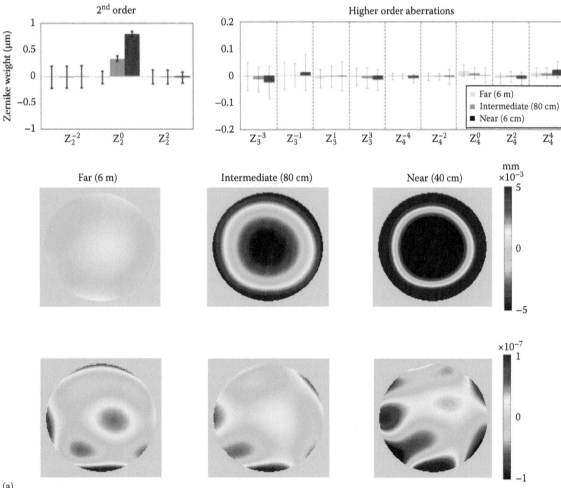

(a)

Figure 20.4 (a) Average Zernike terms in a population of 17 eyes for different accommodative demands, up to 2.5 D (measured with the laser ray tracing, for 4 mm pupils). (a: Adapted from the young control group (28 ± 4 years) in Pérez-Merino, P. et al., *Am. J. Ophthalmol.*, 157, 1077, 2014a.)   *(Continued)*

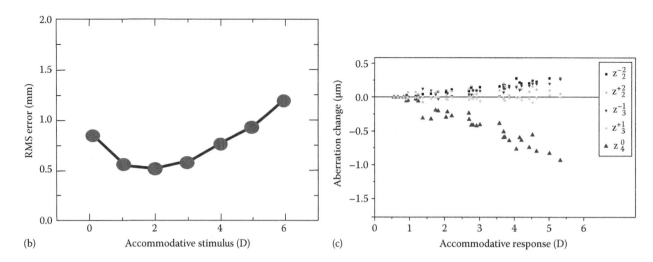

Figure 20.4 *(Continued)* (b) Root-mean-square wavefront error as a function of accommodative stimulus, average of eight eyes (measured with the spatially resolved refractometer, for 6 mm pupil). (c) Example in one subject of the change in several aberration terms (astigmatism, coma, and spherical aberration) as a function of accommodative response, for natural pupils. (b: Adapted from He, J.C. et al., *Vision Res.*, 40, 41, 2000; c: Adapted from Gambra, E. et al., *J. Vis.*, 9, 1, 2009.)

aberration measures should take into account interactions between low-order aberrations and HOA, as well as the actual pupil diameter, which are implicit when using retinal image quality metrics. It was concluded that fluctuations of accommodation may serve to preserve retinal image quality when errors of accommodation are moderate, by temporarily searching for the best focus.

Gambra et al. (2009) used dynamic Shack–Hartmann aberrometry to measure the temporal changes with accommodation upon steady fixation and increased accommodative effort (from 0 to 6 D), Figure 20.4c. In the absence of HOA (corrected using adaptive optics), the accommodative response was more accurate (smaller accommodative lag) than under natural aberrations and upon induced spherical aberration and/or coma.

Aberrometry appears therefore as an excellent tool to track fluctuations of accommodation and as an alternative tool for dynamic photoretinoscopy to measure objectively the accommodative response.

20.3.4 MONOCHROMATIC ABERRATIONS AND REFRACTIVE ERROR

Study of monochromatic aberrations in eyes with different refractive errors, and in myopic eyes in particular, has interest to shed light on potential cause/effect relationships between retinal image quality and myopia development, and to better understand the optical properties of eyes that need to be provided with optical or surgical corrections. Some population studies did not find a correlation between aberrations and refractive error (Porter et al. 2001, Cheng et al. 2003) or differences in the amount of aberrations across refractive groups (Cheng et al. 2003). However, an increasing number of studies have reported higher amounts of aberrations in myopes when compared to emmetropes (Collins et al. 1995, He et al. 2002, Marcos et al. 2002, Paquin et al. 2002). For the spherical aberrations specifically, some authors find significant correlation between spherical aberration and myopia (Collins et al. 1995) or significant differences across high myopes with respect to low myopes, emmetropes, or hyperopes (Carkeet

et al. 2002), whereas others did not find a significant correlation between spherical aberration and myopic refractive error, but did between coma and myopia (Marcos et al. 2002). Different authors found biometric and optical differences between hyperopic and myopic eyes (Collins et al. 1995, Marcos et al. 2000, 2001a). A study comparing two groups of hyperopic and myopic eyes (matched in age and in absolute refractive error) showed higher amounts of corneal asphericity and corneal spherical aberration in the hyperopic group (see Figure 20.5), apart from longer axial lengths (Llorente et al. 2004a).

The differences across studies may be due to several reasons: different age groups, refractive error ranges, and populations and ethnicities, differences in the statistical power of the studies. In an attempt to make groups more comparable, Llorente et al. (2004a) studied corneal and total aberrations (using LRT aberrometry) as well as biometry and corneal topography in two groups of age-matched (30 years of age), absolute refractive error-matched (~3 D) young myopes and hyperopes eyes (see Figure 20.5 for wave aberration examples in one hyperopic and one myopic eye). Hyperopic eyes tended to have higher total spherical aberration and third-order RMS than myopes: 0.22 ± 0.17 μm and 0.10 ± 0.13 μm spherical aberration for hyperopes and myopes, respectively, and 0.34 ± 0.13 μm (Figure 20.5b) and 0.24 ± 0.13 μm third-order RMS for hyperopes and myopes, respectively, for 6 mm pupils. Internal aberrations were not significantly different between the myopic and hyperopic groups, although internal spherical aberration showed a significant age-related shift toward less negative values in the hyperopic group. The results were also indicative of presbyopic changes occurring earlier in hyperopes than in myopes.

These results are confirmed by larger population studies. Philip et al. (2012) measured ocular higher-order aberrations (using a Shack–Hartmann-based aberrometer) and corneal topography on myopic, emmetropic, and hyperopic eyes of 675 adolescents (16.9 ± 0.7 years). Hyperopic eyes (+0.083 ± 0.05 μm) had more positive total ocular primary spherical aberration compared to emmetropic (+0.036 ± 0.04 μm) and myopic eyes

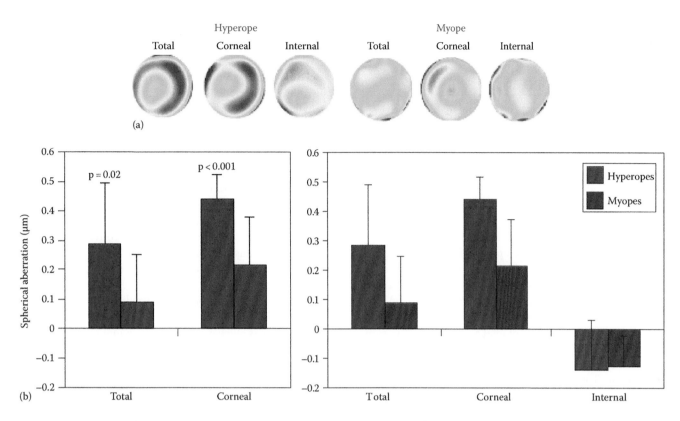

Figure 20.5 (a) Examples of total, corneal, and internal (total corneal) wave aberrations, for a hyperopic eye and for a myopic eye (measured using laser ray tracing, for 6.5 mm pupils). (b) Average total and corneal spherical aberrations in two groups of age-matched and absolute refractive error-matched (22 hyperopes and myopes eyes). (Adapted from Llorente, L. et al., *J. Vis.*, 4, 288, 2004a.)

(low myopia = +0.038 ± 0.05 μm, moderate myopia = +0.026 ± 0.06 μm) (p < 0.05), for 5 mm pupils. Similarly to previous studies, no difference was observed for the anterior corneal spherical aberration. These findings suggest the existence of differences in the characteristics of the crystalline lens (asphericity, curvature, and gradient refractive index) of hyperopic eyes versus other eyes.

The study of the variation of the ocular aberrations with refractive error is generally based on cross-sectional studies, where results may be masked by intersubject variability in the ocular structure. Also, cross-sectional studies are unable to determine whether the association between increased HOA and myopia is a consequence or a cause for myopia development. A longitudinal study (French et al. 2013), part of the Sydney Adolescent Vascular and Eye Study, measured refractive error and HOA in 166 emmetropic participants at age 12, and 5 years later, at age 17, of whom 25% experienced a myopic change. More positive change in spherical aberration was associated with lesser myopic change in refraction. However, significant association was observed between HOA and retinal image quality at baseline and development and progression of myopia among emmetropic eyes, indicating that the change in spherical aberration with myopic change is possibly associated with changes occurring in crystalline lens during ocular growth.

20.3.5 MONOCHROMATIC ABERRATIONS AND WAVELENGTH

Monochromatic aberrations are measured with monochromatic light. For patient's comfort, most aberrometers use IR illumination, although for many applications, including predictions of visual performance or refraction estimation, aberration data in visible light are required. It has been shown experimentally that, in general, HOA measurements in IR (780 nm) and in green light (543 nm) are equivalent (Marcos et al. 1999, Llorente et al. 2003, Pérez-Merino et al. 2013), with only the defocus Zernike term being different (due to chromatic aberration of the eye). Measurements in different wavelengths in the visible and IR spectral ranges reveal that the measured HOA of the eye are equivalent regardless the measurement wavelength (Fernandez et al. 2006, Vinas et al. 2015).

20.4 MONOCHROMATIC ABERRATIONS IN PATHOLOGY AND TREATMENT

While characterizing the ocular aberrations in the normal eye is important to understand the spatial limits of vision, it is in the clinical applications in the ophthalmology practice where aberrometry has seen the largest impact. Initially, aberrometers were commercialized as part of laser refractive surgery units with the aim of guiding corneal ablation in corneal refractive surgery procedures. Their applicability has since then expanded into other clinical areas such as cataract surgery (helping in the selection of intraocular lenses), contactology (to improve fitting of contact lenses [CLs]), and corneal treatments (such as the management of keratoconus). Improvements in ocular treatments (new corneal laser ablation algorithms, new lens designs, etc.) have paralleled the increased understanding of the effects of those treatments in the optical quality if the eye, provided by the evaluations obtained from aberrometry.

20.4.1 MONOCHROMATIC ABERRATIONS AND REFRACTIVE SURGERY

Corneal refractive surgery has become a popular alternative for correction of refractive errors. The use of aberrometry in the early 1990s revealed that while defocus or astigmatism is generally successfully corrected, refractive surgery (RK, PRK, and LASIK) increased the amount of HOA (Applegate et al. 1998, Oshika et al. 1999). Seiler et al. (2000) in standard myopic PRK (15 eyes, mean pre-op spherical error = −4.8 D) and Moreno-Barriuso et al. (2001b) for LASIK (22 eyes, mean pre-op spherical error = −6.5 D) measured for the first time the changes in the total aberration pattern induced by surgery (Figure 20.6a left panel; Figure 20.6b red symbols). Both studies found a significant increase in third- and higher-order aberrations (by a factor of 4.2 and 1.9 in the RMS, respectively). The larger increase occurred for spherical and third-order aberrations. Figure 20.6a shows an example of total and corneal aberrations (third and higher order) for one patient before and after myopic LASIK, and the Figure 20.6b, red symbols, the relationship between induced aberrations and preoperative myopia. Marcos et al. (2001b) found that the decrease in the MTF computed from wave aberrations agreed with the decrease in contrast sensitivity measured psychophysically in those subjects. The changes of total spherical aberrations are not fully accounted by changes in the anterior corneal surface (Marcos et al. 2001b). In all eyes, total spherical aberration increased slightly less than corneal aberrations, likely due to significant changes in the posterior corneal shape (shifting toward more negative values of spherical aberration). The increase in the total spherical aberration is highly correlated to the amount of spherical error corrected,

and it is associated to an increase in corneal asphericity (Figure 20.6b). Computer simulations and experiments on corneal models in plastic revealed that the increased spherical aberration (associated to an increase in corneal asphericity) was not caused by the theoretical ablation profile, but to changes in the laser efficiency across the cornea (Gatinel et al. 2001, Anera et al. 2003, Jiménez et al. 2003, Cano et al. 2004). While spherical aberration becomes more positive after myopic LASIK, it shifts toward negative values after hyperopic LASIK (see Figure 20.6a, right panel, for an example in one eye pre- and poststandard hyperopic LASIK). For the same absolute amount of correction, the absolute increase of corneal spherical aberration is larger with hyperopic LASIK (Llorente et al. 2004) (Figure 20.6b, blue symbols). These studies revealed that correction of the laser ablation algorithms to incorporate laser efficiency losses and a better understanding of the corneal biomechanical changes are needed to avoid induction of HOA. Since these early studies, ablation algorithms have been optimized, and the applications of aberrometry in refractive surgery have increased exponentially. Aberrometry has allowed to deploy wavefront-guided refractive surgery procedures, which adjust the ablation pattern to optimize the asphericity of the cornea, or even to correct the HOA of the eye, and it is also used as a reference as well as to evaluate the outcomes of new refractive surgery procedures. Aberrometry has been applied in the evaluation of presbyLASIK (aiming at increasing the depth of focus of the eye to compensate the accommodative loss in presbyopia) (Gifford et al. 2014), phakic lens treatments (refractive surgery procedures involving the implantation of an intraocular lens [IOL] while preserving the natural crystalline lens), and femtosecond laser refractive surgery, such as the SMILE procedure (Kamiya et al. 2014), where a lenticule is removed from the cornea using femtosecond laser, rather than corneal tissue ablation, among others.

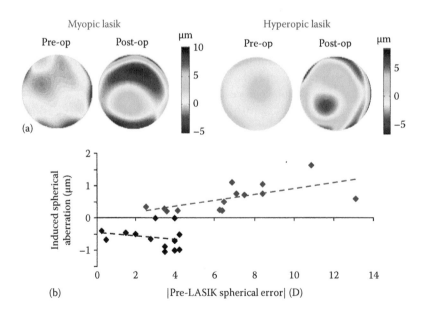

Figure 20.6 (a) Examples of total wave aberrations pre- and poststandard LASIK for myopia and hyperopia correction (measured with laser ray tracing, for 6.5 mm pupils). (b) Induced spherical aberration following surgery, as a function of preoperative spherical error (in absolute values). (Adapted from Marcos S. et al., *Invest. Ophthalmol. Vis. Sci.*, 42, 3349, 2001a; Marcos, S. et al., Why high myopic eyes tend to be more aberrated?, Presented at the *Optical Society of America Technical Digest*, Long Beach, CA, 2001b; Llorente, L. et al., *J. Refract. Surg.*, 20, 203, 2004b.)

20.4.2 MONOCHROMATIC ABERRATIONS AND CATARACT SURGERY

In virtually all cataract procedures, the natural crystalline lens is replaced by an artificial IOL. IOL manufacturers typically assess the optical quality of these lenses by measuring the MTF and the resolution *ex vivo*, in an optical bench. With the advent of aberrometry, the aberrations of the IOL were measured for the first time *in vivo* by Barbero et al. (2003). Corneal and total aberrations were measured using videokeratoscopy and LRT, respectively, in patients implanted with standard spherical IOLs. The aberrations of the IOL were estimated as the total minus corneal aberrations. Figure 20.7a shows average total, corneal, and internal spherical aberrations in one patient before and after cataract surgery from Barbero et al. study (Barbero et al. 2003). While scattering is removed with the extraction of the cataract and replacement by IOL, aberrations are not reduced.

Corneal aberrations tend to increase with surgery, due to the incision (Marcos et al. 2007). Pre- and postcataract surgery aberrations are significantly larger than in a young control group. Measurements *ex vivo* showed that the IOL is not aberration-free. The amount of aberrations (particularly astigmatism and third order) increases after implantation, likely due to tilt and decentration of the lens. Similarly to what Artal et al. (2002) reported for old eyes, postsurgical eyes do not show a good balance of corneal and internal IOL aberrations. This is particularly due to the fact that the spherical aberration of the IOL tends to be positive. Aberrometry has been instrumental in the development of further improvements of the surgical procedures and, in particular, the development of new lens aspheric designs aiming at canceling the spherical aberration of the cornea may result in better optical outcomes of cataract surgery (Piers et al. 2007). Indeed, patients implanted with aspheric IOLs show negative internal aberration

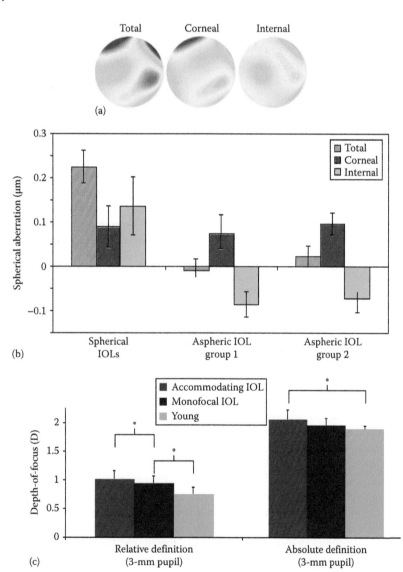

Figure 20.7 (a) Example of total, corneal, and internal wave aberrations in one patient after cataract surgery implanted with a spherical intraocular lens (IOL) (measured with laser ray tracing, for 6.5 mm pupils). (b) Average total, corneal, and internal, spherical aberration in patients implanted with spherical IOLs, and two groups implanted with two types of aspheric IOL. (c) Through-focus optical quality computed from total wave aberrations measured in eyes implanted with monofocal IOLs and with accommodating IOLs (3 mm pupil diameter). (a: Adapted from Barbero, S. et al., *J. Opt. Soc. Am. A*, 20, 1841, 2003; b: Adapted from Marcos, S. et al., *J. Refract. Surg.*, 21, 223, 2005; c: Adapted from Pérez-Merino, P. et al., *Am. J. Ophthalmol.*, 157, 1077, 2014a.)

and significantly lower spherical aberration than patients with spherical IOLs (Figure 20.7b), resulting in better optical quality in focus (Marcos et al. 2004). As a drawback, depth of field (both referred to relative and absolute criteria) was decreased in patients with aspheric IOLs.

Aberrometry is being used to evaluate the outcomes of premium IOLs, occasionally attempting to even measure the optical quality of multifocal diffractive IOLs (for which aberrometry is not well suited). Very recently, aberrometry has provided insights into the mechanisms of accommodating IOLs, which aim at restoring the accommodative ability of the eye by changing power in response to an accommodating stimulus. A study by Pérez-Merino et al. (2014a) measured optical aberrations for different accommodative demands (from 0 to 2.5 D) in 22 eyes implanted with axially shifting accommodating IOLs (Crystalens). Vertical trefoil was the predominant higher-order aberration in the Crystalens group and significantly higher than in a young group, but similar to a monofocal IOL group, consistent with the effects of corneal incision on the aberration pattern of pseudophakic eyes. Coma RMS was higher in the Crystalens group than in the young group. On average, the defocus term, astigmatism, or higher-order aberrations did not change systematically with accommodative demand in Crystalens eyes. The fact that defocus does not change with accommodation supports previous evidence that in fact these lenses do not shift sufficiently to produce a relevant change in power. Depth of focus, computed from the through-focus visual Strehl ratio from the measured aberrations, was statistically significantly higher in the Crystalens eyes than in the control groups (Figure 20.7c), indicating that these lenses may produce some functional near vision by increased depth of focus, rather than by dynamic accommodation.

20.4.3 MONOCHROMATIC ABERRATIONS IN KERATOCONUS AND ITS TREATMENT

Progressive distortion of the cornea in keratoconus leads to abnormal corneal topography and decreased visual performance in keratoconus patients (Chan 1999). Astigmatism is typically high in these patients. Additionally, coma (normally in the vertical direction) is much higher than in normal subjects (Schwiegerling 1997, Barbero et al. 2002b, Maeda et al. 2002), approximately 3.7 times on average, in one of the first studies assessing ocular aberrations in keratoconus (Barbero et al. 2002) (see one example in one patient in Figure 20.8a). Numerous studies have used aberrometry to investigate HOA in keratoconic eyes in different stages of the disease (Maeda et al. 2002, Bühren et al. 2007, Jafri et al. 2007, Kosaki et al. 2007, Pantanelli et al. 2007, Schlegel et al. 2009, Gordon-Shaag et al. 2012), as well the relative contribution to the aberrations of the anterior and posterior corneal surface. Nakagawa et al. (2009) reported increased magnitude of trefoil, coma, and spherical aberration in keratoconus patients, compared to control eyes, and a partial compensation of coma by the posterior corneal surface.

Aberrometry therefore has shown potential to diagnose and grade keratoconus, for which multiple strategies have been proposed based on aberration data, such as those based on the amount of coma (Alió and Shabayek 2006), or more complex neural network-based approaches (Smolek and Klyce 1997).

An increasingly used treatment for keratoconus are intracorneal ring segments (ICRS), implanted in the corneal stroma with the aim of flattening and increasing symmetry of the anterior corneal surface (therefore decreasing astigmatism and HOA). In a recent study, Pérez-Merino et al. (2014b) measured corneal (from quantitative anterior segment OCT) and total (using LRT) aberrations in keratoconus patients before and after implantation of ICRS was highly correlated (Figure 20.8b). Average RMS of total HOA was 0.57 ± 0.39 μm preoperatively and 0.53 ± 0.24 μm postoperatively (4 mm pupils). The anterior corneal surface aberrations were partially compensated by the posterior corneal surface aberrations (by 8.3% preoperatively and 4.1% postoperatively). Astigmatism was 2.03 ± 1.11 μm preoperatively and 1.60 ± 0.94 μm postoperatively. The dominant HOA both pre- and postoperatively were vertical coma, vertical trefoil, and secondary astigmatism. ICRS decreased corneal astigmatism by 27% and corneal coma by 5%, but on average, the overall amount of HOA did not decrease significantly with ICRS treatment. Other studies using aberrometry to evaluate the outcomes of ICRS also show a consistent flattening of the cornea postoperatively, but less systematic reduction of HOA, particularly in patients with mechanically assisted (instead of femtosecond laser-assisted) tunnel creation (Piñero et al. 2009).

Other treatments for keratoconus, depending on the stage of the disease, involve rigid gas-permeable (RGP) CLs (see Section 20.3.4), corneal cross-linking (aiming at stiffening the cornea through the stimulation of covalent bond formations between collagen fibers by means of a photosensitizer and UVA light) and corneal keratoplasty (corneal transplant). Aberrometry has been used to evaluate the outcomes of these treatments (Vinciguerra et al. 2009, Greenstein et al. 2012, Salvetat et al. 2013, Ghanem et al 2014).

20.4.4 MONOCHROMATIC ABERRATIONS AND CONTACT LENSES

CLs are frequently used to correct for refractive errors, although some designs also aim at correcting presbyopia (multifocal CLs) and at customized correction of HOA (in patients with pathological aberrations, such as those occurring in keratoconus).

Several studies investigate the interactions between the optics of the CL and that of the eye through computer simulations (Atchison 1995, Martin and Roorda 2003, De Brabander et al. 2003), using computer models to predict the interaction of the lens with the corneal surface and to estimate the optical contribution of the tear medium between the cornea and the CL. However, full prediction of optical performance of CLs is difficult, as effects such as lens flexure and lens conformity to the cornea, dehydration, and CL decentration produce larger variability in the on-eye than in off-eye (optical bench measurements) (Collins et al. 2001), and the simulated performance may differ substantially from that obtained assuming the nominal design of the lens.

Unlike soft CLs, RGP CLs conform less to the cornea, and it is widely recognized that the smooth lens surface align

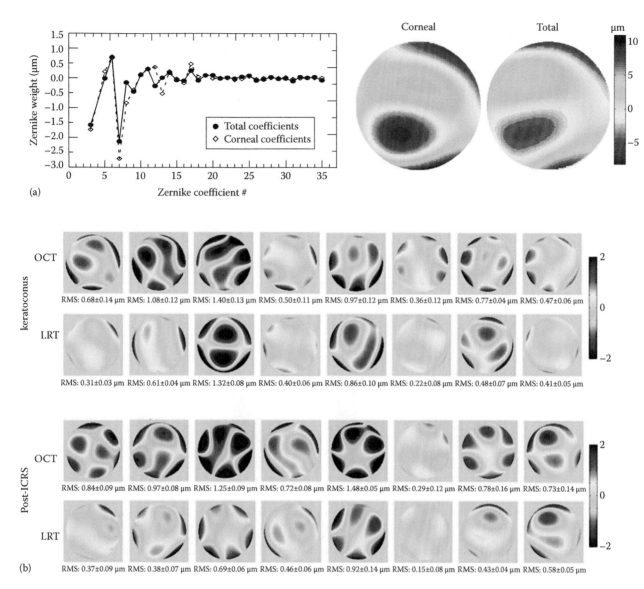

Figure 20.8 (a) Total and corneal high-order aberrations in a keratoconic eye (left panel, Zernike coefficients; right panel, wave aberration maps), for 6.5 mm pupil. (b) Corneal wave aberration maps from optical coherence tomography and total aberration maps from laser ray tracing in keratoconic eyes and upon intracorneal ring segment treatment. Data are for 4 mm pupil diameter. Top, preoperative data (keratoconus); bottom, postoperative data (3 months post intracorneal ring segment implantation). (a: Adapted from Barbero, S. et al., *J. Opt. Soc. Am. A*, 20, 1841, 2003. b: Adapted from Pérez-Merino et al., *Am. J. Ophthalmol.*, 157, 116, 2014b.)

with the tear film meniscus formed between the posterior lens surface and the anterior cornea (tear lens), which can neutralize the irregular aberrations of the anterior corneal surface and even correct modest amounts of astigmatism. Hong and colleagues (Hong et al. 2001) measured aberrations in subjects wearing RGP CLs and found that in three out of four subjects, fitting of RGP CL resulted in lower aberrations than soft CLs and spectacle lenses. Dorronsoro et al. (2003) provided a direct comparison of the optical changes produced by RGP CLs on the anterior surface of the cornea and on the total optical system. They demonstrated that rigid lenses provide passive aberration compensation of anterior corneal aberrations, as shown in Figure 20.9. A more recent study showed that RGP CLs reduced 66% of the severe HOA found in postsurgical corneas (Gemoules and Morris 2007). Conversely, several studies have shown that soft CLs, for

myopia, increase HOA (Roberts et al. 2006) to different extent depending on the manufacturing method (Jiang et al. 2006), power (Awwad et al. 2008), and thickness (Dietze and Cox 2004).

There have been few attempts to explore the on-eye behavior of multifocal contact using aberrometry and customize their prescription (Charman 2005). Conclusions are limited by the fact that designs with optical power alternating between distances and near corrections and diffractive designs are either challenging or not suited for aberrometry measurements. Computer simulations of optical performance of bifocal soft diffractive CLs on real eyes show that simulated soft bifocal CLs combined with real ocular aberrations do not always prove bifocal vision (Martin and Roorda 2003). Further discrepancies are expected if lens flexure and conformity were considered.

Optical properties of the eye

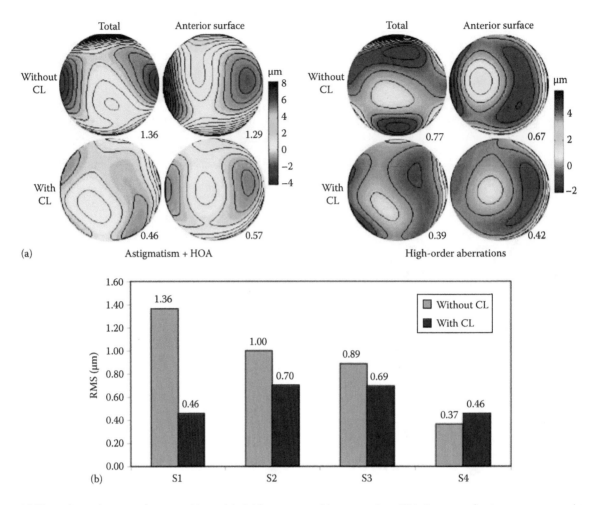

Figure 20.9 (a) Wave aberration maps for one subject with rigid gas-permeable contact lens (CL). Four wavefronts were measured: total (left column) and anterior surface (right column), with and without rigid gas-permeable (CL). Anterior surface aberrations stand for aberrations of the anterior corneal surface for the natural eye and aberrations of the anterior surface of the CL when the eye is wearing the CL. For each subject, the four upper maps include all aberrations except tilt and defocus, and in the four lower maps astigmatism has also been removed. Contours have been plotted at 1 μm interval. The root-mean-square (in microns) is indicated for each wavefront. The rigid gas-permeable CL reduces the aberrations of the anterior surface, producing a partial compensation of the aberrations of the eye. (b) Corresponding ocular RMS without CL and with CL. Pupil diameter is 5 mm in all cases. (Adapted from Dorronsoro, C. et al., *Optom. Vis. Sci.*, 80, 115, 2003.)

Advances in free-form lens manufacturing open the possibilities suggested by the first practitioner of aberrometry, Smirnov (1961), who suggested in 1961 that "in principle it is possible to manufacture a lens compensating the wave aberration of the eye …" and noted that as spectacle lenses do not move with the eye as it rotates, "… the lenses must obviously be contact ones." Although full correction of aberrations with CLs is probably not possible, due to dynamic changes associated with tears, accommodation (Gambra et al. 2009), lens rotations and translations (Guirao et al. 2001), and chromatic aberration, customized correction of HOA can undoubtedly improve the optical performance of eyes with abnormally high aberrations, as those with keratoconic or postsurgical corneas. Preliminary results with customized CLs show a decrease in the HOA by a factor of 3 in normal and keratoconic eyes (Lopez-Gil et al. 2002, Sabesan et al. 2007).

Wavefront aberrometry offers exciting new opportunities to advance our understanding of the ways in which CLs interact with both the normal and abnormal eye and will likely become a common tool in the contactology practice.

20.5 SOURCES OF MONOCHROMATIC ABERRATIONS

20.5.1 INTERACTIONS BETWEEN OCULAR ABERRATIONS

The cornea and the crystalline lens are the major refractive components in the eye, and aberrations of the individual components contribute to overall image quality. It has been shown that for young eyes, a proportion of the corneal aberrations are compensated by aberrations of the crystalline lens. A partial compensation of corneal astigmatism by the crystalline lens is well known from clinical optometry literature (Java's rule) (Keller et al. 1996). Recent studies of crystalline lens topography *in vivo* suggest that compensation of astigmatism between the crystalline lens surfaces may also occur (Ortiz et al. 2012). The spherical aberration of the cornea is typically positive, while the spherical aberration of the crystalline lens tends to be negative in the young lens (Millodot and Sivak 1979, Artal et al. 2002). Figure 20.10a shows an interesting case of a monolateral aphakic young patient, revealing corneal/lens compensation in the normal eye and total aberration dominated by the aberrations of the cornea (primarily positive spherical aberration). Interestingly, a partial compensation of asymmetric aberrations, such as coma, also seems to occur, at least in young, low myopic eyes (Artal et al. 2002). With aging, the spherical aberration of the crystalline lens shifts toward positive (or less negative) values, therefore, compromising the balance with the corneal aberrations, resulting in an increase of the ocular aberrations with aging (see Section 20.2.2).

Interactions between the aberrations terms in the Zernike polynomial series that describe a given wave aberration also occur. Several studies have demonstrated the interactions between low-order aberrations and HOA (Applegate et al. 2003, Thibos et al. 2004). In particular, adding spherical aberration to defocus can improve retinal quality over defocus alone, indicating that canceling defocus in the wave aberration Zernike polynomial expansion does not necessarily produce the best optical quality (Figure 20.10b). For these reason, the use of retinal image quality metrics (such as visual Strehl ratio) accounts for a better descriptor of optical quality than RMS, where the interactions between terms are not taken into account. Aberrometers are increasingly used as autorefractometers. In this case, the interactions between the Zernike defocus term (sometimes simply taken to derive the spherical error) and other higher-order terms need to be considered to estimate accurate refraction (Guirao and Williams 2003, Cheng et al. 2004b).

Favorable interactions between other HOA have also been shown. McLellan et al. (2006) showed that the actual combination of HOA found in eyes produced typically better MTFs than most combinations of equal amounts of aberrations and random signs (Figure 20.10c).

It has also been shown that certain combinations of nonrotationally symmetric aberrations (coma and astigmatism) can improve retinal image quality over the condition with the same amount of astigmatism alone. For example, for a 6 mm pupil, in the presence of 0.5 D of astigmatism, adding 0.23 μm of coma produced (for best focus) a peak improvement in Strehl ratio by a factor of 1.7, over having 0.5 D of astigmatism alone (De Gracia et al. 2010) (Figure 20.10d).

Chromatic and monochromatic aberrations also interact favorably: the relative degradation produced by longitudinal and transverse chromatic aberration of the eye on the MTF at short wavelengths with respect to the MTF at higher wavelengths is much higher in diffraction-limited eyes than in eyes with natural monochromatic aberrations (McLellan et al. 2002).

The presence of interactions between different aberrations is important in the management of refractive error corrections. Similarly, it should be considered in refractive error procedures where correction of given monochromatic aberrations is attempted. Conversely, certain presbyopic corrections, such as PresbyLASIK, multifocal contact, and intraocular lenses, attempt to increase depth of focus by adding aberrations, in

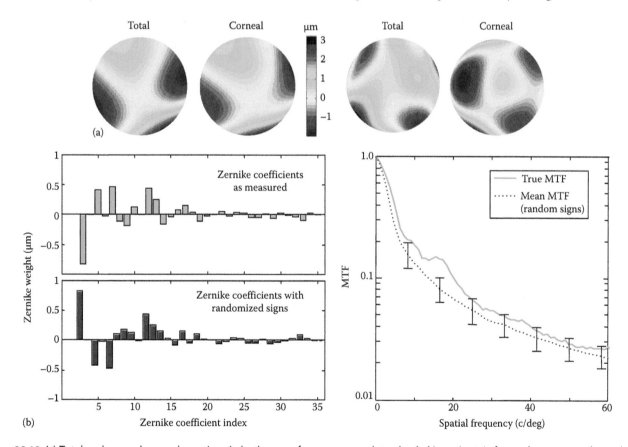

Figure 20.10 (a) Total and corneal wave aberrations in both eyes of a young monolateral aphakic patient. Left panels correspond to aphakic eye (no lens). Right panels to the normal eye, where the compensatory role of the crystalline lens can be observed; data are from laser ray tracing and 6.5 mm pupil. (b) Estimations of modulation transfer function from real set of Zernike coefficients and same coefficients with random signs; left panels are examples of Zernike sets. *(Continued)*

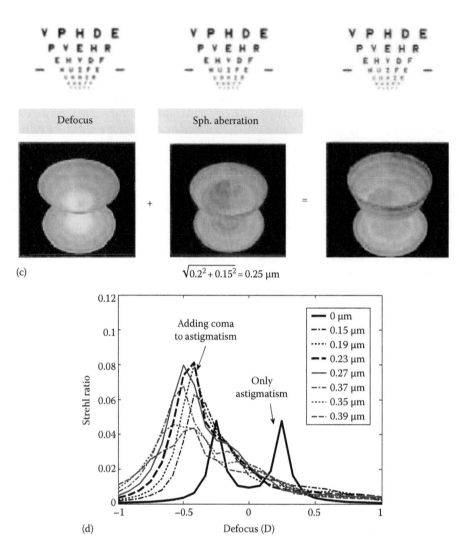

Figure 20.10 (*Continued*) (c) Optical quality of combined spherical aberration and defocus is better than aberration and defocus alone, even if the combined RMS is higher; (d) Through-focus Strehl ratio of combined 0.5 D of astigmatism with different amounts of coma (blue curves) as opposed to astigmatism alone (black curve): combined aberrations result in better quality than astigmatism alone. (a: Adapted from Barbero, S. et al., *J. Refract. Surg.*, 18, 263, 2002a; Barbero, S. et al., *J. Cataract Refract. Surg.*, 28, 1594, 2002b; b: Adapted from McLellan, J.S. et al., *Vision Res.*, 46, 300, 2006; c: Adapted from Applegate, R.A. et al., *Optom. Vis. Sci.*, 80, 97, 2003; d: Adapted from de Gracia et al., 2013.)

certain cases spherical aberration. Considering the specific interactions with the aberration pattern at the patient level will help to customize correcting solutions.

20.5.2 RELATING OCULAR STRUCTURE AND MONOCHROMATIC ABERRATIONS

Monochromatic aberrations result from the geometrical and structural properties of the ocular components as well as the alignment between the cornea and the lens. The factors contributing to aberrations can be best assessed from customized computer eye models using anatomical data from each patient. Quantitative anatomical data for these customized eye models (Tabernero et al. 2006, Rosales and Marcos 2007) can be gathered from corneal topography, ocular biometry, and Purkinje imaging. More recently, this information, including anterior and posterior corneal elevation map, anterior chamber depth, lens geometry, lens tilt

and decentration, foveal position, can be obtained in three dimensions from quantitative OCT (Ortiz et al. 2012) (see Figure 20.11).

While the contribution of the crystalline lens geometry, a structure to the aberrations of the eye, starts to be unraveled (see Section 20.4.5), customized eye models have been most used to evaluate pseudophakic eyes, in which the crystalline lens has been replaced by an artificial IOL (with known geometry and constant refractive index). Simulated aberrations by ray tracing on customized eye models match the measured aberrations in the same patient's eyes by close to 90% (Figure 20.11). In eyes implanted with aspheric IOLs, it has been shown that this IOL geometry cancels out a large proportion of the corneal spherical aberration (and horizontal coma), while the typical amounts of IOL tilt and decentration do not induce further aberrations, but actually contributes to further compensation of coma in more than 70% of the patients

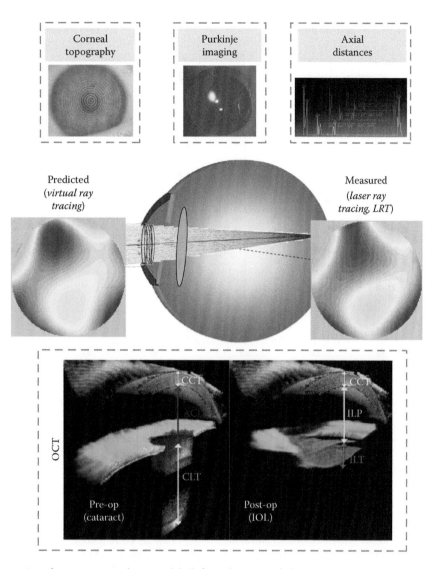

Figure 20.11 Predicted aberrations from customized eye models (left) and measured aberrations (laser ray tracing) on the same pseudopha-kic patient, implanted with a monofocal aspheric intraocular lens (IOL). Anatomical data come from Placido disk corneal topography, Purkinje imaging, and IOLmaster (top). (From Rosales, P. and Marcos, S., *Opt. Express*, 15, 2204, 2007.) Customized model eyes can also be built using quantitative anterior segment optical coherence tomography data (bottom). (From Ortiz, S. et al., *Biomed. Opt. Express*, 4, 387, 2013.)

studied (Rosales and Marcos 2007). Predictions of postopera-tive aberrations can therefore be undertaken through the use of custom eye models, which allow linking ocular geometry with aberrations.

20.5.3 IMPACT OF GRADIENT REFRACTIVE INDEX ON MONOCHROMATIC ABERRATIONS

Both corneal and lens geometry contribute to ocular aberra-tions (see Section 20.4.2). Furthermore, the crystalline lens of the eye shows a gradient refractive index (GRIN), which contributes to the ocular aberration. Most of the informa-tion available on the gradient index distribution comes from measurements on lenses *ex vivo*. Recently, the GRIN in mam-mals and humans has been measured in three dimensions, and the contributions of both lens shape and GRIN to the spherical aberration of the crystalline lens have been revealed (Birkenfeld et al. 2013).

Virtual ray tracing estimations of the crystalline lens from direct measurements of the lens geometry and GRIN (Figure 20.12a) on human donor lenses of different ages confirm that the spherical aberration becomes more positive with age (see Section 20.2.2). While the changes in lens shape with age con-tribute to this shift, the presence of a more distributed GRIN in young lenses, and a flatter GRIN in old lenses, plays a primary role in this changes (Birkenfeld et al. 2014) (Figure 20.12b). Similarly, the change of spherical aberration toward more nega-tive values with accommodation results from both a contribution of changes in lens shape with accommodation and the pres-ence of GRIN, which further shifts spherical aberration toward more negative values (de Castro et al. 2013) (Figure 20.12c). Interestingly, the spherical aberration predicted by virtual ray tracing on crystalline lenses of known shape and GRIN matches well, at the individual level, the spherical aberration on the same lenses measured by laser ray tracing (Maceo et al. 2014).

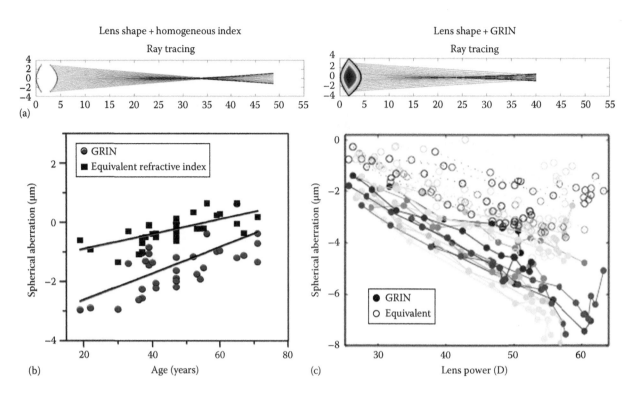

Figure 20.12 (a) Estimation of spherical aberration of the crystalline lens with measured lens shape and a homogeneous equivalent index, and measured lens shape and measured gradient refractive index (GRIN), by virtual ray tracing. (b) Estimated spherical aberration as a function of aging in human lenses *ex vivo* from lens shape and equivalent homogeneous index or lens shape and GRIN. (c) Estimated spherical aberration as a function of accommodation in nonhuman primate lenses *ex vivo* from lens shape and equivalent homogeneous index or lens shape and GRIN. (a: Adapted from de Castro et al., 2014; b: Adapted from Bierkenfeld, J. et al., *Vision Res.*, 86, 27, 2013; c: Adapted from de Castro, A. et al., *Invest. Ophthalmol. Vis. Sci.*, 54, 6197, 2013.)

REFERENCES

Alió, J.L., Shabayek, M.H., Corneal higher order aberrations: A method to grade keratoconus, *J Refract Surg* 22 (2006): 539–545.

Amano, S., Amano, Y., Yamagami, S. et al., Age-related changes in corneal and ocular higher-order wavefront aberrations, *Am J Ophthalmol* 137 (2004): 988–992.

Anera, R., Jiménez, J., Jiménez del Barco, L., Hita, E., Changes in corneal asphericity after laser refractive surgery, including reflection losses and nonnormal incidence upon the anterior cornea, *Opt Lett* 28 (2003): 417–419.

Applegate, R., Hilmante, I.G., Howland, H., Tu, E., Starck, T., Zayac, E., Corneal first surface optical aberrations and visual performance, *J Refract Surg* 16 (2000): 507–514.

Applegate, R.A., Ballentine, C., Gross, H., Sarver, E.J., Sarver, C.A., Visual acuity as a function of Zernike mode and level of root mean square error, *Optom Vis Sci* 80 (2003): 97–105.

Applegate, R.A., Howland, H.C., Sharp, R.P., Cottingham, A.J., Yee, R.W., Corneal aberrations and visual performance after radial keratotomy, *J Refract Surg* 14 (1998): 397–407.

Artal, P., Berrio, E., Guirao, A., Piers, P., Contribution of the cornea and internal surfaces to the change of ocular aberrations with age, *J Opt Soc Am A Opt Image Sci Vis* 19 (2002): 137–143.

Artal, P., Guirao, A., Contributions of the cornea and the lens to the aberrations of the human eye, *Opt Lett* 23 (1998): 1713–1715.

Atchison, D., Aberrations associated with rigid contact lenses, *J Opt Soc Am A Opt Image Sci Vis* 12 (1995): 2267–2273.

Awwad, S.T., Sanchez, P., Sanchez, A., McCulley, J.P., Cavanagh, H.D., A preliminary in vivo assessment of higher-order aberrations induced by a silicone hydrogel monofocal contact lens, *Eye Contact Lens Sci Clin Pract* 34 (2008): 2–5.

Barbero, S., Marcos, S., Jimenez-Alfaro, I., Optical aberrations of intraocular lenses measured *in vivo* and *in vitro*, *J Opt Soc Am A Opt Image Sci Vis* 20 (2003): 1841–1851.

Barbero, S., Marcos, S., Merayo-Lloves, J.M., Total and corneal aberrations in an unilateral aphakic subject, *J Cataract Refract Surg* 28 (2002a): 1594–1600.

Barbero, S., Marcos, S., Merayo-Lloves, J.M., Moreno-Barriuso, E., Validation of the estimation of corneal aberrations from videokeratography in keratoconus, *J Refract Surg* 18 (2002b): 263–270.

Birkenfeld, J., de Castro, A., Marcos, S., Contribution of shape and gradient refractive index to the spherical aberration of isolated human lenses, *Invest Ophthalmol Vis Sci* 55 (2014): 2599–2607.

Birkenfeld, J., de Castro, A., Ortiz, S., Pascual, D., Marcos, S., Contribution of the gradient index and shape to the crystalline lens spherical aberration and astigmatism, *Vision Res* 86 (2013): 27–34.

Born, M., Wolf, E., *Principles of Optics*, 6th ed. (Pergamon Press, Oxford, U.K., 1993).

Bühren, J., Kühne, C., Kohnen, T., Defining subclinical keratoconus using corneal first-surface higher-order aberrations, *Am J Ophthalmol* 143 (2007): 381–389.

Burns, S.A., Marcos, S., Measurement of the image quality of the eye with the spatially resolved refractometer, in *Customized Corneal Ablations*, S. McRae, R. Krueger, and R. Applegate, eds. (Slack. Thorofare, NJ, 2000).

Calver, R.I., Cox, M.J., Elliott, D.B., Effect of aging on the monochromatic aberrations of the human eye, *J Opt Soc Am A Opt Image Sci Vis* 16 (1999): 2069–2078.

Cano, D., Barbero, S., Marcos, S., Comparison of real and computer-simulated outcomes of LASIK refractive surgery, *J Opt Soc Am A Opt Image Sci Vis* 21 (2004): 926–936.

Carkeet, A., Luo, H.D., Tong, L., Saw, S.M., Tan, D.T., Refractive error and monochromatic aberrations in Singaporean children, *Vision Res* 42 (2002): 1809–1824.

Castejon-Mochon, F.J., Lopez-Gil, N., Benito, A., Artal, P., Ocular wave-front aberration statistics in a normal young population, *Vision Res* 42 (2002): 1611–1617.

Chan, D., Bilateral circumscribed posterior keratoconus, *J Am Optom Assoc* 70 (1999): 581–586.

Charman, W., Wavefront technology: Past, present and future, *Cont Lens Ant Eye* 28 (2005): 75–92.

Cheng, H., Barnett, J.K., Vilupuru, A.S. et al., A population study on changes in wave aberration with accommodation, *J Vis* 16 (2004a): 272–280.

Cheng, X., Bradley, A., Hong, X., Thibos, L.N., Relationship between refractive error and monochromatic aberrations of the eye, *Optom Vis Sci* 80 (2003): 43–49.

Cheng, X., Bradley, A., Thibos, L.N., Predicting subjective judgment of best focus with objective image quality metrics, *J Vis* 4 (2004b): 310–321.

Collins, M., Franklin, R., Carney, L., Bergiel, C., Lagos, P., Chebib, D., Flexure of thin rigid contact lenses. *Contact Lens Ant Eye* 24 (2001): 59–64.

Collins, M.J., Wildsoet, C.F., Atchinson, D.A., Monochromatic aberrations and myopia, *Vision Res* 35 (1995): 1157–1163.

De Brabander, J., Chateau, N., Marin, G., Lopez-Gil, N., van der Worp, E., Benito, A., Simulated optical performance of custom wavefront soft contact lenses for keratoconus, *Optom Vis Sci* 80 (2003): 637–643.

de Castro, A., Birkenfeld, J., Maceo, B. et al., Influence of shape and gradient index in the accommodative changes of spherical aberration in nonhuman primate crystalline lenses, *Invest Ophthalmol Vis Sci* 54 (2013): 6197–6207.

de Gracia, P., Dorronsoro, C., Gambra, E., Marin, G., Hernández, M., Marcos, S., Combining coma with astigmatism can improve retinal image over astigmatism alone, *Vision Res* 50 (2010): 2008–2014.

de Gracia, P., Dorronsoro, C., Sánchez-González, Á. et al., Experimental simulation of simultaneous vision, *Invest Ophthalmol Vis* Sci 54 (2013): 415–422.

Dietze, H.H., Cox, M.J., Correcting ocular spherical aberration with soft contact lenses, *J Opt Soc Am A Opt Image Sci Vis* 21 (2004): 473–485.

Dorronsoro, C., Barbero, S., Llorente, L., Marcos, S., On-eye measurement of optical performance of Rigid Gas Permeable contact lenses based on ocular and corneal aberrometry, *Optom Vis Sci* 80 (2003): 115–125.

Fernandez, E.J., Vabre, L., Hermann, B., Unterhuber, A., Pavazay, B., Drexler, W., Chromatic aberration correction of the human eye for retinal imaging in the near infrared, *Opt Express* 14 (2006): 6213–6225.

French, A.N., Morgan, I.G., Mitchell, P., Rosem, K.A., Risk factors for incident myopia in Australian schoolchildren: The Sydney adolescent vascular and eye study, *Ophthalmology* 120 (2013): 2100–2108.

Gambra, E., Sawides, L., Dorronsoro, C., Marcos, S., Accommodative lag and fluctuations when optical aberrations are manipulated, *J Vis* 9 (2009): 1–15.

Gatinel, D., Hoang-Xuan, T., Azar, D., Determination of corneal asphericity after myopia surgery with the excimer laser: A mathematical model, *Invest Ophthalmol Vis Sci* 42 (2001): 1736–1742.

Gemoules, G., Morris, K.M., Rigid gas-permeable contact lenses and severe higher-order aberrations in postsurgical corneas, *Eye Contact Lens* 33 (2007): 304–307.

Ghanem, R.C., Santhiago, M.R., Berti, T., Netto, M.V., Ghanem, V.C., Topographic, corneal wavefront, and refractive outcomes 2 years after collagen crosslinking for progressive keratoconus, *Cornea* 33 (2014): 43–48.

Gifford, P., Kang, P., Swarbrick, H., Versace, P., Changes to corneal aberrations and vision after PresbyLASIK refractive surgery using the MEL platform, *J Refract Surg* 30 (2014): 598–603.

Glasser, A., Campbell, M., Presbyopia and the optical changes in the human crystalline lens with age, *Vision Res* 38 (1998): 209–229.

Goodman, J.W., *Introduction to Fourier Optics*, 2nd ed., Electrical Engineering Series (McGraw-Hill International Editions, New York, 1996).

Gordon-Shaag, A., Millodot, M., Ifrah, R., Shneor, E., Aberrations and topography in normal, keratoconus-suspect, and keratoconic eyes, *Optom Vis Sci* 89 (2012): 411–418.

Greenstein, S.A., Fry, K.L., Hersh, M.J., Hersh, P.S., Higher-order aberrations after corneal collagen crosslinking for keratoconus and corneal ectasia, *J Cataract Refract Surg* 38 (2012): 292–302.

Guirao, A., Artal, P., Corneal wave aberration from videokeratography: Accuracy and limitations of the procedure, *J Opt Soc Am A Opt Image Sci Vis* 17 (2000): 955–965.

Guirao, A., Williams, D.R., A method to predict refractive errors from wave aberration data, *Optom Vis Sci* 80 (2003): 36–42.

Guirao, A., Williams, D.R., Cox, I.G., Effect of rotation and translation on the expected benefit of an ideal method to correct the eye's higher-order aberrations, *J Opt Soc Am A Opt Image Sci Vis* 18 (2001): 1003–1015.

He, J.C., Burns, S.A., Marcos, S., Monochromatic aberrations in the accommodated human eye, *Vision Res* 40 (2000): 41–48.

He, J.C., Marcos, S., Webb, R.H., Burns, S.A., Measurement of the wave-front aberration of the eye by a fast psychophysical procedure, *J Opt Soc Am A Opt Image Sci Vis* 15 (1998): 2449–2456.

He, J.C., Sun, P., Held, R., Thorn, F., Sun, X., Gwiazda, J.E., Wavefront aberrations in eyes of emmetropic and moderately myopic school children and young adults, *Vision Res* 42 (2002): 1063–1070.

Helmholtz, von H., *Physiological Optics* (J.P.C. Southall, Dover, New York, 1986).

Hofer, H., Artal, P., Singer, B., Aragon, J., Williams, D., Dynamics of the eye's wave aberration, *J Opt Soc Am A Opt Image Sci Vis Opt Image Sci Vis* 18 (2001): 497–506.

Hong, X., Himebaugh, N., Thibos, L., On-eye evaluation of optical performance of rigid and soft contact lenses, *Optom Vis Sci* 78 (2001): 872–880.

Jafri, B., Li, X., Yang, H., Rabinowitz, Y.S., Higher-order aberrations and topography in early and suspected keratoconus, *J Refract Surg* 23 (2007): 774–781.

Jahnke, M., Wirbelauer, C., Pham, D.T., Influence of age on optical aberrations of the human eye, *Ophthalmologe* 103 (2006): 596–604.

Jiang, H.J., Wang, D., Yang, L.N., Xie, P., He, J.C., A comparison of wavefront aberrations in eyes wearing different types of soft contact lenses, *Optom Vis Sci* 83 (2006): 769–774.

Jiménez, J., Anera, R., Jiménez del Barco, L., Equation for corneal asphericity after corneal refractive surgery, *J Refract Surg* 19 (2003): 65–69.

Kamiya, K., Shimizu, K., Igarashi, A., Kobashi, H., Visual and refractive outcomes of femtosecond lenticule extraction and small-incision lenticule extraction for myopia, *Am J Ophthalmol* 157 (2014): 128–134.

Keller, P., Collins, M., Carney, L., Davis, B., van Saarloos, P.P., The relation between corneal and total astigmatism, *Optom Vis Sci* 73 (1996): 86–91.

Kosaki, R., Maeda, N., Bessho, K. et al., Magnitude and orientation of Zernike terms in patients with keratoconus, *Invest Ophthalmol Vis Sci* 48 (2007): 3062–3068.

Liang, J., Grimm, B., Goelz, S., Bille, J.F., Objective measurement of wave aberrations of the human eye with the use of a Hartmann-Shack wave-front sensor, *J Opt Soc Am A Opt Image Sci Vis* 11 (1994): 1949–1957.

Lopez-Gil, N., Castejon-Mochon, J.F., Benito, A. et al., Aberration generation by contact lenses with aspheric and asymmetric surfaces. *Third International Congress of Wavefront Sensing and Aberration-free Refractive Correction.* Interlaken, Switzerland, Slack Inc. (2002).

Llorente, L., Barbero, S., Cano, D., Dorronsoro, C., Marcos, S., Myopic versus hyperopic eyes: Axial length, corneal shape and optical aberrations, *J Vis* 4 (2004a): 288–298.

Llorente, L., Barbero, S., Merayo, J., Marcos, S., Changes in corneal and total aberrations induced by LASIK surgery for hyperopia, *J Refract Surg* 20 (2004b): 203–216.

Llorente, L., Diaz-Santana, L., Lara-Saucedo, D., Marcos, S., Aberrations of the human eye in visible and near infrared illumination, *Optom Vis Sci* 80 (2003): 26–35.

Lyall, D.A., Srinivasan, S., Gray, L.S., Changes in ocular monochromatic higher-order aberrations in the aging eye, *Optom Vis Sci* 90 (2013): 996–1003.

Maceo-Heilman, B., Manns, F., de Castro, A. et al., Changes in monkey crystalline lens spherical aberration during simulated accommodation in a lens stretcher, *Invest Ophthalmol Vis Sci* 56 (2015): 1743–1750.

Maeda, N., Fujikado, T., Kuroda, T. et al., Wavefront aberrations measured with Hartmann-Shack sensor in patients with keratoconus, *Ophthalmology* 109 (2002): 1996–2003.

Mahajan, V.N., Zernike circle polynomials and optical aberrations of systems with circular pupils, *Appl Opt* 33 (1994): 8121–8124.

Malacara, D., *Optical Shop Testing*, 2nd ed. (John Wiley & Sons, Inc., New York, 1992).

Marcos, S., Aberrations and visual performance following standard laser vision correction, *J Refract Surg* 17 (2001): 596–601.

Marcos, S., Barbero, S., Jiménez-Alfaro, I., Optical quality and depth-of-field of eyes implanted with spherical and aspheric intraocular lenses, *J Refract Surg* 21 (2005): 223–235.

Marcos, S., Barbero, S., Llorente, L., Why high myopic eyes tend to be more aberrated?, Presented at the *Optical Society of America Technical Digest*, Long Beach, CA, 2001a.

Marcos S., Barbero, S., Llorente, L., Merayo-Lloves, J., Optical response to LASIK for myopia from total and corneal aberrations, *Invest Ophthalmol Vis Sci* 42 (2001b): 3349–3356.

Marcos, S., Burns, S.A., On the symmetry between eyes of wavefront aberration and cone directionality, *Vision Res* 40 (2000): 2437–2447.

Marcos, S., Burns, S.A., Moreno-Barriuso, E., Navarro, R., A new approach to the study of ocular chromatic aberrations, *Vision Res* 39 (1999): 4309–4323.

Marcos, S., Díaz-Santana, L., Llorente, L., Dainty, C., Ocular aberrations with ray tracing and Shack-Hartmann wavefront sensors: Does polarization play a role?, *J Opt Soc Am A Opt Image Sci Vis* 19 (2002): 1063–1072.

Marcos, S., Moreno-Barriuso, E., Llorente, L., Navarro, R., Barbero, S., Do myopic eyes suffer from larger amount of aberrations?, Presented at the *Myopia 200. Proceedings of the Eighth International Conference on Myopia*, Boston, MA, 2000.

Marcos, S., Rosales, P., Llorente, L., Jiménez-Alfaro, I., Change in corneal aberrations after cataract surgery with 2 types of aspherical intraocular lenses, *J Cataract Refract Surg* 33 (2007): 217–226.

Martin, J.A., Roorda, A., Predicting and assessing visual performance with multizone bifocal contact lenses, *Optom Vis Sci* 80 (2003): 812–819.

McLellan, J., Marcos, S., Burns, S., Age-related changes in monochromatic wave aberrations in the human eye, *Invest Ophtalmol Vis Sci* 42 (2001): 1390–1395.

McLellan, J.S., Marcos, S., Prieto, P.M., Burns, S.A., Imperfect optics may be the eye's defence against chromatic blur, *Nature* 417 (2002): 174–176.

McLellan, J.S., Prieto, P.M., Marcos, S., Burns, S.A., Effects of interactions among wave aberrations on optical image quality, *Vision Res* 46 (2006): 3009–3016.

Mierdel, P., Krinke, H.E., Wiegand, W., Kaemmerer, M., Seiler, T., Measuring device for determining monochromatic aberration of the human eye, *Ophthalmologe* 94 (1997): 441–445.

Millodot, M., Sivak, J., Contribution of the cornea and lens to the spherical aberration of the eye, *Vision Res* 19 (1979): 685–687.

Moreno-Barriuso, E., Marcos, S., Navarro, R., Burns, S.A., Comparing laser ray tracing, spatially resolved refractometer and Hartmann-Shack sensor to measure the ocular wavefront aberration, *Optom Vis Sci* 78 (2001a): 152–156.

Moreno-Barriuso, E., Merayo-Lloves, J., Marcos, S., Navarro, R., Llorente, L., Barbero, S., Ocular aberrations before and after myopic corneal refractive surgery: LASIK-induced changes measured with Laser Ray Tracing, *Invest Ophthalmol Vis Sci* 42 (2001b): 1396–1403.

Nakagawa, T., Maeda, N., Kosaki, R. et al., Higher-order aberrations due to the posterior corneal surface in patients with keratoconus, *Invest Ophthalmol Vis Sci* 50 (2009): 2660–2665.

Oberts, B., Athappilly, G., Tinio, B., Naikoo, H., Asbell, P., Higher order aberrations induced by soft contact lenses in normal eyes with myopia, *Eye Contact Lens* 32 (2006): 138–142.

Ortiz, S., Pérez-Merino, P., Durán, S. et al., Full OCT anterior segment biometry: An application in cataract surgery, *Biomed Opt Express* 4 (2013): 387–396.

Ortiz, S., Pérez-Merino, P., Gambra, E., de Castro, A., Marcos, S., *In vivo* crystalline lens topography, *Biomed Opt Express* 3 (2012): 2471–2488.

Oshika, T., Klyce, S.D., Applegate, R.A., Howland, H.C., El Danasoury, M.A., Comparison of corneal wavefront aberrations after photorefractive keratectomy and laser in situ keratomileusis, *Am J Ophthalmol* 127 (1999): 1–7.

Pantanelli, S., MacRae, S., Ieong, T.M., Yoon, G., Characterizing the wave aberrations in eyes with keratoconus or penetrating keratoplasty using a high-dynamic range wavefront sensor, *Ophthalmology* 114 (2007): 2013–2021.

Paquin, M.P., Hamam, H., Simonet, P., Objective measurement of optical aberrations in myopic eyes, *Optom Vis Sci* 79 (2002): 285–291.

Pérez-Merino, P., Birkenfeld, J., Dorronsoro, C. et al., Aberrometry in patients implanted with accommodative intraocular lenses, *Am J Ophthalmol* 157 (2014a): 1077–1089.

Pérez-Merino, P., Dorronsoro, C., Llorente, L., Durán, S., Jiménez-Alfaro, I., Marcos, S., In vivo chromatic aberration in eyes implanted with intraocular lenses, *Invest Ophthalmol Vis Sci* 54 (2013): 2654–2661.

Pérez-Merino, P., Ortiz, S., Alejandre, N., de Castro, A., Jiménez-Alfaro, I., Marcos, S., Ocular and optical coherence tomography-based corneal aberrometry in keratoconic eyes treated by intracorneal ring segments, *Am J Ophthalmol* 157 (2014b): 116–127.

Philip, K., Martinez, A., Ho, A. et al., Total ocular, anterior corneal and lenticular higher order aberrations in hyperopic, myopic and emmetropic eyes, *Vision Res* 52 (2012): 31–37.

Piers, P.A., Weeber, H.A., Artal, P., Norrby, S., Theoretical comparison of aberration-correcting customized and aspheric intraocular lenses, *J Refract Surg* 23 (2007): 374–384.

Piñero, D.P., Alió, J.L., El Kady, B. et al., Refractive and aberrometric outcomes of intracorneal ring segments for keratoconus: Mechanical *versus* femtosecond-assisted procedures, *Ophthalmology* 116 (2009): 1675–1687.

Plainis, S., Ginis, H.S., Pallikaris, A., The effect of ocular aberrations on steady-state errors of accommodative response, *J Vis* 5 (2005): 466–477.

Plainis, S., Pallikaris, I.G., Ocular monochromatic aberration statistics in a large emmetropic population, *J Modern Optics* 55 (2008): 759–772.

Porter, J., Guirao, A., Cox, I.G., Williams, D.R., Monochromatic aberrations of the human eye in a large population, *J Opt Soc Am A Opt Image Sci Vis* 18 (2001): 1793–1803.

Rosales, P., Marcos, S., Customized computer models of eyes with intraocular lenses, *Opt Express* 15 (2007): 2204–2218.

Sabesan, R., Jeong, T.M., Carvalho, L., Cox, I.G., Williams, D.R., Yoon, G., Vision improvement by correcting higher-order aberrations with customized soft contact lenses in keratoconic eyes, *Opt Lett* 32 (2007): 1000–1002.

Salvetat, M.L., Brusini, P., Pedrotti, E. et al., Higher order aberrations after keratoplasty for keratoconus, *Optom Vis Sci* 90 (2013): 293–301.

Schlegel, Z., Lteif, Y., Bains, H.S., Gatinel, D., Total, corneal, and internal ocular optical aberrations in patients with keratoconus, *J Refract Surg* 25 (2009): S951–S957.

Schwiegerling, J., Cone dimensions in keratoconus using zernike polynomials, *Optom Vis Sci* 74 (1997): 963–969.

Seiler, T., Kaemmerer, M., Mierdel, P., Krinke, H.E., Ocular optical aberrations after photorefractive keratectomy for myopia and myopic astigmatism, *Arch Ophthalmol* 118 (2000): 17–21.

Smirnov, M.S., Measurement of the wave aberration of the human eye, *Biofizika* 6 (1961): 776–795.

Smith, G., Cox, M.J., Calver, R. et al., The spherical aberration of the crystalline lens of the human eye, *Vision Res* 41 (2001): 235–243.

Smolek, M.K., Klyce, S.D., Current keratoconus detection methods compared with a neural network approach, *Invest Ophthalmol Vis Sci* 38 (1997): 2290–2299.

Tabernero, J., Piers, P., Benito, A., Redondo, M., Artal, P., Predicting the optical performance of eyes implanted with IOLs to correct spherical aberration, *Invest Ophthalmol Vis Sci* 47 (2006): 4651–4658.

Thibos, L.N., Applegate, R.A., Schwiegerling, J.T., Webb, R.H., Members, V.S.T., Standards for reporting the optical aberrations of eyes, *Sci Appl OSA Trends Opt Photon* 35 (2000): 110–130.

Thibos, L.N., Hong, X., Bradley, A., Applegate, R.A., Accuracy and precision of objective refraction from wavefront aberrations, *J Vis* 4 (2004): 329–351.

Tscherning, M., Die monochromatishen aberrationen des menschlichen auges, *Z Psycho Physiol Sinn* 6 (1894): 456–471.

Vinas, M., Dorronsoro, C., Sawides, L., Cortés, D., Pascual, D., Radhakrishnan, A., Marcos, S., Longitudinal Chromatic Aberration of the human eye in the visible and near infrared from Hartmann-Shack wavefront sensing, double-pass and psychophysics, *Assoc Res Vis Ophthalmol* 6 (2015): 948–962.

Vinciguerra, P., Albè, E., Trazza, S. et al., Refractive, topographic, tomographic, and aberrometric analysis of keratoconic eyes undergoing corneal cross-linking, *Ophthalmology* 116 (2009): 369–378.

Webb, R.H., Penney, C.M., Thompson, K.P., Measurement of ocular wavefront distortion with a spatially resolved refractometer, *Appl Opt* 31 (1992): 3678–3686.

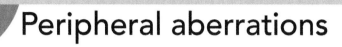

21 Peripheral aberrations

Linda Lundström and Robert Rosén

Contents

21.1 INTRODUCTION

Our peripheral vision is essential for many daily tasks, such as locomotion and detection. However, the human eye is optimized for central vision, both neurally and optically, and similar to man-made optics, the optical errors of the eye generally increase with incident angle. Peripheral visual function therefore depends on a combination of optical and neural factors. The peripheral visual field is defined as the area in object space that is not imaged to the macula lutea of the eye, that is, typically angles equal to or larger than 7°–10° off the visual axis. The monocular visual field stretches out to around 90° temporally, 60° nasally, 70° inferiorly, and 50° superiorly (Millodot 1997). This chapter starts by presenting the different types of optical errors that are often found in the periphery. An overview will then be given on how the optics can be measured and described as well as population data. The last part of this chapter discusses how these off-axis errors can affect our visual performance.

21.2 PERIPHERAL OPTICAL ERRORS

The largest optical errors of the peripheral eye, that is, the deviations from perfect imaging of light incident in oblique angles, can be understood qualitatively from a simple eye model through the Seidel aberration theory (Born and Wolf 1999, Kidger 2002, Freeman and Hull 2003). However, the current section will focus less on the mathematics and more on understanding the origin of the deviations and their effect on image quality. In Figure 21.1 the image quality on the foveal and peripheral retina of a schematic emmetropic eye is illustrated. The largest aberrations present when the object is located on the optical axis are spherical aberration and longitudinal chromatic aberration (inset A); these aberrations remain the same for objects away from the optical axis. When going to off-axis angles, coma and transverse chromatic aberration (TCA) will become apparent, increasing linearly with the angle (inset B). Further out in the peripheral field, the oblique (off-axis) astigmatism, with a close to quadratic dependency on angle, will dominate the image quality (inset C).

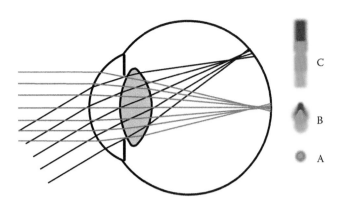

Figure 21.1 Illustration of the imaging and the image quality on the foveal and peripheral retina of a schematic emmetropic eye. The eye is here drawn with aligned spherical surfaces and thereby a clearly defined optical axis coinciding with the visual axis. The gray rays originate from an object point on the optical axis and show positive spherical aberration. The black rays correspond to the tangential rays from an object located in the peripheral field and demonstrate field curvature as well as coma (the sagittal rays would in this case place its best image closer to or behind the retina). The insets are depicturing the image quality on different off-axis locations on the superior retina (the eye is seen from the side): (A) foveal image quality with spherical and longitudinal chromatic aberration present, (B) peripheral image quality at moderate angles with coma and transverse chromatic aberration, and (C) image quality at large angles with oblique astigmatism (sagittal line focus is shown). These figures are not drawn to scale and the reader is referred to Figure 21.2 for comparison to the natural eye.

The schematic of Figure 21.1 is highly simplified, assuming spherical and aligned surfaces. Nevertheless, it demonstrates two important facts regarding the optical errors of the human eye: that the blur increases with the off-axis angle to the object and that it depends on the size of the pupil. For instance, spherical aberration has a cubic dependence on pupil size and coma has a quadratic dependence, whereas TCA is independent. To exemplify, Figure 21.2 shows the variation in monochromatic image

quality over the visual field measured for one subject. The following paragraphs will briefly explain the origin of the four largest peripheral aberrations, namely, astigmatism, field curvature, coma, and TCA.

21.2.1 OBLIQUE ASTIGMATISM

Light from an object in our peripheral visual field will be refracted asymmetrically by the cornea, depending on the meridian of the incident light. Two planes can then be defined: the tangential (also known as the meridional) plane and the sagittal plane. The tangential plane includes both the chief ray from the object to the vertex of the cornea and the optical axis of the eye. The sagittal plane is perpendicular to the tangential and includes the chief ray but not the optical axis. For example, for a point object located horizontally to the right in the temporal visual field of the right eye, the tangential plane is horizontal, whereas the sagittal plane is vertical, connecting the point object and the vertex of the cornea. This means that rays in the tangential plane will be incident on the cornea at a larger angle and therefore travel a longer distance through the cornea than rays in the sagittal plane. The net effect will be that tangential rays are refracted more, creating an astigmatic error with two separate line foci. The tangential line focus will be a line perpendicular to the tangential plane and located closer to the anterior part of the eye than the sagittal line focus, which forms a line in the tangential plane. The power of a correcting negative cylinder will be oriented along the tangential plane. In the example earlier the axis of a negative correcting cylinder would then be 90°. The lines in Figure 21.3 show the orientation of the tangential plane (and hence the sagittal line foci) for eight different field locations. The amount of off-axis astigmatism can be approximated by the Coddington equations (see Section 7.4.1 of Freeman and Hull 2003).

21.2.2 DEFOCUS DUE TO FIELD CURVATURE

Defocus means that the circle of least confusion is not located on the retina. It depends on the optical power, the length of the

Figure 21.2 Monochromatic point-spread functions over the peripheral retina for an elliptic pupil with major diameter of 4 mm using central refractive correction (example for one subject). The center dot is the foveal point-spread function and the horizontal meridian out to ±40° and vertical meridian out to ±30° are sampled in steps of 10°. Nasal retina is shown toward the right, temporal to the left, superior retina up, and inferior down. The inset shows the corresponding image size on the retina of a letter E with total height of 5 arcmin, that is, visual acuity 1.0.

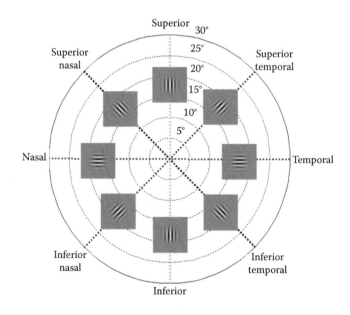

Figure 21.3 The visual field as seen by the right eye. The lines show the orientation of the tangential plane (and hence the sagittal line foci) for eight different field locations.

eye, and the distance to the object. If an object is moved perpendicularly further away from the optical axis of a system in a flat plane, the distance from the system to the object will increase. For a positive system, such as the eye, this means that the further the object is moved off-axis, the closer the image will be to the system, resulting in a curved image plane. This aberration is called field curvature and would mean that the eye becomes more myopic in oblique angles. However, as the retina of the eye is curved, it is the offset between the image plane and the retina that matters. In Section 21.4.1, it is shown that the circle of least confusion of the peripheral eye can be located both in front of and behind the retina depending on the shape of the eyeball, which in turn depends on the central refractive error of the eye.

21.2.3 COMA

The off-axis aberration coma has a similar origin as spherical aberration: it is caused by that fact that rays entering the pupil further away from the optical axis (i.e., the marginal rays) are refracted more. As illustrated by the black rays in Figure 21.1, the marginal rays are converging not only at another distance from the retina but also at a different height than the paraxial rays through the center of the pupil. The rays therefore will have an effective difference in image magnification and retinal blur, which will result in a coma-like point-spread function. In the human eye, off-axis coma is often oriented tangentially, with its blurred tail pointing toward the optical axis (see Figure 21.1 and the lines of Figure 21.3).

21.2.4 TRANSVERSE CHROMATIC ABERRATION

The dispersion of the optical media of the eye is similar to that of water and the refractive index for blue wavelengths is higher than for longer red wavelengths. This means that the power of the eye is more positive for shorter wavelengths, leading to longitudinal chromatic aberration with a more myopic refractive error for blue light and more hyperopic for red light. For off-axis objects, this difference in refractive power will have an additional effect, as the refraction of the principal ray will depend on the wavelength. The principal ray is the ray from the object point going through the center of the entrance pupil of the eye. This ray defines the location of the center of the image on the retina also when the image is out of focus. Hence, a wavelength-dependent deviation of the principal ray will lead to a difference in magnification, with blue light being less magnified than red light. This phenomenon is known as transverse or lateral chromatic aberration (see the insets of Figure 21.1). The further off-axis the object is located, the larger is the TCA becomes. Theory predicts a linear increase with the angle and no dependence on the size of the pupil (Thibos 1987). As the principal ray gives the amount of TCA, the aberration will change if the pupil is decentered, for example, using an artificial pupil, which is one technique used for measuring foveal longitudinal chromatic aberrations (Atchison and Smith 2000). Many eyes have TCA in the fovea, due to the naturally decentered surfaces.

21.3 MEASURING PERIPHERAL OPTICS

Measuring the peripheral refractive errors with the traditional subjective method is difficult due to reduced retinal function and large aberrations. Additionally, objective techniques to measure optical errors, well calibrated for foveal measurements, can be hampered by the large aberrations and elliptical pupils. The first reports on the optical errors over the visual field were published by Ferree, Rand, and Hardy in the early 1930s (Ferree et al. 1931, 1932, Ferree and Rand 1933). They used a manual optometer, a modified Zeiss parallax refractometer, to measure the peripheral refraction over the ±60° horizontal field. Forty years later, in 1971, Rempt, Hoogerheide, and Hoogenboom published peripheral refraction data estimated through retinoscopy at eccentric angles (Hoogerheide et al. 1971, Rempt et al. 1971). Since the shift of the millennium, the number of studies on peripheral refractive errors has increased rapidly, to a large extent due to renewed interest in the hypothesis of a connection between peripheral image quality and myopia development (see Section 21.5.3). The most common methods nowadays are wavefront sensors and different types of autorefractors. This section gives a brief review of techniques used for assessing the peripheral optical errors and discusses the challenges with peripheral measurements (the reader is referred to other chapters of this Handbook for the basic foveal methodology). Table 21.1 summarizes studies that have compared the outcome of two or more methods for peripheral optical measurements; for a more extensive list of studies performing eccentric refraction, see the review of Fedtke et al. (2009).

21.3.1 TRADITIONAL TECHNIQUES: SUBJECTIVE REFRACTION AND RETINOSCOPY

The gold standard for foveal refraction is subjective refraction through optimizing central vision with spherical and cylindrical trial lenses. Attempts have been made to find a similar technique for off-axis angles. Such an eccentric refraction is more challenging than foveal refraction due to the reduced retinal function and the large aberrations in the periphery. Millodot and Lamont used a bracketing technique with spherical lenses to find the lens power that gave subjective optimum clarity of a Landolt C optotype for each meridian separately (Millodot and Lamont 1974). The task of the subject can thereby be classified as resolution acuity, probably to a high contrast (HC) optotype. In the periphery, the lower sampling density of the retina can limit HC resolution, which implies that a HC resolution test will be less sensitive to refractive errors. As explained more in Section 21.5.1, suprathreshold peripheral detection tasks can be performed utilizing aliasing, which is limited by optical blur even when the Nyquist sampling limit of the retina is exceeded. Utilizing this phenomenon, two studies from the University of Indiana found the eccentric refraction through a detection acuity task. Thibos et al. (1996) used retinoscopy as a starting point but then refined the peripheral refraction in a two-alternative procedure, in which the subject was asked to identify the trial lens that maximized the apparent contrast of high-frequency stimuli, as seen through aliasing. Wang et al. (1996) implemented a task in which the detection acuity of horizontal and vertical gratings was maximized by spherical lenses. Both studies were implemented with forced-choice routines to control the subjective bias. The refractive correction obtained through these subjective refractions was found to give better visual outcome than when obtained with retinoscopy, which strengthens the argument that a subjective refraction should be considered as a gold standard also for peripheral vision.

Although the earlier mentioned studies are reporting successful subjective refraction, it should be noted that the measurement

Optical properties of the eye

Table 21.1 Studies comparing different methods for peripheral optical measurements over the temporal (T), nasal (N), and inferior (I) visual fields

STUDY	NO. OF SUBJECTS	SUBJECTIVE REFRACTION	RETINOSCOPY	ZEISS REFRACTOMETER	AUTOREFRACTOR	PHOTOREFRACTION	LAB DOUBLE-PASS	HARTMANN–SHACK SENSORS	COMPARISON BETWEEN METHODS
Millodot and Lamont (1974)	3	10–60° T (clarity of Landolt C)	10–50° T	10–60° T					Good agreement on astigmatism, proximal accommodation in refractometer.
Dunne et al. (1993)	34			0–40° N & T	Canon R-1 0–30° N & T				Not specifically mentioned, graph shows higher astigmatism with Canon R-1.
Wang et al. (1996)	4	20–40° N (detection acuity for horizontal and vertical gratings)	20–40° N (average of 3 optometrists)		20–40° N Canon R-1				Good agreement in meridian of axis 180°. The prescription determined by subjective refraction gave better detection acuity than that from retinoscopy. More astigmatism found with retinoscopy than subjectively, even more found with Canon R-1.
Seidemann et al. (2002)	6					PowerRefractor 15–25° N	0–45° N		Average absolute differences (nonsignificant) were 0.78 D in defocus and 0.85 D in astigmatism. All cylinder axes very close to 90°.
Atchison (2003)	5				Shin-Nippon: 0–35° N & T Canon R-1: 0–40° N & T			Lab-based 0–40° N & T	Good agreement between Hartmann–Shack and Shin-Nippon autorefractor (0.3 D–0.7 D average difference). Canon R-1 did not agree.

(Continued)

Table 21.1 (Continued) Studies comparing different methods for peripheral optical measurements over the temporal (T), nasal (N), and inferior (I) visual fields

STUDY	NO. OF SUBJECTS	SUBJECTIVE REFRACTION	RETINOSCOPY	ZEISS REFRACTOMETER	AUTOREFRACTOR	PHOTOREFRACTION	LAB DOUBLE-PASS	HARTMANN–SHACK SENSORS	COMPARISON BETWEEN METHODS
Lundström et al. (2005a)	50	30° N (detection contrast sensitivity for gratings of different orientations)	20° N, 30° N			PowerRefractor 20° N, 30° N		Lab-based 20° N, 30° N Metric: Strehl ratio	Average differences between methods were 0.7 D in sphere. PowerRefractor underestimates high myopia and not able to measure all in 30°. Relatively larger spread for subjective refraction. Retinoscopy less accurate for cylinder axis. Wavefront senor myopic shift of 0.5 D.
Lundström et al. (2005b) + Lundström et al. (2007)	7					PowerRefractor 10° I, 20° T, 35° N		Lab-based 10° I, 20° T, 35° N Metric: Strehl ratio	Differences in astigmatism; wavefront sensor gave more oblique cylinder axis (larger J45°), but PowerRefractor gave larger total cylinder for 5 subjects. Refraction by the wavefront sensor gave better visual outcome.
Berntsen et al. (2008)	30				Grand Seiko WR-5100K (also called Shin-Nippon) 0°, 30° N & T			COAS 0°, 30° N & T	No significant differences for relative peripheral refraction ($p = 0.34$). COAS gave -0.41 D ± 0.61 D myopic shift.
Jaecken et al. (2012)	35					Lab-based scanning out to 45° N & T		Lab-based scanning out to 45° N & T Refraction from second-order Zernikes	Significant differences in mean spherical equivalent: $M_{PR} = 1.421 \times M_{HS} + 0.513$.

time was long and the studies were only performed on a few subjects, the authors themselves, who can be considered to be highly motivated and well-trained subjects. To perform eccentric subjective refraction also on inexperienced subjects, Lundström et al. (2005a) implemented a quicker routine for peripheral refraction. The method of limits was used to estimate the detection contrast sensitivity of two cycles/degree gratings for different trial lenses, and the sphere–cylinder–axis combination that gave optimum sensitivity was chosen. The routine was tested on 50 subjects, who found it to be tiresome, and when compared to other refraction techniques, this subjective refraction gave a larger spread.

Retinoscopy has the advantage of being a rapid and objective method in the sense that no judgment is made by the subject. However, it requires an experienced examiner to correctly judge the reflex. Similar to subjective refraction, peripheral retinoscopy also suffers from the large off-axis aberrations, leading to a larger depth of field and a less well-defined far point. The point of neutralization is therefore difficult to assess, and Rempt et al. (1971) adopted a new technique "the sliding door effect" to interpret the distorted reflex. Jackson et al. (2004) concluded that off-axis retinoscopy becomes more difficult to perform when the eccentric angle increases. Three studies have compared retinoscopy to other refraction techniques (see Table 21.1). Millodot and Lamont (1974) found retinoscopy to be reliable out to 50°. Wang et al. (1996) reported that the results from three independent examiners were repeatable but gave a too myopic refraction in the meridian of axis 180°. Lundström et al. (2005a) also found inconsistencies with the astigmatism, especially to assess the cylinder axis.

21.3.2 FOVEAL REFRACTOMETERS: ADDITIONAL REQUIREMENTS FOR PERIPHERAL MEASUREMENTS

Many of the objective techniques used for assessing the optical quality of the eye on-axis can also be used for off-axis measurements provided that three requirements are fulfilled. First of all, the optical errors in the peripheral field can be very high, especially astigmatism. The instrument therefore needs a large dynamic range. However, it can be challenging to combine a large dynamic range with high measurement accuracy. It is therefore common to introduce additional optics that partially compensate for the largest optical errors. An elegant way, which is often implemented also for foveal measurements, is a telescope with a Badal lens as the first component and an adjustable distance to the second lens to compensate for defocus. One solution to also compensate for high astigmatism is to place spherical and cylindrical trial lenses in front of the eye until the remaining optical errors are within the dynamic range of the instrument. When using trial lenses in this way, it is important to take the effective power of the lens in the plane of the pupil into consideration as well as the magnification effects: for example, a +10 D lens mounted 15 mm from the eye will be equivalent to a +11.75 D in the plane of the pupil and will image the pupil with a magnification of 1.176. To avoid optical side effects, the trial lenses can instead be placed in conjugating plane with the pupil. Often, it is sufficient to compensate for defocus and astigmatism. If correction is also needed for the high-order aberrations, for example, coma, the system will require some type of adaptive optics.

The second requirement for peripheral measurements is that the instrumental design allows for the mounting of suitable fixation targets to control the viewing angle. This can often be achieved in a laboratory instrument but many commercial instruments restrict the visual field, and only a few have an open field of view. An instrument for peripheral measurements can either include scanning components to analyze parts of the visual field with the subject keeping stable fixation at a central target or include a movable fixation target or a number of targets at different angles that the eye is turned toward. Additionally, if a subject without foveal vision is to be measured, even more specialized fixation targets may be needed (Gustafsson 2001, Gustafsson and Unsbo 2003, Lundström et al. 2005b). A majority of the instruments used today are stationary and therefore require the subject to turn relative to the instrument, which means that the alignment between the instrument and the subject has to be adjusted when the angle is changed. The turning of the subject can be made (1) by only allowing the eye to turn, thus keeping the head fixated in a chin rest or a bite bar; (2) by turning the chin rest/bite bar the required angle, leaving the eye in its primary position; or (3) by using a combination of eye and head turn. Turning the eye in a large angle implies changes in the muscular stress on the eyeball, in the eyelid pressure on the cornea, and in the gravitational effects on the crystalline lens compared to the primary position. This effect of eye turn was first reported by Ferree et al. (1931), who noted that prolonged oblique fixation in angles out to 60° horizontally gave an increasing myopic peripheral refraction with as much as up to 2.5 D. Further studies by Seidemann et al. (2002) indicated a similar trend of increased myopia in 40°, although denoted as small compared with the measured myopia. In a similar study also including high-order aberrations, Lundström et al. (2009a) did not note any such trend in the Zernike coefficient for defocus (c_2^0), although the spherical aberration (c_4^0) changed significantly to more negative values. However, when measuring in more moderate off-axis angles around 30°–34°, no significant changes have been found between eye and head turn, regarding neither the peripheral refractive errors nor the high-order aberrations (Radhakrishnan and Charman 2008, Mathur et al. 2009a). In this context, it should be noted that the earlier mentioned studies have all only investigated horizontal eye turns. Foveal studies have found more significant changes in the optical errors when the eye is in downward gaze (Buehren et al. 2003, Ghosh et al. 2011) compared to horizontal eye turns (Radhakrishnan and Charman 2007, Prado et al. 2009). To conclude, although the effect of moderate eye turn on the measured optical errors has been found to be small, it is recommended to use a head-turn procedure to avoid artifacts.

The third and final additional requirement for peripheral measurements concerns how the instrument treats the pupil. In oblique angles, the pupil will appear elliptic in shape, which will affect the quality of the image on the retina. Ideally, the instrument should measure the optical errors over the entire elliptical pupil and report those errors together with the shape of the measured pupil. Many instruments for on-axis measurements are instead measuring over a circular pupil, either by having a physical aperture mounted in the system or by software. Such systems will be of limited use in large off-axis angles but may be used in more moderate angles. In that case, the elliptical shape of the pupil has to be assumed to be able to estimate the image quality. A simple assumption is that the minor axis of the elliptic pupil is equal to the radius of the circular pupil multiplied with the cosine of the off-axis angle; with some minor adjustments, this has been shown to be a good estimate of the shape of the natural pupil viewed obliquely (Mathur et al. 2013).

Note that the foveal instruments listed in Table 21.1 are no longer commercially available in the tested versions. However, the instruments are still in use in many clinics and labs. As can be seen from Table 21.1, the foveal refractometers have been reported to compare well also for peripheral measurements, at least over the horizontal visual field, with the exception of the Canon R-1 autorefractor. One reason to why different techniques estimate the location of the far point differently may be the large depth of field in the periphery (as discussed in Section 21.4.3).

21.3.3 LAB-BASED SYSTEMS 1: THE DOUBLE-PASS TECHNIQUE

The first studies of the peripheral ocular optics were focused on the largest optical errors: defocus and astigmatism. Although retinoscopy gives the examiner some indication also of the high-order aberrations, the double-pass technique for the first time offered quantifiable measurements of the off-axis image quality. The double-pass technique has the advantage of having a large dynamic range, which can be further extended by trial lenses. The first study was conducted by Jennings and Charman (1978) with a system very similar to those measuring central image quality generating a double-pass line-spread function. Compared to using the instrument foveally, they chose to measure through the natural undilated pupil and avoided using a circular artificial pupil (note the error in table 5 of the review by Fedtke et al. (2009); no slit aperture was imaged to the plan of the pupil). This choice was based on the fact that the retinal image quality recorded with the double-pass method depends on the size and shape of the pupil at the time of the measurement and cannot be analytically reconstructed for other pupils' postmeasurement. As the natural pupil appear elliptic in off-axis angles, a true measure of image quality on the peripheral retina has to include this elliptic shape. However, measurements without paralyzing accommodation are easily degraded by accommodation errors, and in a subsequent study, the same authors did use cycloplegia (Jennings and Charman 1981). Navarro et al. (1993) chose to measure through natural pupils but used a separate camera to record the size and shape of the pupil and varied the surrounding luminance to achieve the desired size. In that study, a point object was used instead of a line and the point-spread function as well as the modulation transfer function (MTF) was computed. The methodology was further developed by Artal et al. (1995a) by introducing an artificial pupil that was rotated the same angle as the eye, thus rendering an elliptical pupil with the same major axis for all eccentricities and a minor axis = major axis × cos(eccentric angle). The same methodology was also used by Williams et al. (1996).

A disadvantage with using a double-pass methodology is that aberrations with odd symmetry will be cancelled upon the second traverse through the optics of the eye. As the largest high-order

aberrations of the peripheral eye coma and transverse (lateral) chromatic aberration both have an odd symmetry, the double-pass technique provides too optimistic estimates of the peripheral point-spread function (Artal et al. 1995b). Although the authors emphasized that the MTF of the peripheral eye can still be correctly estimated by the double-pass method, the odd symmetry aberrations cannot be quantified. The first quantification of the peripheral coma aberration was made with a procedure with separate and unequal apertures for the incoming and the outgoing light (Guirao and Artal 1999).

The double-pass technique can be used to assess the peripheral refractive errors by simply introducing trial lenses in front of the eye to find the optimum image quality. The refractive correction obtained with the double-pass method has been compared to photorefraction in six subjects and no significant differences were found (see Table 21.1 and Seidemann et al. 2002). The measurement of off-axis astigmatism has also been found to be repeatable with a standard deviation of 0.60 D (Gustafsson et al. 2001).

21.3.4 LAB-BASED SYSTEMS 2: WAVEFRONT SENSING

The most common lab-based technique today to measure the peripheral optical quality is wavefront sensing. Compared to the double-pass technique, wavefront measurements have the advantage of being (close to) single-pass measurements in which the monochromatic aberrations with odd symmetry can be directly measured. Additionally, wavefront measurements give access to the phase information, which is difficult to retrieve from double-pass measurements. Furthermore, the technique to sample the wavefront aberrations in the plane of the pupil makes the measurements less dependent on the size of the pupil at the time of the measurement, as the optical errors within a smaller pupil can be computed postmeasurement.

There are two different objective techniques to assess the wavefront errors of the peripheral eye: the Hartmann–Shack sensor and laser ray tracing. The first wavefront technique used to evaluate the peripheral optical errors was based on laser ray tracing. It was first presented by Navarro et al. (1998) and further developed by Mazzaferri and Navarro (2012). The laser ray tracing setup is more complicated than that of the Hartmann–Shack sensor as it requires scanning elements and hence a longer time to sample the complete pupil. However, it has two main advantages of special interest to peripheral measurements. First, each pupil location, and hence each local wavefront tilt, is recorded separately in a sequential manner, which drastically increases the dynamic range of the sensor without effecting the sampling density. Second, it provides a flexible sampling pattern of the pupil and can therefore easily be adapted to sufficiently sample pupils of different sizes and shapes.

The currently most popular technique for wavefront sensing is based on the Hartmann–Shack principle, which measures the peripheral optical quality in a fraction of a second. It compares favorably to other methodologies of peripheral refraction (see Table 21.1 and Fedtke et al. 2009) and gives repeatable results also for peripheral aberrations (Baskaran et al. 2010). However, one disadvantage is the limited dynamic range due to cross talking between light entering through different parts of the pupil, the so-called unwrapping problem. For this reason peripheral Hartmann–Shack sensors usually include separate correction of the refractive errors. However, the design of the system is still a compromise between sampling density (resolution) and dynamic range. The unwrapping problem can be reduced by choosing lenslets with a lower *f*-number, that is, a shorter focal length and larger diameter. This will enable the measurement of larger aberrations but also decrease the accuracy because of the lower number of sampling points and the smaller displacements of the spots. There are also more advanced optical solutions to the unwrapping problem. Some approaches are a moveable aperture transmitting the light from different lenslets at a time (i.e., similar to laser ray tracing) (Olivier et al. 2000, Yoon et al. 2006), astigmatic lenslets for easier identification of the spots (Lindlein and Pfund 2002), or an additional measurement of the spot positions in a plane between the lenslets and the detector. However, as these methods are introducing a larger complexity, there are also many software-based suggestions to the unwrapping problem (see Bedggood and Metha 2010 for a review).

Atchison and Scott (2002) presented the first lab-based Hartmann–Shack sensor specifically developed for peripheral measurements. In this system a movable Badal lens was used for compensating for defocus and the unwrapping problem was handled postmeasurement in a graphics user interface. Lundström et al. (2005b) built a similar setup for peripheral measurements, but implemented a different software-based automatic routine for solving the unwrapping problem (Lundström and Unsbo 2004). Another important difference between the two lab-based systems is how the peripheral wavefront aberrations are quantified. As the off-axis pupil will appear elliptic in shape, the standard for central measurements with Zernike polynomials defined over a circular pupil cannot be directly applied. There are three possible ways to quantify the aberrations with Zernike coefficients (see Figure 21.4): over a small circular aperture located within the elliptic pupil, over a large circular aperture encircling the elliptic pupil, or by stretching the elliptic pupil into a circle. Atchison and Scott (2002) choose to stretch the wavefront, whereas Lundström et al. (2005b) used a large circular aperture encircling the pupil and then removed the extrapolated part for image quality evaluation.

There are advantages and disadvantages with all three methods to handle the elliptic pupil. A circular subaperture of the real elliptic pupil can be realized either with software, which removes spots outside the circle in the measurement image, or with a circular aperture placed in an image plane of the pupil during the measurement. The aperture should be centered and its radius equal to, or smaller than, the minor axis of the elliptic pupil (second graph in Figure 21.4). This representation is less well correlated with the peripheral image quality in large oblique angles,

because parts of the natural wavefront are omitted. Many HS sensors calculate the Zernike coefficients over a circular aperture that encircles all spots in the measurement image, that is, for peripheral wavefronts they would automatically use the second approach. The resulting coefficients will describe a wavefront that is extrapolated outside the borders of the elliptic pupil (third graph in Figure 21.4). This means that the root-mean-square (RMS) value of the real wavefront cannot be calculated directly from the Zernike coefficients (Thibos et al. 2002). However, together with information on the true elliptic shape of the pupil, these coefficients give a full description of the original wavefront and the peripheral image quality. Alternatively, the pupil can be scaled down analytically to instead represent the first version with a circular aperture within the pupil.

The last alternative for representing peripheral wavefront aberrations with Zernike polynomials stretches the elliptic pupil into a circular shape (fourth graph in Figure 21.4). Atchison and Scott (2002) implemented this method by stretching the spot pattern into a circle before fitting Zernike coefficients. It was also used for laser ray tracing by Navarro et al. (1998), with denser sampling of the wavefront along the minor axis of the pupil than along the major axis. In both cases the Zernike coefficients were then fitted to a stretched version of these sampling points, now spaced uniformly. The stretching of the wavefront is performed so that the height of the wavefront is conserved (i.e., the slope of the wavefront is decreased). The advantage with stretching the wavefront is that the RMS value of the wavefront is given directly by the Zernike coefficients. However, the drawback is that a Zernike polynomial in this stretched version will describe a different wavefront shape than the same polynomial would for the case with circular pupils. For example, spherical aberration, c_4^0, will in the stretched version not be rotationally symmetric but transform to second- and fourth-order terms (c_2^0, c_2^2, c_4^0, c_4^2, and c_4^4). Furthermore, extra manipulations are needed to find the retinal image quality. If peripheral wavefront aberrations are represented by Zernike coefficients for a stretched wavefront, these can be transformed into coefficients for the two other approaches, with circular pupils, using the theory described by Lundström and Unsbo (2007) and Lundström et al. (2009b).

21.3.5 FUTURE SYSTEMS FOR COMPLETE IMAGE QUALITY EVALUATION

Currently, the wavefront technique gives the most complete description of the peripheral optical errors. The wavefront sensors described in the previous paragraph are quick to use for one eccentric angle, but rely on the subject to shift the angle of gaze if

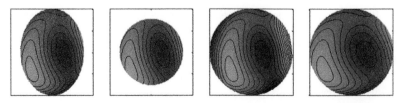

Figure 21.4 Wavefront aberrations with an elliptic pupil reconstructed with Zernike coefficients in three alternative ways: the leftmost graph shows the original wavefront as it emerges from the eye; the second graph presents the version with Zernike coefficients describing a circular part of the original wavefront; the third graph is the alternative, in which the Zernike coefficients also contain extrapolated wavefront data; and the last graph shows the original wavefront stretched out into a circle.

more angles are to be measured. Apart from making a full mapping of the visual field very time-consuming, a shift of gaze also includes a risk of changes in the accommodative state of the eye. It is therefore advantageous to have a scanning system that allows the subject to keep stable fixation during the measurement. Three such systems have been presented. One system is built on custom-design lenses and scanning mirrors, which can scan the ±15° visual field in a spiral pattern within 7 seconds (Wei and Thibos 2010). The second system rotates the whole sensor around the subject and can thereby scan the ±40° horizontal visual field in 1.8 seconds (Jaeken et al. 2011a). The third system consists of 11 different optical paths covering horizontal off-axis angles ±50° in 10° steps, which are scanned through in less than half a second. All three systems enable population studies in a large number of field angles.

The systems described here so far are all assessing the monochromatic aberrations of the peripheral eye. For the chromatic aberrations to be measured more than one wavelength is needed. Two studies have implemented multiple wavelengths into an objective technique for assessing peripheral image quality. Rynders et al. (1998) used a double-pass technique and Jaeken et al. (2011b) a Hartmann–Shack sensor to measure the longitudinal chromatic aberration. Both studies found the longitudinal chromatic aberration to be relatively constant with eccentricity. Although theory predicts TCA to be one of the larger optical aberrations off-axis (Thibos 1987), only two studies on the topic have been published. This is because existing instruments are based on evaluating the reflex from the retina and they can therefore not assess TCA with its odd symmetry. Centrally, the TCA can be assessed through subjective techniques, but a similar technique for eccentric angles gave large variations (Ogboso and Bedell 1987). To overcome these difficulties, a recent study used an adaptive optics scanning laser ophthalmoscope, in which the recording of the retinal cell structure at different wavelengths was used to compute the peripheral TCA (Winter et al. 2016).

21.4 POPULATION DATA ON PERIPHERAL OPTICAL ERRORS

This section will mainly report on the results of studies on peripheral optical errors that have been performed on larger sets of subjects. A summary of the peripheral refractive errors and their relation to the central refractive error will first be given. We will then focus on the peripheral aberrations and image quality. In Section 21.4.3 further processing of the data from the so far largest population study of peripheral wavefront aberrations (Jaeken and Artal 2012) will be presented. Finally, variations with age and accommodation will be discussed.

21.4.1 REFRACTIVE ERRORS OVER THE PERIPHERAL FIELD AND FOR DIFFERENT TYPES OF AMETROPIA

When the peripheral refractive errors were first studied, it was early noted that the astigmatism tended to increase with the off-axis angle and that the variation of the mean spherical equivalent (also known as M or the best sphere that is equal to the sphere plus half of the cylinder) with eccentricity seemed to depend on the central refractive state of the eye. In Figure 21.5 the off-axis astigmatism over the horizontal visual field is shown, with the data taken from nine studies using different measurement methodologies (see Table 21.2 for details on the individual studies). The astigmatism shown in Figure 21.5 is given as the Jackson cross-cylinder with axis 90/180 (J_0 as described in Rabbetts 2007). As the axis of the negative cylinder over the horizontal field is very close to 90, the J_0 values can be converted to cylindrical values, or the interval of Sturm, in diopters simply by multiplying J_0 by 2. The data shown have been averaged over all central refractive errors and ages. The astigmatism shows a clear increase with the off-axis angle, with a close to quadratic behavior. However, the increase is not symmetric around the fovea

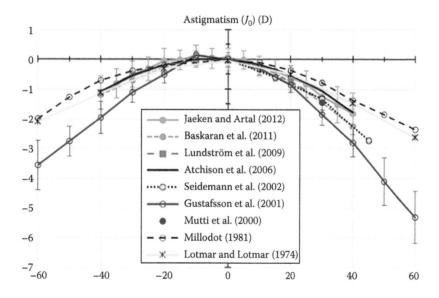

Figure 21.5 Off-axis astigmatism plotted over the horizontal visual field, where negative angles indicate temporal visual field (nasal retina) and positive angles nasal visual field (temporal retina). The astigmatism is given as J_0, the Jackson cross-cylinder with axis 90/180 in diopters. Negative J_0 values mean that the axis of the corresponding negative cylinder is 90. The results from nine studies are given as the average over the complete sample, irrespective of central refractive error or age. The error bars give the standard deviation of the measured sample (not given for all studies). See Table 21.2 for details about the separate studies.

Table 21.2 Details on population studies on peripheral optical errors, the data of which have been used to create Figures 21.5 through 21.16 and 21.19

STUDY	METHODOLOGY	SUBJECTS	VISUAL FIELD	COMMENT
Lotmar and Lotmar (1974) and Rempt et al. (1971)	Retinoscopy of right and left eyes	363 (20–50 year): all emmetropes	±20°, ±40°, ±60° horizontal	J_0 calculated from the interval of Sturm assuming $J_{45} = 0$; no variation given
Millodot (1981)	Topcon refractometer on right and left eyes	32 (18–57 year): 13 emmetropes 30 myopes 19 hyperopes	±10°, ±20°, ±30°, ±40°, ±50°, ±60° horizontal	J_0 calculated from the interval of Sturm assuming $J_{45} = 0$; no variation given
Mutti et al. (2000)	Canon R-1 autorefractor on right eyes	822 (5–14 year): 83.6% emmetropes 7.1% myopes 9.4% hyperopes	30° nasal visual field	J_0 calculated from the cylinder assuming $J_{45} = 0$
Gustafsson et al. (2001)	Lab-based double-pass technique	20 (20–45 year): all emmetropes	±10°, ±20°, ±30°, ±40°, ±50°, ±60° horizontal	One eye measured per subject, either the right or the left
Seidemann et al. (2002)	Lab-based double-pass technique on right and left eyes	35 (21–28 year): 11 emmetropes 9 myopes 5 hyperopes	0°, 15°, (20°) 30°, (40°) 45° nasal visual field	J_0 calculated from the interval of Sturm assuming $J_{45} = 0$; variation only given for sphere at 45° off-axis
Atchison et al. (2006)	Shin-Nippon SRW5000 autorefractor on right eyes (left eye in nine cases)	116 (18–35 year): 32 emmetropes 84 myopes	±0°, ±5°, ±10°, ±15°, ±20°, ±25°, ±30°, ±35° horizontal and vertical	Only polynomial fit used (averaged for the refractive error groups without any variation given)
Lundström et al. (2009b)	Lab-based Hartmann–Shack sensor on right eyes	43 (19–66 year): 15 emmetropes 19 myopes 9 hyperopes	0°, 20°, 30° nasal visual field	J_0 calculated from coefficient c_2^2 for 4 mm circular pupil
Baskaran et al. 2011	COAS wavefront sensor on right eyes	30 (20–29 year)+ 30 (52–67 year): all emmetropes	±0°, ±10°, ±20°, ±30°, ±40° horizontal	J_0 calculated from coefficient c_2^2 for 4 mm circular pupil
Jaeken and Artal (2012)	Lab-based Hartmann–Shack sensor on right and left eyes	101 (15–49 year): 100 emmetropes 90 myopes 12 hyperopes	±40° horizontal with continuous scanning (but plotted at discrete angles)	J_0 calculated from coefficient c_2^2 for 4 mm circular pupil for the right eye

(denoted as 0° in the graph). All of the studies presented here that have measured both the nasal and the temporal field found that the astigmatism is larger in the nasal visual field (temporal retina); on average the astigmatism was 1.2 D at 40° in the temporal and 1.9 D in the nasal visual field (Lotmar and Lotmar 1974, Millodot 1981, Gustafsson et al. 2001, Atchison et al. 2006, Baskaran et al. 2011, Jaeken and Artal 2012). Larger astigmatism in the nasal visual field has also been seen in other studies on peripheral astigmatism (Dunne et al. 1993, Berntsen et al. 2008, Lundström et al. 2009a, Mathur et al. 2009b,c, 2010). One explanation to why nasal visual field astigmatism is larger than temporal is that the fovea is not located on the optical axis of the eye, but around 5° more nasally (Rabbetts 2007). However, as discussed by Dunne et al. (1993), it may also be partly caused by a tilt of the crystalline lens, although that was not confirmed in

the study by Atchison et al. (2006). The astigmatism shows similar quadratic increases also for nonhorizontal meridians, with the axis of the negative cylinder oriented close to perpendicular to the measured meridian (Seidemann et al. 2002, Atchison et al. 2006, Lundström et al. 2009a, Mathur et al. 2009b,c, 2010).

The position of the two astigmatic line foci, and hence the circle of least confusion, relative to the retina shows more individual variations over the visual field than off-axis astigmatism. It was found early on that the peripheral mean spherical equivalent (M_{per}) depends on the central mean spherical equivalent of the eye (M_{fov}). Figures 21.6 through 21.8 therefore give the relative peripheral refraction (RPR) in diopters defined as $RPR = M_{per} - M_{fov}$ for emmetropic, myopic, and hyperopic eyes separately. A positive RPR means that the eye is more hyperopic, that is, that the circle of least confusion is located more behind the retina, in the

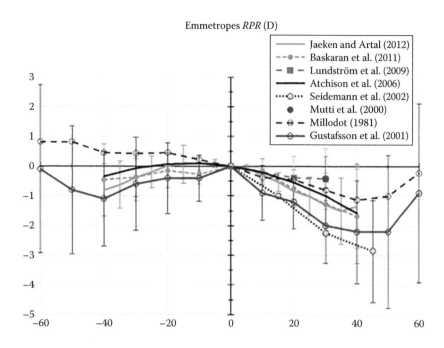

Figure 21.6 Relative peripheral refraction (*RPR*), defined as the difference in mean spherical equivalent between peripheral and central field, for emmetropic eyes given in diopters. A negative *RPR* means that the eye is more myopic, that is, that the circle of least confusion is located more in front of the retina, in the periphery compared to the fovea. The *RPR* is plotted as a function of the off-axis angle in degrees over the horizontal visual field, where negative angles indicate temporal visual field (nasal retina) and positive angles nasal visual field (temporal retina). The results from eight studies are given as the average over the complete sample, irrespective of age, and can be compared to the *RPR* for myopic and hyperopic eyes in Figures 21.7 and 21.8 (the scale of which cover the same amount of diopters for easy comparison). The error bars give the standard deviation of the measured sample (not given for all studies). See Table 21.2 for details about the separate studies.

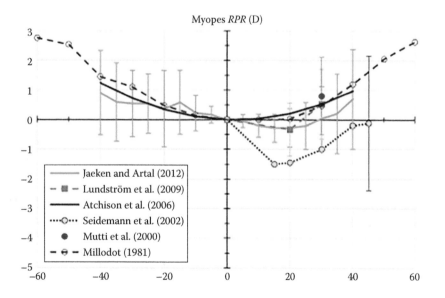

Figure 21.7 Relative peripheral refraction for myopic eyes plotted over the horizontal visual field. See Figure 21.6 for more explanations.

periphery compared to the fovea. Ferree et al. (1931) classified the peripheral refraction pattern of 21 subjects into three categories:
1. Mixed astigmatism increasing with angle (*RPR* close to zero)
2. Hyperopic relative to the fovea (positive *RPR*)
3. Nasal–temporal asymmetry
This classification was further refined by Rempt et al. (1971), who divided the skiagrams of 442 subjects into five different types and compared them to the central refractive error. Mathur and

Atchison (2013) later added a sixth class, IV/I, which they found to be the most common in their study:
I.   Less myopic or even hyperopic toward the periphery with quite small astigmatism (positive *RPR*), most common for myopes
II. One line foci behind and one on the retina (positive *RPR*), shown by some emmetropic eyes
III. Asymmetric off-axis astigmatism, quite uncommon

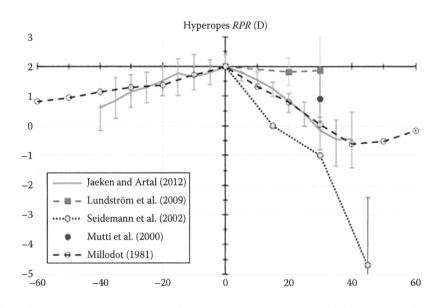

Figure 21.8 Relative peripheral refraction for hyperopic eyes plotted over the horizontal visual field. See Figure 21.6 for more explanations.

IV.   Large astigmatism with one line foci in front of and one behind the retina (*RPR* close to zero), most common skiagram especially for emmetropic and hyperopic eyes

IV/I. Large astigmatism with *RPR* close to zero out to 40°–50° and then a hyperopic shift for the largest angles (positive *RPR*)

V.   One line foci in front of and one on the retina (negative *RPR*), shown by some hyperopic eyes

The correlation between central refraction and the profile of the *RPR* can also be seen in the studies presented in Table 21.2, which are plotted in Figures 21.6 through 21.8. Figure 21.6 shows that emmetropic eyes generally are myopic in the periphery with the circle of least confusion located in front of the retina, that is, negative *RPR*. Hyperopic eyes have a yet clearer myopic shift toward the periphery; the periphery may experience myopic defocus even when the central hyperopic error is left uncompensated (Figure 21.8). However, for myopic eyes the *RPR* tends to go toward positive values, as can be seen in Figure 21.7, which means that the peripheral retina will experience less myopia than the fovea. The difference in peripheral profiles between ametropia can, at least to a large extent, be explained by a difference in the shape of the eye globe, myopic eyes having a more prolate shape than emmetropic and hyperopic eyes (see, e.g., the MRI study by Atchison et al. (2004)).

21.4.2 WAVEFRONT ABERRATIONS OVER THE PERIPHERAL VISUAL FIELD

The largest optical error in the periphery is astigmatism, even when no central astigmatism is present. However, when the astigmatism is corrected, it is evident that high-order aberrations also increase with the off-axis angle. Three of the studies in Table 21.2 also presented high-order aberration data in the form of Zernike coefficients for several peripheral angles (Lundström et al. 2009, Baskaran et al. 2011, Jaeken and Artal 2012). From these studies the individual signed Zernike coefficients for a 4 mm inscribed pupil of the right eye over the horizontal meridian have been averaged over each population sample. Note that

the value of Zernike coefficients is highly dependent on the size of the pupil and therefore has to be recalculated to the same pupil size before comparison (Lundström and Unsbo 2007). For central vision spherical aberration (c_4^0) is dominating with an average of +0.02 μm. Average foveal coma and trefoil are close to zero as it is equally common to have negative and positive signs of these aberrations; if we would instead look at the absolute value of the coefficients, the average amount of foveal coma and trefoil would be similar to that of spherical aberration (although the data by Jaeken and Artal (2012) indicate similar amount of c_4^0 and c_3^1 even in the signed version). In the periphery, the spherical aberration stays relatively unchanged with increasing angle, as shown by Figure 21.9. However, coma grows rapidly toward the periphery and already at 10° in the nasal and at 15° in the temporal visual field the horizontal coma (c_3^1) is three times the size of the spherical aberration (c_3^1 around ±0.05 μm in Jaeken and Artal 2012). Figure 21.10 shows that the coma coefficient c_3^1 becomes increasingly more positive in the temporal and more negative in the nasal visual field, with an angular dependence that can be described either as linear or to the power of three.

In Figure 21.11 an approximate measure of the total amount of high-order aberrations is plotted over the horizontal visual field, in the form of the RMS wavefront error of the coefficients of the third and higher orders (Thibos et al. 2002). Similar to astigmatism, the high-order RMS error shows a quadratic variation with the off-axis angle and is larger in the nasal than in the temporal visual field, with an average of 0.53 compared to 0.41 μm in ±40°. The results of Figures 21.9 through 21.11 are in agreement with the findings of other reports on the peripheral optical aberrations (Guirao and Artal 1999, Atchison and Scott 2002, Mathur et al. 2008, 2009b,c, 2010, Lundström et al. 2009a). Similar aberration profiles can be expected for the left eye, as it tends to be mirror symmetric with that of the right eye for defocus, astigmatism, and coma (Lundström et al. 2011). Interestingly, and similar to the central field, the peripheral aberrations of the eye are mainly generated by the passage through the cornea and partly compensated for by the internal optics of the eye (Atchison 2004).

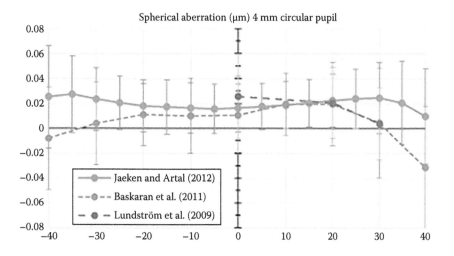

Figure 21.9 Zernike coefficient c_4^0 (spherical aberration) in μm for a 4 mm circular pupil as a function of the off-axis angle in degrees over the horizontal visual field, where negative angles indicate temporal visual field (nasal retina) and positive angles nasal visual field (temporal retina). The results from three studies are given as the average over the complete sample, irrespective of central refractive error or age. The error bars give the standard deviation of the measured sample. See Table 21.2 for details about the separate studies.

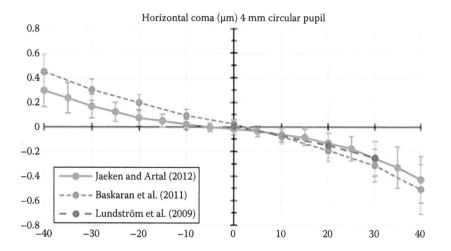

Figure 21.10 Zernike coefficient c_3^1 (horizontal coma) in μm for a 4 mm circular pupil plotted over the horizontal visual field. See Figure 21.9 for more explanations.

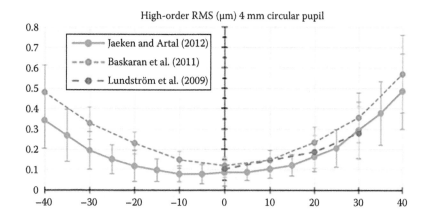

Figure 21.11 The high-order root-mean-square wavefront error in μm for a 4 mm circular pupil plotted over the horizontal visual field. See Figure 21.9 for more explanations.

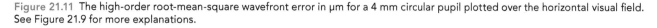

21.4.3 IMAGE QUALITY OVER THE PERIPHERAL VISUAL FIELD

Large aberrations make it difficult to predict the quality of the image on the peripheral retina from the refractive errors alone and even harder to estimate the effect of additional optical errors, for example, peripheral defocus. Such predictions require a full estimation of the peripheral refractive errors as well as high-order aberrations. So far it is the wavefront sensing techniques that offer the most complete picture of the off-axis image quality. There are only a few population studies available on the complete monochromatic image quality on the peripheral retina. In this section further image quality computation with MTF, visual Strehl ratio (VSOTF), and depth of field is presented from the largest study (Jaeken and Artal 2012; presented in Table 21.2). In all computations the central defocus, that is, the c_2^0 Zernike coefficient, has been subtracted from the measurements, thus imitating a foveal spherical correction with the same effect irrespective of measurement angle. The remaining coefficients are given for the original measurement wavelength of 780 nm.

Naturally, larger optical errors in off-axis angles also mean that the MTF will be worse. Figure 21.12 shows the monochromatic MTF as a function of spatial frequency in cycles per degree plotted for nine different horizontal eccentricities out to 40° in steps of 10°, with dashed lines indicating the temporal visual field (negative angles, nasal retina) and solid the nasal field. The MTF curves have been calculated from the wavefront aberrations of the right eye over an elliptical pupil with a vertical diameter of 4 mm and a horizontal diameter of 4 mm multiplied with the cosine of the off-axis angle. The shown values are the population average of all subjects of the radial average of the two-dimensional MTF. The gray shaded area in Figure 21.12 represents the foveal MTF ± one standard deviation of the population sample and the red shaded area the 40° nasal visual field MTF ± one standard deviation. Note that the spread in MTF over the population sample is larger for foveal than for peripheral vision, most likely due to some remaining foveal astigmatism. As can be seen, the nasal visual field (solid lines) has lower MTF values indicating larger optical errors compared to the temporal side. The spatial frequency at which the MTF value has dropped to 0.2 is approximately 4, 6, 10, and 16 cycles/degree in the 40°–10° nasal visual field, 22 cycles/degree foveally, and 7, 11, 15, and 19 cycles/degree in the 40°–10° temporal visual field. Similar curves have been presented also in earlier studies on fewer subjects, although comparison is complicated by differences in pupil size (Navarro et al. 1993, Williams et al. 1996, Jennings and Charman 1997, Guirao and Artal 1999, Yamaguchi et al. 2013).

To make the comparison between different off-axis angles clearer, a condensed measure of the image quality, areaMTF, can be used (Thibos et al. 2004). The areaMTF is calculated from the area under the MTF curve from zero cycles/degree out to a maximum spatial frequency of 50 cycles/degree and expressed in percentage of the area under the diffraction-limited MTF. Figure 21.13 shows the area MTF for the left and right eye over the horizontal visual field; here, the red curves have been calculated with an elliptical pupil as for Figure 21.12, whereas the blue lines used a 4 mm circular pupil. As can be seen the difference in areaMTF between the different pupil shapes increases with angle, with the elliptical pupil giving better image quality. The areaMTF provides an overview of the monochromatic optical quality of the eye. However, it has equal weight for all spatial frequencies, even though the high frequencies cannot be detected by the peripheral retina. The results of Figure 21.13 have therefore been weighted with the neural contrast sensitivity at different off-axis angles. The weighting is performed as for the foveal metric known as the visual Strehl ratio (VSOTF) (Thibos et al. 2004, Iskander 2006). The contrast sensitivities used here for the weighting are from the right eye of one subject measured foveally, 10°, 20°, and 30° in the nasal visual field under full monochromatic correction using specialized adaptive optics and quick psychophysical routines (Rosén et al. 2012d, 2014). These contrast sensitivity curves are presented in Figure 21.14. The VSOTF is plotted over the horizontal field for the right and left eye in Figure 21.15 in a similar way as for the areaMTF in Figure 21.13. Note that the drop in VSOTF over the temporal visual field is much less dramatic than the corresponding drop in areaMTF as well as the drop in VSOTF for the nasal visual field.

Another clinically relevant measure of the retinal image quality over the visual field is the depth of field (also known as the depth of focus). Depth of field is defined as the amount of spherical defocus in diopters that can be added to an image before it appears blurred to the subject. High-order aberrations interact with defocus and make the depth of field larger. In other words, the subject is less sensitive to induced defocus because the aberrations make the best image (the circle of least confusion) more blurred already without defocus. The effect on the retinal image quality of a certain amount of defocus will therefore be less dramatic, and it will hence be more difficult to judge the position

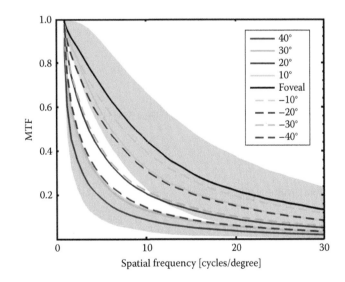

Figure 21.12 The average modulation transfer function (MTF) for an elliptic pupil (4 mm major diameter, 4 mm × cos(off-axis angle) minor diameter) plotted for different angles in the horizontal visual field: black line is the foveal MTF, light blue the 10° field, dark blue the 20° field, green the 30° field, and red the 40° field. Negative angles (dashed lines) indicate temporal visual field (nasal retina) and positive angles (solid lines) nasal visual field (temporal retina). (The data shown are the average over the right eye of all subjects in the study by Jaeken and Artal (2012) (see Table 21.2 and the text for more details). Shaded gray and red areas indicate the ± one standard deviation for the foveal MTF and the 40° nasal MTF, respectively.)

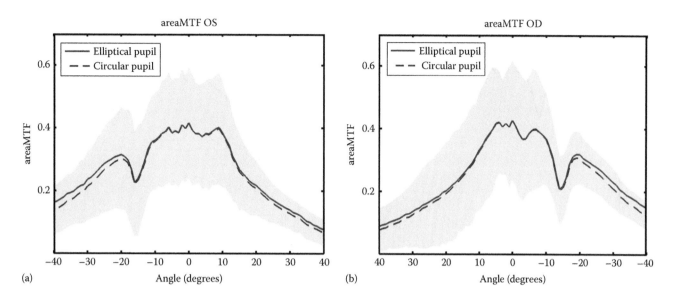

Figure 21.13 The area under the modulation transfer function, areaMTF, as a function of the off-axis angle in degrees over the horizontal field; negative angles for the temporal and positive angles for the nasal visual field (the dip in the temporal field corresponds to the blind spot). The graphs show (a) the average of the left eyes (OS) and (b) the average of the right eyes (OD) in the study by Jaeken and Artal (2012). Blue dashed lines, using a circular pupil of 4 mm diameter; red solid lines, using an elliptical pupil (4 mm major diameter, 4 mm × cos(off-axis angle) minor diameter). The shaded areas indicate the ± one standard deviation for the areaMTF for an elliptic pupil. See Table 21.2 and the text for more details.

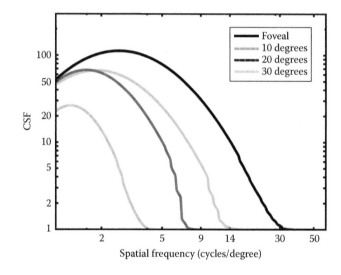

Figure 21.14 The contrast sensitivity for the resolution of gratings for one subject measured foveally, 10°, 20°, and 30° in the nasal visual field under full monochromatic correction. (From Rosén, R., Lundström, L., Venkataraman, A.P., Winter, S., Unsbo, P., *J. Vis.*, 14(8), 3, 2014.) These curves together with the modulation transfer function of Figure 21.13 were used to compute the visual Strehl ratio of Figure 21.15.

of the far point. Note that the effect of high-order aberrations can be different depending on the sign of the induced defocus. One such famous interaction is made by spherical aberration, which is used in many lens designs to induce multifocality, that is, larger depth of field, to foveal vision. For the peripheral eye, there is a similar interaction between the large off-axis astigmatism and coma, which extends the depth of field asymmetrically for positive and negative defocus (Rosén et al. 2012b). This asymmetric depth of field can also be found from the data of Jaeken and

Artal (2012); in Figure 21.16 the depth of field is plotted over the horizontal visual field of the right eyes. The blue line represents the border of the depth of field when positive defocus is added to the eye and the red line the border for inducing negative defocus. The limits of the depth of field are here defined as the amount of defocus needed to reduce the individual areaMTF by 20%, that is, to a level that is 80% of the maximum areaMTF for that angle and subject (same definition as used by Marcos et al. 1999, 2005). This was done computationally for each individual and angle by calculating the areaMTF with varying amounts of induced defocus (from +4 D to –4 D in steps of 0.1 D). The absolute value of the depth of field depends on the criteria used (in this case 20% reduction), but increases in depth of field with the off-axis angle have been found also in other studies (Ronchi and Molesini 1975, Wang and Ciuffreda 2004).

21.4.4 PERIPHERAL VARIATIONS WITH AGE AND ACCOMMODATION

The optical properties of the human eye are not static with time; there are both long- and short-term variations. The long-term changes are mainly due to axial growth of the eye during childhood and adolescent and during adulthood the main reason is the continuous growth of the crystalline lens. The short-term changes are related to changes in accommodative state and thereby also to changes in the crystalline lens. The foveal effects of these changes are discussed in other chapters of this Handbook as well as by Atchison and Smith (2000) and by Rabbetts (2007).

Cross-sectional studies on age-related differences in peripheral refractive errors are complicated by the fact that the peripheral refraction depends on the central refraction, which in turn is known to vary with age. Atchison et al. (2005) handled this by dividing 55 young (24 ± 3 years) and 41 old

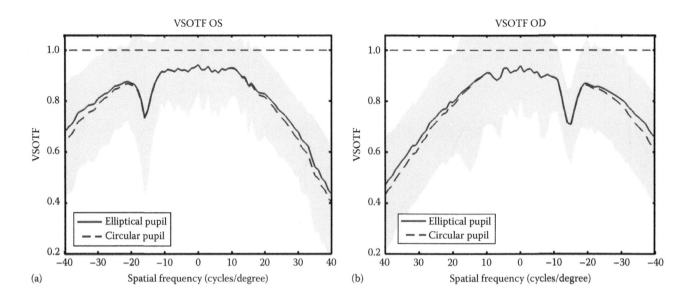

Figure 21.15 The visual Strehl ratio plotted as a function of the off-axis angle in degrees over the horizontal field for the (a) left (OS) and (b) right (OD) eye. See the text and Figure 21.13 for more explanations.

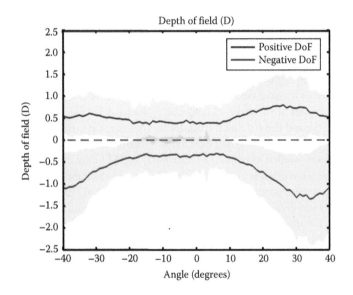

Figure 21.16 The depth of field for an elliptic pupil (4 mm major diameter, 4 mm · cos(off-axis angle) minor diameter) plotted as a function of the off-axis angle in degrees over the horizontal field; negative angles for the temporal and positive angles for the nasal visual field. The blue upper line represents the border of the depth of field when positive defocus is added and the red lower line the border for inducing negative defocus. (The data have been calculated from the areaMTF in Figure 21.13 of the right eye of all subjects in the study by Jaeken and Artal (2012) (see Table 21.2 and the text for more details). Shaded blue and red areas indicate the ± one standard deviation for the respective limits of the depth of field.)

(59 ± 3 years) subjects into four subgroups depending on the central refraction (limits of +1.5 D, +0.5 D, −0.5 D, −1.5 D, −2.5 D). The profile of the peripheral refractive errors over the ±35° horizontal field was then compared between subgroups. They found that the peripheral refraction profile was similar for young and old subjects with similar central refraction. Three subsequent cross-sectional studies have concentrated on emmetropic subjects and come to the same result that the peripheral refraction is relatively unaffected by age (Atchison

and Markwell 2008, Mathur et al. 2010, Baskaran et al. 2011). The same conclusion was also reached by the only longitudinal study of peripheral refractive errors; two subjects who were centrally nearly emmetropic in their thirties shifted about 1 D toward foveal hyperopia over 26 years at the same time as their peripheral refraction became more myopic (Charman and Jennings 2006). The only significant changes in peripheral refraction with age were reported by Atchison and Markwell (2008), who found a nasal shift in the turning point of the mean spherical equivalent and the astigmatism with age. Two studies using wavefront sensing also concluded that the peripheral spherical aberration (c_4^0) became more positive with age, similar to the central field, but remained relatively constant over the visual field (Mathur et al. 2010, Baskaran et al. 2011). Furthermore, the horizontal coma (c_3^1) increased faster with the off-axis angle for older subjects (0.018 vs 0.006 μm/degree for a 5 mm pupil and 0.015 vs 0.007 μm/degree for a 4 mm pupil).

The first measurement of the peripheral refractive errors under different accommodative demands was performed by Smith et al. (1988); eleven young emmetropes were measured with the Topcon refractometer out to 60° in the nasal visual field under up to 5 D of accommodation. They found that accommodation demands of 3 D or more gave a myopic shift in the *RPR* as well as larger astigmatism for visual field angles of 50° and 60°. For lower accommodation and smaller angles, the peripheral refraction profile was very similar to the unaccommodated situation. This first study is also the hitherto only study of the peripheral optics that have measured on accommodation levels above 4 D and angles larger than 45°. Several subsequent studies have confirmed the findings of Smith et al. (1988); for five studies the in total 66 emmetropic subjects measured kept the same negative RPR profile over the horizontal field during accommodation (Calver et al. 2007, Davies and Mallen 2009, Mathur et al. 2009c, Tabernero and Schaeffel 2009, Liu et al. 2016). However, a myopic shift in the *RPR* of up to 0.76 D at 40° temporal visual field was noted by Lundström et al. (2009a) for 4 D accommodation in 5 emmetropic eyes. Myopic subjects have also been measured with three studies on 34 myopes in total finding no difference with accommodation (Calver et al. 2007, Davies and Mallen 2009, Lundström et al. 2009a) and one study on 20 myopes finding a myopic shift in the *RPR* of up to 0.74 D at 40° temporal visual field with accommodation (Whatham et al. 2009). From these studies it seems like both emmetropic and myopic eyes either do not change their peripheral profile or go toward more myopic *RPR* with accommodation, which is also what simple eye models predict when the optical power is increased. However, there is one study on 22 emmetropic and myopic subjects combined that report a hyperopic shift in *RPR* of 0.4 D in the 30° nasal visual field with 3 D accommodation (Walker and Mutti 2002). This hyperopic shift disappeared after prolonged near work and the authors explained it as a change in the ocular shape due to tension on the choroid. From these studies it is not possible to conclude what effects accommodation have on the *RPR* nor what the cause of any possible changes might be.

21.5 EFFECT OF PERIPHERAL OPTICAL ERRORS ON VISION

For the normal healthy eye the peripheral vision is degraded compared to central vision, due to both neural and optical factors. In daily life, we mainly use the peripheral vision for tasks that do not require high resolution, such as locomotion, motion perception, scene recognition, and driving (Wood 2002, Lemmink et al. 2005, Larson and Loschky 2009). Therefore, peripheral optical errors are not routinely measured or corrected. However, there are some areas of vision research where the peripheral optical quality is of special interest, namely, for patients with central visual field loss (CFL) and for research on myopia development. This section will discuss these research areas together with an overview of how normal peripheral vision is affected by the peripheral image quality.

A third clinical application where peripheral optical errors can be of interest is visual field testing with perimetry. In one of the most common instruments, the Humphrey perimeter, the task of the subject is to detect a small white target located on a uniform background at different places in the visual field. Lower than normal light detection ability may indicate a reduction in retinal function and a need for medical treatment. However, a loss in detection may also result from large optical errors and the psychophysical method as well as the stimulus has to be chosen to be as insensitive to optical errors as possible. This application is beyond the scope of this text and the reader is referred to, for example, the review by Bosworth et al. (2000).

21.5.1 OPTICAL EFFECTS ON PERIPHERAL VISION

Peripheral spatial vision is limited by coarse sampling of the retina. While the cone density will decrease an order of magnitude just a few degrees outside the fovea (Curcio et al. 1990), the major limitation is imposed by the decrease in ganglion cell density (Curcio and Allen 1990), as the retinal structure shifts from having a single ganglion cell per cone to several cones being connected to the same ganglion cell (Banks et al. 1991). The decrease in visual acuity as a function of retinal eccentricity for a few studies can be seen in Figure 21.17 (Wertheim 1894, Kerr 1971, Thibos et al. 1987, Anderson et al. 1991,

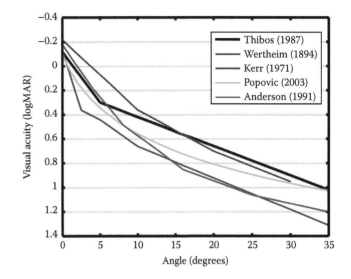

Figure 21.17 Resolution acuity as a function of retinal eccentricity (see text for more details).

Popovic 2003). This psychophysical limit of high contrast (HC) resolution has been found to correspond to the theoretically calculated limit for ganglion cell data (Thibos et al. 1987, Popovic 2003, Watson 2014).

As HC resolution acuity in the periphery is sampling limited, decreasing retinal contrast through defocus has limited impact compared to the case for foveal vision. Several studies have systematically induced defocus at various locations in the peripheral field and measured the visual outcome as a function of defocus as shown in Figure 21.18 (Anderson 1996, Wang et al. 1997, Rosén et al. 2011, 2012c, Lewis et al. 2014). Apart from HC resolution, these studies have also investigated low contrast (LC) resolution and HC detection. It should be noted that resolution and detection tasks often have different acuities in the periphery because of the sampling limitation. Patterns of spatial frequencies above the Nyquist limit of the ganglion cells are perceived as aliased if their retinal image is not too blurred (Thibos et al. 1996, Ennis and Anderson 2000). An aliased image is possible to detect, as the sampled sharp retinal image contrast creates shifting patterns, but impossible to resolve.

Three main conclusions can be drawn from the studies of Figure 21.18:

1. HC resolution acuity is unaffected by optical defocus, with a flat through-focus curve until high levels of defocus.
2. HC detection acuity is affected by optical defocus, with a peak above the resolution acuity. This difference between detection and resolution is due to detection by aliasing. At higher levels of defocus, the detection curve levels off and stays flat, until the defocus becomes high enough (e.g., 4 D in the 20° field) to sufficiently degrade both resolution and detection.
3. LC resolution and LC detection acuity (detection not shown in Figure 21.18) are identical and both affected by defocus. As the contrast is low to begin with, contrast loss due to defocus will affect both resolution and detection acuity (e.g., 0.10–0.20 logMAR/D at 20° eccentricity). It should be noted that for pure defocus, the impact also depends on the sign of defocus as some individuals can exhibit a substantial asymmetry, with a higher sensitivity to positive defocus (Rosén et al. 2012c).

Optical properties of the eye

As HC visual acuity is unaffected by defocus, most studies argue that the peripheral field is specialized for motion perception, detection, low spatial frequencies, and LC (cf. Brown 1972, Millodot et al. 1975, Wood 2002, Brooks et al. 2005, Schieber et al. 2009, Atchison et al. 2013). The benefit of correcting the peripheral optical errors described in Section 21.4 should therefore be illustrated for the relevant spatial frequencies, that is, using the VSOTF of Figure 21.14. This is done in Figure 21.19, which shows the peripheral VSOTF (area under MTF weighted by the neural CSF) of the right eye for the subjects in Jaeken and Artal (2012) and the potential benefits of different corrections. The red line shows the average VSOTF for uncorrected peripheral optics (i.e., the same as the blue dashed line in Figure 21.15b) and the black line shows the average VSOTF if each subject was given individual full correction of defocus, astigmatism, and coma. "Average" correction refers to giving the same peripheral correction to all subjects and "full" refers to individual correction. It can be seen that both the refractive errors and the aberrations, primarily coma, of an average subject substantially degrade the peripheral visual quality. This has been corroborated in the studies referred to earlier, as well as when correcting higher-order aberrations in the periphery (Rosén et al. 2012d). It should be noted that most of these studies of peripheral vision have used gratings as stimuli, and that grating acuity is somewhat higher than letter acuity (Anderson and Thibos 1999). Additionally, resolution acuity is higher for radially oriented gratings than for tangentially oriented gratings (Rovamo et al. 1982, Anderson et al. 1992, Wang et al. 1997, Venkataraman et al. 2016).

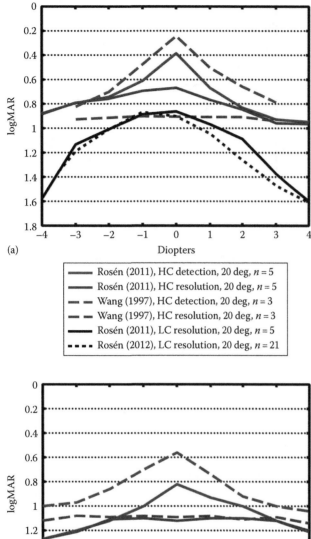

(a)

Legend:
— Rosén (2011), HC detection, 20 deg, $n = 5$
— Rosén (2011), HC resolution, 20 deg, $n = 5$
-- Wang (1997), HC detection, 20 deg, $n = 3$
-- Wang (1997), HC resolution, 20 deg, $n = 3$
— Rosén (2011), LC resolution, 20 deg, $n = 5$
···· Rosén (2012), LC resolution, 20 deg, $n = 21$

(b)

Legend:
— Anderson (1996), HC detection, 30 deg, $n = 1$
— Anderson (1996), HC resolution, 30 deg, $n = 1$
-- Wang (1997), HC detection, 30 deg, $n = 1$
-- Wang (1997), HC resolution, 30 deg, $n = 1$

Figure 21.18 High contrast (HC) resolution and detection acuity and low contrast (LC) resolution acuity from several studies in the 20° (a) and 30° (b) visual field. While studies differ in absolute levels due to methodology, the overall trends of HC resolution being unaffected and HC detection as well as LC resolution being affected by the optical defocus are consistent. Also given is n, the number of subjects in the different studies. (Data adapted from Anderson, R.S., *Curr. Eye Res.*, 15(3), 351, 1996; Wang, Y.Z., Thibos, L.N., Bradley, A., *Invest. Ophthalmol. Vis. Sci.*, 38(10), 2134, 1997; Rosén, R., Lundström, L., Unsbo, P., *Invest. Ophthalmol. Vis. Sci.*, 53, 7176, 2012c; Lewis, P., Baskaran, K., Rosén, R., Lundström, L., Unsbo, P., Gustafsson, J., *Optom. Vis. Sci.*, 91(7), 740, 2014.)

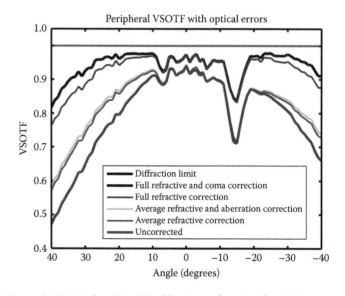

Legend:
— Diffraction limit
— Full refractive and coma correction
— Full refractive correction
— Average refractive and aberration correction
— Average refractive correction
— Uncorrected

Figure 21.19 Peripheral visual Strehl ratio as a function of retinal eccentricity for the average uncorrected individual (red) and with various amounts of correction. Negative angles denote the temporal and positive angles the nasal visual field. For the green and magenta curves, the population average of refractive and full aberration correction has been applied to illustrate the benefits of such an average approach. For the blue and black curves, individual correction of refractive errors and refractive errors plus coma, respectively, has been simulated. (Aberration data from Jaeken, B., Artal, P., *Invest. Ophthalmol. Vis. Sci.*, 53(7), 3405, 2012.)

21.5.2 APPLICATION 1: CENTRAL VISUAL FIELD LOSS

Peripheral vision is especially important for patients suffering from CFL. A number of diseases can cause CFL, including age-related macular degeneration, Stargardt's disease, Leber's hereditary optic neuropathy, and Best disease. Of these diseases, macular degeneration is the most common with more than 100 million persons suffering from it globally, disproportionately affecting elderly and persons of European ethnicity (Wong et al. 2014). Patients with CFL find it harder to recognize faces, read, watch TV, walk, cook, move outdoors, shop, and participate in leisure activities (Williams et al. 1998, Brody et al. 2001, Lamoureux et al. 2004, Hassel et al. 2006).

As the diseases progress and the CFL extends, patients find themselves unable to utilize their foveal vision. If the CFL is severe enough, most patients will develop one or more preferred retinal loci (PRL), a peripheral retinal area that they preferentially use for tasks such as reading or face recognition (Crossland et al. 2005). The eccentricity of these PRL vary from a few degrees up to 30° for some patients (White and Bedell 1990, Gustafsson 2001, Gustafsson and Unsbo 2003, Lundström et al. 2005b, 2007) and can be located in any field direction, although using the inferior field tends to be most beneficial for the patient (Petre et al. 2000). Using the PRL is challenging, but training can facilitate better use of the remaining peripheral vision (Nilsson et al. 2003). Current extraocular optical aids for these patients include magnifying telescopes for improved distance acuity and strong loupes to allow reading at very close distances (Markowitz 2006). Besides the magnification offered by these visual aids, patients with macular degeneration also benefit from improved retinal image contrast, as can be achieved with correction of peripheral optical errors (Gustafsson and Unsbo 2003, Lundström et al. 2007, Baskaran et al. 2012). It is noteworthy that the benefits of optical correction for patients with CFL are different from the situation for healthy individuals described in Section 21.4.1. Peripheral HC resolution acuity for healthy individuals was unaffected by optical defocus. On the other hand, the HC resolution acuity of patients with CFL increased when their peripheral optical errors were corrected. This improvement comes from a worse starting acuity than the healthy subjects at corresponding retinal location. Furthermore, Eisenbarth et al. (2008) reported severe impairment in dynamic processing properties well beyond the visible retinal scar. It therefore seems that the retinal areas used as PRL, outside the scar and with no sign of retinal degeneration, still suffer from a decreased sensitivity and therefore have an increased need of retinal image contrast. If patients with CFL have an elevated need for retinal image contrast, it is not surprising that substantial improvements of HC resolution acuity in the order of 0.2 logMAR have been reported when cataract surgery has been performed on these patients (Armbrecht et al. 2000, El Mallah et al. 2001, Huynh et al. 2013). These improvements come despite the limitations of retinal sampling and the fact that current IOL technology worsens the peripheral optical errors (Jaeken et al. 2013).

21.5.3 APPLICATION 2: MYOPIA DEVELOPMENT

The main reason for the increasing interest in the peripheral optical quality of the eye today is because of its link with myopia development. The prevalence of myopia is increasing fast in the world and many research projects attempt to understand the reasons for this increased prevalence as well as on how to slow down or completely prevent the development of myopia (Holden et al. 2014).

The most common cause of myopia is that the eye grows too long. Even though the disposition of developing myopia is partly inheritable, the increasing prevalence of myopia is triggered by the environment (Morgan and Rose 2005, Goldschmidt and Jacobsen 2014). A leading environmental hypothesis is that the blur in the image on the retina controls the ocular growth and thereby the progression of myopia. This has been shown in a number of studies on animals, for example, that it is possible to provoke myopia development by fitting a negative lens in front of the eye. Doing so will place the plane of the circle of least confusion behind the retina and the eye will grow to shift the retina to that plane (this was shown by Schaeffel et al. (1988) and Irving et al. (1992) for chickens, Hung et al. (1995) and Whatham and Judge (2001) for monkeys, and McFadden et al. (2004) for guinea pigs). The progression of myopia might therefore be due to insufficient accommodation during near work, and studies have been performed in which schoolchildren are fitted with reading spectacles to ensure that the sharpest image is placed on the retina also during near tasks. But the success of these studies has varied; Leung and Brown (1999) only found a small reduction in the progression of myopia, as did Fulk et al. (2000) and Gwiazda et al. (2003), whereas Shih et al. (2001) did not find any effect at all. One reason to why the reading spectacles have not worked as expected from the animal studies could be that the image quality outside of the fovea was not taken into consideration.

Today, a commonly held view is that the image quality in our peripheral field of view affects the growth of the eye and the course of myopia development (Wallman and Winawer 2004, Charman 2005, Smith et al. 2005, Charman and Radhakrishnan 2010, Benavente-Pérez et al. 2014). Compared to emmetropic eyes, the majority of myopic eyes have been found to be elongated, with a more prolate shape (Atchison et al. 2004). A prolate eye shape will result in a more positive *RPR*, which is indeed what was found in the studies presented in Section 21.4.1. Already in 1971 Hoogerheide et al. suggested that there is an increased risk for humans to become myopic if the peripheral refractive errors place the clearest image behind the peripheral retina, that is, a positive *RPR*, although it is not clear whether they had both pre- and postonset data to support this statement (Rosén et al. 2012a). Furthermore, there are other findings suggesting that the prolate eye shape is a consequence of and not a cause to the growth of the myopic eye (Mutti et al. 2007, 2011). Nevertheless, several studies on children have been carried out to evaluate the effect of optical corrections that shift the peripheral image more in front of the retina at the same time as the central image is focused on the fovea (see Koffler and Sears 2013, Smith 2013, Si et al. 2015 for some recent reviews). In one such study the increase in axial length at 2 years follow-up was as low as 43% of that of the control group (Cho and Cheung 2012, Huang et al. 2016). Although the results are encouraging, it should be noted that there are no studies on humans that have managed to completely avoid any further myopia development. The fact that it is not yet possible to get the same treatment effects on children as found in the animal studies demonstrates the need of better understanding of the underlying mechanisms for the optical control of eye growth in humans (Smith et al. 2013).

Several different optical correction designs have been suggested to induce negative *RPR* to prevent myopia progression, both soft and rigid contact lenses with bi- or multifocal designs as well as orthokeratology lenses and laser surgery. It is necessary with a full quantification of the optical quality to judge the usefulness of a certain correction. To rely on autorefractor sphere and cylinder readings is not sufficient as the aberrations in off-axis angles through multifocal optics are large and affect the measurement (see Section 21.4.3 and Bakaraju et al. 2015). For example, the so far most successful myopia prevention technique, orthokeratology lenses (corneal reshaping rigid contact lenses worn overnight) even changes the sign of the peripheral coma (Mathur and Atchison 2009). Also soft contact lenses and laser surgery are increasing the peripheral aberrations (Ma et al. 2005, Rosén et al. 2012b, Wagner et al. 2015). Larger peripheral aberrations give a larger depth of field that makes the desired shift in *RPR* less clear (Rosén et al. 2012b). Furthermore, the change in depth of field itself may be important as there are studies indicating that myopic subjects tend to be less sensitive to negative than to positive peripheral defocus (Rosén et al. 2011, 2012b,c).

REFERENCES

Anderson R.S. The selective effect of optical defocus on detection and resolution acuity in peripheral vision. *Curr Eye Res* 15.3 (1996): 351–353.

Anderson R.S., Thibos L.N. Relationship between acuity for gratings and for tumbling-E letters in peripheral vision. *J Opt Soc Am A Opt Image Sci Vis* 16.10 (1999): 2321–2333.

Anderson R.S., Wilkinson M.O., Thibos L.N. Psychophysical localization of the human visual streak. *Optom Vis Sci* 69.3 (1992): 171–174.

Anderson S.J., Mullen K.T., Hess R.F. Human peripheral spatial resolution for achromatic and chromatic stimuli: Limits imposed by optical and retinal factors. *J Physiol* 442.1 (1991): 47–64.

Armbrecht A.M., Findlay C., Kaushal S., Aspinall P., Hill A.R., Dhillon B. Is cataract surgery justified in patients with age related macular degeneration? A visual function and quality of life assessment. *Br J Ophthalmol* 84.12 (2000): 1343–1348.

Artal P., Derrington A.M., Colombo E. Refraction, aliasing, and the absence of motion reversals in peripheral vision. *Vision Res* 35 (1995a): 939–947.

Artal P., Marcos S., Navarro R., Williams D.R. Odd aberrations and double-pass measurements of retinal image quality. *J Opt Soc Am A Opt Image Sci Vis* 12 (1995b): 195–201.

Atchison D., Scott D. Monochromatic aberrations of human eyes in the horizontal visual field. *J Opt Soc Am A Opt Image Sci Vis* 19.11 (2002): 2180–2184.

Atchison D.A. Comparison of peripheral refractions determined by different instruments. *Optom Vis Sci* 80 (2003): 655–660.

Atchison D.A. Anterior corneal and internal contributions to peripheral aberrations of human eyes. *J Opt Soc Am A Opt Image Sci Vis* 21 (2004): 355–359.

Atchison D.A., Jones C.E., Schmid K.L., Pritchard N., Pope J.M., Strugnell W.E., Riley R.A. Eye shape in emmetropia and myopia. *Invest Ophthalmol Vis Sci* 45.10 (2004): 3380–3386.

Atchison D.A., Markwell E.L. Aberrations of emmetropic subjects at different ages. *Vision Res* 48 (2008): 2224–2231.

Atchison D.A., Mathur A., Varnas S.R. Visual performance with lenses correcting peripheral refractive errors. *Optom Vis Sci* 90.11 (2013): 1304–1311.

Atchison D.A., Pritchard N., Schmid K.L. Peripheral refraction along the horizontal and vertical visual fields in myopia. *Vision Res* 46 (2006): 1450–1458.

Atchison D.A., Pritchard N., White S.D., Griffiths A.M. Influence of age on peripheral refraction. *Vision Res* 45 (2005): 715–720.

Atchison D.A., Smith G. *Optics of the Human Eye* (Oxford, U.K.: Butterworth-Heinemann, 2000).

Bakaraju R.C., Fedtke C., Ehrmann K., Ho A. Comparing the relative peripheral refraction effect of single vision and multifocal contact lenses measured using an autorefractor and an aberrometer: A pilot study. *J Optom* 8.3 (2015): 206–218.

Banks M.S., Sekuler A.B., Anderson S.J. Peripheral spatial vision: Limits imposed by optics, photoreceptors, and receptor pooling. *J Opt Soc Am A Opt Image Sci Vis* 8.11(1991): 1775–1787.

Baskaran K., Rosén R., Lewis P., Unsbo P., Gustafsson J. Benefit of adaptive optics aberration correction at preferred retinal locus. *Optom Vis Sci* 89 (2012): 1417–1423.

Baskaran K., Theagarayan B., Carius S., Gustafsson J. Repeatability of peripheral aberrations in young emmetropes. *Optom Vis Sci* 87 (2010): 751–759.

Baskaran K., Unsbo P., Gustafsson J. Influence of age on peripheral ocular aberrations. *Optom Vis Sci* 88.9 (2011): 1088–1098.

Bedggood P., Metha A. Comparison of sorting algorithms to increase the range of Hartmann-Shack aberrometry. *J Biomed Opt* 15.6 (2010): 067004.

Benavente-Pérez A., Nour A., Troilo D. Axial eye growth and refractive error development can be modified by exposing the peripheral retina to relative myopic or hyperopic defocus. *Invest Ophthalmol Vis Sci* 55.10 (2014): 6765–6773.

Berntsen D.A., Mutti D.O., Zadnik K. Validation of aberrometry based relative peripheral refraction measurements. *Ophthalmic Physiol Opt* 28(2008): 83–90.

Born M., Wolf E. Chapter V: Geometrical theory of aberrations, in *Principles of Optics* (Cambridge, U.K.: Cambridge University Press, 7th edn., 1999), pp. 228–260.

Bosworth C.F., Sample P.A., Johnson C.A., Weinreb R.N. Current practice with standard automated perimetry. *Semin Ophthalmol* 15.4 (2000): 172–181.

Brody B.L., Gamst A.C., Williams R.A., Smith A.R., Lau P.W., Dolnak D., Rapaport M.H., Kaplan R.M., Brown S.I. Depression, visual acuity, comorbidity, and disability associated with age-related macular degeneration. *Ophthalmology* 108.10 (2001): 1893–1900.

Brooks J.O., Tyrrell R.A., Frank T.A. The effects of severe visual challenges on steering performance in visually healthy young drivers. *Optom Vis Sci* 82.8 (2005): 689–697.

Brown B. Resolution thresholds for moving targets at the fovea and in the peripheral retina. *Vision Res* 12.2 (1972): 293–304.

Buehren T., Collins M.J., Carney L. Corneal aberrations and reading. *Optom Vis Sci* 80 (2003): 159–166.

Calver R., Radhakrishnan H., Osuobeni E., O'Leary D. Peripheral refraction for distance and near vision in emmetropes and myopes. *Ophthalmic Physiol Opt* 27 (2007): 584–593.

Charman W.N. Aberrations and myopia. *Ophthalmic Physiol Opt* 25 (2005): 285–301.

Charman W.N., Jennings J.A. Longitudinal changes in peripheral refraction with age. *Ophthalmic Physiol Opt* 26 (2006): 447–455.

Charman W.N., Radhakrishnan H. Peripheral refraction and the development of refractive error: A review. *Ophthalmic Physiol Opt* 30.4 (2010): 321–338.

Cho P., Cheung S.W. Retardation of myopia in orthokeratology (ROMIO) study: A 2-year randomized clinical trial. *Invest Ophthalmol Vis Sci* 53 (2012): 7077–7085.

Crossland M.D., Culham L.E., Kabanarou S.A., Rubin G.S. Preferred retinal locus development in patients with macular disease. *Ophthalmology* 112.9 (2005): 1579–1585.

Curcio C.A., Allen K.A. Topography of ganglion cells in human retina. *J Comp Neurol* 300.1 (1990): 5–25.

Curcio C.A., Sloan K.R., Kalina R.E., Hendrickson A.E. Human photoreceptor topography. *J Comp Neurol* 292.4 (1990): 497–523.

Davies L.N., Mallen E.A.H. Influence of accommodation and refractive status on the peripheral refractive profile. *Br J Ophthalmol* 93 (2009): 1186–1190.

Dunne M.C.M., Misson G.P., White E.K., Barnes D.A. Peripheral astigmatic asymmetry and angle alpha. *Ophthalmic Physiol Opt* 13 (1993): 303–305.

Eisenbarth W., Mackeben M., Poggel D.A., Strasburger H. Characteristics of dynamic processing in the visual field of patients with age-related maculopathy. *Graefes Arch Clin Exp Ophthalmol* 246.1 (2008): 27–37.

Ennis F.A., Anderson R.S. Aliasing in peripheral vision for flickering gratings under different levels of illumination. *Curr Eye Res* 20.5 (2000): 413–419.

Fedtke C., Ehrmann K., Falk D., Bakaraju R.C., Holden B.A. The BHVI-EyeMapper: Peripheral refraction and aberration profiles. *Optom Vis Sci* 91.10 (2014): 1199–1207.

Fedtke C., Ehrmann K., Holden B.A. A review of peripheral refraction techniques. *Optom Vis Sci* 86.5 (2009): 429–446.

Ferree C.E., Rand G. Interpretation of refractive conditions in the peripheral field of vision. *Arch Ophthalmol* 9 (1933): 925–938.

Ferree C.E., Rand G., Hardy C. Refraction for the peripheral field of vision. *Arch Ophthalmol* 5 (1931): 717–731.

Ferree C.E., Rand G., Hardy C. Refractive asymmetry in the temporal and nasal halves of the visual field. *Am J Ophthalmol* 15 (1932): 513–522.

Freeman M.H., Hull C.C. *Optics* (Eastbourne, England: Butterworth-Heinemann, 11th edn., 2003).

Fulk G.W., Cyert L.A., Parker D.E. A randomized trial of the effect of single-vision vs. bifocal lenses on myopia progression in children with esophoria. *Optom Vis Sci* 77 (2000): 395–401.

Ghosh A., Collins M.J., Read S.A., Davis B.A., Iskander D.R. The influence of downward gaze and accommodation on ocular aberrations over time. *J Vis* 11.10 (2011): 17.

Goldschmidt E., Jacobsen N. Genetic and environmental effects on myopia development and progression. *Eye* 28 (2014): 126–133.

Guirao A., Artal P. Off-axis monochromatic aberrations estimated from double pass measurements in the human eye. *Vis Res* 39 (1999): 207–217.

Gustafsson J. The first successful eccentric correction. *Vis Impair Res* 3.3 (2001): 147–155.

Gustafsson J., Terenius E., Buchheister J., Unsbo P. Peripheral astigmatism in emmetropic eyes. *Ophthalmic Physiol Opt* 21.5 (2001): 393–400, 2001. Erratum in *Ophthalmic Physiol Opt* 21.6 (2001): 491.

Gustafsson J., Unsbo P. Eccentric correction for off-axis vision in central visual field loss, *Optom Vis Sci* 80.7 (2003): 535–541.

Gwiazda J., Hyman L., Hussein M., Everett D., Norton T.T., Kurtz D., Leske M.C. et al. A randomized clinical trial of progressive addition lenses versus single vision lenses on the progression of myopia in children. *Invest Ophthalmol Vis Sci* 44 (2003): 1492–1500.

Hassell J.B., Lamoureux E.L., Keeffe J.E. Impact of age related macular degeneration on quality of life. *Br J Ophthalmol* 90.5 (2006): 593–596.

Holden B., Sankaridurg P., Smith E., Aller T., Jong M., He M. Myopia, an underrated global challenge to vision: Where the current data takes us on myopia control. *Eye* 28 (2014): 142–146.

Hoogerheide J., Rempt F., Hoogenboom W.P.H. Acquired myopia in young pilots. *Ophthalmologica* 163 (1971): 209–215.

Huang J., Wen D., Wang Q., McAlinden C., Flitcroft I., Chen H., Saw S.M. et al. Efficacy comparison of 16 interventions for myopia control in children: A network meta-analysis. *Ophthalmology* 123.4 (2016): 697–708.

Hung L.F., Crawford M.L., Smith E.L. Spectacle lenses alter eye growth and the refractive status of young monkeys. *Nat Med* 1 (1995): 761–765.

Huynh N., Nicholson B.P., Agrón E., Clemons T.E., Bressler S.B., Rosenfeld P.J., Chew E.Y. Age-Related Eye Disease Study 2 Research Group. Visual acuity outcomes after cataract surgery in patients with age-related macular degeneration in the Age-Related Eye Disease Study 2. *Invest Ophthalmol Vis Sci* 54.6 (2013): 243.

Irving E.L., Sivak J.G., Callender M.G. Refractive plasticity of the developing chick eye. *Ophthalmic Physiol Opt* 12 (1992): 448–456.

Iskander D.R. Computational aspects of the visual strehl ratio. *Optom Vis Sci* 83.1 (2006): 57–59.

Jackson D.W., Paysse E.A., Wilhelmus K.R., Hussein M.A., Rosby G., Coats D.K. The effect of off-the-visual-axis retinoscopy on objective refractive measurement. *Am J Ophthalmol* 137 (2004): 1101–1104.

Jaeken B., Artal P. Optical quality of emmetropic and myopic eyes in the periphery measured with high-angular resolution. *Invest Ophthalmol Vis Sci* 53.7 (2012): 3405–3413.

Jaeken B., Lundström L., Artal P. Fast scanning peripheral wavefront sensor for the human eye. *Opt Express* 19 (2011a): 7903–7913.

Jaeken B., Lundström L., Artal P. Peripheral aberrations in the human eye for different wavelengths: Off-axis chromatic aberration. *J Opt Soc Am A Opt Image Sci Vis* 28 (2011b): 1871–1879.

Jaeken B., Mirabet S., Marín J.M., Artal P. Comparison of the optical image quality in the periphery of phakic and pseudophakic eyes. *Invest Ophthalmol Vis Sci* 54.5 (2013): 3594–3599.

Jaeken B., Tabernero J., Frank Schaeffel F., Artal P. Comparison of two scanning instruments to measure peripheral refraction in the human eye. *J Opt Soc Am A Opt Image Sci Vis* 29.3 (2012): 258–264.

Jennings J.A., Charman W.N. Analytic approximation of the off-axis modulation transfer function of the eye. *Vision Res* 37(1997): 697–704.

Jennings J.A.M., Charman W.N. Optical image quality in the peripheral retina. *Am J Optom Physiol Opt* 55.8 (1978): 582–590.

Jennings J.A.M., Charman W.N. Off-axis image quality in the human eye. *Vision Res* 21.4 (1981): 445–455.

Kerr J.L. Visual resolution in the periphery. *Percept Psychophys* 9.3B (1971): 375–378.

Kidger M.J. *Fundamental Optical Design* (Bellingham, WA: SPIE Press, 2002).

Koffler B.H., Sears J.J. Myopia control in children through refractive therapy gas permeable contact lenses: Is it for real? *Am J Ophthalmol* 156.6 (2013): 1076–1081.

Lamoureux E.L., Hassell J.B., Keeffe J.E. The determinants of participation in activities of daily living in people with impaired vision. *Am J Ophthalmol* 137.2 (2004): 265–270.

Larson A.M., Loschky L.C. The contributions of central versus peripheral vision to scene gist recognition. *J Vis* 9.10 (2009): 6.

Lemmink K.A., Dijkstra B., Visscher C. Effects of limited peripheral vision on shuttle sprint performance of soccer players. *Percept Mot Skills* 100.1 (2005): 167–175.

Leung J., Brown B. Progression of myopia in Hong Kong Chinese schoolchildren is slowed by wearing progressive lenses. *Optom Vis Sci* 76 (1999): 346–354.

Lewis P., Baskaran K., Rosén R., Lundström L., Unsbo P., Gustafsson J. Objectively determined refraction improves peripheral vision. *Optom Vis Sci* 91.7 (2014): 740–746.

Optical properties of the eye

Lindlein N., Pfund J. Experimental results for expanding the dynamic range of a Shack-Hartmann sensor using astigmatic microlenses. *Opt Eng* 41 (2002): 529–533.

Liu T., Sreenivasan V., Thibos L.N. Uniformity of accommodation across the visual field. *J Vis* 16.3 (2016): 6.

Lotmar W., Lotmar T. Peripheral astigmatism in the human eye: Experimental data and theoretical predictions. *J Opt Soc Am* 64 (1974): 510–513.

Lundström L., Gustafsson J., Svensson I., Unsbo P. Assessment of objective and subjective eccentric refraction. *Optom Vis Sci* 82 (2005a): 298–306.

Lundström L., Gustafsson J., Unsbo P. Vision evaluation of eccentric refractive correction. *Optom Vis Sci* 84 (2007): 1046–1052.

Lundström L., Gustafsson J., Unsbo P. Population distribution of wavefront aberrations in the peripheral human eye. *J Opt Soc Am A Opt Image Sci Vis* 26 (2009b): 2192–2198.

Lundström L., Mira-Agudelo A., Artal P. Peripheral optical errors and their change with accommodation differ between emmetropic and myopic eyes. *J Vis* 9 (2009a): 17.

Lundström L., Rosén R., Baskaran K., Jaeken B., Gustafsson J., Artal P., Unsbo P. Symmetries in peripheral ocular aberrations. *J Mod Opt* 58 (2011): 1690–1695.

Lundström L., Unsbo P. Unwrapping Hartmann-Shack images from highly aberrated eyes using an iterative B-spline based extrapolation method. *Optom Vis Sci* 81 (2004): 383–388.

Lundström L., Unsbo P. Transformation of Zernike coefficients: Scaled, translated, and rotated wavefronts with circular and elliptical pupils. *J Opt Soc Am A Opt Image Sci Vis* 24 (2007): 569–577.

Lundström L., Unsbo P., Gustafsson J. Off-axis wave front measurements for optical correction in eccentric viewing. *J Biomed Opt* 10 (2005b): 034002.

Ma L., Atchison D.A., Charman W.N. Off-axis refraction and aberrations following conventional laser in situ keratomileusis. *J Cataract Refract Surg* 31 (2005): 489–498.

Mallah M.K., Hart P.M., McClure M., Stevenson M.R., Silvestri G., White S.T., Chakravarthy U. Improvements in measures of vision and self-reported visual function after cataract extraction in patients with late-stage age-related maculopathy. *Optom Vis Sci* 78.9 (2001): 683–688.

Marcos S., Barbero S., Jiménez-Alfaro I. Optical quality and depth-of-field of eyes implanted with spherical and aspheric intraocular lenses. *J Refract Surg* 21(2005): 223–235.

Marcos S., Moreno E., Navarro R. The depth-of-field of the human eye from objective and subjective measurements. *Vision Res* 39 (1999): 2039–2049.

Markowitz S.N. Principles of modern low vision rehabilitation. *Can J Ophthalmol* 41.3 (2006): 289–312.

Mathur A., Atchison D.A. Effect of orthokeratology on peripheral aberrations of the eye. *Optom Vis Sci* 86.5 (2009): E476–E484.

Mathur A., Atchison D.A. Peripheral refraction patterns out to large field angles. *Optom Vis Sci* 90.2 (2013): 140–147.

Mathur A., Atchison D.A., Charman W.N. Myopia and peripheral ocular aberrations. *J Vis* 9.10 (2009b): 15.

Mathur A., Atchison D.A., Charman W.N. Effect of accommodation on peripheral ocular aberrations. *J Vis* 9.12 (2009c): 20.

Mathur A., Atchison D.A., Charman W.N. Effects of age on peripheral ocular aberrations. *Opt Express* 18.6 (2010): 5840–5853.

Mathur A., Atchison D.A., Kasthurirangan S., Dietz N.A., Luong S., Chin S.P., Lin W.L., Hoo S.W. The influence of oblique viewing on axial and peripheral refraction for emmetropes and myopes. *Ophthalmic Physiol Opt* 29 (2009a): 155–161.

Mathur A., Atchison D.A., Scott D.H. Ocular aberrations in the peripheral visual field. *Opt Lett* 33.8 (2008): 863–865.

Mathur A., Gehrmann J., Atchison D.A. Pupil shape as viewed along the horizontal visual field. *J Vis* 13.6 (2013): 3.

Mazzaferri J., Navarro R. Wide two-dimensional field laser ray-tracing aberrometer. *J Vis* 12.2 (2012): 2.

McFadden S.A., Howlett M.H., Mertz J.R. Retinoic acid signals the direction of ocular elongation in the guinea pig eye. *Vision Res* 44 (2004): 643–653.

Millodot M. Effect of ametropia on peripheral refraction. *Am J Optom Physiol Opt* 58 (1981): 691–695.

Millodot M. *Dictionary of Optometry and Visual Science* (Oxford, U.K.: Butterworth Heinemann, 4th edn., 1997).

Millodot M., Johnson C.A., Lamont A., Leibowitz H.W. Effect of dioptrics on peripheral visual acuity. *Vision Res* 15.12 (1975): 1357–1362.

Millodot M., Lamont A. Letter: Refraction of the periphery of the eye. *J Opt Soc Am* 64 (1974): 110–111.

Morgan I., Rose K. How genetic is school myopia? *Prog Retin Eye Res* 24 (2005): 1–38.

Mutti D.O., Hayes J.R., Mitchell G.L., Jones L.A., Moeschberger M.L., Cotter S.A., Kleinstein R.N., Manny R.E., Twelker J.D., Zadnik K. Refractive error, axial length, and relative peripheral refractive error before and after the onset of myopia. *Invest Ophthalmol Vis Sci* 48 (2007): 2510–2519.

Mutti D.O., Sinnott L.T., Mitchell G.L., Jones-Jordan L.A., Moeschberger M.L., Cotter S.A., Kleinstein R.N. et al. Relative peripheral refractive error and the risk of onset and progression of myopia in children. *Invest Ophthalmol Vis Sci* 52.1 (2011): 199–205.

Navarro R., Artal P., Williams D.R. Modulation transfer of the human eye as a function of retinal eccentricity. *J Opt Soc Am A Opt Image Sci Vis* 10.2 (1993): 201–212.

Navarro R., Moreno E., Dorronsoro C. Monochromatic aberrations and point-spread functions of the human eye across the visual field. *J Opt Soc Am A Opt Image Sci Vis* 15.9 (1998): 2522–2529.

Nilsson U.L., Frennesson C., Nilsson S.E.G. Patients with AMD and a large absolute central scotoma can be trained successfully to use eccentric viewing, as demonstrated in a scanning laser ophthalmoscope. *Vision Res* 43.16 (2003): 1777–1787.

Ogboso Y.U., Bedell H.E. Magnitude of lateral chromatic aberration across the retina of the human eye. *J Opt Soc Am A Opt Image Sci Vis* 4 (1987): 1666–1672.

Olivier S., Laude V., Huignard J.P. Liquid-crystal Hartmann wave-front scanner. *Appl Optics* 39 (2000): 3838–3846.

Petre K.L., Hazel C.A., Fine E.M., Rubin G.S. Reading with eccentric fixation is faster in inferior visual field than in left visual field. *Optom Vis Sci* 77.1 (2000): 34–39.

Popovic Z. Neural limits of visual resolution, PhD thesis (2003), University of Gothenburg, Sweden, ISBN 91-628-5708-8.

Prado P., Arines J., Bará S., Manzanera S., Mira-Agudelo A., Artal P. Changes of ocular aberrations with gaze. *Ophthalmic Physiol Opt* 29 (2009): 264–271.

Rabbetts R.B. *Clinical Visual Optics* (Philadelphia, PA: Butterworth-Heinemann, 4th edn., 2007).

Radhakrishnan H., Charman W.N. Refractive changes associated with oblique viewing and reading in myopes and emmetropes. *J Vis* 7.8 (2007): 5.

Radhakrishnan H., Charman W.N. Peripheral refraction measurement: Does it matter if one turns the eye or the head? *Ophthalmic Physiol Opt* 28 (2008): 73–82.

Rempt F., Hoogerheide J., Hoogenboom W.P.H. Peripheral retinoscopy and the skiagram. *Ophthalmologica* 162 (1971): 1–10.

Ronchi L., Molesini G. Depth of focus in peripheral vision. *Ophthalmic Res* 7 (1975): 152–157.

Rosén R., Jaeken B., Lindskoog Pettersson A., Artal P., Unsbo P., Lundström L. Evaluating the peripheral optical effect of multifocal contact lenses. *Ophthalmic Physiol Opt* 32 (2012b): 527–534.

Rosén R., Lundström L., Unsbo P. Influence of optical defocus on peripheral vision. *Invest Ophthalmol Vis Sci* 52 (2011): 318–323.

Rosén R., Lundström L., Unsbo P. Sign-dependent sensitivity to peripheral defocus for myopes due to aberrations. *Invest Ophthalmol Vis Sci* 53 (2012c): 7176–7182.

Rosén R., Lundström L., Unsbo P. Adaptive optics for peripheral vision. *J Mod Opt* 59 (2012d): 1064–1070.

Rosén R., Lundström L., Unsbo P., Atchison D.A. Have we misinterpreted the study of Hoogerheide et al. (1971)? *Optom Vis Sci* 89 (2012a): 1235–1237.

Rosén R., Lundström L., Venkataraman A.P., Winter S., Unsbo P. Quick contrast sensitivity measurements in the periphery. *J Vis* 14.8 (2014): 3.

Rovamo J., Virsu V., Laurinen P., Hyvärinen L. Resolution of gratings oriented along and across meridians in peripheral vision. *Invest Ophthalmol Vis Sci* 23.5 (1982): 666–670.

Rynders M.C., Navarro R., Losada M.A. Objective measurement of the off-axis longitudinal chromatic aberration in the human eye. *Vision Res* 38 (1998): 513–522.

Schaeffel F., Glasser A., Howland H.C. Accommodation, refractive error and eye growth in chickens. *Vision Res* 28 (1988): 639–657.

Schieber F., Schlorholtz B., McCall R. Visual requirements of vehicular guidance, in *Human Factors of Visual and Cognitive Performance in Driving* (Taylor & Francis, 2009), pp. 31–50.

Seidemann A., Schaeffel F., Guirao A., Lopez-Gil N., Artal P. Peripheral refractive errors in myopic, emmetropic, and hyperopic young subjects. *J Opt Soc Am A Opt Image Sci Vis* 19.12 (2002): 2363–2373.

Shih Y.F., Hsiao C.K., Chen C.J., Chang C.W., Hung P.T., Lin L.L. An intervention trial on efficacy of atropine and multi-focal glasses in controlling myopic progression. *Acta Ophthalmol Scand* 79.3 (2001): 233–236.

Si J.K., Tang K., Bi H.S., Guo D.D., Guo J.G., Wang X.R. Orthokeratology for myopia control: A meta-analysis. *Optom Vis Sci* 92.3 (2015): 252–257.

Smith III E.L. Optical treatment strategies to slow myopia progression: Effects of the visual extent of the optical treatment zone. *Exp Eye Res* 114 (2013): 77–88.

Smith III E.L., Campbell M.C.W., Irving E. Does peripheral retinal input explain the promising myopia control effects of corneal reshaping therapy (CRT or ortho-K) & multifocal soft contact lenses? *Ophthalmic Physiol Opt* 33.3 (2013): 379–384.

Smith III E.L., Kee C.S., Ramamirtham R., Qiao-Grider Y., Hung L.F. Peripheral vision can influence eye growth and refractive development in infant monkeys. *Invest Ophthalmol Vis Sci* 46 (2005): 3965–3972.

Smith G., Millodot M., McBrien N. The effect of accommodation on oblique astigmatism and field curvature of the human eye. *Clin Exp Optom* 71 (1988): 119–125.

Tabernero J., Schaeffel F. Fast scanning photoretinoscope for measuring peripheral refraction as a function of accommodation. *J Opt Soc Am A Opt Image Sci Vis* 26.10 (2009): 2206–2210.

Thibos L.N. Calculation of the influence of lateral chromatic aberration on image quality across the visual field. *J Opt Soc Am A Opt Image Sci Vis* 4 (1987): 1673–1680.

Thibos L.N., Applegate R.A., Schwiegerling J.T., Webb R., VSIA Standards Taskforce Members, Standards for reporting the optical aberrations of eyes. *J Refract Surg* 18 (2002): 652–660.

Thibos L.N., Cheney F.E., Walsh D.J. Retinal limits to the detection and resolution of gratings. *J Opt Soc Am A Opt Image Sci Vis* 4.8 (1987): 1524–1529.

Thibos L.N., Hong X., Bradley A., Applegate R.A. Accuracy and precision of objective refraction from wavefront aberrations. *J Vis* 4 (2004): 329–351.

Thibos L.N., Still D.L., Bradley A. Characterization of spatial aliasing and contrast sensitivity in peripheral vision. *Vision Res* 36 (1996): 249–258.

Venkataraman A.P., Winter S., Rosén R., Lundström L. Choice of grating orientation for evaluation of peripheral vision. *Optom Vis Sci* 93.6 (2016): 567–574.

Wagner S., Conrad F., Bakaraju R.C., Fedtke C., Ehrmann K., Holden B.A. Power profiles of single vision and multifocal soft contact lenses. *Cont Lens Anterior Eye* 38.1 (2015): 2–14.

Walker T.W., Mutti D.O. The effect of accommodation on ocular shape. *Optom Vis Sci* 79 (2002): 424–430.

Wallman J., Winawer J. Homeostasis of eye growth and the question of myopia. *Neuron* 43 (2004): 447–468.

Wang B., Ciuffreda K.J. Depth-of-focus of the human eye in the near retinal periphery. *Vision Res* 44 (2004): 1115–1125.

Wang Y.Z., Thibos L.N., Bradley A. Effects of refractive error on detection acuity and resolution acuity in peripheral vision. *Invest Ophthalmol Vis Sci* 38.10 (1997): 2134–2143.

Wang Y.Z., Thibos L.N., Lopez N., Salmon T., Bradley A. Subjective refraction of the peripheral field using contrast detection acuity. *J Am Optom Assoc* 67 (1996): 584–589.

Watson A.B. A formula for human retinal ganglion cell receptive field density as a function of visual field location. *J Vis* 14.7 (2014): 15.

Wei X., Thibos L. Design and validation of a scanning Shack Hartmann aberrometer for measurements of the eye over a wide field of view. *Opt Express* 18.2 (2010): 1134–1143.

Wertheim T. Über die indirekte Sehschärfe (peripheral visual acuity). *Z Psychol Physiol Sinnesorgane* 7 (1894): 172–87. Translated by Dunsky I.L. *Am J Optom Physiol Opt* 57 (1980): 915–924.

Whatham A., Zimmermann F., Martinez A., Delgado S., de la Jara P.L., Sankaridurg P., Ho A. Influence of accommodation on off-axis refractive errors in myopic eyes. *J Vis* 9.3 (2009): 14.

Whatham A.R., Judge S.J. Compensatory changes in eye growth and refraction induced by daily wear of soft contact lenses in young marmosets. *Vision Res* 41 (2001): 267–273.

White J.M., Bedell H.E. The oculomotor reference in humans with bilateral macular disease. *Invest Ophthalmol Vis Sci* 31.6 (1990): 1149–1161.

Williams D.R., Artal P., Navarro R., McMahon R.N., Brainard D. Off-axis optical quality and retinal sampling in the human eye. *Vision Res* 36 (1996): 1103–1114.

Williams R.A., Brody B.L., Thomas R.G., Kaplan R.M., Brown S.I. The psychosocial impact of macular degeneration. *Arch Ophthalmol* 116.4 (1998): 514–520.

Winter S., Sabesan R., Tiruveedhula P.N., Privitera C., Unsbo P., Lundström L., Roorda A. Transverse chromatic aberration across the visual field of the human eye. *J Vis* (2016): in press.

Wood J.M. Age and visual impairment decrease driving performance as measured on a closed-road circuit. *Hum Factors* 44.3 (2002): 482–494.

Wong W.L., Su X., Li X., Cheung C.M.G., Klein R., Cheng C.Y., Wong T.Y. Global prevalence of age-related macular degeneration and disease burden projection for 2020 and 2040: A systematic review and meta-analysis. *Lancet Glob Health* 2.2 (2014): e106–e116.

Yamaguchi T., Ohnuma K., Konomi K., Satake Y., Shimazaki J., Negishi K. Peripheral optical quality and myopia progression in children. *Graefes Arch Clin Exp Ophthalmol* 251 (2013): 2451–2461.

Yoon G., Pantanelli S., Nagy L.J. Large-dynamic-range Shack-Hartmann wavefront sensor for highly aberrated eyes. *J Biomed Opt* 11.3 (2006): 30502.

Optical properties of the eye

22 Customized eye models

Juan Tabernero

Contents

22.1 INTRODUCTION

Classical schematic eye models of the human eye provide a useful understanding of some of the optical characteristics of the *average* human eye. They are mostly based on *average* values of the geometrical parameters that constitute the human eye, like the corneal and lens radius of curvature, the asphericity of the surfaces, and the axial distances between the surface components (see Chapter 16).

This chapter introduces the idea of a customized eye model to complement the classical concept of the schematic eye. The goal of such models is to provide *individual* predictions of the full optical quality as accurate as possible (beyond the paraxial approximation). They can be used to anticipate the effects of any refractive procedure that might potentially modify the geometry of the elements of a specific eye model from a *particular* subject.

In order to build up individual eye models, a precise knowledge of the geometry and positioning of the cornea and the lens is typically required. Accurate corneal data can be obtained from corneal topography providing a nearly point-by-point description of the first corneal surface. However, accurate in vivo posterior corneal data are still scarce. Although the cornea can be imaged on a meridian using Scheimpflug lamps (Dubbelman et al. 2002, 2006), the reconstruction of a 3D corneal surface still has some technical limitations. There is not a lens topography system, either, to reproduce the exact geometry of the in vivo lens surfaces. For these reasons a 100% customization of the full eye geometry cannot be achieved and some assumptions must be taken.

The development of customized eye modeling requires an exact ray-tracing engine to calculate the complete optical quality of any eye model. Living in an advanced computer age simplifies greatly this task since ray-tracing software has become relatively popular in optics labs today. Back on the 1960s, early computers were able to trace 1 ray per surface per second (Turner 2000). Nowadays, an ordinary PC can trace some tens of millions of ray-surface per second. From a technical point of view, this is no longer a problem, but still a proper knowledge of how a ray-tracing engine works can be educationally relevant and it may help to improve creative thinking to the user end of the software.

The possibilities opened by ray-tracing software applied to customized eye models are enormous. A huge range of optical properties and metrics can be examined within these eye models. Calculations can easily go way beyond paraxial approximations (cardinal point locations, retinal size magnification, refractive errors, etc.) to include aberrations and any other optical parameter/metric that can be derived from them.

Practical applications of customized eye models are, for instance, the possibility to perform personalized predictions on the optical outcome of many refractive procedures, like the implantation of different types of intraocular lenses (IOLs) with spherical or aspheric surfaces (Tabernero et al. 2006a; Rosales and Marcos 2007) or the optical effects of a particular corneal ablation profile performed on each subject's corneal surface (Cano et al. 2004). Other examples include the individual prediction of the IOL dioptric power (Canovas and Artal 2011), predictions of depth of focus using any ophthalmic device designed for this task (Tabernero and Artal 2012) or even to test the individual effects of optical devices designed to modify the optical quality not only in the central area of the retina but also in the eccentric retina (Tabernero et al. 2012; Jaeken et al. 2013). Customized eye models have also been used to explain the mechanism of compensation of aberrations between the cornea and the lens and their potential changes with age and accommodation (Tabernero et al. 2007, 2011a; Artal and Tabernero 2008).

This chapter first introduces the reader to the mathematical techniques used by common ray-tracing software (the fundamentals of exact ray-tracing theory) followed by an explanation of how cornea, lens, and other customized parameters are typically introduced in the software to build the models. Finally, the chapter ends with a short compilation of typical examples where a customized model is applied. All parameters required to reproduce the model using ray-tracing software are included.

22.2 EXACT RAY TRACING

Ray-tracing programs use the laws of geometrical optics to calculate the exact trajectory of a "ray" of light when it travels through different optical medium. This is equivalent to consider the nature of light as plane waves. Theoretically, any wave can be considered as planar in a particular environment as long as that environment is *small* enough. Maxwell's theory tells us that this is strictly true in the limit when the wavelength of light tends to be zero. In practical terms, that means that the dimensions of the obstacles and apertures of the physical environment where light travels must be very large compared to the wavelength of light. If the eye is considered as the physical *environment* where light travels, then the dimensionality assumption holds true. The sizes of the apertures are in the order of several millimeters compared to the less than a micrometer wavelength of the visible range. Therefore, ray tracing is a useful and valuable tool to calculate the propagation of light in the eye.

A ray-tracing procedure can be split in two sequences. The first one involves propagation of the ray until it hits a surface (it requires the calculation of the intersection point between the ray and the surface), while the second one involves the refraction of the ray through the surface at that particular point (it requires to solve the Snell law for any ray hitting the surface). In an optical system made of many surfaces, the ray-tracing procedure would be a concatenated series of alternated propagation and refractive sequences.

Typically, a light ray travelling in a homogeneous medium is defined by the director cosines $\bar{r}$ (the cosines of the angles that

the vector makes with the x-, y-, and z-axes, respectively) and the starting point for the ray tracing, P_0:

$$\bar{r} = (L, M, N) \tag{22.1}$$

$$P_0 = (x_0, y_0, z_0) \tag{22.2}$$

A refractive surface that separates two optical medium of different refractive index can be defined implicitly as

$$S(x, y, z) = 0 \tag{22.3}$$

The mathematical problem of finding the intersection point between the ray and the surface can only be solved analytically in a reduced number of cases where the surface S has the particular shape of a sphere or a conicoid (Welford 1986). In the general case where S is a nonrotationally symmetrical surface, the problem has to be solved numerically using an iterative procedure. Figure 22.1 illustrates the first four iterations of a typical algorithm to solve this problem. Initially, the intersection point between the tangent plane to the surface vertex and the ray is found (I_0). This point is projected to the surface (I_1) and then the intersection between the ray and the coplanar tangent line is found (I_2). Again, this point is projected to the surface (I_3) and the procedure continues until the distance between two consecutive iterative solutions converge to an established threshold.

Once the intersection point of the ray and the surface is found, the second step in a ray-tracing procedure requires to solve the exact Snell's law at the intersection point. Given the director cosines of the incident and refracted ray as $\bar{r}$ and $\bar{r}'$ and the normal unit vector to the surface at the intersection point as $\overline{N} = (\alpha, \beta, \gamma)$, Snell's law in vector form can be expressed as

$$n(\bar{r} \times \overline{N}) = n'(\bar{r}' \times \overline{N}) \tag{22.4}$$

where n and n' are the refractive index of the optical medium. Taking the magnitude of the cross products, the well-known scalar form is derived as

$$n \operatorname{sen} I = n' \operatorname{sen} I' \tag{22.5}$$

where I and I' are the angles of incidence and refraction between the ray and the normal vector. Equation 22.4 can be transformed

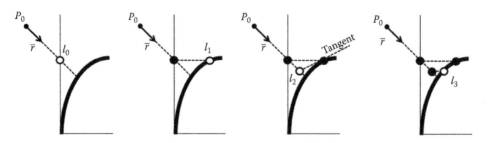

Figure 22.1 Iterative process to calculate the intersection of a ray with a general surface (Welford 2002). The iterative solutions are denoted as *Ii*.

multiplying on both sides by the cross product $\times \overline{N}$ and using the triple product expansion:

$$n(\overline{r} - \overline{N}(\overline{r} \cdot \overline{N})) = n'(\overline{r}' - \overline{N}(\overline{r}' \cdot \overline{N})) \qquad (22.6)$$

Using the scalar form of (22.6), the following set of algebraic equations is derived:

$$n'L' - nL = k\alpha$$
$$n'M' - nM = k\beta \qquad (22.7)$$
$$n'N' - nN = k\gamma$$

where k is

$$k = n'(\overline{r'N}) - n(\overline{rN}) = n' \cos I' - n \cos I \qquad (22.8)$$

Given an incidence ray with director cosine $\overline{r} = (L, M, N)$ that intersect a refractive surface at a certain known point, the following sequence of operations can be used to obtain the director cosine $\overline{r}' = (L', M', N')$ of the refracted ray at that surface (Welford 2002):

1. Calculate the normal unit vector to the surface ($\overline{N} = (\alpha, \beta, \gamma)$) at the intersection point with the incidence ray.
2. Get the cosine of the angle of incidence ($\cos I$) as the dot product $\overline{r} \cdot \overline{N}$.
3. Get the cosine of the angle of refraction ($\cos I'$) using the scalar form of Snell's law $n' \cos I' = \sqrt{n'^2 - n^2(1 - \cos^2 I)}$.
4. Calculate the magnitude k from Equation 22.8.
5. Finally, using Equation 22.7, $\overline{r}' = (L', M', N')$ can be calculated.

After a refractive sequence is finished, a propagation sequence follows the procedure by finding the intersection point of the light ray with the next surface in the optical system and alternates with the next refraction, repeating the sequences until the final image surface is reached.

22.3 BUILDING A CUSTOMIZED MODEL

The customized model incorporates as many details of a particular subject as it can be available from individual measurements of the patient's eye. That include the situation where the cornea is modeled as a one-surface or a two-surface element, and the crystalline is optimized to reproduce certain optical characteristics of the eye (like wavefront aberrations) in an adaptive modeling procedure. This section reviews these procedures.

22.3.1 THE CORNEA

The most common tool to obtain quantitative information of the anterior corneal shape is corneal topography. It is mostly based on Placido disk reflections although other alternatives aiming to correct the "skew ray error" are starting to show up, like color-LED topography (Klijn et al. 2015). Using data from corneal topography, anterior corneal aberrations have been calculated

using analytical approaches (e.g., Schwiegerling and Greivenkamp 1997; Guirao and Artal 2000; Sicam et al. 2004) or ray-tracing techniques (Barbero et al. 2002; Tabernero et al. 2006a).

To incorporate the *full* anterior corneal surface to the ray-tracing package, it is necessary to transform the elevation (or height) data from the topographer to a suitable type of surface accepted by a particular ray-tracing software. Typically, Placido disk topographers provide the elevation data as a polar grid of data (slit-scanning topography that used a Cartesian geometry is also possible but not as common as Placido disk topography, e.g., Orbscan II topographer). One option is to fit the numerically obtained elevation data to an analytical surface available within the ray-tracing software set of surface types. Zernike or Cartesian polynomial surfaces are typically accepted and can be uploaded in the software as a set of polynomial coefficients.

Another possibility is to directly upload a numerical point-by-point description of the cornea and then let the ray-tracing software package to interpolate a surface. However, this situation typically requires the transformation of the polar grid of data to a Cartesian grid. Again, it is necessary to fit first the surface data to a polynomial and, then, resample this polynomial surface to obtain a Cartesian grid of data. These data can be now uploaded into the ray-tracing package where the surface can be interpolated. Cubic spline functions that exactly reproduce the surface at those points provided by the grid and smoothly interpolate the surface between points are typically used.

Customization of the alignment of the cornea is also relevant to measure corneal optical quality (Salmon and Thibos 2002). The elevation data from corneal topography are typically taken with the center of coordinates located at the center of the reflected Placido disks. This is the consequence (common to most corneal topographers) of taking the keratoscope axis as the reference axis (the line that connects the fixation stimulus coaligned with the Placido rings to the center of the corneal reflected rings). However, corneal topographers also typically provide the users with the distance between the center of the reflected Placido rings and the center of the entrance pupil of the eye. Since the natural alignment of an eye model must be along the line of sight (the line that connects the fixation stimulus to the center of the entrance pupil), the corneal elevation data must be evaluated and recentered to the location of the entrance pupil center. The pupil is typically observed in the temporal direction with respect to the coaxially sighted corneal reflex (Artal et al. 2006).

It is also possible to include some personalized data from the posterior corneal surface. Some corneal topographers and corneal imaging instruments (rotating Scheimpflug lamps) can estimate data on the second surface of the cornea. However, because of the rotating scanning system of Scheimpflug lamps, the interpolation required to collect 3D data between corneal meridians might generate reconstruction artifacts as the spatial resolution decreases with the radial distance. In some cases, an equivalent refractive index that accounts for the refractive change induced by the second surface of the cornea is considered as an alternative to add some uncertainty to the model (Hemenger et al. 1995; Guirao and Artal 2000). Another possibility is to use a generic model for the posterior corneal surface (Navarro et al. 2006) taken from average values of the population. In any case, since the dioptric power of the second corneal surface is much less than the power of the anterior surface (about −6 D vs 48 D), the expected changes in optical quality using a one-surface cornea model or a two-surface cornea with a personalized or generic posterior surface are small. This is supported by data measured with corrected Scheimpflug lamps where only a modest contribution of the posterior corneal surface to the total corneal spherical aberration and coma was found (Sicam et al. 2006; Dubbelman et al. 2007).

22.3.2 THE LENS

Because of the technical difficulties to visualize the in vivo crystalline lens, the personalization of the optics of the lens cannot be performed with as much detail as the corneal surface permits. The possibility to measure the geometry of the lens is still very much constrained to certain experimental prototypes in the lab. While Scheimpflug lamps are becoming popular in eye clinics claiming the possibility of visualizing the structure of the lens, to obtain an accurate quantitative measurement of the geometry of the lens is still complicated with commercial instruments. The lens is visualized through the previous optical surfaces of the eye that requires a personalized correction algorithm to avoid optical distortions from the light travelling through each particular preceding surface. Besides, the refractive index from average eye models must be used to derive such distortion that also adds some uncertainty to the potential "customized" surface corrections. In some cameras, parts of these correction algorithms have been implemented providing valuable information on the geometry of the lens (Dubbelman and Van der Heijde 2001; Dubbelman et al. 2001). Another technique that can potentially became a good choice to visualize lens surfaces is anterior chamber optical coherence tomography imaging, although it faces similar challenges as Scheimpflug lamps to provide accurate data. Recently, some advances to obtain quantitative data from OCT lens images have been done (Ortiz et al. 2012; Sun et al. 2014).

The customization of the optics of the crystalline lens requires also dealing with the gradient index structure of the element. To measure this characteristic of the lens in vivo is still complicated and requires, for instance, magnetic resonance imaging of the eye (Kasthurirangan et al. 2008). Most of the data available have been collected from ex vivo lenses (e.g., Pierscionek and Chan 1989; Birkenfeld et al. 2014). For this reason, the choice of using an equivalent constant refractive index for the lens of a customized model can be adopted. Other possibilities exist in the field of *average* eye models where average data can be incorporated to predict mean values of the aberrations in a population and its changes with age (Navarro 2014). But concerning customized eye models, the only possibility is to leave the gradient index as a fitting variable in the model or as a fixed constant equivalent value.

In practice, several options to introduce the geometry of the lens are available. Navarro et al. (2006) proposed the use of "generic" lens models. Similar to the posterior corneal surface, data taken from an average lens model representing the mean of a population can be introduced. Such a model itself would fail to give an accurate individualized optical quality prediction. The solution was to set some parameters of the model as variables and to run an optimization procedure where the individual

optical quality of the subject (expressed as Zernike polynomial terms) acts as the optimization target. After setting constraints and weights to the different parameters to force them to keep physiological values and running the optimization procedure from starting generic parameters, an anatomically valid solution can be obtained. Potential risks of this procedure are the multivariate space of solutions and the dependence on the starting point. It should be noticed, however, that the aim of the procedure is not the calculation of the real parameters of the human lens but instead to obtain a physiologically plausible eye model that reproduces the individual's eye quality.

The choice of the variables to optimize depends on how many parameters of the lens model must be modified. While Navarro et al. modified a substantial number of parameters, including surface curvatures and deformations together with lens alignment and the refractive gradient index parameters to fit a full set of optical quality Zernike aberrations, Tabernero et al. (2007) chose to adjust only for the spherical aberration of the eye, setting the asphericity of the lens surfaces and the gradient refractive index quadratic parameters as variables. The first approach (Navarro et al. 2006) rendered a more accurate eye model, but on the other hand, it was also more artificial in the sense that many unknown variables are introduced. The second choice in Tabernero et al.'s approach included a further step in customization. Measurements of lens alignment, tilt, and decentration with respect to the line of sight were incorporated to the modeling procedure. This step was important to naturally reproduced coma-like aberrations of the eye without further optimization. However, other aberrations, like trefoil, were not considered by this model. Those aberrations intrinsic to the surfaces of the lens were defined as "structural" lens aberrations.

22.3.3 AXIAL DISTANCES

Customized axial distances in the eye can be obtained using a classical variety of methods very common in the clinical practice. They are typically used to estimate the IOL dioptric power required for cataract surgery. Ultrasound biometry is the oldest method (Franken 1961) used to obtain axial distances of the eye (anterior chamber depth, lens thickness, and axial length of the eye). It is based on recording the temporal echoes of ultrasonic pulses coming back from the different ocular surfaces. The time of flight is then converted to axial distances using an average speed of sound in the eye. The need for mechanical contact between the transducer and the eye is the main drawback with respect to new optical and more accurate methods. Those are based on partial coherence interferometry (Fercher et al. 1988; Hitzenberger 1991) achieving high accuracy (about ±30 μ) and more comfort for the subject since they do not need to contact the eye surface. The optical method originally measures optical path differences that need to be converted to geometrical distances using the refractive index from eye models.

Commercial optical biometers are very common and successful today in clinics around the world, like the IOL Master (Carl Zeiss Meditec, Oberkochen, Germany) that provides optical measurements of the axial length or the Lenstar (Haag-Streit, Koeniz, Switzerland) that similarly calculates a full set of axial distances in the eye that can be directly incorporated to the customized eye models.

22.3.4 OCULAR ALIGNMENT

As briefly mentioned in Section 22.3.1, the optical surfaces in a customized eye model should be naturally aligned along the line of sight that connects the fixation stimulus with the center of the entrance pupil. This axis is not typically coaligned with the most accurate (if any) optical axis of the eye. In particular, there is a tendency in the optics of human eyes to point toward a temporal position with respect to the location of the fixation stimulus (Le Grand and El Hage 1980). The origin of the misalignments in the eye is related to the foveal position at the retina. Because the position of maximum retinal resolution does not coincide with the intersection of the optical axis at the retinal plane, the eye forces to stay in a slightly tilted position to display an image from the central object field of view over this retinal point.

To customize the alignment characteristics of the eye, the corneal surface must be recentered from the topographical coordinates (origin on the corneal reflex/videokeratoscope axis) to the pupil center data using the relative pupil position (typically available from the corneal topographer). However, the customization of the position of the lens with respect to the line of sight requires more specific instrumentation. Purkinje-based instruments that record the specular reflections of light coming back from the ocular surfaces can be used to calculate the angle that the lens subtends with respect to the line of sight and to measure the position of the apex of the lens with respect to the center of the entrance pupil (Rosales and Marcos 2006; Tabernero et al. 2006b; Schaeffel 2008). These measurements are based on the location of the overlapping position of the anterior (Purkinje III) and posterior (Purkinje IV) lens surface reflections. The overlapping position with respect to the pupil center is used to calculate lens decentration, and the angular position of the fixation stimulus at the overlapping point is used to estimate lens tilt with respect to the line of sight.

While the effect of the alignment parameters is relatively small to induce changes on the paraxial properties of the eye (typically, lens tilt does not exceed 10°, while lens decentration does not goes beyond 0.3 mm), it was showed that the inclusion of these parameters in the models had a very strong effect on the prediction of off-axis aberrations like coma (Tabernero et al. 2007). The clearest effect was observed when lens tilt is incorporated to the eye models. Without this parameter, coma aberrations could not be predicted in a customized eye, reflecting the very unique nature of the ocular alignment in the human eye, very different to any other man-made optical instrument.

In addition to the estimation of the lens alignment parameters, Purkinje-based instruments have been used to obtain alignment parameters from pseudophakic customized eye models. In this case, reflections with higher intensity and less noisy signals than in normal phakic eyes can be recorded back from the IOL surfaces, since they are manufactured with a constant refractive index that generates a clear mirror reflection image from each optical surface.

22.3.5 VALIDATION

Several approaches have been used to validate the prediction from customized eye models. In general, in order to cross-check the procedure, the optical quality of the subject should be measured (typically using an aberrometer that allows quantification of the Zernike aberration modes and any other optical quality metric) and be compared to the prediction extracted from the eye model. As mentioned in the lens section, not all the optical characteristics of the crystalline lens can be measured and incorporated to an eye model. For that reason, in order to validate the computational procedure, some specific groups of populations can be used. In particular, pseudophakic eyes where lens parameters are exchanged by an IOL with nominal geometries have been studied.

Pseudophakic eyes have an intrinsic and an additional interest since (1) it is possible to test a full customization procedure where the crystalline lens is exchange by a well-known lens and (2) the study of this group of subjects has very interesting clinical implications that go beyond the checking procedure alone. For instance, it can be predicted what kind of IOL would work better in combination with the optics of every particular subject. In a pseudophakic eye model, the anterior corneal surface can be incorporated from topography as in any customized model while the geometrical parameters of the IOL are potentially known. Interocular axial distances and IOL misalignments can be also incorporated, which makes a fully completed optical eye modeling without missing parameters. The agreement between the predicted values and those measured with a wavefront sensor should be very close for this particular group of subjects (Tabernero et al. 2006a; Rosales and Marcos 2007; Rosales et al. 2010).

In those customized models built as a result of an inverse optical problem (Navarro et al. 2006; Goncharov et al. 2008; Sakamoto et al. 2008; Wei and Thibos 2008), validation should be performed in a different way since the physiological parameters are not measured but adjusted to fully reproduce the optical quality of the subject. The question here is how accurate are those ocular parameters obtained as the result of the optimization process. Validation is often performed using simulated computational data as an input for the reconstruction algorithms, for example, generating Hartmann–Shack spot patterns from an eye model with well-known optical parameters and then using them in a similar way as real data. A good agreement between the "reconstructed" geometry of the eye and the "nominal" geometry can be obtained (Sakamoto et al. 2008; Wei and Thibos 2008). Another possibility to cross-check the model and procedure is to obtain wavefront data from a physically built eye where geometries and materials are known to try to calculate them with the fitting procedure (Wei and Thibos 2008).

22.4 EXAMPLES

The final section of this chapter presents examples of typical uses of a customized eye model, providing the reader with real data to test, practice, and get familiar with the procedures introduced in this chapter. In particular, data from a pseudophakic customized model implanted with a specifically designed IOL are given here to implement some of the techniques detailed in the previous sections. The model can be built step-by-step and the optical quality can be analyzed. A pseudophakic eye model has been chosen in view of the potential clinical applications. As an educational purpose, Table 22.1 contains all the geometry of the

Table 22.1 **Geometry of the eye model used in Section 22.4**

CORNEAL SURFACE ZERNIKE COEFFICIENTS (mm−1; 5 mm PUPIL DIAMETER)		OCULAR BIOMETRY (mm)		INTRAOCULAR LENS GEOMETRY AND ALIGNMENT		REFRACTIVE INDEXES	
Z_1^{-1}	−0.019	Anterior chamber depth	3.7	Anterior radius (mm)	16	Corneal and ACD refractive index	1.3375
Z_1^{1}	-2.738×10^{-3}	Pupil to IOL distance	1.0	Posterior radius (mm)	−18.551	IOL refractive index	1.492
Z_2^{-2}	-8.68×10^{-4}	Axial length	23.2	Thickness (mm)	1.2	Vitreous refractive index	1.336
Z_2^{0}	0.124			Tilt X (°)	4.5	Wavelength	589.3 nm
Z_2^{-2}	-1.228×10^{-3}			Tilt Y(°)	1		
Z_3^{-3}	5.18×10^{-4}			Decentration X (mm)	−0.150		
Z_3^{-1}	-4.29×10^{-4}			Decentration Y (mm)	0.11		
Z_3^{1}	-3.77×10^{-4}						
Z_3^{3}	4.18×10^{-4}						
Z_4^{-4}	-6.9×10^{-5}						
Z_4^{-2}	1.4×10^{-5}						
Z_4^{0}	9.73×10^{-4}						
Z_4^{2}	-3.1×10^{-5}						
Z_4^{4}	-8.3×10^{-5}						

model that is required to replicate the examples here presented using any potentially available ray-tracing software. All data in Table 22.1 were collected from the same particular subject that was selected from a published study, except the IOL parameters that have been specifically design for this example. The Zernike surface coefficients on this table were calculated from a least squares fitting procedure that mathematically adjusted a 5 mm pupil diameter anterior corneal elevation map to a seventh-order Zernike polynomial. For simplicity, only coefficients up to the fourth-order term are shown in the table and are used here in the simulations. Also, the cornea is simply model as a one-surface element with a refractive index of 1.3375 for the anterior chamber. Using the optical design ray-tracing software Zemax, the corneal parameters shown in Table 22.1 can be directly uploaded to a "Zernike Standard Sag" surface in the Zemax "lens data editor." The IOL implanted with this model is a custom-designed spherical lens made of PMMA (refractive index is 1.492 at 589.3 nm) with a dioptric power of 18 D inside the capsular bag.

Figure 22.2 shows the optical layouts of the cornea model alone and the full customized model together with the corresponding wavefront aberration maps. In this pseudophakic eye, the cornea is the optical element that dominates the wavefront aberration function, while implanting a spherical IOL does not improve the optics of the eye and in fact it slightly deteriorates the optical quality. The first two columns in Table 22.2 displays the wavefront aberration terms expressed as a Zernike polynomial expansion (entrance pupil diameter was set to 5 mm) for the

cornea and the total eye (with the spherical IOL). It shows how spherically designed IOLs induce positive spherical aberration that overall increments the typically positive corneal spherical aberration. Small changes in astigmatism and coma can be also observed as a result of IOL tilt and decentration although, in this subject, IOL misalignments were small. It can be noted that the surface geometry of the IOL did not induce changes in other corneal higher-order aberrations (trefoil and higher-order astigmatisms). Additionally, the astigmatism in diopters of this eye model can be calculated from the Zernike coefficients in Table 22.2 using the conversion equation (e.g., Guirao and Williams 2003):

$$C = \frac{-4\sqrt{6}}{r^2} \sqrt{\left(z_2^{-2}\right)^2 + \left(z_2^{2}\right)^2} \qquad (22.9)$$

where r is the entrance pupil radius (2.5 mm). For this particular subject, the residual astigmatism is 0.77 D.

Once the model is fully established and analyzed, it is possible to elaborate some predictions from potential "virtual" modifications in the optics of this subject. Some of these possibilities are the following:

22.4.1 THE EFFECTS OF CORRECTING THE RESIDUAL ASTIGMATISM

Correction of residual astigmatism can be surgically implemented in different ways. Toric lenses are typically used to compensate for precataract refractive errors, and more recently, it has been

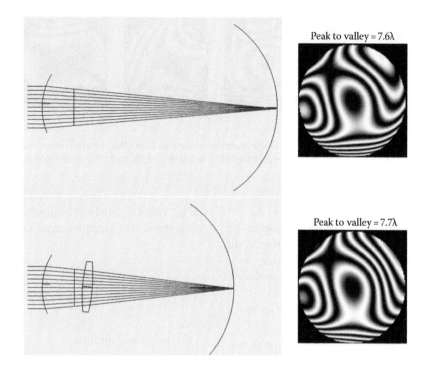

Figure 22.2 Ray tracing depicted through the cornea and the pseudophakic eye model that corresponds to the example of Section 22.4 (Table 22.1). Corresponding wavefront aberration maps are plotted to the right.

Table 22.2 Zernike polynomial coefficients (5 mm pupil diameter) for the cornea and the total eye model example of Section 22.4

	CORNEAL ABERRATION COEFFICIENTS (µM)	TOTAL EYE (SPHERICAL IOL) ABERRATION COEFFICIENTS (µM)	TOTAL EYE (ASPHERIC IOL) ABERRATION COEFFICIENTS (µM)
Z_2^{-2}	−0.29	−0.33	−0.31
Z_2^{-2}	−0.41	−0.37	−0.38
Z_3^{-3}	0.17	0.17	0.17
Z_3^{-1}	−0.14	−0.17	−0.10
Z_3^{1}	−0.09	0.00	−0.09
Z_3^{3}	0.14	0.14	0.14
Z_4^{-4}	−0.02	−0.02	−0.02
Z_4^{-2}	0.00	0.00	0.00
Z_4^{0}	0.18	0.23	0.00
Z_4^{2}	0.00	0.00	−0.01
Z_4^{4}	−0.03	−0.03	−0.03

shown that light adjustable lenses can be used to compensate for surgically induced astigmatic errors. Ultraviolet light is used to radiate the IOL surface (after cataract surgery) that responds to the treatment with changes in curvature that compensate the residual refractive errors (Villegas et al. 2014). Additionally, it is also possible to perform an astigmatic LASIK treatment on the corneal surface to eliminate cylindrical errors. The benefits of the astigmatic correction in this particular subject can be further explored within this model. Corneal surface Zernike coefficients Z_2^{-2}, Z_2^{0}, and Z_2^{2} can be modified to correct astigmatism and keep the light well in focus at the retina. In particular for this subject, if those Zernike terms of the corneal surface are modified correspondingly to the values of 1.203×10^{-4} mm^{-1}, 0.125 mm^{-1}, and -1.74×10^{-4} mm^{-1}, the eye model would be free of residual spherocylindrical errors. These coefficients are calculated from a least squares optimization procedure to correct astigmatism

Optical properties of the eye

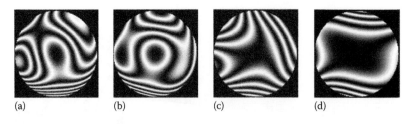

Figure 22.3 Wavefront aberration maps corresponding to the eye model of Section 22.4. The full aberration map is presented (a) together with the situation where astigmatism is corrected (b), spherical aberration is corrected (c), and both astigmatism and spherical aberration are simultaneously corrected (d).

and defocus. Figure 22.3 shows the wavefront maps from the original model (a) and the model after correcting astigmatism (b). Figure 22.3 shows the effects of correcting astigmatism in the point spread function (PSF) with the corresponding increase in the Strehl ratio (the peak of the PSF). Additionally, a convolution of the PSF with an extended object (an image that subtends 1° vertically) shows some of the potential visual benefits of correcting this residual astigmatism.

22.4.2 THE EFFECTS OF CORRECTING SPHERICAL ABERRATION

Average corneal spherical aberration can be corrected with aspheric IOLs implanted after cataract extraction (Guirao et al. 2002; Holladay et al. 2002; Packer et al. 2004). Using customized eye models, it is also possible to explore the potential benefits of a full individualized correction (different to an average correction) of spherical aberration (Piers et al. 2004; Tabernero et al. 2006a; Piers et al. 2007). For this particular case, the anterior surface sag from the original IOL model has been redesigned to include an aspheric term and a second-order radial polynomial term. The conic (aspheric) term is used to fully correct spherical aberration

and the radial squared term helped to keep the light well focused on the retina. The anterior surface sag (z) is described as

$$z = \frac{cr}{1 + \sqrt{1 - (1+k)c^2 r^2}} + \alpha r^2 \tag{22.10}$$

where
  r is the radial coordinate
  c is the curvature of the surface (the inverse of the radius of curvature R; $c = 1/R$)
  k is the conic constant (k)
  α is the coefficient of the squared term

After running an optimization procedure to fully correct spherical aberration in this subject and keep the image well focused on the retina, the variables k and α must be −48.134 and 5.666×10^{-3} mm^{-1}. The radius of curvature, thickness, and refractive index remained exactly the same as in the previous model.

Figures 22.3 and 22.4 show the wavefront map, PSF, Strehl ratio, and the extended retinal image obtained when this IOL

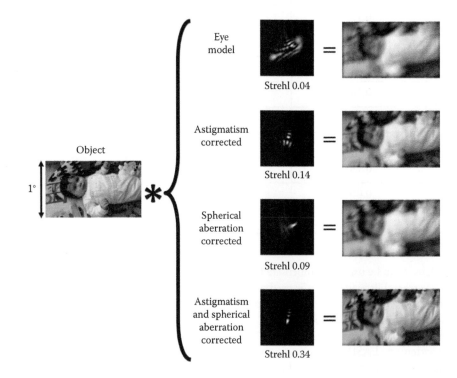

Figure 22.4 Monochromatic point spread functions and extended retinal images obtained from the eye model example in Section 22.4.

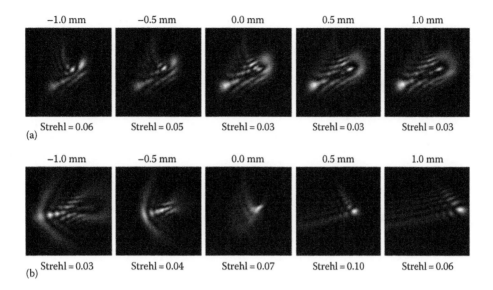

Figure 22.5 Monochromatic point spread functions and Strehl ratios as a function of intraocular lens (IOL) decentration for the pseudophakic eye model of Section 22.4. Two different IOL models were studied: a spherical IOL (a) and a aspheric IOL (b) that fully correct corneal spherical aberration.

is "virtually" implanted in the eye model of this example. Also shown in these figures is the case where the residual 0.77 D of cylinder is corrected together with spherical aberration. This is the more favorable case as the Strehl ratio is higher, the energy is more concentrated in a small retinal area, and the extended image has consequently better contrast and visibility of details.

A potential drawback of this aspheric design might be an increase in the sensitivity to misalignments. Similar to commercially available aspheric IOLs, incorporating a conic constant to the surface sag could reduce the tolerance to displacements of the IOL. While in this particular case IOL decentration was moderately small, it is interesting to explore what situation would be expected if larger values of misalignments were presented. Figure 22.5 shows the PSFs when the aspheric and the spherical IOLs are decentered from –1 to 1 mm horizontally (from temporal to nasal side) together with the Strehl ratios. As expected, the PSFs from the aspheric IOL with the conic term in the surface sag changed this shape as a function of IOL decentration more dramatically than in the case of the spherical IOL. A clearly visible comet-shaped PSF due to the induced coma aberration can be clearly observed at the extremes of the decentration intervals.

22.4.3 EXTENDING DEPTH OF FOCUS

The final example of this section shows a very simple method to increase the tolerance to the defocus generated by the lack of accommodation in this particular subject. Different surgical procedures, in diffractive IOLs, accommodating IOLs, or specific LASIK profiles to generate multifocal patters (presbyLASIK), are currently presented as clinical solutions. All of them can be individually simulated for each particular subject (as long as the nominal IOL surfaces or corneal ablation parameters are known). Another possibility is to calculate a customized solution for every particular subject in terms of a specific multifocal pattern that interacts with the subject's own optical aberrations. In general, these kinds of solutions are

effective but at the expense of less optical quality in the primary (far) focus that is transferred to secondary (near) and eventually intermediary focus. Alternatively, a very simple solution that can be directly tested with this model is to reduce the pupil size of the subject to approximately 1.6 mm of diameter, increasing the tolerance to depth of focus compared to a natural mesopic pupil. This is the approach used by the commercialized intracorneal inlay (ACI-7000; AcuFocus Inc., Irvine, CA). The surgical implantation is monocular to avoid problems with the reduction of light intensity at the retina as much as possible (Seyeddain et al. 2010; Yilmaz et al. 2011). It has been shown that the small aperture effect can potentially increase binocular depth of focus (Tabernero et al. 2011b), but testing the optical quality of this approach using customized optical models is simple and very illustrative. Selecting an entrance pupil diameter of 1.6 mm, it is possible to compare any optical quality metric under this small aperture to the case where a mesopic pupil diameter has been used. In Figure 22.6 the size of the spots diagrams at the retina with the small aperture (1.6 mm diameter) pupil and with a mesopic pupil (a 4 mm pupil diameter has been considered) is plotted for far vision (infinity), at 1 m (1 diopter), and at 0.5 m (2 diopters). In Figure 22.7, a picture that contains some printed text was used as an extended object (subtending 5° of height) and convolved with the PSF to obtain a retinal image for different distances and the two aperture conditions. From both figures, it is clear that the small aperture (1.6 mm) can effectively increase the range of defocus distances that the subject might tolerate. However, some drawbacks exist like the chromatic sensitivity to the location where the aperture is implanted. This issue, together with the optimal residual refractive errors of the procedure, has been also analyzed with the use of personalized optical models (Tabernero and Artal 2012). Typically, the small aperture is surgically centered on the coaxially viewed corneal reflex in order to minimize transversal chromatic aberration, but in some cases there might be individual differences with the best centration location (Manzanera et al. 2015).

Optical properties of the eye

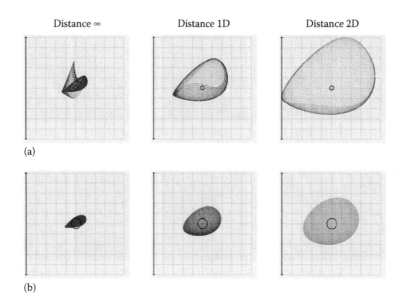

Distance ∞   Distance 1D   Distance 2D

(a)

(b)

Figure 22.6 Monochromatic spots diagram at the retinal surface of the pseudophakic eye model of Section 22.4 as function of the distance from the object (point source) to the eye. Two different sizes of the entrance pupil diameter are considered: 4 mm (a) and 1.6 mm (b).

Distance ∞   Distance 1D   Distance 2D

4. DISCUSSION

The experimental data and me paper provided several importan stand the mechanisms of compe tions between cornea and intern

(a)

4. DISCUSSION

The experimental data and me paper provided several importan stand the mechanisms of compe tions between cornea and intern

(b)

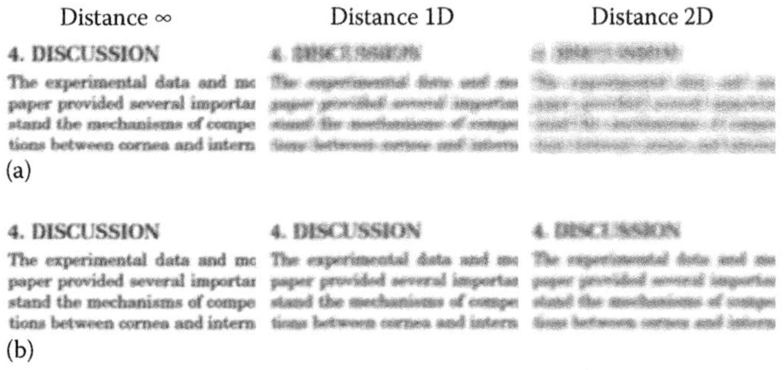

Figure 22.7 Extended retinal images (the object subtends 5° of height) calculated for the eye model example of Section 22.4 as function of the distance from the object to the eye. Two different sizes of the entrance pupil diameter are considered: 4 mm (a) and 1.6 mm (b).

REFERENCES

Artal, P., Benito, A., and Tabernero, J., The human eye is an example of robust optical design, *Journal of Vision* 10 (2006): 1–7.

Artal, P. and Tabernero, J., The eye's aplanatic answer, *Nature Photonics* 2 (2008): 586–589.

Barbero, S., Marcos, S., Merayo-Lloves, J., and Moreno-Barriuso, E., Validation of the estimation of corneal aberrations from videokeratography in keratoconus, *Journal of Refractive Surgery* 18 (2002): 263–270.

Birkenfeld, J., de Castro, A., and Marcos, S., Contribution of shape and gradient refractive index to the spherical aberration of isolated human lenses, *Investigative Ophthalmology and Visual Science* 55 (2014): 2599–2607.

Cano, D., Barbero, S., and Marcos, S., Comparison of real and computer-simulated outcomes of LASIK refractive surgery, *Journal of the Optical Society of America A* 21 (2004): 926–936.

Canovas, C. and Artal, P., Customized eye models for determining optimized intraocular lenses power, *Biomedical Optics Express* 2 (2011): 1649–1662.

Dubbelman, M., Sicam, V. A., and Van der Heijde, G. L., The shape of the anterior and posterior surface of the aging human cornea, *Vision Research* 46 (2006): 993–1001.

Dubbelman, M., Sicam, V. A., and van der Heijde, R. G., The contribution of the posterior surface to the coma aberration of the human cornea, *Journal of Vision* 30 (2007): 1–8.

Dubbelman, M. and Van der Heijde, G. L., The shape of the aging human lens: Curvature, equivalent refractive index and the lens paradox, *Vision Research* 41 (2001): 1867–1877.

Dubbelman, M., van der Heijde, G. L., and Weeber, H. A., The thickness of the aging human lens obtained from corrected Scheimpflug images, *Optometry and Vision Science* 78 (2001): 411–416.

Dubbelman, M., Weeber, H. A., Van der Heijde, R. G., and Völker-Dieben, H. J., Radius and asphericity of the posterior corneal surface determined by corrected Scheimpflug photography, *Acta Ophthalmologica Scandinavica* 80 (2002): 379–383.

Fercher, A. F., Mengedoht, K., and Werner, W., Eye-length measurement by interferometry with partially coherent light, *Optics Letters* 13 (1988): 186–188.

Franken, S., Length measurement of the eye with the aid of an ultrasonic echo, *Nederlands tijdschrift voor geneeskunde* 105 (1961): 1803–1804.

Goncharov, A. V., Nowakowski, M., Sheehan, M. T., and Dainty, C., Reconstruction of the optical system of the human eye with reverse ray-tracing, *Optics Express* 16 (2008): 1692–1703.

Guirao, A. and Artal, P., Corneal wave aberration from videokeratography: Accuracy and limitations of the procedure, *Journal of the Optical Society of America A* 17 (2000): 955–965.

Guirao, A., Redondo, M., Geraghty, E., Piers, P., Norrby, S., and Artal, P., Corneal optical aberrations and retinal image quality in patients in whom monofocal intraocular lenses were implanted, *Archives of Ophthalmology* 120 (2002): 1143–1151.

Guirao, A. and Williams, D., A method to predict refractive errors from wave aberration data, *Optometry and Vision Science* 80 (2003): 36–42.

Hemenger, R. P., Tomlinson, A., and Oliver, K., Corneal optics from videokeratographs, *Ophthalmic and Physiological Optics* 15 (1995): 63–68.

Hitzenberger, C. K., Optical measurement of the axial eye length by laser Doppler interferometry, *Investigative Ophthalmology and Visual Science* 32 (1991): 616–624.

Holladay, J. T., Piers, P. A., Koranyi, G., van der Mooren, M., and Norrby, N. E., A new intraocular lens design to reduce spherical aberration of pseudophakic eyes, *Journal of Refractive Surgery* 18 (2002): 683–691.

Jaeken, B., Mirabet, S., Marín, J. M., and Artal, P., Comparison of the optical image quality in the periphery of phakic and pseudophakic eyes, *Investigative Ophthalmology and Visual Science* 54 (2013): 3594–3599.

Kasthurirangan, S., Markwell, E. L., Atchison, D. A., and Pope, J. M., In vivo study of changes in refractive index distribution in the human crystalline lens with age and accommodation, *Investigative Ophthalmology and Visual Science* 49 (2008): 2531–2540.

Klijn, S., Reus, N. J., and Sicam, V.A., Evaluation of keratometry with a novel Color-LED corneal topographer, *Journal of Refractive Surgery* 31 (2015): 249–256.

Le Grand, Y. and El Hage, S. G., *Physiological Optics* (Berlin, Germany: Springer-Verlag, 1980).

Manzanera, S., Prieto, P. M., Benito, A., Tabernero, J., and Artal, P., Location of achromatizing pupil position and first Purkinje reflection in a normal population, *Investigative Ophthalmology and Visual Science* 56 (2015): 962–966.

Navarro, R., Adaptive model of the aging emmetropic eye and its changes with accommodation, *Journal of Vision* 14 (2014), 21: 1–17.

Navarro, R., González, L., and Hernández-Matamoros, J. L., On the prediction of optical aberrations by personalized eye models, *Optometry and Vision Science* 83 (2006): 371–381.

Ortiz, S., Pérez-Merino, P., Gambra, E., de Castro, A., and Marcos, S., In vivo human crystalline lens topography, *Biomedical Optics Express* 3 (2012): 2471–2488.

Packer, M., Fine, I. H., Hoffman, R. S., and Piers, P. A., Improved functional vision with a modified prolate intraocular lens, *Journal of Cataract and Refractive Surgery* 30 (2004): 986–992.

Piers, P. A., Norrby, N. E., and Mester, U., Eye models for the prediction of contrast vision in patients with new intraocular lens designs, *Optics Letters* 29 (2004): 733–735.

Piers, P. A., Weeber, H. A., Artal, P., and Norrby, S., Theoretical comparison of aberration-correcting customized and aspheric intraocular lenses, *Journal of Refractive Surgery* 23 (2007): 374–384.

Pierscionek, B. K. and Chan, D. Y., Refractive index gradient of human lenses, *Optometry and Vision Science* 66 (1989): 822–829.

Rosales, P., De Castro, A., Jiménez-Alfaro, I., and Marcos, S., Intraocular lens alignment from Purkinje and Scheimpflug imaging, *Clinical and Experimental Optometry* 93 (2010): 400–408.

Rosales, P. and Marcos, S., Customized computer models of eyes with intraocular lenses, *Optics Express* 15 (2007): 2204–2218.

Rosales, P. and Marcos, S., Phakometry and lens tilt and decentration using a custom-developed Purkinje imaging apparatus: Validation and measurements, *Journal of the Optical Society of America A* 23 (2006): 509–520.

Sakamoto, J. A., Barrett, H. H., and Goncharov, A. V., Inverse optical design of the human eye using likelihood methods and wavefront sensing, *Optics Express* 16 (2008): 304–314.

Salmon, T. O. and Thibos, L. N., Videokeratoscope-line-of-sight misalignment and its effect on measurements of corneal and internal ocular aberrations, *Journal of the Optical Society of America A* 19 (2002): 657–669.

Schaeffel, F., Binocular lens tilt and decentration measurements in healthy subjects with phakic eyes, *Investigative Ophthalmology and Visual Science* 49 (2008): 2216–2222.

Schwiegerling, J. and Greivenkamp, J. E., Using corneal height maps and polynomial decomposition to determine corneal aberrations, *Optometry and Vision Science* 74 (1997): 906–916.

Seyeddain, O., Riha, W., Hohensinn, M., Nix, G., Dexl, A. K., and Grabner, G., Refractive surgical correction of presbyopia with the AcuFocus small aperture corneal inlay: Two-year follow-up, *Journal of Refractive Surgery* 26 (2010): 707–715.

Sicam, V. A., Coppens, J., van den Berg, T. J., and van der Heijde, R. G., Corneal surface reconstruction algorithm that uses Zernike polynomial representation, *Journal of the Optical Society of America A* 21 (2004): 1300–1306.

Sicam, V. A., Dubbelman, M., and van der Heijde, R. G., Spherical aberration of the anterior and posterior surfaces of the human cornea, *Journal of the Optical Society of America A* 23 (2006): 544–549.

Sun, M., Birkenfeld, J., de Castro, A., Ortiz, S., and Marcos, S., OCT 3-D surface topography of isolated human crystalline lenses, *Biomedical Optics Express* 5 (2014): 3547–3561.

Tabernero, J. and Artal, P., Optical modeling of a corneal inlay in real eyes to increase depth of focus: Optimum centration and residual defocus, *Journal of Cataract and Refractive Surgery* 38 (2012): 270–277.

Tabernero, J., Benito, A., Alcón, E., and Artal, P., Mechanism of compensation of aberrations in the human eye, *Journal of the Optical Society of America A* 24 (2007): 3274–3283.

Tabernero, J., Benito, A., Nourrit, V., and Artal, P., Instrument for measuring the misalignments of ocular surfaces, *Optics Express* 14 (2006b): 10945–10956.

Tabernero, J., Berrio, E., and Artal, P., Modeling the mechanism of compensation of aberrations in the human eye for accommodation and aging, *Journal of the Optical Society of America A* 28 (2011a): 1889–1895.

Tabernero, J., Ohlendorf, A., Fischer, M. D., Bruckmann, A. R., Schiefer, U., and Schaeffel, F., Peripheral refraction in pseudophakic eyes measured by infrared scanning photoretinoscopy, *Journal of Cataract and Refractive Surgery* 38 (2012): 807–815.

Tabernero, J., Piers, P., Benito, A., Redondo, M., and Artal, P., Predicting the optical performance of eyes implanted with IOLs to correct spherical aberration, *Investigative Ophthalmology and Visual Science* 47 (2006a): 4651–4658.

Tabernero, J., Schwarz, C., Fernández, E. J., and Artal, P., Binocular visual simulation of a corneal inlay to increase depth of focus, *Investigative Ophthalmology and Visual Science* 52 (2011b): 5273–5277.

Turner, M. G., Pushing the envelope in optical design software, *Optical Engineering* 39 (2000): 1735–1736.

Villegas, E. A., Alcon, E., Rubio, E., Marín, J. M., and Artal, P., Refractive accuracy with light-adjustable intraocular lenses, *Journal of Cataract and Refractive Surgery* 40 (2014): 1075–1084.

Wei, X. and Thibos, L., Modeling the eye's optical system by ocular wavefront tomography, *Optics Express* 16 (2008): 20490–20502.

Welford, W. T., *Aberrations of Optical Systems* (Bristol, England: Adam Hilger, 1986).

Yılmaz, O. F., Alagöz, N., Pekel, G., Azman, E., Aksoy, E. F., Cakır, H., Bozkurt, E., and Demirok, A., Intracorneal inlay to correct presbyopia: Long-term results, *Journal of Cataract and Refractive Surgery* 37 (2011): 1275–1281.

23 Scattering, straylight, and glare

Thomas J.T.P. van den Berg

Contents

23.1 INTRODUCTION

The effect of light scattering in the optical eye media, with its visual counterpart called "straylight," and the resulting glare problem have since long drawn the attention of scientists. This can be understood as everybody experiences straylight and glare, also people with the best eyes. Goethe discusses the "subjective Höfe" seen around light sources in his "Zur Farbenlehre" (von Goethe 1810), assuming them to be of neuronal origin. Purkinje on the other hand already guesses what we now know to be right that light scattering by small particles is the cause (Purkinje 1823). Helmholtz discusses the matter thoroughly, assuming again small particles to be important and stating that scientific experiments must be conducted to decide whether neuronal factors play a role also (von Helmholtz 1852). An important step forward is made by Cobb introducing the concept of an equivalent light, that is, that the light veil, as seen around a light source, can in all its effects be replaced by a real background light (Cobb 1911). We now call the luminance of the replacement light the

"equivalent luminance" of the straylight. Holladay follows Cobb and establishes how the equivalent luminance (Leq) declines with distance to the light source: Leq = Ebl × k/θ², with Ebl the illuminance caused by the (bl = blinding) light source on the pupil of the subject (Holladay 1926, 1927). The proportionality between Leq and Ebl corresponds of course with the physical nature of light scattering as believed to be the case by Holladay. Stiles and Crawford perform a series of thorough studies, confirming in essence Holladay's results (Stiles and Crawford 1935, 1937). The previous formula for Leq has been called the "Stiles–Holladay approximation," often with constant k = 10.

Although, nowadays, the physical (light scattering) explanation of straylight and glare is almost self-evident, around WWII, a great many studies were devoted to the question whether maybe (partly) lateral neuronal interactions played a role too. The thesis work of Vos (1963) can be considered an endpoint in this discussion, in favor of the notion that light scattering is the sole determinant of straylight and glare. Later, Vos played an important role (e.g., disability glare review (Vos 1984)) in the international

Figure 23.1 Cartoon of glare disturbance. (Courtesy of Dr. Houdijn Beekhuis.)

normalization committee Commission International d'Éclairage (CIE), well known among others for the definition of the spectral sensitivity function of the human eye Vλ, and the chromaticity diagram. This committee established that straylight must be used as norm for quantification of disability glare, defined using the equivalent luminance principle. Straylight is part of the functional point spread function psf of the human eye. It constitutes the periphery from say 1° to 100°, whereas the center up till say 10′ is dominated by errors of refraction and by diffraction. It is efficient to express straylight with the "straylight parameter" defined as

$$\text{Straylight parameter } s = \theta^2 psf = \theta^2 L_{eq}/E_{bl},$$

among others because s is relatively insensitive to differences in measurement angle because of the approximate Stiles–Holladay rule (van den Berg 1995). In normal, young eyes log(s) = 0.9, but increases with age and cataract formation. An increase of a factor of 3 (log(s) = 1.4) is considered a serious visual handicap and reason for cataract surgery. As limit factor for safe driving an increase of a factor of 4 (log(s) = 1.5) and for demanding professions such as pilots, a factor of 2 (log(s) = 1.2) is proposed (Figure 23.1).

23.1.1 DISABILITY GLARE AND DISCOMFORT GLARE

While disability glare can be measured directly and is well defined in the previous explanation, discomfort glare is more difficult to define and measure. Discomfort glare is typically identified when observers complain of annoyance by a bright light source in their field of view. Discomfort glare is not necessarily accompanied by loss of visual capacity. A discomfort glare rating system, called the de Boer rating scale, was developed in the 1960s and consists of a scale from 1 to 9 that can be used to rate the discomfort glare from a light source. The de Boer rating scale runs from 1 = unbearable, via 3 = disturbing, 5 = just acceptable, and 7 = satisfactory, to 9 = unnoticeable. Several authors have proposed formulae to calculate a de Boer rate from specifications of the disturbing lights, and the CIE has adopted standards for specific applications. An example is de Boer grade $W = 5 - 2\log\Sigma E_i / 0.020\theta_i^{0.46}\left(1 + \sqrt{\left(L_{b_i}/0.04\right)}\right)$ where E_i is the illuminance (in lux) at the eye from the ith glare source, θ_i is its eccentricity (in degrees), and L_{b_i} is its background luminance (in cd/m²) (Schmidt-Clausen and Bindels 1974). Discomfort glare is not considered further in this chapter.

23.2 ASSESSMENT

23.2.1 EQUIVALENT LUMINANCE APPROACHES

Several techniques have been developed to assess equivalent luminance values. In the older literature, most often an adaptation paradigm was used. Some visual threshold, like visual acuity (e.g., Cobb 1911), brightness increment threshold (e.g., Holladay 1927; Stiles and Crawford 1937; Vos 1963; Wooten and Geri 1987), or contrast sensitivity (Paulsson and Sjöstrand 1980 simplified by assuming constancy of retinal contrast sensitivity) is determined in two conditions, in the presence of the glare source (at a distance θ) and in the presence of a veiling background. This is done for a series of strengths of the glare source and/or the background. The conditions for which the two thresholds are equal give values for the equivalent luminance at the distance θ from the glare source. Le Grand published a critical test of this approach (le Grand 1937). Le Grand also introduced an approach to more directly equate the glare veil to a real veil. He alternated in quick succession the glare source and a real veil. Like with visual photometry, the point of identity between the two stimuli is arrived when the perception of flicker is at a minimum. The strong inhomogeneities in the field made the task difficult, although the presented veil was given an intensity distribution somewhat adapted to the glare veil.

This use of counterphase flicker to assess the Leq value can be much improved by using an annulus instead of a (point) glare source, with a small counterphase flickering test field in the center. This way, the test is much more like the well-known technique of visual photometry called "heterochromatic flicker photometry," which is a very precise measurement technique. The approach to measure straylight with counterphase flicker was called the "direct compensation" approach (van den Berg 1986). A further development was the "compensation comparison" approach. The central test field is now divided in two-halves, one with and one without counterphase flicker. This enables the use of the more rigid two-alternative forced choice or 2AFC psychophysical technique, including estimates of accuracy of the outcome. Moreover, this made it possible that the outcome cannot be influenced by the individual being tested as both half fields appear interchangeably. On this approach, a commercial instrument is based (C-Quant straylight meter, from Oculus) and a survey has appeared of the basic and clinical studies (van den Berg et al. 2013; Figure 23.2).

23.2.2 GLARE TESTING

In general, glare testing involves assessment of a visual function aspect under the influence of glare. Often, the visual function outcome in the presence of glare is considered sufficient as glare test. However, as many patients suffer from decreased visual function already without glare, this is no selective test for glare. Sometimes, the glare test *per se* is considered as the difference between the visual function results with and without the glare source present. This would indeed be selective for glare. The glare source can be a small (point) source, often intended to mimic the situation with driving at night. The disadvantage is that misfixations give afterimages interfering with the test outcome. Sometimes the glare source is extended, most often of

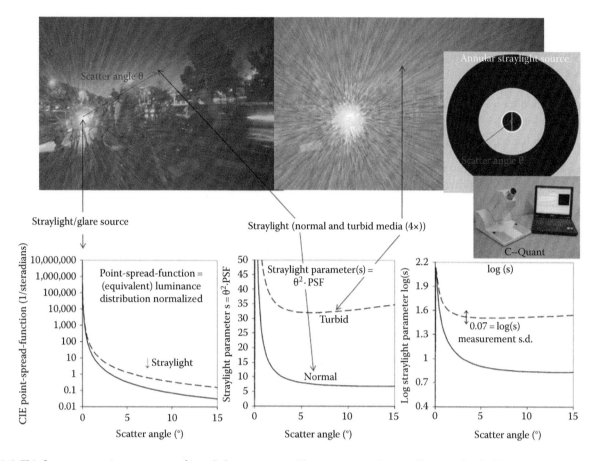

Figure 23.2 This figure summarizes concepts of straylight assessment. The street scene images illustrate that (left) a young normal eye does experience some radiation emanating from a bright light source, but this can be much worse to the level that vision is impossible (right) if the eye is older or has turbidity. This light as seen by the eye is called "straylight," and the intensity of this light as seen by the eye is measured with its equivalent luminance value Leq. The upper-right image shows the field of view in the C-Quant instrument. The annulus is the straylight source, which is presented flickering. The two half fields in the center are the test fields, one of which flickers in counterphase. For details, see Franssen et al. (2006). The lower-left graph shows the straylight part of the point spread function for the normal and turbid eye; the lower-middle graph shows the same expressed linearly with the straylight parameter s and the lower-right graph shows s logarithmically. (Reproduced from van den Berg, T.J.T.P. et al., *Z. Med. Phys.*, 23, 6, 2013. With permission.)

annular shape. A great number of glare tests, commercial and noncommercial, have been designed, differing not only in design (with/without glare), and in glare source (shape, size, location, brightness), but also in visual function tested (visual acuity, contrast sensitivity, increment threshold). The most simple glare test employs a simple pen torch, directed to the eye of the patient while the patient reads a vision chart. One step more advanced is the brightness acuity tester, with an illuminated semisphere, put over the eye, with a hole in the center to observe a vision chart. Then there are several instruments that combine normal visual function testing with glare testing. The reader is referred to a critical review of these instruments given by Aslam et al. (2007).

23.2.3 OPTICAL MEASUREMENTS

Early attempts on optical measurement were undertaken by Boynton and coworkers (Boynton et al. 1954; De Mott and Boynton 1958). They excised the posterior of cadaver eyes and placed a photocell to collect scattered light. The amounts recorded were very high, leading the authors to conclude that there is more than sufficient scattered light to explain the in vivo straylight/ glare data. Later researchers assumed postmortem changes might have occurred. Since later measurements on isolated human eye

lenses (see Section 23.3) showed proper correspondence with the in vivo results of straylight, probably the cornea was to blame. Westheimer and Liang (1994) proposed to use the peripheral part of a double-pass recording as measure for light scatter, but this is hampered by optical artifacts, especially if infrared light is used (Pinero et al. 2010; van den Berg 2010). Artal and Ginis carefully adapted the double-pass approach to avoid the problems associated with forward scatter assessment (Ginis et al. 2012). Problems include interference from backscatter in the eye media, recording limitations because of the dynamic range involved, and interference from light coming back from the depth of the fundus (Williams et al. 1994). Another approach uses the spot patterns from a Shack–Hartmann wavefront aberrometer (Donnelly et al. 2004). In this case, one records a great many double-pass images all over the pupillary plane. In case of disturbances in the optical media, some of these spots are more widened as compared to other spots, and this could be analyzed to give a measure of local light scatter. However, as these recordings include only very small angles, it is questionable whether (volume) light scatter as discussed in this chapter is involved. Possibly mainly aberrations are involved. It must be realized that aberrations could extend to scales much smaller than can be resolved with customary aberrometers. Indeed Thibos and coworkers coined the name "microaberrations" for aberrations of scale smaller than the lenslet of a Shack–Hartmann lens set (Nam et al. 2011). As with customary aberrations of large extent ("macroaberrations") do, microaberrations deflect light over small angles. In this chapter, the word scatter is used for the light spreading over relatively large angles, order 1° or more.

23.2.4 FORWARD VERSUS BACKWARD SCATTERING

The scattering discussed in the present chapter is forward scattering. Part of the light entering the eye deviates from the straight path, but this deviation must be smaller than 90° to bother vision, otherwise it would not reach the retina. Light scattered over more than 90°, does not reach the retina, but leaves the eye again. This is called backward scatter or backscatter. The backward scattered light is typically observed with the biomicroscope with slit lamp or with a Scheimpflug camera. It could be thought that these techniques can be used to assess straylight/glare because both are based on scattering. That would be the case if the same light scattering processes dominate in forward and backward directions. For the human eye lens, this was however found not to be the case (van den Berg and Spekreijse 1999). Indeed studies comparing backscattered light from the eye lens to straylight found rather low correlations (de Waard et al. 1992; Yager et al. 1993; Patel et al. 2009; Rozema et al. 2011).

23.2.5 LIGHT SCATTERING ⇒ STRAYLIGHT ⇒ GLARE AND QUALITY OF VISION

Although scattering, straylight, and disability glare are intimately related, it is good to discuss the essential differences. Scattering is a physical optical process of interaction between light and matter. When light hits matter, it is scattered because the molecules are excited by the electromagnetic waves of the light. Each molecule acts as a point source emitting light in all directions. Depending on interference with the light waves emitted by molecules in the surrounding, the composite scattering is more or less directional. In case the relevant molecules are clustered in very small particles, the scattered rays do not interfere destructively, but only constructively, leading to the Rayleigh type of scattering as well known from the blue of the sky. In the human eye, backward scattering from the eye lens is for an important part of this type, with the protein molecules being the particles of the lens. But the more relevant type of light scattering originates from larger particles, dominating in the forward scattering.

The second stage in the process is that the scattered light impinges on the retina. It excites the photoreceptors and degrades vision. The visual effect in case of a point source of light is that we see a radiation of light around the point source. This is called "straylight." To be precise, the word straylight is used for the visual effect, as quantified with the equivalent luminance; see Sections 23.1 and 23.2.1. However, one might presume a close one-to-one correspondence between the light distribution impinging on the retina and the equivalent luminance distribution. Hence, often the adjective "retinal" is added, but this may not be precisely correct as we must consider important differences between true retinal light distribution and the visual effect. The (scattered/nonfocused) retinal light distribution is composed of (at least) three rather distinct parts: (1) light scattered forward by the eye media such as the lens and cornea, (2) light scattered by the eye wall toward

the retina, and (3) light penetrating the tissue behind the retina and scattered backward toward the retina. The photoreceptors are maximally sensitive to component (1), but much less (a factor 10 or more) sensitive to component (2) because of the Stiles–Crawford effect. See the discussion section in van den Berg et al. (1991). With respect to component (3), the situation is further complicated because its direction is reversed. If we consider the relative intensities as seen ophthalmoscopically (double pass), the photoreceptors are even (1000×) less sensitive to component (3) because apart from the Stiles–Crawford effect, this component is ophthalmoscopically (double pass) observed with about 100× more sensitivity than the components impinging in forward direction on the retina. This is caused by the fact that the retina has very low reflectance causing ophthalmoscopy and double pass to "see" the visually relevant light very ineffectively as opposed to the light coming back from the deeper layers.

The step toward glare and more in general quality of vision depends of course on the visually effective stimulus, that is, on straylight, as defined with the equivalent luminance. This holds true for both discomfort glare and disability glare, the two well-established concepts according to CIE definitions. However, in literature, often the expressions glare, and glare disability, and even disability glare are used more loosely to indicate some chosen disturbance from a glare source. The disturbance is, for example, the effect of glare on visual acuity or light sensitivity; see Section 23.2.2. These effects originate from the equivalent luminance of the respective glare sources used, the equivalent luminance defining the desensitization of the eye through light adaptation. However, in most studies, the connection between glare effect and equivalent luminance is not elaborated. There is one notable exception. Sjöstrand and coworkers used the effect of a glare source on the contrast threshold (Paulsson and Sjöstrand 1980). If it can be assumed that the retinal contrast threshold M does not change with retinal illuminance (Weber's law), equivalent luminance can be calculated as (M(w glare) – M(wo glare))/M(wo glare) × luminance of the contrast test. It has been reported that the outcome of this calculation can give a likeness to true equivalent luminance (Whitaker et al. 1994). However, one has to be careful with such elaboration of glare effects because, in practice, also significant errors were found (de Waard et al. 1992; Yager et al. 1992), as Weber's law does not always apply.

Straylight does not only affect the eye's integrity because of glare, but also in a more general sense. Straylight is part of the eye's functional point spread function psf; it is its outer skirt. In formula, $psf(\theta) = Leq(\theta)/Ebl = s(\theta)/\theta^2$ (for definitions, see the section 23.1). The central part of the psf is dominated by refraction type errors and diffraction. The retinal image is the result of convolution of the scene visualized, with the psf. So, different aspects of a scene can be degraded by straylight. Figure 23.3 illustrates some effects straylight can have on different aspects of quality of vision. Those include, apart from glare, also face recognition difficulty, spatial orientation problems, contrast, and color loss. On the other hand, scattering/straylight does basically hardly affect visual acuity. This can be understood from consideration of the optics involved (van den Berg et al. 2010) and is found experimentally (McLaren and Patel 2012). An association between visual acuity and straylight could however result from the processes affecting the optics of the eye, but, for example, in the case of cataract, the association is weak; see Section 23.5.1.

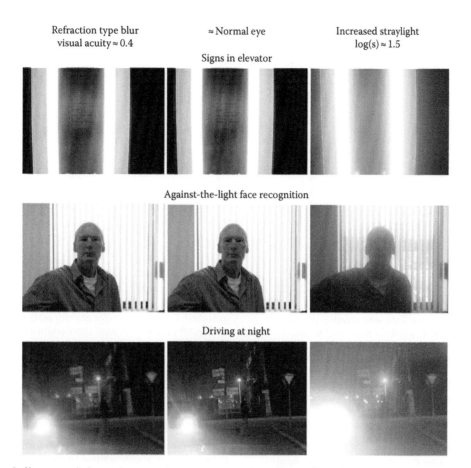

Refraction type blur
visual acuity ≈ 0.4 ≈ Normal eye Increased straylight
log(s) ≈ 1.5

Signs in elevator

Against-the-light face recognition

Driving at night

Figure 23.3 Examples of effects straylight can have on vision, as compared to effects of visual acuity loss. In the middle column is depicted how a normal young eye different scenes would see. To the left, the same scenes are shown, but "seen" with an eye with reduced visual acuity. In the examples given, there is not much difference if visual acuity is down to 0.4 decimal (logMAR 0.4), except that the street signs with text can no longer be read. Visual acuity 0.4 is about a factor of 4 loss compared to the young normal eye. A straylight change by a factor of 4 (log(s) = 1.5) has however dramatic effects on vision in these examples. (Reproduced from van den Berg, T.J.T.P. et al., Ocular media clarity and straylight, in *Encyclopedia of the Eye*, ed. D.A. Dart, Academic Press, Oxford, U.K., 2010, pp. 173–183. With permission.)

23.3 PHYSICS

23.3.1 BASIC OPTICS OF SCATTERING

Historically, the physics of light scattering was studied from the perspective of the astronomers and meteorologists. For overview, see, for example, van de Hulst (1981). Well known is the name of Lord Rayleigh, who explained why light scattering in the clear sky is so blue dominant. For particles much smaller than wavelength λ, light scattering is proportional to λ^{-4}. It is moreover "isotropic," that is, equal in all directions (apart from a "natural light" correction $(1 + \cos(\theta)^2)/2$ if the incident light contains all polarization directions). Subsequently, light scattering characteristics for many different particle types were resolved. In the human eye lens, the most relevant light scattering is from particles in the "Rayleigh–Gans" domain, that is, from particles that are not small compared to wavelength but have a small index of refraction difference with the surrounding medium. Later, Mie solved the mathematics for spherical particles of any size and index of refraction. Those equations are accessible through several sources on the internet. For Figures 23.4 through 23.6, we used the code of Christian Mätzler from the Universität Bern (http://www.iap.unibe.ch/) and checked the results with outcomes from Scott Prahl (http://omlc.org/calc/mie_calc.html).

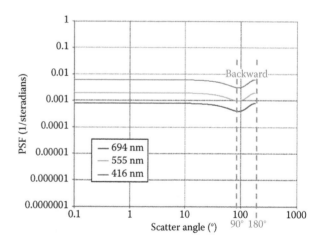

Figure 23.4 Light scattering from particles much smaller than wavelength, from natural light, and from wavelengths λ = 694 nm (red), 555 nm (green), and 416 nm (blue). For such particles light scattering is very dominant at the short wavelength side based on the Rayleigh law λ^{-4}.

Figure 23.4 shows the results for particles much smaller than wavelength (Rayleigh scattering). It shows clearly the strong blue dominance. In all three figures, the horizontal scale is logarithmic, making the portion of backward scattering that in reality covers of course half the angular range to

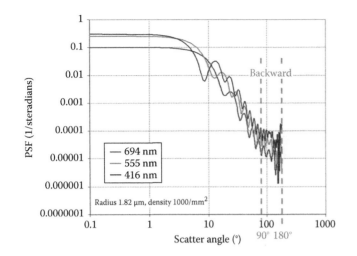

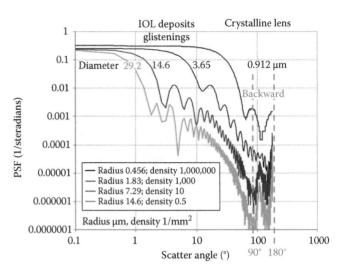

Figure 23.5 Light scattering from particles of size of the order of wavelength and with index of refraction close to that of the embedding medium. Calculations were performed based on the rigorous Mie theory, for natural light with wavelengths in air λ = 694 nm (red), 555 nm (green), and 416 nm (blue). For such particles, light scattering is less strongly dominant on the short wavelength side, and strong forward directionality appears. In this example, particle radius is 1.82 μm, embedding medium is water (n = 1.336), and relative refractive index is 1.16, as would be valid for protein particles in the eye (lens). The figure can be compared in absolute sense to the psf as depicted in Figure 23.2.

Figure 23.6 Light scattering from particles with index of refraction close to that of the embedding medium and different sizes as indicated. Calculations were performed based on the rigorous Mie theory, for natural light with wavelength in air λ = 555 nm (green). The forward directionality is shown to be strongly dependent on size. The particle densities are chosen such that the scatter in the zero direction is approximately the same. It is clear that those densities are vastly different. In this example, embedding medium is water (n = 1.336), and relative refractive index is 1.16, as would be valid for protein particles in the eye (lens). For "glistenings" (water inclusions in lens implants) medium has somewhat higher refractive index, and relative refractive index is the inverse, but that does not make much of a difference in these scatter characteristics. The figure can be compared in an absolute sense to the psf as depicted in Figure 23.2.

seem narrow. At 90°, one sees a relative minimum of a factor 2 because of the natural light correction. In all three figures, the vertical scaling corresponds to proper normalization of the point spread function, such that these results can be applied in absolute sense to the eye. Figure 23.5 shows the result for Rayleigh–Gans particles. The wavelength dependence has weakened, and a strong directionality appears, with dominance of the forward directions, as seen in the human eye. Because the calculations are performed for one precise size of particle, a lobular structure appears from the Mie functions. In practice the particles are not of one size but have a distribution of sizes that causes the lobular structure to disappear. Figure 23.6 shows the results for four differently sized particles, each for wavelength 555 nm (in air).

Another issue to consider is whether the scattering is single or multiple. If the light scattered from one particle would hit for most part other particles, the distributions shown in the figures would be altered. This, however, is not an important issue in practice. In the human eye, normally only about 10% of the light is scattered, that is, only about 1% would be scattered twice. This holds true for the scattering in the eye media *per se*, cornea and lens. This does not hold true for the light scattered from the fundus, but in that case the distributions shown in the figures do not apply.

23.3.2 SOURCES OF LIGHT SCATTERING IN THE EYE

In principle all matter scatters light, but in the eye, some parts scatter so little as to be insignificant compared to other parts. This applies for young healthy eyes to the aqueous and vitreous (Vos 1984). However, pathological conditions can make these structures to scatter light to significant amounts, such as cells in the aqueous in the case of inflammation or vitreous

opacities or floaters (Mura et al. 2011). Vos (1984) concluded that cornea, lens, and reflectance from the fundus contribute to about equal amounts to straylight as he measured it in the young eye. Later it was found that the eye wall is somewhat translucent, also in normal eyes, which contributes a rather isotropic component to straylight (van den Berg et al. 1991). Both the reflectance from the fundus and the transmittance through the eye wall depend strongly on pigmentation of the eye, thus making straylight dependent on the state of pigmentation. The angular distributions of these straylight components have been elaborated for the normal eye to some detail (Vos and van den Berg 1997), showing them to be much less dependent on angle as compared to the Stiles–Holladay approximation.

The crystalline lens has received most attention as the source of straylight because the aging process of the lens makes overall straylight of the human eye to increase strongly with age, also in noncataractous eyes, and makes the lens the dominating factor in most older eyes (Vos 1984; van den Berg et al. 2007). Details on angle and wavelength dependence of scatter/straylight will be given in Section 23.3.3 for the lens.

The contribution of the cornea to straylight seems to be rather stable with age, and the wavelength dependence is strongly blue dominant, according to the Rayleigh distribution (McCally and Farrell 1988; van den Berg and Tan 1994). The angular distribution does not comply with the Rayleigh type of scattering though (Vos 1984). Figure 23.7 shows a schematic overview of the four most important sources of straylight in the normal eye.

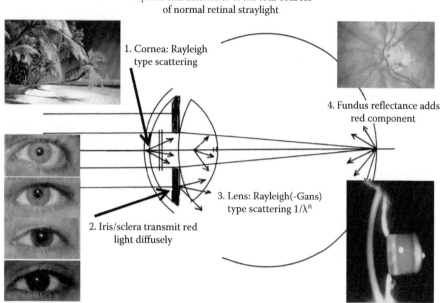

Physical characteristics of the four sources
of normal retinal straylight

1. Cornea: Rayleigh type scattering

4. Fundus reflectance adds red component

3. Lens: Rayleigh(-Gans) type scattering $1/\lambda^n$

2. Iris/sclera transmit red light diffusely

Figure 23.7 Overview of the four most important sources of straylight in the normal eye. (1) The cornea contributes a component with strong blue dominance, (almost) as strong as the λ^{-4} Rayleigh dependence, but with forward directionality. (2) Translucency of the eye wall contributes a diffuse red component. (3) The lens contributes a component with relatively weak wavelength dependence in forward direction, which has strong directionality, but exhibits pure Rayleigh behavior in backward direction. (4) Fundus reflectance adds a red dominant component with weak directionality. (Reproduced from van den Berg, T.J.T.P. et al., Ocular media clarity and straylight, in *Encyclopedia of the Eye*, ed. D.A. Dart, Academic Press, Oxford, U.K., 2010, pp. 173–183. With permission.)

23.3.3 PHYSICS OF LENTICULAR LIGHT SCATTERING

As explained earlier, good knowledge exists about the basic physics of scattering. If the characteristics of particles are known, for many cases, the mathematics is resolved to calculate wavelength and angular dependence of light scattering. Reversibly, if angular and wavelength dependence of the light scattering are known, one can infer much about the underlying light scattering process. Figure 23.8 shows these characteristics for an isolated lens from the eye of a deceased human donor (van den Berg and Spekreijse 1999). These measurements were done for the middle of the lens, but essentially the same pattern was obtained throughout the lens. Only the relative strengths of the different components differ as function of location in the body of the lens. From the fact that light scattering occurs more or less uniformly throughout the lens, a conclusion can be drawn that light scattering derives from a dense distribution of small-sized irregularities (molecules, cells, etc.) throughout the lens. For angles >30° including backward angles a particularly well-known pattern is seen. It is the Rayleigh pattern, as discussed in Section 23.3.1 (Figure 23.4). Please note how precisely the λ^{-4} dependence and the isotropic behavior apply, including the "natural light" correction. This leads to the conclusion that for these directions, light scattering is dominated by particles much smaller than wavelength. It can be assumed that these are the protein molecules giving the lens its relatively high index of refraction to fulfill its optical function. However, if reversibly one calculates from the known density of proteins the light scattering to be expected, one arrives at much stronger light scattering values, and the eye would effectively be blind. This issue was already realized long ago, and a theory was developed to explain this seeming incongruity on the basis of so-called

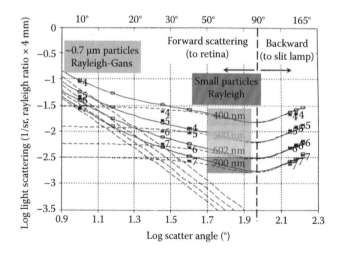

Figure 23.8 Light scattering by a crystalline lens isolated from a human donor eye, from an area 1 × 0.1 mm at 50% depth in the middle of the lens. Four wavelengths were used, 400 nm (blue), 500 nm (green), 602 nm (orange), and 700 nm (red). For larger angles the Rayleigh pattern of scattering is seen, being isotropic, and having the strong λ^{-4} wavelength dependence, corresponding to scattering by particles much smaller than wavelength, presumably the protein molecules. For smaller angles, the forward-direction, weaker wavelength dependence is seen, and nonisotropic scattering, corresponding to light scattering by particles with sizes around 0.7 µm radius. Assuming these particles to have a relative refractive index close to one (1.13 would correspond to protein of refractive index 1.51 in water of refractive index 1.336), their scattering characteristics can be calculated with the Rayleigh–Gans approximation (van de Hulst 1981). (Adapted from van den Berg, T.J.T.P. and Spekreijse, H., *Vision Res.*, 39, 1437, 1999.)

short range order of the molecules (Trokel 1962). The proteins are ordered in a more or less strict crystalline structure leading to destructive interference of the scattered light. This lowers the light scattering efficiency with a factor of 10 or more based on the actual measurements, illustrated in Figure 23.8 (van den Berg and Spekreijse 1999). Maybe surprisingly, this Rayleigh behavior was found for all lenses, independent of cataractous state. Only, in the case of cataract, the scattering was much stronger, corresponding to the degree of ordering being (much) less, in the worst case to the degree of almost no ordering. It must be added though that for the largest scattering angles used (165°), an extra component appeared, interpreted to originate from "zones of discontinuity" or "water clefts." Also those were stronger in the case of cataract (van den Berg and Spekreijse 1999).

For small scattering angles, Figure 23.8 shows an increase in scattering, with less strong wavelength dependence. This could be fitted on the basis of the assumption that a distribution of larger particles is present in the lens. If we assume these particles to be protein aggregates, from the found light scattering characteristic, a median size of 1.4 μm diameter can be estimated (van den Berg and Spekreijse 1999). The amount needed for the level of light scattering found proved to be very low: only a volume fraction of 0.000006 needs to be occupied by these particles. However, Costello and coworkers also found multilamellar bodies that might explain these data (Gilliland et al. 2001).

23.3.4 CILIARY CORONA

One might expect the straylight pattern that we see with our own eyes around a point source of light to be comparable to the light scattering pattern measured optically as described earlier and corresponding with the mentioned physical theory. For single particles, those patterns are somewhat like the well-known Airy disc pattern, with concentric rings around a central disc. In Figures 23.5 and 23.6, the lobules correspond to the rings. However, as the particles are not all of the same size and have a distribution of diameters, the rings/lobules disappear, and one would expect to see a uniform field of scattered light, declining in intensity with eccentricity. However, what we actually see around a point source of light is a radiation of very sharp needles with local colorations. This visual phenomenon is called "ciliary corona." Cilia is eyelashes in Latin. This phenomenon is seen by everybody, including normal subjects and subjects with cataract. The explanation for this seeming discrepancy was found in physical theory as follows. It must be realized that the light waves being scattered from the particles present in the eye lens or cornea are coherent because the light hitting the particles is coherent, also in the case of a natural light source, such as a halogen light, if the light source is small enough. The scattered waves interfere at the retina in a complicated way, and this process results in the pattern of fine needles. This was proven with a simulation of this physical process, using parameters as valid for the human eye, resulting in Figure 23.9 (van den Berg et al. 2005). So the seeming discrepancy proved to result in a confirmation that the light scattering in the human eye is dominated by a distribution of small particles.

The fine needles are seen only for small-sized point sources. For extended sources, the fine needles disappear, and a uniform light field is seen to surround the glare source. The "ray formation angle" is the limiting angular size and was found to be around 19′

Figure 23.9 Mathematical simulation of the ciliary corona, as seen by human eyes when observing a small light source against a dark background. For the simulation, a random distribution of small light scattering particles of sizes around 0.7 μm radius was assumed to be present in the eye lens. As light source, a small point source of white (equal energy) light was assumed. The patterns of scattered light from all the particles superimpose and interfere at the retina, leading to this figure. (Reproduced from van den Berg, T.J.T.P. et al., *Invest. Ophthalmol. Vis. Sci.*, 46, 2627, 2005. With permission.)

in one subject (Simpson 1953). It must be noted that many light sources are highly inhomogeneous, such as headlights having a bright spot in the center, and the needles may be seen from light sources larger than 19′ as a result.

23.3.5 PHYSICS OF CORNEAL LIGHT SCATTERING

The physics of light scattering in the cornea has been studied to a high degree of physical detail by the group of Hart, Farrell, McCally, Freund, and others (McCally and Farrell 1988; Freund et al. 1995), but more groups had a considerable impact in this area, such as those from Maurice (1957) and Benedek (1971). The cornea derives its mechanical strength from a special ultrastructural arrangement. It is somewhat like plywood. The cornea is built up from thin layers, parallel to its surface, each about 2 μm thick, containing a dense packing of parallel collagen fibrils. In subsequent lamellae, the fibrils are oriented at large angles with respect to the neighboring lamellae. The fibrils have a diameter of around 30 nm, which is much smaller than wavelength. Maurice already realized that if those fibrils were to act as independent scatterers, the cornea would be opaque. However, as the fibrils are much closer to each other as compared to the size of one wavelength, their centers are around 50 nm apart, with a relatively regular ordered arrangement, the scattered waves interfere destructively to a large degree, except in the very forward direction. The residual scattering depends on the statistics of the fibril distances, it being essential that the index of refraction does not fluctuate over distances comparable or larger than the wavelength of light. It was found for rabbit corneas that the scattering is strongly dependent on wavelength, as λ^{-3} (McCally and Farrell 1988). On the other hand, Feuk found experimentally a λ^{-5} dependence in those corneas (Feuk 1971). For human corneas, in vivo measurements suggested a λ^{-4} dependence (van den Berg and Tan 1994). In the case of edematous swelling of the cornea, light scattering increases while the wavelength dependence decreases, to about λ^{-2}, supposedly as the result of "voids" or "lakes" between the fibrils (McCally and Farrell 1988).

23.4 STRAYLIGHT IN NORMAL EYES

23.4.1 NORMAL POPULATION DATA COLLECTIONS

Vos has reviewed the older literature with the view to establish a CIE reference norm for straylight as a function of angle and age (Vos 1984). He lists 13 papers giving functions for the dependence of straylight on angle and 5 papers for the dependence of straylight on age. The angular dependencies vary somewhat around the Stiles–Holladay approximation, depending on the range of angles studied. For smaller angles, a steeper dependence is found, up till about θ^{-3}, with an angular minimum of 10′ studied, whereas for larger angles, a shallower dependence is found, down to about $\theta^{-1.5}$. In fact, of course the angular dependence changes gradually over angle, having the approximate θ^{-2} dependence in the most studied range between 1° and 30°. All authors find clear age dependencies, also in the best eyes, in reasonable agreement with each other. Figure 23.10 shows recent results on the European driver population with the more accurate technique of "compensation comparison," mentioned earlier (van den Berg et al. 2007). The red dots are for eyes with good visual acuity and no cataract, showing a clear straylight increase with age. The curved line has equation $\log(s) = 0.90 + \log(1 + (age/65)^4)$, more or less in accordance with the older literature. The open circles are for pseudophakic eyes, and the crosses for eyes, with some degree of cataract based on the slit lamp observation (LOCS III).

As mentioned in Section 23.1, the CIE has established straylight as the norm for the peripheral part of the functional point spread function psf based on the equivalent luminance principle. The CIE has defined normal reference functions for many different aspects of the human eye, such as the spectral sensitivity Vλ, and has also established functions for straylight (Vos and van den Berg 1999). In fact several norm functions were adopted with different levels of complexity, valid in different angular ranges, including as parameters age and pigmentation dependence. These norm functions are valid for best eyes, as based on population studies for eyes with good acuity and no cataract. The simplest norm function is the "age-adapted Stiles–Holladay equation" psf = $(1 + (age/70)^4)10/\theta^2$ with validity range 3°–30°. The next simplest is the "simplified glare equation" psf = $10/\theta^3 + (1 + (age/62.5)^4)5/\theta^2$ with validity range 0.1°–30°. There are two more functions with increasing complexity and validity (Vos and van den Berg 1999). Figure 23.11 shows all functions expressed as straylight parameter s (van den Berg et al. 2013), illustrating the validity of the approximate θ^{-2} course for the psf and the corresponding constancy of s over 1°–30°.

23.4.2 WAVELENGTH DEPENDENCE ISSUE

As explained earlier, light scattering from volumes of matter shows wavelength dependence. In the early days, it was assumed that the molecules of the eye would scatter based on the very strong wavelength dependence of Rayleigh scattering. However, straylight studies failed to find such wavelength dependence; they even failed to find any wavelength dependence at all (Wooten and Geri 1987; Whitaker et al. 1993). Some authors concluded that the light scattering in the eye must originate from particles very large compared to wavelength. However, such particles should have been detectable with biomicroscopy, and no such particles are known. Previously, the evidence is explained that light scattering in cornea and lens is dominated by small particles, of sizes maximally a few times wavelength, and that the light scattered has wavelength dependence. So how can the failure to find in vivo wavelength dependence be understood? The explanation is that the other components of light scattering, that is, fundal reflectance and eye wall transmittance, have a wavelength dependence

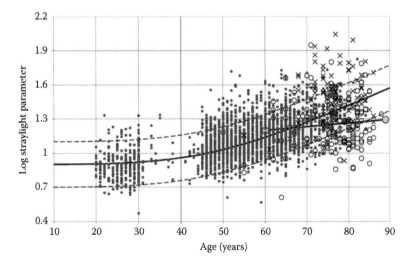

Figure 23.10 Straylight in log(s) as function of age in a population of European drivers. The red dots and curved lines are for the best eyes, with good visual acuity and no cataract based on the slit lamp observation (LOCS III). The dashed lines indicate the normal interval of ±0.20 log units. The open circles with linear regression line are for pseudophakic eyes. The crosses are for eyes with cataract based on the slit lamp observation (LOCS III). (Reproduced from van den Berg, T.J.T.P. et al., *Am. J. Ophthalmol.*, 144, 358, 2007. With permission.)

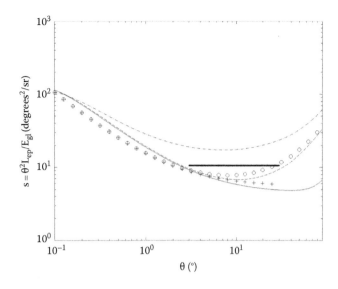

support the hypothesis that red dominant sources of straylight counteract the blue dominant sources so as to almost quench the wavelength dependence (Coppens et al. 2006).

23.5 CLINICAL STRAYLIGHT DATA

23.5.1 LENS AGING AND CATARACT

It has long been realized that straylight must be considered for clinical assessment of patient problems. Also in case visual acuity is no problem, straylight/disability glare can be. Cataract is the area where this has received most attention and is accepted for clinical management of patients (Koch 1989; Masket 1989; Elliott et al. 1991; Rubin et al. 1993). Figure 23.12 illustrates that visual acuity and straylight are quite independent in the normal aging population. This figure is from the same study as Figure 23.10, with red dots indicating best normal eyes, circles indicating pseudophakic eyes, and crosses indicating eyes with a degree of cataract. It can be seen that quite often straylight is considerably increased, also when visual acuity is normal. The reverse is also true, indicating that the one measure of visual function cannot be inferred from the other, and that both must be assessed and considered separately. Although many older clinical studies already concluded straylight/disability glare to be essential for proper management of cataract patients, a quantitative study on the relative importance of straylight for overall visual functioning was performed only rather recently. This was done by comparing the outcome of a visual function questionnaire to the values of visual acuity and straylight in a population of cataract patients (van der Meulen et al. 2012). It proved that straylight was almost as important for the overall questionnaire outcome as visual acuity, confirming the earlier conclusions. Figure 23.13 shows that also in the cataract study visual acuity and straylight were quite independent both extending to much worse values of course as compared to the driver study.

Figure 23.11 The four CIE-defined norm functions for straylight, expressed as straylight parameter s. (1) Thick straight line is based on the age-adapted Stiles–Holladay equation for age 35, having validity from 3° to 30°. (2) Plusses are based on the simplified glare equation for age 35, having validity from 0.1° to 30°. (3) Open circles are based on the "general glare equation" for age 35 and blue-green eye color, having validity from 0.1° to 100°. (4) Three curves are given corresponding to the "total glare function," having validity from 0° to 100°. The continuous line is for a Negroid eye aged 35. The dashed line is for a blue-green eye aged 35. The dash-dotted line is for a blue-green eye aged 80. It can be seen that for all conditions the angular dependence is close to θ^{-2} around 10°. (Reproduced from van den Berg, T.J.T.P. et al., Z. Med. Phys., 23, 6, 2013. With permission.)

that is opposite to that of cornea and lens. Fundal reflectance has been studied by many authors and has been shown to have strong red dominance because of the blood and pigments in the choroid and pigment epithelium (Van Norren and Tiemeijer 1986; Delori and Pflibsen 1989). Relatively recently, new spectrally resolved straylight measurements were done showing spectral effects that

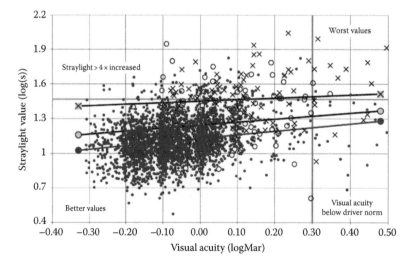

Figure 23.12 This figure shows that visual acuity and straylight are not strongly related. The red dots are for best normal eyes, open circles for pseudophakic eyes, and crosses for eyes with some degree of cataract. The lines are linear regression lines for the respective subpopulations. The relative independence of visual acuity and straylight can be understood because they depend on different aspects of disturbances to the eye's optics. Visual acuity depends on errors in refraction (aberrations) and the center of the psf; straylight depends on light scattering and the periphery of the psf. (Reproduced from van den Berg, T.J.T.P. et al., Am. J. Ophthalmol., 144, 358, 2007. With permission.)

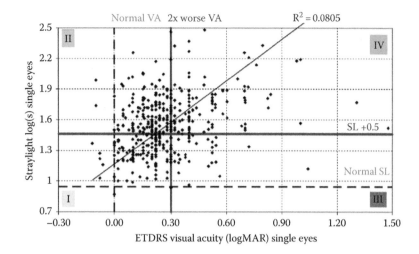

Figure 23.13 This figure shows that also for cataract eyes, scheduled to undergo surgery, visual acuity and straylight are not strongly related. The same argument applies that both aspects of visual function are based on very different optical aspects (aberrations vs scattering). (Reproduced from van der Meulen, I.J.E. et al., *J. Cataract Refract. Surg.*, 38, 840, 2012. With permission.)

23.5.2 PSEUDOPHAKIA

If we compare the straylight values for pseudophakic eyes in Figure 23.12, to the straylight values for cataractous eyes in Figure 23.13, it is clear that cataract surgery is very effective to reduce straylight. In fact, if one considers the straylight values as function of age in Figure 23.10, after cataract surgery on average most patients have straylight better than the normal reference for phakic eyes with good acuity. As a surprise, it was even found that in case of "clear lens exchange" straylight improved as result of the surgery (Lapid-Gortzak et al. 2014). Clear lens exchange is an elective procedure as one of the options for refractive surgery, performed as a rule in good eyes, apart from the refraction. Straylight improvement may contribute to the postoperative satisfaction of these patients. It corresponds with the finding of straylight increase with aging in the best eyes. This increase can be attributed to the normal aging of the eye lens and is remedied upon replacement by an artificial lens IOL. However, looking at Figure 23.12, it appears that the improvement is not as big as one might have expected. If the straylight increase with age is completely due to the eye lens, replacement of the eye lens should bring the straylight level back to the values of the youth. It was a surprise that this is not always the case and it was considered that posterior capsule opacification (PCO) was to blame (Witmer et al. 1989; Hard et al. 1993). However, later studies showed that also in case PCO was excluded, straylight could be elevated. Several reports have shown that IOLs can give increased straylight (de Vries et al. 2008; Ferrer-Blasco et al. 2009; Blundell et al. 2010; Ehmer et al. 2011), but in general it is unclear at this moment in how far the IOLs, the surgery, or preexisting disturbances to the ocular structures are the reason for the widespread occurrence of straylight elevation in pseudophakic eyes.

23.5.3 CORNEAL CONDITIONS AND REFRACTIVE SURGERY

The cornea seems to be particularly liable to cause straylight elevation, and in this section, only a limited overview can be given. The groups of Elliott, Applegate, McLaren/Patel, Nuijts,

Delleman, van der Meulen, Montes-Mico/Cervino, and others have studied many different conditions, including contact lens effects (Applegate and Wolf 1987; Elliott et al. 1993; Cervino et al. 2008; van der Meulen et al. 2010), dystrophies (van den Berg et al. 1993; van der Meulen et al. 2011a), and transplantation (Ahmed et al. 2010; Cheng et al. 2011; Seery et al. 2011; van der Meulen et al. 2011b), to reference just a few studies.

Because of this corneal sensitivity, it can be understood that the potential effect of corneal refractive surgery has gained much interest. As refractive surgery has the potential to induce haze in the cornea, and other disturbances, such as folds or wrinkling of the flap, interface debris, or ingrowth, it may be expected that in such cases straylight can increase. Indeed, starting from the 1990s, many studies have reported straylight increase correlated with such morphological findings, but with the more advanced recent surgery techniques, the occurrence of straylight elevations seems to have reduced to incidental cases (Rocha et al. 2009; Barreto et al. 2010; Nieto-Bona et al. 2010; Cervino et al. 2011; Li and Wang 2011). In recent years, several studies have suggested that refractive surgery can also have the opposite effect, that is, straylight improvement. Population averages showed no change or slight improvement upon refractive surgery, which is surprising in view of the fact that most studies also include incidental cases with deterioration (Lapid-Gortzak et al. 2010; Nieto-Bona et al. 2010; Rozema et al. 2010). It has been speculated that this effect may be caused by relief of the eyes from ill-tolerated contact lenses. This would be consistent with the result that also in case of the use of a phakic implant lens for refractive correction improvement is found (Paarlberg et al. 2011), but this speculation is in need of verification.

23.5.4 PIGMENTATION INSUFFICIENCIES

As explained in Section 23.3.2, not only the optical media *per se*, in particular cornea and lens, contribute to straylight but also reflectance from the fundus and transmittance through the anterior eye wall. Both these components depend on the state of pigmentation of the eye, and indeed several studies have appeared on this relationship. Differences in pigmentation already occur in the normal population, and part of the differences found between normal young individuals can be attributed to pigmentation differences (Nischler et al. 2013). Blue-eyed individuals have on average more straylight as compared to brown-eyed individuals, but in the normal population the effects are small. More important effects are found in individuals with pathological conditions, in particular in case of Fuchs heterochromic cyclitis or some hereditary conditions such as albinism, or X-linked megalocornea. Not much can be done to reduce the straylight effect of hyperreflectivity of the fundus, except maybe the use of color filters to reduce the amount of red light. The effect of hypertransmittance (clinically called diaphany) of the iris can effectively be counteracted with "iris print" contact lenses (Kruijt et al. 2011).

REFERENCES

Ahmed, K. A., J. W. McLaren, K. H. Baratz, L. J. Maguire, K. M. Kittleson, and S. V. Patel, Host and graft thickness after Descemet stripping endothelial keratoplasty for Fuchs endothelial dystrophy. *Am J Ophthalmol* 150 (2010): 490–497.

Applegate, R. A. and M. Wolf, Disability glare increased by hydrogel lens wear. *Am J Optom Physiol Opt* 64 (1987): 309–312.

Aslam, T. M., D. Haider, and I. J. Murray, Principles of disability glare measurement: An ophthalmological perspective. *Acta Ophthalmol Scand* 85 (2007): 354–360.

Barreto, J., Jr., M. T. Barboni, C. Feitosa-Santana, J. R. Sato, S. J. Bechara, D. F. Ventura, and M. R. Alves, Intraocular straylight and contrast sensitivity after contralateral wavefront-guided LASIK and wavefront-guided PRK for myopia. *J Refract Surg* 26 (2010): 588–593.

Benedek, G. B., Theory of transparency of the eye. *Appl Opt* 10 (1971): 459–473.

Blundell, M. S., E. J. Mayer, N. E. Knox Cartwright, L. P. Hunt, D. M. Tole, and A. D. Dick, The effect on visual function of Hydroview intraocular lens opacification: A cross-sectional study. *Eye (Lond)* 24 (2010): 1590–1598.

Boynton, R. M., J. M. Enoch, and W. Bush, Physical measures of stray light in excised eyes. *J Opt Soc Am* 44 (1954): 879–886.

Cervino, A., J. M. Gonzalez-Meijome, J. M. Linhares, S. L. Hosking, and R. Montes-Mico, Effect of sport-tinted contact lenses for contrast enhancement on retinal straylight measurements. *Ophthalmic Physiol Opt* 28 (2008): 151–156.

Cervino, A., C. Villa-Collar, J. M. Gonzalez-Meijome, T. Ferrer-Blasco, and S. Garcia-Lazaro, Retinal straylight and light distortion phenomena in normal and post-LASIK eyes. *Graefes Arch Clin Exp Ophthalmol* 249 (2011): 1561–1566.

Cheng, Y. Y., T. J. van den Berg, J. S. Schouten, E. Pels, R. J. Wijdh, H. van Cleynenbreugel, C. A. Eggink, W. J. Rijneveld, and R. M. Nuijts, Quality of vision after femtosecond laser-assisted descemet stripping endothelial keratoplasty and penetrating keratoplasty: A randomized, multicenter clinical trial. *Am J Ophthalmol* 152 (2011): 556–566.

Cobb, P. W., The influence of illumination of the eye on visual acuity. *Am J Physiol* 29 (1911): 76–99.

Coppens, J. E., L. Franssen, and T. J. T. P. van den Berg, Wavelength dependence of intraocular straylight. *Exp Eye Res* 82 (2006): 688–692.

De Mott, D. W. and R. M. Boynton, Sources of entoptic stray light. *J Opt Soc Am* 48 (1958): 120–125.

de Vries, N. E., L. Franssen, C. A. B. Webers, N. G. Tahzib, Y. Y. Y. Cheng, F. Hendrikse, K. F. Tjia, T. J. T. P. van den Berg, and R. M. M. A. Nuijts, Intraocular straylight after implantation of the multifocal AcrySof ReSTOR SA60D3 diffractive intraocular lens. *J Cataract Refract Surg* 34 (2008): 957–962.

de Waard, P. W., J. K. IJspeert, T. J. T. P. van den Berg, and P. T. de Jong, Intraocular light scattering in age-related cataracts. *Invest Ophthalmol Vis Sci* 33 (1992): 618–625.

Delori, F. C. and K. P. Pflibsen, Spectral reflectance of the human ocular fundus. *Appl Opt* 28 (1989): 1061–1077.

Donnelly, W. J., III, K. Pesudovs, J. D. Marsack, E. J. Sarver, and R. A. Applegate, Quantifying scatter in Shack-Hartmann images to evaluate nuclear cataract. *J Refract Surg* 20 (2004): S515–S522.

Ehmer, A., T. M. Rabsilber, A. Mannsfeld, M. J. Sanchez, M. P. Holzer, and G. U. Auffarth, Influence of different multifocal intraocular lens concepts on retinal stray light parameters. *Ophthalmologe* 108 (2011): 952–956.

Elliott, D. B., D. Fonn, J. Flanagan, and M. Doughty, Relative sensitivity of clinical tests to hydrophilic lens-induced corneal thickness changes. *Optom Vis Sci* 70 (1993): 1044–1048.

Elliott, D. B., M. A. Hurst, and J. Weatherill, Comparing clinical tests of visual loss in cataract patients using a quantification of forward light scatter. *Eye* 5 (Pt 5) (1991): 601–606.

Ferrer-Blasco, T., R. Montes-Mico, A. Cervino, and J. F. Alfonso, Light scatter and disability glare after intraocular lens implantation. *Arch Ophthalmol* 127 (2009): 576–577.

Feuk, T., The wavelength dependence of scattered light intensity in rabbit corneas. *IEEE Trans Biomed Eng* 18 (1971): 92–96.

Franssen, L., J. E. Coppens, and T. J. T. P. van den Berg, Compensation comparison method for assessment of retinal straylight. *Invest Ophthalmol Vis Sci* 47 (2006): 768–776.

Freund, D. E., R. L. McCally, R. A. Farrell, S. M. Cristol, N. L. L'Hernault, and H. F. Edelhauser, Ultrastructure in anterior and posterior stroma of perfused human and rabbit corneas. Relation to transparency. *Invest Ophthalmol Vis Sci* 36 (1995): 1508–1523.

Gilliland, K. O., C. D. Freel, C. W. Lane, W. C. Fowler, and M. J. Costello, Multilamellar bodies as potential scattering particles in human age-related nuclear cataracts. *Mol Vis* 7 (2001): 120–130.

Ginis, H., G. M. Perez, J. M. Bueno, and P. Artal, The wide-angle point spread function of the human eye reconstructed by a new optical method. *J Vis* 12 (2012): 1–10.

Hard, A. L., C. Beckman, and J. Sjöstrand, Glare measurements before and after cataract surgery. *Acta Ophthalmol (Copenh)* 71 (1993): 471–476.

Holladay, L. L., The fundamentals of glare and visibility. *J Opt Soc Am* 12 (1926): 271–319.

Holladay, L. L., Action of a light source in the field of view in lowering visibility. *J Opt Soc Am* 14 (1927): 1–15.

Koch, D. D., Glare and contrast sensitivity testing in cataract patients. *J Cataract Refract Surg* 15 (1989): 158–164.

Kruijt, B., L. Franssen, L. J. Prick, J. M. van Vliet, and T. J. van den Berg, Ocular straylight in albinism. *Optom Vis Sci* 88 (2011): E585–E592.

Lapid-Gortzak, R., J. W. van der Linden, I. van der Meulen, C. Nieuwendaal, and T. van den Berg, Straylight measurements in laser in situ keratomileusis and laser-assisted subepithelial keratectomy for myopia. *J Cataract Refract Surg* 36 (2010): 465–471.

Lapid-Gortzak, R., I. J. van der Meulen, J. W. van der Linden, M. P. Mourits, and T. J. van den Berg, Straylight before and after phacoemulsification in eyes with preoperative corrected distance visual acuity better than 0.1 logMAR. *J Cataract Refract Surg* 40 (2014): 748–755.

le Grand, Y., Recherches sur la diffusion de la lumiere dans loeil humain. Deuxieme partie. *Reveu d'Optique* 16 (1937): 241–266.

Li, J. and Y. Wang, Characteristics of straylight in normal young myopia and changes before and after LASIK. *Invest Ophthalmol Vis Sci* 52 (2011): 3069–3073.

Masket, S., Reversal of glare disability after cataract surgery. *J Cataract Refract Surg* 15 (1989): 165–168.

Maurice, D. M. The structure and transparency of the cornea. *J Physiol* 136 (1957): 263–286.

McCally, R. L. and R. A. Farrell, Interaction of light and the cornea: Light scattering versus transparency. In *The Cornea: Transactions of the World Congress on the Cornea III*, ed. H.D. Cavanagh, 1988, pp. 165–171 (New York: Raven Press Ltd.).

McLaren, J. W. and S. V. Patel, Modeling the effect of forward scatter and aberrations on visual acuity after endothelial keratoplasty. *Invest Ophthalmol Vis Sci* 53 (2012): 5545–5551.

Mura, M., L. A. Engelbrecht, M. D. de Smet, T. G. Papadaki, T. J. van den Berg, and H. S. Tan, Surgery for floaters. *Ophthalmology* 118 (2011): 1894.

Nam, J., L. N. Thibos, A. Bradley, N. Himebaugh, and H. Liu, Forward light scatter analysis of the eye in a spatially-resolved double-pass optical system. *Opt Express* 19 (2011): 7417–7438.

Nieto-Bona, A., A. Lorente-Velazquez, C. V. Collar, P. Nieto-Bona, and A. G. Mesa, Intraocular straylight and corneal morphology six months after LASIK. *Curr Eye Res* 35 (2010): 212–219.

Nischler, C., R. Michael, C. Wintersteller, P. Marvan, L. J. van Rijn, J. E. Coppens, T. J. van den Berg, M. Emesz, and G. Grabner, Iris color and visual functions. *Graefes Arch Clin Exp Ophthalmol* 251 (2013): 195–202.

Paarlberg, J. C., M. Doors, C. A. Webers, T. T. Berendschot, T. J. van den Berg, and R. M. Nuijts, The effect of iris-fixated foldable phakic intraocular lenses on retinal straylight. *Am J Ophthalmol* 152 (2011): 969–975.

Patel, S. V., K. H. Baratz, D. O. Hodge, L. J. Maguire, and J. W. McLaren, The effect of corneal light scatter on vision after descemet stripping with endothelial keratoplasty. *Arch Ophthalmol* 127 (2009): 153–160.

Paulsson, L. E. and J. Sjöstrand, Contrast sensitivity in the presence of a glare light. Theoretical concepts and preliminary clinical studies. *Invest Ophthalmol Vis Sci* 19 (1980): 401–406.

Pinero, D. P., D. Ortiz, and J. L. Alio, Ocular scattering. *Optom Vis Sci* 87 (2010): E682–E696.

Purkinje, J. E., *Beobachtungen und Versuche zur Physiologie der Sinnesorgane* (Prague, Czech Republic: Calve, 1823).

Rocha, K. M., R. Kagan, S. D. Smith, and R. R. Krueger, Thresholds for interface haze formation after thin-flap femtosecond laser in situ keratomileusis for myopia. *Am J Ophthalmol* 147 (2009): 966–972.

Rozema, J. J., T. Coeckelbergh, T. J. van den Berg, R. Trau, N. C. Duchateau, S. Lemmens, and M. J. Tassignon, Straylight before and after LASEK in myopia: Changes in retinal straylight. *Invest Ophthalmol Vis Sci* 51 (2010): 2800–2804.

Rozema, J. J., R. Trau, K. H. Verbruggen, and M. J. Tassignon, Backscattered light from the cornea before and after laser-assisted subepithelial keratectomy for myopia. *J Cataract Refract Surg* 37 (2011): 1648–1654.

Rubin, G. S., I. A. Adamsons, and W. J. Stark, Comparison of acuity, contrast sensitivity, and disability glare before and after cataract surgery. *Arch Ophthalmol* 111 (1993): 56–61.

Schmidt-Clausen, H. J. and J. Th. H. Bindels, Assessment of discomfort glare in motor vehicle lighting. *Lighting Res Technol* 6 (1974): 79–87.

Seery, L. S., J. W. McLaren, K. M. Kittleson, and S. V. Patel, Retinal point-spread function after corneal transplantation for Fuchs' dystrophy. *Invest Ophthalmol Vis Sci* 52 (2011): 1003–1008.

Simpson, G. C., Ocular haloes and coronas. *Br J Ophthalmol* 37 (1953): 450–486.

Stiles, W. S. and B. H. Crawford, The liminal brightness increment for white light for different conditions of the foveal and parafoveal retina. *Proc R Soc Lond* 116B (1935): 55.

Stiles, W. S. and B. H. Crawford, The effect of a glaring light source on extrafoveal vision. *Proc R Soc* 122B (1937): 255–280.

Trokel, S., The physical basis for transparency of the crystalline lens. *Invest Ophthalmol* 1 (1962): 493–501.

van de Hulst, H. C., *Light Scattering by Small Particles* (New York: Dover Publications Inc.), 1981.

van den Berg, T. J. T. P., Analysis of intraocular straylight, especially in relation to age. *Optom Vis Sci* 72 (1995): 52–59.

van den Berg, T. J. T. P., To the editor: Intra- and intersession repeatability of a double-pass instrument. *Optom Vis Sci* 87 (2010): 920–921.

van den Berg, T. J. T. P., J. E. Coppens, and L. Franssen, Ocular media clarity and straylight. In *Encyclopedia of the Eye*, ed. D.A. Dart, 2010. pp. 173–183 (Oxford, U.K.: Academic Press).

van den Berg, T. J. T. P., L. Franssen, B. Kruijt, and J. E. Coppens, History of ocular straylight measurement: A review. *Z Med Phys* 23 (2013): 6–20.

van den Berg, T. J. T. P., M. P. J. Hagenouw, and J. E. Coppens, The ciliary corona: Physical model and simulation of the fine needles radiating from point light sources. *Invest Ophthalmol Vis Sci* 46 (2005): 2627–2632.

van den Berg, T. J. T. P., B. S. Hwan, and J. W. Delleman, The intraocular straylight function in some hereditary corneal dystrophies. *Doc Ophthalmol* 85 (1993): 13–19.

van den Berg, T. J. T. P., J. K. IJspeert, and P. W. de Waard, Dependence of intraocular straylight on pigmentation and light transmission through the ocular wall. *Vision Res* 31 (1991): 1361–1367.

van den Berg, T. J. T. P. and H. Spekreijse, Light scattering model for donor lenses as a function of depth. *Vision Res* 39 (1999): 1437–1445.

van den Berg, T. J. T. P. and K. E. Tan, Light transmittance of the human cornea from 320 to 700 nm for different ages. *Vision Res* 34 (1994): 1453–1456.

van den Berg, T. J. T. P., L. J. van Rijn, R. Michael, C. Heine, T. Coeckelbergh, C. Nischler, H. Wilhelm et al., Straylight effects with aging and lens extraction. *Am J Ophthalmol* 144 (2007): 358–363.

van der Meulen, I., L. A. Engelbrecht, J. M. van Vliet, R. Lapid-Gortzak, C. P. Nieuwendaal, M. P. Mourits, R. O. Schlingemann, and T. J. van den Berg, Straylight measurements in contact lens wear. *Cornea* 29 (2010): 516–522.

van der Meulen, I. J., S. V. Patel, R. Lapid-Gortzak, C. P. Nieuwendaal, J. W. McLaren, and T. J. van den Berg, Quality of vision in patients with fuchs endothelial dystrophy and after descemet stripping endothelial keratoplasty. *Arch Ophthalmol* 129 (2011a): 1537–1542.

van der Meulen, I. J., T. C. Van Riet, R. Lapid-Gortzak, C. P. Nieuwendaal, and T. J. van den Berg, Correlation of straylight and visual acuity in long-term follow-up of manual descemet stripping endothelial keratoplasty. *Cornea* 31 (2011b): 380–386.

van der Meulen, I. J. E., J. Gjertsen, B. Kruijt, J. P. Witmer, A. Rulo, R. O. Schlingemann, and T. J. T. P. van den Berg, Straylight measurements as an indication for cataract surgery. *J Cataract Refract Surg* 38 (2012): 840–848.

Van Norren, D. and L. F. Tiemeijer, Spectral reflectance of the human eye. *Vision Res* 26 (1986): 313–320.

von Goethe, J. W. *Zur Farbenlehre* (Tübingen, Germany: 2 Bde. Cotta, 1810).

von Helmholtz, H., Ueber Hern Brewsters neue Analyse des Sonnenlichts. *Ann der Physik und Chemi* 86 (1852): 501.

Vos, J. J. 1963. On mechanisms of glare. (Utrecht, the Netherlands: v/h Kemink en Zoon N.V.)

Vos, J. J., Disability glare—A state of the art report. *Comm Int de l'Eclairag J* 3/2 (1984): 39–53.

Vos, J. J. and T. J. T. P. van den Berg, On the course of the disability glare function and its attribution to components of ocular scatter. *CIE Collection* 124 (1997): 11–29.

Vos, J. J. and T. J. T. P. van den Berg, Report on disability glare. *CIE Collection* 135 (1999): 1–9.

Westheimer, G. and J. Liang, Evaluating diffusion of light in the eye by objective means. *Invest Ophthalmol Vis Sci* 35 (1994): 2652–2657.

Whitaker, D., D. B. Elliott, and R. Steen, Confirmation of the validity of the psychophysical light scattering factor. *Invest Ophthalmol Vis Sci* 35 (1994): 317–321.

Whitaker, D., R. Steen, and D. B. Elliott, Light scatter in the normal young, elderly, and cataractous eye demonstrates little wavelength dependency. *Optom Vis Sci* 70 (1993): 963–968.

Williams, D. R., D. H. Brainard, M. J. McMahon, and R. Navarro, Double-pass and interferometric measures of the optical quality of the eye. *J Opt Soc Am A Opt Image Sci Vis* 11 (1994): 3123–3135.

Witmer, F. K., H. J. van den Brom, A. C. Kooijman, and L. J. Blanksma, Intra-ocular light scatter in pseudophakia. *Doc Ophthalmol* 72 (1989): 335–340.

Wooten, B. R. and G. A. Geri, Psychophysical determination of intra-ocular light scatter as a function of wavelength. *Vision Res* 27 (1987): 1291–1298.

Yager, D., C. L. Liu, N. Kapoor, and R. Yuan, Relations between three measures of ocular forward light scatter and two measures of backward light scatter. *OSA Tech Dig Ser* 3 (1993): 174–177.

Yager, D., R. Yuan, and S. Mathews, What is the utility of the psychophysical 'light scattering factor'? *Invest Ophthalmol Vis Sci* 33 (1992): 688–690.

24 Accommodation mechanisms*

* *Funding source:* Fast Track Young Scientist scheme from Department of Science and Technology, Government of India and Ramalingaswami fellowship from Department of Biotechnology, Government of India.

Shrikant R. Bharadwaj

Contents

24.1 INTRODUCTION

Clarity of vision at different viewing distances is critical for optimal visual performance in routine day-to-day tasks. This is achieved in different species through a variety of mechanisms, all of which are broadly classified as ocular accommodation. The overall goal of these mechanisms is to adjust the optics of the eye such that light rays entering the eye become optically conjugate with the image-forming plane, the retina. Majority of these mechanisms therefore involve dynamically changing the optical power of one or both refractive elements of the eye—the cornea and the crystalline lens—either by changing their curvature or by adjusting their refractive indices or by translating the two positive optical elements toward each other. The readers are referred to the classic textbook by Gordon Walls for a detailed description of the mechanisms of ocular accommodation in various animal, avian, Piscean, and reptilian species.[1] This chapter will restrict itself to the accommodative mechanism employed by humans to achieve clear vision across a range of viewing distances.

Human ocular accommodation functions as a "closed-loop" negative feedback controlled mechanism where sensory information falling on the retina is processed in the cortex to generate neural commands that will alter the biomechanics of the accommodative apparatus to change the optical power of the eye.[2] The resultant change in image quality brought about the optical power change is fed back to the cortex for further processing and for generating a new set of neural commands to the accommodative apparatus. This process continues on in an iterative fashion until the quality of image formed on the retina reaches a certain desirable state. With this background, this chapter will first briefly describe the anatomy of the accommodative apparatus, the neural control of accommodation, and controls engineering

models of the negative feedback control of accommodation. The chapter shall then discuss the sensory cues that stimulate the accommodative process and how they may be weighted during habitual viewing conditions. The discussion shall then tend toward how this weighting of cues may need recalibration in order to maintain optimal performance in the face of changing biomechanical and optical demands imposed during visual development and aging. By and large, this chapter shall take an engineer's approach to describing the accommodative mechanism and, wherever possible, a connection will be made to our current clinical understanding of this mechanism. This chapter is written with the more specialized reader of this topic in mind. Readers are also referred to the excellent textbook chapters on this topic by Adrian Glasser,[3] Kenneth Ciuffreda,[4] and others.

24.2 ANATOMY AND NEUROPHYSIOLOGY OF ACCOMMODATION

The anatomy of the human accommodative apparatus is rather complex involving four different structures that bring about a change in optical power of the eye. These structures include the ciliary muscle, the choroid, the suspensory zonules, and the crystalline lens (Figure 24.1a).

A change in fixation from a distant to a near target requires the optical power of the eye to increase in order to achieve clear vision while a change in fixation from a near to a distant target requires the optical power to decrease in order to achieve clear vision. Such alternations to the optical power of the eye are brought about by changing the shape of the crystalline lens in response to constriction of the ciliary muscle. The most widely

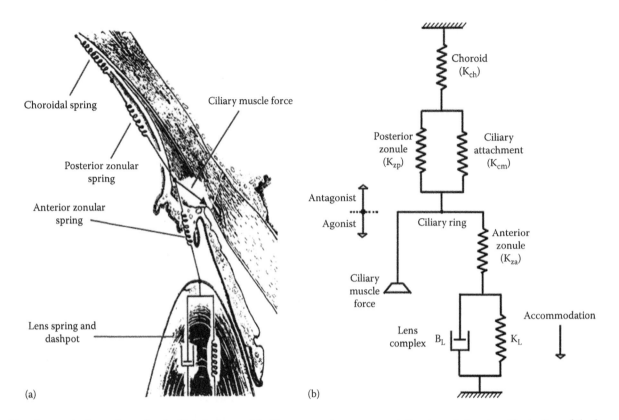

Figure 24.1 Anatomical correlates for the Helmholtz model of human ocular accommodation (panel a) and an illustration of the lumped biomechanical model of the passive plant (panel b). (Adapted from Beers, A.P. and van der Heijde, G.L., *Optom. Vis. Sci.*, 73, 235, 1996; Beers, A.P. and Van Der Heijde, G.L., *Vision Res.*, 34, 2897, 1994.).[5,6] In both panels, the ciliary muscle innervation acting on the choroid, zonules, and the crystalline lens is indicated as an outward directed force, while the choroid and the zonules are represented as pure spring elements. The crystalline lens is represented as a dashpot containing both viscous and elastic elements. Antagonist components in the lumped biomechanical model include the choroid spring, ciliary muscle attachment to the choroid, and posterior zonules while the agonist components include the anterior zonules and the lens–capsule complex. (Adapted from Schor, C.M. and Bharadwaj, S.R., *Vision Res.*, 45, 1237, 2005.)[67]

accepted Helmholtz theory of accommodation posits that the forces applied to the crystalline lens are changed during accommodation by agonist–antagonist interactions between the ciliary muscle and choroid.[2] These forces are transferred via the suspensory zonules to the crystalline lens and its capsule, as indicated in the biomechanical schematic adapted from Beers and van der Heijde[5,6] (Figure 24.1b). When changing fixation from a distant to a near target, the ciliary muscle constricts leading to anterior movement of the choroid and a relaxation of the zonules that suspend the crystalline lens in the anterior chamber of the eye. The zonular relaxation allows the crystalline lens to become more curved and thicker, thereby increasing the optical power of the eye. The reverse happens when fixation changes from near to distance (i.e., during the process of disaccommodation) wherein the ciliary muscle relaxes, leading to a posterior retraction of the choroid, a tightening of the zonules and the flattening of the crystalline lens curvature. This general scheme of the accommodative apparatus has been elaborated upon in static[7–10] and dynamic[5,6] biomechanical models of accommodation. These models have also been used to describe how age-related changes in biomechanics of each structure contribute to the reduction in the amplitude of accommodation in presbyopia.

Of the four structures in the human accommodative apparatus, the ciliary muscle is the only "active" structure to receive neural innervation from the midbrain to change the optical power of the eye. The remaining three structures are "passive" in that they act upon the command received from the midbrain via the ciliary muscle. Neural innervation to accommodation is largely parasympathetic in nature, although a small component of sympathetic innervation has also been reported in the literature.[11] The parasympathetic innervation is fast and robust, and it regulates most of the dynamic changes in accommodation that happen under naturalistic viewing conditions. Sympathetic innervation, on the other hand, is slow and appears to set only the tone of accommodation during periods of extended near viewing.[11–13] Although a significant body of literature exists on the parasympathetic neural innervation to accommodation, all cortical and subcortical areas involved in this process are yet to be completely mapped out. Areas far away from the visual occipital cortex such as the frontal eye fields have been shown to be activated during the process of accommodation and associated binocular near vergence responses.[14–16] The final common pathway for accommodation is more thoroughly mapped out with midbrain areas of the Edinger–Westphal nucleus, pretectal nucleus, and the III (oculomotor) cranial nerve complex actively involved in transmitting parasympathetic neural commands to the ciliary muscle for driving the accommodative response. The review chapters by Paul Gamlin and colleagues provide a more detailed discussion of the cortical and subcortical areas involved in the process of accommodation.[14,15]

24.3 SENSORY CUES FOR ACCOMMODATION

When fixation is changed from one viewing distance to another, the accommodative system has to solve two challenges in order to generate a response that will reinstate clear vision on the retina. The first challenge is to determine the direction of accommodation—that is, should the optical power of the eye increase (i.e., generate accommodation) or should it decrease (i.e., generate disaccommodation) in order to restore clear vision. The second challenge is to determine the magnitude of the accommodative response that will be needed to reinstate clear vision. Once the former challenge is solved, the latter challenge is partly solved by the negative feedback control nature of the accommodative system, although a relative accurate calculation of the magnitude is still needed to generate the aforementioned preprogrammed pulse innervation that overcomes the viscoelasticity of the accommodative apparatus. Sensory stimuli (or cues) that provide information to the visual system on only the magnitude are typically referred to as "even-error cues" to accommodation, while those that provide information on both magnitude and direction are typically referred to as "odd-error" cues to accommodation. A pure defocus blur on the retina is a typical example of "even-error cue" to accommodation. Such a blur is characterized by a progressive loss of contrast in the image that remains identical irrespective of whether the blur is hyperopic (i.e., requiring the eye to accommodate) or myopic (i.e., requiring the eye to disaccommodate) in nature (Figure 24.4).[17]

The contrast loss increases with the magnitude of blur, more so for higher than for lower spatial frequencies. However, when the defocus term is combined with other higher-order wavefront aberrations terms (e.g., spherical aberrations or coma), the form of the resultant point spread function changes depending on whether the defocus is myopic or hyperopic, thereby allowing the visual system to decipher the direction of accommodation (Figure 24.4).[17] Thus, a pure defocus *even-error* cue is converted into *odd-error* cue by the addition of blur from higher-order wavefront aberrations.[17] Other examples of *odd-error* cues to accommodation include retinal image disparity, longitudinal chromatic aberrations, and the microfluctuations of accommodation (see, however, Metlapally et al. for the interaction between the eye's higher-order wavefront aberrations and accommodative microfluctuations[18]).[19–22] Early studies on this topic by Stark and colleagues suggested that the visual system might use blur as a trial-and-error mechanism to determine the direction of accommodation.[23] The evidence for such a behavior came from the number of initial direction errors that the accommodative system made in response to a monocular step change in blur generated using an optometer before embarking

on a course correction in the correct direction.[23] However, such a behavior may not be extrapolated to accommodative responses generated under habitual viewing conditions that contain a combination of blur and other odd-error cues that may provide the desired direction information to the accommodative system. Blur generated using an optometer, as was the case in the study by Phillips and Stark,[23] are devoid of odd-error cues (e.g., retinal disparity, retinal image size), and hence the accommodative system may have resorted to the trial-and-error behavior.

A different way to categorize the sensory cues to accommodation is based on the reference frame used by the cue to drive accommodation. Retinal image blur and binocular retinal image disparity that use the eye as the reference are referred to as retinotopic cues to accommodation, while target proximity that uses the entire body as the reference is referred to as a spatiotopic cue to accommodation.[24] These cues are also routinely described in clinical texts as the three Maddox components of accommodation—blur-driven accommodation, retinal disparity (or vergence)-driven accommodation, and proximity-driven accommodation.[25] The fourth Maddox component is tonic accommodation that represents the baseline state of accommodation when there is no stimulus to focus (e.g., a Ganzfeld screen[26,27] or complete darkness[28,29]).[12] Discussion of tonic accommodation is however outside the purview of this chapter and will not be discussed any further here (readers are referred to the excellent reviews by Rosenfield and colleagues on this topic[12,13]). Extensive literature exists on how the two retinotopic cues of retinal blur and disparity interact with each other and also with the spatiotopic cue of proximity to generate accommodative responses.[20–22,24,30–32] All three cues operate in a negative feedback control scheme to generate the accommodative response, with their individual contributions to the total accommodative response varying with the operating range of each cue.[24] The retinotopic and spatiotopic cues are shown to be effective over complementary operating ranges, with the retinotopic retinal blur cue having a smaller but finer operating range (2D ± 2D) than the spatiotopic proximity cue that has a larger but coarser operating range.[24] The former cue therefore functions as an optimal stimulus for large changes in near demand, while the latter provides precise quantitative information about small fixation errors. Large spatiotopic errors are sampled intermittently at the beginning of the near response, thereby providing critical information for response initiation, whereas small retinotopic errors are sampled continuously to fine-tune the near response.[24] In general, the spatiotopic and retinotopic cues tend to work in a supplementary fashion to improve the efficiency and reduce variability of the accommodative response. Kruger and Pola elegantly showed that the accommodative gain increased and the phase lag reduced when spatiotopic size cues were added to sinusoidal changes in blur.[19,20] The positive impact of size cues were greater for moderate to high temporal frequencies of the sinusoid, indicating that size cue may be more effective for dynamic changes than for steady-state accommodation (Figure 24.2).[19,20]

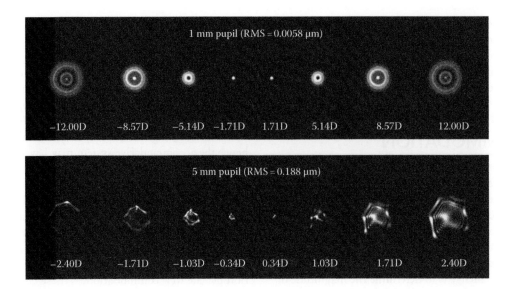

Figure 24.2 Simulated monochromatic point spread functions (PSFs) for a range of myopic and hyperopic defocus values for one subject in the study by Wilson et al.[17] PSFs are shown for 1 mm and 5 mm pupil diameter, with the effect of higher-order monochromatic aberrations being minimal for the former pupil diameter than for the latter. The similarity in the pattern of myopic and hyperopic defoci can be appreciated for the 1 mm pupil diameter (i.e., in the absence of monochromatic higher-order aberrations). PSFs for myopic and hyperopic defoci assume different shapes for the 5 mm pupil diameter suggesting that this information could function as "odd-error" cue for determining the direction of accommodation.

24.4 MODELING THE INTERACTION BETWEEN BLUR AND DISPARITY CUES TO ACCOMMODATION

Extensive literature is available on the independent functioning and the interaction of the two-retinotopic cues of blur and disparity to drive the accommodative response. Such an analysis also forms the basis for the management of nonstrabismic binocular vision disorders in the clinic, with the primary goal there to establish or expand the zone of clear and single binocular vision of the patient.[33] Controls engineering models have especially been useful in this endeavor, with separate models being developed to understand how blur and disparity cues interact with each other to determine the static and dynamic aspects of the accommodative response.[34–43] The static models will be discussed first followed by a discussion of the dynamic models of accommodation. In the static models,[36,38,41] retinal blur is considered as the primary cue for accommodation, while retinal disparity is deemed as the primary cue for binocular vergence (Figure 24.3).

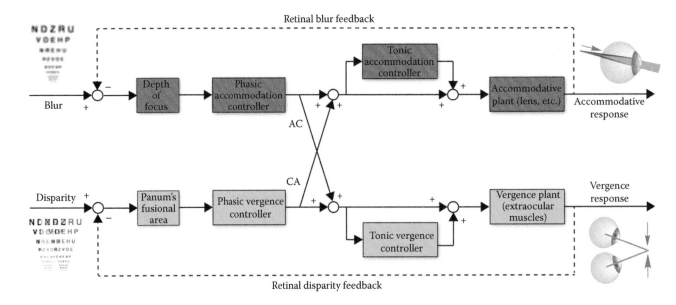

Figure 24.3 Cross-coupling model of accommodation and vergence proposed by Schor (1992).[41] Accommodative and vergence responses are stimulated by retinal blur and retinal disparity cues, respectively. These cues are processed by phasic and tonic controllers that feed their output into the respective biomechanical plants to generate the accommodative and vergence motor responses. The two motor responses are also coupled through the accommodation–convergence (AC) and convergence–accommodation (CA) cross-links such that a change in blur-driven accommodation also stimulates vergence through the AC cross-link, while a change in disparity-driven vergence stimulates accommodation through the CA cross-link. Blur and disparity are processed in a negative feedback loop until both cues reach below their respective thresholds (i.e., the depth of focus and Panum's fusional area, respectively).

A step change in retinal blur and retinal disparity above the visual system's detection threshold (i.e., above the depth of focus for blur and above the Panum's fusional area for disparity) initiates the cortical negative feedback loop for accommodation and vergence, and these continue on until the blur and disparity cues are reduced below their respective threshold values. Blur and disparity are processed by two types of controllers in the cortex—the phasic controller and the tonic controller—to convert the sensory cues into motor commands that then feed the respective biomechanical plants for generating the motor responses (crystalline lens complex for accommodation and extraocular muscles for vergence). The phasic and tonic controllers are usually modeled as first-order "leaky integrators" containing a gain element and time constant that determine the controller's integrated output and the sustenance of this output, respectively.[36,38,41] Properties of these two controllers are therefore of interest to understand how rapid versus sustained changes in the near response are brought about. The phasic controllers are modeled to have a large gain in order to generate large changes in neural output for accommodation and vergence within a short period of time, thereby generating rapid step changes in the motor response until the new desired steady state is achieved. However, these controllers have a relatively short time constant indicating that their output is ill sustained and cannot maintain the response in the desired steady state for extended periods of time.[36,38,41] The tonic controllers, on the other hand, have a smaller gain but longer time constant; they therefore respond more sluggishly to the step change in blur and disparity, but their output is far more sustained when compared to that of the phasic controller. The tonic controllers therefore aid the sustenance of steady-state accommodation and vergence for an extended period of time and, as their names suggest, play a key role in determining the tonic state of these motor responses.[36,38,41] Said another way, a step change in accommodative or vergence demand that is within the operating range of the retinotopic cues of blur or disparity will stimulate a response initially from the phasic accommodation/vergence controllers until a desired feedback-driven steady state is achieved. Sustenance of this steady state for extended periods of time is ensured by the activity of the tonic accommodation/vergence controllers.

Any change in blur-driven accommodation also stimulates the vergence system through the accommodative–vergence cross-coupling (AC cross-link), while any change in disparity-driven vergence stimulates the accommodative system through the vergence–accommodation cross-coupling (CA cross-link) (Figure 24.3). The strength of the AC and CA cross-links determine how much of coupled vergence and accommodative responses are generated per unit change in blur-driven accommodation and disparity-driven vergence, respectively. The strengths of these cross-links are also measured in the clinic during an orthoptic evaluation as the accommodative–convergence-to-accommodation ratio (AC/A ratio recorded in units of prism diopters per diopter or degrees per diopter or meter angles per diopter) and the convergence–accommodation-to-convergence ratio (CA/C ratio recorded in units of diopters per prism diopter or diopters per prism degree or diopters per meter angle).[44] AC/A ratio may be measured by making the disparity-driven vergence feedback open loop (e.g., by occluding one eye), while CA/C ratio may be measured by making the blur-driven accommodation feedback open loop (e.g., using pinholes[45–47] or a very low spatial frequency difference-of-Gaussian-type targets[48,49]).[44] There seems to exist a small but

significant reciprocal relationship between the AC/A and CA/C ratios, with individuals with high AC/A ratio showing lower CA/C ratio and vice versa.[50,51] The strength of this reciprocity is however stronger in patient with accommodative–vergence anomalies[51] than in those with normal binocular vision status.[50] Neural correlates for the cross-links have also been demonstrated in the form of neurons in the near-response complex in the supraoculomotor area of the midbrain (comprising the III cranial nerve nucleus and the Edinger–Westphal nucleus) systematically increasing their firing rate for a near stimulus and also responding to the target even when either the blur- or disparity feedback is made open loop.[14,52–54] AC/A ratio is more routinely measured in the clinic than the CA/C ratio because of the relative ease in removing disparity feedback than removing the blur feedback. Nonstrabismic binocular vision anomalies in the clinic are therefore often times attributed to nonoptimized AC cross-link, while the contribution of the CA cross-link is generally ignored. Ironically, however, many laboratory-based studies investigating the relative contributions of the cross-links to total accommodation and vergence response have indicated that the CA cross-link contributes equally or more to the total accommodative response than does the AC cross-link to the total vergence response.[55,56] The population average AC/A ratio is 4.6ΔD/D, indicating that the AC cross-link contributes to about 77% of the total vergence response in a subject with 6 cm interpupillary distance (IPD) (i.e., with a vergence demand of 6Δ at 100 cm viewing distance or 12Δ at 50 cm viewing distance; Δ = IPD/viewing distance in meters). The average CA/C ratio ranges between 0.58 and 0.9D/MA in previous studies, indicating that the CA cross-link may contribute anywhere between 58% and 90% of the total accommodative response in subjects with similar IPDs.[50,55,57] Note, however, that the contribution of the cross-link responses to the total near response is less than 100% indicating that the remainder of the contribution needs to be fulfilled by direct blur- and disparity-driven responses. Such an analysis, of course, assumes a linear interaction between the different Maddox components of accommodation and vergence, which has been contested in the past by Judge.[58]

The location of the phasic and tonic elements of the direct response with respect to the AC and CA cross-links has been a matter of debate with conflicting models proposed by Schor[41] and Hung and Semmlow[38] to explain the results of empirical studies conducted to understand the interaction between the accommodation and vergence motor responses. The cross-link model proposed by Schor places the tonic element after the phasic element and the cross-link, indicating that the output of both the direct phasic controller and the cross-link can influence the output of the tonic controller (Figure 24.3).[41] On the other hand, the model proposed by Hung and Semmlow places the tonic element at a summing junction along with the output of the phasic and the cross-link element.[38] While the purpose of this discussion is not to delve into the nuances of these models (the readers are referred to the textbook chapter by Jiang for such a discussion[39]), it suffices to say that both models sufficiently explain the empirical phenomenon of tonic accommodation and tonic vergence adaptation through a buildup in activity of their tonic controllers.[59–65] According to the Schor model, the buildup in the output of the tonic controller occurs with sustained activity of the phasic controller or that of the cross-link, and the tonic system eventually takes control of the accommodation and vergence motor output.[41] In this scheme, the AC and CA cross-links are primarily stimulated by the output of the phasic controller, while the tonic controller exerts an influence on the cross-links only through blur and disparity feedback (Figure 24.3).[66] Hung and colleagues explain this buildup as separate A_{Bias} and V_{Bias} (for accommodation and vergence bias, respectively) elements in their model.[38]

The first- and second-order dynamics (velocity and acceleration, respectively) of accommodation to a step change in near visual demand have also been modeled more recently as a combination of two innervations that are optimized to overcome the biomechanical viscoelasticity of the accommodative plant (Figure 24.4).[42,67] The organization of these models are very similar to those proposed for the saccadic[68,69] and vergence[70–72] eye movements as the overarching goal of all three oculomotor systems are the same—to generate a response as rapidly as possible by overcoming the viscoelasticity of their respective biomechanical plants (extraocular muscles in the case of saccades and vergence and crystalline lens complex in the case of accommodation). The pulse innervation for accommodation that lasts for a very short period of time (50–150 ms) is thought to be fed back independently and preprogrammed based on the near vision demand, and it primarily exists to overcome the initial viscoelasticity of the elements in the accommodative apparatus (largely the crystalline lens).[42,67] The step innervation, on the other hand, is feedback controlled and regulates the magnitude of the accommodative response to achieve the desired retinal image quality (Figure 24.4).[42,67] Such a dual innervation scheme explains the empirical trends in the first- and second-order dynamics of accommodative step responses (i.e., velocity and acceleration characteristics, respectively) and its changes with presbyopia better than a single-step innervation scheme.[73–78] A similar dual-innervation model has also been proposed to explain the dynamics of the accommodation and vergence cross-link responses[79] and also for slow ramp changes in accommodative demand (i.e., while following a target that changes smoothly from one distance to another).[40] For ramp stimuli, the sustained neural innervation maintains clarity of the target as it changes viewing distances while a pulse-like innervation acts as a preprogrammed catch-up mechanism when retinal image quality falls below a desired level of clarity.[40]

24.5 CUE CONFLICTS AND ITS INFLUENCE ON ACCOMMODATION

Under naturalistic viewing conditions, the blur cue to accommodation remains consistent in the two eyes and the blur cue is also consistent with the disparity and proximity cues available in the stimulus. There are however several conditions where the consistency between the cues is broken down, and it is of interest to understand how the visual system handles such a cue-conflict environment. Two such situations will be discussed here. The first situation is anisometropia where the refractive powers of the two eyes are dissimilar, and therefore the accommodative demands in the two eyes are also dissimilar.[80,81] In myopic anisometropia, the accommodative demand of the myopic eye is relatively smaller

when compared to the fellow eye with lesser magnitude of myopia or emmetropia. Alternatively, in hyperopic anisometropia, the accommodative demand of the hyperopic eye is relatively larger when compared to the fellow eye with lesser magnitude of hyperopia or emmetropia. Physiologically, anisometropia-like viewing experience is encountered during off-axis viewing wherein a target situated away from the midline creates dissimilar accommodative demands in the two eyes.[82] The eye with the target at a closer viewing distance experiences greater accommodative demand than the eye where the target is relatively farther away.[82] The visual system could, in theory, be adopting several strategies to overcome this challenge. First, the visual system could be generating dissimilar accommodative responses to cater to the individual demands of each other (i.e., aniso-accommodation). Accommodative responses of humans are however centrally controlled as described earlier, and they are typically consensual, with the accommodative performance deteriorating in the presence of significantly dissimilar interocular retinal image qualities.[81,83–86] Marran and Schor studied the ability of humans to generate aniso-accommodative responses and found that only a maximum of 0.5D of aniso-accommodation could be generated and, that too, after significant amount of training.[85,86] Such small magnitudes of aniso-accommodation are likely to be little practical consequence in handling anisometropia. In a related experiment, Flitcroft and colleagues stimulated accommodation of their subjects in a haploscope setup with the optical blur changing in counterphase in the two eyes.[81] They found the resulted accommodation of the two eyes to be an average of the optical demands in the eye—that is, in a counterphase scenario, the binocular accommodative response was close to zero diopters.[81] Second, the visual system could be adopting a monovision-type strategy with one eye being used for near viewing while the fellow eye is used for distance viewing.[87,88] In this case, minimal modulation of accommodation might be expected in the two eyes in response to a change in target position. This strategy may work well for myopic anisometropes, but it is difficult to conceive how such a strategy would function for hyperopic anisometropes. Nevertheless, Bharadwaj and Candy found little evidence for such a strategy being used by the adult and infant visual system for different magnitudes of induced myopic and hyperopic anisometropia (Figure 24.5).[80]

Results of their study indicated that the accommodative response in one eye (typically the less ametropic eye) maintained focus on targets at different distances, while accommodation in the fellow eye consensually followed the tracking eye, resulting in chronic retinal defocus in that eye.[80] This strategy was independent of whether the target was slowly ramped before the subjects or if it was stepped before them between two viewing distances (Figure 24.6).

The change in target position in these experiments happened in real space, and therefore disparity-driven vergence response was also present along with the blur cue. The extent to which the accommodative responses generated in the presence of anisometropia reflected a cross-link vergence response (through the CA/C cross-link) remains unknown. Further, these experiments were performed on subjects with induced anisometropia—what the accommodative response of habitual anisometropes would be also remains to be explored.

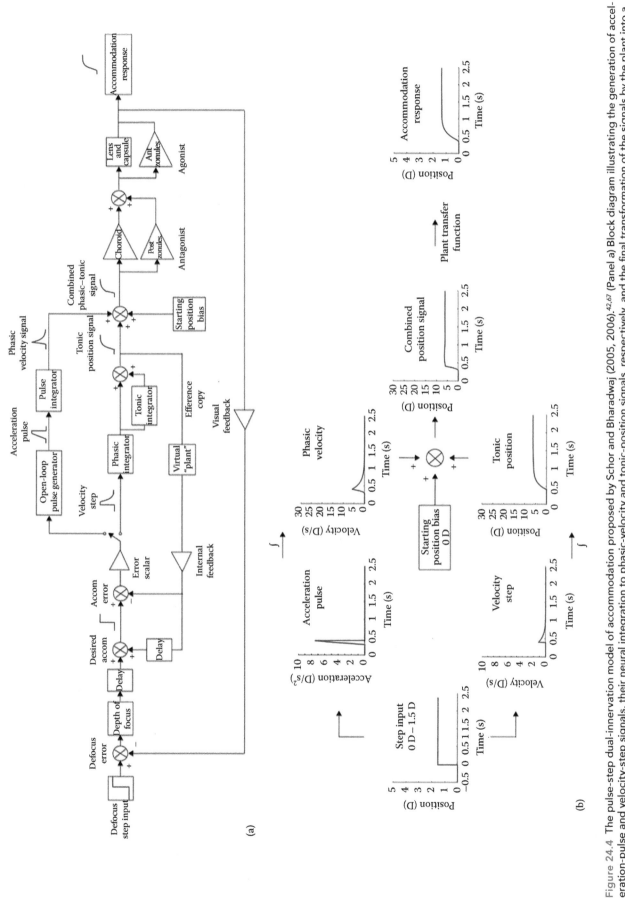

Figure 24.4 The pulse-step dual-innervation model of accommodation proposed by Schor and Bharadwaj (2005, 2006).[42,67] (Panel a) Block diagram illustrating the generation of acceleration-pulse and velocity-step signals, their neural integration to phasic-velocity and tonic-position signals, respectively, and the final transformation of the signals by the plant into a disaccommodation step response. Insets in figure a show signal profiles at each stage of the models. The block diagram for the pulse-step model of disaccommodation (i.e., near to far accommodation) is identical to the block diagram of the accommodation shown here. (Panel b) Signal processing flow charts for accommodation illustrating the neural integration of independent acceleration-pulse and velocity-step signals into phasic-velocity and tonic-position signals, respectively, which is transformed by the plant into accommodation step responses. The evolution of the neural signal for disaccommodation step responses is identical to what is shown here for accommodation. The readers are referred to Schor and Bharadwaj (2005, 2006) for details of the pulse-step model of accommodation.[42,67] (Adapted from Schor, C.M. and Bharadwaj, S.R., *Vision Res.*, 46, 242, 2006.)[42]

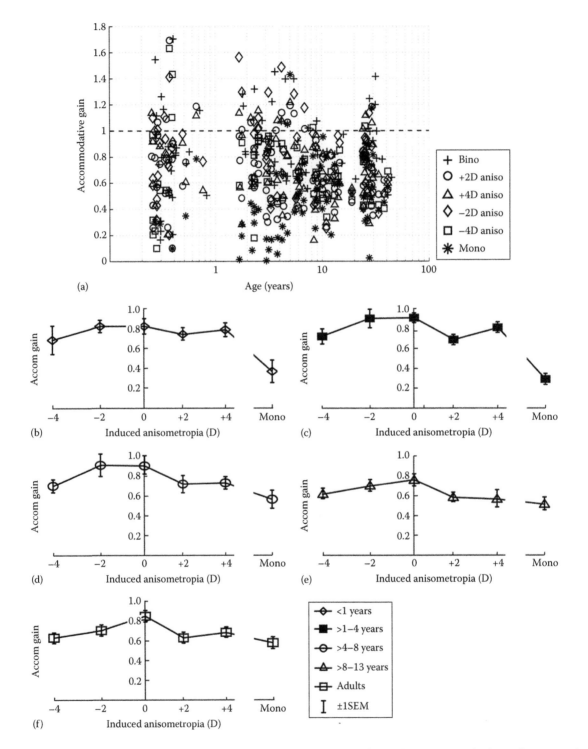

Figure 24.5 (Panel a) Scatter plot of accommodative gain in the left eye as a function of the subject's age under binocular, monocular, and the four induced myopic (+2D and +4D) and hyperopic (−2D and −4D) anisometropia conditions. Mean (±1-SEM) accommodative gain of the left eye plotted as a function of the level of induced anisometropia for <1-year-olds (Panel b), 2- to 4-year-olds (Panel c), 4- to 8-year-olds (Panel d), 8- to 13-year-olds (Panel e), and adults (Panel f). A value of unity on the abscissa scale indicates the accommodative response equaled the stimulus, while a value of zero on this scale indicates that there was zero accommodative response. All responses were obtained when the target ramped between 80 and 33 cm before the subject. (Adapted from Bharadwaj, S.R. and Candy, T.R., J. Vis., 9(4), 1, 2009.)[80]

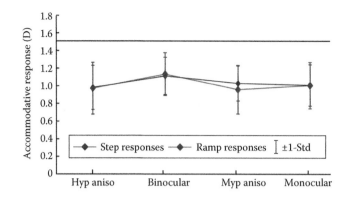

Figure 24.6 Mean ± 1SD accommodative response amplitude plotted in the step and ramp stimulus paradigm. Hyp Aniso and Myp Aniso indicate 3D of induced hyperopic and induced myopic anisometropia, respectively. The target either ramped or stepped (electronically between two LCD screens placed at two different viewing distances) between 67 and 33 cm before the subject. (Unpublished data from the author's laboratory at LVPEI, Hyderabad, India.)

Overall, it appears that the accommodative system may not be well suited to handle interocular differences in retinal image quality. These results, especially from children, may therefore have important implications for developmental anomalies like amblyopia that are strongly associated with refractive anisometropia.[89,90]

Conflicts between blur and disparity cues may be experienced by the visual system in the presence of uncorrected refractive errors or uncorrected phorias. For instance, the demand on accommodation for a given target viewing distance is greater than that of vergence (in units of meter angles) in individuals with uncorrected hyperopia. Similarly, for the same target viewing distance, the demand on vergence is greater than that of accommodation in individuals with uncorrected exophoria. Individuals with uncorrected myopia or uncorrected exophoria will experience the opposite. Such dissimilar demands on accommodation and vergence will challenge the simultaneous achievement of clear and single vision and may abnormally constrict the zone of clear and single binocular vision of an individual.[91] The developing visual system may encounter such a conflict between accommodative and vergence demands on a routine basis if they are born with adultlike cross-link gains between the two systems.[92,93] Infants are typically born hyperopic (mean ± 1SD: 2D ± 2SD vs. emmetropia of adults) and with an IPD that is smaller than those of adults (44 mm of an infant vs. 65 mm of an adult).[94–97] For a given target viewing distance, the accommodative demand of an infant is therefore greater and the vergence demand is therefore smaller than that of an emmetropic adult (Figure 24.7).

If adultlike interactions between blur-driven accommodation and disparity-driven vergence were to exist from infancy, then the developing visual system is in a continuous state of conflict, challenging the achievement of simultaneous clear and single binocular vision.[92,93] Furthermore, the hyperopia declines and the IPD widens with the age of the child, suggesting that the relative demands on blur-driven accommodation and disparity-driven vergence vary during development.[92,93]

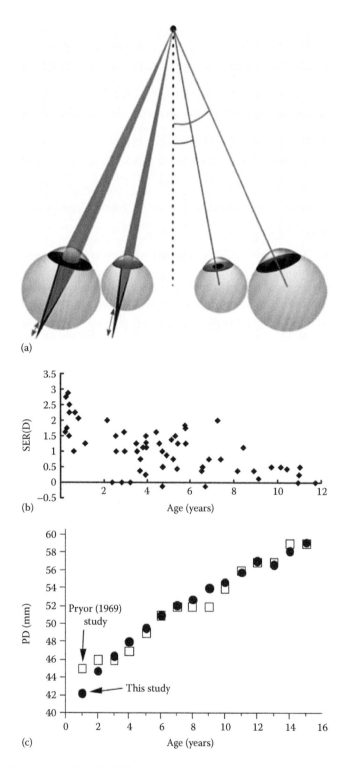

Figure 24.7 (Panel a) Schematic representation of the hyperopic refractive error and the narrow interpupillary distance of infants relative to that of adults. The accommodative demands of infants are therefore larger than that of adults, while the vergence demands of infants are smaller than those of adults. (Panel b) Scatter diagram of the spherical equivalent refractive error plotted as a function of age of the subject. (Adapted from Bharadwaj, S.R. and Candy, T.R., J. Vis., 9(4), 1, 2009.)[93] (Panel c) Interpupillary distance of subjects from <1 year old to adults obtained by MacLachlan and Howland (2002)[96] and comparative data from Pryor (1969).[135] (Adapted from MacLachlan, C. and Howland, H.C., Ophthalmic Physiol. Opt., 22, 175, 2002.)[96]

The ability of the adult and the developing accommodative and vergence system to handle and compensate such cue conflicts has been explored well in the past.[57,98] Interactions between accommodation and vergence in the presence of such conflicting dual closed-loop viewing conditions (i.e., both blur and disparity feedback are functional but conflicting in nature) are likely to be complex as the final responses could consist of combinations of responses driven by the primary cues, plus the coupling, and negative feedback used to correct any remaining errors.[56,99–101] In the clinic, the ability to handle such conflicts is routinely assessed by stimulating either blur-driven accommodation ("relative accommodation") or disparity-driven vergence ("relative vergence"), while keeping the other cue clamped to a constant value and the subject instructed to maintain a clear and single percept of a visual target.[91,102–104] Using subjective estimates of the first perception of target blur or diplopia at the testing distance, the typical adult ranges of relative accommodation and vergence have been estimated to extend from −3.00D to +2.25D and from 24ΔD base-out to 23ΔD base-in, respectively.[105] This indicates that the visual system is capable of handling up to 5.0–5.5D and up to ±13°–14° (degrees = atan(ΔD/100); the equivalent degree value for a 24ΔD stimulus is atan(24/100) = 13.5°) of conflicting demands between the two cues. Objective measurements of accommodative and vergence responses to such "conflicting" stimuli confirm that the motor responses are indeed dissociated, with accommodation responding more to the lens stimulation and vergence responding more to the prism stimulation (Figure 24.8).[57,98,106,107]

Similar responses were obtained by Bharadwaj and Candy[93] who explored the ability of the developing visual system to handle such cue conflicts by placing either a pair of negative lenses or a pair of base-out prisms (of equivalent magnitude to the negative lens) before the child's eye while the child maintained binocular fixation on a near target (Figure 24.9a and b).

The accommodative response generated when placing negative lenses was somewhat smaller than the vergence response generated when placing base-out prisms (Figure 24.9a and b).[93] This is somewhat expected from the larger sensory threshold for blur (depth of focus = ~0.5D[108]) when compared to that of disparity (Panum's fusional area = 0.1MA[109]). The response amplitude of the stimulated pathway (i.e., blur-driven accommodation when stimulated with negative lenses and disparity-driven vergence when stimulated with base-out prisms) also increased gradually with age (more for accommodation than for vergence), indicating a gradual maturation of the visual system to handle such conflicting scenarios (Figure 24.9b).[93] Interestingly, there was no strong bias toward either blur or disparity cue while generating these responses, as indicated from the frequency of responses being similar to both lens and prism stimulation (Figure 24.9c).[93] Taken together, the results of all these experiments indicate that the developing and the adult human visual system has built-in buffer mechanisms to handle conflicting demands of blur and disparity and achieve simultaneous clear and single binocular vision. The compromise of single vision at the cost of blurred vision or clear vision at the cost of diplopia appears only when the conflict

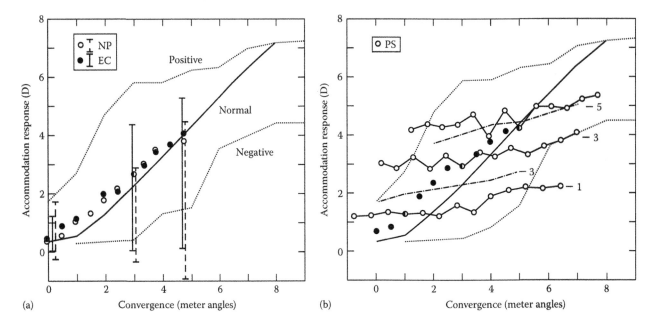

Figure 24.8 Data showing the ability of the adult visual system to handle conflicts between blur and disparity cues. (Adapted from Ramsdale, C. and Charman, W.N., *Ophthalmic Physiol. Opt.*, 8, 43, 1988[98] with comparison to the data obtained from their study and by Fincham and Walton (1957).[57]) (Panel a) Positive and negative relative accommodation obtained from a 32-year-old subject for various values of fixed convergence stimuli. The continuous line shows Fincham and Walton's result of normal accommodation (i.e., in the absence of cue conflict), and the dotted lines show the limits of relative accommodation. Open and filled circles show normal right-eye responses obtained from two subjects in the study by Ramsdale and Charman,[95] and the vertical bars show the amplitudes of relative accommodation of these individuals at three different object distances (0.25MA, 3MA, and ~4.8MA).[98] The limits of relative accommodation were taken as the points where the Snellen acuity appeared first blurred. (Panel b) Effect of changing convergence on accommodation to targets at a fixed viewing distance. The heavy and dotted lines are as in panel a. The filled circles show the responses with normal convergence (i.e., in the absence of cue conflict). The open circles show responses when convergence was altered by prism, while the target was set at −1D, −3D, and −5D viewing distances. Data are plotted over the convergence range within which the subject reported the target clear and single.

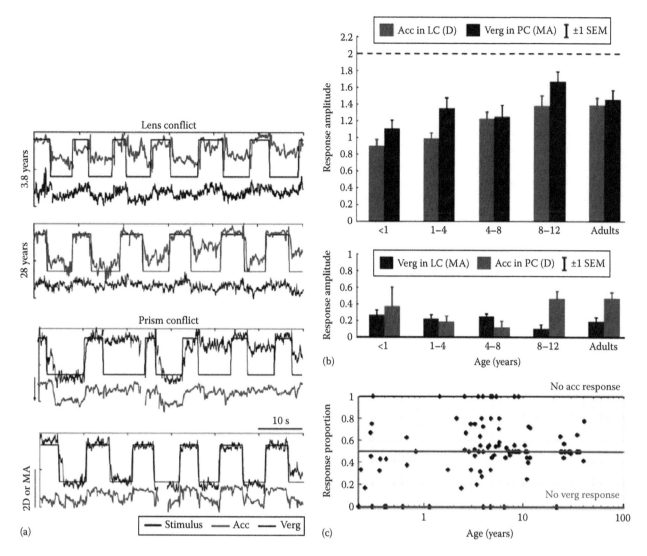

Figure 24.9 Data showing the ability of the developing visual system to handle conflicts between blur and disparity cues. (Adapted from Bharadwaj, S.R. and Candy, T.R., *J. Vis.*, 9(4), 1, 2009.)[93] (Panel a) Representative raw traces of accommodation and vergence from a 3.8- and a 28-year-old when stimulated with negative lenses (indicated as lens conflict [LC] in this figure) (top panel) and with base-out prisms (indicated as prism conflict [PC] in this figure) (bottom panel). The raw traces have been vertically shifted for clarity, with no shifting of the horizontal (time) axis. Downward on the ordinate indicates an increasing stimulus or response. (Panel b) Mean (±1 SEM) amplitude of the stimulated response (top panel) and the coupled response (bottom panel) in the negative lens and base-out prism conditions. The dashed line in the top panel indicates the amplitude of lens (in diopters) and prisms (in meter angles) used in this study. (Panel c) Response proportion (i.e., ratio of the number of valid vergence responses in the PC condition to the total number of valid responses in the LC and PC conditions combined) of individual subjects plotted as a function of their age. Subjects with no valid response in both LC and PC conditions were not included in these analyses. The purple line in the figure indicates equal proportion of responses to the lens and prism stimulation.

between the two cues is large (e.g., large hyperopia or phorias) or when the buffering mechanisms have limited bandwidth (e.g., low relative accommodation or relative vergence values). Optical aids like spectacles or prisms, in effect therefore, reduce this conflict by making the demands of blur and disparity more similar, while orthoptic exercises train the visual system to handle such conflicts by increasing the bandwidth of these buffering mechanisms.

24.6 CONCLUSION

This chapter has intended to provide an overview of the accommodative mechanism employed by the mammalian (primarily, human) visual system to achieve clear vision at a range of different viewing distances. The chapter initially discussed the complex anatomy of the accommodative plant and the neural strategies employed to control this plant for generating rapid and sustained changes in the optical power of the eye. The discussion then shifted toward understanding the behavioral aspects of accommodation, with specific reference to how sensory cues of blur, disparity, and proximity are utilized by the visual system to generate a given accommodative response. The aforementioned discussion of this fascinating autofocus mechanism of the eye is by no means exhaustive. Even while Fick observed in 1879 that "there is no need to discuss sundry theories of accommodation now that the problem has been solved in its entirety [by Helmholtz]," the field of accommodation research continues to evolve rapidly by attracting the attention of clinicians, psychophysicists, neuroscientists, bioengineers, and optical engineers.[110] Three recent developments in this field that were not covered in this chapter include (1) how the accommodative response of the eye modulates[111–118] and gets modulated[119–123] by the higher-order wavefront aberrations of the eye, (2) how the waning accommodative ability of presbyopes may be restored through optical and surgical means (e.g., accommodative intraocular lenses) thereby providing freedom by conventional bifocal and progressive addition lenses,[124–127] and (3) how the accommodative mechanism adapts to the changing optical and biomechanical demands imposed by these wavefront aberrations and presbyopia, respectively.[128–134] Peer-reviewed scientific literature is accumulating on these topics at a very rapid rate, and a detailed discussion of these topics is worth a separate chapter in itself. For sure, this body of literature is changing our fundamental understanding of the refraction of light through the optical media and taking us closer than ever toward the restoration of focusing ability in the aging eye.

REFERENCES

1. Walls GL. Accommodation and its substitutes. In: Walls GL (ed.), *The Vertebrate Eye and Its Adaptive Radiation*. New York: Hafner Publishing Company; 1963: pp. 247–283.
2. Helmholtz von H. *Handbuch der Physiologishen Optik* (Translated by Southall JPC, 1962). New York: Dover; 1909.
3. Glasser A, Kaufman PL. Accommodation and presbyopia. In: Kaufman PL, Alm A (eds.), *Adler's Physiology of the Eye*. St. Louis, MO: Mosby; 2003: pp. 197–236.
4. Ciuffreda KJ. Accommodation and its anomalies. In: Charman WN (ed.), *Vision and Visual Dysfunction: Visual Optics and Instrumentation*. London, U.K.: MacMillan; 1991: pp. 231–279.
5. Beers AP, Van Der Heijde GL. In vivo determination of the biomechanical properties of the component elements of the accommodation mechanism. *Vis Res* 1994;34:2897–2905.
6. Beers AP, van der Heijde GL. Age-related changes in the accommodation mechanism. *Optom Vis Sci* 1996;73:235–242.
7. Stark L. Presbyopia in light of accommodation. In: Stark L, Obrecht G (eds.), *Presbyopia: Recent Research and Reviews from the Third International Symposium*. New York: Professional Press Books; 1987: pp. 264–274.
8. Judge SJ, Burd HJ. Modelling the mechanics of accommodation and presbyopia. *Ophthalmic Physiol Opt* 2002;22:397–400.
9. Wyatt HJ. Application of a simple mechanical model of accommodation to the aging eye. *Vis Res* 1993;33:731–738.
10. Wyatt HJ. Some aspects of the mechanics of accommodation. *Vis Res* 1988;28:75–86.
11. Gilmartin B. A review of the role of sympathetic innervation of the ciliary muscle in ocular accommodation. *Ophthalmic Physiol Opt* 1986;6:23–37.
12. Rosenfield M, Ciuffreda KJ, Hung GK, Gilmartin B. Tonic accommodation: A review. I. Basic aspects. *Ophthalmic Physiol Opt* 1993;13:266–284.
13. Rosenfield M, Ciuffreda KJ, Hung GK, Gilmartin B. Tonic accommodation: A review. II. Accommodative adaptation and clinical aspects. *Ophthalmic Physiol Opt* 1994;14:265–277.
14. Gamlin PD. Subcortical neural circuits for ocular accommodation and vergence in primates. *Ophthalmic Physiol Opt* 1999;19:81–89.
15. Gamlin PD. Neural mechanisms for the control of vergence eye movements. *Ann N Y Acad Sci* 2002;956:264–272.
16. Gamlin PD, Yoon K. An area for vergence eye movement in primate frontal cortex. *Nature* 2000;407:1003–1007.
17. Wilson BJ, Decker KE, Roorda A. Monochromatic aberrations provide an odd-error cue to focus direction. *J Opt Soc Am A Opt Image Sci Vis* 2002;19:833–839.
18. Metlapally S, Tong JL, Tahir HJ, Schor CM. The impact of higher-order aberrations on the strength of directional signals produced by accommodative microfluctuations. *J Vis* 2014;14.
19. Kotulak JC, Schor CM. A computational model of the error detector of human visual accommodation. *Biol Cybern* 1986;54:189–194.
20. Kruger PB, Pola J. Stimuli for accommodation: Blur, chromatic aberration and size. *Vis Res* 1986;26:957–971.
21. Kruger PB, Pola J. Dioptric and non-dioptric stimuli for accommodation: Target size alone and with blur and chromatic aberration. *Vis Res* 1987;27:555–567.
22. Kruger PB, Pola J. Accommodation to size and blur changing in counterphase. *Optom Vis Sci* 1989;66:455–458.
23. Phillips S, Stark L. Blur: A sufficient accommodative stimulus. *Doc Ophthalmol* 1977;43:65–89.
24. Schor CM, Alexander J, Cormack L, Stevenson S. Negative feedback control model of proximal convergence and accommodation. *Ophthalmic Physiol Opt* 1992;12:307–318.
25. Morgan MW. The Maddox classification of vergence eye movements. *Am J Optom Physiol Opt* 1980;57:537–539.
26. Westheimer G. Accommodation measurements in empty visual fields. *J Opt Soc Am* 1957;47:714–718.
27. Whiteside TC. Accommodation of the human eye in a bright and empty visual field. *J Physiol* 1952;118:65P–66P.
28. Leibowitz HW, Owens DA. Anomalous myopias and the intermediate dark focus of accommodation. *Science* 1975;189:646–648.
29. Leibowitz HW, Owens DA. Night myopia and the intermediate dark focus of accommodation. *J Opt Soc Am* 1975;65:1121–1128.
30. Hung GK, Ciuffreda KJ, Rosenfield M. Proximal contribution to a linear static model of accommodation and vergence. *Ophthalmic Physiol Opt* 1996;16:31–41.
31. Joubert C, Bedell HE. Proximal vergence and perceived distance. *Optom Vis Sci* 1990;67:29–35.
32. Rosenfield M, Ciuffreda KJ, Hung GK. The linearity of proximally induced accommodation and vergence. *Invest Ophthalmol Vis Sci* 1991;32:2985–2991.
33. Scheiman M, Wick B. *Clinical Management of Binocular Vision*, 3rd ed. Philadelphia, PA: Lippincott Williams & Wilkins; 2008.
34. Carroll JP. Control theory approach to accommodation and vergence. *Am J Optom Physiol Opt* 1982;59:658–669.
35. Krishnan VV, Stark L. Integral control in accommodation. *Comput Programs Biomed* 1975;4:237–245.
36. Eadie AS, Carlin PJ. Evolution of control system models of ocular accommodation, vergence and their interaction. *Med Biol Eng Comput* 1995;33:517–524.
37. Hung GK, Ciuffreda KJ, Khosroyani M, Jiang B-C. Models of accommodation. In: Hung GK, Ciuffreda KJ (eds.), *Models of the Visual System*. New York: Kluwer Academic/Plenum Publishers; 2002: pp. 287–340.
38. Hung GK, Semmlow JL. Static behavior of accommodation and vergence: Computer simulation of an interactive dual-feedback system. *IEEE Trans Biomed Eng* 1980;27:439–447.
39. Jiang B-C, Hung GK, Ciuffreda KJ. Models of vergence and accommodation–vergence interactions. In: Hung GK, Ciuffreda KJ (eds.), *Models of the Visual System*. New York: Kluwer Academic/Plenum Publishers; 2002: pp. 341–384.
40. Khosroyani M, Hung GK. A dual-mode dynamic model of the human accommodation system. *Bull Math Biol* 2002;64:285–299.
41. Schor CM. A dynamic model of cross-coupling between accommodation and convergence: Simulations of step and frequency responses. *Optom Vis Sci* 1992;69:258–269.
42. Schor CM, Bharadwaj SR. Pulse-step models of control strategies for dynamic ocular accommodation and disaccommodation. *Vis Res* 2006;46:242–258.

43. Toates FM. The accommodation control system of the human eye. *Exp Eye Res* 1971;11:142–143.

44. Goss DA. *Ocular Accommodation, Convergence and* Fixation *Disparity*, 3rd ed. Santa Ana, CA: Optometric Extension Program Foundation Press; 2009: pp. 286.

45. Hennessy RT, Iida T, Shina K, Leibowitz HW. The effect of pupil size on accommodation. *Vis Res* 1976;16:587–589.

46. Ripps H, Chin NB, Siegel IM, Breinin GM. The effect of pupil size on accommodation, convergence, and the AC/A ratio. *Invest Ophthalmol* 1962;1:127–135.

47. Ward PA, Charman WN. Effect of pupil size on steady state accommodation. *Vis Res* 1985;25:1317–1326.

48. Bobier WR, Guinta A, Kurtz S, Howland HC. Prism induced accommodation in infants 3 to 6 months of age. *Vis Res* 2000;40:529–537.

49. Tsuetaki TK, Schor CM. Clinical method for measuring adaptation of tonic accommodation and vergence accommodation. *Am J Optom Physiol Opt* 1987;64:437–449.

50. Bruce AS, Atchison DA, Bhoola H. Accommodation-convergence relationships and age. *Invest Ophthalmol Vis Sci* 1995;36:406–413.

51. Schor C, Horner D. Adaptive disorders of accommodation and vergence in binocular dysfunction. *Ophthalmic Physiol Opt* 1989;9:264–268.

52. Judge SJ, Cumming BG. Neurons in the monkey midbrain with activity related to vergence eye movement and accommodation. *J Neurophysiol* 1986;55:915–930.

53. Cumming BG, Judge SJ. Disparity-induced and blur-induced convergence eye movement and accommodation in the monkey. *J Neurophysiol* 1986;55:896–914.

54. Gamlin PD, Zhang Y, Clendaniel RA, Mays LE. Behavior of identified Edinger–Westphal neurons during ocular accommodation. *J Neurophysiol* 1994;72:2368–2382.

55. Kersten D, Legge GE. Convergence accommodation. *J Opt Soc Am* 1983;73:332–338.

56. Semmlow J, Heerema D. The synkinetic interaction of convergence accommodation and accommodative convergence. *Vis Res* 1979;19:1237–1242.

57. Fincham EF, Walton J. The reciprocal actions of accommodation and convergence. *J Physiol* 1957;137:488–508.

58. Judge SJ. Do target angular size-change and blur cues interact linearly in the control of human accommodation? *Vis Res* 1988;28:263–268.

59. Bobier WR, Sivak JG. Orthoptic treatment of subjects showing slow accommodative responses. *Am J Optom Physiol Opt* 1983;60:678–687.

60. Judge SJ, Miles FA. Changes in the coupling between accommodation and vergence eye movements induced in human subjects by altering the effective interocular separation. *Perception* 1985;14:617–629.

61. Liu JS, Lee M, Jang J et al. Objective assessment of accommodation orthoptics. I. Dynamic insufficiency. *Am J Optom Physiol Opt* 1979;56:285–294.

62. Miles FA, Judge SJ, Optican LM. Optically induced changes in the couplings between vergence and accommodation. *J Neurosci* 1987;7:2576–2589.

63. Sreenivasan V, Bobier WR. Reduced vergence adaptation is associated with a prolonged output of convergence accommodation in convergence insufficiency. *Vis Res* 2014;100:99–104.

64. Sreenivasan V, Bobier WR, Irving E, Lakshminarayanan V. Effect of vergence adaptation on convergence accommodation: Model simulations. *IEEE Trans Biomed Eng* 2010;57.

65. Sreenivasan V, Irving EL, Bobier WR. Effect of heterophoria type and myopia on accommodative and vergence responses during sustained near activity in children. *Vis Res* 2012;57:9–17.

66. Jiang BC. Accommodative vergence is driven by the phasic component of the accommodative controller. *Vis Res* 1996;36:97–102.

67. Schor CM, Bharadwaj SR. A pulse-step model of accommodation dynamics in the aging eye. *Vis Res* 2005;45:1237–1254.

68. Robinson DA. Models of the saccadic eye movement control system. *Kybernetik* 1973;14:71–83.

69. Robinson DA. The systems approach to the oculomotor system. *Vis Res* 1986;26:91–99.

70. Hung GK. Dynamic model of the vergence eye movement system: Simulations using MATLAB/SIMULINK. *Comput Methods Programs Biomed* 1998;55:59–68.

71. Yuan W, Semmlow JL, Alvarez TL, Munoz P. Dynamics of the disparity vergence step response: A model-based analysis. *IEEE Trans Biomed Eng* 1999;46:1191–1198.

72. Yuan W, Semmlow JL, Muller-Munoz P. Model–based analysis of dynamics in vergence adaptation. *IEEE Trans Biomed Eng* 2001;48:1402–1411.

73. Bharadwaj SR, Schor CM. Acceleration characteristics of human ocular accommodation. *Vis Res* 2005;45:17–28.

74. Bharadwaj SR, Schor CM. Dynamic control of ocular disaccommodation: First and second-order dynamics. *Vis Res* 2006;46:1019–1037.

75. Bharadwaj SR, Schor CM. Initial destination of the disaccommodation step response. *Vis Res* 2006;46:1959–1972.

76. Kasthurirangan S, Glasser A. Age related changes in accommodative dynamics in humans. *Vis Res* 2006;46:1507–1519.

77. Kasthurirangan S, Vilupuru AS, Glasser A. Amplitude dependent accommodative dynamics in humans. *Vis Res* 2003;43:2945–2956.

78. Suryakumar R, Meyers JP, Irving EL, Bobier WR. Vergence accommodation and monocular closed loop blur accommodation have similar dynamic characteristics. *Vis Res* 2007;47:327–337.

79. Maxwell J, Tong J, Schor CM. The first and second order dynamics of accommodative convergence and disparity convergence. *Vis Res* 2010;50:1728–1739.

80. Bharadwaj SR, Candy TR. The effect of lens–induced anisometropia on accommodation and vergence during human visual development. *Invest Ophthalmol Vis Sci* 2011;52:3595–3603.

81. Flitcroft DI, Judge SJ, Morley JW. Binocular interactions in accommodation control: Effects of anisometropic stimuli. *J Neurosci* 1992;12:188–203.

82. Charman WN. Aniso-accommodation as a possible factor in myopia development. *Ophthalmic Physiol Opt* 2004;24:471–479.

83. Ball EA. A study of consensual accommodation. *Am J Optom Arch Am Acad Optom* 1952;29:561–574.

84. Campbell FW. Correlation of accommodation between the two eyes. *J Opt Soc Am* 1960;50:738.

85. Marran L, Schor CM. Lens induced aniso-accommodation. *Vis Res* 1998;38:3601–3619.

86. Marran L, Schor CM. The effect of target proximity on the aniso-accommodative response. *Ophthalmic Physiol Opt* 1999;19:376–392.

87. Evans BJ. Monovision: A review. *Ophthalmic Physiol* Opt 2007;27:417–439.

88. Josephson JE, Erickson P, Back A et al. Monovision. *J* Am *Optom Assoc* 1990;61:820–826.

89. Donahue SP. The relationship between anisometropia, patient age, and the development of amblyopia. *Trans Am Ophthalmol Soc* 2005;103:313–336.

90. Fielder AR, Moseley MJ. Anisometropia and amblyopia—Chicken or egg? *Br J Ophthalmol* 1996;80:857–858.

91. Fry GA. Basic concepts underlying graphical analysis. In: Schor CM, Ciuffreda K (eds.), *Vergence Eye Movements: Basic and Clinical Aspects*. Boston, MA: Butterworths; 1983: pp. 605–646.

92. Aslin RN, Jackson RW. Accommodative–convergence in young infants: Development of a synergistic sensory–motor system. *Can J Psychol* 1979;33:222–231.

93. Bharadwaj SR, Candy TR. Accommodative and vergence responses to conflicting blur and disparity stimuli during development. *J Vis* 2009;9(4):1–18.

94. Atkinson J, Braddick OJ, Durden K, Watson PG, Atkinson S. Screening for refractive errors in 6–9 month old infants by photorefraction. *Br J Ophthalmol* 1984;68:105–112.

95. Ingram RM. Refraction of 1-year-old children after atropine cycloplegia. *Br J Ophthalmol* 1979;63:343–347.

96. MacLachlan C, Howland HC. Normal values and standard deviations for pupil diameter and interpupillary distance in subjects aged 1 month to 19 years. *Ophthalmic Physiol Opt* 2002;22:175–182.

97. Mayer DL, Hansen RM, Moore BD, Kim S, Fulton AB. Cycloplegic refractions in healthy children aged 1 through 48 months. *Arch Ophthalmol* 2001;119:1625–1628.

98. Ramsdale C, Charman WN. Accommodation and convergence: Effects of lenses and prisms in 'closed-loop' conditions. *Ophthalmic Physiol Opt* 1988;8:43–52.

99. Schor CM, Narayan V. Graphical analysis of prism adaptation, convergence accommodation, and accommodative convergence. *Am J Optom Physiol Opt* 1982;59:774–784.

100. Semmlow J, Venkiteswaran N. Dynamic accommodative vergence components in binocular vision. *Vis Res* 1976;16:403–410.

101. Semmlow JL, Hung G. Accommodative and fusional components of fixation disparity. *Invest Ophthalmol Vis Sci* 1979;18:1082–1086.

102. Morgan MW. Analysis of clinical data. *Am J Optom Arch Am Acad Optom* 1944;21.

103. Morgan MW. The clinical aspects of accommodation and convergence. *Am J Optom Arch Am Acad Optom* 1944;21:301–313.

104. Morgan MW. Accommodation and vergence. *Am J Optom Arch Am Acad Optom* 1968;45:417–454.

105. Goss DA. Morgan's norms and clinical analysis. In: Goss DA (ed.), Ocular *Accommodation, Convergence, and Fixation Disparity: A Manual of Clinical Analysis.* New York: Professional Press Books; 1986: pp. 107–110.

106. Jaschinski W. Fixation disparity and accommodation as a function of viewing distance and prism load. *Ophthalmic Physiol Opt* 1997;17:324–339.

107. Ogle KN, Mussey F, Prangen AD. Fixation disparity and the fusional processes in binocular single vision. *Am J Ophthalmol* 1949;32:1069–1087.

108. Charman WN, Whitefoot H. Pupil diameter and the depth-of-field of the human eye as measured by laser speckle. *Optica Acta* 1977;24:1211–1216.

109. Schor C, Wood I, Ogawa J. Binocular sensory fusion is limited by spatial resolution. *Vis Res* 1984;24:661–665.

110. Weale RA. New light on old eyes. *Nature* 1963;198:944–946.

111. Charman WN. The Charles F. Prentice Award Lecture 2005: Optics of the human eye: Progress and problems. *Optom Vis Sci* 2006;83:335–345.

112. Cheng H, Barnett JK, Vilupuru AS et al. A population study on changes in wave aberrations with accommodation. *J Vis* 2004;4:272–280.

113. Ghosh A, Collins MJ, Read SA, Davis BA, Iskander DR. The influence of downward gaze and accommodation on ocular aberrations over time. *J Vis* 2011;11:17.

114. Lopez-Gil N, Fernandez-Sanchez V. The change of spherical aberration during accommodation and its effect on the accommodation response. *J Vis* 2010;10:12.

115. Lopez-Gil N, Fernandez-Sanchez V, Legras R, Montes-Mico R, Lara F, Nguyen-Khoa JL. Accommodation-related changes in monochromatic aberrations of the human eye as a function of age. *Invest Ophthalmol Vis Sci* 2008;49:1736–1743.

116. Mathur A, Atchison DA, Charman WN. Effect of accommodation on peripheral ocular aberrations. *J Vis* 2009;9:20 21–11.

117. Radhakrishnan H, Charman WN. Age-related changes in ocular aberrations with accommodation. *J Vis* 2007;7:11 11–21.

118. Yuan Y, Shao Y, Tao A et al. Ocular anterior segment biometry and high-order wavefront aberrations during accommodation. *Invest Ophthalmol Vis Sci* 2013;54:7028–7037.

119. Buehren T, Collins MJ. Accommodation stimulus-response function and retinal image quality. *Vis Res* 2006;46:1633–1645.

120. Tarrant J, Roorda A, Wildsoet CF. Determining the accommodative response from wavefront aberrations. *J Vis* 2010;10:4.

121. Gambra E, Sawides L, Dorronsoro C, Marcos S. Accommodative lag and fluctuations when optical aberrations are manipulated. *J Vis* 2009;9:4 1–15.

122. Plainis S, Ginis HS, Pallikaris A. The effect of ocular aberrations on steady-state errors of accommodative response. *J Vis* 2005;5:466–477.

123. Thibos LN, Bradley A, Lopez-Gil N. Modelling the impact of spherical aberration on accommodation. *Ophthalmic Physiol Opt* 2013;33:482–496.

124. Buznego C, Trattler WB. Presbyopia-correcting intraocular lenses. *Curr Opin Ophthalmol* 2009;20:13–18.

125. Charman WN. Developments in the correction of presbyopia II: Surgical approaches. *Ophthalmic Physiol Opt* 2014;34:397–426.

126. Charman WN. Developments in the correction of presbyopia I: Spectacle and contact lenses. *Ophthalmic Physiol Opt* 2014;34:8–29.

127. Ong HS, Evans JR, Allan BD. Accommodative intraocular lens versus standard monofocal intraocular lens implantation in cataract surgery. *Cochrane Database Syst Rev* 2014;5:CD009667.

128. Bharadwaj SR, Vedamurthy I, Schor CM. Short-term adaptive modification of dynamic ocular accommodation. *Invest Ophthalmol Vis Sci* 2009;50:3520–3528.

129. Schor CM. Charles F. Prentice award lecture 2008: Surgical correction of presbyopia with intraocular lenses designed to accommodate. *Optom Vis Sci* 2009;86:E1028–1041.

130. Cufflin MP, Mallen EA. Dynamic accommodation responses following adaptation to defocus. *Optom Vis Sci* 2008;85:982–991.

131. Jiang BC, White JM. Effect of accommodative adaptation on static and dynamic accommodation in emmetropia and late-onset myopia. *Optom Vis Sci* 1999;76:295–302.

132. Pepin SM. Neuroadaptation of presbyopia-correcting intraocular lenses. *Curr Opin Ophthalmol* 2008;19:10–12.

133. Schor CM, Bharadwaj SR. Adaptive calibration of dynamic accommodation—Implications for accommodating intraocular lenses. *J Refract Surg* 2008;24:984–990.

134. Schor CM, Bharadwaj SR, Burns CD. Dynamic performance of accommodating intraocular lenses in a negative feedback control system: A simulation-based study. *Comput Biol Med* 2007;37:1020–1035.

135. Pryor HB. Objective measurement of interpupillary distance. *Pediatrics* 1969;44:973–977.

25 Accommodation dynamics

Lyle S. Gray and Barry Winn

Contents

25.1 INTRODUCTION

Ocular accommodation describes the neuromuscular process that allows the eye to focus clearly objects located at distances from infinity to a near point determined by the amplitude of accommodation and is a reflex response that appears to occur instantaneously in prepresbyopic individuals.[1] Objects continue to appear clear when attention is altered from one object to another even when a significant increase in ocular accommodation is required.[2] Increases in accommodation for near focusing are produced by an increase in the optical power of the crystalline lens, first identified by Thomas Young (1801).[3] Young was able to identify through a series of elegant experiments that the radius of curvature of both surfaces of the crystalline lens decreased during accommodation, and the theory of decreased tension was subsequently developed by Helmholtz (1865) to describe the physical changes in the crystalline lens during accommodation.[4]

Until recently, the techniques available for measurement of the accommodation response precluded the investigation of dynamic responses. The advent of continuously recording infrared, objective optometers, which did not interfere with experimental viewing conditions, allowed the dynamics of the accommodation response to be investigated systematically across a range of controlled stimulus conditions.[5,6] Collins' (1937)[6] early pioneering work using an ingenious electronic refractometer was the forerunner of laboratory-based infrared optometers and subsequently modern clinical autorefractors.[7]

25.1.1 CONTROL OF THE ACCOMMODATION RESPONSE

The accommodation response can be understood to have two primary functions: the ability to alter ocular power rapidly in response to a change in object or fixation distance and the ability to maintain a steady level of focus at a chosen fixation distance[2]; the neuromuscular system of accommodation is ideally suited to these functions.[2] Rapid alterations in response occur reflexively and are known as step changes in accommodation due to the characteristic trace obtained when recording such responses continuously.[2,5] Perceptually, there is no effect upon vision during a step response despite significant dioptric change in accommodation level,[8] and the response acts in a reflex manner producing the required alterations in ocular power without any conscious input.[8,9]

The second function of the accommodation system, which has important evolutionary consequences by allowing humans to work at close distances with fine tools, is the ability to maintain accurate steady-state focus at a fixed distance.[2] Again, the steady-state response occurs without conscious control, and the accommodation system makes continual, fine adjustments necessary to maintain clear focus, which can be maintained for considerable periods of time.[2]

The primary stimulus to accommodation has received considerable attention over time. Maddox (1893) described the four classes of accommodation response as reflex, tonic, psychic, and cross-stimulation from vergence eye movements.[10] The reflex response in Maddox' classification was taken to be the response to blur of the retinal image, while the psychic response represented mainly proximal stimulation arising from a knowledge of the spatiotopic relationship between the subject and the object being viewed.[8]

Early experiments by Fincham (1951)[11] and Allen (1955)[12] using coincidence optometers showed that the accommodation response to blur stimuli with a magnitude of ~1.50D or less occurred in the correct direction 99% of the time, suggesting that not only was blur the primary stimulus to accommodation, but the accommodation error detection system had the ability to extract odd-error directional information from an even-error stimulus.[11,12]

Neurological control of accommodation is via the autonomic system, primarily parasympathetic, arising from the

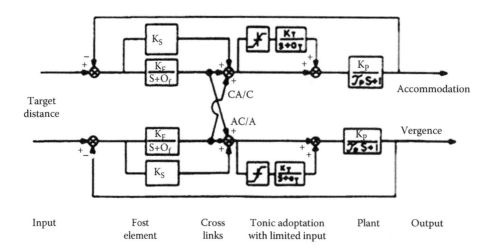

Figure 25.1 Schor and Kotulak (1986)[9] system model representation of accommodation and vergence control and the interactions between the systems. The key elements are negative feedback subtracted from the desired stimulus level to produce an error signal. If the error signal is greater than the system dead space (for accommodation this would be ocular depth of focus), then a phasic response is initiated to shift the response level. Upon achieving this response level if fixation is sustained, the output from the adaptive control element will increase to maintain the response. Note that the cross-link interaction between accommodation and vergence occurs after the phasic component but prior to the adaptive control output. For further description, see Reference 9.

Edinger–Westphal nucleus of the III cranial nerve.[2] There is a small (2% of nerve fibers) β-adrenergic input to the ciliary muscle arising from the superior cervical ganglion.[13] To understand the mechanisms of accommodation control and to characterize the response, a number of investigators have used modeling techniques from the engineering domain.[9,14,15] Schor and Kotulak's (1986) dual mode systems control model described the components necessary to produce both dynamic step changes and steady-state responses in accommodation when blur is the primary stimulus for accommodation (Figure 25.1).[9]

The model contains a phasic element with an integral controller to produce rapid step changes in accommodation response. The output of this controller is fed forward to an adaptive element with a slower time constant, which produces the ongoing output necessary to maintain a steady-state response at that level.[9] A key component of all models was negative feedback of the response level, which was then subtracted from the required stimulus level in order to determine the need for further phasic response.[9] The presence of negative feedback is essential in any feed-forward system to maintain response stability.[9,14,15] Subtraction of negative feedback from the required stimulus level results in an accommodative error signal that must be greater than the system dead space (in this case the ocular depth of focus) in order to initiate a new response.[9,14,15] With some minor alterations, these models describe well the accommodation response to optical blur stimuli.[9,14,15]

Difficulties arise in our understanding of the accommodation system when we consider the response to pure blur stimuli greater than ~1.50D. A number of early studies showed clearly that the accommodation response to pure blur stimuli of 2D or larger demonstrated even error behavior, suggesting that the accommodation error detector could not be a directional signal from the larger blur stimulus.[16,17] This inconsistency was addressed by recent studies demonstrating how potent perceived proximity was as a cue for the accommodation response.[8,18,19] Consideration of the complementary but distinct operating

ranges for the stimuli of blur and proximity strongly suggested that large accommodation responses would be initiated by spatiotopic, proximal stimuli, until the accommodation response reduced the blur error signal to levels within the retinotopic blur stimulus range that could then refine the accuracy of the accommodation response to the optically required level.[8] Schor et al. (1992) developed a model based on that described previously that summarized this retinotopic/spatiotopic division of the accommodation response and how such a system would enable accurate accommodation responses to any change in object or fixation distance across the full range of the accommodation system.[8]

25.1.2 ACCOMMODATION STEP RESPONSES

Empirical observation shows that the response time for accommodation begins to increase concurrently with decreases in the amplitude of accommodation response around the age of 45 years in Caucasian patients.[1,2] This deterioration of the accommodation system leads to a functional inability to focus near distances, which progresses until no useful accommodation response is found at around 60 years of age, and has been termed presbyopia.[1,2] The majority of studies investigating dynamic accommodation responses have therefore been conducted on prepresbyopic adults.

Campbell and Westheimer (1960) conducted a groundbreaking series of experiments to investigate the dynamic accommodation response using both step and sinusoidal changes in stimulus vergence.[20] They used a newly constructed infrared optometer to measure continuously the accommodation response during an abrupt 2D change in stimulus vergence under monocular conditions (Figure 25.2).[20]

They reported reaction times typically between 300 and 400 ms with a maximum velocity of around 10D/s. The average response time for far-to-near accommodation was 0.64 s and for near-to-far accommodation, 0.56 s. Therefore, the total time for an accommodation response to a dynamic step stimulus of 2D

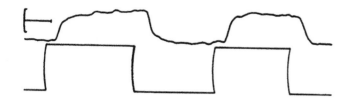

Figure 25.2 Record of accommodation responses to a 2D step stimulus and return to zero level of accommodation (subject F.W.C.). Allowance should be made for the arc of the pen. Top line, accommodation (length of horizontal line, 1 s; height of arc, 1D): upward movement represents far-to-near accommodation. Bottom line, stimulus signal, the same scale. This record is an example of single-sweep accommodation responses. (After Campbell, F.W. and Westheimer, G., *J. Physiol.*, 151(2), 285, 1960.)

was approximately 1 s. However, they did report variability in responses between observers and viewing conditions.[20]

The dynamic accommodation response to abrupt changes in stimulus level has been evaluated comprehensively by a number of groups around the world following these initial experiments by Campbell and Westheimer (1960). A number of studies have shown that near-to-far accommodation has a longer response time than that found for far-to-near responses,[5,21–25] and continuous accommodation recordings show that the near-to-far response has a different pattern with a more gradual reduction in response compared to the abrupt increase found for far-to-near responses (Figure 25.3).[5,20]

The majority of these studies have been conducted monocularly to avoid input from vergence eye movements via the cross-links between the two systems described previously.[8,9] When dynamic step responses of accommodation are measured binocularly, then reaction and response times have been shown to be the same as those found in monocular conditions,[22] with the responses showing a high degree of symmetry between the two eyes (Figure 25.3).[5,20] The addition of binocular vergence eye

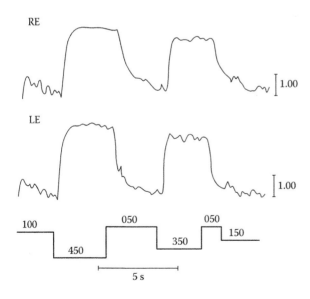

Figure 25.3 A typical binocular accommodation response to a step change in target distance. The lower trace indicates the change in target vergence. The near-to-far response gradually approaches its final level in comparison with the far-to-near response. (After Heron, G. and Winn, B., *Ophthal. Physiol. Opt.*, 9(2), 176, 1898.)

movements does provide the accommodation–vergence complex with an odd-error signal in the form of binocular retinal image disparity that allows both systems to respond in the correct direction.[26] This interaction between the accommodation and vergence systems has been shown previously to be active primarily during the phasic element of the response,[9] meaning that during the steady-state response there is a requirement for odd-error modulation of retinal image blur in order to maintain accurate steady-state focus.[8]

Abrupt step changes in accommodation show a temporal, dual mode pattern of response with an initial preprogrammed component (ref), which generates a ballistic response not influenced by negative feedback of retinal image blur.[27,28] Studies have shown that during this period of the step response the interposition of a further blur stimulus does not affect the completion of this preprogrammed component.[27,28] Once the preprogrammed response is completed, the response level will be within the range of the retinotopic negative feedback control system, which then completes the response bringing it within the ocular depth of focus. This dual mode control of step responses of accommodation has been modeled using the same engineering tools as described previously.[27,28]

With accommodation responses taking up to 1 s to complete for a large dioptric change, it is interesting to note that objects rarely appear blurred during an accommodation step response,[2] although there would be ample time for the sensory visual system to recognize and process this information.[29] Saccadic suppression is a well-documented process that prevents the visual system from becoming perceptually aware of the motion of the images across the retina during a saccadic movement.[30] For large saccadic movements, this motion can reach the speed of up to 500°/s.[31] Saccadic suppression has been shown to increase thresholds primarily for the detection of low spatial frequency information.[30] In contrast, the accommodation system detects and responds to image blur, which affects high spatial frequency information to a greater extent.[8,14] Recent studies have demonstrated a suppression mechanism, which suppresses the sensory visual response to retinal image blur during abrupt step changes in accommodation response, by selective suppression of high spatial frequency content in the target.[32,33]

25.1.3 STEADY-STATE RESPONSE

An important characteristic of the accommodation response is the ability to maintain stable, clear focus upon an object of regard for relatively long periods of time.[1,2] The accommodation system shows a characteristic pattern of accommodative lead for targets at a distance of ≥1 m and a lag of accommodation for closer targets.[1,2]

Intriguingly, when the steady-state accommodation response is measured continuously, it demonstrates a continual variation in response level with an amplitude of ~0.5D and temporal frequencies up to ~5 Hz.[34] This temporal instability in the steady-state accommodation response has attracted the interest of a number of investigators over the last 80 years since the first direct observation of these accommodative microfluctuations by Collins (1937).[6] A number of investigations of steady-state accommodation dynamics suggest that rather than being an extraneous characteristic of the steady-state accommodation response, microfluctuations could provide the odd-error cue required to maintain

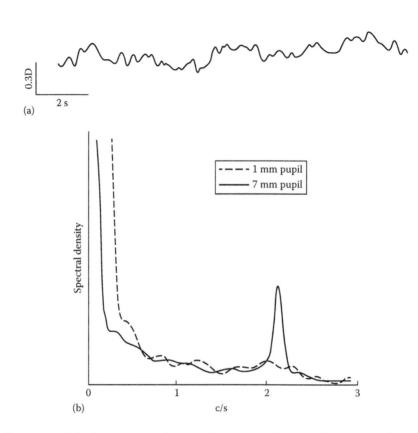

Figure 25.4 (a) Accommodation record of subject J.G.R. under normal viewing conditions with a 7 mm pupil (upper) and with a 1 mm effective entrance pupil of the eye (lower). The records have the same average accommodation level. (b) Frequency spectra of the two records shown in (a) (linear ordinates). (After Campbell, F.W. et al., *J. Physiol.*, 145(3), 579, 1959.)

an optimum accommodation response by providing subthreshold changes in retinal image contrast that could be detected by the sensory error detection mechanism.[35,36]

The first systematic investigation of the magnitude and temporal characteristics of the steady-state response concluded that microfluctuations must play a role in sensory feedback.[34] They reported temporal frequencies up to 3 Hz with dominant components occurring under 0.5 Hz and between 1.3 and 2.2 Hz (Figure 25.4).[34]

A number of groups have confirmed the observation that the waveform of the microfluctuations exhibits temporal variations characterized by two dominant regions of activity: a low-frequency component (LFC) typically broadband with frequencies up to 0.6 Hz and a high-frequency component (HFC) typically narrowband and occurring in the range of frequencies between 1.0 and 2.3 Hz.[35,36] The microfluctuations typically occur with a root-mean-square (rms) magnitude of approximately 0.02D–0.25D, which has been shown to be positively correlated with increases in the level of accommodation response (Figure 25.5).[37,38]

Functionally, microfluctuations offer a means by which an odd-error, directional cue can be elicited from the primary, even-error stimulus of retinal image blur.[8] By monitoring variations in retinal image contrast and correlating these with the small variations in dioptric power resulting from the micro-fluctuations, the accommodation error detection mechanism can maintain an accurate steady-state accommodation response within the ocular depth of focus for a given stimulus.[39,40] Previous work suggests that this modulation of retinal image contrast by accommodation microfluctuations is available to the accommodation system.[41–43]

As described previously, when all cues other than blur are removed, the initial accommodation response to large step stimuli (>2D) is even error in nature.[16,17] This suggests that the microfluctuations are unlikely to play a role in guiding the initial response as the magnitude of change is beyond the range for retinotopic information.[8] Consideration of the latency of accommodation step responses (300–400 ms) also suggests that a frequency component of ~2.5–3.3 Hz would be required to provide the required directional information.[44] In contrast, the presence of odd-error cues to the accommodation step response has been established for small (<1.5D) changes in the stimulus to accommodation where the accommodation response shows a directional accuracy of 99%.[11,12]

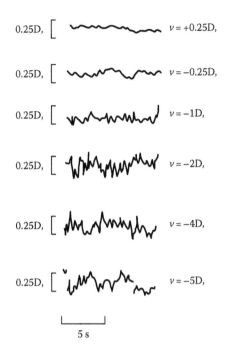

Figure 25.5 Records of accommodation microfluctuations as a function of target vergence (*V*). Observer M.B. (After Denieul, P., *Vision Res.*, 22(5), 561, 1982.)

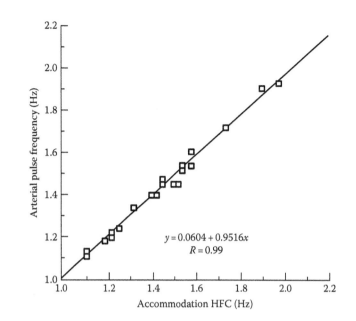

Figure 25.6 Correlation between arterial pulse frequency and high-frequency component for group data (*n* = 20: *r* = 0.99, *P* < 0.001). The regression line is *y* = 0.0604 + 0.9516*x*. (After Winn, B. et al., *Curr. Eye Res.*, 9(10), 971, 1990.)

A number of studies have examined the contribution of the two dominant frequency components within the microfluctuations to the negative feedback control mechanism of the steady-state accommodation response.[45–50] The source of the HFCs was of particular interest as their characteristics did not appear to be related to changes in stimulus parameters suggesting they may simply represent "plant noise" derived from the mechanical and elastic properties of the crystalline lens and its support structures.[35,36,45,46] Studies showed that while there is very little intrasubject variation in the peak frequency of the HFC, it was evident that there was significant intersubject variation.[34,45,46] The significant intersubject variability led to consideration of the relationship between the HFC and other physiological systems that create rhythmic intraocular variation. Simultaneous measurements of ocular accommodation and systemic arterial pulse on 20 subjects demonstrated that the location of the HFC peak frequency was significantly correlated with arterial pulse frequency (Figure 25.6).[46]

Subsequent studies revealed that the magnitude of the LFCs alters with changes in stimulus parameters in a manner that suggests they play a role in accommodation control.[34,47–50] Reductions in pupil size are known to increase the ocular depth of focus[51–53] with increases in the magnitude of the microfluctuations reported.[34,47–50] Campbell et al.,[34] in their original paper from 1959, showed that microfluctuations were larger through a 1 mm pupil compared to a 7 mm pupil (Figure 25.4a and b).

A systematic study of the relationship between pupil diameter and accommodation microfluctuations was conducted with the stimulus placed at the subjects' tonic position to ensure that the mean accommodation response level remained constant throughout the study.[47] This was an important design feature of the study as it is known that accommodation microfluctuations are larger at higher levels of accommodation response.[37,38] For pupil diameters >2 mm, the fluctuations remained approximately constant with an rms magnitude of ~0.20D. A significant increase in the rms magnitude to ~0.31D was observed for pupil sizes ≤2 mm, and these changes were found to be due primarily to an increase in the LFC (Figure 25.7).[47]

Power spectrum analysis for a typical observer highlighted the increase in magnitude of the LFC for smaller pupil sizes while the HFC remains fairly constant in magnitude and frequency (Figure 25.8).[47] The power of the LFC is approximately constant (0.05D²/Hz) for pupil sizes above 2 mm but increases to 0.12D²/Hz for the 2 mm pupil, 0.13D²/Hz for the 1 mm pupil, and 0.22D²/Hz for the 0.5 mm pupil. Clearly, the pupil diameters producing increases in the microfluctuations correspond with those that produce substantial increases in the ocular depth of focus.[47] The increase in magnitude of the microfluctuations has the potential to provide the accommodation error detector with consistent feedback as the size of the depth of focus increases.

It has been proposed that the presence of an inherent accommodative "lag" or "lead" (steady-state error) would enhance the effectiveness of microfluctuations[54] as it is known that the sensitivity to blur is increased when the retinal image is slightly defocused.[55] A computer simulation of an accommodative feedback control system identified the frequency of oscillation permissible in the response before the loop becomes unstable to be 0.45 Hz, which is consistent with a typical LFC.[44] Opening the accommodation loop causes the response to regress to a tonic position[56] with large drifts in accommodative level occurring at low frequencies.[48,57–59]

The overall profile of accommodative microfluctuations was thought to be the result of a combination of both neurological control and localized plant noise.[35,36] The HFC is not under direct neurological control but may still be utilized as part of the overall waveform in conjunction with the LFC by the accommodation error detector.[35,36] A functional role for the microfluctuations as

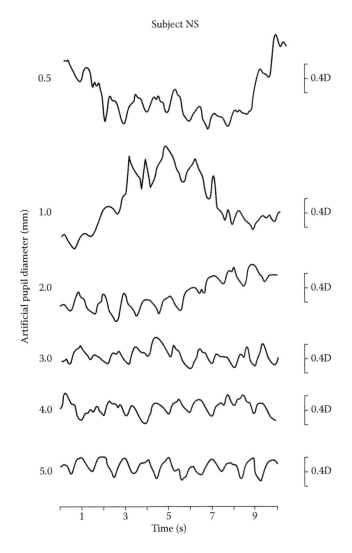

Figure 25.7 Accommodation traces for one observer (subject NS) for each artificial pupil diameter. Each trace is of 10 s duration and has been smoothed to 10 Hz. Note the high incidence of low-frequency components, which can be identified for the smallest pupil diameters (0.5 and 1.0 mm) and which decreases for the larger pupil diameters. The incidence of high-frequency components is approximately the same for all artificial pupil diameters. (From Gray, L.S. et al., *Vision Res.*, 33(15), 2083, 1993a.)

an error detector is probably related to the maintenance of focus on a stationary stimulus, as the neurologically controlled component is too slow to provide the necessary information to optimize the response to rapid step changes in stimulus vergence.[35,36,47]

25.1.4 DETECTABILITY OF ACCOMMODATION MICROFLUCTUATIONS

The microfluctuations introduce a blur stimulus that is smaller than the ocular depth of focus hence below the perceptual blur threshold, yet it is apparently of sufficient magnitude to provide the required odd-error signal to the accommodation control system allowing maintenance of the steady-state response via negative feedback of retinal image blur.[8] A model was described to explain how the accommodation error detector could extract the required information from subperceptual threshold stimuli.[39] Although the model was originally conceived using the

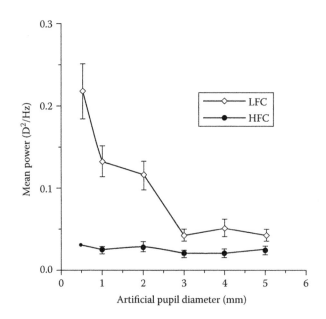

Figure 25.8 Mean power in the low-frequency and high-frequency components of the microfluctuations as a function of pupil diameter for the three subjects. Each point represents the mean of 15 power spectra and is calculated for three frequency bins. For further details, see Reference 47.

high-frequency (2 Hz) component of the fluctuations, it is equally applicable to any frequency of oscillation including those found within the LFC range.[47] Magnitude and directional information can be determined from subperceptual stimuli by calculating and comparing the first derivatives of temporal changes in retinal image contrast and temporal changes in ocular lens power.[39] Directional information is extracted by comparing the signs of the first derivatives of these two time functions: an overaccommodated eye will have the lens power function out of phase with the retinal contrast function and an underaccommodated eye will have these two functions in phase.[39] The retinal image contrast is directly related to the instantaneous focus error present hence magnitude information can be extracted from the model by direct comparison of the two first derivatives.[39]

As stated previously, low-frequency drifts in the accommodation response are found when the system is placed under open-loop conditions.[48,57–59] The slope of the accommodation stimulus/response curve decreases when the luminance of the target is reduced,[60] and reducing stimulus luminance has been shown to result in an increase in the magnitude of the microfluctuations and the magnitude of the LFCs in the waveform.[48] The magnitude of accommodation microfluctuations was constant for target luminances >0.010 cd/m² but increased and became more variable for luminances ≤0.010 cd/m². The increase in magnitude of the fluctuations was attributable to changes in the LFC (Figure 25.9).[48]

Reduction in stimulus luminance does not alter the stimulus contrast per se but causes the higher spatial frequency content within the target to fall below threshold producing a shallower contrast gradient in the cortical image and, consequently, increasing the ocular depth of focus.[50] Day et al. (2009) showed that the contrast gradient and alterations in depth of focus resulting from reductions in target luminance remain relatively constant until

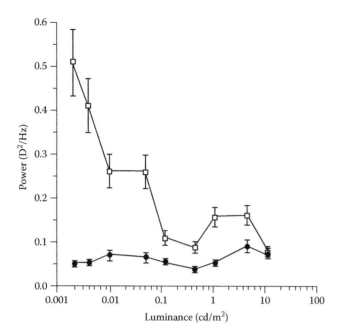

Figure 25.9 Mean power in the low-frequency (open circles) and high-frequency components (closed circles) as a function of target luminance for three subjects. Each point represents the mean of 15 power spectra and is calculated for three frequency bins; error bars represent ±1 SD. (After Gray, L.S. et al., *Ophthal. Physiol. Opt.*, 13(3), 258, 1993b.)

the luminance is reduced to 0.002 cd/m², which corresponds to the levels of luminance found to produce increases in the magnitude of the microfluctuations.[48]

The relationship between ocular depth of focus and accommodation microfluctuations has been used as the basis for investigation of differences between refractive groups. Several studies have reported that late-onset myopes (LOM: onset after the age of 15 years) demonstrate a larger depth of focus than emmetropes.[61,62] Significantly, larger microfluctuations have been reported in LOM compared to emmetropes suggesting that the larger depth of focus found in LOM leads to a higher threshold for retinal image blur.[21,22,49,50]

25.1.5 AGE-RELATED CHANGES IN ACCOMMODATION RESPONSE

The ability to accommodate diminishes with increasing age resulting in the need for spectacles to read in patients over the age of 45 years, and the underlying decline in the amplitude of accommodation with age has been well documented.[63] There is a lack of consensus regarding the exact anatomical and physiological changes that underlie the onset of presbyopia,[64] although, regardless of the mechanism, presbyopia affects 100% of the population.[65]

Although the decrease in amplitude with increasing age is well established, there has been less work on the age-related changes in dynamics of the accommodation response. Measuring dynamic accommodation requires intrusive conditions for subjects including a head restraint and often the use of a dental bite to control head position.[36,46] The calibration procedures are time consuming and repeated trials under these conditions can be challenging.[66]

This has inevitably led to the use of small sample sizes, resulting in studies with equivocal results. This is especially important when looking for subtle age-related changes in the response, as significant interindividual variation in accommodation response characteristics is known to occur.[23,54]

Kasthurirangan and Glasser (2005) attempted to resolve this lack of consensus by conducting a study of accommodation dynamics in a large group (*n* = 66) of subjects between the ages of 14 and 45 years.[64] The experiment used a number of accommodation stimulus amplitudes over repeated trials, allowing a comprehensive analysis of data across the age range. They were able to confirm a linear decrease in accommodative amplitude with increasing age at a rate of 0.26D per year when measured objectively and 0.35D per year for subjective observations although significant interindividual variability was apparent. Extrapolation of the data showed that any functional accommodation response was completely lost at 50 years of age.[64]

Measurement of accommodation step responses showed no differences in response latency with increasing age.[67] Time constants for accommodation step responses are known to increase linearly with increasing stimulus amplitude,[67] and this relationship was shown to increase with increasing age at a rate of 0.01 s/D/year.[64] Interestingly, this relationship occurred only for far-to-near responses and no systematic change in the relationship between time constants and response amplitude could be found for near-to-far (disaccommodation) responses.[64,67]

The amplitude of accommodation has an impact on response dynamics with the saturation level reducing with increasing age.[68] Saturation of the peak velocity occurs at lower response amplitudes with increasing age.[67] By evaluating responses well within the total amplitude, it is possible to get a clearer view of age-related changes, and peak velocity was shown to be invariant with increasing response amplitude in older subjects although it increases with response amplitude in younger subjects.[64]

There appears to be a general consensus that the speed of the accommodation response declines with increasing age.[69] However, the effect of increasing age on disaccommodation remains equivocal.[64,67] There have been several studies that report that disaccommodation does not change with age.[23,24,64] Other studies have suggested a reduction in response dynamics[69] although this finding may be the result of not adjusting for reduced response amplitudes in older subjects.

REFERENCES

1. Rosenfield, M. and Logan, N. (2009). *Optometry: Science, Techniques and Clinical Management*. Butterworth-Heinemann, London, U.K.
2. Glasser, A. (2011). Accommodation. In *Adler's Physiology of the Eye* (ed. 11), Levin, L. A. et al. (Eds.), pp. 40–70, Saunders, London, U.K.
3. Young, T. (1801). The Bakerian Lecture: On the mechanism of the eye. *Philosophical Transactions of the Royal Society of London*, 91, 23–88.
4. Helmholtz, H. L. F. (1909–1911). *Treatise on Physiological Optics* (Translated from the third German edition, 1924), Southall, J. P. C. (Ed.), Optical Society of America, New York.
5. Heron, G. and Winn, B. (1989). Binocular accommodation reaction and response times for normal observers. *Ophthalmic and Physiological Optics*, 9(2), 176–183.

6. Collins, G. (1937). The electronic refractionometer. *British Journal of Physiological Optics*, 11, 30–42.

7. Wood, I. C. J. (1987). A review of autorefractors. *Eye*, 1(4), 529–535.

8. Schor, C. M., Alexander, J., Cormack, L., and Stevenson, S. (1992). Negative feedback control model of proximal convergence and accommodation. *Ophthalmic and Physiological Optics*, 12(3), 307–318.

9. Schor, C. M. and Kotulak, J. C. (1986). Dynamic interactions between accommodation and convergence are velocity sensitive. *Vision Research*, 26(6), 927–942.

10. Maddox, E. E. (1893). *The Clinical Use of Prisms and the Decentering of Lenses*. John Wright & Sons, Bristol, England.

11. Fincham, E. F. (1951). The accommodation reflex and its stimulus. *The British Journal of Ophthalmology*, 35(7), 381.

12. Allen, M. J. (1955). The stimulus to accommodation. *American Journal of Optometry and Archives of American Academy of Optometry*, 32(8), 422–431.

13. Gilmartin, B. (1986). A review of the role of sympathetic innervation of the ciliary muscle in ocular accommodation. *Ophthalmic and Physiological Optics*, 6(1), 23–37.

14. Hung, G. K. and Semmlow, J. L. (1980). Static behavior of accommodation and vergence: Computer simulation of an interactive dual-feedback system. *IEEE Transactions on Biomedical Engineering*, BME-27 (8), 439–447.

15. Toates, F. M. (1970). A model of accommodation. *Vision Research*, 10(10), 1069–1076.

16. Stark, L. and Takahashi, Y. (1965). Absence of an odd-error signal mechanism in human accommodation. *IEEE Transactions on Biomedical Engineering*, BME-12 (3 and 4), 138–146.

17. Smithline, L. M. (1974). Accommodative response to blur. *Journal of the Optical Society of America*, 64(11), 1512–1516.

18. Rosenfield, M., Ciuffreda, K. J., and Hung, G. K. (1991). The linearity of proximally induced accommodation and vergence. *Investigative Ophthalmology and Visual Science*, 32(11), 2985–2991.

19. Morrison, K. A., Seidel, D., Strang, N. C., and Gray, L. S. (2010). The effect of proximity on open-loop accommodation responses measured with pinholes. *Ophthalmic and Physiological Optics*, 30(4), 365–370.

20. Campbell, F. W. and Westheimer, G. (1960). Dynamics of accommodation responses of the human eye. *The Journal of Physiology*, 151(2), 285–295.

21. Seidel, D., Gray, L. S., and Heron, G. (2003). Retinotopic accommodation responses in myopia. *Investigative Ophthalmology and Visual Science*, 44(3), 1035–1041.

22. Seidel, D., Gray, L. S., and Heron, G. (2005). The effect of monocular and binocular viewing on the accommodation response to real targets in emmetropia and myopia. *Optometry and Vision Science*, 82(4), 279–285.

23. Heron, G., Charman, W. N., and Gray, L. S. (1999). Accommodation responses and ageing. *Investigative Ophthalmology and Visual Science*, 40, 2872–2883.

24. Heron, G., Charman, W. N., and Schor, C. (2001). Dynamics of the accommodation response to abrupt changes in target vergence as a function of age. *Vision Research*, 41(4), 507–519.

25. Mordi, J. A. and Ciuffreda, K. J. (2004). Dynamic aspects of accommodation: Age and presbyopia. *Vision Research*, 44(6), 591–601.

26. Tyler, C. W. (1983). Sensory processing of binocular disparity. In *Vergence Eye Movements: Basic and Clinical Aspects*, Schor C. M. and Ciuffreda, K. J. (Eds.), pp. 199–295. Butterworth, Boston, MA.

27. Hung, G. K. and Ciuffreda, K. J. (1988). Dual-mode behaviour in the human accommodation system. *Ophthalmic and Physiological Optics*, 8(3), 327–332.

28. Schor, C. M. and Bharadwaj, S. R. (2006). Pulse-step models of control strategies for dynamic ocular accommodation and disaccommodation. *Vision Research*, 46(1), 242–258.

29. Wurtz, R. H. (2008). Neuronal mechanisms of visual stability. *Vision Research*, 48, 2070–2089.

30. Burr, D. C., Morrone, M. C., and Ross, J. (1994). Selective suppression of the magnocellular visual pathway during saccadic eye movements. *Nature*, 371, 511–513.

31. Alhazmi, M., Seidel, D., and Gray, L. S. (2014). The effect of ocular rigidity upon the characteristics of saccadic eye movements. *Investigative Ophthalmology and Visual Science*, 55(3), 1251–1258.

32. Mucke, S., Manahilov, V., Strang, N. C., Seidel, D., and Gray, L. S. (2008). New type of perceptual suppression during dynamic ocular accommodation. *Current Biology*, 18(13), R555–R556.

33. Mucke, S., Manahilov, V., Strang, N. C., Seidel, D., Gray, L. S., and Shahani, U. (2010). Investigating the mechanisms that may underlie the reduction in contrast sensitivity during dynamic accommodation. *Journal of Vision*, 10(5), 5.

34. Campbell, F. W., Robson, J. G., and Westheimer, G. (1959). Fluctuations of accommodation under steady viewing conditions. *The Journal of Physiology*, 145(3), 579–594.

35. Charman, W. N. and Heron, G. (2015). Microfluctuations in accommodation: An update on their characteristics and possible role. *Ophthalmic and Physiological Optics*, 35(5), 476–499.

36. Winn, B. and Gilmartin, B. (1992). Current perspective on microfluctuations of accommodation. *Ophthalmic and Physiological Optics*, 12(2), 252–256.

37. Denieul, P. (1982). Effects of stimulus vergence on mean accommodation response, microfluctuations of accommodation and optical quality of the human eye. *Vision Research*, 22(5), 561–569.

38. Kotulak, J. C. and Schor, C. M. (1986). Temporal variations in accommodation during steady-state conditions. *Journal of the Optical Society of America A*, 3(2), 223–227.

39. Kotulak, J. C. and Schor, C. M. (1986). A computational model of the error detector of human visual accommodation. *Biological Cybernetics*, 54(3), 189–194.

40. Hung, G. K., Ciuffreda, K. J., Khosroyani, M., and Jiang, B. C. (2002). Models of accommodation. In *Models of the Visual System*, pp. 287–339, Springer, New York.

41. Ludlam, W. M., Wittenberg, S., Giglio, E. J., and Rosenberg, R. (1968). Accommodative responses to small changes in dioptric stimulus. *American Journal of Optometry. Archives of the American Academy of Optometry*, 45, 483–506.

42. Winn, B., Charman, W. N., Pugh, J. R., Heron, G., and Eadie, A. S. (1989). Perceptual detectability of ocular accommodation microfluctuations. *Journal of the Optical Society of America A*, 6(3), 459–462.

43. Metlapally, S., Tong, J. L., Tahir, H. J., and Schor, C. M. (2014). The impact of higher-order aberrations on the strength of directional signals produced by accommodative microfluctuations. *Journal of Vision*, 14(12), 25. doi:10.1167/14.12.25.

44. Hung, G. K., Semmlow, J. L., and Ciuffreda, K. J. (1982). Accommodative oscillation can enhance average accommodative response: A simulation study. *IEEE Transactions on Systems, Man, and Cybernetics*, 12(4), 594–598.

45. Winn, B., Pugh, J. R., Gilmartin, B., and Owens, H. (1990). The frequency characteristics of accommodative microfluctuations for central and peripheral zones of the human crystalline lens. *Vision Research*, 30(7), 1093–1099.

46. Winn, B., Pugh, J. R., Gilmartin, B., and Owens, H. (1990). Arterial pulse modulates steady-state ocular accommodation. *Current Eye Research*, 9(10), 971–975.

47. Gray, L. S., Winn, B., and Gilmartin, B. (1993). Accommodative microfluctuations and pupil diameter. *Vision Research*, 33(15), 2083–2090.

48. Gray, L. S., Winn, B., and Gilmartin, B. (1993). Effect of target luminance on microfluctuations of accommodation. *Ophthalmic and Physiological Optics*, 13(3), 258–265.

49. Day, M., Strang, N. C., Seidel, D., Gray, L. S., and Mallen, E. A. (2006). Refractive group differences in accommodation microfluctuations with changing accommodation stimulus. *Ophthalmic and Physiological Optics*, 26(1), 88–96.

50. Day, M., Seidel, D., Gray, L. S., and Strang, N. C. (2009). The effect of modulating ocular depth of focus upon accommodation microfluctuations in myopic and emmetropic subjects. *Vision Research*, 49(2), 211–218.

51. Campbell, F. W. (1957). The depth of field of the human eye. *Journal of Modern Optics*, 4(4), 157–164.

52. Charman, W. N. and Whitefoot, H. (1977). Pupil diameter and the depth-of-field of the human eye as measured by laser speckle. *Journal of Modern Optics*, 24(12), 1211–1216.

53. Atchison, D. A., Charman, W. N., and Woods, R. L. (1997). Subjective depth-of-focus of the eye. *Optometry & Vision Science*, 74(7), 511–520.

54. Charman, W. N. and Tucker, J. (1978). Accommodation as a function of object form. *American Journal of Optometry and Physiological Optics*, 55(2), 84–92.

55. Campbell, F. W., Westheimer, G., and Robson, J. G. (1958). Significance of fluctuations of accommodation. *Journal of the Optical Society of America*, 48(9), 669.

56. Gilmartin, B. and Hogan, R. E. (1985). The relationship between tonic accommodation and ciliary muscle innervation. *Investigative Ophthalmology and Visual Science*, 26(7), 1024–1028.

57. Westheimer, G. (1957). Accommodation measurements in empty visual fields. *Journal of the Optical Society of America*, 47(8), 714–718.

58. Alpern, M. (1958). Variability of accommodation during steady fixation at various levels of illuminance. *Journal of the Optical Society of America*, 48(3), 193–197.

59. Baker, R., Brown, B., and Garner, L. (1983). Time course and variability of dark focus. *Investigative Ophthalmology and Visual Science*, 24(11), 1528–1531.

60. Johnson, C. A. (1976). Effects of luminance and stimulus distance on accommodation and visual resolution. *Journal of the Optical Society of America*, 66(2), 138–142.

61. Rosenfield, M. and Abraham-Cohen, J. A. (1999). Blur sensitivity in myopes. *Optometry and Vision Science*, 76(5), 303–307.

62. Vasudevan, B., Ciuffreda, K. J., and Wang, B. (2006). Objective blur thresholds in free space for different refractive groups. *Current Eye Research*, 31(2), 111–118.

63. Duane, A. (1912). Normal values of the accommodation at all ages. *Journal of the American Medical Association*, 59(12), 1010–1013.

64. Kasthurirangan, S. and Glasser, A. (2006). Age related changes in accommodative dynamics in humans. *Vision Research*, 46(8), 1507–1519.

65. Weale, R. A. (2003). Epidemiology of refractive errors and presbyopia. *Survey of Ophthalmology*, 48(5), 515–543.

66. Pugh, J. R. and Winn, B. (1988). Modification of the Canon Auto Ref R1 for use as a continuously recording infra-red optometer. *Ophthalmic and Physiological Optics*, 8(4), 460–464.

67. Kasthurirangan, S., Vilupuru, A. S., and Glasser, A. (2003). Amplitude dependent accommodative dynamics in humans. *Vision Research*, 43(27), 2945–2956.

68. Ciuffreda, K. J. and Kruger, P. B. (1988). Dynamics of human voluntary accommodation. *American Journal of Optometry and Physiological Optics*, 65(5), 365–370.

69. Schaeffel, F., Wilhelm, H., and Zrenner, E. (1993). Interindividual variability in the dynamics of natural accommodation in humans: Relation to age and refractive errors. *The Journal of Physiology*, 461(1), 301–320.

26

Eye movements*

* Support: Australian Research Council Discovery Project DP120100651.

Andrew J. Anderson

Contents

26.1 INTRODUCTION

A key aspect of the optical components of the human eye described in previous chapters is that they are contained within a package that can *move*. Although we typically pay little attention to these movements, a moment's reflection will find that our eyes are rarely still and that eye movements determine how we obtain visual information from our environment. This chapter outlines some of the reasons why our eyes move, the mechanisms that bring about these movements, and the relationship between the different types of eye movements and the stimuli that produce them. It is also instructive to consider the evolutionary history of eye movements, although this is beyond the scope of the current chapter: interested readers may wish to refer to the work of Walls (1962).

Eye movements can be broadly classified as either *gaze-holding* or *gaze-shifting*. Gaze-holding movements are designed to prevent the image of the world slipping across the retina, whereas gaze-shifting are designed to direct our gaze to an appropriate point of interest.

26.2 THE NEED FOR EYE MOVEMENTS

26.2.1 MINIMIZING VELOCITY BLUR: GAZE-HOLDING EYE MOVEMENTS

Relative motion between a visual target and the eye produces an image that sweeps across the retina. As the eye integrates visual information over time (Hildreth 1973), such retinal image motion can "smear" the perceived retinal image (although not necessarily by as much as a simple analysis of visual integration times might predict [Hildreth 1973]) and potentially decrease visual performance. Eye movements involved in reducing this velocity-induced smearing are called "gaze-holding" eye movements, and they achieve this by moving the eyes with a velocity that approximately matches the retinal image motion, thereby stabilizing the image.

What are the sources of this relative motion? Clearly the scene itself can move, such as when lying in a field and watching the clouds sail by on a windy day. More commonly, however, it is caused by the eye movement relative to the scene as a result of head and body movements: such movements may be actively generated by the observer or passively induced as when riding on a train. In general, the visual effects of either scene or observer motion are equivalent as it is the velocity of the image on the retina that determines visual performance, rather than the cause of this image velocity (Murphy 1978) (although see also Steinman et al. [1985]).

It may be initially thought that gaze-holding eye movements should ideally act to reduce image velocity to zero to maximize visual performance. However, when images are completely stabilized on the retina using appropriate laboratory techniques, perceptually fading occurs (Ditchburn and Ginsborg 1952) and contrast thresholds are increased markedly across a broad range of spatial frequencies (Kelly 1979a) (Figure 26.1). A small amount of retinal motion is therefore necessary to prevent such fading. Introducing a drift of around 0.15°/s to an otherwise stabilized grating target markedly improves sensitivity and produces a contrast sensitivity function similar to that obtained when an eye fixates a static grating target that has not been stabilized on the retina (Kelly 1979b). Of course, in the natural world, the head is not stabilized as it is in the laboratory, and so head movements—when combined with the imperfect nature of gaze-holding eye movements—means retinal image velocity is slightly higher at around 2°/s: fortunately, this small increase has little influence on the broadly shaped contrast sensitivity function (Carpenter 1992). Introducing progressively faster image velocities largely preserves the overall shape of the contrast sensitivity function but shifts it to progressively lower spatial frequencies (Kelly 1979b). As a result, high spatial resolution vision (visual acuity) is dramatically impaired by this retinal motion, although sensitivity to a large, low spatial frequency targets can be substantially *improved* if the peak of the contrast sensitivity shifts toward the spatial frequency of the target.

26.2.2 DIRECTING THE FOVEA: GAZE-SHIFTING EYE MOVEMENTS

The eye does not have an equal resolving capability across its entire field of view, but rather has its most acute vision limited to a small area centered on the fovea (see Chapter 12). Therefore,

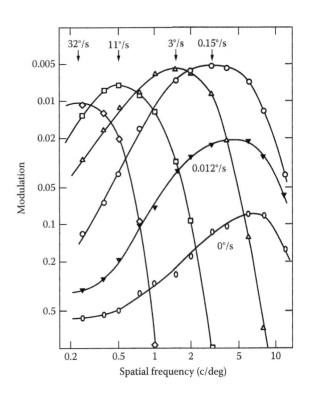

Figure 26.1 Contrast sensitivity curves for drifting grating targets presented using a retinally stabilized display. The curve for a drift rate of 0.15°/s (circles) is almost identical to that obtained when a nondrifting grating is presented on a nonstabilized display (Kelly 1985). A drift rate of 0.012°/s (filled triangles) is approximately equivalent to a retinal velocity of two cones per second. (From Kelly, D.H., *J. Opt. Soc. Am.*, 69, 1340, 1979b.)

there is a need to direct the fovea to whichever part of the visual environment that requires viewing in critical detail, along with the need to keep the fovea directed at the selected target should it move. *Gaze-shifting* eye movements are those responsible for achieving these tasks.

Simple observation of a person's eye can reveal the different types of gaze-shifting eye movements made. Visual inspection of a static scene is characterized by fast, darting eye movements (saccades), whereas observing a moving target can elicit smooth, tracking eye movements (smooth pursuit movements). Eye movements can also be made to ensure that both foveae remain fixed on a target despite changes in how far away the target is: an observer becomes "cross-eyed" when looked at an object brought near the observer's nose (vergence eye movements).

Although eye movements offer the fastest mechanism through which redirection of the fovea can take place, it should be remembered that the eyes can also be redirected by appropriate changes to either the head or body. Indeed, larger eye movements (above around 40°) are often accompanied by a relatively slower head movement in the same direction (Bartz 1966). Such head motion allows the eye to return to a more central position of gaze relative to the head while keeping the fovea directed at the object of interest.

26.2.3 EXPANDING THE EYE'S LIMITED FIELD OF VIEW

With steady fixation, the field of vision of the human eye extends approximately 60° nasally, 100° laterally, 60° superiorly, and 75° inferiorly (Anderson and Patella 1999). Binocularly the horizontal field of vision therefore extends approximately 200° and so allows a person to see slightly behind the plane of their eyes. While impressive, these limits mean that only approximately 40% of the visual environment can be monitored at a given instant.

Eye movements—particularly saccades (see Section 26.4.1)—extend the effective field of view of the eye. Although eye movements of 45° or more are possible, in natural viewing, such eye movements are extremely uncommon: excluding very tiny eye movements (<1°), nearly 90% of saccadic eye movements have amplitudes of 15° or less, with similar patterns observed in the horizontal, vertical, and oblique directions (Bahill et al. 1975). This additional 15° allows approximately 60% of the visual environment to be monitored, if the role of anatomical restrictions to the field of view (e.g., brows) is ignored. A much greater increase is achieved with appropriate movements of the head and body, however, allowing 100% of the environment to be viewed.

26.3 GAZE-HOLDING EYE MOVEMENTS

26.3.1 OPTOKINESIS

Imagine sitting on a moving train and looking out the window with a motionless eye (and with your head held still). As the world outside moves past, the image of the world moves across your retina: this movement is known as "retinal slip" (i.e., the difference between the eye's angular velocity and the target's angular velocity, being equivalent to the angular velocity of the image on the retina) and has the potential to compromise visual performance as described earlier (Section 26.2.1). Such retinal slip is a stimulus for optokinetic eye movements.

It is therefore not surprising that in real life, the eye is not motionless when staring out a train window, but rather engaged in a characteristic sawtooth pattern of eye movements known as "optokinetic nystagmus" (OKN) (Figure 26.2) or sometimes—and rather appropriately!—"train nystagmus" (Rademaker and ter Braak 1948). The *slow phases* of the sawtooth reflects the eyes moving in the same direction as that of the environment and so reducing the retinal slip velocity and stabilizing the retinal image, with the *fast phases* reflecting saccadic eye movements

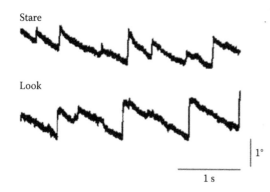

Figure 26.2 Horizontal eye movement trace demonstrating optokinetic nystagmus. The participant viewed a computer monitor (40° horizontal, 30° vertical) displaying a 0.5 cycle/° sine wave grating that drifted horizontally at 1°/s. For the upper trace, the participant attempted to keep staring straight ahead, whereas in the lower trace the participant looked at the stripes in the grating.

made in the opposite direction to avoid the eye reaching a too extreme position of gaze. OKN can be divided into *look OKN* (where the subject actively tracks the moving field) and *stare OKN* (where no effort is made to track the field). Both result in the characteristic sawtooth trace, although the gain (i.e., the ratio of the eye velocity to the target velocity) and amplitude of the slow phase is reduced for the latter (Knapp et al. 2009). In the laboratory or clinical setting, OKN can be invoked by moving a coarsely striped target in close proximity to the observer's eye so that the target occupies a large portion of their visual field. In the absence of a stationary fixation target in front of the striped stimulus, OKN cannot be suppressed and so such testing is useful in assessing vision in young infants (Naegele and Held 1982) or in suspected malingerers claiming to be blind. Although OKN can faithfully match lower target velocities, beyond around 100°/s the reflex breaks down (Carpenter 1988). This velocity limit is less than the retinal image velocity induced by a saccade (see Section 26.4.1.2), hence the large retinal slip induced by saccadic eye movements is too fast to act as a stimulus for OKN.

The latency for the onset of the slow phase of OKN is reasonably slow at approximately 80 ms (Kröller and Behrens 1997), which is not surprising given that it is driven by visual feedback about the retinal slip velocity. When an OKN stimulus is terminated abruptly by plunging the observer into darkness, it is found that OKN does not abruptly cease but rather progressively decays (Rademaker and ter Braak 1948). OKN is, therefore, not simply based on the immediate feedback about slip velocity, but also on a temporary storage of this velocity information. "Charging" this velocity store takes time, and it does not appear to occur if OKN is suppressed by staring at a fixation target (Fletcher et al. 1990).

26.3.2 VESTIBULAR

Most retinal slip is not caused by movement of the objects being viewed, but rather by movements of our heads and bodies. The vestibular system provides information about head movements and position, which is also used to drive correcting eye movements.

For example, when a colleague shakes their head in disagreement, it is observed that their eyes remain fixed on yours through rotations of the eyes that are equal and opposite to the rotation of their head. As mentioned previously (Section 26.2.2), eye movements toward a peripheral target are often followed by a head movement that allows the eye to return to a more central position of gaze relative to the head. In order to maintain fixation on this peripheral target, the angular rotation of the head must be accompanied by an equal and opposite angular rotation of the eyes. Eye movements resulting from such head rotations are driven by signals from the vestibular apparatus of the inner ear via the *vestibulo-ocular reflex* (VOR). Other head rotations, tremors, and translations in head position (such as when walking) must also be compensated for.

There are three fluid-filled semicircular canals (Figure 26.3) located in the inner ear on either side of the head to sense head rotation. They are oriented in largely perpendicular planes, thereby allowing rotations in all directions to be sensed. Head rotations rotate the canal but leave the fluid inside lagging behind (Figure 26.3b). The lagging fluid pushes on the *cupula* within the canal and so signals that a head rotation is occurring.

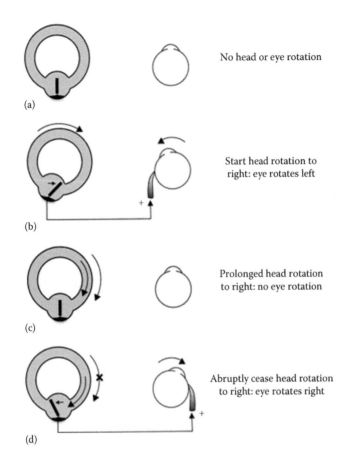

Figure 26.3 Schematic representation of vestibular eye movements. A fluid-filled semicircular canal is represented on the left of each panel, with the flap-like *cupula* (thick straight line) being displaced from its resting position when there is a relative difference in rotational velocities between the canal wall and the fluid. Stimulation arising from the semicircular canal produces a smooth eye movement (e.g., panels b and d): in addition to the excitation of the muscles shown here, the appropriate antagonist muscle must be inhibited.

These signals from the inner ear then pass to structures in the brainstem to generate an appropriate eye movement in the opposite direction to the head rotation by exciting and inhibition the appropriate antagonistic pairs of extraocular muscles (see Section 26.9). The magnitude of the movement must also appropriately match that of the head and in general the gain of the VOR is close to unity for horizontal and vertical movements. For torsional movements of the eyes, however (e.g., when tilting your head from one side to the other), the gain of the VOR is rather less than unity and so the rotation of the eyes does not fully compensate for the rotation of the head (Kushner and Kraft 1983). Indeed, some experimenters have found that no compensatory torsional movements occur at all in response to head tilts (Jampel and Shi 2002).

VOR therefore provides a mechanism to reduce retinal slip velocity that is not dependent upon visual feedback. Provided that vestibular information is rapidly transmitted, the possibility exists to generate compensating eye movements *prior* to the arrival of visual feedback about the true retinal slip velocity as required by optokinetic eye movements. This is indeed what happens: the latency of VOR is approximately 7 ms (Crane and Demer 1998), and so is approximately an order of magnitude faster than the latency for optokinetic eye movements (Kröller and Behrens

1997). The comparative speed of these two gaze-holding mechanisms can be demonstrated by holding a page at arm's length and moving it from side to side a few degrees at around 4 Hz (Robinson 1968): optokinesis on its own is not rapid enough to stabilize the moving image of the page, and so blurring of the text results. In contrast, shaking your head with a similar frequency and angular magnitude while keeping the page still will result in sufficiently stabilized retinal image to allow you to read. Even when attempting to keep your head still, measurable head rotations still occur, and these are only partly compensated by VOR and optokinesis (Skavenski et al. 1979): full compensation is, of course, undesirable given that a small amount of retinal image velocity is required for maintaining optimum contrast sensitivity (Figure 26.1).

With extended head rotation, friction from the canal wall will mean that the fluid in the canal eventually rotates at the same rate as the canal itself, thereby abolishing the stimulus to continued vestibular eye movement (Figure 26.3c). When the head (and canal) is suddenly stopped, however, the fluid continues rotating for a period and so pushes against the cupula, but now in the *opposite* direction to that when the head rotation commenced (Figure 26.3d). It is this sensory signal that creates the familiar sense of dizziness when we abruptly cease spinning around. This sensory signal equally generates a vestibular eye movement although now in the direction of the previous head rotation and one that—rather than stabilizing a visual target—draws it away from the fovea. Saccadic eye movements act to reposition the target on the fovea, thereby creating the characteristic alternation between slow vestibular movements and rapid corrective saccades characteristic of *vestibular nystagmus* (the waveform of the nystagmus is similar to that of OKN, shown in Figure 26.2).

Sensing linear accelerations of the head (e.g., the up–down translation of the head that accompanies walking) requires the action of different sense organs in the middle ear—the *otolith organs*. Here the tips of sensory fibers are embedded in a dense mass (the otolith) that moves in response to head tilts and linear accelerations, thereby bending the sensory fibers and signaling the direction of the tilt or acceleration. Indeed, from the standpoint of the sensory fibers, the stimulation produced by a tilt or an appropriate linear acceleration are identical. The latency for compensatory eye movements generated by translational VOR as also quick, being a few tens of milliseconds (Bronstein and Gresty 1988). The magnitude of the compensatory eye movement required to stabilize retinal imagery in the presence of translational head movement is slightly more complex than for rotational eye movements, as the amount the retinal image moves will depend upon how far away the object being viewed is. In angular terms, an object close to an observer will move appreciably during the vertical translation of the head that accompanies walking, whereas an object on the horizon will be effectively stationary. It is therefore not surprising that the magnitude of translational VOR alters inversely with viewing distance (Busettini et al. 1994). That the magnitude of VOR is modified by factors other than simple rotations and translations of the head means the response is somewhat more complex than the simple reflex arc commonly presented (Crane and Demer 1998).

26.4 GAZE-SHIFTING EYE MOVEMENTS

26.4.1 SACCADES

26.4.1.1 Ballistic nature

Saccadic eye movements—or saccades—are rapid, darting eye movements that quickly direct the fovea to a new position of gaze. They are said to be ballistic, in that the characteristics of a particular saccade is preprogrammed, and once the eye begins to move, the course of the eye movement is not guided to the target via visual feedback (Carpenter 1988). The need to preprogram saccades is due to their rapid nature: there is simply not sufficient time for visual information to be fed back to the appropriate control centers in the brain to modify the eye movement as it occurs. The rapid nature of saccadic eye movements is seen in Figure 26.4, which shows traces of the horizontal position of the eye over time when saccades of various amplitudes are made. A characteristic steplike function results, wherein the eye typically takes only a few tens of milliseconds to reach its target once it begins moving.

26.4.1.2 Amplitude and velocity characteristics

The amplitude of saccadic eye movements varies widely from a fraction of a degree (see also *microsaccades* in Section 26.5.1) to over 40° although, as noted previously, the frequency of large saccades (>15°) is low in natural viewing conditions. Once the eye starts moving, the time taken to complete the saccade increases with its amplitude as can be seen in Figure 26.5 (upper panel) (Bahill et al. 1981). The linearity of this relationship appears to hold even when very large (> 60°) saccades are made (Carpenter 1988). It should be noted that the slope of this relationship is such that when the amplitude of a saccade doubles the duration of movement is less than double, and so it

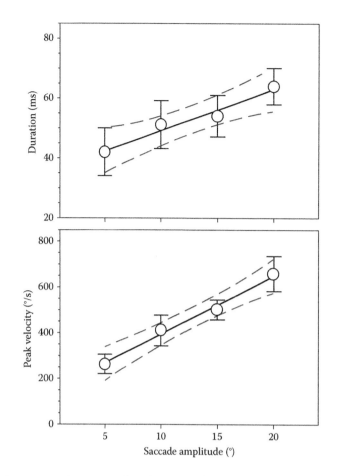

Figure 26.5 Duration and peak velocity of saccades, as a function of the amplitude of the saccade, determined for 13 young, normal observers. Error bars gives ± one standard deviation for the mean data, with solid lines showing linear regression to these data and dashed lines the corresponding 95% confidence bands. (Data from Bahill, A.T. et al., *Invest. Ophthalmol. Visual Sci.*, 21, 116, 1981.)

would be anticipated that the velocity of a saccade would also increase with the amplitude of the saccade: this is indeed the case (Figure 26.5, lower panel: note, however, that peak velocity saturates for saccadic amplitudes above those shown here (Bahill et al. 1975)). The well-defined relationship between a saccade's amplitude, duration, and peak velocity has been termed the *main sequence* and has been used to define what eye movements are, in fact, saccades (see Section 26.5.1).

26.4.1.3 Latency and the notion of oculomotor procrastination

Although the duration of a saccade is short once the eye starts moving, a substantial time elapses (~200 ms) between the onset of the target and the beginning of the saccade made toward it: this interval is called the "saccadic latency" (Figure 26.4). This delay is initially puzzling, as it takes only around 30–60 ms for visual information about the target to reach low-level control centers in the brain capable of initiating saccades, such as the superior colliculus (Sparks 1986). Why don't we respond as quickly as possible? The key to understanding this oculomotor procrastination is that while low-level centers are capable of rapidly determining *where* a target is located, they do not have direct access to appropriate information to determine *whether* or not the target is of potential interest. This idea is developed more in the following

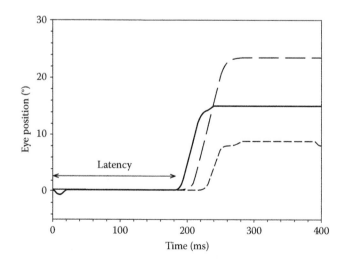

Figure 26.4 Saccadic eye movements. Traces show the horizontal eye position when a target abruptly jumps 8° (short dashes), 16° (solid line), or 24° (long dashes) to the right at time zero. Of note is that while the shape of the eye movement trace is a highly stereotyped step function, the saccadic latency (i.e., the time between the appearance of the target and the commencement of the eye movement) is not: even for saccades of the same amplitude, latency randomly varies from trial to trial.

section (Section 26.4.1.4), although, in brief, saccadic latency can be thought of reflecting—at least in part—the time taken to decide whether a particular target should be looked at next.

An additional key feature of saccadic latencies is that they are highly variable from trial to trial. It may be thought that this delay reflects variability in the time to confidently detect a visual target in the presence of visual noise, except that this variability persists even for high-contrast, highly visible targets. Rather, it has been suggested that this variability arises from the decision-making process itself. Although many models exist, the linear approach to threshold with ergodic rate (LATER) model (Reddi and Carpenter 2000) is probably the simplest principled model to describe this decision-making process. In the LATER model, an internal decision signal S begins to rise linearly from a base level S_0 in response to the appearance of a stimulus, with a saccade being generated once a threshold S_T is reached. The rate of rise is randomized on a trial-to-trial basis and follows a normal distribution, and so it follows that the time S_T is reached—and, therefore, saccadic latency—will be distributed as the reciprocal of a normal distribution. The reciprocal of latencies are indeed essentially normally distributed, as predicted (Carpenter 1981, 1988). Furthermore, neurophysiological evidence shows that reaction time variability for highly visible targets reflects randomness in the rate of rise of signals, rather than variability in the threshold activation required to generate a saccade (Hanes and Schall 1996). Deciding between multiple response options (e.g., several competing targets of interest in the field of vision) can then be viewed as a race to threshold between competing LATER units, with the unit reaching threshold first winning the race and canceling the rate of rise of the remaining units.

The competitors in this race may, however, be handicapped. The starting level S_0 reflects the likelihood of making a particular decision, and so when a target is highly likely to appear in a particular location (e.g., to the left), S_0 rises and latencies correspondingly shorten (Carpenter and Williams 1995). Rates of rise are independently randomized for each LATER unit (Leach and Carpenter 2001) however, ensuring that the most likely decision does not always win the race. This active randomization in our decision-making process means that our responses are not completely determined by our sensory input, which—when considered in relation to decision-making more generally—is probably important for survival. One does not want the pursuing lion—who is also able to see the same sensory input as we can—to be able to perfectly predict our next move. Such partial randomization of responses may also help in discovering new and potentially beneficial behaviors (Carpenter 1999).

Of course, the role of simple target detection on latency cannot be ignored. It is well established that saccadic latencies increase as target contrast decreases (reviewed in Taylor, Carpenter and Anderson [2006]), reflecting that an increasing amount of time is required to distinguish a low-contrast target from background noise prior to any decision about that target being made. This behavior is well captured in a two-stage model for decision-making, which consists of a contrast-dependent target detection stage followed by a LATER decision-making stage (Carpenter et al. 2009).

Saccadic latency therefore reflects the decision-making time and so has been the focus of many studies investigating decision-making processes in the brain. Multiple factors have been shown to influence saccadic latency, including target appearance probability (Carpenter and Williams 1995), contrast (Carpenter 2004a; Taylor et al. 2006), the urgency of responding (Reddi and Carpenter 2000), and whether targets appears as part of a predictable sequence (Anderson and Carpenter 2010; Anderson and Stainer 2014).

26.4.1.4 Relationship between higher centers, superior colliculus, and brainstem

Although the details of the control centers describing how we shift our gaze via saccades are potentially complex, it is instructive to consider the general layout reduced to three hierarchical stages concerned with—in turn—*how* appropriate neural impulses are generated to move the eye, determining *where* potential targets of interest are located, and deciding *which* is the particular target we will look at next (Carpenter 2000) (Figure 26.6).

How the eyes are moved involves two types of neural activity arising in the brainstem. *Burst units* produce a short flurry of neural pulses that is of the precise length in order to ballistically launch the eye in to its new position. In contrast, *tonic units* provide a steady stream of pulses required to hold the eye in its new position. As the size of a saccade increases, the *duration* of the flurry from burst units and the *rate* of pulses from the tonic units correspondingly increase (Luschei and Fuchs 1972). Indeed, the rate of tonic pulses is itself derived from integrating the activity of burst cells in the brainstem: when this integrator is lesioned, the eye can be flung into position but cannot be held there, and so gradually drifts back (Cannon and Robinson 1987). Appropriate action of the integrator is aided by the cerebellum (Carpenter 1972; Cannon and Robinson 1987). The activity of these burst and tonic units is then combined to form the motor neuron signal that is delivered to the extraocular muscle. The timing of the flurry is controlled by a further set of neurons—omnipause neurons—whose suppressive effect on the burst units is transiently lifted for the precise time needed to produce the appropriate duration burst for a given amplitude saccade.

The superior colliculus appears critically involved in calculating *where* the eye might move. The superior colliculus contains a "motor map" in retinotopic coordinates (i.e., positions relative to the fovea), with electrical stimulation of a particular location sending signals to the brainstem to produce a saccade of corresponding direction and amplitude (Sparks 1988). Neurons at the rostral pole of the superior colliculus represent the fovea on this map and are active when the eye is fixating and are suppressed during a saccade (Munoz and Wurtz 1993): unsurprisingly, their output projects to omnipause neurons (Büttner-Ennever et al. 1999). The superior colliculus receives direct input from the eye (Schiller and Malpeli 1977) and from other cortical areas (Krauzlis 2005) and so can produce an area of activation on the map corresponding to where a visual target is (van Opstal and van Gisbergen 1989). What it *cannot* do makes much sense of what these targets are, as the inputs to the colliculus carry only a limited information (e.g., no information about color or fine spatial form) (Schiller and Malpeli 1977; Schiller et al. 1979). In a laboratory setting, targets may obligingly appear in isolation and so the required saccadic eye movement is unambiguous, although in the natural world there are typically many potential targets competing for our gaze.

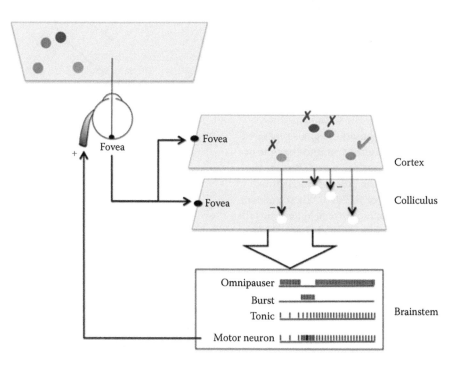

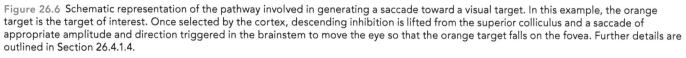

Figure 26.6 Schematic representation of the pathway involved in generating a saccade toward a visual target. In this example, the orange target is the target of interest. Once selected by the cortex, descending inhibition is lifted from the superior colliculus and a saccade of appropriate amplitude and direction triggered in the brainstem to move the eye so that the orange target falls on the fovea. Further details are outlined in Section 26.4.1.4.

The superior colliculus is therefore held in check until higher levels of the brain—with access to important information such as color, form, expectation, and urgency—can decide *which* particular target in our field of vision is the next target of interest. When a decision is made, inhibition of the superior colliculus is momentarily lifted for the appropriate location on the map so that a saccade can be produced. Many areas are involved in this decision-making process, although three of the key cortical areas involved in eye movements are the frontal eye fields (FEF), the lateral intraparietal area (LIP), and the supplementary eye fields (SEF). The FEF is involved in attention and the generation of voluntary saccades (Schall 2004), whereas the role of SEF is more complex but includes involvement in encoding learnt sequences (Gaymard et al. 1993; Tanji and Shima 1994) and monitoring for response errors (Carpenter 2004b). LIP appears to be involved in target selection like the FEF but is also affected by the likely rewards associated with selecting one target over the other (Platt and Glimcher 1999).

Of course, the highly simplified model outlined earlier should not be taken too prescriptively. Based on Figure 26.6, one would predict that lesions to the cortex or to the superior colliculus would mean saccades could no longer be made to visual targets although this is in fact not the case (Sparks 1986). Also, perturbing the eye position through electrical stimulation just after a visual target is briefly flashed, but before a saccade commences, might be expected to result in an error in the final eye position equal to the perturbation. Instead the perturbation is compensated for in the final saccade, suggesting that saccade metrics cannot be driven purely by visual stimulation alone (Sparks and Mays 1983). Certain parts of the superior colliculus may themselves be involved in calculating the final motor error (i.e., the difference between the current position of the eyes and the desired

position) during such perturbation experiments (Sparks and Porter 1983), and indeed it has been suggested that the superior colliculus is more a map of *motor goals* rather than the specific movement required to achieve the goal (Krauzlis 2005). The simplified model presented here also necessarily ignores the multiple other areas of the brain involved in shifting our gaze, along with their high degree of interconnectedness (Carpenter 1988).

26.4.1.5 Saccadic omission, suppression, timing reversal, and postsaccadic enhancement

Given the rapid velocities of saccades outlined earlier, the image of the world correspondingly sweeps across the retina when the eye is in flight. We are not perceptually aware of this motion, as can be readily demonstrated by observing one's eyes in a mirror, where it is impossible to catch one's eyes in flight when looking from one eye to the other (Dodge 1900). Neither is there any perception of image motion or displacement or, indeed, any sense of a *gap* in visual sensation during eye movement.

Our lack of perception can be explained, in part, because the contrast sensitivity of the visual system is severely compromised by the high image velocities during a saccade (see Section 26.2.1 and Figure 26.1). This reduced sensitivity would be expected to manifest as a perceived reduction in contrast (a "grayout") as the eyes move. The comparatively high-contrast, sharp imagery that immediately precedes and follows a saccade acts to mask the perception of this grayout, a process that has been called "saccadic omission" (Campbell and Wurtz 1978). By arranging for a scene to be illuminated only while the eyes are in flight, it can be demonstrated that a smeared, grayed-out version of the world is indeed readily visible (even when examining your own eyes in a mirror) (Campbell and Wurtz 1978).

Optical properties of the eye

While the lack of perception during a saccade is primarily explained by saccadic omission, perception is also decreased during the movement of the eyes through a process called "saccadic suppression." This suppression is not absolute and appears to be selective for the magnocellular visual pathway (Burr et al. 1994) that is critically involved in motion detection. At least part of this suppression appears actively induced by the saccadic eye movement, rather than passively induced by motion signals generated on the retina (Thiele et al. 2002). Curiously, around the time of a saccade perceptual distortions of space can occur (Ross et al. 1997; Zimmermann et al. 2014) and the order of closely timed events can appear reversed (Morrone et al. 2005). These phenomena may be due to predictive shifts in receptive field locations that occur in anticipation of a saccade (Duhamel et al. 1992; Walker et al. 1995).

In contrast to saccadic suppression, in monkeys there is a significant increase in neural activity after a saccade, commencing at the level of the lateral geniculate nucleus (the location where retinal ganglion cell axons first synapse after leaving the eye), which has been argued may be the cause of a postsaccadic enhancement of vision (Ibbotson et al. 2008). Chromatic contrast sensitivity is indeed slightly enhanced immediately after a saccade, although sensitivity to luminance-modulated stimuli—of the sort that have demonstrated enhanced neural responses in monkeys—do not appear to be (Burr, Morrone, and Ross 1994; Knöll et al. 2011). As such, the precise relationship between postsaccadic enhancement of neural responses in monkeys and perceptual responses in humans is unclear.

26.4.2 SMOOTH PURSUIT MOVEMENTS

Smooth pursuit movements are gaze-shifting movements used to track moving object, typically with the aim of keeping the image of the target on the fovea. The velocity of smooth pursuit movements therefore will, ideally, match the velocity of the target. This is indeed the case over a wide range of image velocities for linearly moving targets (Meyer et al. 1985). The maximum velocity of reliable pursuit eye movements is extremely rapid, at approximately 90°/s (Meyer et al. 1985). Not surprisingly, pursuit eye movements are absent in animals without a fovea. For example, rabbits have an extended visual streak containing increased ganglion cell density, rather than a discrete fovea, and so only show optokinetic eye movements (and even then, only for motions covering more than half the visual field) (Collewijn 1969).

When a target abruptly springs in to motion, there will be a short latency before the pursuit movement can commence (in the order of 80 ms) (Lisberger and Westbrook 1985), and so simply producing a pursuit that matches the target velocity will result in a fixed offset between the target and the fovea. Furthermore, the eyes have a fixed acceleration for the first 20 ms or so of the pursuit that is independent of the final pursuit velocity (Lisberger and Westbrook 1985), potentially further widening this offset. Because of this, there is typically a saccadic eye movement shortly after the commencement of a smooth pursuit to close this offset and so place the target on the fovea (Figure 26.7). Such a saccade appears only to occur when the target velocity is sufficiently high (approximately 3°/s) such that the offset reaches a threshold level for generating a visually driven saccade (around a quarter of a degree) (Rashbass 1961).

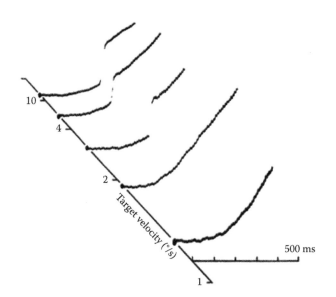

Figure 26.7 Horizontal eye movement traces demonstrating smooth pursuit eye movements in response to a target that abruptly commences moving at a uniform velocity. Time is on the horizontal axis, eye position on the vertical axis, and target velocity on a third axis at 45°. The gain of the recording was made inversely proportional to the target velocity, and so if the velocity of smooth pursuit eye movements is linearly proportional to the target velocity, then the smooth portion of each trace would appear largely the same: this is indeed the case, save for some flattening of smooth pursuit movement for the fastest target investigated (12.5°/s). Note also the intrusion of "catch-up" saccadic eye movements for faster target velocities: for the trace in response to the 3°/s target, the magnitude of the saccade is 0.6°. For target velocities much lower than this, there is little lag produced between the fovea and the target as the smooth pursuit eye movement commences, and so the threshold for a saccade is not reached. The saccade can also be eliminated for faster target velocities by having the target abruptly jump an appropriate amount in the opposite direction to the target velocity just as the target starts to move. (From Rashbass, C., *J. Physiol.*, 159, 326, 1961.)

The onset of the smooth pursuit movement and the saccade are independent (Rashbass 1961).

Generating an acceleration of the eyes in response to the retinal slip velocity of the target to be pursued is a simple way to generate a smooth pursuit movement that precisely matches the velocity of the target (Rashbass 1961). Several models have been proposed to explain the dynamics of smooth pursuit eye movements, although unsurprisingly they share a common input signal of retinal slip velocity (Lencer and Trillenberg 2008). The middle temporal (MT) visual area and medial superior temporal (MST) visual area are highly involved in processing motion signals arising from the retina and traveling through the visual pathway. These areas are also activated during pursuit (Barton et al. 1996) and MST neurons are also able to signal motion differences between their receptive field center and the surround as may occur between a moving target and its background (Eifuku and Wurtz 1998). Signals then pass forward to frontal areas of the cortex, including to the FEF, where smooth pursuit signals can be generated (Tanaka and Lisberger 2001; Lencer and Trillenberg 2008).

As the final velocity of a smooth pursuit movement matches that of a linearly moving target (Meyer et al. 1985), it follows that the retinal slip velocity for the target will be zero. Therefore, the continued tracking of a constant velocity target is not driven by retinal slip velocity *per se* but rather by some calculation by the oculomotor system of the target's velocity. Consistent with this, smooth pursuit eye movements can continue for a brief period when a target disappears momentarily (Becker and Fuchs 1985), which would be of great benefit in the case where a pursued target is momentarily obscured by foreground features.

Although pursuit eye movements are typically in response to image motion, other stimuli are also able to generate smooth pursuits. For example, smooth pursuit eye movements can be generated when a subject is asked to follow a moving hand (their own!) in the dark (Jordan 1970). Similarly, smooth pursuits can be used to track a moving auditory tone (Carpenter 1988). Some form of stimulus is, however, required: attempts to change gaze smoothly in the absence of a moving stimulus generate a series of saccades, rather than a smooth pursuit eye movement.

26.4.3 VERGENCE

So far we have dealt primarily with *conjugate* eye movements, where each eye moves by equal amounts in the same direction. Conjugate eye movements result when we change our gaze between, or track motion within, a plane that is a fixed distance away from the observer. However, objects may appear at different distances away from the observer. As our eyes are separated horizontally in space, each eye must move by a different amount in order to shift our gaze between two objects at different distances (Figure 26.8). Such *disjunctive* gaze-shifting movements are given the special name of *vergence* eye movements. The most common vergence eye movements are *convergence* and *divergence*, describing the gaze of each eye moving either toward or away from each other, respectively, in a horizontal plane.

26.4.3.1 Stimuli for vergence

A key stimulus for vergence eye movements is the *retinal disparity* that arises from an object located at a distance other than the current plane of regard. In Figure 26.8 (panel a), the object positioned at location "b" is further away from where the eyes are currently looking (location "a"), and so the image of the object falls on different retinal locations relative to the fovea in each eye. This retinal disparity is the stimulus that drives, at least in part, a divergent eye movement if we wished to look at object "b." Conversely, an object placed at position "c" again creates a retinal disparity although in the opposite direction and so would drive a convergent eye movement. In contrast, the object located at "d"—although not imaged at the fovea—creates no retinal disparity as the image of the object in each eye is located at the same eccentricity and direction relative to the fovea. In this case, a conjugate eye movement (a saccade) is required to look at this object, rather than a vergence eye movement.

In addition to requiring a change in gaze position, the accommodative state for the eye also needs to alter as object depth changes in order to keep the retinal image in focus. This change in accommodation can itself drive vergence eye movements through the *accommodation/convergence synkinesis* (Schor 1979). On its own, the amount of vergence driven by accommodation is typically insufficient to bring both foveae in to alignment on the new object of regard.

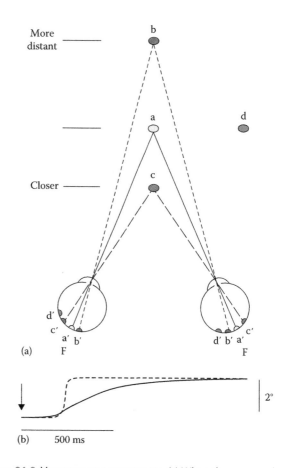

Figure 26.8 Vergence eye movements. (a) When the eyes are looking at object "a," the images of the object (a') are located at the fovea. Objects in either closer or more distant planes cause *retinal disparity* (i.e., the images of these objects are located at different locations relative to the fovea in each eye). Such disparity is a stimulus for a vergence eye movement, being convergence when wanting to look at object "c" and divergence when wanting to look at object "b." In contrast, an object in the same plane (object "d") stimulates corresponding points in each eye (i.e., points with the same location relative to the fovea, thereby producing no disparity) and so do not stimulate a vergence eye movement. Note that looking at "b" or "c" also produces a change in accommodation, which itself can drive vergence eye movements. (b) Schematic eye movement trace (solid curve) when a disparity of 2° is presented abruptly at the downward pointing arrow, in the absence of any accommodative change: time course is based on the data of Rashbass and Westheimer (1961). A schematic representation of a saccadic eye movement of the same magnitude is shown for comparison (dashed curve).

For many people, convergence may also be driven purely voluntarily, with no regard to retinal disparity (the ability to go "cross-eyed").

26.4.3.2 Dynamics and limits

Figure 26.8 (panel b) shows the horizontal eye position for an eye performing a vergence eye movement of 2° degrees in the absence of any accommodative change (achieved in the laboratory by independently controlling the images presented to each eye). The movement is substantially slower than a saccadic eye movement (around 15°/s initially for a 5° disparity), with the velocity of the eye gradually slowing in a roughly exponential manner as the target is approached (Rashbass and Westheimer 1961). This is what would be expected when an eye movement is guided by feedback from a visual signal, in this case retinal disparity: as the target is approached, the retinal disparity decreases, and the drive for the eye movement correspondingly reduces. Consistent with this idea, when an experiment is set up such that a constant disparity is maintained, the velocity of the vergence eye movement is proportional to the disparity (Rashbass and Westheimer 1961). When a step change in disparity is presented, there is a latency of around 160 ms before the eyes commence moving, similar to that seen in saccadic latency, with the new position of the eyes achieved over the course of approximately a second (Rashbass and Westheimer 1961). Under more natural conditions—where multiple cues to object depth exist—the temporal responsiveness of the vergence system improves (Erkelens et al. 1989). Furthermore, when gaze is directed to objects located not simply a different distance away but also to the left or right of current gaze, the bulk of the required convergence or divergence is met by making saccades whose amplitudes differ between the eyes (i.e., disjunctive saccades) (Erkelens et al. 1989).

Retinal disparity can be easily induced by placing a prism in front of one eye to displace the image. Therefore, the magnitude of vergence eye movements is sometimes expressed in terms of prism diopters (Δ) where one prism diopter describes an angular displacement equivalent to a 1 cm lateral shift in an object located 1 meter away. When viewing in the natural world with the naked eye, there is a fixed linear relationship between the accommodative demand in diopters (or reciprocal meters) and the convergence demand in prism diopters. The introduction of various optical instruments (e.g., lenses, prisms, magnification, or binocular instruments) can alter this relationship (Smith and Atchison 1997), with excessive discrepancies between the vergence and accommodative demand meaning that single, clear, binocular images cannot be formed: typically either the images are made clear but appear double, or the image is fused into a single percept but with noticeable image blur. Vertical image disparities do not normally occur in the natural world, and so it is not surprising that our ability to make vertical vergence eye movements in response to a suitably oriented prism is dramatically less than in the horizontal plane (Bennett and Rabbetts 1989).

26.5 OTHER

26.5.1 MINIATURE EYE MOVEMENTS

In between eye movements, such as saccades, pursuits, and OKN, it would be tempting to assume that the eye is at rest. We have already seen (Section 26.3.2) that head rotations cannot be prevented despite a subject's best efforts, and so compensatory VOR and optokinetic responses are constantly occurring (Skavenski et al. 1979). Furthermore, even when the head is mechanically held in place in a laboratory setting (e.g., through the use of a bite-bar), the apparently steadily fixating eye is subject to miniature (small amplitude) eye movements, which can be classified in to three broad categories, outlined as follows.

Drift described comparatively slow, meandering eye movements with velocities of approximately 4'/s and amplitudes around 2–5' (Carpenter 1988) and are believed to reflect instability within the oculomotor system (Cornsweet 1956). Drift appears similar to a "random walk" (Skavenski and Steinman 1970) and

so suggests the action of a noise source existing prior to an integrator of some kind (Carpenter 1988). Based on their frequency components, it has been argued that the neural activity responsible for drift arises in the brainstem (Spauschus et al. 1999). The movements appear coordinated between the two eyes (Riggs and Ratliff 1951), although other works suggests movements between the eyes are not correlated (Krauskopf et al. 1960).

Microsaccades are small saccadic eye movement, although their limits are somewhat ill-defined: typically they are substantially less than a degree in amplitude (Schulz 1984), binocular (Schulz 1984), and with peak velocities up to around 70°/s (Engbert and Kliegl 2003). They share many of the characteristics of regular saccades and fall on the main sequence for saccades described previously (see Section 26.4.1.2) (Zuber et al. 1965). Indeed it has been argued by some that the distinction between saccades and microsaccades is "fundamentally arbitrary" (Otero-Millan et al. 2008). One suggested role for microsaccades is to correct for when drift eye movements shift the image of a target too far away from its desired position on the retina (Cornsweet 1956), although others have argued that other, slow control mechanisms achieve this role (Steinman et al. 1973; Kowler and Steinman 1980). Microsaccades can be voluntarily suppressed (Steinman et al. 1967) and have been argued to be useless "busy work" (Steinman et al. 1973): the over-learnt gaze shifting habit we use when viewing the world unconsciously continues (albeit on a much smaller scale) when we attempt to fixate a target. While some evidence suggests microsaccades are not important for fine visual tasks and indeed tend to be suppressed during such tasks (Kowler and Steinman 1980), other work suggests that microsaccades are visually useful eye movements whose frequency can increase as a result of increasing task demand (Otero-Millan et al. 2008) and are directed to appropriate regions of interest in spatially detailed tasks (Ko et al. 2010). Microsaccades also appear to counteract the effect of perceptual fading that can occur when fixating both foveal and peripherally presented images (McCamy et al. 2013), although others have argued that no special role exists for microsaccades in maintaining the visibility of foveal images (Kowler 2011). Furthermore, it has been argued that one of the primary role of microsaccades may be to improve the eye's sensitivity to very low spatial frequency objects, given the leftward shift in the peak of the contrast sensitivity function during rapid retinal image velocities (Figure 26.1) (Carpenter 1992). For stimuli other than simple gratings, however, things may be more complicated. Fixational eye movements can improve sensitivity to high spatial frequencies for stimuli containing low and high spatial frequencies combined in proportions similar to those seen in natural images (Rucci et al. 2007).

Tremors (or oculomotor microtremor) are the smallest of the miniature eye movements, being less than 0.5 arc minutes (the approximate width of a cone photoreceptor in the fovea). They occur simultaneously with drift. Tremors are rapid, nonperiodic oscillations with dominant frequencies in the order of 70–100 Hz (Bolger et al. 1999) and have been argued to reflect neural activity in the brainstem (Spauschus et al. 1999). The movements are independent for the two eyes (Riggs and Ratliff 1951). In contrast to microsaccades, the frequency of tremors is not related to perceptual fading during fixation (McCamy et al. 2013).

Figure 26.9 Using afterimages to demonstrate small fixational eye movements. By staring at the white dot on the left for approximately 20 s, an afterimage of the rings is produced when fixation is then transferred to the black dot on the right. Small drifts (slow) and microsaccades (rapid) should become apparent as relative motion between the black dot (whose image location depends upon eye position) and the afterimages (whose image location if fixed on the retina and so is independent of eye position). At a viewing distance of 40 cm, the inner and outer rings have radii of approximately 0.5° and 1.0°, thereby allowing the magnitude of the fixation eye movements to be estimated. By resting your chin in your hands while your elbows are on a table—thereby achieving some stabilization of the head—it will be noticed that the nature of the observed image motion does not change appreciably, consistent with the notion that the motion is primarily that of the eye.

Despite these miniature eye movements constantly shifting the position of the eye during fixation, normally we are perceptually unaware of such shifts. It may be thought that they are simply too small to appreciate, although Verheijen has shown that it is easy to visualize one's own drift and microsaccadic eye movements using an appropriately arranged retinal afterimage (Verheijen 1961) (Figure 26.9). How then is perceptual stability achieved under most viewing conditions? There is evidence that the oculomotor system exerts active control over at least a component of drift eye movements (Nachmias 1961), and so efference copy (i.e., using knowledge of where the eyes are being instructed to move) could therefore be assumed to achieved part of this perceptual stability. However, it appears that correction for miniature eye movements is predominantly due to motion signals arising from retinal mechanisms. Murakami and Cavanagh (1998) proposed that, at any given moment, the smallest retinal motion signal across a scene is assumed by the visual system to be the result of miniature eye movements about an object being fixated. Retinal motion signals across the scene can be corrected by this smallest motion signal and perceptual stability thereby achieved. Subsequent experiments employing gaze-contingent display stabilization have provided further insight into these mechanisms (Poletti et al. 2010; Arathorn et al. 2013), although these experiments confirm Murakami and Cavanagh's basic idea of a correction mechanism employing retinal-based motion signals.

26.6 COORDINATING DIFFERENT EYE MOVEMENTS

Although largely treated separately, it should be borne in mind that in natural environments, the net movement of the eyes typically reflects a combined effect of several different types of eye movements. For example, an object placed far off to the right might initiate a saccade to the right, followed by a slower vestibular eye movement to the left to compensate for when the head rotates toward the object: the timing of these movements may

overlap (Morasso et al. 1973) and so their net effect needs to be appropriately coordinated. The complexities quickly escalate if the observer is also walking (producing an up and down translation of the head, and therefore translational VOR), and the object is relatively closer to the observer than where they were previously looking, thereby requiring a vergence eye movement. Not uncommonly the object of regard may also be *moving* and so require that addition of a smooth pursuit eye movement. If coordinating all of these actions was not impressive enough, overlying everything are ceaseless miniature eye movements. It is therefore unsurprising that eye movement control pathways are highly interconnected (Krauzlis 2005) to allow this coordination to take place.

In some situations, coordination may be relatively simple given that different eye movements represent responses to different aspects of the stimulus. For example, the action of smooth pursuit eye movements and catch-up saccades when initially tracking a rapidly moving target are driven by target velocity and target displacement from the fovea, respectively (Section 26.4.2). Although the final eye movement appears as a coordinated whole (e.g., an accelerating smooth pursuit movement in the direction of the target, along with a saccade in the same direction), laboratory experiments that displace the target at the moment it starts moving reveal the large degree of independence of the two underlying control mechanisms (Rashbass 1961). Therefore, the apparently coordinated eye movement pattern we normally see may be more a reflection of how objects typically move in the natural world. The saccadic and smooth pursuit systems do, however, appear to share a common mechanism for selecting the appropriate target on which to act (Krauzlis and Dill 2002; Erkelens 2006).

For other situations, there is a clear need for a more active interaction between control systems. For example, the type of large-field stabilization produced by optokinesis commonly will be at odds with that large-field retinal slip produced by smooth pursuit eye movements. Carpenter (1988) highlights this conflict with his example of a cat watching a mouse that is scurrying through the undergrowth. Smooth pursuit eye movements allow the tiny image of the mouse to be kept on the cat's fovea, but this means that the expansive, textured image of undergrowth slips across the retina, creating a strong, coherent motion signal that is a salient stimulus for optokinetic stabilization. Under such circumstances, the desire to perform a pursuit eye movement requires that the optokinetic stabilization reflex be suppressed, and it appears that appropriate allocation of visual attention is involved in effecting this suppression (Rubinstein and Abel 2011). Image motion at the retina arising from natural scenes is potentially complex, and so optokinesis and pursuit eye movements need to be flexibly engaged to ensure that the correct object or objects are stabilized.

26.7 DECIDING WHERE TO DIRECT OUR GAZE

26.7.1 ROLE OF SALIENCY

From the brief description of the mechanisms controlling saccadic eye movements (Section 26.4.1), it is clear that some form of decision-making process is involved in selecting what target we will look at next. Determining where next to direct our gaze is a complex task, reflecting a combination of both the nature of the features existing in the field of vision (so-called bottom-up control) as well as the nature of the task the observer is performing (top-down control). One target attribute that could influence this process is the visual *saliency* or conspicuity of a target, with this saliency being dependent upon low-level (bottom-up) features. One might expect targets of greater contrast to be more salient and so more commonly selected for visual inspection, and indeed, in the two-stage decision-making mechanism described in Section 26.4.1.3, the highest-contrast target should most often win the race and so be looked toward, all other factors being equal.

More sophisticated models for saliency exist, with conspicuity calculated for individual component features (e.g., color and intensity) at several spatial scales and then used to create feature maps, which are then combined to form a single saliency map (Koch and Ullman 1985; Itti and Koch 2000). The location of greatest activation on the map is where gaze is then drawn to, with additional measures applied to prevent gaze becoming immediately redirected to this most salient location or its immediate surround. At small amplitudes (<8°), saccades indeed appear to be directed in part by the salience of features within natural scene images although larger saccades are less dependent on such features (Tatler et al. 2006). Of course, simply directing the eyes to those locations that are already highly visible does appear to ignore the primary reason we have saccades—to shift our gaze to those locations that currently are not adequately visualized with our peripheral vision (Kowler 2011). Visual search models that select the next point of gaze in order to maximize the likelihood of finding the desired target (Najemnik and Geisler 2005) may better address this issue. What drives the eye movements in a visual search will ultimately be dependent on exactly how the search task is arranged: it has been argued that top-down control dominates behavior when searching for previously viewed natural objects (Chen and Zelinsky 2006).

The time over which information is accumulated and used to select an appropriate target for a saccade has also been the subject of investigation. Because saccades are so rapid, and because the transmission of visual information from the eye to those centers controlling eye movements is not instantaneous, there is a small "dead period" of around 80 ms in which incoming visual information cannot influence an upcoming saccade (Becker 1991). For rapid visual searches, target selection for a saccade depends upon integration of information occurring within a small (hundreds of milliseconds) window prior to this dead period (Caspi et al. 2004). Curiously, information in the dead-period of one saccade does appear to be used for target selection and programming of a subsequent saccade. Indeed, there is evidence for some form of parallel programming of saccades, as information prior to the dead period is utilized both by the upcoming saccade and the subsequent one (Caspi et al. 2004). For simple decision tasks, there is evidence that the visual information may be integrated over an even shorter period (Ludwig et al. 2005).

26.7.2 ROLE OF PATTERN AND EXPECTATION

The influence of top-down control when deciding where to look is probably most dramatically demonstrated in the anti-saccade task (Hallett 1978). Here, subjects are asked to look in the direction opposite to that of an abruptly appearing target, and so to a location completely at odds with what would be expected from a response driven by target saliency alone. More commonly the influence of top-down factors is somewhat less dramatic, where they are used to steer our responses to stimulus features in a way that is appropriate for the task at hand (Tatler and Vincent 2008).

Activities that are performed as part of a habitual, set sequence can also influence our eye movements. For example, making a cup of tea involves a highly learnt sequence of actions, and it has been shown that eye movements in such circumstances also involve patterns (such as locating objects for the task, guiding the movements of the hands or objects) that are far from random and largely unrelated to the comparative visual saliency of objects within the field of vision (Land et al. 1999). The pattern of eye movements used in sequential tasks can also systematically change as a sequential task is learnt and automated (Foerster et al. 2011). Furthermore, eye movement patterns can systematically alter as a result of learning even when the task itself does not have an explicit temporal pattern. For example, gaze patterns in expert versus trainee radiographers differ significantly when hunting for cancer in mammograms (Kundel et al. 2007), as do trainee versus expert ophthalmologists inspecting photographs of the retinal fundus (O'Neill et al. 2011).

The instructions provided to a person can also influence the selection of objects within a visual scene. Arguably, the most well-known demonstration of this is that from Yarbus (Tatler et al. 2010) who found the gaze patterns for examining a painting depended upon what question was asked of the observer (e.g., to estimate the ages of the people of the picture or to surmise what the people in the picture were doing immediately prior to the event depicted). More recent investigations have indeed found that a statistical analysis of gaze patterns can be used to determine what an observer's task was, albeit imperfectly (Borji and Itti 2014).

26.8 EYE MOVEMENTS IN VISION SCIENCE

26.8.1 ALLOWING THE EYES TO MOVE: OVERT ATTENTION

The amount of visual information available in natural visual scenes is vast, and so attentional mechanisms allow us to selectively process a subset of this information. *Overt attention* describes the process whereby a spatial subset of information is selected by directing our gaze toward the appropriate area of interest. In contrast, it is possible to attend to a location in our visual field without directing our gaze toward it: this is known as "covert attention."

In studies where overt attention is explored, the eyes are therefore free to move. For example, in classical visual search tasks the eyes move about a display so that overt attention can be used to hunt for the appropriate target (Treisman and Gelade 1980). It is often the case that eye movements are not measured in such experiments as other behavioral measures (such as the time to complete the search, as a function of search difficulty) are used to track performance (Verghese et al. 2014). In other experiments, knowing where the eye—and, therefore, overt attention—is

Figure 26.10 Eye movement scan path (arrows) for an observer examining an image of a city street. Circles show fixations, with the diameter of the circle being proportional to the duration of the fixation. (For further methodological details, see Stainer, M.J. et al., *Front. Psychol.*, 4, 624, 2013; Data courtesy of Dr. Matthew Stainer.)

directed during the task is critical. For example, gaze tracking experiments can be used to determine what information, and in what order, is extracted from complicated scenes (Figure 26.10), both in normal observers (Borji and Itti 2014) and in those whose visual field is compromised by disease (Crabb et al. 2010).

Experiments in which eye movements are allowed do not necessarily involve overt attention, however. For example, in experiments measuring saccadic latency (Carpenter and Williams 1995) or saccadic programming (Komoda et al. 1973), eye movements are made as a response toward targets presented away from the fovea. Assuming that the mechanisms subserving gaze shifting and the orienting of overt attention are highly related, such experiments do provide insights into what factors influence where our overt attention is to be next directed, however. They may not necessarily give insights into where *covert* attention was directed at the time of the stimulus, as preparing the eyes to move to a location does not necessarily involve covertly allocating attention to that location (Hunt and Kingstone 2003). Changes in the allocation of covert attention has been reported to bias the orientation of microsaccades during fixation, however (Engbert and Kliegl 2003).

It is worth briefly mentioning the eye movements involved in reading. Reading a horizontal line of text produces a horizontal series of saccades (probably first measured by Hering in 1979 [Wade and Tatler 2009]) and so is a clear example of the directing of overt attention. However, it is proposed that during each fixation, covert attention is then used to sweep across the letters in the fixated word (Vidyasagar 2013). Of note is that this specialist task of reading is comparatively recent in human history and so the neural mechanisms underpinning it are apparently "recycled" from other mechanisms evolved for nonreading tasks (Vidyasagar 2013).

Multiple technologies exist to record the position of the eyes (e.g., infrared reflection oculography, video-based eye tracking, electrooculography, and retinal imaging). The performance characteristics of each method differ, although a simple comparison

of these can be potentially misleading. Sometimes a method's strength in one situation becomes an insuperable liability in another. The selection of a suitable method cannot be made without considering what aspect of eye motion needs measuring and, as such, there is no single technology that is optimal for all circumstances. For example, piezoelectric strain gauge recording have been used to capture tiny, rapid tremor eye movements (Bolger et al. 1999), although such methods would be unsuited to measuring the sequence of saccades used when examining a photograph.

26.8.2 KEEPING THE EYES STILL

It is commonly the case that an experimenter wants an observer to maintain a fixed gaze position. This is particularly the case when visual functions away from the fovea are being assessed. For many experiments, participants are simply instructed to maintain fixation on an appropriate marker, and it is assumed that they are able to comply with this instruction. Sometimes greater assurances are desired that fixation is being maintained, however.

Monitoring eye movement provides one way to ensure subjects maintain fixation throughout an experiment. Of course, even the fixating eye is never truly at rest (see Sections 26.2.1 and 26.5.1) and so suitable criteria must be selected as to what tolerance on gaze position is acceptable for a particular experiment (and, correspondingly, what approach is best suited to measure the position of the eyes [see Section 26.8.1]).

Careful design of an experimental protocol can also aid in reducing the likelihood of disruptive shifts in gaze. For example, if peripheral vision is to be assessed, having targets appear randomly at balanced eccentricities either side of the fixation point (Anderson and McKendrick 2007) can reduce the temptation for the observer's gaze to drift away from the fixation target over the course of the experiment as no net performance benefit should result. Using brief stimuli that are shorter than the typical saccadic latency is also a way to reduce the likelihood a participant will make an eye movement toward a target and directly view it

(Anderson and Patella 1999). Short durations can also be used to try to minimize the likelihood of scanning eye movements being employed when an extended visual target is presented (Anderson and Wassnig 2012). Maintaining a fixed position of gaze for an extended period is a rather unnatural behavior, and so recruiting participants experienced in such a task is likely to be of benefit particularly where maintaining accurate fixation is crucial: at the very least, some initial practice at the task is desirable.

If monitoring fixation during an experiment is not possible, it may be useful to demonstrate that participants are able to maintain appropriate fixation upon instruction in a related task. Perimeters are clinical devices used to measure the integrity of a person's visual field and so require responses to stimuli presented in the retinal periphery while maintaining fixation on a central marker. Perimeters commonly contain several measures to monitor fixation (e.g., simple observation of the eye via a video camera, infrared gaze tracking, and Heijl-Krakau blind-spot location monitoring [Heijl and Krakau 1977]). As a bonus, such devices give confidence that the visual field of the observer is, in fact, normal—something that cannot be necessarily assumed, particularly in older participants (Weih et al. 2001).

Of course, the final aim of having observers steadily fixate is normally so the experimenter knows where on the retina a stimulus is located. Gaze-contingent displays allows this to occur *despite* the presence of eye movements, at least over a limited range, by moving the stimulus by the same amount that the eye has moved (e.g., Ditchburn and Ginsborg [(1952)]; Kelly [(1979a)]; Poletti et al. [(2010)]). Although such devices had previously been confined to laboratory environments, commercially available perimeters now exist that compensate for eye movements (Acton and Greenstein 2013), which is likely to be of particular value in diseases such as age-related macular degeneration where fixation stability can be poor (Longhin et al. 2013). By combining gaze-contingent display methods with adaptive optics to correct of optical imperfections in the eye, it is now possible to deliver stimuli to a selected cone photoreceptor (Arathorn et al. 2013).

26.9 MOVING THE EYE: THE EXTRAOCULAR MUSCLES

The oculomotor control systems outlined in previous sections all share a final aim of producing appropriate neural signals in the nerves controlling the six extraocular muscles that move the eye. Four rectus muscles—superior rectus, inferior rectus, lateral rectus, and medial rectus—arise from the back of the orbit that houses the eye and attach on the upper, lower, temporal (lateral), and nasal aspects of the eye globe, respectively (Figure 26.11a). Although the precise movement generated by each muscle is complex and depends upon the eye's direction of gaze relative to the orbit (Figure 26.11d and e), in primary gaze (i.e., with the eyes looking straight ahead relative to the head and observing an object place at eye level but nominally an infinite distance away), the primary actions of the rectus muscles is to move the eye up, down, away from the nose, and toward the nose, respectively. Two oblique muscles—superior oblique and inferior oblique—attach above and below the globe at slightly more posterior locations and pull the eye toward the nasal aspect of the front of the orbit.

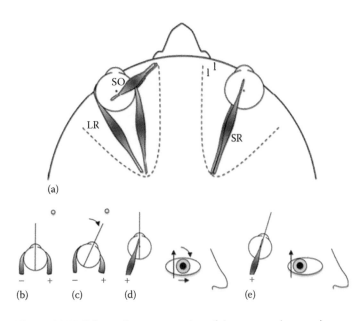

Figure 26.11 Schematic representation of the extraocular muscles and their actions. (a) Approximate positions of the extraocular muscles (red) in relation to the eye and orbit (dashed lines), as viewed from above. For clarity, the superior rectus (SR) is not shown for the left eye, and the lateral and medial recti (LR and MR) and superior oblique (SO) muscles not shown in the right eye. The inferior rectus (IR) and inferior oblique (IO) muscles (not directly shown, but partially visible at the edge of SR and SO, respectively) insert underneath the eyeball in a manner similar to the SR and SO muscles, respectively. The angle between the medial and lateral walls of each orbit is approximately 45°, with each lateral wall pointing approximately straight ahead. Note that the sizes of the muscles have been reduced for clarity: in particular, their tendons are much broader when they insert into the eye. Furthermore, the SO muscle is somewhat more complex than shown here and operates via a pulley near the upper nasal portion of the orbital opening, with the body of the muscle traveling down the length of the orbit: its plane of action is approximately as demonstrated here, however. More detailed anatomical descriptions of the muscles and their pathways can be found in texts such as Snell and Lemp (1989). Note that some texts may refer LR and SO as "abducens" and "trochlear," respectively. (b) In response to the appearance of a target to the right (yellow), increased stimulation is delivered to one muscle (e.g., LR) and decreased stimulation the other muscle (e.g., MR) in the horizontally acting antagonist pair. (c) Contraction of one muscle, and corresponding relaxation of the other muscle (Sherrington 1893), in the antagonist pair results in the eye moving to the right. (d) Due to the angle of the orbits, SR and IR travel approximately 23° to the visual axis when the eye is pointing straight ahead. Because of its attachment in front of the nominal center of rotation of the eye (small black dot), contraction of SR results in the eye's elevation, along with some intorsion (rotation of the top of the eye toward the nose) and adduction (i.e., movement toward the nose). (e) When the eye is turned out (abduction) by 23°, the visual axis of the eye and the SR muscle are aligned, and so contraction of SR now results almost purely in elevation of the eye. A similar principle occurs with the oblique muscles, which approximately align with the visual axis when the eye is adducted by 54° (SO) and 51° (IO).

In primary gaze, the principal action of these muscles is intorsion and extorsion (rotation of the eye, such that the top of the eye rotates either toward or away from the nose, respectively).

Of note is that the extraocular muscles exist in antagonistic pairs (e.g., superior and inferior rectus, lateral and medial rectus), and so an eye movement requires not only the contraction of at least one muscle, but the simultaneous relaxation of its antagonist partner. This can be seen in Figure 26.11b and c where in order to make a horizontal saccadic eye movement to a target appearing to the right, the right eye must simultaneously contract the lateral rectus and relax the medial rectus. The eye movement is not actively stopped, but rather the eye passively decelerates due to the viscosity of structures within the orbit (Robinson 1964).

The mechanical properties of the extraocular muscles and the eye have been well investigated, and reviews of this work can be found elsewhere (Carpenter 1988). Only by understanding these mechanical properties can the nature of the neurological control signals supplied to the muscles to achieve accurate movements be fully appreciated. One clear advantage that the oculomotor system has when trying to accurately direct our gaze is that the eye is not required to carry any loads, unlike some muscle systems in the body, thereby allowing the oculomotor system to reliably learn the relationship between a particular set of neural commands and their eventual motor consequences: a *sine qua non* for the ballistic control of saccades.

REFERENCES

Acton, J H and V C Greenstein. 2013. Fundus-driven perimetry (microperimetry) compared to conventional static automated perimetry: Similarities, differences, and clinical applications. *Canadian Journal of Ophthalmology* 48(5):358–363.

Anderson, A J and R H S Carpenter. 2010. Saccadic latency in deterministic environments: Getting back on track after the unexpected happens. *Journal of Vision* 10(14):1–10.

Anderson, A J and A M McKendrick. 2007. Quantifying adaptation and fatigue effects in frequency doubling perimetry. *Investigative Ophthalmology and Visual Science* 48(2):943–948.

Anderson, A J and M J Stainer. 2014. Target direction rather than position determines oculomotor expectation in repeating sequences. *Experimental Brain Research* 232(7):2187–2195.

Anderson, A J and S E Wassnig. 2012. The role of local separation in spatial frequency discrimination. *Vision Research* 53(1):15–20.

Anderson, D R and V M Patella. 1999. *Automated Static Perimetry*. St. Louis, MO: C V Mosby.

Arathorn, D W, S B Stevenson, Q Yang, P Tiruveedhula, and A Roorda. 2013. How the unstable eye sees a stable and moving world. *Journal of Vision* 13(10):1–19.

Bahill, A T, D Adler, and L Stark. 1975. Most naturally occurring human saccades have magnitudes of 15 degrees or less. *Investigative Ophthalmology* 14(6):468–469.

Bahill, A T, A Brockenbrough, and B T Troost. 1981. Variability and development of a normative data base for saccadic eye movements. *Investigative Ophthalmology and Visual Science* 21:116–125.

Bahill, A T, M R Clark, and L Stark. 1975. Glissades—Eye movements generated by mismatched components of the saccadic motoneuronal control signal. *Mathematical Biosciences* 26:303–318.

Barton, J J S, T Simpson, E Kiriakopoulos, C Stewart, A Crawley, B Guthrie, M Wood, and D Mikulis. 1996. Functional MRI of lateral occipitotemporal cortex during pursuit and motion perception. *Annals of Neurology* 40(3):387–398.

Bartz, A E. 1966. Eye and head movement in peripheral vision: Nature of compensatory eye movements. *Science* 152:1644–1645.

Becker, W. 1991. Saccades. In *Eye Movements*, ed. R H S Carpenter, pp. 95–137. London, U.K.: Macmillan.

Becker, W and A F Fuchs. 1985. Prediction in the oculomotor system: Smooth pursuit during transient disappearance of a visual target. *Experimental Brain Research* 57(3):562–575.

Bennett, A G and R B Rabbetts. 1989. *Clinical Visual Optics*, 2nd edn. Oxford, U.K.: Butterworth-Heinemann Ltd.

Bolger, C, S Bojanic, N F Sheahan, D Coakley, and J F Malone. 1999. Dominant frequency content of ocular microtremor from normal subjects. *Vision Research* 39:1911–1915.

Borji, A and L Itti. 2014. Defending Yarbus: Eye movements reveal observers' task. *Journal of Vision* 14(3):1–22.

Bronstein, A M and M A Gresty. 1988. Short latency compensatory eye movement responses to transient linear head acceleration: A specific function of the otolith-ocular reflex. *Experimental Brain Research* 71(2):406–410.

Burr, D C, M C Morrone, and J Ross. 1994. Selective suppression of the magnocellular visual pathway during saccadic eye movements. *Nature* 371(6497):511–513.

Busettini, C, F A Miles, U Schwartz, and J R Carl. 1994. Human ocular responses to translation of the observer and of the scene: Dependence on viewing distance. *Experimental Brain Research* 100(3):484–494.

Büttner-Ennever, J A, A K E Horn, V Henn, and B Cohen. 1999. Projections from the superior colliculus motor map to omnipause neurons in monkey. *The Journal of Comparative Neurology* 413:55–67.

Campbell, F W and R H Wurtz. 1978. Saccadic omission: Why we do not see a grey-out during a saccadic eye movement. *Vision Research* 18(10):1297–1303.

Cannon, S C and D A Robinson. 1987. Loss of the neural integrator of the oculomotor system from brain stem lesions in monkey. *Journal of Neurophysiology* 57(5):1383–1409.

Carpenter, R H S. 1972. Cerebellectomy and the transfer function of the vestibulo-ocular reflex in the decerebrate cat. *Proceedings of the Royal Society London B Biological Sciences* 181:353–374.

Carpenter, R H S. 1981. Oculomotor procrastination. In *Eye Movements: Cognition and Visual Perception*, eds. D F Fisher, R A Monty, and J W Senders, pp. 237–246. Hillsdale, NJ: Lawrence Erlbaum Associates.

Carpenter, R H S. 1988. *Movements of the Eyes*. London, U.K.: Pion.

Carpenter, R H S. 1992. The visual origins of ocular motility. In *Eye Movements*, ed. R H S Carpenter, pp. 1–10. London, U.K.: MacMillan.

Carpenter, R H S. 1999. A neural mechanism that randomises behaviour. *Journal of Consciousness Studies* 6:13–22.

Carpenter, R H S. 2000. The neural control of looking. *Current Biology* 10(8):R291–R293.

Carpenter, R H S. 2004a. Contrast, probability and saccadic latency: Evidence for independence of detection and decision. *Current Biology* 14(17):1576–1580.

Carpenter, R H S. 2004b. Supplementary eye field: Keeping an eye on eye movement. *Current Biology* 14(11):R416–R418.

Carpenter, R H S, B A J Reddi, and A J Anderson. 2009. A simple two-stage model predicts response time distributions. *Journal of Physiology* 587(16):4051–4062.

Carpenter, R H S and M L L Williams. 1995. Neural computation of log likelihood in control of eye movements. *Nature* 377(7):59–62.

Caspi, A, B R Beutter, and M P Eckstein. 2004. The time course of visual information accrual guiding eye movement decisions. *Proceedings of the National Academy of Sciences of the United States of America* 101(35):13086–13090.

Chen, X and G J Zelinsky. 2006. Real-world visual search is dominated by top-down guidance. *Vision Research* 46:4118–4133.

Collewijn, H. 1969. Optokinetic eye movements in the rabbit: Input-output relations. *Vision Research* 9:117–132.

Cornsweet, T N. 1956. Determination of the stimuli for involuntary drifts and saccadic eye movements. *Journal of the Optical Society of America* 46(11):987–993.

Crabb, D P, N D Smith, F G Rauscher, C M Chisholm, J L Barbur, D F Edgar, and D F Garway-Heath. 2010. Exploring eye movements in patients with glaucoma when viewing a driving scene. *PLoS ONE* 5(3):e9710.

Crane, B T and J L Demer. 1998. Human horizontal vestibulo-ocular reflex initiation: Effects of acceleration, target distance, and unilateral deafferentation. *Journal of Neurophysiology* 80(3):1151–1166.

Ditchburn, R W and B L Ginsborg. 1952. Vision with a stabilized retinal image. *Nature* 170(4314):36–37.

Dodge, R. 1900. Visual perception du eye movement. *Psychological Review* 7:454–465.

Duhamel, J R, C L Colby, and M E Goldberg. 1992. The updating of the representation of visual space in parietal cortex by intended eye movements. *Science* 255(5040):90–92.

Eifuku, S and R H Wurtz. 1998. Response to motion in extrastriate area MST1: Centre-surround interactions. *Journal of Neurophysiology* 80(1):282–296.

Engbert, R and R Kliegl. 2003. Microsaccades uncover the orientation of covert attention. *Vision Research* 43:1035–1045.

Erkelens, C J. 2006. Coordination of smooth pursuit and saccades. *Vision Research* 46:163–170.

Erkelens, C J, R M Steinman, and H Collewijn. 1989. Ocular vergence under natural conditions. II. Gaze shifts between real targets differing in distance and direction. *Proceedings of the Royal Society London B* 236:441–465.

Erkelens, C J, J van der Steen, R M Steinman, and H Collewijn. 1989. Ocular vergence under natural conditions. I. Continuous changes of target distance along the median plane. *Proceedings of the Royal Society London B* 236:417–440.

Fletcher, W A, T C Hain, and D S Zee. 1990. Optokinetic nystagmus and after nystagmus in human beings: Relationship to nonlinear processing of information about retinal slip. *Experimental Brain Research* 81:46–52.

Foerster, R M, E Carbone, H Koesling, and W X Schneider. 2011. Saccadic eye movements in a high-speed bimanaul stacking task: Changes of attentional control during learning and automatization. *Journal of Vision* 11(7):1–16.

Gaymard, B, S Rivaud, and C Pierrot-Deseilligny. 1993. Role of the left and right supplementary motor areas in memory-guided saccade sequences. *Annals of Neurology* 34(3):404–406.

Hallett, P E. 1978. Primary and secondary saccades to goals defined by instructions. *Vision Research* 18:1279–1296.

Hanes, D P and J D Schall. 1996. Neural control of voluntary movement initiation. *Science* 274(5286):427–430.

Heijl, A and C E Krakau. 1977. A note on fixation during perimetry. *Acta Ophthalmol (Copenh)* 55(5):854–861.

Hildreth, J D. 1973. Bloch's law and a temporal integration model for simple reaction time to light. *Perception and Psychophysics* 14:421–432.

Hunt, A M and A Kingstone. 2003. Covert and overt voluntary attention: Linked or independent? *Cognitive Brain Research* 18:102–105.

Ibbotson, M R, N A Crowder, S L Cloherty, N S Price, and M J Mustari. 2008. Saccadic modulation of neural responses: Possible roles in saccadic suppression, enhancement, and time compression. *Journal of Neuroscience* 28(43):10952–10960.

Itti, L and C Koch. 2000. A saliency-based search mechanism for overt and covert shifts of visual attention. *Vision Research* 40:1498–1506.

Jampel, R S and D X Shi. 2002. The absence of so-called compensatory ocular countertorsion: The response of the eyes to head tilt. *Archives of Ophthalmology* 120(10):1331–1340.

Jordan, S. 1970. Ocular pursuit movement as a function of visual and proprioceptive stimulation. *Vision Research* 10(8):775–780.

Kelly, D H. 1979a. Motion and vision. I. Stabilized images of stationary gratings. *Journal of the Optical Society of America* 69(9):1266–1274.

Kelly, D H. 1979b. Motion and vision. II. Stabilized spatio-temporal threshold surface. *Journal of the Optical Society of America* 69(10):1340–1349.

Kelly, D H. 1985. Visual processing of moving stimuli. *Journal of the Optical Society of America A* 2(2):216–225.

Knapp, C M, I Gottlob, R J McLean, S Rafelt, and F A Proudlock. 2009. Effect of distance upon horizontal and vertical look and stare OKN. *Journal of Vision* 9(12):1–9.

Knöll, J, P Binda, M C Morrone, and F Bremmer. 2011. Spatiotemporal profile of peri-saccadic contrast sensitivity. *Journal of Vision* 11(14):1–12.

Ko, H K, M Poletti, and M Rucci. 2010. Microsaccades precisely relocate gaze in a high visual acuity task. *Nature Neuroscience* 13(12):1549–1553.

Koch, C and S Ullman. 1985. Shifts in selective visual attention: Towards the underlying neural circuitry. *Human Neurobiology* 4:219–227.

Komoda, M K, L Festinger, L J Phillips, R H Duckman, and R A Young. 1973. Some observations concerning saccadic eye movements. *Vision Research* 13(6):1009–1020.

Kowler, E. 2011. Eye movements: The past 25 years. *Vision Research* 51:1457–1483.

Kowler, E and R M Steinman. 1980. Small saccades serve no useful purpose: Reply to a letter by R. W. Ditchburn. *Vision Research* 20(3):273–276.

Krauskopf, J, T N Cornsweet, and L A Riggs. 1960. Analysis of eye movements during monocular and binocular fixation. *Journal of the Optical Society of America* 50(6):572–578.

Krauzlis, R J. 2005. The control of voluntary eye movements: New perspectives. *The Neuroscientist* 11(2):124–137.

Krauzlis, R J and N Dill. 2002. Neural correlates of target choice for pursuit and saccades in the primate superior colliculus. *Neuron* 35(2):355–363.

Kröller, J and F Behrens. 1997. On the optokinetic response during stepwise changes in stimulus velocity in squirrel monkeys. *Journal of Vestibular Research* 7(1):35–44.

Kundel, H L, C F Nodine, E F Conant, and S P Weinstein. 2007. Holistic component of image perception in mammogram interpretation: Gaze-tracking study. *Radiology* 242(2):396–402.

Kushner, B J and S Kraft. 1983. Ocular torsional movements in normal humans. *American Journal of Ophthalmology* 95(6):752–762.

Land, M F, N R Mennie, and J Rusted. 1999. The roles of vision and eye movements in the control of activities of daily living. *Perception* 28(11):1311–1328.

Leach, J C D and R H S Carpenter. 2001. Saccadic choice with asynchronous targets: Evidence for independent randomisation. *Vision Research* 41:3437–3445.

Lencer, R and P Trillenberg. 2008. Neurophysiology and neuroanatomy of smooth pursuit in humans. *Brain and Cognition* 68:219–228.

Lisberger, S G and L E Westbrook. 1985. Properties of visual inputs that initiate horizontal smooth pursuit eye movements in monkeys. *Journal of Neuroscience* 5(6):1662–1673.

Longhin, E, E Convento, E Pilotto, G Bonin, S Vujosevic, O Kotsafti, and E Midena. 2013. Static and dynamic retinal fixation stability in microperimetry. *Canadian Journal of Ophthalmology* 48(5):375–380.

Ludwig, C J H, I D Gilchrist, E McSorley, and R J Baddeley. 2005. The temporal impulse response underlying saccadic decisions. *Journal of Neuroscience* 25(43):9907–9912.

Luschei, E S and A F Fuchs. 1972. Activity of brain stem neurons during eye movements of alert monkeys. *Journal of Neurophysiology* 35(4):445–461.

McCamy, M B, N Collins, J Otero-Millan, M Al-Kalbani, S L Macknik, D Coakley, X G Troncoso et al. 2013. Simultaneous recordings of ocular microtremor and microsaccades with a piezoelectric sensor and a video-oculography system. *Peer Journal* 1(e14). doi:10.7717/peerj.14.

Meyer, C H, A G Lasker, and D A Robinson. 1985. The upper limit of human smooth pursuit velocity. *Vision Research* 25(4):561–563.

Morasso, P, E Bizzi, and J Dichgans. 1973. Adjustment of saccade characteristics during head movements. *Experimental Brain Research* 16:492–500.

Morrone, M C, J Ross, and D Burr. 2005. Saccadic eye movements cause compression of time as well as space. *Nature Neuroscience* 8(7):950–954.

Munoz, D P and R H Wurtz. 1993. Fixation cells in monkey superior colliculus I. Characteristics of cell discharge. *Journal of Neurophysiology* 70(2):559–575.

Murakami, I and P Cavanagh. 1998. A jitter after-effect reveals motion-based stabilization of vision. *Nature* 395:798–801.

Murphy, B J. 1978. Pattern thresholds for moving and stationary gratings during smooth eye movement. *Vision Research* 18:521–530.

Nachmias, J. 1961. Determiners of the drift of the eye during monocular fixation. *Journal of the Optical Society of America* 51:761–766.

Naegele, J R and R Held. 1982. The postnatal development of monocular optokinetic nystagmus in infants. *Vision Research* 22:341–346.

Najemnik, J and W S Geisler. 2005. Optimal eye movement strategies in visual search. *Nature* 434(7031):387–391.

O'Neill, E C, Y X Kong, P P Connell, D N Ong, S A Haymes, M A Coote, and J G Crowston. 2011. Gaze behavior among experts and trainees during optic disc examination: Does how we look affect what we see? *Investigative Ophthalmology and Visual Science* 52(7):3976–3983.

Otero-Millan, J, X G Troncoso, S L Macknik, I Serrano-Pedraza, and S Martinez-Conde. 2008. Saccades and microsaccades during visual fixation, exploration, and search: Foundations for a common saccadic generator. *Journal of Vision* 8(14):1–18.

Platt, M L and P W Glimcher. 1999. Neural correlates of decision variables in parietal cortex. *Nature* 400(6741):233–238.

Poletti, M, C Listorti, and M Rucci. 2010. Stability of the visual world during eye drift. *The Journal of Neuroscience* 30(33):11143–11150.

Rademaker, G G J and J W G ter Braak. 1948. On the central mechanism of some optic reactions. *Brain* 71(1):48–76.

Rashbass, C. 1961. The relationship between saccadic and smooth tracking eye movements. *Journal of Physiology* 159:326–338.

Rashbass, C and G Westheimer. 1961. Disjunctive eye movements. *Journal of Physiology* 159:339–360.

Reddi, B A J and R H S Carpenter. 2000. The influence of urgency on decision time. *Nature Neuroscience* 3(8):827–830.

Riggs, L A and F Ratliff. 1951. Visual acuity and the normal tremor of the eyes. *Science* 114(2949):17–18.

Robinson, D A. 1964. The mechanics of human saccadic eye movement. *Journal of Physiology* 174:245–264.

Robinson, D A. 1968. The oculomotor control system: A review. *Proceedings of the IEE* 56(6):1032–1049.

Ross, J, M C Morrone, and D C Burr. 1997. Compression of visual space before saccades. *Nature* 386(6625):598–601.

Rubinstein, N J and L A Abel. 2011. Optokinetic nystagmus suppression as an index of the allocation of visual attention. *Investigative Ophthalmology and Visual Science* 52(1):462–467.

Rucci, M, R Iovin, M Poletti, and F Santini. 2007. Miniature eye movements enhance fine spatial details. *Nature* 447(7146):851–854.

Schall, J D. 2004. On the role of frontal eye field in guiding attention and saccades. *Vision Research* 44:1453–1467.

Schiller, P H and J G Malpeli. 1977. Properties and tectal projections of the monkey retinal ganglion cells. *Journal of Neurophysiology* 40:428–445.

Schiller, P H, J G Malpeli, and S J Schein. 1979. Composition of geniculostriate input to superior colliculus of the rhesus monkey. *Journal of Neurophysiology* 42(4):1124–1133.

Schor, C M. 1979. The relationship between fusional vergence eye movements and fixation disparity. *Vision Research* 19:1359–1367.

Schulz, E. 1984. Binocular micromovements in normal persons. *Graefe's Archive for Clinical and Experimental Ophthalmology* 222:95–100.

Sherrington, C S. 1893. Further experimental note on the correlation of action of antagonistic muscles. *The British Medical Journal* 1(1693):1218.

Skavenski, A A, R M Hansen, R M Steinman, and B J Winterson. 1979. Quality of retinal image stabilization during small natural and artificial body rotations in man. *Vision Research* 19:675–683.

Skavenski, A A and R M Steinman. 1970. Control of eye position in the dark. *Vision Research* 10:193–203.

Smith, G and D A Atchison. 1997. *The Eye and Visual Optical Instruments*. Cambridge, U.K.: Cambridge University Press.

Snell, R S and M A Lemp. 1989. *Clinical Anatomy of the Eye*. Boston, MA: Blackwell Scientific Publications.

Sparks, D L. 1986. Translation of sensory signals into commands for control of saccadic eye movements: Role of primate superior colliculus. *Physiological Reviews* 66(1):118–171.

Sparks, D L. 1988. Neural cartography: Sensory and motor maps in the superior colliculus. *Brain, Behavior and Evolution* 31:49–56.

Sparks, D L and L E Mays. 1983. Spatial localization of saccade targets. I. Compensation for stimulation-induced perturbations in eye position. *Journal of Neurophysiology* 49(1):45–63.

Sparks, D L and J D Porter. 1983. Spatial localization of saccade targets. II. Activity of superior colliculus neurons preceding compensatory saccades. 49(1):64–74.

Spauschus, A, J Marsden, D M Halliday, J R Rosenberg, and P Brown. 1999. The origin of ocular microtremor in man. *Experimental Brain Research* 126:556–562.

Stainer, M J, K C Scott-Brown, and B W Tatler. 2013. Behavioral biases when viewing multiplexed scenes: Scene structure and frames of reference for inspection. *Frontiers in Psychology* 4:624.

Steinman, R M, R J Cunitz, G T Timberlake, and M Herman. 1967. Voluntary control of microsaccades during maintained monocular fixation. *Science* 155(3769):1577–1579.

Steinman, R M, G M Haddad, A A Skavenski, and D Wyman. 1973. Miniature eye movement. *Science* 181(4102):810–819.

Steinman, R M, J Z Levinson, H Collewijn, and J van der Steen. 1985. Vision in the presence of known natural retinal image motion. *Journal of the Optical Society of America A* 2(2):226–233.

Tanaka, M and S G Lisberger. 2001. Regulation of the gain of visually guided smooth-pursuit eye movements by frontal cortex. *Nature* 409(6817):191–194.

Tanji, J and K Shima. 1994. Role for supplementary motor area cells in planning several movements ahead. *Nature* 371:413–416.

Tatler, B W, R J Baddeley, and B T Vincent. 2006. The long and the short of it: Spatial statistics at fixation vary with saccade amplitude and task. *Vision Research* 46:1857–1862.

Tatler, B W and B T Vincent. 2008. Systematic tendencies in scene viewing. *Journal of Eye Movement Research* 2(2):1–18.

Tatler, B W, N J Wade, H Kwan, J M Findlay, and B M Velichkovsky. 2010. Yarbus, eye movements, and vision. *i-Perception* 1:7–27.

Taylor, M J, R H Carpenter, and A J Anderson. 2006. A noisy transform predicts saccadic and manual reaction times to changes in contrast. *Journal of Physiology* 15(573):741–751.

Thiele, A, P Henning, M Kubishik, and K P Hoffmanh. 2002. Neural mechanisms of saccadic suppression. *Science* 295(5564):2460–2462.

Treisman, A M and G Gelade. 1980. A feature-integration theory of attention. *Cognitive Psychology* 12:97–136.

van Opstal, A J and J A M van Gisbergen. 1989. Scatter in the metrics of saccades and properties of the collicular motor map. *Vision Research* 29(9):1183–1196.

Verghese, A, S C Kolbe, A J Anderson, G F Egan, and T R Vidyasagar. 2014. Functional size of human visual area V1: A neural correlate of top-down attention. *NeuroImage* 93(Part 1):47–52. doi: 10.1016/j.neuroimage.2014.02.023.

Verheijen, F J. 1961. A simple after image method demonstrating the involuntary multidirectional eye movements during fixation. *Optica Acta* 8:309–311.

Vidyasagar, T R. 2013. Reading into neuronal oscillations in the visual system: Implications for developmental dyslexia. *Frontiers in Human Neuroscience* 7:811.

Wade, N J and B W Tatler. 2009. Did Javal measure eye movements during reading? *Journal of Eye Movement Research* 2(5):1–7.

Walker, M F, E J Fitzgibbon, and M E Goldberg. 1995. Neurons in the monkey superior colliculus predict the visual result of impending saccadic eye movements. *Journal of Neurophysiology* 73(5):1988–2003.

Walls, G L. 1962. The evolutionary history of eye movements. *Vision Research* 1962(2):69–80.

Weih, L M, M Nanjan, C A McCarty, and H R Taylor. 2001. Prevalence and predictors of open-angle glaucoma: Results from the visual impairment project. *Ophthalmology* 108(11):1966–1972.

Zimmermann, E, M C Morrone, and D C Burr. 2014. The visual component to saccadic compression. *Journal of Vision* 14(12):1–9.

Zuber, B L, L Stark, and G Cook. 1965. Microsaccades and the velocity-amplitude relationship for saccadic eye movements. *Science* 150(3702):1459–1460.

Aging and the eye's optics

W. Neil Charman

Contents

27.1 INTRODUCTION

Although the most rapid period of growth in the eye occurs in early childhood, changes in its optical characteristics and performance continue throughout life.

The optical development of the eye is governed by both genetic and environmental factors (e.g., Lyhne et al., 2001; Morgan and Rose, 2005; Young, 2009). Changes in dimensions are most rapid in the first few years of life, when the axial length increases markedly, from about 17 to 24 mm. Accompanying this size increase, the relative thicknesses of the cornea and lens decrease and the lens surfaces become flatter (Weale, 1982). Postnatal visual experience appears to have an important role in coordinating these changes to produce an overall refractive condition that approximates to emmetropia or slight hyperopia. Moreover, although the individual optical parameters are each normally distributed, the distribution of overall refraction becomes leptokurtic and shows a marked near-emmetropic peak. This coordinating process is known as emmetropization. Figure 27.1 compares typical data for the distributions of refractive error at birth, when there is a wide initial range of refractions, and at ages 6–8 years, by which time the coordination of growth of the different ocular optical parameters has produced a state of near emmetropia for most children.

Although the most obvious dimensional changes occur in early life, slow geometric changes continue throughout adulthood, particularly in the lens (e.g., Atchison et al., 2008). These dimensional changes are accompanied by changes in the mechanical and optical characteristics of the lens that, together with changes in the rest of the accommodation apparatus, affect its ability to change shape and power to allow the eye to focus near objects. This review will concentrate on these and other relevant changes through adulthood and their impacts on visual performance.

27.2 EFFECTS OF AGE ON THE OPTICAL COMPONENTS OF THE EYE

27.2.1 TEAR FILM

The tear film is important in maintaining the smoothness of the anterior surface of the eye. In the absence of any disease process, it appears that a tear film of satisfactory optical quality is usually

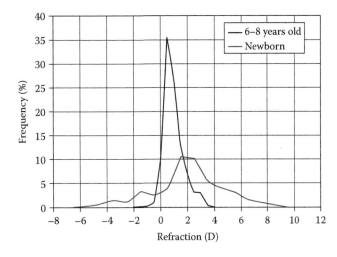

Figure 27.1 The distribution of refractive error in neonates and at age 6–8 years, showing the narrowing of the distribution (emmetropization) that occurs in the first few years of life. (Based on Cook, R.C. and Glasscock, R.E., *Am. J. Ophthalmol.*, 34, 1407, 1951; Kemph, G.A. et al., Public Health Bulletin No. 182, U.S. Government Printing Office, Washington, DC, 1928.)

maintained in the older eye (Tomlinson and Craig, 2002), although tear production is reduced and evaporation rates are higher (Guillon and Maissa, 2010).

27.2.2 CORNEA

The central cornea maintains reasonable optical quality throughout life, although there are some structural and other changes with age, particularly a reduction in endothelial cell density (Abib and Barreto, 2001) and in sensitivity (Millodot, 1977). There is little change in axial direct spectral transmittance, which shows a sharp cutoff below about 320 nm (Beems and Van Best, 1994; van den Berg and Tam, 1990), but there is often some loss of transparency in the peripheral cornea, near the limbus, where fatty and other deposits may build up to form *arcus senilis*. Corneal light scattering normally increases only slightly with age (Allen and Vos, 1967; Olsen, 1982). Corneal power and aberrations change slowly (Oshika et al., 1999; Guirao et al., 2000). There is a gradual shift from generally with-the-rule (vertical meridian more myopic) toward against-the-rule (inverse) astigmatism with age (Fledelius, 1984; Saunders, 1988; Hayashi et al., 1995), due largely to a steepening of the cornea in the horizontal meridian (Baldwin and Mills, 1981). More recent studies using Scheimpflug imaging (Dubbelman et al., 2006; Navarro et al., 2013a,b) have examined the shape of the anterior and posterior surfaces, and their effects on aberration, in more detail. Navarro et al. (2013b) find a small increase in corneal power, amounting to about 0.75D between the ages of 20 and 80 years: corneal aberrations appear to increase with age at a similar rate to total eye aberrations (see in the following text).

27.2.3 AQUEOUS

Due to its constant circulation and replacement, the aqueous maintains its clarity and refractive index throughout life, scattering only becoming a problem as a result of pathology, such as uveitis.

27.2.4 PUPIL

The pupil forms the aperture stop of the eye. For any given level of illumination, the diameter of the pupil reduces with age (*senile miosis* [see Figure 27.2, after Winn et al., 1994]). This has the advantages of increasing the depth-of-focus, hence partly compensating for losses in the amplitude of accommodation and reducing the impact of age-increased monochromatic aberration. On the other hand, the retinal illuminance is diminished, adversely affecting vision in mesopic and scotopic conditions. Pupil diameters reduce as the diameter of the adapting field is increased (Atchison et al., 2011) and the pupil center may change with dilation by up to around 0.5 mm (Yang et al., 2002).

27.2.5 LENS

The lens increases in thickness throughout life as a result of the growth of new lens fibers, but shows at most only minor changes in diameter. Accompanying the thickness changes are changes in surface curvatures and in the internal gradients of refractive index. Representative regression-line fits to *in vivo* data for the dimensions of the unaccommodated lens are given in Table 27.1.

These external changes are accompanied by internal changes, particularly in the gradients of refractive index within the lens. While there are still disagreements about the exact nature of the

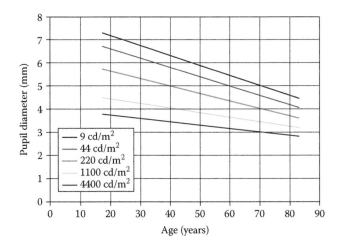

Figure 27.2 Pupil diameter as a function of age and adapting luminance (monocular observation, 10° diameter adapting field). (After Winn, B. et al., *Invest. Ophthalmol. Vis. Sci.*, 35, 1132, 1994.)

Table 27.1 Regression-line data for the changes with age in the dimensions (in mm) of the in vivo, unaccommodated, adult crystalline lens

PARAMETER	REGRESSION
Equatorial diameter (mm)	$D = 9.2$
Sagittal thickness (mm)	$T = 0.0238 * age + 2.93$
Anterior radius of curvature (mm)	$r_a = 12.9 - 0.057 * age$
Posterior radius of curvature (mm)	$r_p = -6.2 + 0.012 * age$

Note: The data for diameter and thickness come from magnetic resonance studies by Strenk et al. (1999), those for surface radii from Scheimpflug measurements by Dubbelman et al. (2001).

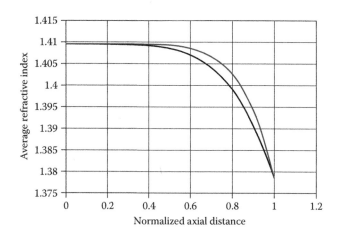

Figure 27.3 Models for the axial distributions of refractive index of the unaccommodated crystalline lens at mean ages of 23 (black) and 64 (red) years, as derived from, in vivo, MRI-based data. The axial distance from the center of the lens has been normalized. The true mean lens thicknesses are 3.78 and 4.75 mm for the young and old lenses, respectively. (After Kasthurirangan, S. et al., *Invest. Ophthalmol. Vis. Sci.*, 49, 2531, 2008.)

index distribution, it is generally thought that in early adulthood there is a small central volume, with an index about 1.41, from which the index decreases fairly gradually but at an increasing rate to about 1.38 at the surface. In contrast, in old age the central volume of near-uniform index is much larger: the fall-in index occurs more rapidly and almost entirely in the surface layers (Jones et al., 2005; Kasthurirangan et al., 2008; de Castro et al., 2011; Birkenfield et al., 2013) (Figure 27.3). This reduces the contribution that the gradient index makes to the overall lens power and tends to compensate for the increase in surface power with age, so that overall lens power remains almost constant (Dubbelman et al., 2001; Jones et al., 2005).

The non-cataractous lens suffers from a gradual loss in transmittance with age, caused by increases in thickness, absorption, and scattering (Sample et al., 1991). The effects differ in the nucleus and cortex (Mellerio, 1987). The loss occurs at all wavelengths but is most severe at the blue end of the spectrum (Figure 27.4). It starts from birth and continues throughout life (Weale, 1988; Delori and Burns, 1996).

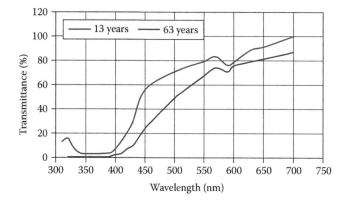

Figure 27.4 Spectral transmittance of lenses from the eyes of a 13- and a 63-year-old. (Based on data compiled by Weale, R.A., *J. Physiol.*, 395, 577, 1988.)

27.2.6 VITREOUS

Unlike the aqueous, the vitreous is not regularly renewed. The major ageing effect is that the vitreous gel is gradually replaced by unbound water, so that more movement within the vitreous becomes possible. Optically the chief effect is that the number and movement of "floaters" and *muscae volitantes* that are seen tends to increase, these being the shadows on the retina of clusters of fibers and other debris within the vitreous chamber. However, the overall vitreous transmittance is little changed.

27.2.7 AXIAL LENGTH

Grosvenor (1987) speculated, on the basis of data from Sorsby et al. (1957, 1962), that in order for the older eye to remain approximately emmetropic in spite of lenticular changes in power, the axial length might decrease through adult life by about 0.6 mm (equivalent to a hyperopic shift of about 2D). However, in the transverse study of emmetropes by Atchison et al. (2008), ultrasound measurements showed an increase, rather than a decrease of about this magnitude. The same study found no change with age in the vitreous length. Thus at the present time the question of age-dependent changes in axial length remains unresolved.

27.3 EFFECTS IN THE COMPLETE EYE

27.3.1 ABERRATIONS

27.3.1.1 Chromatic aberrations

Longitudinal chromatic aberration varies little with age or with the individual (Howarth et al., 1988; Morrell et al., 1991), but little is known about the effects of age on the small, subject-dependent amounts of transverse chromatic aberration that are found on the line of sight (Rynders et al., 1995; Marcos et al., 2001).

27.3.1.2 Monochromatic aberrations

Axial higher-order monochromatic aberrations at constant pupil diameter increase markedly with age (e.g., Guirao et al., 1999; McLellan et al., 2001; Amano et al., 2004). This is because lenticular growth and corneal change destroy the balance between the corneal and lenticular aberrations that applies in younger life (e.g., Artal et al., 2002; Kelly et al., 2004; Tabernero et al., 2007; Sharma et al., 2008; Navarro et al., 2013b). Representative data (Applegate et al., 2007) for the total higher-order root-mean-square (RMS) wavefront aberration as a function of age and pupil diameter in normal eyes are shown in Figure 27.5, together with the levels of aberration corresponding to errors of focus of 0.125D, 0.25D, and 0.50D. Although the RMS errors in the older eye appear likely to cause substantial blur, particularly at larger pupil diameters, it must be remembered that the older pupil is of smaller diameter at any fixed light level. For example, Figure 27.2 suggests that a light level that results in a pupil diameter of 5 mm for a 20-year-old will give a 4 mm pupil diameter for a 60-year-old, so that levels of wavefront aberration in the two eyes become almost equal (Figure 27.5). Overall aberration levels will, of course, rise in individuals who develop conditions such as cataract or keratoconus, due to large increases in the contributions made

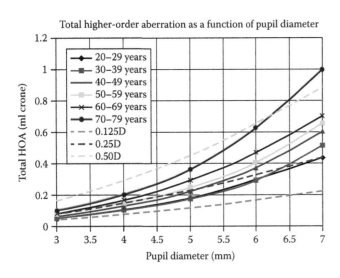

Figure 27.5 Plots of mean total, axial, root-mean-square, higher-order, wavefront aberration as a function of pupil diameter for the age groups indicated. For comparison, the dashed curves show the wavefront aberrations associated with pure spherical defocus levels of 0.125D, 0.25D, and 0.50D. (Based on Applegate, R.A. et al., *J. Opt. Soc. Am. A*, 24, 578, 2007.)

by lenticular and corneal aberrations, respectively (Kuroda et al., 2002; Maeda et al., 2002).

Although peripheral higher-order aberrations are higher in older eyes when measured at constant pupil diameter, in practice their impact on peripheral vision is again lessened by the reduction in pupil diameter found under normal viewing conditions. The blur caused by higher-order aberrations is also masked by the larger defocus blur produced by second-order peripheral refractive errors, which appear to change little with age (Mathur et al., 2010).

27.3.2 SCATTERED LIGHT

Relatively wide-angle scattered light arises from a variety of sources within the eye, including the cornea, lens, and retina, and its angular characteristics depend upon a variety of factors, including iris pigmentation (Ijspeert et al., 1990). It remains relatively constant in earlier adult life but rises rapidly after the age of about 45 (van den Berg, 1995; Hennelly et al., 1998), resulting in a marked loss of contrast in retinal images and disability glare problems with strong light sources, such as the sun at low elevations or car headlamps at night.

27.3.3 OVERALL OCULAR TRANSMITTANCE

The overall transmittance to the retinal receptors is largely governed by the changes in lenticular transmittance, although macular pigment contributes to blue absorption in the central few degrees of visual field (Baptista and Nascimento, 2014). Useful summaries of experimental data and models for the transmittance as a function of age have been given by van der Kraats and van Norren (2007). Figure 27.6 plots the variation in overall optical density ($\log_{10}(1/\text{transmittance})$) as a function of wavelength at several different ages for a small field, using the summary equation given by van der Kraats and van Norren. The spectral selectivity of the absorption, with more light being lost at shorter wavelengths, adversely affects various aspects of color

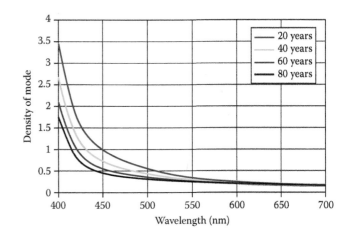

Figure 27.6 Optical density spectra for the pre-receptoral ocular media at four ages, as derived from a formula given by van der Kraats and van Norren (2007).

perception (e.g., Sagawa and Takahashi, 2001; Nguyen-Tri et al., 2003; Paramel and Oakley, 2014) and may also disrupt circadian rhythms in the elderly (Charman, 2003; Brøndsted et al., 2013).

27.3.4 CHANGES IN REFRACTION

Although distance refraction in individual eyes may follow a variety of time courses, with a subset developing myopia in childhood, a general pattern of lifelong refractive change emerges from the combined results of available transverse and longitudinal studies (Figure 27.7). Overall mean spherical refraction tends to move in a myopic direction up to the age of about 30 and then to slowly drift back in a hyperopic direction by a total of 1D–2D up to the age of

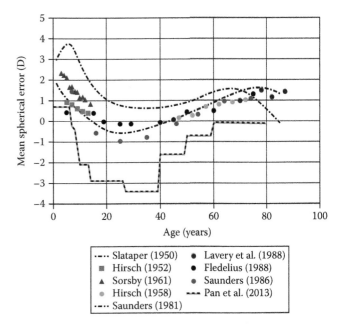

Figure 27.7 Average mean sphere refractive error as a function of age, as found in various mainly transverse (cross-sectional) studies. Most of the data are for western populations of subjects, with varying results somewhat depending upon such factors as the selection of the subject groups and the use of means or modes. The black stepped continuous line is a compilation by Pan et al. (2013) of several studies of Asian (mostly ethnic Chinese) subjects.

about 60. Thereafter, the trend appears to reverse, to move in the myopic direction again. Most of these changes are probably associated with lenticular changes, particularly in its index gradients, with the change after 60 being influenced by the increasing development of cataract. However, there may also be contributions from other factors, such as changes in axial length (Atchison et al., 2008) and corneal power (Navarro et al., 2013b). Most of the data of Figure 27.7 are for western populations. Recent Asian data, shown by the continuous black line, suggest that mean levels of refraction there are significantly more myopic than in the west at any age, although the pattern of change is broadly similar (Pan et al., 2013).

Off axis, the pattern of absolute peripheral refraction varies with the axial ametropia (Millodot, 1981), but if peripheral error is expressed relative to the axial value, it varies little between individuals and does not appear to change appreciably with age in adulthood (Charman and Jennings, 2006; Atchison and Markwell, 2008; Charman and Radhakrishnan, 2010).

27.3.5 ACCOMMODATION

As discussed in Chapters 3 and 4 of *Handbook of Visual Optics: Instrumentation and Vision Correction, Volume Two*, the objective amplitude accommodation of accommodation (i.e., the maximum achievable change in power of the eye) declines steadily, to reach zero at the age of about 50, the exact age varying both with the individual and their environment (Weale, 1981). The decline in objective dioptric amplitude (OA) is almost linear with age (years), a typical relationship being (Charman, 1989)

$$OA = 12.7 - 0.27 \times age$$

While lenticular changes play an important role in this loss in accommodation, other parts of the accommodation system are also involved (Charman, 2008). Depth-of-focus extends subjective amplitudes, so that they typically exceed objectively measured amplitudes by about 1D. The onset of presbyopia, when accommodation is no longer sufficient to allow easy performance of near tasks, is often taken as the age at which the subjective amplitude falls to 3D. Presbyopic symptoms typically appear around the age of 40, the exact age depending upon such factors as arm length and posture, as well as the amplitude. There is a progressive decline in the slope of the accommodation response/stimulus curve as presbyopia is approached (Kalsi et al., 2001). Although the bandwidth of the temporal frequency response of accommodation dynamics within the available accommodation amplitude is not affected by age, remaining at about 1.7 Hz, gain at any frequency reduces and the phase lags increases (Heron and Charman, 2004).

Fortunately, just as distance refractive error can be corrected by a variety of types of spectacles, contact lenses, intraocular lenses, or refractive surgery, variants of these approaches can yield reasonable distance and near vision (see Volume Two, Section IV, Vision Correction). As yet, however, it has not proved possible to restore or duplicate the flexibility and efficiency of the young accommodative system (Charman, 2014a,b).

27.3.6 OVERALL ON-AXIS RETINAL IMAGE QUALITY

This is affected by both wavefront aberration and scattering (Diaz-Doutón et al., 2006). Since both of these increase with age, it would be expected that image quality at constant pupil

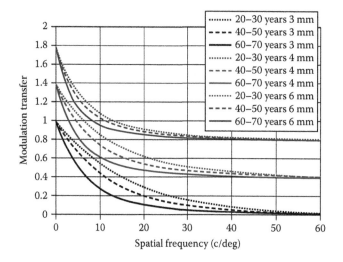

Figure 27.8 In-focus ocular modulation transfer functions for different age groups and pupil diameters. For clarity, the results for the 4 and 6 mm pupils have been moved upward by 0.4 and 0.8 units of modulation transfer, respectively. (Based on formulae given by Guirao, A. et al., *Invest. Ophthalmol. Vis. Sci.*, 40, 203, 1999.)

diameter would decline with age, as is found in typical direct measurements using the double-pass ophthalmoscopic technique (Artal et al., 1993; Guirao et al., 1999). Figure 27.8 illustrates that, for pupil diameters of 3, 4, and 6 mm and age groups 20–30, 40–50, and 60–70 years, the mean modulation transfer function and Strehl ratio get worse as the pupil diameter and age increase. However, as discussed in the context of aberrations, age has less impact on the retinal image quality if results are compared when the pupil diameter of each age group is that which is appropriate for the same fixed adapting luminance (see Figure 27.2, Guirao et al., 1999). Nevertheless, the older eye is still at a disadvantage, since the smaller pupil, combined with the lower ocular transmittance, reduces the retinal illuminance by a factor of about 3 between the ages of 20 and 60 (Weale, 1961).

27.4 OVERALL VISUAL PERFORMANCE

Overall visual performance depends upon both optical and neural factors, both of which show deterioration with age. However, it appears that long-term adaptation to the changes in aberration, scattering, and blur helps to minimize any deterioration in vision (see Volume 2, Section V, chapters 66 and 67). Similarly, although the spectrally selective changes in ocular transmittance discussed earlier might be expected to have large effects on color vision, color perception as assessed by color-naming remains surprisingly stable across the life span due to long-term adaptation effects (e.g., Werner and Webster, 2012). The relative stability of spatial vision under photopic conditions is illustrated in Figure 27.9, which shows high-contrast acuity as a function of age in healthy eyes: mean acuity remains better than zero logMAR (6/6) throughout life (Elliott et al., 1994). Nevertheless, acuity and contrast sensitivity do decline after the age of about 30 and adaptive optics studies in which aberrations are corrected suggest that this is not purely due to increased aberration but must involve other optical or neural factors (Elliott et al., 2009).

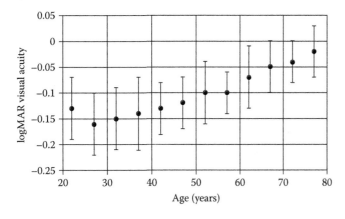

Figure 27.9 Mean and standard deviation of the high-contrast logMAR acuity of healthy eyes as a function of age. Each point refers to eyes within a 5-year interval. Chart luminance is 160 cd/m² (i.e., recommended clinical test chart luminance). (After Elliott, D.B. et al., *Optom. Vision Sci.*, 72, 186, 1994.)

Under mesopic and scotopic conditions, the older eye is at a real disadvantage, due to the lower retinal illuminance caused by the smaller pupil area and reduced ocular transmittance associated with aging and to a decline in neural performance. Increased intraocular scattering may also cause glare problems in situations such as night driving and such that the older driver may have major difficulties in seeing pedestrians at night (Wood et al., 2014). These problems are, of course, exacerbated if cataract or other pathology develops, although man-made, implantable intraocular lenses are now highly successful in restoring good quality distance vision to cataract patients and, in the case of some designs, near vision (see Volume Two, Section IV, Chapters 53–56).

REFERENCES

Abib, F.C. and Barreto, J., Behavior of corneal endothelial cell density over a lifetime, *Journal of Cataract and Refractive Surgery* 27 (2001): 1574–1578.

Allen, M.J. and Vos, J.J., Ocular scattered light and visual performance as a function of age, *American Journal of Optometry and Archives of the American Academy of Optometry* 44 (1967): 717–727.

Amano, S., Amano, Y., Yamagami, S., Miyai, T., Miyata, K., Samejima, T., and Oshika, T., Age-related changes in corneal and ocular higher-order wavefront aberrations, *American Journal of Ophthalmology* 137 (2004): 988–992.

Applegate, R.A., Donnelly, W.J., Marsack, J.D., and Koenig, D.E., Three-dimensional relationship between higher-order root-mean-square wavefront error, pupil diameter, and aging, *Journal of the Optical Society of America A* 24 (2007): 578–587.

Artal, P., Berrio, E., Guirao, A., and Piers, P., Contribution of the cornea and internal surfaces to the change of ocular aberrations with age, *Journal of the Optical Society of America A* 19 (2002): 137–143.

Artal, P., Ferro, M., Miranda, I., and Navarro, R., Effects of aging in retinal image quality, *Journal of the Optical Society of America A* 10 (1993): 1656–1662.

Atchison, D.A., Campbell, G.M., Dodds, J.P., Byrnes, T.M., and Zele, A.J., Influence of field size on pupil diameter under photopic and mesopic light levels, *Clinical and Experimental Optometry* 94 (2011): 545–548.

Atchison, D.A. and Markwell, E.L., Aberrations of emmetropic subjects at different ages, *Vision Research* 48 (2008): 2224–2231.

Atchison, D.A., Markwell, E.L., Kasthurirangan, S., Pope, J.M., Smith, G., and Swann, P.G., Age-related changes in optical and biometric characteristics of emmetropic eyes, *Journal of Vision* 8(4) (2008): 29.1–29.20.

Baldwin, W.R. and Mills, D., A longitudinal study of corneal astigmatism and total astigmatism, *American Journal of Optometry and Physiological Optics* 58 (1981): 206–211.

Baptista, A.M.G. and Nascimento, S.M.C., Changes in spatial extent and peak double optical density of human macular pigment with age, *Journal of the Optical Society of America A* 31 (2014): A87–A92.

Beems, E.M. and Van Best, J.A., Light transmission of the human cornea in whole human eyes, *Experimental Eye Research* 50 (1990): 393–395.

Birkenfeld, J., de Castro, A., Ortiz, S., Pascual, D., and Marcos, S., Contribution of the gradient refractive index and shape to the crystalline lens spherical aberration and astigmatism, *Vision Research* 86 (2013): 27–34.

Brøndsted, A.E., Lundeman, J.H., and Kessel, L., Short wavelength light filtering by the natural human lens and IOLs—Implications for entrainment of circadian rhythm, *Acta Ophthalmologica* 91 (2013): 52–57.

Charman, W.N., The path to presbyopia. Straight or crooked? *Ophthalmic and Physiological Optics* 9 (1989): 424–430.

Charman, W.N., Age, lens transmittance, and the possible effects of light on melatonin suppression, *Ophthalmic and Physiological Optics* 23 (2003): 181–187.

Charman, W.N., The eye in focus: Accommodation and presbyopia, *Clinical and Experimental Optometry* 91 (2008): 207–225.

Charman, W.N., Developments in the correction of presbyopia I: Spectacle and contact lenses, *Ophthalmic and Physiological Optics* 34 (2014a): 8–29.

Charman, W.N., Developments in the correction of presbyopia II: Surgical approaches, *Ophthalmic and Physiological Optics* 34 (2014b): 397–426.

Charman, W.N. and Jennings, J.A.M., Longitudinal changes in peripheral refraction, *Ophthalmic and Physiological Optics* 26 (2006): 447–455.

Charman, W.N. and Radhakrishnan, H., Peripheral refraction and the development of refractive error: A review, *Ophthalmic and Physiological Optics* 30 (2010): 321–328.

Cook, R.C. and Glasscock, R.E., Refractive and ocular findings in the newborn, *American Journal of Ophthalmology* 34 (1951): 1407–1413.

De Castro, A., Siedlecki, D., Borja, D., Uhlhorn, S., Parel, J.-M., Manns, F., and Marcos, S., Age-dependent variation in the gradient index profile in human crystalline lenses, *Journal of Modern Optics* 58 (2011): 19–20, 1781–1787.

Delori, F.C. and Burns, S.A., Fundus reflectance and the measurement of crystalline lens density, *Journal of the Optical Society of America A* 13 (1996): 215–226.

Diaz-Doutón, F., Benito, A., Pujol, J., Arjona, M., Güell, J.L., and Artal, P., Comparison of retinal image quality with a Hartmann-Shack wavefront sensor and a double-pass instrument, *Investigative Ophthalmology and Visual Science* 47 (2006): 1710–1716.

Dubbelman, M., Sicam, V.A.D.P., and Van der Heijde, G.L., The shape of the aging human lens: Curvature, equivalent refractive index and the lens paradox, *Vision Research* 41 (2001): 1867–1877.

Dubbelman, M., Sicam, V.A.D.P., and Van der Heijde, G.L., The shape of the anterior and posterior surface of the aging human cornea, *Vision Research* 46 (2006): 993–1001.

Elliott, D.B., Yang, K.C.H., and Whitaker, D., Visual acuity throughout adulthood in normal, healthy eyes: Seeing beyond 6/6, *Optometry and Vision Science* 72 (1994): 186–191.

Elliott, S.L., Choi, S.S., Doble, N., Hardy, J.L., Evans, J.W., and Werner, J.S., Role of high-order aberrations in senescent changes in spatial vision, *Journal of Vision* 9(2) (2009): 24, 1–16.

Fledelius, H.C., Prevalences of astigmatism and anisometropia in adult Danes, *Acta Ophthalmologica* 62 (1984): 391–400.

Fledelius, H.C., Refraction and eye size in the elderly. A review based on literature, including own results, *Acta Ophthalmologica (Copenhagen)* 66 (1988): 241–248.

Grosvenor, T., Reduction in axial length with age: An emmetropizing mechanism for the adult eye? *American Journal of Optometry and Physiological Optics* 64 (1987): 657–663.

Guillon, M. and Maissa, C., Tear film evaporation—Effect of age and gender, *Contact Lens and Anterior Eye* 33 (2010): 171–175.

Guirao, A., González, C., Redondo, M., Geraghty, E., Norrby, S., and Artal, P., Average optical performance of the human eye as a function of age in a normal population, *Investigative Ophthalmology and Visual Science* 40 (1999): 203–213.

Guirao, A., Redondo, M., and Artal, P., Optical aberrations of the human cornea as a function of age, *Journal of the Optical Society of America A* 17 (2000): 1697–1702.

Hayashi, K., Hayashi, H., and Hayashi, F., Topographic analysis of the changes in corneal shape due to aging, *Cornea* 14 (1995): 527–532.

Hennelly, M.L., Barbur, J.L., Edgar, D.F., and Woodward, E.G., The effect of age on the light scattering characteristics of the eye, *Ophthalmic and Physiological Optics* 18 (1998): 197–203.

Heron, G. and Charman, W.N., Accommodation as a function of age and the linearity of the response dynamics, *Vision Research* 44 (2004): 3119–3130.

Hirsch, M.J., The changes in refraction between the ages of 5 and 14—Theoretical and practical considerations, *American Journal of Optometry and Archives of the American Academy of Optometry* 29 (1952): 445–459.

Hirsch, M.J., Changes in refractive state after the age of forty-five, *American Journal of Optometry and Archives of the American Academy of Optometry* 35(1958): 229–237.

Howarth, P.A., Zhang, X.X., Bradley, A., Still, D.I., and Thibos, L.N., Does the chromatic aberration of the eye vary with age? *Journal of the Optical Society of America A* 5 (1988): 2087–2092.

Ijspeert, J.K., de Waard, P.W.T., van den Berg, T.J.T.P., and de Jong, P.T.V.M., The healthy straylight function in 129 healthy volunteers: Dependence on angle, age and pigmentation, *Vision Research* 30 (1990): 699–707.

Jones, C.E., Atchison, D.A., Meder, R., and Pope, J.M., Refractive index distribution and optical properties of the isolated human lens measured using magnetic resonance imaging (MRI), *Vision Research* 45 (2005): 2352–2366.

Kalsi, M., Heron, G., and Charman, W.N., Changes in the static accommodation response with age, *Ophthalmic and Physiological Optics* 21 (2001): 77–84.

Kasthurirangan, S., Markwell, E.L., Atchison, D.A., and Pope, J.SM., In vivo study of changes in refractive index distribution in the human crystalline lens with age and accommodation, *Investigative Ophthalmology and Visual Science* 49 (2008): 2531–2540.

Kelly, J.E., Mihashi, T., and Howland, H.C., Compensation of corneal horizontal/vertical astigmatism, lateral coma and spherical aberration by internal optics of the eye, *Journal of Vision* 4(4) (2004): 2, 262–271.

Kemph, G.A., Collins, S.D., and Jarman, B.L., Refractive errors in the eyes of children as determined by retinoscopic examination with a cycloplegic, Public Health Bulletin No. 182, U.S. Government Printing Office, Washington, DC, 1928.

Kuroda, T., Fujikado, T., Maeda, N., Oshika, T., Hirohara, Y., and Mihashi, T., Wavefront analysis in eyes with nuclear or cortical cataract, *American Journal of Ophthalmology* 134 (2002): 1–9.

Lavery, J.R., Gibson, J.M., Shaw, D.E., and Rosenthal, A.R., Refraction and refractive errors in an elderly population. *Ophthalmic and Physiological Optics* 8 (1988): 394–396.

Lyhne, N., Sjolie, A.K., Kyvik, K.O., and Green, A., The importance of genes and environment for ocular refraction and its determiners: A population based study among 20–45 year old twins, *British Journal of Ophthalmology* 85 (2001): 1470–1476.

Maeda, N., Fujikado, T., Kuroda, T., Mihashi, T., Hirohara, Y., Nishida, K., Watanabe, H., and Tano, Y., Wavefront aberrations measured with Hartmann-Shack sensor in patients with keratoconus, *Ophthalmology* 109 (2002): 1996–2003.

Marcos, S., Burns, S.A., Prieto, P.M., Navarro, R., and Baraibar, B., Investigating sources of variability of monochromatic and transverse chromatic aberrations across eyes, *Vision Research* 41 (2001): 3861–3871.

Mathur, A., Atchison, D.A., and Charman, W.N., Effects of age on peripheral ocular aberrations, *Optics Express* 18(6) (2010): 5840–5853.

McLellan, J.S., Marcos, S., and Burns, S.A., Age-related changes in monochromatic aberrations of the human eye, *Investigative Ophthalmology and Visual Science* 42 (2001): 1390–1395.

Mellerio, J., Yellowing of the human lens: Nuclear and cortical contributions, *Vision Research* 27 (1987): 1581–1587.

Millodot, M., The influence of age on the sensitivity of the cornea, *Investigative Ophthalmology and Visual Science* 16 (1977): 240–242.

Millodot, M., Effect of ametropia on peripheral refraction, *American Journal of Optometry and Physiological Optics* 58 (1981): 691–695.

Morgan, I. and Rose, K., How genetic is school myopia? *Progress in Retinal and Eye Research* 24 (2005): 1–38.

Morrell, A., Whitefoot, H.D., and Charman, W.N., Ocular chromatic aberration and age, *Ophthalmic and Physiological Optics* 11 (1991): 385–390.

Navarro, R., Rozema, J.J., and Tassignon, M.-J., Orientation changes of the main corneal axes as a function of age, *Optometry and Vision Science* 90 (2013a): 23–30.

Navarro, R., Rozema, J.J., and Tassignon, M.-J., Optical changes of the human cornea as a function of age, *Optometry and Vision Science* 90 (2013b): 587–598.

Nguyen-Tri, D., Overbury, O., and Faubert, J., The role of lenticular senescence in age-related color vision changes, *Investigative Ophthalmology and Visual Science* 44 (2003): 3698–3704.

Olsen, T., Light scattering from the human cornea, *Investigative Ophthalmology and Visual Science* 23 (1982): 81–86.

Oshika, T., Klyce, S.D., Applegate, R.A., and Howland, H.C., Changes in corneal wavefront aberrations with aging, *Investigative Ophthalmology and Visual Science* 40 (1999): 1351–1355.

Pan, C.-W., Zheng, Y.-F., Anuar, A., Chew, M., Gazzard, G., Aung, T., Cheng, C.-Y., Wong, T.Y., and Saw, S.-M., Prevalence of refractive errors in a multi-ethnic Asian population: The Singapore Epidemiology of Eye Disease Study, *Investigative Ophthalmology and Visual Science* 54 (2013): 2590–2598.

Paramel, G.V. and Oakley, B., Variation of colour discrimination across the life span, *Journal of the Optical Society of America A* 31 (2014): A375–A384.

Rynders, M., Lidkea, B., Chisholm, W., and Thibos, L.N., Statistical distribution of foveal transverse chromatic aberration, pupil centration, and angle ψ in a population of young adult eyes, *Journal of the Optical Society of America A* 12 (1995): 2348–2357.

Sagawa, K. and Takahashi, Y., Spectral luminous efficiency as a function of age, *Journal of the Optical Society of America A* 18 (2001): 2659–2667.

Sample, P.A., Quirante, J.S., and Weinreb, R.N., Age-related changes in the human lens, *Acta Ophthalmologica* 69 (1991): 310–314.

Saunders, H., Age-dependence of human refractive errors, *Ophthalmic and Physiological Optics* 1 (1981): 159–174.

Saunders, H., A longitudinal study of the age-dependence of human ocular refraction. I. Age-dependent changes in the equivalent sphere, *Ophthalmic and Physiological Optics* 6 (1986): 39–46.

Saunders, H., Changes in the axis of astigmatism, *Ophthalmic and Physiological Optics* 8 (1988): 37–42.

Sharma, R., Mihashi, T., and Howland, H.C., Compensation of monochromatic aberrations in older eyes, *Journal of Modern Optics* 55 (2008): 773–781.

Slataper, F.J., Age norms of refraction and vision, *Archives of Ophthalmology* 43 (1950): 468–481.

Sorsby, A., Benjamin, J.B., Davey, M., Sheridan, M., and Tanner, J.M., *Emmetropia and Its Aberrations* (London, U.K.: Her Majesty's Stationery Office, 1957).

Sorsby, A., Benjamin, B., and Sheridan, M., Refraction and its components during growth of the eye from the age of three, *Special Reports Series of the Medical Research Council No 301* (London, U.K.: Her Majesty's Stationery Office, 1961).

Sorsby, A., Sheridan, M., and Leary, G.A., *Refraction and Its Components in Twins* (London, U.K.: Her Majesty's Stationery Office, 1962).

Strenk, S.A., Semmlow, J.L., Strenk, L.M., Munoz, P., Grunlund-Jacob, J., and DeMarco, J.K., Age-related changes in human ciliary muscle and lens: A magnetic resonance imaging study, *Investigative Opthalmology and Visual Science* 40 (1999): 1162–1169.

Tabernero, J., Alcón, A.B.E., and Artal, P., Mechanism of compensation of aberrations in the human eye, *Journal of the Optical Society of America A* 24 (2007): 3274–3283.

Tomlinson, A. and Craig, J., Time and the tear film, in *The Tear Film*, eds. D.R. Korb, J. Craig, M. Doughty, J.-P. Guillon, G. Smith, and A. Tomlinson (London, U.K.: Butterworth-Heinemann, 2002), pp. 83–103.

van den Berg, T.J.T.P., Analysis of intraocular stray light, especially in relation to age, *Optometry and Vision Science* 72 (1995): 52–59.

van den Berg, T.J.T.P. and Tan, K.E.W.P., Light transmittance of the human cornea from 320 to 700 nm for different ages, *Vision Research* 34 (1994): 1453–1456.

van der Kraats, J. and van Norren, D., Optical density of the aging human ocular media in the visible and the UV, *Journal of the Optical Society of America A* 24 (2007): 1842–1857.

Weale, R.A., Retinal illumination and age, *Transactions of the illuminating Engineering Society* 26 (1961): 95–100.

Weale, R.A., Human ocular aging and ambient temperature, *British Journal of Ophthalmology* 65 (1981): 869–870.

Weale, R.A., *A Biography of the Eye* (London, U.K.: Lewis, 1982), pp. 94–120.

Weale, R.A., Age and transmittance of the human crystalline lens, *Journal of Physiology* 395 (1988): 577–587.

Werner, J.S. and Webster, M., Neural changes in vision affecting the presbyopic eye, in *Presbyopia: Origins, Effects and Treatment*, eds. I.G. Pallikaris, S. Plainis, and W.N. Charman (Thorofare, NJ: Slack, 2012) pp. 85–92.

Winn, B., Whitaker, D., Elliott, D.B., and Phillips, N.J., Factors affecting light-adapted pupil size in normal human subjects, *Investigative Ophthalmology and Visual Science* 35 (1994): 1132–1137.

Wood, J.M., Lacherez, P., and Tyrell, R.A., Seeing pedestrians at night: Effect of driver age and visual abilities, *Ophthalmic and Physiological Optics* 34 (2014): 452–458.

Yang, Y., Thompson, K., and Burns, S.A., Pupil location under mesopic, photopic and pharmacologically dilated conditions, *Investigative Ophthalmology and Visual Science* 43 (2002): 2508–2512.

Young, T.L., Molecular genetics of human myopia: An update, *Optometry and Visual Science* 86 (2009): E8–E22.

Polarization properties

Juan M. Bueno

Contents

28.1 OVERVIEW OF POLARIZATION CONCEPTS: DEFINITIONS AND FORMALISM

Along with intensity and wavelength, polarization is one of the fundamental characteristics of light. This is related to the vectorial nature of light early suggested by Christian Huygens (Kliger et al. 1990). In this sense, when a light beam is propagating along the Z direction, the electrical field vector E (with components E_x and E_y and phases δ_x and δ_y) describes, in general, an ellipse in the plane XY (Figure 28.1):

$$\left(\frac{E_x}{A_x}\right)^2 + \left(\frac{E_y}{A_y}\right)^2 - 2 \cdot \frac{E_x \cdot E_y}{A_x \cdot A_y} \cdot \cos\delta = \sin^2\delta, \quad (28.1)$$

with A_x and A_y being, respectively, the amplitudes of E_x and E_y.

The angle of the major axis and the direction X is called azimuth χ ($0 \leq \chi \leq \pi$) and can be computed from:

$$tg\,2\chi = 2 \cdot \frac{A_x \cdot A_y}{A_x^2 - A_y^2} \cdot \cos\delta. \quad (28.2)$$

The ratio of the length of the semiminor axis to that of the semimajor axis is the ellipticity $\varphi = \pm b/a$ ($-1 \leq tg\varphi \leq 1$). If the tip of the electrical field vector maps out the ellipse in the clockwise direction, the polarization form is right-handed ($tg\varphi > 0$). If the direction of tracing is counterclockwise, the form is left-handed. Then, any polarization state is defined by χ, φ and the handedness of the polarization ellipse.

Elliptically polarized light might turn into particular polarization states, such as circular and linear. In the former $A_x = A_y$ and $\delta = \pi/2$. The light will be right circular if $tg\varphi = 1$ (left circular

for $tg\varphi = -1$). For the latter, $\delta = 0$ or π. Moreover, if one of the components of the electric field vector is null, the light will be also linearly polarized (vertical if $E_x = 0$, horizontal if $E_y = 0$). The azimuth of the linearly polarized light can be easily computed by means of Equation 28.2.

Unlike (totally) polarized light, when the electric field vector vibrates in all directions with no preferential orientation, the light is depolarized. Light containing both polarized and depolarized components is known as partially polarized.

In 1852 G. G. Stokes reported that the state of polarization of a light beam can be represented by four measurable quantities grouped in a 4×1 vector known as the Stokes vector (Shurcliff 1962). Whereas this vector can describe any polarization state, the Jones vector is only useful for totally polarized beams (Jones 1941). As a function of the parameters of the ellipse of polarization, this vector is expressed as:

$$S = \begin{pmatrix} S_0 \\ S_1 \\ S_2 \\ S_3 \end{pmatrix} = \begin{pmatrix} A_x^2 + A_y^2 \\ A_x^2 - A_y^2 \\ 2A_xA_y\cos\delta \\ 2A_xA_y\sin\delta \end{pmatrix} = \begin{pmatrix} I \\ I \cdot \cos 2\chi \cdot \cos 2\varphi \\ I \cdot \sin 2\chi \cdot \cos 2\varphi \\ I \cdot \sin 2\varphi \end{pmatrix}, \quad (28.3)$$

where the intensity of the light beam always verifies:

$$I^2 = S_0^2 \geq S_1^2 + S_2^2 + S_3^2. \quad (28.4)$$

The degree of polarization (DOP) of a light beam is defined as the ratio of polarized-component intensity to total intensity:

$$DOP = \frac{\left(S_1^2 + S_2^2 + S_3^2\right)^{1/2}}{S_0}, \quad (28.5)$$

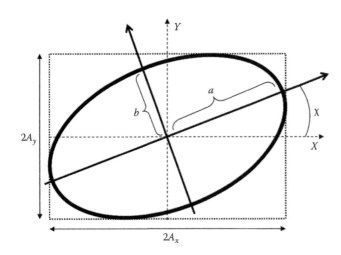

Figure 28.1 Ellipse of polarization.

which ranges from zero (depolarized light) to unity (totally polarized light). If $0 < DOP < 1$, the beam is partially polarized.

Moreover, the Stokes parameters of Equation 28.3 are easily identified as spherical coordinates. That is, the polarization state of a light beam is described by a four-element column vector that can be located on the surface of the so-called Poincaré sphere (Figure 28.2) (Jerrard 1954).

As Equation 28.6 shows, a change in the polarization state of an incident beam produced by a system is a linear transformation in a 4D space. When a light beam with a Stokes vector S_{in} passes an optical medium, this turns into S_{out} due to the effect of the medium, represented by a matrix M called Mueller matrix (Chipman 1995). This is a 4×4 matrix with real valued elements M_{ij} ($i, j = 0, 1, 2, 3$) containing information about all polarization properties of the medium:

$$S_{out} = M \cdot S_{in} = \begin{pmatrix} M_{00} & M_{01} & M_{02} & M_{03} \\ M_{10} & M_{11} & M_{12} & M_{13} \\ M_{20} & M_{21} & M_{22} & M_{23} \\ M_{30} & M_{31} & M_{32} & M_{33} \end{pmatrix} \cdot S_{in}. \quad (28.6)$$

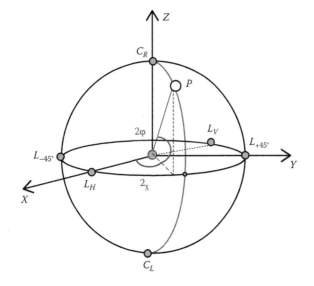

Figure 28.2 Poincaré sphere. L and C indicate linear and circular states.

However, the parameters characterizing the polarizing properties of an optical system do not appear explicitly in its Mueller matrix. To obtain useful information theorems of polar decomposition are required (Gil and Bernabeu 1987, Lu and Chipman 1996, Gil et al. 2013).

In particular, M_{00} represents the intensity of the emergent beam when nonpolarized light is entering the system. Elements M_{01}, M_{02}, and M_{03} describe the diattenuation (D) (intensity transmittance as a function of the incident polarization state). M_{10}, M_{20}, and M_{30} characterize the polarizance (P) (DOP of the transmitted light when nonpolarized light is incident). All elements contribute to the calculation of the DOP of the system (G_T). Depolarization ($Dep = 1 - G_T$) is understood as a process coupling polarized light into depolarized light. Evidently, if $Dep = 0$, the sample does not depolarize the totally polarized incident light. G_T, P, and D range from 0 to 1 and are defined, respectively, as (Chipman 1995)

$$D = \frac{\sqrt{M_{01}^2 + M_{02}^2 + M_{03}^2}}{M_{00}^2} \quad P = \frac{\sqrt{M_{10}^2 + M_{20}^2 + M_{30}^2}}{M_{00}^2}$$

$$G_T = \frac{\sqrt{\left(\sum_{i,j=0}^{3} M_{ij}^2\right) - M_{00}^2}}{\sqrt{3} \cdot M_{00}}. \quad (28.7)$$

The lower 3×3 submatrix contains information on the retardation introduced by birefringent structures. This represents an equivalent retarder characterized by its retardation (δ), its azimuth (α, i.e., orientation of its fast or slow axis), and its ellipticity (ω). Retardation, birefringence (Δn), and thickness (d) of a medium are related as

$$\delta = \frac{2\pi}{\lambda} \cdot \Delta n \cdot d. \quad (28.8)$$

Birefringence (or double refraction) is associated with optically anisotropic materials exhibiting different indices of refraction. In such media, the velocity of light depends on its direction of travel. Changes in the state of polarization caused by birefringent structures can be represented on the Poincaré sphere as a rotation of an angle δ (i.e., retardation value) around an axis or eigenvector. Half of the value of the longitude and the latitude of this eigenvector indicate, respectively, the azimuth and the ellipticity of the equivalent retarder. If the latter is null, this equivalent retarder is linear, and the axis lies on the equatorial plane of the Poincaré sphere.

Although birefringence is thought as an inherent property of anisotropic media, it can also arise from factors, such as structural ordering, physical stress, deformation, and strain, among others. When the asymmetry in refractive index occurs naturally, this is known as intrinsic birefringence. When a medium is composed of elongated parallel structures (fibers, microtubules, etc.) whose thickness and separation are much smaller than the wavelength of the incident light, it presents form birefringence. Moreover, if a material is not naturally birefringent but it is under external forces and/or deformation, stress- and/or strain-induced birefringence occurs.

In a birefringent material, the orthogonal directions X, Y, and Z are called principal directions. The simplest (and most common) birefringence is described as uniaxial, with a direction governing the optical anisotropy and the rest being optically equivalent. Two different indices of refraction are present, and then the propagation velocity of light is the same along X and Y, but different from that along the Z direction. If the velocity differs among the three principal directions, the material is called biaxial.

28.2 IMAGING POLARIMETRY: BASIS AND APPLICATIONS

Polarimetry is a technique used to determine the polarization properties of samples and light beams. It has been used in different fields, ranging from basic research (see Chipman 1995, as a general review) to clinical applications (Greenfield et al. 2000, Novikova et al. 2012). When analyses involve changes in the polarization state of the light beam upon reflection in a surface or interface, the usual term is ellipsometry (Azzam 1992). Imaging polarimetry is understood as a spatially resolved technique where a point-by-point detection is required.

A polarimeter is composed of a light source, a polarization state generator (PSG), a polarization state analyzer (PSA), and a detection unit. The typical configuration of a polarimeter is in transmission mode. However, to explore the polarization properties of the living eye, a double-pass polarimetric configuration is required (i.e., the light beam passes twice the ocular media and suffers reflection at the retina as shown in Figure 28.3).

The objective of a polarimeter is to determine the 16 elements of the Mueller matrix. For this aim, four independent polarization states in both the PSG and the PSA are required. To compute the elements of a Stokes vector, just four independent PSA states are necessary (and a fixed polarization state in the PSG). When

the full Mueller matrix or the complete Stokes vector cannot be reconstructed, the experimental system is known as an incomplete polarimeter (Chipman 1995), polariscope, or linear polariscope (when incorporating just linear polarizers) (Bueno 2002). These setups are often used when one polarization property of the system dominates and then the calculation of the complete Mueller matrix is not required.

28.3 OCULAR POLARIZATION PROPERTIES

Unlike many vertebrates (Able 1982, Hawryshyn and McFarland 1987), the human visual system cannot detect polarized light. However, this fact contrasts with the early entopic phenomenon described in 1844 and nowadays known as Haidinger's brushes (Haidinger 1844). This consists on observing for a few seconds an obscure pale-yellow figure against blue background when looking at a uniform field illuminated by plane polarized light. The search of an explanation for this opened the doors to the study of the ocular polarization properties. Since the eye is composed of complex structures, each contributes in a particular way to the "total ocular" polarization properties.

The analysis of the polarization properties of the eye uses the light that travels twice the ocular media and is reflected at the retina. Early studies were based on a polarizer–analyzer technique (Weale 1966, Charman 1980). However, since a set polarizer–analyzer might identify totally elliptical polarized light as partially polarized, a complete characterization of the eye requires the calculation of the Mueller matrix. Using different polarimetric systems, the Mueller matrix of the living human has been measured (Van Blokland 1985, Bueno and Artal 1999, Bueno 2000, Twietmeyer et al. 2008). Their results showed the normal healthy eye as a linear birefringent optical system with low depolarization effects (~10%–20%). The contribution of the different ocular elements to these properties will be treated in the following sections.

28.3.1 THE CORNEA

The cornea is the clear, curved "window" at the front of the ocular globe that protects from external injuries. It is the main refractive ocular component (~2/3 of the total power). The stroma occupies about 90% of the corneal thickness (which increases from central area to the periphery), and it is composed of 250–300 layers of collagen fibers, called lamellae. An individual lamella is composed of uniformly spaced fibrils and can transverse the cornea from limbus to limbus. Lamellar distribution is not always parallel to each other. They can also be arranged in interwoven and more complex patterns (Komai and Ushiki 1991, Bueno et al. 2011). The collagen arrangement depends on the stoma location, is particular for every species, and may be altered by pathologies, surgery, or injuries.

The corneal anisotropy has its origin in the birefringent nature of the stroma, which was early recognized by David Brewster (1815). As a result of this birefringence, a cross-shaped pattern appears when the cornea is imaged between crossed linear polarizers (see left panel in Figure 28.4). Nowadays, it is well known that the birefringence of each individual lamella arises from two different sources: (1) the intrinsic birefringence due to each

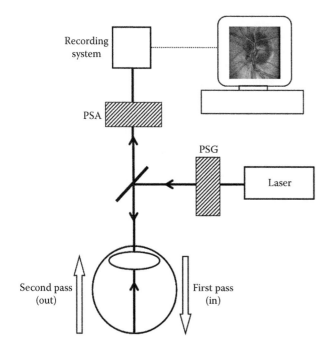

Figure 28.3 Schematic diagram of a double-pass polarimetric device for the living human eye.

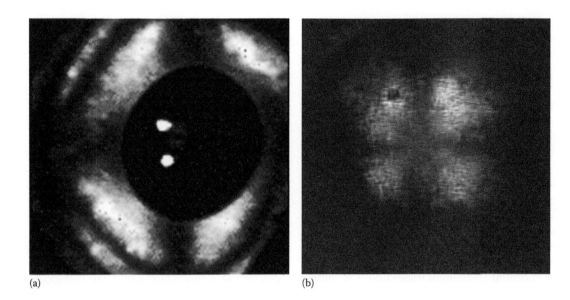

Figure 28.4 Cross-like patterns observed when the cornea (a) and the lens (b) are imaged under crossed linear polarizers. (a: Courtesy of Bille, J. and Pelz, B., University of Heidelberg, Heidelberg, Germany.)

individual fibril and (2) the form birefringence due to the parallel stack formed by the fibrils (since they are embedded in an optically homogeneous substance of different refractive index). The total birefringence of a lamella is equal to the sum of both.

For almost 200 years, numerous (*in vivo* and *in vitro*) studies have tried to understand corneal birefringence. However, the models are different and sometimes contradictory. Moreover, the question on the uniaxial or biaxial birefringence nature of the cornea is still open. Early qualitative analyses of corneal sections by using polarized light showed that lamellae exhibited double refraction (i.e., birefringence) with the optic axis along their length, lying on the cornea plane (His 1856). This uniaxial birefringence contrasted with a spherical-curved sheet of uniaxial crystal with its optical axis along the radius of curvature, reported some years later (Valentin 1861).

The first quantitative report on ocular birefringence was carried out by Boehm (1940). He observed the change in Haidinger's brush appearance using blue light (480 nm), concluding that the eye behaved as a retardation plate with fix retardation (45° on average) and a slow axis along the nasal-downward direction. This result was confirmed by de Vries and coworkers, who measured retardations between 30° and 90° (for light of 460 nm) in living human corneas and a slow axis ranging from 10° to 30° (nasal downward). These retardation values implied that, depending on the subject, corneal birefringence ranged between 0.0001 and 0.00015 (de Vries et al. 1953).

In an isolated cat cornea and for normal incidence, Stanworth and Naylor found a rapid increase in retardation with corneal eccentricity, from close to zero up to 75° at 3 mm (Stanworth and Naylor 1950). They reported a cornea behaving like a bent uniaxial crystal plate with the optic axis perpendicular to the surface (i.e., perpendicular to the fibers). Based on the hypothesis of a random distribution of lamellae, a birefringence of 0.0014 was later on derived (Stanworth and Nayor 1953). This value might change as a result of variations in the intraocular pressure (Stanworth 1953).

By that time (i.e., middle of the last century), two opposite models describing the corneal retardation had been already

reported in the literature. Whereas one considered the cornea as a fixed retarder and a slow axis mainly along the lower-nasal direction, the other showed an increase with corneal eccentricity and an optic axis along the radius of curvature and perpendicular to the surface. The corneal birefringence value and the type of lamellar arrangement (random or preferential) were issues still to be solved.

By analyzing *in vitro* rabbit corneas for different hydration states, Maurice was able to separate the contribution of both intrinsic and form birefringence (Maurice 1957). The latter was found to be 74% of the total. The former remains in a dehydrated cornea without ground substance. The birefringence of a lamella was found to be 0.0027 (0.00135 for total). A scattered-light method was used with the cornea of an enucleated cat eye providing an average birefringence of 0.0017 for the central cornea (Post and Gurland 1966). Since this value reduced from the anterior to the posterior corneal surface, the authors supposed the existence of local regions where fibers were essentially parallel. Changes in intraocular pressure (up to 30 mmHg) hardly had influence (<3%) in the measured birefringence. Similar results were found for *in vitro* bovine corneas (Kaplan and Bettelheim 1972). An increase in retardation from 80° at the center to 200° at 1 cm (for light of 633 nm) was measured. However, it was supposed that the corneal thickness was uniform, what is extremely unlikely. Based on qualitative observations, other authors also proposed a preferential orientation of human corneal lamellae (Shute 1974, Cope et al. 1978).

Spatially resolved quantitative measurements in living human corneas using a subjective technique reported that ocular retardation increased from zero at the center of the pupil to a maximum at the margins (50°–100° for 633 nm) (Bour and Lopes Cardozo 1981). The corresponding average birefringence for two subjects was 0.0020. This finding only fits the behavior of a bent uniaxial plate with an axis perpendicular to the surface. This model was difficult to reconcile with a fixed retardation and an axis parallel to the corneal surface earlier reported. The only way of matching both models is supposing a biaxial cornea with two slow axes.

This implies that corneal lamellae are not randomly distributed, but they present a dominant orientation. This was partially supported by McCally and Farell after measuring scattering patterns in rabbit and bovine corneas. They suggested either one or two (orthogonal) preferred directions, although some lamellae are evidently oriented in all directions (McCally and Farell 1982).

In that sense, van Blokland and Verhelst used Mueller-matrix polarimetry to measure the birefringence of living human corneas (van Blokland and Verhelst 1987). Despite a large variability among subjects, spatially resolved results showed an approximately constant retardation (55° on average, i.e., greater than zero) at the central pupil area that increased with eccentricity (for some subjects larger than 200°). The axes changed from a nasal-downward direction at the central zone to a tangential direction at large eccentricities. Then, they proposed that the cornea could be treated as a biaxial crystal with a slow Z-axis perpendicular to the surface and a second slow axis (Y-axis) associated with the preferential direction of the stromal lamellae at the central cornea. Birefringence values with respect to the X-axis (parallel to the surface) were 0.00159 and 0.00014, respectively.

This biaxial anisotropy has been explored more recently in 14 donor corneas with a polarizing microscope used for optical mineralogy studies. The images showed hyperbolic isogyres, characteristic of a negative biaxial material. Both the angle between the optic axes and the retardation were fairly constant among most samples, what suggested a uniform corneal structure (Bone and Draper 2007). Knighton and coworkers used a commercially available scanning laser (SL) polarimeter to compute the retardation and slow axis of the cornea at many points using images of the bow-tie pattern formed by the macular radial birefringence (see Section 28.3.3 for details on this). Results were compared to data of retardation produced by a curved biaxial material between two spherical surfaces. They found areas where the retardation

patterns could be mimicked (but not accurately) by a biaxial model and others in which behavior closely resembled that of a uniaxial model. This indicated that corneal birefringence varies greatly among people and, within a single cornea, significantly with position (Knighton et al. 2008).

The comparison of simulated and measured data by Donohue and colleagues confirmed that lamellar orientation is not entirely random, but rather a significant fraction are oriented in a fixed, preferred direction (Donohue et al. 1995). This confirms that a fixed-retarder model is only appropriate for the central cornea, but not for peripheral areas. Experimental data of *in vitro* human corneas obtained with phase-stepping imaging polarimetry partially confirmed that funding, showing that the distribution of retardation is nearly constant in the center of the cornea but highly nonhomogenous over the periphery (Jaronski and Kasprzak 1999).

The use of Mueller-matrix imaging in normal human eyes allowed to get more information on ocular polarization properties (mainly from the central cornea) apart from retardation data due to birefringence (van Blokland 1985, Bueno 2000). Results showed that central corneal birefringence was linear (i.e., the ellipticity of the equivalent retarder is close to zero), the slow axis was directed along the upper-temporal to lower-nasal line, and retardation values varied across subjects. Moreover, properties of depolarization, diattenuation, and polarizance are less important (Louis-Dorr et al. 2004, Bueno and Artal 2008). This agreed with spatially resolved analyses in porcine, bovine, and human corneas in transmission mode computed from the corresponding Mueller matrices (Figure 28.5) (Bueno and Jaronski 2001, Jaronski and Kasprzak 2003). The retardation was found to be approximately constant at the center, but it increased when going toward the periphery. The azimuth angle remained also constant at the center but varied significantly out of this area.

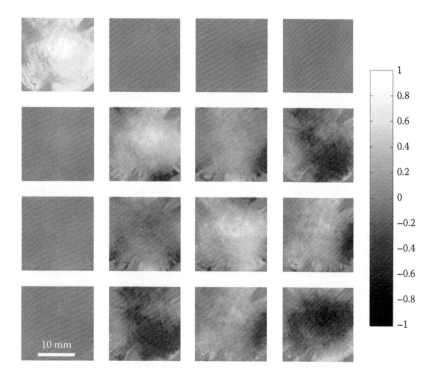

Figure 28.5 Spatially resolved Mueller matrix of a porcine cornea.

Ocular surgery and pathologies might also alter corneal optical anisotropies. In this sense, Bueno and coworkers explored changes in corneal polarization properties after refractive surgery. Spatially resolved polarization properties were compared in young normal and post-LASIK eyes (Bueno et al. 2006). Results showed that in normal eyes the retardation increased along the radius; however, this pattern became irregular after LASIK refractive surgery. The maps of slow axis also differ in normal and post-surgery eyes, with a larger disorder in post-LASIK eyes. Adrovani and colleagues analyzed polarization properties in corneas of diabetic and healthy mice. Corneal retardation was found to be higher in pathological eyes that in control ones (Adrovani et al. 2007).

However, during the last decade, the main interest in determining the corneal retardation and azimuth has been supported by clinical environments involved in glaucoma diagnosis using SL polarimetry. In particular, the corneal (or anterior segment) influence has to be correctly compensated to obtain accurate peripapillary retardation measurements (Zhou and Weinreb 2002) (see Section 28.4 for more details). Nowadays, this clinical corneal compensation is variable since polarimetric experiments carried out in sets of normal adult eyes with different experimental systems have shown a large variability in central corneal retardation, but a (slow) axis lying mainly along the nasally downward direction ($10°–20°$) (Knighton and Huang 2002, Bueno 2011). Recent measurements in children with a clinical SL polarimeter provided similar results. This indicates that there are not significant age-related differences in central corneal birefringence between children and adults (Irsch and Shah 2012).

In summary, the cornea is an ocular component with a noticeable birefringent nature. Whether its structure is uniaxial or a biaxial is a question still pending on. However, most results indicate that corneal retardation increases from the center to the periphery. At the central cornea, the slow axis lies along the nasal-downward direction. Other polarization effects such as depolarization or diattenuation are less important.

28.3.2 THE LENS

The lens is the refractive component of the eye responsible for accommodation. It grows and increases its weight and thickness throughout life. Unlike the cornea, the studies on the polarization properties of the lens are less numerous and not fully understood.

The lens presents a gradient of refractive index distribution, where the cortex has a lower refractive index than the core. Whereas the latter does not exhibit a regular pattern, the fibers of the former are distributed as the layers of an onion (Charman 1991). Since these are thicker (~ 8 µm) than the wavelength of visible light, a certain amount of form birefringence is expected (Bour 1991). This form birefringence together with intrinsic birefringence, originated from the internal distribution of molecules in the fibers, gives rise to the optical anisotropy of the lens. Apart from birefringence, the analyses of other lenticular polarization properties are scarce in the literature.

Although Brewster was the first to qualitatively analyze the behavior of an isolated lens placed between crossed polarizers (Brewster 1816) (see Figure 28.4), it was Valentin who reported the human lens as uniaxial birefringent (Valentin 1859). If the contribution of a birefringent lens to the total ocular retardation is important, differences between accommodated and nonaccommodated eyes, or between normal and aphakic eyes, must be significant. However, Boehm compared the mean retardation of normal and aphakic eyes and did not find differences (Boehm 1940).

Kirschenbaum demonstrated the capacity of optical rotation of polarized light by isolated lenses of different vertebrate species, what might be associated to a certain amount of (linear) birefringence (Kirschenbaum 1962). Later Takeguchi and Nakagary corroborated this in bovine lenses for different wavelengths, reporting a maximum rotation value of $-0.91°$ (for the lowest wavelength used) (Takeguchi and Nakagary 1968). This optical activity was much later analyzed using the Purkinje images (Pierscionek and Weale 1998).

In bovine freshly cut lens sections, Bettelheim found small birefringence values (in the range $10^{-7}–10^{-6}$) and no differences between the core and the cortex of the lens (Bettelheim 1975). Using a swelling mechanism, the contribution of intrinsic and form birefringence were separated, showing an opposite sign and leading to the fact that "the lens has practically no birefringence." Values of form and intrinsic birefringence varied between 10^{-5} and 10^{-3}.

Some years later, Weale investigated the changes in the birefringence of *in vitro* human lenses with age, sex, and stress (Weale 1979). Values were one order of magnitude higher than those reported by Bettelheim. Neither sex nor age had significant effect on birefringence. On the opposite, stress led to a reduction in birefringence.

Klein Brink carried out a study in living human eyes using Mueller-matrix polarimetry (klein Brink 1991). The total ocular retardation was found not to change when comparing accommodated and unaccommodated eyes. This technique was also used by Bueno and Campbell to explore for the first time the spatially resolved polarization properties of *in vitro* nonaccommodated human lenses (Figure 28.6) (Bueno and Campbell 2003). Their results showed a small overall retardation and a decrease from the center of the lens to the periphery. Birefringence was found to be linear, although it reduced outward along the radius (values reduced from 4.3×10^{-6} to 2.5×10^{-6}). Diattenuation values were also small, but depolarization was about 35%, probably due to the fact that some lenses came from old donors (older than 88 years) and were affected by the loss of transparency.

Since depolarization is intrinsically associated with scattering (Chipman 1995, Bueno et al. 2004), an increase in lenticular depolarization can directly be related to the presence of cataracts or posterior capsular opacification. This fact as well as the birefringence of intraocular lenses (Miura et al. 2004) must be taken into account during polarimetric measurements since they might introduce artifacts and lead to erroneous diagnoses (García-Medina et al. 2006, Bueno 2004).

To summarize, most experiments indicate that the lens, unlike the cornea, does not contribute substantially to the total optical retardation of the eye. However, when the effects of depolarization (mainly due to cataractous processes) are noticeable, the reliability of ocular retardation measurements might be affected.

28.3.3 THE RETINA

The retina contains complex structures that may present different polarization properties including birefringence, depolarization, and diattenuation (or alternatively dichroism). Probably the

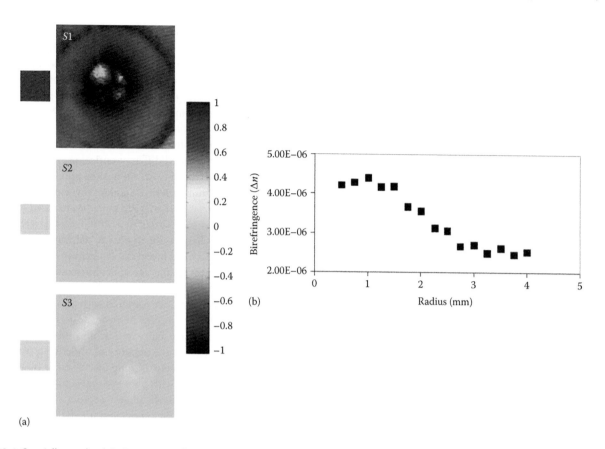

Figure 28.6 Spatially resolved Stokes vector of the light emerging from an isolated nonaccommodated human lens for incident linear vertical polarized light (a). The elements of the incoming vector are presented by each image for direct comparison. Values of birefringence as a function of the distance to the lens center (b).

discovery of the ability to perceive polarized light (Haidinger's brushes, see previous text) was the starting point of an increasing interest in exploring the polarization properties of the retina. Although the initial attention was centered on the explanation of this phenomenon, the main interest when analyzing these retinal polarization characteristics has been their association with changes produced with pathologies (i.e., healthy and diseased eyes might present significantly different polarization maps). Details and a number of explanations on the Haidinger's brushes can be found in Bour (1991) and Zhevandrov (1995) and references therein.

Boehm and de Vries et al. reported that the ocular media were double refracting although at that time it was not possible to localize that anisotropic structure (Boehm 1940, de Vries et al. 1953). Now it is well known that birefringence is mainly associated with the cornea, but it is not straightforward separate out the individual contributions of the cornea and the retina from this total ocular birefringence. Summer et al. suggested a model with the retina presenting properties of an anisotropic absorbing material that combined dichroism and birefringence (Summers et al. 1970). This was reinforced by the experiment carried out by Hochheimer and Kues (1982). They recorded retinal photographs with crossed linear polarizers in the input and output pathways. The cross or brush pattern overlying the macular was attributed to the birefringence of the cone photoreceptor outer layer. These exhibit negative form birefringence due to their stratified structure (Hárosi 1981). Since the visual pigment was found to present

positive intrinsic birefringence of similar magnitude (Weale 1971), both types of birefringence cancel out and cone outer segment birefringence seemed to play a minor role.

The first authors to report values of retinal retardation were Delori and coworkers. They demonstrated the presence in the macula of a radial birefringent network with the slow axis oriented radially and a retardation of 13°–52° (Delori et al. 1979). This demonstrated that macular birefringence arises from the Henle fiber layer.

Klein Brink and van Blokland used *in vivo* Mueller-matrix polarimetry to assess the retinal birefringence at the foveal region (Klein Brink and van Blokland 1988). Keeping constant the point of incidence over the corneal surface, the ocular retardation along an annular radius around the fovea presented a double-hump pattern. This was understood as a result of the combination of corneal and Henle fiber layer birefringence patterns. The former behaves as a fixed linear retarder and the latter is another linear retarder with its slow axis radial oriented from the fovea. This combination enabled to separate retinal and corneal component. The global ocular retardation will be maximum (minimum) where both corneal and retinal azimuths are parallel (perpendicular). Moreover, from the maximum and the minimum values, the corneal retardation can be easily computed and subtracted from the total. For a single pass, Henle fiber layer retardation values were 13.5° and 12.5° for 514 and 568 nm, respectively.

Changes in the appearance of the retinal nerve fiber layer (RNFL) covering the retina seen under polarized light were attributed to the presence of birefringence (Sommer et al. 1984). The RNFL covering the retina contains microtubules, cylindrical parallel intracellular organelles with diameter much smaller than the wavelength of the illuminating light. RNFL form birefringence is due to the preferential orientation of these cellular organelles (Wiener 1912, Sato et al. 1975). A general expression for the form birefringence of a medium of parallel cylinders of arbitrary sizes and separations was developed by Hemenger and evaluated for a model of RNFL (Hemenger 1989). This birefringence was shown to be linear and exhibited a slow axis parallel to the direction of nerve fiber bundles (Dreher et al. 1988).

Dreher and colleagues developed a polarimeter to assess polarization changes around the optic nerve head (ONH) of postmortem human eyes computed from the Mueller matrices of the retina. The area showed a substantial amount of linear (uniaxial) birefringence. The optic axis was found to be perpendicular to the incident laser beam and at a radial symmetrical direction, closely correlated with the physical orientation of the RNFL axons arrangement around the ONH. Moreover, the retardation local distribution showed two maxima that coincided with the thickest areas of RNFL (Dreher et al. 1992).

A multispectral imaging micropolarimeter was used to measure the retardation of the RNFL in the rat retina. As expected, the RNFL behaved as a linear retarder with the slow axis along the bundle. The retardation hardly changed between 440 and 830 nm (average ~1.8° and $\Delta n = 0.13°/\mu m$ in living eyes). This is consistent with the mechanism of form birefringence (Huang and Knighton 2002). The contribution of the microtubules of ganglion cell axons to the RNFL birefringence was also explored in rat retinas with an imaging polarimeter by analyzing the changes in the contrast over time. Results showed that microtubules make a significant contribution to RNFL birefringence. This indicates that a decrease in RNFL birefringence with glaucoma may indicate a loss of microtubules (Huang and Knighton 2005). A SL polarimeter and an optical coherence tomography (OCT) device were combined to measure the retinal retardation and the thickness, respectively, in a set of normal subjects (Huang et al. 2004). In most subjects Δn varied significantly along a circular path around the ONH, with maxima in superior and inferior bundles, and minima temporally and nasally (mean value = 0.15°/μm). However, Δn profile values were similar on circles of different diameter (i.e., Δn did not vary along the bundles). Retardation values varied from 10° to 50° with a mean of 33°.

The contribution of the foveal Henle fiber layer, particularly cone axons, was explored with a commercially available SL polarimeter in 20 normal subjects by Elsner and coworkers (2008). As the bow-tie pattern appears in normal macular retardation maps, this was modelled as rings increasing in radius around the fovea, using a sine wave of two periods (2*f*). The amplitude of the sine wave increased linearly with eccentricity and a good linear fit implied regular cone distribution and radial symmetry. This macular pattern imaged with SLP was later found to be affected by normal aging, that is, structural alterations in central cone photoreceptor morphology (either fewer cone photoreceptors or a change in the orientation of their axons) occur as a function of age (VanNasdale et al. 2011).

Moreover, form birefringence exhibited by the Henle fiber layer (since it is made up cylindrical and radially orientated

photoreceptor axons centered on the fovea) and computed by SLP was used to determine the foveal location in healthy patients with accuracy sufficient for clinical purposes (see further details on foveal fixation tracking in Section 28.4) (VanNasdale et al. 2009). In eyes suffering from nonexudative age-related macular degeneration, the disruption in Henle fiber birefringence was evident but nevertheless sufficient to help in foveal localization despite macular pathology (VanNasdale et al. 2012).

Sardar et al. compared different optical properties (transmission, reflection, scattering, etc.) in healthy and diseased (neovascularized) intact human retinal tissues. The study showed that the polarization characteristics are more pronounced in diseased than in healthy tissues (Sardar et al. 2005). These were attributed to the structural changes due to neovascularization. In particular, the relationship between foveal birefringence and visual acuity in patients with neovascular age-related macular degeneration was studied by Weber and colleagues (2007). They evaluated the macular bow-tie appearance acquired with a commercial SLP (see Figure 28.7). Regular bow ties were associated with subjects with higher mean visual acuity indicating a substantially intact Henle fiber layer despite the presence of underlying pathology. Conversely, disrupted bow-tie group (or no visible bow tie) corresponded to the group with poorer visual acuity values. This may indicate severe damage to the photoreceptors.

Clinical available SLPs do not compute the Mueller matrices of the analyzed fundus regions. They use a combination of sets of images acquired with linear incident light and two detection channels (parallel and crossed with respect to the entry) to extract only the information of the ocular retardation. This instrument was modified in 2008 to provide spatially resolved Mueller matrices of healthy living human retinas (Twietmeyer et al. 2008). From these, all polarimetric information (retardation, diattenuation, depolarization) could be obtained (see Figure 28.8). The cornea (or anterior segment) influence was eliminated by using a screen method (Knighton and Huang 2002). This assumes the corneal retardation may be approximated by averages over a large area of the macular region (i.e., assumes a negligible macular retardation). Results showed that the magnitude of the retinal retardation (both macular area and ONH) was particular for each subject but always presented the typical distribution previously highlighted. Mean values were 28.4° for ONH (ranging from 10° to 60°) and 17.3° for macula. The slow axis was radially symmetric around the fovea or the central ONH.

The number of images required to reconstruct polarimetric maps of the living human retina was recently reduced to a unique image by Fukuma et al. They combined a polarization analysis system and a fundus camera to measure the retardation distribution of the RNFL from a single image, correcting numerically the corneal influence and the retinal background scattering (Fukuma et al. 2011). Burns and colleagues combined speed and high-resolution polarimetry by implementing electro-optical polarization elements into an adaptive optics SL ophthalmoscope (Song et al. 2010). The system was able to reconstruct the Stokes vector of light returning from retinal structures at cellular level in living human eyes. These allowed improving the retinal image contrast of retinal structures based on their polarization properties.

To conclude, retinal birefringence together with that of the cornea are the main contributors to ocular polarization

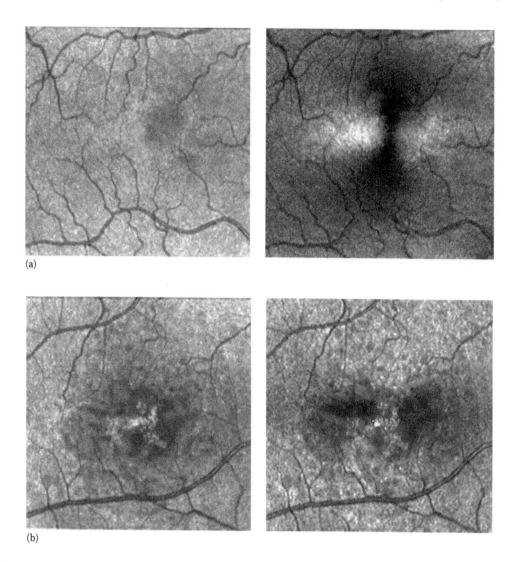

Figure 28.7 Regular scanning laser images and retardation maps showing the typical bow-tie pattern for a 56-year-old normal subject (a) and an irregular bow tie for a 68-year-old subject suffering from dry age-related macular degeneration (b). (Images courtesy of Elsner, A.E., Indiana University, Bloomington, IN.)

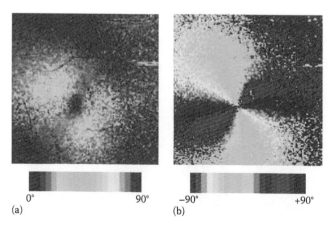

Figure 28.8 Macular retardation (a) and azimuth (b) maps in a normal eye after performing corneal compensation. (Reproduced from Twietmeyer, K.M. et al., Opt. Express, 16, 21339, 2008. With permission.)

properties. Changes in retinal retardation patterns might be associated to pathologies and might help in clinical diagnoses. Apart from birefringence, depolarization and diattenuation could also play a role (van Blokland and van Norren 1986, Dreher et al. 1992, Bueno 2001, Naoun et al. 2005, Bueno and Artal 2008, Twietmeyer et al. 2008, Baumann et al. 2009, Bueno et al. 2010, Bueno and Perez 2010) (Figure 28.9) and have some nonnegligible influence in the performance of clinically oriented devices (Burns et al. 2003, Bueno 2004, Elsner et al. 2007).

28.4 CLINICAL-ORIENTED IMAGING POLARIMETRY FOR OCULAR DIAGNOSIS

The eyes of many insects, fishes, and birds are basically analyzers of polarized light used as a compass for navigation and migration (Wehner 1976, Able 1982, Hawryshyn and McFarland 1987). Only a few humans can detect different types of polarized light (Bour 1991). Despite the complicated polarization properties of

Optical properties of the eye

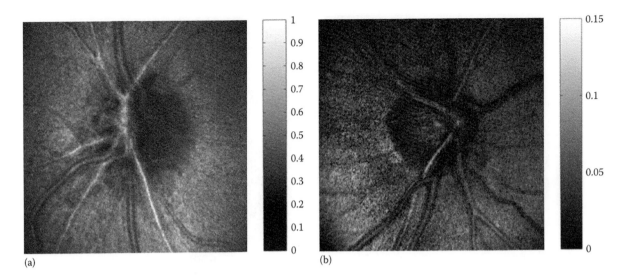

(a)

(b)

Figure 28.9 Degree of polarization (a) and diattenuation (b) maps in two normal eyes.

the human eye, the retinal image quality (or, alternatively, the ocular aberrations) is nearly independent on the state of polarization of the incident light (Bueno and Artal 2001, Prieto et al. 2002). However, these polarization properties have been used to obtain quantitative or qualitative ocular information useful for clinical applications. In particular, elderly eyes and other undergoing surgery or suffering from some ocular pathology might benefit from this.

In retinal imaging, the use of crossed polarizers has been reported to enhance the visibility of the arcuate bundles of the RNFL (Sommer et al. 1984). Moreover, a cross-like or brushlike pattern appears when imaging the macular area. Changes in this pattern can be a useful tool for diagnosing diseases affecting the macula (Hochheimer and Kues 1982). This configuration has also been used to observe the corneal and the lenticular crosses (see Figure 28.4) (Cope 1978). The distortions in the corneal pattern appearing with this configuration of crossed polarizers were analyzed to explore the *in vivo* corneal stress patterns (Ichihashi et al. 1995). Lenticular features and corneal endothelial cells hardly discernible when using standard illumination can be seen when

linear or circular polarizers are incorporated into a biomicroscope (Peli 1985, Pierscionek and Weale 1995). Scattering (or alternatively depolarization) might also reduce the visibility of retinal features in clinical fundus imaging (Hunter et al. 2007) as shown in Figure 28.10.

Retinal birefringence has also been used as a marker for foveal fixation and to detect amblyopia. Both the radial distribution of the nerve fibers around the fovea and the fixed central corneal birefringence give place to a particular polarization distribution (see Section 28.3.3). The clinical device, named as retinal birefringence scanning, scans an annulus around the central fovea with polarized light. The polarization state of the returning light is analyzed and the retardation computed. For a central fixation, the retardation pattern around the fovea provides two maximum and two minimum values. Out of this area (i.e., parafoveal fixation) the patterns clearly changes (Hunter et al. 1999, 2003).

Polarimetry has also been proposed as a noninvasive for glucose monitoring in diabetic patients. This will be an alternative *in vivo* test to the regular invasive blood glucose readings (Baba et al. 2002, Ansari et al. 2004, Rawer et al. 2004, Purvinis et al. 2011).

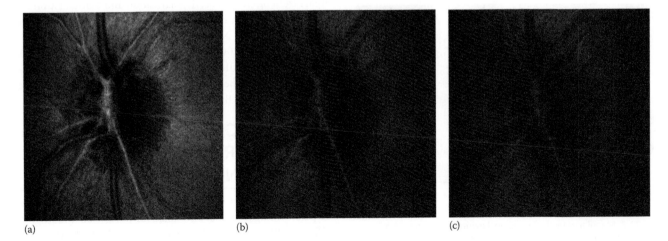

(a)

(b)

(c)

Figure 28.10 Optic nerve head images acquired with incident linear (a) and elliptical (b) polarized light. (c) It corresponds to linear polarization with induced scattering.

The aqueous humor (i.e., the clear fluid located in the anterior chamber between the cornea and the lens) has been reported to be a blood glucose sensor. As the glucose molecules within the humor present optical activity, these modify the polarization of the incident light. This "rotation" of the incident polarization state can be associated with the glucose concentration and correlated with actual blood glucose levels. This noncontact polarimetric-based technique may be more comfortable for the patient and would allow fast noninvasive glucose readings.

However, the main clinical application based on ocular polarimetry has been oriented to the diagnosis of glaucoma. This ocular pathology is caused by progressive apoptosis of retinal ganglion cells and a reduction in the RNFL thickness. Since the RNFL is birefringent, there is a linear relationship between the anatomical thickness and the optical retardation introduced. Then, if the RNFL retardation can be measured, the retinal thickness can be estimated.

Taking this into account, Weinreb and colleagues implemented a polarimeter into a laser tomographic scanner (Weinreb et al. 1990). They compared retardation data from fixed monkey retinas with the RNFL thickness at particular imaged points. An excellent correlation was found ($\Delta n = 0.14°/\mu m$). For the measured locations, retardation and thickness values ranged between 0.9° and 23.7° and 20.4 and 213.9 μm, respectively. Since then, this histopathologic validation of the RNFL has helped to enhance the discrimination between glaucomatous and normal eyes and has opened the door to the development of clinical devices for glaucoma diagnosis. An improved device turned into the first clinical instrument suitable for spatially resolved (i.e., imaging) polarimetry of the living human retina (Dreher et al. 1992, Dreher and Reiter 1992, Dreher and Baile 1994). This was later able to distinguish normal from glaucomatous eyes by measuring the retinal retardation map around the ONH (Weinreb et al. 1995, Greenfield 2002). Results show that pathological eyes present lower retardation since the RNFL thickness is reduced due to the loss of the axons of the ganglion cells.

Since both the cornea and the retina are birefringent structures, the influence of the former must be subtracted from the total ocular information. Early clinical instruments included a fixed corneal compensator (Greenfield et al. 2000). However, due to the considerable interindividual variation in the corneal axis (Knighton and Huang 2002), individual corneal compensation was required in order to obtain reliable clinical diagnoses in glaucomatous eyes (Zhou and Weinreb 2002).

Pelz and coworkers also worked in another version of a clinical instrument, imaging a retinal area covering not only the ONH but also the macular zone (25° × 12.5°) (Pelz et al. 1996). They compensated for the corneal influence using the specular reflection from the retinal blood vessels. Macular retardation maps showed a homogenous radial distribution, and the area around the ONH showed two typical maxima associated with the arcuate bundles (thickest RNFL areas). As expected, the azimuth around both the fovea and the ONH was radially distributed, following the direction of the nerve fibers. The mean retardation around the ONH decreased from 16° in healthy eyes to about 10° in glaucomatous eyes. Moreover, the latter did not show the two retardation peaks. As an example, Figure 28.11 shows the retinal retardation map in the left eye of this author before and after

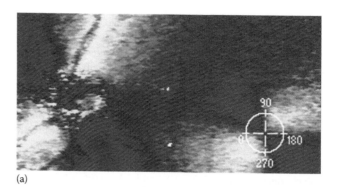

(a)

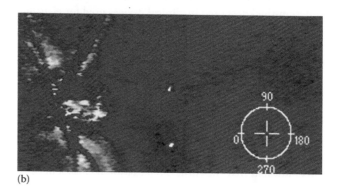

(b)

Figure 28.11 Retardation maps (double pass) of the left eye of this author without (a) and with (b) corneal compensation. Hot colors indicate higher retardation values. Images subtend 25° × 12.5°. (Courtesy of Bille, J. and Pelz, B., University of Heidelberg, Heidelberg, Germany.)

compensating for the corneal influence. A bow-tie pattern overlying the macula can be observed before the corneal retardation was eliminated. After compensation, this turns into a homogeneous pattern around the fovea and the typical retardation pattern around the ONH appears.

Since glaucoma leads to an irreversible damage of the retinal tissue, it is crucial an early detection of the pathology to initiate the appropriate treatment. In that sense, the main efforts in the field of ocular polarimetry have been centered on the development of clinical instruments to accurately assess the retinal thickness.

28.5 POLARIZATION-SENSITIVE OPTICAL COHERENCE TOMOGRAPHY

Based on the principle of low coherence interferometry, OCT provides depth-resolved images of biological samples (Huang et al. 1991). The combination with polarization led to the imaging technique known as polarization-sensitive OCT (PS-OCT) (Hee et al. 1992). This gives additional and quantitative information by combining the birefringence (and depolarizing) properties of the tissue under analysis and the advantages of OCT. First developed by Hee and coworkers (1992) and later improved by others (de Boer et al. 1997, Everett et al. 1998), PS-OCT instruments have been reported as powerful tools to explore the polarization properties of the cornea and the retina, both in depth and en face.

Early PS-OCT experiments in *ex vivo* rabbit and primate eyes (Ducros et al. 1999, 2001) reported the RNFL as a birefringent structure. Depth-resolved images of the Stokes parameters were used to compute the retardation. The associated retinal thickness showed a good correlation with histology. *In vivo* depth-resolved RNFL measurements reported for the first time by Cense and colleagues provided a double-pass phase retardation of 39° ± 6°/100 μm (Cense et al. 2002). Later they demonstrated that around the ONH the retardation is constant with radius but varies with retinal location (between 0.10° and 0.35°/μm) (Cense et al. 2004a,b). Moreover, superior and inferior regions were found to be more birefringent than temporal and nasal areas. These changes (between 1.2×10^{-4} and 4.1×10^{-4} at 840 nm wavelength) must be taken into account in the equivalence retardation-thickness used in clinical devices used for glaucoma diagnosis. Cucu et al. reported similar values of retardation (0.29°/μm) estimated from en face PS-OCT images of the living human retina at the ONH region (Cucu et al. 2003).

In 2004, Hitzenberger and his coworkers in Vienna centered their PS-OCT research in measuring and imaging the birefringent properties of the cornea (Götzinger et al. 2004, Pircher et al. 2004). They presented maps of retardation and slow axis orientation of human donor corneas in longitudinal cross sections and en face images at the posterior corneal surface. Their results indicated that the corneal retardation increases in the radial direction and with depth. The slow axis changes in the transversal direction (Figure 28.12). When comparing normal and keratoconic corneas, heavy distortions in retardation and slow axis orientation patterns were found. In particular, pathological corneas showed large regions of decreased and increased retardation and areas where the slow axis markedly changed with depth (Götzinger et al. 2007).

PS-OCT devices have also been used to improve previous results on retinal polarization properties. In particular, a research instrument was developed to show *in vivo* 2D (regular tomography and en face) and 3D data sets of retardation and azimuth of the fovea and the ONH of the human eye (Pircher et al. 2006). They corroborated early results obtained by different techniques, including SL polarimetry (klein Brink and van Blokland 1988, Pelz et al. 1996, Zhou and Weinreb 2002,

Twietmeyer et al. 2008). An hourglass-shaped pattern appears in the en face retardation maps. As previously explained, this is a result of the combination of two retarders, the anterior segment (mainly the cornea) and the Henle fibers. The former has a fixed axis and the latter has a radial orientation. The retardation is maximum when both axes are aligned and minimum when orthogonal. At the ONH, the pattern is similar due to the combination of the cornea and the RNFL. However, unlike the fovea, the retardation introduced by the RNFL is not constant along a circle around the ONH.

However, the most important finding was an abrupt change in polarization properties at the location of the retinal pigment epithelium (RPE)/Bruch's membrane. Random retardation and azimuth values were found and associated with depolarization effects. This polarization-scrambling layer was identified as the RPE by imaging the retina of patients with neurosensory retinal detachment and RPE atrophy (Figure 28.13) (Pircher et al. 2006).

Spectral-domain (SD) technology was also included in PS-OCT devices. Compared to time-domain instruments, these shorten the acquisition time and improve the signal-to-noise ratio of the images (Götzinger et al. 2005). SD-PS-OCT was originally used to measure the RNFL retardation in a healthy eye, and results were shown to be similar to those obtained with time-domain systems (between 0.18°/μm and 0.40°/μm) (Cense et al. 2007).

Since retinal retardation maps are distorted by the effects of the corneal birefringence, these must be compensated to obtain reliable distributions of polarization parameters of the retina that can be used in glaucoma diagnosis (Dreher et al. 1992). This problem also arises in SLP where a variable retarder is used (Choplin et al. 2003). As an alternative to this hardware-based solution, Pircher and colleagues reported a numerical method to numerically compensate the corneal birefringence using the retinal surface polarization state as input (Pircher et al. 2007). Moreover, the physical origin of atypical polarimetric patterns was analyzed by comparing images provided by the GDx VCC (commercially available instrument) and a research PS-OCT

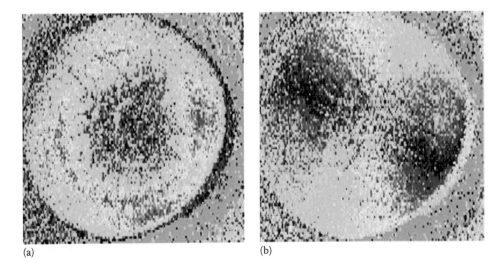

(a) (b)

Figure 28.12 En face optical coherence tomography images at the posterior surface of an *in vitro* human cornea. Retardation (a) and fast axis orientation (b). Image size: 8 × 8 mm². (Reproduced from Pircher, M. et al., *Phys. Med. Biol.*, 49, 1257, 2004. With permission.)

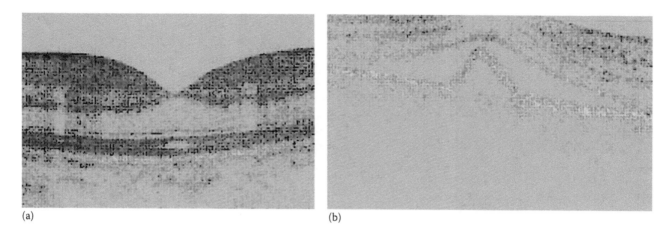

Figure 28.13 Retardation maps computed from polarization-sensitive optical coherence tomography images of the foveal region in a healthy subject (a) and in a patient with neurosensory retinal and retinal pigment epithelium detachment (b). The depolarizing layer can easily be observed. (Modified from Pircher, M. et al., *Invest. Ophthalmol. Vis. Sci.*, 47, 5487, 2006.)

instrument (Götzinger et al. 2008). Artifacts were found to be due to an increased penetration of the probing beam into the birefringent sclera.

In the last years, PS-OCT has enabled direct identification of retinal layers and corneal structure. It has provided useful information on ocular diseases (Michels et al. 2008, Ahlers et al. 2010, Schlanitz et al. 2011) in such a way that future implementations in clinical environments may help to understand certain pathologies and to improve diagnosis.

28.6 IMPROVEMENT OF OCULAR IMAGING THROUGH POLARIZATION

The reliability of a clinical diagnosis strongly depends on the quality of the acquired images. In the eye this is limited by a number of factors including aberrations, scattering, and the properties of the light reflected from the structures of interest. Moreover, since the ocular media and the retina change the polarization state of the incident light, ocular imaging techniques might be potentially affected by polarization.

With the classical configuration of two linear polarizers, the light returning from the retina has often been interpreted in terms of two components: one maintaining the polarization (directional component or guided by the photoreceptors) and other becoming depolarized (diffuse or scattered component) (Weale 1966, Charman 1980, van Blokland and van Norren 1986, Burns et al. 1995, Bueno 2001). However, intraocular scattering (directly associated with depolarizing properties) might change the polarization state in a complicated way, and this configuration is not necessarily complete or even correct, leading to erroneous outcomes (see, for instance, Bueno 2001, 2002). This portion of depolarized light might increase with age, in particular when the crystalline lens starts to lose its transparency in the early stages of a cataract (Bueno and Campbell 2003, Bueno et al. 2009). At the plane of the pupil, the directional component is a useful tool to measure directional properties of human photoreceptors (Burns et al. 1995).

Although the retinal image quality estimates strongly depend on the combinations of in-and-out polarization states (Bueno and Artal 2001), the use of simple polarizing elements has been shown to enhance the visibility of different structures in the living eye. A pair of crossed linear polarizers (in the incoming and outgoing pathways) was used to observe the brushlike pattern overlying the macula. This polarized-light retinal pattern was reported to be a useful tool for diagnosing diseases affecting the macula (Hochheimer and Kues 1982) as shown many years later (Weber et al. 2007). The implementation of this simple polarization configuration into a fundus camera to observe the changes in the appearance of the RNFL provided an enhanced visualization of the arcuate bundles (Sommer et al. 1984). When using circular polarized light as illumination in a biomicroscope, the contrast of lenticular discontinuities was increased (Weale 1986) and the visibility or corneal endothelial cells enhanced (Peli 1985).

Despite linear and circular polarized light seemed to improve the observation of some ocular structures, some fundus images corresponding to other polarization states showed details of the retinal structures that could not be observed in the rest of the images. This is due to the fact that changes in the polarization state depend on the retinal location. In particular, the use of the Stokes–Mueller algorithm allowed the reconstruction of SL ophthalmoscope images with the highest and lowest quality (according to different metrics). These improved the visualization of structures of clinical importance, such as vessels and the ONH features as shown in Figure 28.14 (Bueno and Campbell 2002, Bueno and Vohnsen 2005, Bueno et al. 2007, 2008).

Since diseases might change the retinal structures in a non-controlled manner, pathological eyes might present additional changes in polarization (in particular, increased depolarization effects). This was used to show that depolarized light images produced higher contrast of drusen and subretinal changes in subjects with age-related macular degeneration (Burns et al. 2003, Elsner et al. 2007). These depolarized images provided also a better contrast for the profiles of retinal arteries and veins that helps to their characterization when affected by some pathologies, such as epiretinal membrane (Weber et al. 2004, Miura et al. 2007). In eyes with central serous chorioretinopathy, depolarized

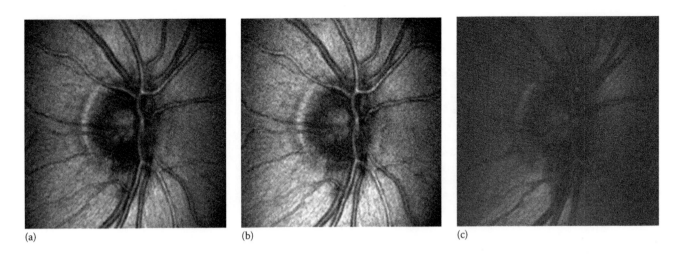

(a)　　　　　　　　　(b)　　　　　　　　　(c)

Figure 28.14 Polarimetric images of the optic nerve head in a normal healthy subject. It can be observed how the visualization of retinal features is readily better in S_0 (a) and degree of polarization (b) images than in the original one (c, acquired with circular polarized incident light).

light images visualized leakage points and areas of fluid (Miura et al. 2005). These images showed increased contrast in subretinal tissues with peripapillary hyperpigmentation in patients with glaucoma (Mellem-Kairala et al. 2005).

To summarize, polarization-sensitive imaging can be useful to improve the visualization of clinically important fundus structures that could aid in the detection, localization, and tracking of ocular disease.

REFERENCES

Able, K. P., Skylight polarization patterns at dusk influence migratory orientation in birds, *Nature* 299 (1982): 550–551.

Ahlers, C., Götzinger, E., Pircher, M., Golbaz, I., Prager, F., Schütze, C., Baumann, B., Hitzenberger, C. K., and Schmidt-Erfurth, U., Imaging of the retinal pigment epithelium in age-related macular degeneration using polarization-sensitive optical coherence tomography, *Invest. Ophthalmol. Vis. Sci.* 51 (2010): 2149–2157.

Aldrovani, M., Guaraldo, A. M. A., and Vidal, B. C., Optical anisotropies in corneal stroma collagen fibers from diabetic spontaneous mice, *Vision Res.* 47 (2007): 3229–3237.

Ansari, R. R., Böckle, S., and Rovati, L., New optical scheme for a polarimetric-based glucose sensor, *J. Biomed. Opt.* 9 (2004): 103–115.

Azzam, R. M. A. and Bashara, N. M., *Ellipsometry and Polarized Light* (New York: North-Holland, 1992).

Baba, J. S., Cameron, B. D., Theru, S., and Coté, G. L., Effect of temperature, pH, and corneal birefringence on polarimetric glucose monitoring in the eye, *J. Biomed. Opt.* 7 (2002): 321–328.

Baumann, B., Götzinger, E., Pircher, M., and Hitzenberger, C., Measurements of depolarization distribution in the healthy human macula by polarization sensitive OCT, *J. Biophotonics* 2 (2009): 426–434.

Bernabeu, E. and Gil, J. J., An experimental device for the dynamic determination of Mueller matrices, *J. Optics* (Paris) 16 (1985): 139–141.

Bettelheim, F. A., On the optical anisotropy of lens fiber cells, *Exp. Eye Res.* 21 (1975): 231–234.

Boehm, G., Ueber maculare (Haidingersche) Polarisations buschel und ueber einen polarisationoptischen Fehler des Auges, *Acta Ophthalmol.* 18 (1940): 109–169.

Bone, R. A. and Draper, G., Optical anisotropy of the human cornea determined with a polarizing microscope, *Appl. Opt.* 46 (2007): 8351–8357.

Bour, L. J., Polarized light and the eye. In *Visual Optics and Instrumentation*, Chap. 13, ed. W. N. Charman (New York: Macmillan Press, 1991).

Bour, L. J. and Lopes Cardozo, N. J., On the birefringence of the living human eye, *Vision Res.* 21 (1981): 1413–1421.

Brewster, D., Experiments on the de-polarization of light as exhibited by various minerals, animal and vegetable bodies with a reference of the phenomena to the general principles of polarization, *Philos. Trans. R. Soc. Lond.* 2 (1815–1830): 4–6.

Brewster, D., On the structure of the crystalline lens in fishes and quadrupeds, as ascertained by its action on polarized light, *Philos. Trans. R. Soc. Lond. Ser. B* 106 (1816): 311–317.

Bueno, J. M., Measurement of parameters of polarization in the living human eye using imaging polarimetry, *Vision Res.* 40 (2000): 3791–3799.

Bueno, J. M., Depolarization effects in the human eye, *Vision Res.* 41 (2001): 2687–2696.

Bueno, J. M., Polarimetry in the human eye using an imaging linear polariscope, *J. Opt. A: Pure Appl. Opt.* 4 (2002): 553–561.

Bueno, J. M., The influence of depolarization and corneal birefringence on ocular polarization, *J. Opt. A: Pure Appl. Opt.* 6 (2004): S91–S99.

Bueno, J. M., Analysis of the central corneal birefringence with double-pass polarimetric images, *J. Modern Opt.* 58 (2011): 1864–1870.

Bueno, J. M. and Artal, P., Double-pass imaging polarimetry in the human eye, *Opt. Lett.* 24 (1999): 64–66.

Bueno, J. M. and Artal, P., Polarization and retinal image quality estimates in the human eye, *J. Opt. Soc. Am. A Opt. Image Sci. Vis.* 18 (2001): 489–496.

Bueno, J. M. and Artal, P., Average double-pass ocular diattenuation using foveal fixation, *J. Mod. Opt.* 55 (2008): 849–859.

Bueno, J. M., Berrio, E., and Artal, P., Corneal polarimetry after LASIK refractive surgery, *J. Biomed. Opt.* 11 (2006): 014001.

Bueno, J. M., Berrio, E., Ozolinsh, M., and Artal, P., Degree of polarization as an objective method of estimating scattering, *J. Opt. Soc. Am. A Opt. Image Sci. Vis.* 21 (2004): 1316–1321.

Bueno, J. M. and Campbell, M. C. W., Confocal scanning laser ophthalmoscopy improvement by use of Mueller-matrix polarimetry, *Opt. Lett.* 27 (2002): 830–832.

Bueno, J. M. and Campbell, M. C. W., Polarization properties for in vitro old human crystalline lens, *Ophthal. Physiol. Opt.* 23 (2003): 109–118.

Bueno, J. M., Cookson, C. J., Hunter, J. J., Kisilak, M. L., and Campbell, M. C. W., Depolarization properties of the optic nerve head: The effect of age, *Ophthal. Physiol. Opt.* 29 (2009): 247–255.

Bueno, J. M., Cookson, C. J., Hunter, J. J., Kisilak, M. L., and Campbell, M. C. W., Imaging the fundus of the eye through polarization: Dependence with age, 2008 Annual Meeting Abstract and Program Planner, www.arvo.org, abstract #3209.

Bueno, J. M., Gualda, E. J., and Artal, P., Analysis of corneal stroma organization with wavefront optimized nonlinear microscopy, *Cornea* 30 (2011): 692–701.

Bueno, J. M., Hunter, J. J., Cookson, C. J., Kisilak, M. L., and Campbell, M. C. W., Improved scanning laser fundus imaging using polarimetry, *J. Opt. Soc. Am. A Opt. Image Sci. Vis.* 24 (2007): 1337–1348.

Bueno, J. M. and Jaronski, J. W., Spatially resolved polarization properties for in vitro corneas, *Ophthal. Physiol. Opt.* 21 (2001): 384–392.

Bueno, J. M. and Pérez, G. M., Combined effect of wavelength and polarization in double-pass retinal images in the human eye, *Vision Res.* 50 (2010): 2439–2444.

Bueno, J. M. and Vohnsen, B., Polarimetric high-resolution confocal scanning laser ophthalmoscope, *Vision Res.* 45 (2005): 3526–3534.

Burns, S. A., Elsner, A. E., Mellem-Kairala, M. B., and Simmons, R. B., Improved contrast of subretinal structures using polarization analysis, *Invest. Ophthalmol. Vis. Sci.* 44 (2003): 4061–4068.

Burns, S. A., Wu, S., Delori, F. C., and Elsner, A. E., Direct measurement of human-cone-photoreceptor alignment, *J. Opt. Soc. Am. A Opt. Image Sci. Vis.* 12 (1995): 2329–2338.

Cense, B., Chen, T. C., Park, B. H., Pierce, M. C., and de Boer, J. F., In vivo depth-resolved birefringence measurements of the human retinal nerve fiber layer by polarization-sensitive optical coherence tomography, *Opt. Lett.* 27 (2002): 1610–1612.

Cense, B., Chen, T. C., Park, B. H., Pierce, M. C., and de Boer, J. F., In vivo birefringence and thickness measurements of the human retinal nerve fiber layer using polarization-sensitive optical coherence tomography, *J. Biomed. Opt.* 9 (2004a): 121–125.

Cense, B., Chen, T. C., Park, B. H., Pierce, M. C., and de Boer, J. F., Thickness and birefringence of healthy retinal nerve fiber tissue measured with polarization-sensitive optical coherence tomography, *Invest. Ophthalmol. Vis. Sci.* 45 (2004b): 2606–2612.

Cense, B., Mujat, M., Chen, T. C., Park, B. H., and de Boer, J. F., Polarization-sensitive spectral-domain optical coherence tomography using a single line scan camera, *Opt. Express* 15 (2007): 2421–2431.

Charman, W. N., Reflection of plane-polarized light by the retina, *Br. J. Physiol. Opt.* 34 (1980): 34–49.

Charman, W. N., Optics of the human eye. In *Visual Optics and Instrumentation*, Chap. 1, ed. W. N. Charman (New York: Macmillan Press, 1991).

Chipman, R. A., Polarimetry. In *Handbook of Optics*, Bass, M. (ed.), Vol. 2, 2nd edn., Chap. 22 (New York: McGraw-Hill, 1995).

Choplin, N. T., Zhou, Q., and Knighton, R. W., Effect of individualized compensation for anterior segment birefringence on retinal nerve fiber layer assessments as determined by scanning laser polarimetry, *Ophthalmology* 110 (2003): 719–725.

Cope, W. T., Wolbarsht, M. L., and Yamanashi, B. S., The corneal polarization cross, *J. Opt. Soc. Am.* 68 (1978): 1139–1141.

Cucu, R. G., Podoleanu, A. Gh., Rosen, R. B., Boxer, A. B., and Jackson, D. A., En-face polarization-sensitive optical coherence tomography, *Proc. SPIE* 5140 (2003): 113–119.

de Boer, J. F., Milner, T. E., van Gemert, M. J. C., and Nelson, J. S., Two-dimensional birefringence imaging in biological tissue by polarization sensitive optical coherence tomography, *Opt. Lett.* 22 (1997): 934–936.

Delori, F. C., Webb, R. H., and Parker, J. S., Macular birefringence, *Invest. Ophthalmol. Vis. Sci.* (ARVO Suppl.) 19 (1979): 53.

de Vries, H. L., Spoor, A., and Jielof, R., Properties of the eye with respect to polarized light, *Physica* 19 (1953): 419–432.

Donohue, D. J., Stoyanov, B. J., McCally, R. L., and Farrell, R. A., Numerical modeling of the cornea's lamellar structure and birefringence properties, *J. Opt. Soc. Am. A Opt. Image Sci. Vis.* 12 (1995): 1425–1438.

Dreher, A., Reiter, K., and Bille, J., Assessment of nerve fiber layer thickness with the LTS laser tomographic scanner, *Invest. Ophthalmol. Vis. Sci.* (ARVO Suppl.) 29 (1988): 355.

Dreher, A. W. and Reiter, K., Scanning laser polarimetry of the retinal nerve fiber layer, *Proc. SPIE* 1746 (1992): 34–41.

Dreher, A. W., Reiter, K., and Weinreb, R. N., Spatially resolved birefringence of the retinal nerve fiber layer assessed with a retinal laser ellipsometer, *Appl. Opt.* 31 (1992): 3730–3735.

Ducros, M. G., de Boer, J. F., Huang, H.-E., Chao, L. C., Chen, Z., Nelson, J. S., Milner, T. E., and Rylander, III, H. G., Polarization sensitive optical coherence tomography of the rabbit eye, *IEEE* 5 (1999): 1159–1167.

Ducros, M. G., Marsack, J. D., Rylander III, H. G., Thomsen, S. L., and Milner, T. E., Primate retina imaging with polarization-sensitive optical coherence tomography, *J. Opt. Soc. Am. A Opt. Image Sci. Vis.* 18 (2001): 2945–2956.

Elsner, A. E., Weber, A., Cheney, M. C., and VanNasdale, D. A., Spatial distribution of macular birefringence associated with the Henle fibers, *Vision Res.* 48 (2008): 2578–2585.

Elsner, A. E., Weber, A., Cheney, M. C., VanNasdale, D. A., and Miura, M., Imaging polarimetry in patients with neovascular age-related macular degeneration, *J. Opt. Soc. Am. A Opt. Image Sci. Vis.* 24 (2007): 1468–1480.

Everett, M. J., Schoenenberger, K., Colston Jr., B. W., and Da Silva, L. B., Birefringence characterization of biological tissue by use of optical coherence tomography, *Opt. Lett.* 23 (1998): 228–230.

Fukuma, Y., Okazaki, Y., Shioiri, T., Iida, Y., Kikuta, H., Shirakashi, M., Yaoeda, K., Abe, H., and Ohnuma, K., Retinal nerve fiber layer retardation measurements using a polarization-sensitive fundus camera, *J. Biomed. Opt.* 16 (2011): 076017.

García Medina, J. J., García Medina, M., Shahin, M., and Pinazo Durán, M. D., Posterior capsular opacification affects scanning laser polarimetry examination, *Graefe's Arch. Clin. Exp. Ophthalmol.* 244 (2006): 520–523.

Gil, J. J., San José, I., and Ossikovski, R., Serial-parallel decompositions of Mueller matrices, *J. Opt. Soc. Am. A Opt. Image Sci. Vis.* 30 (2013): 32–50.

Götzinger, E., Pircher, M., Dejaco-Ruhswurm, I., Kaminski, S., Skorpik, C., and Hitzenberger, C. K., Imaging of birefringent properties of keratoconus corneas by polarization-sensitive optical coherence tomography, *Invest. Ophthalmol. Vis. Sci.* 48 (2007): 3551–3558.

Götzinger, E., Pircher, M., Baumann, B., Hirn, C., Vass, C., and Hitzenberger, C. K., Analysis of the origin of atypical scanning laser polarimetry patterns by polarization-sensitive optical coherence tomography, *Invest. Ophthalmol. Vis. Sci.* 49 (2008): 5366–5372.

Götzinger, E., Pircher, M., and Hitzenberger, C. K., High speed spectral domain polarization sensitive optical coherence tomography of the human retina, *Opt. Express* 13 (2005): 10217–10229.

Götzinger, E., Pircher, M., Sticker, M., Fercher, A. F., and Hitzenberger, C. K., Measurement and imaging of birefringent properties of the human cornea with phase-resolved, polarization-sensitive optical coherence tomography, *J. Biomed. Opt.* 9 (2004): 94–102.

Greenfield, D. S. Optic nerve and retinal nerve fiber layer analyzers in glaucoma, *Curr. Opin. Ophthalmol.* 13 (2002): 68–76.

Greenfield, D. S., Knighton, R. W., and Huang, X.-R., Effect of corneal polarization axis on assessment of retinal nerve fiber layer thickness by scanning laser polarimetry, *Am. J. Ophthalmol.* 129 (2000): 715–722.

Haidinger, W., Über das direkte Erkennen des polarisierten Lichts und der Lage der Polarisationsebene, *Annal. Phys.* (Leipzig) 63 (1844): 29–39.

Hárosi, F. I., Microspectrophotometry and optical phenomena: Birefringence, dichroism and anomalous dispersion. In *Vertebrate Photoreceptor Optics*, eds. J. M. Enoch and F. L. Tobey (Berlin, Germany: Springer-Verlag, 1981), pp. 337–399.

Hawryshyn, C. W. and McFarland, W. N., Cone photoreceptor mechanisms and the detection of polarized light in fish, *J. Comp. Physiol. A* 160 (1987): 459–465.

Hee, M. R., Huang, D., Swanson, E. A., and Fujimoto, J. G., Polarization-sensitive low-coherence reflectometer for birefringence characterization and ranging, *J. Opt. Soc. Am. B* 9 (1992): 903–908.

Hemenger, R. P., Birefringence of a medium of tenuous parallel cylinders, *Appl. Opt.* 28 (1989): 4030–4034.

His, W., Beiträge Zur Normalen Und Pathologischen: Histologie Der Cornea (Basel, Swizerland, 1856).

Hochheimer, B. F. and Kues, H. A., Retinal polarization effects, *Appl. Opt.* 21 (1982): 3811–3818.

Huang, D., Swanson, E. A., Lin, C. P., Schuman, J. S., Stinson, W. G., Chang, W., Hee, M. R. et al., Optical coherence tomography, *Science* 254 (1991): 1178–1181.

Huang, X.-R., Bagga, H., Greenfield, D. S., and Knighton, R. W., Variation of peripapillary retinal nerve fiber layer birefringence in normal human subjects, *Invest. Ophthalmol. Vis. Sci.* 45 (2004): 3073–3080.

Huang, X.-R. and Knighton, R. W., Linear birefringence of the retinal nerve fiber layer measured in vitro with a multispectral imaging micropolarimeter, *J. Biomed. Opt.* 7 (2002): 199–204.

Huang, X.-R. and Knighton, R. W., Microtubules contribute to the birefringence of the retinal nerve fiber layer, *Invest. Ophthalmol. Vis. Sci.* 46 (2005): 4588–4593.

Hunter, D. G., Shah, A. S., Sau, S., Nassif, D., and Guyton, D. L., Automated detection of ocular alignment with binocular retinal birefringence scanning, *Appl. Opt.* 42 (2003): 3047–3053.

Hunter, D. G., Patel, S. N., and Guyton, D. L., Automated detection of foveal fixation by use of retinal birefringence scanning, *Appl. Opt.* 38 (1999): 1273–1279.

Hunter, J. J., Cookson, C. J., Kisilak, M. L., Bueno, J. M., and Campbell, M. C. W., Characterizing image quality in a scanning laser ophthalmoscope with differing pinholes and induced scattered light, *J. Opt. Soc. Am. A Opt. Image Sci. Vis.* 24 (2007): 1284–1295.

Ichihashi, Y., Khin, M. H., Ishikawa, K., and Hatada, T., Birefringence effect of the in vivo cornea, *Opt. Eng.* 34 (1995): 693–700.

Irsch, K. and Shah, A. A., Birefringence of the central cornea in children assessed with scanning laser polarimetry, *J. Biomed Opt.* 17 (2012): 086001.

Jaronski, J. W. and Kasprzak, H. T., Generalized algorithm for photoelastic measurements based on phase-stepping imaging polarimetry, *Appl. Opt.* 38 (1999): 7018–7025.

Jaronski, J. W. and Kasprzak, H. T., Linear birefringence measurements of the in vitro human, *Ophthal. Physiol. Opt.* 23 (2003): 361–369.

Jerrard, H. G., Transmission of light through birefringent and optically active media: The Poincaré sphere, *J. Opt. Soc. Am.* 44 (1954): 634–640.

Jones, R. C., A new calculus for the treatment of optical systems, I. Description and discussion of the calculus, *J. Opt. Soc. Am.* 31 (1941): 488–493.

Kaplan, D. and Bettelheim, F. A., On the birefringence of bovine cornea, *Exp. Eye Res.* 13 (1972): 219–226.

Kirschenbaum, D. M., Optical rotatory capacity of the lens of the vertebrate eye, *Nature* 193 (1962): 392–393.

Klein Brink, H. B., Birefringence of the human crystalline lens in vivo, *J. Opt. Soc. Am. A* 8 (1991): 1788–1793.

Klein Brink, H. B. and van Blokland, G. J., Birefringence of the human foveal area assessed in vivo with Mueller-matrix ellipsometry, *J. Opt. Soc. Am. A* 5 (1988): 49–57.

Kliger, D. S., Lewis, J. W., and Randall, C. E., *Polarized Light in Optics and Spectroscopy* (San Diego, CA: Academic Press, Inc., 1990).

Knighton, R. W. and Huang, X.-R., Linear birefringence of the central human cornea, *Invest. Ophthalmol. Vis. Sci.* 43 (2002a): 82–86.

Knighton, R. W. and Huang, X.-R., Analytical methods for scanning laser polarimetry, *Opt. Express* 10 (2002b): 1179–1189.

Knighton, R. W., Huang, X. R., and Cavuoto, L. A., Corneal birefringence mapped by scanning laser polarimetry, *Opt. Express* 16 (2008): 13738–13751.

Komai, Y. and Ushiki, T., The three-dimensional organization of collagen fibrils in the human cornea and sclera, *Invest. Ophthalmol. Vis. Sci.* 32 (1991): 2244–2258.

Louis-Dorr, V., Naoun, K., Allé, P., Benoit, A. M., and Raspiller, A., Linear dichroism of the cornea, *Appl. Opt.* 43 (2004): 1515–1521.

Lu, S. and Chipman, R. A., Interpretation of Mueller matrices based on polar decomposition, *J. Opt. Soc. Am. A* 13 (1996): 1106–1113.

Maurice, D. M., The structure and transparency of the cornea, *J. Physiol.* 136 (1957): 263–286.

McCally, R. L. and Farrel, R. A., Structural implications of small-angle light scattering from cornea, *Exp. Eye Res.* 34 (1982): 99–113.

Mellem-Kairala, M. B., Elsner, A. E., Weber, A., Simmons, R. B., and Burns, S. A., Improved contrast of peripapillary hyperpigmentation using polarization analysis, *Invest. Ophthalmol. Vis. Sci.* 46 (2005): 1099–1106.

Michels, S., Pircher, M., Geitzenauer, W., Simader, C., Götzinger, E., Findl, O., Schmidt-Erfurth, U., and Hitzenberger, C. K., Value of polarisation-sensitive optical coherence tomography in diseases affecting the retinal pigment epithelium, *Br. J. Ophthalmol.* 92 (2008): 204–209.

Miura, M., Elsner, A. E., Cheney, M. C., Usui, M., and Iwasaki, T., Imaging polarimetry and retinal blood vessel quantification at the epiretinal membrane, *J. Opt. Soc. Am. A Opt. Image Sci. Vis.* 24 (2007): 1431–1437.

Miura, M., Elsner, A. E., Weber, A., Cheney, M. C., Oshako, M., Usui, M., and Iwasaki, T., Imaging polarimetry in central serous chorioretinopathy, *Am. J. Ophthalmol.* 140 (2005): 1014–1019.

Miura, M., Osako, M., Elsner, A. E., Kajizuka, H., Yamada, K., and Usui, M., Birefringence of intraocular lenses, *J. Cataract Refract. Surg.* 30 (2004): 1549–1555.

Naoun, O. K., Louis-Dorr, V., Allé, P., Sablon, J. C., and Benoit, A. M., Exploration of the retinal nerve fiber layer thickness by measurement of the linear dichroism, *Appl. Opt.* 44 (2005): 7074–7082.

Novikova, T., Pierangelo, A., De Martino, A., Benali, A., and Validire, P., Polarimetric imaging for cancer diagnosis and staging, *Opt. Photon. News* (October 2012): 27–33.

Peli, E., Circular polarizers enhance visibility of endothelium in specular reflection biomicroscopy, *Arch. Ophthalmol.* 103 (1985): 670–672.

Pelz, B. C. E., Weschenmoser, C., Goelz, S., Fischer, P., Burk, R. O. W., and Bille, J. F., In-vivo measurement of the retinal birefringence with regard to corneal effects using an electro-optical ellipsometer, *Proc. SPIE* 2930 (1996): 92–101.

Pierscionek, B. K. and Weale, R. A., Polarising light biomicroscopy and the relation between visual acuity and cataract, *Eye* 9 (1995): 304–308.

Pierscionek, B. K. and Weale, R. A., Investigation of the polarization optics of the living human cornea and lens with Purkinje images, *Appl. Opt.* 37 (1998): 6845–6851.

Pircher, M., Götzinger, E., Baumann, B., and Hitzenberger, C. K., Corneal birefringence compensation for polarization sensitive optical coherence tomography of the human retina, *J. Biomed. Opt.* 12 (2007): 041210.

Pircher, M., Götzinger, E., Findl, O., Michels, S., Geitzenauer, W., Leydolt, C., Schmidt-Erfurth, U., and Hitzenberger, C. K., Human macula investigated in vivo with polarization-sensitive optical coherence tomography, *Invest. Ophthalmol. Vis. Sci.* 47 (2006): 5487–5494.

Pircher, M., Götzinger, E., Leitgeb, R., and Hitzenberger, C. K., Transversal phase resolved polarization sensitive optical coherence tomography, *Phys. Med. Biol.* 49 (2004): 1257–1263.

Post, D. and Gurland, J. E., Birefringence of the cat cornea, *Exp. Eye Res.* 5 (1966): 286–295.

Prieto, P. M., Vargas-Martín, F., McLellan, J. S., and Burns, S. A., The effect of polarization on ocular wave aberration measurements, *J. Opt. Soc. Am. A Opt. Image Sci. Vis.* 19 (2002): 809–814.

Purvinis, G., Cameron, B. D., and Altrogge, D. M., Noninvasive polarimetric-based glucose monitoring: An in vivo study, *J. Diabetes Sci. Technol.* 5 (2011): 380–387.

Rawer, R., Stork, W., and Kreiner, C. F., Non-invasive polarimetric measurement of glucose concentration in the anterior chamber of the eye, *Graefes Arch. Clin. Exp. Ophthalmol.* 242 (2004): 1017–1023.

Sardar, D. K., Yow, R. M., Tsin, A. T., and Sardar, R., Optical scattering, absorption, and polarization of healthy and neovascularized human retinal tissues, *J. Biomed. Opt.* 10 (2005): 051501.

Sato, H., Ellis, G. W., and Inoué, S., Microtubular origin of mitotic spindle form birefringence. Demonstration of the applicability of Wiener's equation, *J. Cell Biol.* 67 (1975): 501–517.

Schlanitz, F. G., Baumann, B., Spalek, T., Schütze, C., Ahlers, C., Pircher, M., Götzinger, E., Hitzenberger, C. K., and Schmidt-Erfurth, U., Performance of automated drusen detection by polarization-sensitive optical coherence tomography, *Invest. Ophthalmol. Vis. Sci.* 52 (2011): 4571–4579.

Shurcliff, W. A., *Polarized Light: Production and Use* (Cambridge, U.K.: Harvard University Press, 1962).

Shute, C. C. D., Haidinger's brushes and predominant orientation of collagen in corneal stroma, *Nature* 250 (1974): 163–164.

Sommer, A., Kues, H. A., D'Anna, S. A., Arkell, S., Robin, A., and Quigley, H. A., Cross-polarization photography of the nerve fiber layer, *Arch. Ophthalmol.* 102 (1984): 864–869.

Song, H., Qi, X., Zou, W., Zhong, Z., and Burns, S. A., Dual electro-optical modulator polarimeter based on adaptive optics scanning laser ophthalmoscope, *Opt. Express* 18 (2010): 21892–21904.

Stanworth, A., Polarized light studies of the cornea. II. The effect of intraocular pressure, *J. Exp. Biol.* 30 (1953): 164–169.

Stanworth, A. and Naylor, E. J., The polarization optics of the isolated cornea, *Br. J. Ophtalmol.* 34 (1950): 201–211.

Stanworth, A. and Naylor, E. J., Polarized light studies of the cornea. I. The isolated cornea, *J. Exp. Biol.* 30 (1953): 160–163.

Summers, D. M., Friedmann, G. B., and Clements, R. M., Physical model for Haidinger's brush, *J. Opt. Soc. Am.* 60 (1970): 271–272.

Takeguchi, N. and Nakagary, M., Determination of small birefringence in the bovine lens capsule by optical rotatory dispersion, *J. Opt. Soc. Am.* 58 (1968): 415–418.

Twietmeyer, K. M., Chipman, R. A., Elsner, A. E., Zhao, Y., and VanNasdal, D., Mueller matrix retinal imager with optimized polarization conditions, *Opt. Express* 16 (2008): 21339–21354.

Valentin, G., Neue Untersuchungen Über die Polarisations-Erscheinungen der Krystall-linsen des Menschen und der Thiere, *Arch. Ophthalmol.* 4 (1859): 227–268.

Valentin, G. G., *Die Untersuchung der Pflanzen-und der Tiergewebe in polarisiertem Lichte* (Lepzig, Germany: Engelmann, 1861).

Van Blokland, G. J., Ellipsometry of the human retina in vivo: Preservation of polarization, *J. Opt. Soc. Am. A* 2 (1985): 72–75.

Van Blokland, G. J. and van Norren, D., Intensity and polarization of light scattered at small angles from the human fovea, *Vision Res.* 26 (1986): 485–494.

Van Blokland, G. J. and Verhelst, S. C., Corneal polarization in the living human eye explained with a biaxial model, *J. Opt. Soc. Am. A* 4 (1987): 82–90.

VanNasdale, D. A., Elsner, A. E., Hobbs, T., and Burns, S. A., Foveal phase retardation changes associated with normal aging, *Vision Res.* 51 (2011): 2263–2272.

VanNasdale, D. A., Elsner, A. E., Kohne, K. D., Peabody, T. D., Malinovsky, V. E., Haggerty, B. P., Weber, A., and Clark, C. A., Foveal localization in non-exudative AMD using scanning laser polarimetry, *Optom. Vis. Sci.* 89 (2012): 667–677.

VanNasdale, D. A., Elsner, A. E., Weber, A., Miura, M., and Haggerty, B. P., Determination of foveal location using scanning laser polarimetry, *J. Vis.* 9 (2009): 1–17, 21.

Weale, R. A., Polarized light and the human fundus oculi, *J. Physiol.* 186 (1966): 925–930.

Weale, R. A., Sex, age and the birefringence of the human crystalline lens, *Exp. Eye Res.* 29 (1979): 449–461.

Weale, R. A., New method for visualising discontinuities in the crystal-line lens, *Br. J. Ophthamol.* 70 (1986): 925–930.

Weber, A., Cheney, M. C., Smithwick, Q. Y. J., and Elsner, A. E., Polarimetric imaging and blood vessel quantification, *Opt. Express* 12 (2004): 5178–5190.

Weber, A., Elsner, A. E., Miura, M., Kompa, S., and Cheney, M. C., Relationship between foveal birefringence and visual acuity in neovascular age-related macular degeneration, *Eye* (London) 21 (2007): 353–361.

Wehner, R., Polarized-light navigation by insects, *Sci. Am.* 235 (1976): 106–115.

Weinreb, R. N., Dreher, A. W., Coleman, A., Quigley, H., Shaw, B., and Reiter, K., Histopathologic validation of Fourier-ellipsometry measurements of retinal nerve fiber layer thickness, *Arch. Ophthalmol.* 108 (1990): 557–560.

Weinreb, R. N., Shakiba, S., and Zangwill, L., Scanning laser polarimetry to measure the nerve fiber layer of normal and glaucomatous eyes, *Am. J. Ophthalmol.* 119 (1995): 627–636.

Wiener, O., *Die Theorie des Mischkörpers für das Feld der Statinären Strömung* (Leipzig, Germany: B. G. Teubner, 1912).

Zhevandrov, N. D., Polarisation physiological optics, *Physics-Uspekhi* 38 (1995): 1147–1167.

Zhou, Q. and Weinreb, R. N., Individualized compensation of anterior segment birefringence during scanning laser polarimetry, *Invest. Ophthalmol. Vis. Sci.* 43 (2002): 2221–2228.

Index